Digital Image Processing

Digital Image Processing

Second Edition

Rafael C. Gonzalez
University of Tennessee

Richard E. Woods
MedData Interactive

Prentice Hall
Upper Saddle River, New Jersey 07458

Library of Congress Cataloging-in-Pubblication Data

Gonzalez, Rafael C.
   Digital Image Processing / Richard E. Woods
     p. cm.
   Includes bibliographical references
   ISBN 0-201-18075-8
   1. Digital Imaging. 2. Digital Techniques. I. Title.

  TA1632.G66   2001
  621.3—dc21   2001035846
                                     CIP

Vice-President and Editorial Director, ECS: *Marcia J. Horton*
Publisher: *Tom Robbins*
Associate Editor: *Alice Dworkin*
Editorial Assistant: *Jody McDonnell*
Vice President and Director of Production and Manufacturing, ESM: *David W. Riccardi*
Executive Managing Editor: *Vince O'Brien*
Managing Editor: *David A. George*
Production Editor: *Rose Kernan*
Composition: *Prepare, Inc.*
Director of Creative Services: *Paul Belfanti*
Creative Director: *Carole Anson*
Art Director and Cover Designer: *Heather Scott*
Art Editor: *Greg Dulles*
Manufacturing Manager: *Trudy Pisciotti*
Manufacturing Buyer: *Lisa McDowell*
Senior Marketing Manager: *Jennie Burger*

Prentice Hall   © 2002 by Prentice-Hall, Inc.
Upper Saddle River, New Jersey 07458

The author and publisher of this book have used their best efforts in preparing this book. These efforts
include the development, research, and testing of the theories and programs to determine their
effectiveness. The author and publisher make no warranty of any kind, expressed or implied, with rega
these programs or the documentation contained in this book. The author and publisher shall not be lia
any event for incidental or consequential damages in connection with, or arising out of, the furnishing,
performance, or use of these programs.

Printed in the United States of America
10 9 8 7 6

ISBN: 0-201-18075-8

Pearson Education Ltd., *London*
Pearson Education Australia Pty., Limited, *Sydney*
Pearson Education Singapore, Pte. Ltd.
Pearson Education North Asia Ltd., *Hong Kong*
Pearson Education Canada, Ltd., *Toronto*
Pearson Education de Mexico, S.A. de C.V.
Pearson Education—Japan, *Tokyo*
Pearson Education Malaysia, Pte. Ltd.
Pearson Education, *Upper Saddle River, New Jersey*

To Connie, Ralph, and Robert
and
To Janice, David, and Jonathan

Contents

3 *Image Enhancement in the Spatial Domain 75*

4 *Image Enhancement in the Frequency Domain 147*

5 *Image Restoration   220*

9 *Morphological Image Processing* 519

Preface

> When something can be read without effort,
> great effort has gone into its writing.
>
> *Enrique Jardiel Poncela*

This edition is the most comprehensive revision of *Digital Image Processing* since the book first appeared in 1977. As the 1977 and 1987 editions by Gonzalez and Wintz, and the 1992 edition by Gonzalez and Woods, the present edition was prepared with students and instructors in mind. Thus, the principal objectives of the book continue to be to provide an introduction to basic concepts and methodologies for digital image processing, and to develop a foundation that can be used as the basis for further study and research in this field. To achieve these objectives, we again focused on material that we believe is fundamental and has a scope of application that is not limited to the solution of specialized problems. The mathematical complexity of the book remains at a level well within the grasp of college seniors and first-year graduate students who have introductory preparation in mathematical analysis, vectors, matrices, probability, statistics, and rudimentary computer programming.

The present edition was influenced significantly by a recent market survey conducted by Prentice Hall. The major findings of this survey were:

1. A need for more motivation in the introductory chapter regarding the spectrum of applications of digital image processing.
2. A simplification and shortening of material in the early chapters in order to "get to the subject matter" as quickly as possible.
3. A more intuitive presentation in some areas, such as image transforms and image restoration.
4. Individual chapter coverage of color image processing, wavelets, and image morphology.
5. An increase in the breadth of problems at the end of each chapter.

The reorganization that resulted in this edition is our attempt at providing a reasonable degree of balance between rigor in the presentation, the findings of the market survey, and suggestions made by students, readers, and colleagues since the last edition of the book. The major changes made in the book are as follows.

Chapter 1 was rewritten completely. The main focus of the current treatment is on examples of areas that use digital image processing. While far from exhaustive, the examples shown will leave little doubt in the reader's mind regarding the breadth of application of digital image processing methodologies. Chapter 2 is totally new also. The focus of the presentation in this chapter is on how digital images are generated, and on the closely related concepts of sampling, aliasing, Moiré patterns, and image zooming and shrinking. The new material and the manner in which these two chapters were reorganized address directly the first two findings in the market survey mentioned above.

Chapters 3 though 6 in the current edition cover the same concepts as Chapters 3 through 5 in the previous edition, but the scope is expanded and the presentation is totally different. In the previous edition, Chapter 3 was devoted exclusively to image transforms. One of the major changes in the book is that image transforms are now introduced when they are needed. This allowed us to begin discussion of image processing techniques much earlier than before, further addressing the second finding of the market survey. Chapters 3 and 4 in the current edition deal with image enhancement, as opposed to a single chapter (Chapter 4) in the previous edition. The new organization of this material does not imply that image enhancement is more important than other areas. Rather, we used it as an avenue to introduce spatial methods for image processing (Chapter 3), as well as the Fourier transform, the frequency domain, and image filtering (Chapter 4). Our purpose for introducing these concepts in the context of image enhancement (a subject particularly appealing to beginners) was to increase the level of intuitiveness in the presentation, thus addressing partially the third major finding in the marketing survey. This organization also gives instructors flexibility in the amount of frequency-domain material they wish to cover.

Chapter 5 also was rewritten completely in a more intuitive manner. The coverage of this topic in earlier editions of the book was based on matrix theory. Although unified and elegant, this type of presentation is difficult to follow, particularly by undergraduates. The new presentation covers essentially the same ground, but the discussion does not rely on matrix theory and is much easier to understand, due in part to numerous new examples. The price paid for this newly gained simplicity is the loss of a unified approach, in the sense that in the earlier treatment a number of restoration results could be derived from one basic formulation. On balance, however, we believe that readers (especially beginners) will find the new treatment much more appealing and easier to follow. Also, as indicated below, the old material is stored in the book Web site for easy access by individuals preferring to follow a matrix-theory formulation.

Chapter 6 dealing with color image processing is new. Interest in this area has increased significantly in the past few years as a result of growth in the use of digital images for Internet applications. Our treatment of this topic represents a significant expansion of the material from previous editions. Similarly Chapter 7, dealing with wavelets, is new. In addition to a number of signal processing applications, interest in this area is motivated by the need for more sophisticated methods for image compression, a topic that in turn is motivated by a increase in the number of images transmitted over the Internet or stored in web servers. Chapter 8 dealing with image compression was updated to include new compression methods and standards, but its fundamental structure remains the same as in the previous edition. Several image transforms, previously covered in Chapter 3 and whose principal use is compression, were moved to this chapter.

Chapter 9, dealing with image morphology, is new. It is based on a significant expansion of the material previously included as a section in the chapter on image representation and description. Chapter 10, dealing with image segmentation, has the same basic structure as before, but numerous new examples were included and a new section on segmentation by morphological watersheds was added. Chapter 11, dealing with image representation and description, was shortened slightly by the removal of the material now included in Chapter 9. New examples were added and the Hotelling transform (description by principal components), previously included in Chapter 3, was moved to this chapter. Chapter 12 dealing with object recognition was shortened by the removal of topics dealing with knowledge-based image analysis, a topic now covered in considerable detail in a number of books which we reference in Chapters 1 and 12. Experience since the last edition of *Digital Image Processing* indicates that the new, shortened coverage of object recognition is a logical place at which to conclude the book.

Although the book is totally self-contained, we have established a companion web site (see inside front cover) designed to provide support to users of the book. For students following a formal course of study or individuals embarked on a program of self study, the site contains a number of tutorial reviews on background material such as probability, statistics, vectors, and matrices, prepared at a basic level and written using the same notation as in the book. Detailed solutions to many of the exercises in the book also are provided. For instruction, the site contains suggested teaching outlines, classroom presentation materials, laboratory experiments, and various image databases (including most images from the book). In addition, part of the material removed from the previous edition is stored in the web site for easy download and classroom use, at the discretion of the instructor. A downloadable instructor's manual containing sample curricula, solutions to sample laboratory experiments, and solutions to all problems in the book is available to instructors who have adopted the book for classroom use.

This edition of *Digital Image Processing* is a reflection of the significant progress that has been made in this field in just the past decade. As is usual in a project such as this, progress continues after work on the manuscript stops. One of the reasons earlier versions of this book have been so well accepted throughout the world is their emphasis on fundamental concepts, an approach that, among other things, attempts to provide a measure of constancy in a rapidly-evolving body of knowledge. We have tried to observe that same principle in preparing this edition of the book.

R.C.G.
R.E.W.

Acknowledgments

We are indebted to a number of individuals in academic circles as well as in industry and government who have contributed to this edition of the book. Their contributions have been important in so many different ways that we find it difficult to acknowledge them in any other manner but alphabetically. In particular, we wish to extend our appreciation to our colleagues Mongi A. Abidi, William E. Blass, Ramiro Jordan, Yongmin Kim, Bryan Morse, Andrew Oldroyd, Ali M. Reza, Edgardo Felipe Riveron, and Jose Ruiz Shulcloper, for their many suggestions on how to improve the presentation and/or the scope of coverage in the book.

Numerous individuals and organizations provided us with valuable assistance during the writing of this edition. Again, we list them alphabetically. We are particularly indebted to Steve Eddins and Naomi Fernandes at The MathWorks for providing us with MATLAB software and support that were important in our ability to create or clarify many of the examples and experimental results included in this edition of the book. A significant percentage of the new images used in this edition (and in some cases their history and interpretation) were obtained through the efforts of individuals whose contributions are sincerely appreciated. In particular, we wish to acknowledge the efforts of Serge Beucher, Melissa D. Binde, James Blankenship, Uwe Boos, Ernesto Bribiesca, Dragana Brzakovic, Michael E. Casey, D. R. Cate, Michael W. Davidson, Thomas R. Gest, Lalit Gupta, Zhong He, Roger Heady, Juan Herrera, John M. Hudak, Michael Hurwitz, Chris J. Johannsen, Rhonda Knighton, Ashley Mohamed, A. Morris, Curtis C. Ober, Joseph E. Pascente, David R. Pickens, Michael Robinson, Barrett A. Schaefer, Michael Shaffer, Pete Sites, Sally Stowe, Craig Watson, and David K. Wehe. We also wish to acknowledge other individuals and organizations cited in the captions of numerous figures throughout the book for their permission to use that material.

Special thanks go to Tom Robbins, Rose Kernan, Alice Dworkin, Vince O'Brien, Jody McDonnell, and Heather Scott at Prentice Hall for their commitment to excellence in all aspects of the production of this edition of the book. Their creativity, assistance, and patience are truly appreciated.

R.C.G
R.E.W

About the Authors

Rafael C. Gonzalez

R. C. Gonzalez received the B.S.E.E. degree from the University of Miami in 1965 and the M.E. and Ph.D. degrees in electrical engineering from the University of Florida, Gainesville, in 1967 and 1970, respectively. He joined the Electrical and Computer Engineering Department at University of Tennessee, Knoxville (UTK) in 1970, where he became Associate Professor in 1973, Professor in 1978, and Distinguished Service Professor in 1984. He served as Chairman of the deparment from 1994 through 1997. He is currently a Professor Emeritus at UTK.

Gonzalez is the founder of the Image & Pattern Analysis Laboratory and the Robotics & Computer Vision Laboratory at the University of Tennessee. He also founded Perceptics Corporation in 1982 and was its president until 1992. The last three years of this period were spent under a full-time employment contract with Westinghouse Corporation, who acquired the company in 1989.

Under his direction, Perceptics became highly successful in image processing, computer vision, and laser disk storage technology. In its initial ten years, Perceptics introduced a series of innovative products, including: The world's first commercially-available computer vision system for automatically reading the license plate on moving vehicles; a series of large-scale image processing and archiving systems used by the U.S. Navy at six different manufacturing sites throughout the country to inspect the rocket motors of missiles in the Trident II Submarine Program; the market leading family of imaging boards for advanced Macintosh computers; and a line of trillion-byte laser disk products.

He is a frequent consultant to industry and government in the areas of pattern recognition, image processing, and machine learning. His academic honors for work in these fields include the 1977 UTK College of Engineering Faculty Achievement Award; the 1978 UTK Chancellor's Research Scholar Award; the 1980 Magnavox Engineering Professor Award; and the 1980 M.E. Brooks Distinguished Professor Award. In 1981 he became an IBM Professor at the University of Tennessee and in 1984 he was named a Distinguished Service Professor there. He was awarded a Distinguished Alumnus Award by the University of Miami in 1985, the Phi Kappa Phi Scholar Award in 1986, and the University of Tennessee's Nathan W. Dougherty Award for Excellence in Engineering in 1992.

Honors for industrial accomplishment include the 1987 IEEE Outstanding Engineer Award for Commercial Development in Tennessee; the 1988 Albert Rose Nat'l Award for Excellence in Commercial Image Processing; the 1989 B. Otto Wheeley Award for Excellence in Technology Transfer; the 1989 Coopers and Lybrand Entrepreneur of the Year Award; the 1992 IEEE Region 3 Outstanding Engineer Award; and the 1993 Automated Imaging Association National Award for Technology Development.

Gonzalez is author or co-author of over 100 technical articles, two edited books, and four textbooks in the fields of pattern recognition, image processing, and robotics. His books are used in over 500 universities and research institutions throughout the world. He is listed in the prestigious Marquis *Who's Who in America*, Marquis *Who's Who in Engineering*, Marquis *Who's Who in the World*, and in 10 other national and international biographical citations. He is the co-holder of two U.S. Patents, and has been an associate editor of the IEEE Transactions on Systems, Man and Cybernetics, and the International Journal of Computer and Information Sciences. He is a member of numerous professional and honorary societies, including Tau Beta Pi, Phi Kappa Phi, Eta Kappa Nu, and Sigma Xi. He is a Fellow of the IEEE.

Richard E. Woods

Richard E. Woods earned his B.S., M.S., and Ph.D. degrees in Electrical Engineering from the University of Tennessee, Knoxville. His professional experiences range from entrepreneurial to the more traditional academic, consulting, governmental, and industrial pursuits. Most recently, he founded MedData Interactive, a high technology company specializing in the development of hand-held computer systems for medical applications. He was also a founder and Vice President of Perceptics Corporation, where he was responsible for the development of many of the company's quantitative image analysis and autonomous decision making products.

Prior to Perceptics and MedData, Dr. Woods was an Assistant Professor of Electrical Engineering and Computer Science at the University of Tennessee and prior to that, a computer applications engineer at Union Carbide Corporation. As a consultant, he has been involved in the development of a number of special-purpose digital processors for a variety of space and military agencies, including NASA, the Ballistic Missile Systems Command, and the Oak Ridge National Laboratory.

Dr. Woods has published numerous articles related to digital signal processing and is a member of several professional societies, including Tau Beta Pi, Phi Kappa Phi, and the IEEE. In 1986, he was recognized as a Distinguished Engineering Alumnus of the University of Tennessee.

Digital Image Processing

Introduction

> One picture is worth more than ten thousand words.
>
> *Anonymous*

Preview

Interest in digital image processing methods stems from two principal application areas: improvement of pictorial information for human interpretation; and processing of image data for storage, transmission, and representation for autonomous machine perception. This chapter has several objectives: (1) to define the scope of the field that we call image processing; (2) to give a historical perspective of the origins of this field; (3) to give an idea of the state of the art in image processing by examining some of the principal areas in which it is applied; (4) to discuss briefly the principal approaches used in digital image processing; (5) to give an overview of the components contained in a typical, general-purpose image processing system; and (6) to provide direction to the books and other literature where image processing work normally is reported.

1.1 What Is Digital Image Processing?

An image may be defined as a two-dimensional function, $f(x, y)$, where x and y are *spatial* (plane) coordinates, and the amplitude of f at any pair of coordinates (x, y) is called the *intensity* or *gray level* of the image at that point. When x, y, and the amplitude values of f are all finite, discrete quantities, we call the image a *digital image*. The field of *digital image processing* refers to processing digital images by means of a digital computer. Note that a digital image is composed of a finite number of elements, each of which has a particular location and value. These elements are referred to as *picture elements*, *image elements*, *pels*, and *pixels*. *Pixel* is the term most widely used to denote the elements of a digital image. We consider these definitions in more formal terms in Chapter 2.

Vision is the most advanced of our senses, so it is not surprising that images play the single most important role in human perception. However, unlike humans, who are limited to the visual band of the electromagnetic (EM) spectrum, imaging machines cover almost the entire EM spectrum, ranging from gamma to radio waves. They can operate on images generated by sources that humans are not accustomed to associating with images. These include ultrasound, electron microscopy, and computer-generated images. Thus, digital image processing encompasses a wide and varied field of applications.

There is no general agreement among authors regarding where image processing stops and other related areas, such as image analysis and computer vision, start. Sometimes a distinction is made by defining image processing as a discipline in which both the input and output of a process are images. We believe this to be a limiting and somewhat artificial boundary. For example, under this definition, even the trivial task of computing the average intensity of an image (which yields a single number) would not be considered an image processing operation. On the other hand, there are fields such as computer vision whose ultimate goal is to use computers to emulate human vision, including learning and being able to make inferences and take actions based on visual inputs. This area itself is a branch of artificial intelligence (AI) whose objective is to emulate human intelligence. The field of AI is in its earliest stages of infancy in terms of development, with progress having been much slower than originally anticipated. The area of image analysis (also called image understanding) is in between image processing and computer vision.

There are no clear-cut boundaries in the continuum from image processing at one end to computer vision at the other. However, one useful paradigm is to consider three types of computerized processes in this continuum: low-, mid-, and high-level processes. Low-level processes involve primitive operations such as image preprocessing to reduce noise, contrast enhancement, and image sharpening. A low-level process is characterized by the fact that both its inputs and outputs are images. Mid-level processing on images involves tasks such as segmentation (partitioning an image into regions or objects), description of those objects to reduce them to a form suitable for computer processing, and classification (recognition) of individual objects. A mid-level process is characterized by the fact that its inputs generally are images, but its outputs are attributes extracted from those images (e.g., edges, contours, and the identity of individual objects). Finally, higher-level processing involves "making sense" of an ensemble of recognized objects, as in image analysis, and, at the far end of the continuum, performing the cognitive functions normally associated with vision.

Based on the preceding comments, we see that a logical place of overlap between image processing and image analysis is the area of recognition of individual regions or objects in an image. Thus, what we call in this book *digital image processing* encompasses processes whose inputs and outputs are images and, in addition, encompasses processes that extract attributes from images, up to and including the recognition of individual objects. As a simple illustration to clarify these concepts, consider the area of automated analysis of text. The processes of acquiring an image of the area containing the text, preprocessing that image, extracting (segmenting) the individual characters, describing the characters in a form suitable for computer processing, and recognizing those individual characters are in the scope of what we call digital image processing in this book. Making sense of the content of the page may be viewed as being in the domain of image analysis and even computer vision, depending on the level of complexity implied by the statement "making sense." As will become evident shortly, digital image processing, as we have defined it, is used successfully in a broad range of areas of exceptional social and economic value. The concepts developed in the following chapters are the foundation for the methods used in those application areas.

1.2   The Origins of Digital Image Processing

One of the first applications of digital images was in the newspaper industry, when pictures were first sent by submarine cable between London and New York. Introduction of the Bartlane cable picture transmission system in the early 1920s reduced the time required to transport a picture across the Atlantic from more than a week to less than three hours. Specialized printing equipment coded pictures for cable transmission and then reconstructed them at the receiving end. Figure 1.1 was transmitted in this way and reproduced on a telegraph printer fitted with typefaces simulating a halftone pattern.

Some of the initial problems in improving the visual quality of these early digital pictures were related to the selection of printing procedures and the distribution of intensity levels. The printing method used to obtain Fig. 1.1 was abandoned toward the end of 1921 in favor of a technique based on photographic reproduction made from tapes perforated at the telegraph receiving terminal. Figure 1.2 shows an image obtained using this method. The improvements over Fig. 1.1 are evident, both in tonal quality and in resolution.

FIGURE 1.1 A digital picture produced in 1921 from a coded tape by a telegraph printer with special type faces. (McFarlane.[†])

[†]References in the Bibliography at the end of the book are listed in alphabetical order by authors' last names.

FIGURE 1.2 A digital picture made in 1922 from a tape punched after the signals had crossed the Atlantic twice. Some errors are visible. (McFarlane.)

The early Bartlane systems were capable of coding images in five distinct levels of gray. This capability was increased to 15 levels in 1929. Figure 1.3 is typical of the type of images that could be obtained using the 15-tone equipment. During this period, introduction of a system for developing a film plate via light beams that were modulated by the coded picture tape improved the reproduction process considerably.

Although the examples just cited involve digital images, they are not considered digital image processing results in the context of our definition because computers were not involved in their creation. Thus, the history of digital image processing is intimately tied to the development of the digital computer. In fact, digital images require so much storage and computational power that progress in the field of digital image processing has been dependent on the development of digital computers and of supporting technologies that include data storage, display, and transmission.

The idea of a computer goes back to the invention of the abacus in Asia Minor, more than 5000 years ago. More recently, there were developments in the past two centuries that are the foundation of what we call a computer today. However, the basis for what we call a *modern* digital computer dates back to only the 1940s with the introduction by John von Neumann of two key concepts: (1) a memory to hold a stored program and data, and (2) conditional branching. These two ideas are the foundation of a central processing unit (CPU), which is at the heart of computers today. Starting with von Neumann, there were

FIGURE 1.3 Unretouched cable picture of Generals Pershing and Foch, transmitted in 1929 from London to New York by 15-tone equipment. (McFarlane.)

a series of key advances that led to computers powerful enough to be used for digital image processing. Briefly, these advances may be summarized as follows: (1) the invention of the transistor by Bell Laboratories in 1948; (2) the development in the 1950s and 1960s of the high-level programming languages COBOL (Common Business-Oriented Language) and FORTRAN (Formula Translator); (3) the invention of the integrated circuit (IC) at Texas Instruments in 1958; (4) the development of operating systems in the early 1960s; (5) the development of the microprocessor (a single chip consisting of the central processing unit, memory, and input and output controls) by Intel in the early 1970s; (6) introduction by IBM of the personal computer in 1981; and (7) progressive miniaturization of components, starting with large scale integration (LI) in the late 1970s, then very large scale integration (VLSI) in the 1980s, to the present use of ultra large scale integration (ULSI). Concurrent with these advances were developments in the areas of mass storage and display systems, both of which are fundamental requirements for digital image processing.

The first computers powerful enough to carry out meaningful image processing tasks appeared in the early 1960s. The birth of what we call digital image processing today can be traced to the availability of those machines and the onset of the space program during that period. It took the combination of those two developments to bring into focus the potential of digital image processing concepts. Work on using computer techniques for improving images from a space probe began at the Jet Propulsion Laboratory (Pasadena, California) in 1964 when pictures of the moon transmitted by *Ranger 7* were processed by a computer to correct various types of image distortion inherent in the on-board television camera. Figure 1.4 shows the first image of the moon taken by *Ranger 7* on July 31, 1964 at 9:09 A.M. Eastern Daylight Time (EDT), about 17 minutes before impacting the lunar surface (the markers, called *reseau* marks, are used for geometric corrections, as discussed in Chapter 5). This also is the first image of the moon taken by a U.S. spacecraft. The imaging lessons learned with *Ranger 7* served as the basis for improved methods used to enhance and restore images from the Surveyor missions to the moon, the Mariner series of flyby missions to Mars, the Apollo manned flights to the moon, and others.

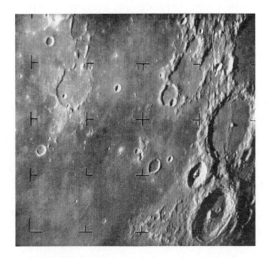

FIGURE 1.4 The first picture of the moon by a U.S. spacecraft. *Ranger 7* took this image on July 31, 1964 at 9:09 A.M. EDT, about 17 minutes before impacting the lunar surface. (Courtesy of NASA.)

In parallel with space applications, digital image processing techniques began in the late 1960s and early 1970s to be used in medical imaging, remote Earth resources observations, and astronomy. The invention in the early 1970s of computerized axial tomography (CAT), also called computerized tomography (CT) for short, is one of the most important events in the application of image processing in medical diagnosis. Computerized axial tomography is a process in which a ring of detectors encircles an object (or patient) and an X-ray source, concentric with the detector ring, rotates about the object. The X-rays pass through the object and are collected at the opposite end by the corresponding detectors in the ring. As the source rotates, this procedure is repeated. Tomography consists of algorithms that use the sensed data to construct an image that represents a "slice" through the object. Motion of the object in a direction perpendicular to the ring of detectors produces a set of such slices, which constitute a three-dimensional (3-D) rendition of the inside of the object. Tomography was invented independently by Sir Godfrey N. Hounsfield and Professor Allan M. Cormack, who shared the 1979 Nobel Prize in Medicine for their invention. It is interesting to note that X-rays were discovered in 1895 by Wilhelm Conrad Roentgen, for which he received the 1901 Nobel Prize for Physics. These two inventions, nearly 100 years apart, led to some of the most active application areas of image processing today.

From the 1960s until the present, the field of image processing has grown vigorously. In addition to applications in medicine and the space program, digital image processing techniques now are used in a broad range of applications. Computer procedures are used to enhance the contrast or code the intensity levels into color for easier interpretation of X-rays and other images used in industry, medicine, and the biological sciences. Geographers use the same or similar techniques to study pollution patterns from aerial and satellite imagery. Image enhancement and restoration procedures are used to process degraded images of unrecoverable objects or experimental results too expensive to duplicate. In archeology, image processing methods have successfully restored blurred pictures that were the only available records of rare artifacts lost or damaged after being photographed. In physics and related fields, computer techniques routinely enhance images of experiments in areas such as high-energy plasmas and electron microscopy. Similarly successful applications of image processing concepts can be found in astronomy, biology, nuclear medicine, law enforcement, defense, and industrial applications.

These examples illustrate processing results intended for human interpretation. The second major area of application of digital image processing techniques mentioned at the beginning of this chapter is in solving problems dealing with machine perception. In this case, interest focuses on procedures for extracting from an image information in a form suitable for computer processing. Often, this information bears little resemblance to visual features that humans use in interpreting the content of an image. Examples of the type of information used in machine perception are statistical moments, Fourier transform coefficients, and multidimensional distance measures. Typical problems in machine perception that routinely utilize image processing techniques are automatic character recognition, industrial machine vision for product assembly and inspection, military recognizance, automatic processing of fingerprints, screening of X-rays and blood samples, and machine processing of aerial and satellite imagery for weather prediction and environmental assessment. The continuing decline in the ratio of computer price to performance and the expansion of networking and communication bandwidth via the World Wide Web and the Internet have created unprecedented opportunities for continued growth of digital image processing. Some of these application areas are illustrated in the following section.

1.3 Examples of Fields that Use Digital Image Processing

Today, there is almost no area of technical endeavor that is not impacted in some way by digital image processing. We can cover only a few of these applications in the context and space of the current discussion. However, limited as it is, the material presented in this section will leave no doubt in the reader's mind regarding the breadth and importance of digital image processing. We show in this section numerous areas of application, each of which routinely utilizes the digital image processing techniques developed in the following chapters. Many of the images shown in this section are used later in one or more of the examples given in the book. All images shown are digital.

The areas of application of digital image processing are so varied that some form of organization is desirable in attempting to capture the breadth of this field. One of the simplest ways to develop a basic understanding of the extent of image processing applications is to categorize images according to their source (e.g., visual, X-ray, and so on). The principal energy source for images in use today is the electromagnetic energy spectrum. Other important sources of energy include acoustic, ultrasonic, and electronic (in the form of electron beams used in electron microscopy). Synthetic images, used for modeling and visualization, are generated by computer. In this section we discuss briefly how images are generated in these various categories and the areas in which they are applied. Methods for converting images into digital form are discussed in the next chapter.

Images based on radiation from the EM spectrum are the most familiar, especially images in the X-ray and visual bands of the spectrum. Electromagnetic waves can be conceptualized as propagating sinusoidal waves of varying wavelengths, or they can be thought of as a stream of massless particles, each traveling in a wavelike pattern and moving at the speed of light. Each massless particle contains a certain amount (or bundle) of energy. Each bundle of energy is called a *photon*. If spectral bands are grouped according to energy per photon, we obtain the spectrum shown in Fig. 1.5, ranging from gamma rays (highest energy) at one end to radio waves (lowest energy) at the other. The bands are shown shaded to convey the fact that bands of the EM spectrum are not distinct but rather transition smoothly from one to the other.

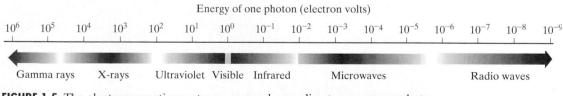

Energy of one photon (electron volts)

10^6 10^5 10^4 10^3 10^2 10^1 10^0 10^{-1} 10^{-2} 10^{-3} 10^{-4} 10^{-5} 10^{-6} 10^{-7} 10^{-8} 10^{-9}

Gamma rays   X-rays   Ultraviolet Visible Infrared   Microwaves   Radio waves

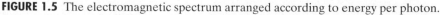

FIGURE 1.5 The electromagnetic spectrum arranged according to energy per photon.

1.3.1 Gamma-Ray Imaging

Major uses of imaging based on gamma rays include nuclear medicine and astronomical observations. In nuclear medicine, the approach is to inject a patient with a radioactive isotope that emits gamma rays as it decays. Images are produced from the emissions collected by gamma ray detectors. Figure 1.6(a) shows an image of a complete bone scan obtained by using gamma-ray imaging. Images of this sort are used to locate sites of bone pathology, such as infections or tumors. Figure 1.6(b) shows another major modality of nuclear imaging called positron emission tomography (PET). The principle is the same

a b
c d

FIGURE 1.6
Examples of gamma-ray imaging. (a) Bone scan. (b) PET image. (c) Cygnus Loop. (d) Gamma radiation (bright spot) from a reactor valve. (Images courtesy of (a) G.E. Medical Systems, (b) Dr. Michael E. Casey, CTI PET Systems, (c) NASA, (d) Professors Zhong He and David K. Wehe, University of Michigan.)

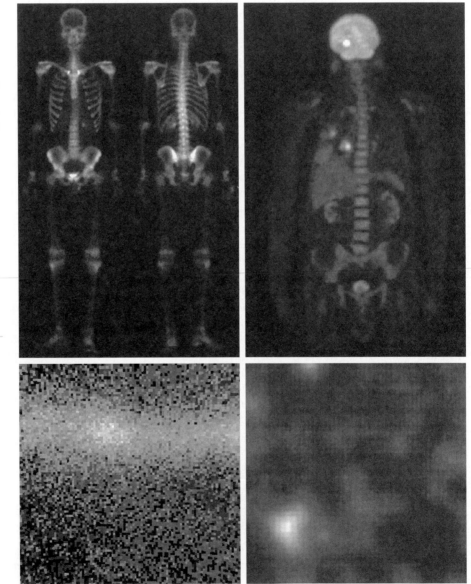

as with X-ray tomography, mentioned briefly in Section 1.2. However, instead of using an external source of X-ray energy, the patient is given a radioactive isotope that emits positrons as it decays. When a positron meets an electron, both are annihilated and two gamma rays are given off. These are detected and a tomographic image is created using the basic principles of tomography. The image shown in Fig. 1.6(b) is one sample of a sequence that constitutes a 3-D rendition of the patient. This image shows a tumor in the brain and one in the lung, easily visible as small white masses.

A star in the constellation of Cygnus exploded about 15,000 years ago, generating a superheated stationary gas cloud (known as the Cygnus Loop) that glows in a spectacular array of colors. Figure 1.6(c) shows the Cygnus Loop imaged in the gamma-ray band. Unlike the two examples shown in Figs. 1.6(a) and (b), this image was obtained using the natural radiation of the object being imaged. Finally, Fig. 1.6(d) shows an image of gamma radiation from a valve in a nuclear reactor. An area of strong radiation is seen in the lower, left side of the image.

1.3.2 X-ray Imaging

X-rays are among the oldest sources of EM radiation used for imaging. The best known use of X-rays is medical diagnostics, but they also are used extensively in industry and other areas, like astronomy. X-rays for medical and industrial imaging are generated using an X-ray tube, which is a vacuum tube with a cathode and anode. The cathode is heated, causing free electrons to be released. These electrons flow at high speed to the positively charged anode. When the electrons strike a nucleus, energy is released in the form of X-ray radiation. The energy (penetrating power) of the X-rays is controlled by a voltage applied across the anode, and the number of X-rays is controlled by a current applied to the filament in the cathode. Figure 1.7(a) shows a familiar chest X ray generated simply by placing the patient between an X-ray source and a film sensitive to X-ray energy. The intensity of the X-rays is modified by absorption as they pass through the patient, and the resulting energy falling on the film develops it, much in the same way that light develops photographic film. In digital radiography, digital images are obtained by one of two methods: (1) by digitizing X-ray films; or (2) by having the X-rays that pass through the patient fall directly onto devices (such as a phosphor screen) that convert X-rays to light. The light signal in turn is captured by a light-sensitive digitizing system. We discuss digitization in detail in Chapter 2.

Angiography is another major application in an area called contrast-enhancement radiography. This procedure is used to obtain images (called *angiograms*) of blood vessels. A catheter (a small, flexible, hollow tube) is inserted, for example, into an artery or vein in the groin. The catheter is threaded into the blood vessel and guided to the area to be studied. When the catheter reaches the site under investigation, an X-ray contrast medium is injected through the catheter. This enhances contrast of the blood vessels and enables the radiologist to see any irregularities or blockages. Figure 1.7(b) shows an example of an aortic angiogram. The catheter can be seen being inserted into the large blood vessel on the lower left of the picture. Note the high contrast of the

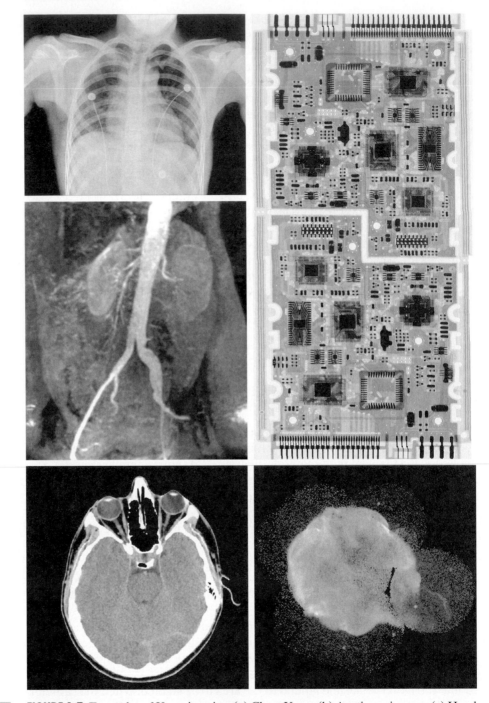

a	d
b	
c	e

FIGURE 1.7 Examples of X-ray imaging. (a) Chest X-ray. (b) Aortic angiogram. (c) Head CT. (d) Circuit boards. (e) Cygnus Loop. (Images courtesy of (a) and (c) Dr. David R. Pickens, Dept. of Radiology & Radiological Sciences, Vanderbilt University Medical Center, (b) Dr. Thomas R. Gest, Division of Anatomical Sciences, University of Michigan Medical School, (d) Mr. Joseph E. Pascente, Lixi, Inc., and (e) NASA.)

large vessel as the contrast medium flows up in the direction of the kidneys, which are also visible in the image. As discussed in Chapter 3, angiography is a major area of digital image processing, where image subtraction is used to enhance further the blood vessels being studied.

Perhaps the best known of all uses of X-rays in medical imaging is computerized axial tomography. Due to their resolution and 3-D capabilities, CAT scans revolutionized medicine from the moment they first became available in the early 1970s. As noted in Section 1.2, each CAT image is a "slice" taken perpendicularly through the patient. Numerous slices are generated as the patient is moved in a longitudinal direction. The ensemble of such images constitutes a 3-D rendition of the inside of the patient, with the longitudinal resolution being proportional to the number of slice images taken. Figure 1.7(c) shows a typical head CAT slice image.

Techniques similar to the ones just discussed, but generally involving higher-energy X-rays, are applicable in industrial processes. Figure 1.7(d) shows an X-ray image of an electronic circuit board. Such images, representative of literally hundreds of industrial applications of X-rays, are used to examine circuit boards for flaws in manufacturing, such as missing components or broken traces. Industrial CAT scans are useful when the parts can be penetrated by X-rays, such as in plastic assemblies, and even large bodies, like solid-propellant rocket motors. Figure 1.7(e) shows an example of X-ray imaging in astronomy. This image is the Cygnus Loop of Fig. 1.6(c), but imaged this time in the X-ray band.

1.3.3 Imaging in the Ultraviolet Band

Applications of ultraviolet "light" are varied. They include lithography, industrial inspection, microscopy, lasers, biological imaging, and astronomical observations. We illustrate imaging in this band with examples from microscopy and astronomy.

Ultraviolet light is used in fluorescence microscopy, one of the fastest growing areas of microscopy. Fluorescence is a phenomenon discovered in the middle of the nineteenth century, when it was first observed that the mineral fluorspar fluoresces when ultraviolet light is directed upon it. The ultraviolet light itself is not visible, but when a photon of ultraviolet radiation collides with an electron in an atom of a fluorescent material, it elevates the electron to a higher energy level. Subsequently, the excited electron relaxes to a lower level and emits light in the form of a lower-energy photon in the visible (red) light region. The basic task of the fluorescence microscope is to use an excitation light to irradiate a prepared specimen and then to separate the much weaker radiating fluorescent light from the brighter excitation light. Thus, only the emission light reaches the eye or other detector. The resulting fluorescing areas shine against a dark background with sufficient contrast to permit detection. The darker the background of the nonfluorescing material, the more efficient the instrument.

Fluorescence microscopy is an excellent method for studying materials that can be made to fluoresce, either in their natural form (primary fluorescence) or when treated with chemicals capable of fluorescing (secondary fluorescence). Figures 1.8(a) and (b) show results typical of the capability of fluorescence

a b
c

FIGURE 1.8
Examples of
ultraviolet
imaging.
(a) Normal corn.
(b) Smut corn.
(c) Cygnus Loop.
(Images courtesy
of (a) and
(b) Dr. Michael
W. Davidson,
Florida State
University,
(c) NASA.)

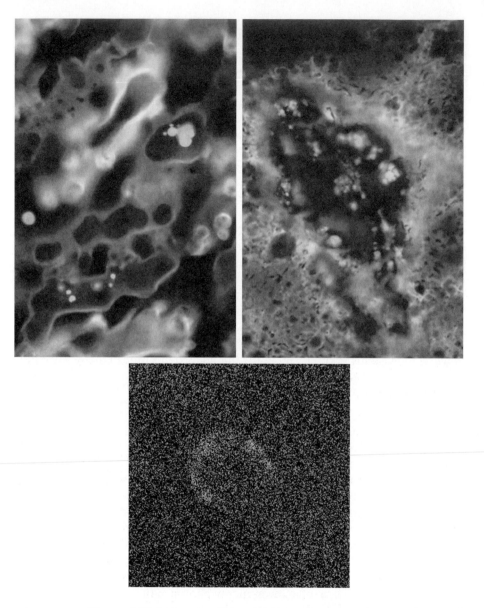

microscopy. Figure 1.8(a) shows a fluorescence microscope image of normal corn, and Fig. 1.8(b) shows corn infected by "smut," a disease of cereals, corn, grasses, onions, and sorghum that can be caused by any of more than 700 species of parasitic fungi. Corn smut is particularly harmful because corn is one of the principal food sources in the world. As another illustration, Fig. 1.8(c) shows the Cygnus Loop imaged in the high-energy region of the ultraviolet band.

1.3.4 Imaging in the Visible and Infrared Bands

Considering that the visual band of the electromagnetic spectrum is the most familiar in all our activities, it is not surprising that imaging in this band outweighs by far all the others in terms of scope of application. The infrared band often is used in conjunction with visual imaging, so we have grouped the visible and infrared bands in this section for the purpose of illustration. We consider in the following discussion applications in light microscopy, astronomy, remote sensing, industry, and law enforcement.

Figure 1.9 shows several examples of images obtained with a light microscope. The examples range from pharmaceuticals and microinspection to materials characterization. Even in just microscopy, the application areas are too numerous to detail here. It is not difficult to conceptualize the types of processes one might apply to these images, ranging from enhancement to measurements.

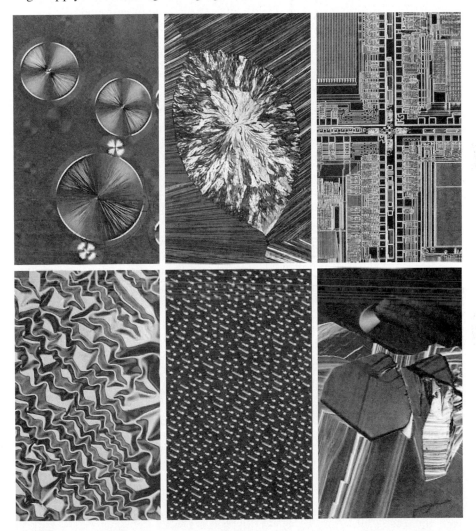

a b c
d e f

FIGURE 1.9 Examples of light microscopy images. (a) Taxol (anticancer agent), magnified 250×. (b) Cholesterol—40×. (c) Microprocessor—60×. (d) Nickel oxide thin film—600 ×. (e) Surface of audio CD—1750×. (f) Organic superconductor—450×. (Images courtesy of Dr. Michael W. Davidson, Florida State University.)

TABLE 1.1
Thematic bands
in NASA's
LANDSAT
satellite.

Band No.	Name	Wavelength (μm)	Characteristics and Uses
1	Visible blue	0.45–0.52	Maximum water penetration
2	Visible green	0.52–0.60	Good for measuring plant vigor
3	Visible red	0.63–0.69	Vegetation discrimination
4	Near infrared	0.76–0.90	Biomass and shoreline mapping
5	Middle infrared	1.55–1.75	Moisture content of soil and vegetation
6	Thermal infrared	10.4–12.5	Soil moisture; thermal mapping
7	Middle infrared	2.08–2.35	Mineral mapping

Another major area of visual processing is remote sensing, which usually includes several bands in the visual and infrared regions of the spectrum. Table 1.1 shows the so-called *thematic bands* in NASA's LANDSAT satellite. The primary function of LANDSAT is to obtain and transmit images of the Earth from space, for purposes of monitoring environmental conditions on the planet. The bands are expressed in terms of wavelength, with 1 μm being equal to 10^{-6} m (we discuss the wavelength regions of the electromagnetic spectrum in more detail in Chapter 2). Note the characteristics and uses of each band.

In order to develop a basic appreciation for the power of this type of *multispectral* imaging, consider Fig. 1.10, which shows one image for each of the spec-

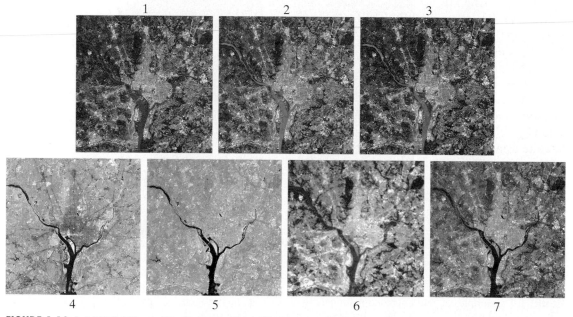

FIGURE 1.10 LANDSAT satellite images of the Washington, D.C. area. The numbers refer to the thematic bands in Table 1.1. (Images courtesy of NASA.)

FIGURE 1.11
Multispectral
image of
Hurricane
Andrew taken by
NOAA GEOS
(Geostationary
Environmental
Operational
Satellite) sensors.
(Courtesy of
NOAA.)

tral bands in Table 1.1. The area imaged is Washington D.C., which includes features such as buildings, roads, vegetation, and a major river (the Potomac) going though the city. Images of population centers are used routinely (over time) to assess population growth and shift patterns, pollution, and other factors harmful to the environment. The differences between visual and infrared image features are quite noticeable in these images. Observe, for example, how well defined the river is from its surroundings in Bands 4 and 5.

Weather observation and prediction also are major applications of multispectral imaging from satellites. For example, Fig. 1.11 is an image of a hurricane taken by a National Oceanographic and Atmospheric Administration (NOAA) satellite using sensors in the visible and infrared bands. The eye of the hurricane is clearly visible in this image.

Figures 1.12 and 1.13 show an application of infrared imaging. These images are part of the *Nighttime Lights of the World* data set, which provides a global inventory of human settlements. The images were generated by the infrared imaging system mounted on a NOAA DMSP (Defense Meteorological Satellite Program) satellite. The infrared imaging system operates in the band 10.0 to 13.4 μm, and has the unique capability to observe faint sources of visible-near infrared emissions present on the Earth's surface, including cities, towns, villages, gas flares, and fires. Even without formal training in image processing, it is not difficult to imagine writing a computer program that would use these images to estimate the percent of total electrical energy used by various regions of the world.

FIGURE 1.12
Infrared satellite
images of the
Americas. The
small gray map is
provided for
reference.
(Courtesy of
NOAA.)

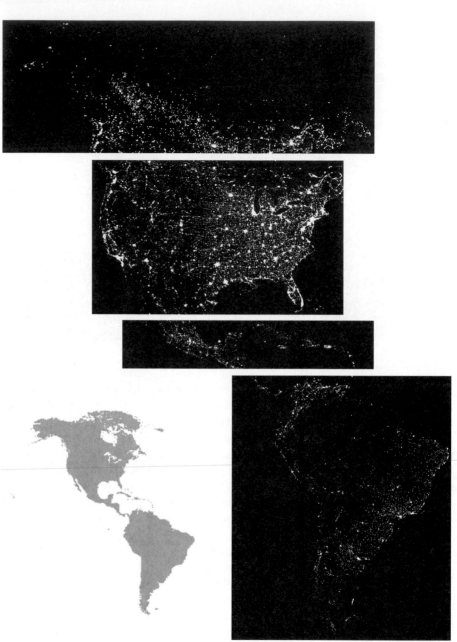

A major area of imaging in the visual spectrum is in automated visual inspection of manufactured goods. Figure 1.14 shows some examples. Figure 1.14(a) is a controller board for a CD-ROM drive. A typical image processing task with products like this is to inspect them for missing parts (the black square on the top, right quadrant of the image is an example of a missing component). Figure 1.14(b) is an imaged pill container. The objective here is to have a machine look for missing pills. Figure 1.14(c) shows an application in which image processing is used to look for bottles that are not filled up to an acceptable level. Figure 1.14(d) shows

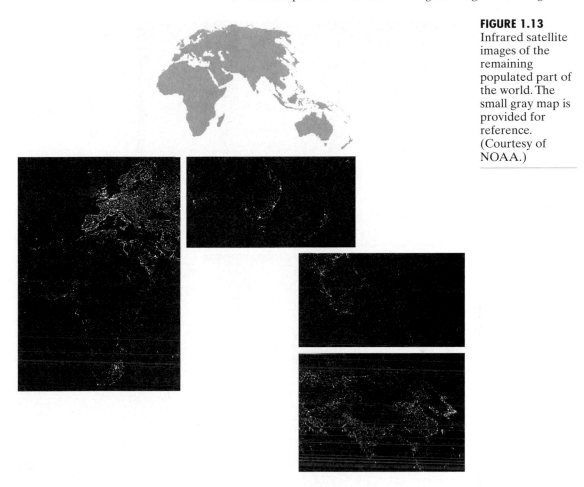

FIGURE 1.13
Infrared satellite images of the remaining populated part of the world. The small gray map is provided for reference. (Courtesy of NOAA.)

a clear-plastic part with an unacceptable number of air pockets in it. Detecting anomalies like these is a major theme of industrial inspection that includes other products such as wood and cloth. Figure 1.14(e) shows a batch of cereal during inspection for color and the presence of anomalies such as burned flakes. Finally, Fig. 1.14(f) shows an image of an intraocular implant (replacement lens for the human eye). A "structured light" illumination technique was used to highlight for easier detection flat lens deformations toward the center of the lens. The markings at 1 o'clock and 5 o'clock are tweezer damage. Most of the other small speckle detail is debris. The objective in this type of inspection is to find damaged or incorrectly manufactured implants automatically, prior to packaging.

As a final illustration of image processing in the visual spectrum, consider Fig. 1.15. Figure 1.15(a) shows a thumb print. Images of fingerprints are routinely processed by computer, either to enhance them or to find features that aid in the automated search of a database for potential matches. Figure 1.15(b) shows an image of paper currency. Applications of digital image processing in this area include automated counting and, in law enforcement, the reading of the serial number for the purpose of tracking and identifying bills. The two vehicle images shown in Figs. 1.15 (c) and (d) are examples of automated license plate reading.

a b
c d
e f

FIGURE 1.14
Some examples of
manufactured
goods often
checked using
digital image
processing. (a) A
circuit board
controller.
(b) Packaged pills.
(c) Bottles.
(d) Bubbles in
clear-plastic
product.
(e) Cereal.
(f) Image of
intraocular
implant.
(Fig. (f) courtesy
of Mr. Pete Sites,
Perceptics
Corporation.)

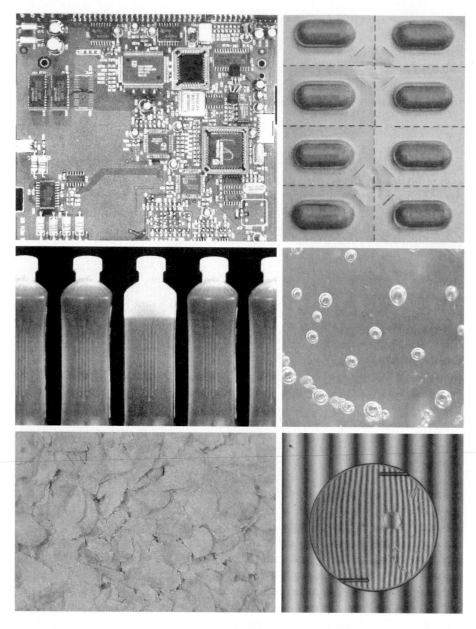

The light rectangles indicate the area in which the imaging system detected the
plate. The black rectangles show the results of automated reading of the plate
content by the system. License plate and other applications of character recognition are used extensively for traffic monitoring and surveillance.

1.3.5 Imaging in the Microwave Band

The dominant application of imaging in the microwave band is radar. The unique
feature of imaging radar is its ability to collect data over virtually any region at
any time, regardless of weather or ambient lighting conditions. Some radar

FIGURE 1.15
Some additional examples of imaging in the visual spectrum. (a) Thumb print. (b) Paper currency. (c) and (d). Automated license plate reading. (Figure (a) courtesy of the National Institute of Standards and Technology. Figures (c) and (d) courtesy of Dr. Juan Herrera, Perceptics Corporation.)

waves can penetrate clouds, and under certain conditions can also see through vegetation, ice, and extremely dry sand. In many cases, radar is the only way to explore inaccessible regions of the Earth's surface. An imaging radar works like a flash camera in that it provides its own illumination (microwave pulses) to illuminate an area on the ground and take a snapshot image. Instead of a camera lens, a radar uses an antenna and digital computer processing to record its images. In a radar image, one can see only the microwave energy that was reflected back toward the radar antenna.

Figure 1.16 shows a spaceborne radar image covering a rugged mountainous area of southeast Tibet, about 90 km east of the city of Lhasa. In the lower right corner is a wide valley of the Lhasa River, which is populated by Tibetan farmers and yak herders and includes the village of Menba. Mountains in this area reach about 5800 m (19,000 ft) above sea level, while the valley floors lie about 4300 m (14,000 ft) above sea level. Note the clarity and detail of the image, unencumbered by clouds or other atmospheric conditions that normally interfere with images in the visual band.

FIGURE 1.16
Spaceborne radar
image of
mountains in
southeast Tibet.
(Courtesy of
NASA.)

1.3.6 Imaging in the Radio Band

As in the case of imaging at the other end of the spectrum (gamma rays), the major applications of imaging in the radio band are in medicine and astronomy. In medicine radio waves are used in magnetic resonance imaging (MRI). This technique places a patient in a powerful magnet and passes radio waves through his or her body in short pulses. Each pulse causes a responding pulse of radio waves to be emitted by the patient's tissues. The location from which these signals originate and their strength are determined by a computer, which produces a two-dimensional picture of a section of the patient. MRI can produce pictures in any plane. Figure 1.17 shows MRI images of a human knee and spine.

The last image to the right in Fig. 1.18 shows an image of the Crab Pulsar in the radio band. Also shown for an interesting comparison are images of the same region but taken in most of the bands discussed earlier. Note that each image gives a totally different "view" of the Pulsar.

1.3.7 Examples in which Other Imaging Modalities Are Used

Although imaging in the electromagnetic spectrum is dominant by far, there are a number of other imaging modalities that also are important. Specifically, we discuss in this section acoustic imaging, electron microscopy, and synthetic (computer-generated) imaging.

Imaging using "sound" finds application in geological exploration, industry, and medicine. Geological applications use sound in the low end of the sound spectrum (hundreds of Hertz) while imaging in other areas use ultrasound (millions of Hertz). The most important commercial applications of image processing in geology are in mineral and oil exploration. For image acquisition over land, one of the main approaches is to use a large truck and a large flat steel plate. The plate is pressed on the ground by the truck, and the truck is vibrated through a fre-

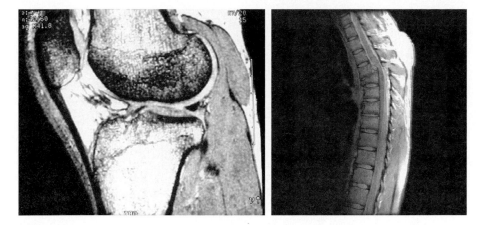

a b

FIGURE 1.17 MRI images of a human (a) knee, and (b) spine. (Image (a) courtesy of Dr. Thomas R. Gest, Division of Anatomical Sciences, University of Michigan Medical School, and (b) Dr. David R. Pickens, Department of Radiology and Radiological Sciences, Vanderbilt University Medical Center.)

quency spectrum up to 100 Hz. The strength and speed of the returning sound waves are determined by the composition of the earth below the surface. These are analyzed by computer, and images are generated from the resulting analysis.

For marine acquisition, the energy source consists usually of two air guns towed behind a ship. Returning sound waves are detected by hydrophones placed in cables that are either towed behind the ship, laid on the bottom of the ocean, or hung from buoys (vertical cables). The two air guns are alternately pressurized to ~2000 psi and then set off. The constant motion of the ship provides a transversal direction of motion that, together with the returning sound waves, is used to generate a 3-D map of the composition of the Earth below the bottom of the ocean.

Figure 1.19 shows a cross-sectional image of a well-known 3-D model against which the performance of seismic imaging algorithms is tested. The arrow points to a hydrocarbon (oil and/or gas) trap. This target is brighter than the surrounding layers because of the change in density in the target region is larger.

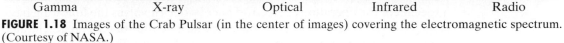

| Gamma | X-ray | Optical | Infrared | Radio |

FIGURE 1.18 Images of the Crab Pulsar (in the center of images) covering the electromagnetic spectrum. (Courtesy of NASA.)

FIGURE 1.19
Cross-sectional
image of a seismic
model. The arrow
points to a
hydrocarbon (oil
and/or gas) trap.
(Courtesy of
Dr. Curtis Ober,
Sandia National
Laboratories.)

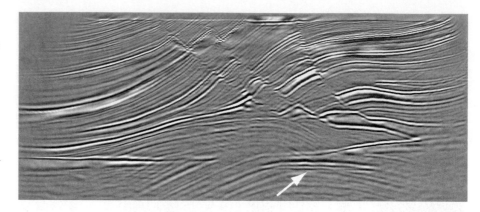

Seismic interpreters look for these "bright spots" to find oil and gas. The layers above also are bright, but their brightness does not vary as strongly across the layers. Many seismic reconstruction algorithms have difficulty imaging this target because of the faults above it.

Although ultrasound imaging is used routinely in manufacturing, the best known applications of this technique are in medicine, especially in obstetrics, where unborn babies are imaged to determine the health of their development. A byproduct of this examination is determining the sex of the baby. Ultrasound images are generated using the following basic procedure:

1. The ultrasound system (a computer, ultrasound probe consisting of a source and receiver, and a display) transmits high-frequency (1 to 5 MHz) sound pulses into the body.
2. The sound waves travel into the body and hit a boundary between tissues (e.g., between fluid and soft tissue, soft tissue and bone). Some of the sound waves are reflected back to the probe, while some travel on further until they reach another boundary and get reflected.
3. The reflected waves are picked up by the probe and relayed to the computer.
4. The machine calculates the distance from the probe to the tissue or organ boundaries using the speed of sound in tissue (1540 m/s) and the time of the each echo's return.
5. The system displays the distances and intensities of the echoes on the screen, forming a two-dimensional image.

In a typical ultrasound image, millions of pulses and echoes are sent and received each second. The probe can be moved along the surface of the body and angled to obtain various views. Figure 1.20 shows several examples.

We continue the discussion on imaging modalities with some examples of electron microscopy. Electron microscopes function as their optical counterparts, except that they use a focused beam of electrons instead of light to image a specimen. The operation of electron microscopes involves the following basic steps: A stream of electrons is produced by an electron source and accelerated toward the specimen using a positive electrical potential. This stream is con-

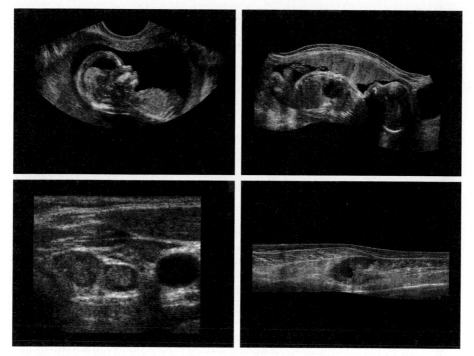

FIGURE 1.20
Examples of
ultrasound
imaging. (a) Baby.
(b) Another view
of baby.
(c) Thyroids.
(d) Muscle layers
showing lesion.
(Courtesy of
Siemens Medical
Systems, Inc.,
Ultrasound
Group.)

fined and focused using metal apertures and magnetic lenses into a thin, focused, monochromatic beam. This beam is focused onto the sample using a magnetic lens. Interactions occur inside the irradiated sample, affecting the electron beam. These interactions and effects are detected and transformed into an image, much in the same way that light is reflected from, or absorbed by, objects in a scene. These basic steps are carried out in all electron microscopes, regardless of type.

A *transmission electron microscope* (TEM) works much like a slide projector. A projector shines (transmits) a beam of light through the slide; as the light passes through the slide, it is affected by the contents of the slide. This transmitted beam is then projected onto the viewing screen, forming an enlarged image of the slide. TEMs work the same way, except that they shine a beam of electrons through a specimen (analogous to the slide). The fraction of the beam transmitted through the specimen is projected onto a phosphor screen. The interaction of the electrons with the phosphor produces light and, therefore, a viewable image. A *scanning electron microscope* (SEM), on the other hand, actually scans the electron beam and records the interaction of beam and sample at each location. This produces one dot on a phosphor screen. A complete image is formed by a raster scan of the bean through the sample, much like a TV camera. The electrons interact with a phosphor screen and produce light. SEMs are suitable for "bulky" samples, while TEMs require very thin samples.

Electron microscopes are capable of very high magnification. While light microscopy is limited to magnifications on the order 1000×, electron microscopes

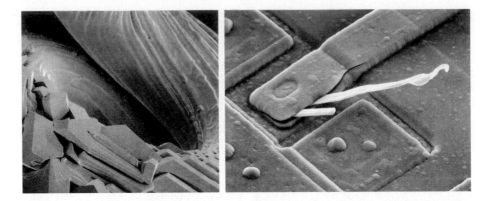

a b

FIGURE 1.21 (a) 250× SEM image of a tungsten filament following thermal failure. (b) 2500× SEM image of damaged integrated circuit. The white fibers are oxides resulting from thermal destruction. (Figure (a) courtesy of Mr. Michael Shaffer, Department of Geological Sciences, University of Oregon, Eugene; (b) courtesy of Dr. J. M. Hudak, McMaster University, Hamilton, Ontario, Canada.)

can achieve magnification of 10,000× or more. Figure 1.21 shows two SEM images of specimen failures due to thermal overload.

We conclude the discussion of imaging modalities by looking briefly at images that are not obtained from physical objects. Instead, they are generated by computer. *Fractals* are striking examples of computer-generated images (Lu [1997]). Basically, a fractal is nothing more than an iterative reproduction of a basic pattern according to some mathematical rules. For instance, *tiling* is one of the simplest ways to generate a fractal image. A square can be subdivided into four square subregions, each of which can be further subdivided into four smaller square regions, and so on. Depending on the complexity of the rules for filling each subsquare, some beautiful tile images can be generated using this method. Of course, the geometry can be arbitrary. For instance, the fractal image could be grown radially out of a center point. Figure 1.22(a) shows a fractal grown in this way. The reader will recognize this image as the theme image used in the beginning page of each chapter in this book, selected because of its artistic simplicity and abstract analogy to a human eye. Figure 1.22(b) shows another fractal (a "moonscape") that provides an interesting analogy to the images of space used as illustrations in some of the preceding sections.

Fractal images tend toward artistic, mathematical formulations of "growth" of subimage elements according to some rules. They are useful sometimes as random textures. A more structured approach to image generation by computer lies in 3-D modeling. This is an area that provides an important intersection between image processing and computer graphics and is the basis for many 3-D visualization systems (e.g., flight simulators). Figures 1.22(c) and (d) show examples of computer-generated images. Since the original object is created in 3-D, images can be generated in any perspective from plane projections of the 3-D volume. Images of this type can be used for medical training and for a host of other applications, such as criminal forensics and special effects.

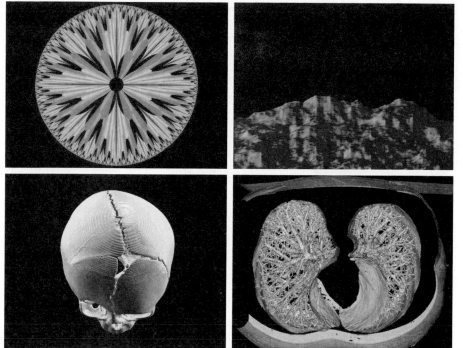

a b
c d

FIGURE 1.22
(a) and (b) Fractal images. (c) and (d) Images generated from 3-D computer models of the objects shown. (Figures (a) and (b) courtesy of Ms. Melissa D. Binde, Swarthmore College, (c) and (d) courtesy of NASA.)

1.4 Fundamental Steps in Digital Image Processing

It is helpful to divide the material covered in the following chapters into the two broad categories defined in Section 1.1: methods whose input and output are images, and methods whose inputs may be images, but whose outputs are attributes extracted from those images. This organization is summarized in Fig. 1.23. The diagram does not imply that every process is applied to an image. Rather, the intention is to convey an idea of all the methodologies that can be applied to images for different purposes and possibly with different objectives. The discussion in this section may be viewed as a brief overview of the material in the remainder of the book.

Image acquisition is the first process shown in Fig. 1.23. The discussion in Section 1.3 gave some hints regarding the origin of digital images. This topic is considered in much more detail in Chapter 2, where we also introduce a number of basic digital image concepts that are used throughout the book. Note that acquisition could be as simple as being given an image that is already in digital form. Generally, the image acquisition stage involves preprocessing, such as scaling.

Image enhancement is among the simplest and most appealing areas of digital image processing. Basically, the idea behind enhancement techniques is to bring out detail that is obscured, or simply to highlight certain features of interest in an image. A familiar example of enhancement is when we increase the contrast of an image because "it looks better." It is important to keep in mind that

FIGURE 1.23
Fundamental
steps in digital
image processing.

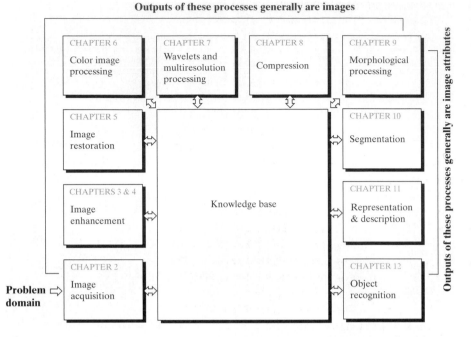

enhancement is a very subjective area of image processing. Two chapters are devoted to enhancement, not because it is more important than the other topics covered in the book but because we use enhancement as an avenue to introduce the reader to techniques that are used in other chapters as well. Thus, rather than having a chapter dedicated to mathematical preliminaries, we introduce a number of needed mathematical concepts by showing how they apply to enhancement. This approach allows the reader to gain familiarity with these concepts in the context of image processing. A good example of this is the Fourier transform, which is introduced in Chapter 4 but is used also in several of the other chapters.

Image restoration is an area that also deals with improving the appearance of an image. However, unlike enhancement, which is subjective, image restoration is objective, in the sense that restoration techniques tend to be based on mathematical or probabilistic models of image degradation. Enhancement, on the other hand, is based on human subjective preferences regarding what constitutes a "good" enhancement result.

Color image processing is an area that has been gaining in importance because of the significant increase in the use of digital images over the Internet. Chapter 5 covers a number of fundamental concepts in color models and basic color processing in a digital domain. Color is used also in later chapters as the basis for extracting features of interest in an image.

Wavelets are the foundation for representing images in various degrees of resolution. In particular, this material is used in this book for image data compression and for pyramidal representation, in which images are subdivided successively into smaller regions.

Compression, as the name implies, deals with techniques for reducing the storage required to save an image, or the bandwidth required to transmit it. Although storage technology has improved significantly over the past decade, the same cannot be said for transmission capacity. This is true particularly in uses of the Internet, which are characterized by significant pictorial content. Image compression is familiar (perhaps inadvertently) to most users of computers in the form of image file extensions, such as the jpg file extension used in the JPEG (Joint Photographic Experts Group) image compression standard.

Morphological processing deals with tools for extracting image components that are useful in the representation and description of shape. The material in this chapter begins a transition from processes that output images to processes that output image attributes, as indicated in Section 1.1.

Segmentation procedures partition an image into its constituent parts or objects. In general, autonomous segmentation is one of the most difficult tasks in digital image processing. A rugged segmentation procedure brings the process a long way toward successful solution of imaging problems that require objects to be identified individually. On the other hand, weak or erratic segmentation algorithms almost always guarantee eventual failure. In general, the more accurate the segmentation, the more likely recognition is to succeed.

Representation and description almost always follow the output of a segmentation stage, which usually is raw pixel data, constituting either the boundary of a region (i.e., the set of pixels separating one image region from another) or all the points in the region itself. In either case, converting the data to a form suitable for computer processing is necessary. The first decision that must be made is whether the data should be represented as a boundary or as a complete region. Boundary representation is appropriate when the focus is on external shape characteristics, such as corners and inflections. Regional representation is appropriate when the focus is on internal properties, such as texture or skeletal shape. In some applications, these representations complement each other. Choosing a representation is only part of the solution for transforming raw data into a form suitable for subsequent computer processing. A method must also be specified for describing the data so that features of interest are highlighted. *Description*, also called *feature selection*, deals with extracting attributes that result in some quantitative information of interest or are basic for differentiating one class of objects from another.

Recognition is the process that assigns a label (e.g., "vehicle") to an object based on its descriptors. As detailed in Section 1.1, we conclude our coverage of digital image processing with the development of methods for recognition of individual objects.

So far we have said nothing about the need for prior knowledge or about the interaction between the *knowledge base* and the processing modules in Fig. 1.23. Knowledge about a problem domain is coded into an image processing system in the form of a knowledge database. This knowledge may be as simple as detailing regions of an image where the information of interest is known to be located, thus limiting the search that has to be conducted in seeking that information. The knowledge base also can be quite complex, such as an interrelated list of all major possible defects in a materials inspection problem or an image database containing high-resolution satellite images of a region in connection with change-detection applications. In addition to guiding the operation of each processing module, the knowledge base also controls the interaction between modules. This distinction is made in Fig. 1.23 by the use of double-headed arrows between the processing modules and the knowledge base, as opposed to single-headed arrows linking the processing modules.

Although we do not discuss image display explicitly at this point, it is important to keep in mind that viewing the results of image processing can take place at the output of any stage in Fig. 1.23. We also note that not all image processing applications require the complexity of interactions implied by Fig. 1.23. In fact, not even all those modules are needed in some cases. For example, image enhancement for human visual interpretation seldom requires use of any of the other stages in Fig. 1.23. In general, however, as the complexity of an image processing task increases, so does the number of processes required to solve the problem.

1.5 Components of an Image Processing System

As recently as the mid-1980s, numerous models of image processing systems being sold throughout the world were rather substantial peripheral devices that attached to equally substantial host computers. Late in the 1980s and early in the 1990s, the market shifted to image processing hardware in the form of single boards designed to be compatible with industry standard buses and to fit into engineering workstation cabinets and personal computers. In addition to lowering costs, this market shift also served as a catalyst for a significant number of new companies whose specialty is the development of software written specifically for image processing.

Although large-scale image processing systems still are being sold for massive imaging applications, such as processing of satellite images, the trend continues toward miniaturizing and blending of general-purpose small computers with specialized image processing hardware. Figure 1.24 shows the basic components comprising a typical *general-purpose* system used for digital image processing. The function of each component is discussed in the following paragraphs, starting with image sensing.

With reference to *sensing*, two elements are required to acquire digital images. The first is a physical device that is sensitive to the energy radiated by the object we wish to image. The second, called a *digitizer*, is a device for converting the output of the physical sensing device into digital form. For instance, in a digital video camera, the sensors produce an electrical output proportional to light intensity. The digitizer converts these outputs to digital data. These topics are covered in some detail in Chapter 2.

Specialized image processing hardware usually consists of the digitizer just mentioned, plus hardware that performs other primitive operations, such as an arithmetic logic unit (ALU), which performs arithmetic and logical operations in parallel on entire images. One example of how an ALU is used is in averaging images as quickly as they are digitized, for the purpose of noise reduction. This type of hardware sometimes is called a *front-end subsystem*, and its most

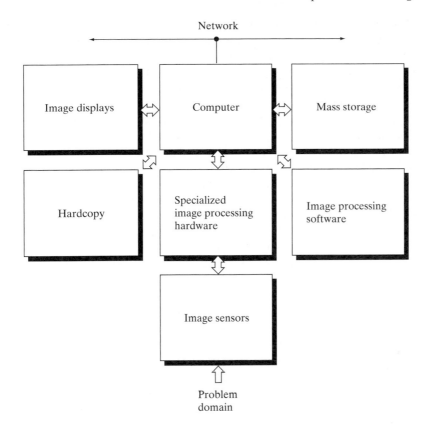

FIGURE 1.24
Components of a
general-purpose
image processing
system.

distinguishing characteristic is speed. In other words, this unit performs functions that require fast data throughputs (e.g., digitizing and averaging video images at 30 frames/s) that the typical main computer cannot handle.

The *computer* in an image processing system is a general-purpose computer and can range from a PC to a supercomputer. In dedicated applications, sometimes specially designed computers are used to achieve a required level of performance, but our interest here is on general-purpose image processing systems. In these systems, almost any well-equipped PC-type machine is suitable for off-line image processing tasks.

Software for image processing consists of specialized modules that perform specific tasks. A well-designed package also includes the capability for the user to write code that, as a minimum, utilizes the specialized modules. More sophisticated software packages allow the integration of those modules and general-purpose software commands from at least one computer language.

Mass storage capability is a must in image processing applications. An image of size 1024 × 1024 pixels, in which the intensity of each pixel is an 8-bit quantity, requires one megabyte of storage space if the image is not compressed. When dealing with thousands, or even millions, of images, providing adequate storage in an image processing system can be a challenge. Digital storage for image processing applications falls into three principal categories: (1) short-term storage for use during processing, (2) on-line storage for relatively fast recall, and (3) archival storage, characterized by infrequent access. Storage is measured in bytes (eight bits), Kbytes (one thousand bytes), Mbytes (one million bytes), Gbytes (meaning giga, or one billion, bytes), and Tbytes (meaning tera, or one trillion, bytes).

One method of providing short-term storage is computer memory. Another is by specialized boards, called *frame buffers*, that store one or more images and can be accessed rapidly, usually at video rates (e.g., at 30 complete images per second). The latter method allows virtually instantaneous image *zoom*, as well as *scroll* (vertical shifts) and *pan* (horizontal shifts). Frame buffers usually are housed in the specialized image processing hardware unit shown in Fig. 1.24. On-line storage generally takes the form of magnetic disks or optical-media storage. The key factor characterizing on-line storage is frequent access to the stored data. Finally, archival storage is characterized by massive storage requirements but infrequent need for access. Magnetic tapes and optical disks housed in "jukeboxes" are the usual media for archival applications.

Image displays in use today are mainly color (preferably flat screen) TV monitors. Monitors are driven by the outputs of image and graphics display cards that are an integral part of the computer system. Seldom are there requirements for image display applications that cannot be met by display cards available commercially as part of the computer system. In some cases, it is necessary to have stereo displays, and these are implemented in the form of headgear containing two small displays embedded in goggles worn by the user.

Hardcopy devices for recording images include laser printers, film cameras, heat-sensitive devices, inkjet units, and digital units, such as optical and CD-ROM disks. Film provides the highest possible resolution, but paper is the obvious medium of choice for written material. For presentations, images are displayed on film transparencies or in a digital medium if image projection equipment is used. The latter approach is gaining acceptance as the standard for image presentations.

Networking is almost a default function in any computer system in use today. Because of the large amount of data inherent in image processing applications, the key consideration in image transmission is bandwidth. In dedicated networks, this typically is not a problem, but communications with remote sites via the Internet are not always as efficient. Fortunately, this situation is improving quickly as a result of optical fiber and other broadband technologies.

Summary

The main purpose of the material presented in this chapter is to provide a sense of perspective about the origins of digital image processing and, more important, about current and future areas of application of this technology. Although the coverage of these topics in this chapter was necessarily incomplete due to space limitations, it should have left the reader with a clear impression of the breadth and practical scope of digital image processing. As we proceed in the following chapters with the development of image processing theory and applications, numerous examples are provided to keep a clear focus on the utility and promise of these techniques. Upon concluding the study of the final chapter, the reader of this book will have arrived at a level of understanding that is the foundation for most of the work currently underway in this field.

References and Further Reading

References at the end of later chapters address specific topics discussed in those chapters, and are keyed to the Bibliography at the end of the book. However, in this chapter we follow a different format in order to summarize in one place a body of journals that publish material on image processing and related topics. We also provide a list of books from which the reader can readily develop a historical and current perspective of activities in this field. Thus, the reference material cited in this chapter is intended as a general-purpose, easily accessible guide to the published literature on image processing.

Major refereed journals that publish articles on image processing and related topics include: *IEEE Transactions on Image Processing; IEEE Transactions on Pattern Analysis and Machine Intelligence; Computer Vision, Graphics, and Image Processing* (prior to 1991); *Computer Vision and Image Understanding; IEEE Transactions on Systems, Man and Cybernetics; Artificial Intelligence; Pattern Recognition; Pattern Recognition Letters; Journal of the Optical Society of America* (prior to 1984); *Journal of the Optical Society of America—A: Optics, Image Science and Vision; Optical Engineering; Applied Optics—Information Processing; IEEE Transactions on Medical Imaging; Journal of Electronic Imaging; IEEE Transactions on Information Theory; IEEE Transactions on Communications; IEEE Transactions on Acoustics, Speech and Signal Processing; Proceedings of the IEEE;* and issues of the *IEEE Transactions on Computers* prior to 1980. Publications of the International Society for Optical Engineering (SPIE) also are of interest.

The following books, listed in reverse chronological order (with the number of books being biased toward more recent publications), contain material that complements our treatment of digital image processing. These books represent an easily accessible overview of the area for the past 30 years and were selected to provide a variety of treatments. They range from textbooks, which cover foundation material; to handbooks, which give an overview of techniques; and finally to edited books, which contain material representative of current research in the field.

Duda, R. O., Hart, P. E., and Stork, D. G. [2001]. *Pattern Classification*, 2nd ed., John Wiley & Sons, NY.

Pratt, W. K. [2001]. *Digital Image Processing*, 3rd ed., John Wiley & Sons, NY.

Ritter, G. X. and Wilson, J. N. [2001]. *Handbook of Computer Vision Algorithms in Image Algebra*, CRC Press, Boca Raton, FL.

Shapiro, L. G. and Stockman, G. C. [2001]. *Computer Vision*, Prentice Hall, Upper Saddle River, NJ.

Dougherty, E. R. (ed.) [2000]. *Random Processes for Image and Signal Processing*, IEEE Press, NY.

Etienne, E. K. and Nachtegael, M. (eds.). [2000]. *Fuzzy Techniques in Image Processing*, Springer-Verlag, NY.

Goutsias, J, Vincent, L., and Bloomberg, D. S. (eds.). [2000]. *Mathematical Morphology and Its Applications to Image and Signal Processing*, Kluwer Academic Publishers, Boston, MA.

Mallot, A. H. [2000]. *Computational Vision*, The MIT Press, Cambridge, MA.

Marchand-Maillet, S. and Sharaiha, Y. M. [2000]. *Binary Digital Image Processing: A Discrete Approach*, Academic Press, NY.

Mitra, S. K. and Sicuranza, G. L. (eds.) [2000]. *Nonlinear Image Processing*, Academic Press, NY.

Edelman, S. [1999]. *Representation and Recognition in Vision*, The MIT Press, Cambridge, MA.

Lillesand, T. M. and Kiefer, R. W. [1999]. *Remote Sensing and Image Interpretation*, John Wiley & Sons, NY.

Mather, P. M. [1999]. *Computer Processing of Remotely Sensed Images: An Introduction*, John Wiley & Sons, NY.

Petrou, M. and Bosdogianni, P. [1999]. *Image Processing: The Fundamentals*, John Wiley & Sons, UK.

Russ, J. C. [1999]. *The Image Processing Handbook*, 3rd ed., CRC Press, Boca Raton, FL.

Smirnov, A. [1999]. *Processing of Multidimensional Signals*, Springer-Verlag, NY.

Sonka, M., Hlavac, V., and Boyle, R. [1999]. *Image Processing, Analysis, and Computer Vision*, PWS Publishing, NY.

Umbaugh, S. E. [1998]. *Computer Vision and Image Processing: A Practical Approach Using CVIPtools*, Prentice Hall, Upper Saddle River, NJ.

Haskell, B. G. and Netravali, A. N. [1997]. *Digital Pictures: Representation, Compression, and Standards*, Perseus Publishing, NY.

Jahne, B. [1997]. *Digital Image Processing: Concepts, Algorithms, and Scientific Applications*, Springer-Verlag, NY.

Castleman, K. R. [1996]. *Digital Image Processing*, 2nd ed., Prentice Hall, Upper Saddle River, NJ.

Geladi, P. and Grahn, H. [1996]. *Multivariate Image Analysis*, John Wiley & Sons, NY.

Bracewell, R. N. [1995]. *Two-Dimensional Imaging*, Prentice Hall, Upper Saddle River, NJ.

Sid-Ahmed, M. A. [1995]. *Image Processing: Theory, Algorithms, and Architectures*, McGraw-Hill, NY.

Jain, R., Rangachar, K., and Schunk, B. [1995]. *Computer Vision*, McGraw-Hill, NY.

Mitiche, A. [1994]. *Computational Analysis of Visual Motion*, Perseus Publishing, NY.

Baxes, G. A. [1994]. *Digital Image Processing: Principles and Applications*, John Wiley & Sons, NY.

Gonzalez, R. C. and Woods, R. E. [1992]. *Digital Image Processing*, Addison-Wesley, Reading, MA.

Haralick, R. M. and Shapiro, L. G. [1992]. *Computer and Robot Vision*, vols. 1 & 2, Addison-Wesley, Reading, MA.

Pratt, W. K. [1991] *Digital Image Processing*, 2nd ed., Wiley-Interscience, NY.

Lim, J. S. [1990]. *Two-Dimensional Signal and Image Processing*, Prentice Hall, Upper Saddle River, NJ.

Jain, A. K. [1989]. *Fundamentals of Digital Image Processing*, Prentice Hall, Upper Saddle River, NJ.

Schalkoff, R. J. [1989]. *Digital Image Processing and Computer Vision*, John Wiley & Sons, NY.

Giardina, C. R. and Dougherty, E. R. [1988]. *Morphological Methods in Image and Signal Processing*, Prentice Hall, Upper Saddle River, NJ.

Levine, M. D. [1985]. *Vision in Man and Machine*, McGraw-Hill, NY.

Serra, J. [1982]. *Image Analysis and Mathematical Morphology*, Academic Press, NY.

Ballard, D. H. and Brown, C. M. [1982]. *Computer Vision*, Prentice Hall, Upper Saddle River, NJ.

Fu, K. S. [1982]. *Syntactic Pattern Recognition and Applications*, Prentice Hall, Upper Saddle River, NJ.

Nevatia, R. [1982]. *Machine Perception*, Prentice Hall, Upper Saddle River, NJ.

Pavlidis, T. [1982]. *Algorithms for Graphics and Image Processing*, Computer Science Press, Rockville, MD.

Rosenfeld, R. and Kak, A. C. [1982]. *Digital Picture Processing*, 2nd ed., vols. 1 & 2, Academic Press, NY.

Hall, E. L. [1979]. *Computer Image Processing and Recognition*, Academic Press, NY.

Gonzalez, R. C. and Thomason, M. G. [1978]. *Syntactic Pattern Recognition: An Introduction*, Addison-Wesley, Reading, MA.

Andrews, H. C. and Hunt, B. R. [1977]. *Digital Image Restoration*, Prentice Hall, Upper Saddle River, NJ.

Pavlidis, T. [1977]. *Structural Pattern Recognition*, Springer-Verlag, NY, 1977.

Tou, J. T. and Gonzalez, R. C. [1974]. *Pattern Recognition Principles*, Addison-Wesley, Reading, MA, 1974.

Andrews, H. C. [1970]. *Computer Techniques in Image Processing*, Academic Press, NY.

2 Digital Image Fundamentals

Those who wish to succeed must ask the right preliminary questions.

Aristotle

Preview

The purpose of this chapter is to introduce several concepts related to digital images and some of the notation used throughout the book. Section 2.1 briefly summarizes the mechanics of the human visual system, including image formation in the eye and its capabilities for brightness adaptation and discrimination. Section 2.2 discusses light, other components of the electromagnetic spectrum, and their imaging characteristics. Section 2.3 discusses imaging sensors and how they are used to generate digital images. Section 2.4 introduces the concepts of uniform image sampling and gray-level quantization. Additional topics discussed in that section include digital image representation, the effects of varying the number of samples and gray levels in an image, some important phenomena associated with sampling, and techniques for image zooming and shrinking. Section 2.5 deals with some basic relationships between pixels that are used throughout the book. Finally, Section 2.6 defines the conditions for linear operations. As noted in that section, linear operators play a central role in the development of image processing techniques.

2.1 Elements of Visual Perception

Although the digital image processing field is built on a foundation of mathematical and probabilistic formulations, human intuition and analysis play a central role in the choice of one technique versus another, and this choice often is made based on subjective, visual judgments. Hence, developing a basic understanding of human visual perception as a first step in our journey through this book is appropriate. Given the complexity and breadth of this topic, we can only aspire to cover the most rudimentary aspects of human vision. In particular, our interest lies in the mechanics and parameters related to how images are formed in the eye. We are interested in learning the physical limitations of human vision in terms of factors that also are used in our work with digital images. Thus, factors such as how human and electronic imaging compare in terms of resolution and ability to adapt to changes in illumination are not only interesting, they also are important from a practical point of view.

2.1.1 Structure of the Human Eye

Figure 2.1 shows a simplified horizontal cross section of the human eye. The eye is nearly a sphere, with an average diameter of approximately 20 mm. Three membranes enclose the eye: the *cornea* and *sclera* outer cover; the *choroid*; and the *retina*. The cornea is a tough, transparent tissue that covers the anterior

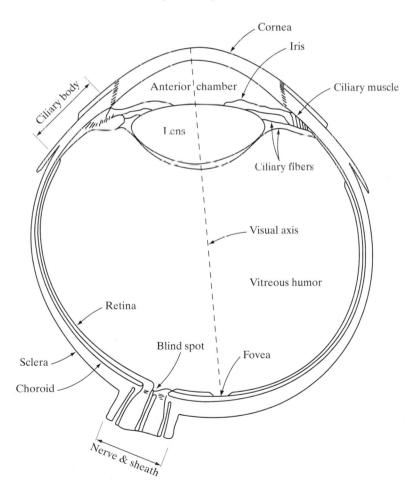

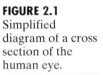

FIGURE 2.1
Simplified diagram of a cross section of the human eye.

surface of the eye. Continuous with the cornea, the sclera is an opaque membrane that encloses the remainder of the optic globe.

The choroid lies directly below the sclera. This membrane contains a network of blood vessels that serve as the major source of nutrition to the eye. Even superficial injury to the choroid, often not deemed serious, can lead to severe eye damage as a result of inflammation that restricts blood flow. The choroid coat is heavily pigmented and hence helps to reduce the amount of extraneous light entering the eye and the backscatter within the optical globe. At its anterior extreme, the choroid is divided into the *ciliary body* and the *iris diaphragm*. The latter contracts or expands to control the amount of light that enters the eye. The central opening of the iris (the *pupil*) varies in diameter from approximately 2 to 8 mm. The front of the iris contains the visible pigment of the eye, whereas the back contains a black pigment.

The *lens* is made up of concentric layers of fibrous cells and is suspended by fibers that attach to the ciliary body. It contains 60 to 70% water, about 6% fat, and more protein than any other tissue in the eye. The lens is colored by a slightly yellow pigmentation that increases with age. In extreme cases, excessive clouding of the lens, caused by the affliction commonly referred to as *cataracts*, can lead to poor color discrimination and loss of clear vision. The lens absorbs approximately 8% of the visible light spectrum, with relatively higher absorption at shorter wavelengths. Both infrared and ultraviolet light are absorbed appreciably by proteins within the lens structure and, in excessive amounts, can damage the eye.

The innermost membrane of the eye is the retina, which lines the inside of the wall's entire posterior portion. When the eye is properly focused, light from an object outside the eye is imaged on the retina. Pattern vision is afforded by the distribution of discrete light receptors over the surface of the retina. There are two classes of receptors: *cones* and *rods*. The cones in each eye number between 6 and 7 million. They are located primarily in the central portion of the retina, called the *fovea*, and are highly sensitive to color. Humans can resolve fine details with these cones largely because each one is connected to its own nerve end. Muscles controlling the eye rotate the eyeball until the image of an object of interest falls on the fovea. Cone vision is called *photopic* or bright-light vision.

The number of rods is much larger: Some 75 to 150 million are distributed over the retinal surface. The larger area of distribution and the fact that several rods are connected to a single nerve end reduce the amount of detail discernible by these receptors. Rods serve to give a general, overall picture of the field of view. They are not involved in color vision and are sensitive to low levels of illumination. For example, objects that appear brightly colored in daylight when seen by moonlight appear as colorless forms because only the rods are stimulated. This phenomenon is known as *scotopic* or dim-light vision.

Figure 2.2 shows the density of rods and cones for a cross section of the right eye passing through the region of emergence of the optic nerve from the eye. The absence of receptors in this area results in the so-called *blind spot* (see Fig. 2.1). Except for this region, the distribution of receptors is radially symmetric about the fovea. Receptor density is measured in degrees from the fovea (that is, in degrees off axis, as measured by the angle formed by the visual axis and a line passing through the center of the lens and intersecting the retina).

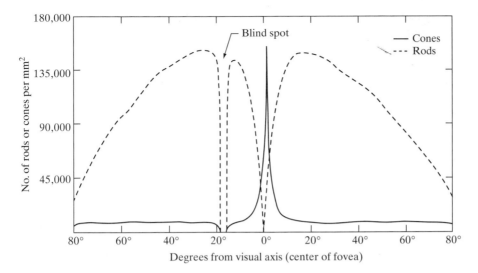

FIGURE 2.2
Distribution of
rods and cones in
the retina.

Note in Fig. 2.2 that cones are most dense in the center of the retina (in the center area of the fovea). Note also that rods increase in density from the center out to approximately 20° off axis and then decrease in density out to the extreme periphery of the retina.

The fovea itself is a circular indentation in the retina of about 1.5 mm in diameter. However, in terms of future discussions, talking about square or rectangular arrays of sensing elements is more useful. Thus, by taking some liberty in interpretation, we can view the fovea as a square sensor array of size 1.5 mm × 1.5 mm. The density of cones in that area of the retina is approximately 150,000 elements per mm^2. Based on these approximations, the number of cones in the region of highest acuity in the eye is about 337,000 elements. Just in terms of raw resolving power, a charge-coupled device (CCD) imaging chip of medium resolution can have this number of elements in a receptor array no larger than 5 mm × 5 mm. While the ability of humans to integrate intelligence and experience with vision makes this type of comparison dangerous. Keep in mind for future discussions that the basic ability of the eye to resolve detail is certainly within the realm of current electronic imaging sensors.

2.1.2 Image Formation in the Eye

The principal difference between the lens of the eye and an ordinary optical lens is that the former is flexible. As illustrated in Fig. 2.1, the radius of curvature of the anterior surface of the lens is greater than the radius of its posterior surface. The shape of the lens is controlled by tension in the fibers of the ciliary body. To focus on distant objects, the controlling muscles cause the lens to be relatively flattened. Similarly, these muscles allow the lens to become thicker in order to focus on objects near the eye.

The distance between the center of the lens and the retina (called the *focal length*) varies from approximately 17 mm to about 14 mm, as the refractive power of the lens increases from its minimum to its maximum. When the eye

FIGURE 2.3
Graphical
representation of
the eye looking at
a palm tree. Point
C is the optical
center of the lens.

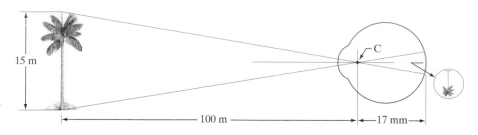

focuses on an object farther away than about 3 m, the lens exhibits its lowest refractive power. When the eye focuses on a nearby object, the lens is most strongly refractive. This information makes it easy to calculate the size of the retinal
image of any object. In Fig. 2.3, for example, the observer is looking at a tree
15 m high at a distance of 100 m. If *h* is the height in mm of that object in the
retinal image, the geometry of Fig. 2.3 yields $15/100 = h/17$ or $h = 2.55$ mm. As
indicated in Section 2.1.1, the retinal image is reflected primarily in the area of
the fovea. Perception then takes place by the relative excitation of light receptors, which transform radiant energy into electrical impulses that are ultimately decoded by the brain.

2.1.3 Brightness Adaptation and Discrimination

Because digital images are displayed as a discrete set of intensities, the eye's
ability to discriminate between different intensity levels is an important consideration in presenting image-processing results. The range of light intensity levels to which the human visual system can adapt is enormous—on the order of
10^{10}—from the scotopic threshold to the glare limit. Experimental evidence indicates that *subjective brightness* (intensity as *perceived* by the human visual
system) is a logarithmic function of the light intensity incident on the eye. Figure 2.4, a plot of light intensity versus subjective brightness, illustrates this char-

FIGURE 2.4
Range of
subjective
brightness
sensations
showing a
particular
adaptation level.

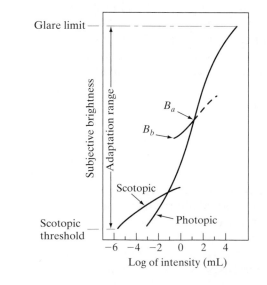

acteristic. The long solid curve represents the range of intensities to which the visual system can adapt. In photopic vision alone, the range is about 10^6. The transition from scotopic to photopic vision is gradual over the approximate range from 0.001 to 0.1 millilambert (-3 to -1 mL in the log scale), as the double branches of the adaptation curve in this range show.

The essential point in interpreting the impressive dynamic range depicted in Fig. 2.4 is that the visual system cannot operate over such a range *simultaneously*. Rather, it accomplishes this large variation by changes in its overall sensitivity, a phenomenon known as *brightness adaptation*. The total range of distinct intensity levels it can discriminate simultaneously is rather small when compared with the total adaptation range. For any given set of conditions, the current sensitivity level of the visual system is called the *brightness adaptation level*, which may correspond, for example, to brightness B_a in Fig. 2.4. The short intersecting curve represents the range of subjective brightness that the eye can perceive when adapted to this level. This range is rather restricted, having a level B_b at and below which all stimuli are perceived as indistinguishable blacks. The upper (dashed) portion of the curve is not actually restricted but, if extended too far, loses its meaning because much higher intensities would simply raise the adaptation level higher than B_a.

The ability of the eye to discriminate between *changes* in light intensity at any specific adaptation level is also of considerable interest. A classic experiment used to determine the capability of the human visual system for brightness discrimination consists of having a subject look at a flat, uniformly illuminated area large enough to occupy the entire field of view. This area typically is a diffuser, such as opaque glass, that is illuminated from behind by a light source whose intensity, I, can be varied. To this field is added an increment of illumination, ΔI, in the form of a short-duration flash that appears as a circle in the center of the uniformly illuminated field, as Fig. 2.5 shows.

If ΔI is not bright enough, the subject says "no," indicating no perceivable change. As ΔI gets stronger, the subject may give a positive response of "yes," indicating a perceived change. Finally, when ΔI is strong enough, the subject will give a response of "yes" all the time. The quantity $\Delta I_c/I$, where ΔI_c is the increment of illumination discriminable 50% of the time with background illumination I, is called the *Weber ratio*. A small value of $\Delta I_c/I$, means that a small percentage change in intensity is discriminable. This represents "good" brightness discrimination. Conversely, a large value of $\Delta I_c/I$, means that a large percentage change in intensity is required. This represents "poor" brightness discrimination.

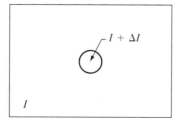

FIGURE 2.5 Basic experimental setup used to characterize brightness discrimination.

FIGURE 2.6
Typical Weber
ratio as a function
of intensity.

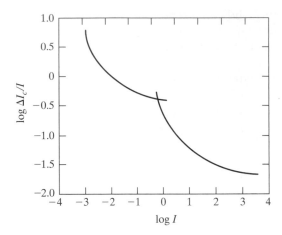

A plot of $\log \Delta I_c / I$, as a function of $\log I$ has the general shape shown in Fig. 2.6. This curve shows that brightness discrimination is poor (the Weber ratio is large) at low levels of illumination, and it improves significantly (the Weber ratio decreases) as background illumination increases. The two branches in the curve reflect the fact that at low levels of illumination vision is carried out by activity of the rods, whereas at high levels (showing better discrimination) vision is the function of cones.

If the background illumination is held constant and the intensity of the other source, instead of flashing, is now allowed to vary incrementally from never being perceived to always being perceived, the typical observer can discern a total of one to two dozen different intensity changes. Roughly, this result is related to the number of different intensities a person can see at any one point in a monochrome image. This result does not mean that an image can be represented by such a small number of intensity values because, as the eye roams about the image, the average background changes, thus allowing a *different* set of incremental changes to be detected at each new adaptation level. The net consequence is that the eye is capable of a much broader range of *overall* intensity discrimination. In fact, we show in Section 2.4.3 that the eye is capable of detecting objectionable contouring effects in monochrome images whose overall intensity is represented by fewer than approximately two dozen levels.

Two phenomena clearly demonstrate that perceived brightness is not a simple function of intensity. The first is based on the fact that the visual system tends to undershoot or overshoot around the boundary of regions of different intensities. Figure 2.7(a) shows a striking example of this phenomenon. Although the intensity of the stripes is constant, we actually perceive a brightness pattern that is strongly scalloped, especially near the boundaries [Fig. 2.7(b)]. These seemingly scalloped bands are called *Mach bands* after Ernst Mach, who first described the phenomenon in 1865.

The second phenomenon, called *simultaneous contrast*, is related to the fact that a region's perceived brightness does not depend simply on its intensity, as Fig. 2.8 demonstrates. All the center squares have exactly the same intensity.

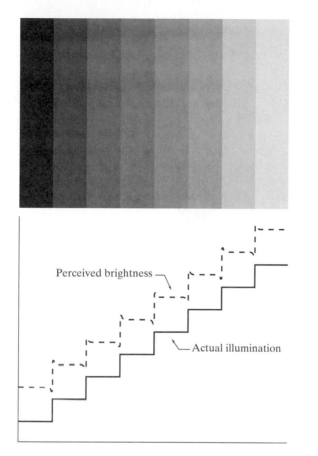

FIGURE 2.7
(a) An example showing that perceived brightness is not a simple function of intensity. The relative vertical positions between the two profiles in (b) have no special significance; they were chosen for clarity.

However, they appear to the eye to become darker as the background gets lighter. A more familiar example is a piece of paper that seems white when lying on a desk, but can appear totally black when used to shield the eyes while looking directly at a bright sky.

a b c

FIGURE 2.8 Examples of simultaneous contrast. All the inner squares have the same intensity, but they appear progressively darker as the background becomes lighter.

a b
c d

FIGURE 2.9 Some
well-known
optical illusions.

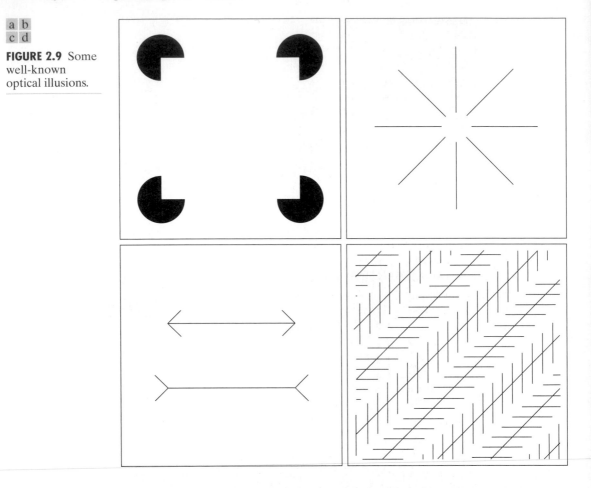

Other examples of human perception phenomena are optical illusions, in which the eye fills in nonexisting information or wrongly perceives geometrical properties of objects. Some examples are shown in Fig. 2.9. In Fig. 2.9(a), the outline of a square is seen clearly, in spite of the fact that no lines defining such a figure are part of the image. The same effect, this time with a circle, can be seen in Fig. 2.9(b); note how just a few lines are sufficient to give the illusion of a complete circle. The two horizontal line segments in Fig. 2.9(c) are of the same length, but one appears shorter than the other. Finally, all lines in Fig. 2.9(d) that are oriented at 45° are equidistant and parallel. Yet the crosshatching creates the illusion that those lines are far from being parallel. Optical illusions are a characteristic of the human visual system that is not fully understood.

2.2   Light and the Electromagnetic Spectrum

The electromagnetic spectrum was introduced in Section 1.3. We now consider this topic in more detail. In 1666, Sir Isaac Newton discovered that when a beam of sunlight is passed through a glass prism, the emerging beam of light is not

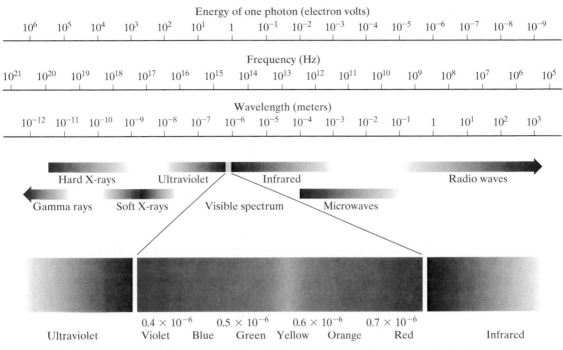

FIGURE 2.10 The electromagnetic spectrum. The visible spectrum is shown zoomed to facilitate explanation, but note that the visible spectrum is a rather narrow portion of the EM spectrum.

white but consists instead of a continuous spectrum of colors ranging from violet at one end to red at the other. As shown in Fig. 2.10, the range of colors we perceive in visible light represents a very small portion of the electromagnetic spectrum. On one end of the spectrum are radio waves with wavelengths billions of times longer than those of visible light. On the other end of the spectrum are gamma rays with wavelengths millions of times smaller than those of visible light. The electromagnetic spectrum can be expressed in terms of wavelength, frequency, or energy. Wavelength (λ) and frequency (ν) are related by the expression

$$\lambda = \frac{c}{\nu} \tag{2.2-1}$$

where c is the speed of light (2.998×10^8 m/s). The energy of the various components of the electromagnetic spectrum is given by the expression

$$E = h\nu \tag{2.2-2}$$

where h is Planck's constant. The units of wavelength are meters, with the terms *microns* (denoted μm and equal to 10^{-6} m) and *nanometers* (10^{-9} m) being used just as frequently. Frequency is measured in Hertz (Hz), with one Hertz being equal to one cycle of a sinusoidal wave per second. A commonly used unit of energy is the electron-volt.

FIGURE 2.11
Graphical
representation of
one wavelength.

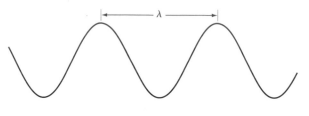

Electromagnetic waves can be visualized as propagating sinusoidal waves with wavelength λ (Fig. 2.11), or they can be thought of as a stream of massless particles, each traveling in a wavelike pattern and moving at the speed of light. Each massless particle contains a certain amount (or bundle) of energy. Each bundle of energy is called a *photon*. We see from Eq. (2.2-2) that energy is proportional to frequency, so the higher-frequency (shorter wavelength) electromagnetic phenomena carry more energy per photon. Thus, radio waves have photons with low energies, microwaves have more energy than radio waves, infrared still more, then visible, ultraviolet, X-rays, and finally gamma rays, the most energetic of all. This is the reason that gamma rays are so dangerous to living organisms.

Light is a particular type of electromagnetic radiation that can be seen and sensed by the human eye. The visible (color) spectrum is shown expanded in Fig. 2.10 for the purpose of discussion (we consider color in much more detail in Chapter 6). The visible band of the electromagnetic spectrum spans the range from approximately 0.43 μm (violet) to about 0.79 μm (red). For convenience, the color spectrum is divided into six broad regions: violet, blue, green, yellow, orange, and red. No color (or other component of the electromagnetic spectrum) ends abruptly, but rather each range blends smoothly into the next, as shown in Fig. 2.10.

The colors that humans perceive in an object are determined by the nature of the light *reflected* from the object. A body that reflects light and is relatively balanced in all visible wavelengths appears white to the observer. However, a body that favors reflectance in a limited range of the visible spectrum exhibits some shades of color. For example, green objects reflect light with wavelengths primarily in the 500 to 570 nm range while absorbing most of the energy at other wavelengths.

Light that is void of color is called *achromatic* or *monochromatic* light. The only attribute of such light is its *intensity*, or amount. The term *gray level* generally is used to describe monochromatic intensity because it ranges from black, to grays, and finally to white. Chromatic light spans the electromagnetic energy spectrum from approximately 0.43 to 0.79 μm, as noted previously. Three basic quantities are used to describe the quality of a chromatic light source: radiance; luminance; and brightness. *Radiance* is the total amount of energy that flows from the light source, and it is usually measured in watts (W). *Luminance*, measured in lumens (lm), gives a measure of the amount of energy an observer *perceives* from a light source. For example, light emitted from a source operating in the far infrared region of the spectrum could have significant energy (radiance), but an observer would hardly perceive it; its luminance would be almost zero. Finally, as discussed in Section 2.1, *brightness* is a subjective descriptor of light perception that is practically impossible to measure. It embodies the achromatic notion of intensity and is one of the key factors in describing color sensation.

Continuing with the discussion of Fig. 2.10, we note that at the short-wavelength end of the electromagnetic spectrum, we have gamma rays and hard X-rays. As discussed in Section 1.3.1, gamma radiation is important for medical and astronomical imaging, and for imaging radiation in nuclear environments. Hard (high-energy) X-rays are used in industrial applications. Chest X-rays are in the high end (shorter wavelength) of the soft X-rays region and dental X-rays are in the lower energy end of that band. The soft X-ray band transitions into the far ultraviolet light region, which in turn blends with the visible spectrum at longer wavelengths. Moving still higher in wavelength, we encounter the infrared band, which radiates heat, a fact that makes it useful in imaging applications that rely on "heat signatures." The part of the infrared band close to the visible spectrum is called the *near-infrared* region. The opposite end of this band is called the *far-infrared* region. This latter region blends with the microwave band. This band is well known as the source of energy in microwave ovens, but it has many other uses, including communication and radar. Finally, the radio wave band encompasses television as well as AM and FM radio. In the higher energies, radio signals emanating from certain stellar bodies are useful in astronomical observations. Examples of images in most of the bands just discussed are given in Section 1.3.

In principle, if a sensor can be developed that is capable of detecting energy radiated by a band of the electromagnetic spectrum, we can image events of interest in that band. It is important to note, however, that the wavelength of an electromagnetic wave required to "see" an object must be of the same size as or smaller than the object. For example, a water molecule has a diameter on the order of 10^{-10} m. Thus, to study molecules, we would need a source capable of emitting in the far ultraviolet or soft X-ray region. This limitation, along with the physical properties of the sensor material, establishes the fundamental limits on the capability of imaging sensors, such as visible, infrared, and other sensors in use today.

Although imaging is based predominantly on energy radiated by electromagnetic waves, this is not the only method for image generation. For example, as discussed in Section 1.3.7, sound reflected from objects can be used to form ultrasonic images. Other major sources of digital images are electron beams for electron microscopy and synthetic images used in graphics and visualization.

2.3 ■ Image Sensing and Acquisition

The types of images in which we are interested are generated by the combination of an "illumination" source and the reflection or absorption of energy from that source by the elements of the "scene" being imaged. We enclose *illumination* and *scene* in quotes to emphasize the fact that they are considerably more general than the familiar situation in which a visible light source illuminates a common everyday 3-D (three-dimensional) scene. For example, the illumination may originate from a source of electromagnetic energy such as radar, infrared, or X-ray energy. But, as noted earlier, it could originate from less traditional sources, such as ultrasound or even a computer-generated illumination pattern. Similarly, the scene elements could be familiar objects, but they can just as easily be molecules, buried rock formations, or a human brain. We could even image a source, such as acquiring images of the sun. Depending on the nature of the source, illumination energy is reflected from, or transmitted through, objects. An example in the first category is light reflected from a planar surface. An example in the second category is when X-rays pass through a patient's body for the purpose of generating a diagnostic X-ray film. In some applications, the reflected or transmitted energy is focused onto a photoconverter (e.g., a phosphor screen), which converts the energy into visible light. Electron microscopy and some applications of gamma imaging use this approach.

Figure 2.12 shows the three principal sensor arrangements used to transform illumination energy into digital images. The idea is simple: Incoming energy is

a
b
c

FIGURE 2.12
(a) Single imaging sensor.
(b) Line sensor.
(c) Array sensor.

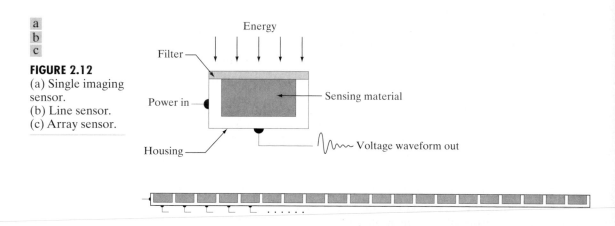

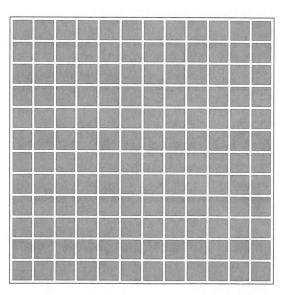

transformed into a voltage by the combination of input electrical power and sensor material that is responsive to the particular type of energy being detected. The output voltage waveform is the response of the sensor(s), and a digital quantity is obtained from each sensor by digitizing its response. In this section, we look at the principal modalities for image sensing and generation. Image digitizing is discussed in Section 2.4.

2.3.1 Image Acquisition Using a Single Sensor

Figure 2.12(a) shows the components of a single sensor. Perhaps the most familiar sensor of this type is the photodiode, which is constructed of silicon materials and whose output voltage waveform is proportional to light. The use of a filter in front of a sensor improves selectivity. For example, a green (pass) filter in front of a light sensor favors light in the green band of the color spectrum. As a consequence, the sensor output will be stronger for green light than for other components in the visible spectrum.

In order to generate a 2-D image using a single sensor, there has to be relative displacements in both the x- and y-directions between the sensor and the area to be imaged. Figure 2.13 shows an arrangement used in high-precision scanning, where a film negative is mounted onto a drum whose mechanical rotation provides displacement in one dimension. The single sensor is mounted on a lead screw that provides motion in the perpendicular direction. Since mechanical motion can be controlled with high precision, this method is an inexpensive (but slow) way to obtain high-resolution images. Other similar mechanical arrangements use a flat bed, with the sensor moving in two linear directions. These types of mechanical digitizers sometimes are referred to as *microdensitometers*.

Another example of imaging with a single sensor places a laser source coincident with the sensor. Moving mirrors are used to control the outgoing beam in a scanning pattern and to direct the reflected laser signal onto the sensor. This arrangement also can be used to acquire images using strip and array sensors, which are discussed in the following two sections.

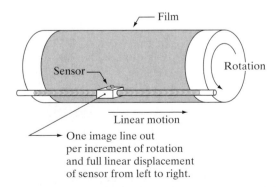

FIGURE 2.13 Combining a single sensor with motion to generate a 2-D image.

2.3.2 Image Acquisition Using Sensor Strips

A geometry that is used much more frequently than single sensors consists of an in-line arrangement of sensors in the form of a sensor strip, as Fig. 2.12(b) shows. The strip provides imaging elements in one direction. Motion perpendicular to the strip provides imaging in the other direction, as shown in Fig. 2.14(a). This is the type of arrangement used in most flat bed scanners. Sensing devices with 4000 or more in-line sensors are possible. In-line sensors are used routinely in airborne imaging applications, in which the imaging system is mounted on an aircraft that flies at a constant altitude and speed over the geographical area to be imaged. One-dimensional imaging sensor strips that respond to various bands of the electromagnetic spectrum are mounted perpendicular to the direction of flight. The imaging strip gives one line of an image at a time, and the motion of the strip completes the other dimension of a two-dimensional image. Lenses or other focusing schemes are used to project the area to be scanned onto the sensors.

Sensor strips mounted in a ring configuration are used in medical and industrial imaging to obtain cross-sectional ("slice") images of 3-D objects, as Fig. 2.14(b) shows. A rotating X-ray source provides illumination and the por-

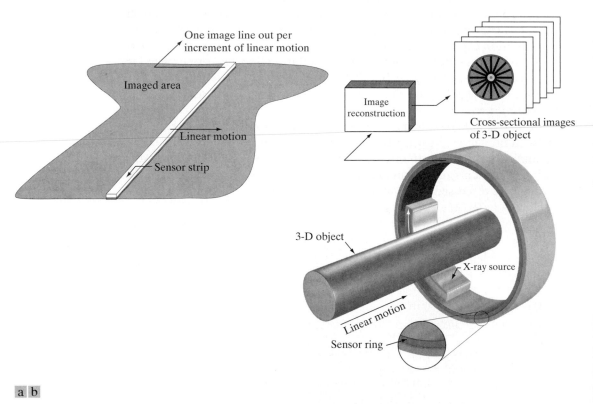

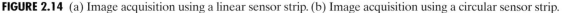

a b

FIGURE 2.14 (a) Image acquisition using a linear sensor strip. (b) Image acquisition using a circular sensor strip.

tion of the sensors opposite the source collect the X-ray energy that pass through the object (the sensors obviously have to be sensitive to X-ray energy). This is the basis for medical and industrial computerized axial tomography (CAT) imaging as indicated in Sections 1.2 and 1.3.2. It is important to note that the output of the sensors must be processed by reconstruction algorithms whose objective is to transform the sensed data into meaningful cross-sectional images. In other words, images are not obtained directly from the sensors by motion alone; they require extensive processing. A 3-D digital volume consisting of stacked images is generated as the object is moved in a direction perpendicular to the sensor ring. Other modalities of imaging based on the CAT principle include magnetic resonance imaging (MRI) and positron emission tomography (PET). The illumination sources, sensors, and types of images are different, but conceptually they are very similar to the basic imaging approach shown in Fig. 2.14(b).

2.3.3 Image Acquisition Using Sensor Arrays

Figure 2.12(c) shows individual sensors arranged in the form of a 2-D array. Numerous electromagnetic and some ultrasonic sensing devices frequently are arranged in an array format. This is also the predominant arrangement found in digital cameras. A typical sensor for these cameras is a CCD array, which can be manufactured with a broad range of sensing properties and can be packaged in rugged arrays of 4000 × 4000 elements or more. CCD sensors are used widely in digital cameras and other light sensing instruments. The response of each sensor is proportional to the integral of the light energy projected onto the surface of the sensor, a property that is used in astronomical and other applications requiring low noise images. Noise reduction is achieved by letting the sensor integrate the input light signal over minutes or even hours (we discuss noise reduction by integration in Chapter 3). Since the sensor array shown in Fig. 2.15(c) is two dimensional, its key advantage is that a complete image can be obtained by focusing the energy pattern onto the surface of the array. Motion obviously is not necessary, as is the case with the sensor arrangements discussed in the preceding two sections.

The principal manner in which array sensors are used is shown in Fig. 2.15. This figure shows the energy from an illumination source being reflected from a scene element, but, as mentioned at the beginning of this section, the energy also could be transmitted through the scene elements. The first function performed by the imaging system shown in Fig. 2.15(c) is to collect the incoming energy and focus it onto an image plane. If the illumination is light, the front end of the imaging system is a lens, which projects the viewed scene onto the lens focal plane, as Fig. 2.15(d) shows. The sensor array, which is coincident with the focal plane, produces outputs proportional to the integral of the light received at each sensor. Digital and analog circuitry sweep these outputs and convert them to a video signal, which is then digitized by another section of the imaging system. The output is a digital image, as shown diagrammatically in Fig. 2.15(e). Conversion of an image into digital form is the topic of Section 2.4.

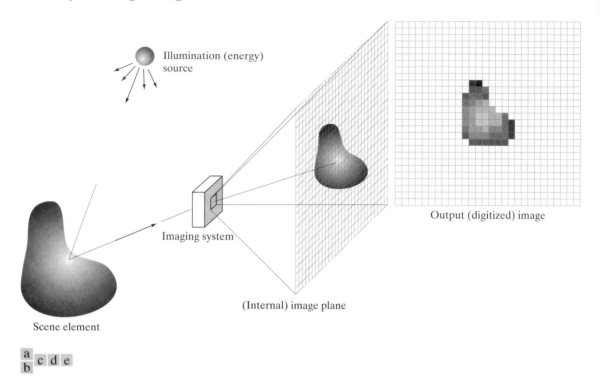

Illumination (energy) source

Imaging system

(Internal) image plane

Output (digitized) image

Scene element

a
b c d e

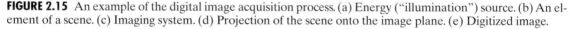

FIGURE 2.15 An example of the digital image acquisition process. (a) Energy ("illumination") source. (b) An element of a scene. (c) Imaging system. (d) Projection of the scene onto the image plane. (e) Digitized image.

2.3.4 A Simple Image Formation Model

As introduced in Section 1.1, we shall denote images by two-dimensional functions of the form $f(x, y)$. The value or amplitude of f at spatial coordinates (x, y) is a positive scalar quantity whose physical meaning is determined by the source of the image. Most of the images in which we are interested in this book are monochromatic images, whose values are said to span the gray scale, as discussed in Section 2.2. When an image is generated from a physical process, its values are proportional to energy radiated by a physical source (e.g., electromagnetic waves). As a consequence, $f(x, y)$ must be nonzero and finite; that is,

$$0 < f(x, y) < \infty. \tag{2.3-1}$$

The function $f(x, y)$ may be characterized by two components: (1) the amount of source illumination incident on the scene being viewed, and (2) the amount of illumination reflected by the objects in the scene. Appropriately, these are called the *illumination* and *reflectance* components and are denoted by $i(x, y)$ and $r(x, y)$, respectively. The two functions combine as a product to form $f(x, y)$:

$$f(x, y) = i(x, y)r(x, y) \qquad\qquad (2.3\text{-}2)$$

where

$$0 < i(x, y) < \infty \qquad\qquad (2.3\text{-}3)$$

and

$$0 < r(x, y) < 1. \qquad\qquad (2.3\text{-}4)$$

Equation (2.3-4) indicates that reflectance is bounded by 0 (total absorption) and 1 (total reflectance). The nature of $i(x, y)$ is determined by the illumination source, and $r(x, y)$ is determined by the characteristics of the imaged objects. It is noted that these expressions also are applicable to images formed via transmission of the illumination through a medium, such as a chest X-ray. In this case, we would deal with a *transmissivity* instead of a *reflectivity* function, but the limits would be the same as in Eq. (2.3-4), and the image function formed would be modeled as the product in Eq. (2.3-2).

■ The values given in Eqs. (2.3-3) and (2.3-4) are theoretical bounds. The following *average* numerical figures illustrate some typical ranges of $i(x, y)$ for visible light. On a clear day, the sun may produce in excess of 90,000 lm/m² of illumination on the surface of the Earth. This figure decreases to less than 10,000 lm/m² on a cloudy day. On a clear evening, a full moon yields about 0.1 lm/m² of illumination. The typical illumination level in a commercial office is about 1000 lm/m². Similarly, the following are some typical values of $r(x, y)$: 0.01 for black velvet, 0.65 for stainless steel, 0.80 for flat-white wall paint, 0.90 for silver-plated metal, and 0.93 for snow. ■

EXAMPLE 2.1: Some typical values of illumination and reflectance.

As noted in Section 2.2, we call the intensity of a monochrome image at any coordinates (x_0, y_0) the *gray level* (ℓ) of the image at that point. That is,

$$\ell = f(x_0, y_0) \qquad\qquad (2.3\text{-}5)$$

From Eqs. (2.3-2) through (2.3-4), it is evident that ℓ lies in the range

$$L_{min} \le \ell \le L_{max} \qquad\qquad (2.3\text{-}6)$$

In theory, the only requirement on L_{min} is that it be positive, and on L_{max} that it be finite. In practice, $L_{min} = i_{min}r_{min}$ and $L_{max} = i_{max}r_{max}$. Using the preceding average office illumination and range of reflectance values as guidelines, we may expect $L_{min} \approx 10$ and $L_{max} \approx 1000$ to be typical limits for indoor values in the absence of additional illumination.

The interval $[L_{min}, L_{max}]$ is called the *gray scale*. Common practice is to shift this interval numerically to the interval $[0, L - 1]$, where $\ell = 0$ is considered black and $\ell = L - 1$ is considered white on the gray scale. All intermediate values are shades of gray varying from black to white.

2.4 Image Sampling and Quantization

From the discussion in the preceding section, we see that there are numerous ways to acquire images, but our objective in all is the same: to generate digital images from sensed data. The output of most sensors is a continuous voltage waveform whose amplitude and spatial behavior are related to the physical phenomenon being sensed. To create a digital image, we need to convert the continuous sensed data into digital form. This involves two processes: *sampling* and *quantization*.

2.4.1 Basic Concepts in Sampling and Quantization

The basic idea behind sampling and quantization is illustrated in Fig. 2.16. Figure 2.16(a) shows a continuous image, $f(x, y)$, that we want to convert to digital form. An image may be continuous with respect to the x- and y-coordinates, and also in amplitude. To convert it to digital form, we have to sample the function in both coordinates and in amplitude. Digitizing the coordinate values is called *sampling*. Digitizing the amplitude values is called *quantization*.

The one-dimensional function shown in Fig. 2.16(b) is a plot of amplitude (gray level) values of the continuous image along the line segment AB in Fig. 2.16(a). The random variations are due to image noise. To sample this function, we take equally spaced samples along line AB, as shown in Fig. 2.16(c). The location of each sample is given by a vertical tick mark in the bottom part of the figure. The samples are shown as small white squares superimposed on the function. The set of these discrete locations gives the sampled function. However, the values of the samples still span (vertically) a continuous range of gray-level values. In order to form a digital function, the gray-level values also must be converted (*quantized*) into discrete quantities. The right side of Fig. 2.16(c) shows the gray-level scale divided into eight discrete levels, ranging from black to white. The vertical tick marks indicate the specific value assigned to each of the eight gray levels. The continuous gray levels are quantized simply by assigning one of the eight discrete gray levels to each sample. The assignment is made depending on the vertical proximity of a sample to a vertical tick mark. The digital samples resulting from both sampling and quantization are shown in Fig. 2.16(d). Starting at the top of the image and carrying out this procedure line by line produces a two-dimensional digital image.

Sampling in the manner just described assumes that we have a continuous image in both coordinate directions as well as in amplitude. In practice, the method of sampling is determined by the sensor arrangement used to generate the image. When an image is generated by a single sensing element combined with mechanical motion, as in Fig. 2.13, the output of the sensor is quantized in the manner described above. However, sampling is accomplished by selecting the number of individual mechanical increments at which we activate the sensor to collect data. Mechanical motion can be made very exact so, in principle, there is almost no limit as to how fine we can sample an image. However, practical limits are established by imperfections in the optics used to focus on the

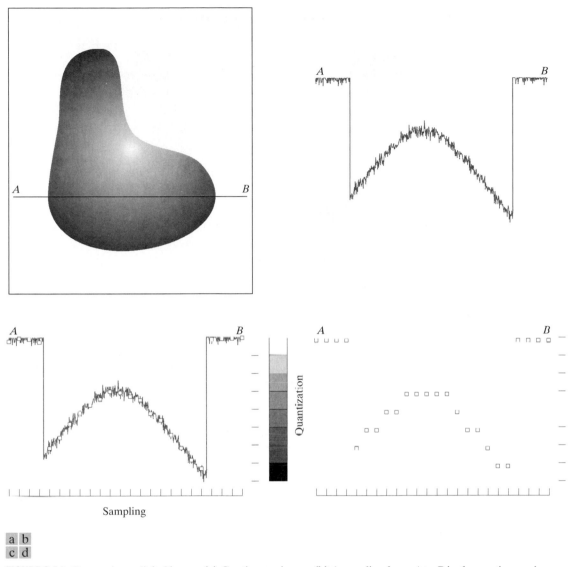

a b
c d

FIGURE 2.16 Generating a digital image. (a) Continuous image. (b) A scan line from A to B in the continuous image, used to illustrate the concepts of sampling and quantization. (c) Sampling and quantization. (d) Digital scan line.

sensor an illumination spot that is inconsistent with the fine resolution achievable with mechanical displacements.

When a sensing strip is used for image acquisition, the number of sensors in the strip establishes the sampling limitations in one image direction. Mechanical motion in the other direction can be controlled more accurately, but it makes little sense to try to achieve sampling density in one direction that exceeds the

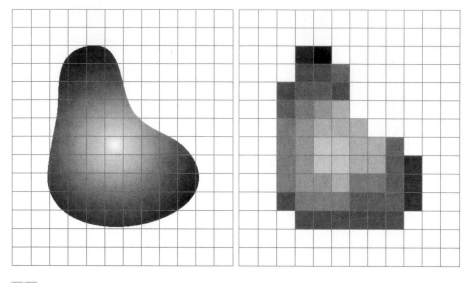

a b

FIGURE 2.17 (a) Continuos image projected onto a sensor array. (b) Result of image sampling and quantization.

sampling limits established by the number of sensors in the other. Quantization of the sensor outputs completes the process of generating a digital image.

When a sensing array is used for image acquisition, there is no motion and the number of sensors in the array establishes the limits of sampling in both directions. Quantization of the sensor outputs is as before. Figure 2.17 illustrates this concept. Figure 2.17(a) shows a continuous image projected onto the plane of an array sensor. Figure 2.17(b) shows the image after sampling and quantization. Clearly, the quality of a digital image is determined to a large degree by the number of samples and discrete gray levels used in sampling and quantization. However, as shown in Section 2.4.3, image content is an important consideration in choosing these parameters.

2.4.2 Representing Digital Images

The result of sampling and quantization is a matrix of real numbers. We will use two principal ways in this book to represent digital images. Assume that an image $f(x, y)$ is sampled so that the resulting digital image has M rows and N columns. The values of the coordinates (x, y) now become *discrete* quantities. For notational clarity and convenience, we shall use integer values for these discrete coordinates. Thus, the values of the coordinates at the origin are $(x, y) = (0, 0)$. The next coordinate values along the first row of the image are represented as $(x, y) = (0, 1)$. It is important to keep in mind that the notation $(0, 1)$ is used to signify the second sample along the first row. It does *not* mean that these are the actual values of physical coordinates when the image was sampled. Figure 2.18 shows the coordinate convention used throughout this book.

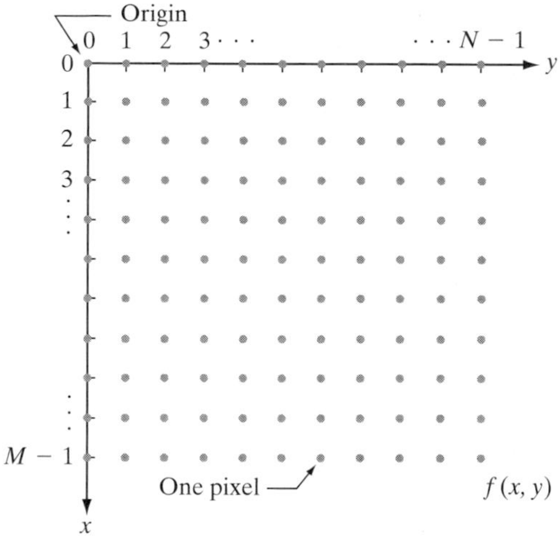

FIGURE 2.18
Coordinate
convention used
in this book to
represent digital
images.

The notation introduced in the preceding paragraph allows us to write the complete $M \times N$ digital image in the following compact matrix form:

$$f(x, y) = \begin{bmatrix} f(0,0) & f(0,1) & \cdots & f(0, N-1) \\ f(1,0) & f(1,1) & \cdots & f(1, N-1) \\ \vdots & \vdots & & \vdots \\ f(M-1,0) & f(M-1,1) & \cdots & f(M-1, N-1) \end{bmatrix}. \quad (2.4\text{-}1)$$

The right side of this equation is by definition a digital image. Each element of this matrix array is called an *image element, picture element, pixel,* or *pel.* The terms *image* and *pixel* will be used throughout the rest of our discussions to denote a digital image and its elements.

In some discussions, it is advantageous to use a more traditional matrix notation to denote a digital image and its elements:

$$\mathbf{A} = \begin{bmatrix} a_{0,0} & a_{0,1} & \cdots & a_{0, N-1} \\ a_{1,0} & a_{1,1} & \cdots & a_{1, N-1} \\ \vdots & \vdots & & \vdots \\ a_{M-1,0} & a_{M-1,1} & \cdots & a_{M-1, N-1} \end{bmatrix}. \quad (2.4\text{-}2)$$

Clearly, $a_{ij} = f(x = i, y = j) = f(i, j)$, so Eqs. (2.4-1) and (2.4-2) are identical matrices.

Expressing sampling and quantization in more formal mathematical terms can be useful at times. Let Z and R denote the set of real integers and the set of real numbers, respectively. The sampling process may be viewed as partitioning the xy plane into a grid, with the coordinates of the center of each grid being a pair of elements from the Cartesian product Z^2, which is the set of all ordered pairs of elements (z_i, z_j), with z_i and z_j being integers from Z. Hence, $f(x, y)$ is a digital image if (x, y) are integers from Z^2 and f is a function that assigns a gray-level value (that is, a real number from the set of real numbers, R) to each distinct pair of coordinates (x, y). This functional assignment obviously is the quantization process described earlier. If the gray levels also are integers (as usually is the case in this and subsequent chapters), Z replaces R, and a digital image then becomes a 2-D function whose coordinates and amplitude values are integers.

This digitization process requires decisions about values for M, N, and for the number, L, of discrete gray levels allowed for each pixel. There are no requirements on M and N, other than that they have to be positive integers. However, due to processing, storage, and sampling hardware considerations, the number of gray levels typically is an integer power of 2:

$$L = 2^k. \tag{2.4-3}$$

We assume that the discrete levels are equally spaced and that they are integers in the interval $[0, L - 1]$. Sometimes the range of values spanned by the gray scale is called the *dynamic range* of an image, and we refer to images whose gray levels span a significant portion of the gray scale as having a high dynamic range. When an appreciable number of pixels exhibit this property, the image will have high contrast. Conversely, an image with low dynamic range tends to have a dull, washed out gray look. This is discussed in much more detail in Section 3.3.

The number, b, of bits required to store a digitized image is

$$b = M \times N \times k. \tag{2.4-4}$$

When $M = N$, this equation becomes

$$b = N^2 k. \tag{2.4-5}$$

Table 2.1 shows the number of bits required to store square images with various values of N and k. The number of gray levels corresponding to each value of k is shown in parentheses. When an image can have 2^k gray levels, it is common practice to refer to the image as a "k-bit image." For example, an image with 256 possible gray-level values is called an 8-bit image. Note that storage requirements for 8-bit images of size 1024×1024 and higher are not insignificant.

TABLE 2.1
Number of storage bits for various values of N and k.

N/k	1 ($L = 2$)	2 ($L = 4$)	3 ($L = 8$)	4 ($L = 16$)	5 ($L = 32$)	6 ($L = 64$)	7 ($L = 128$)	8 ($L = 256$)
32	1,024	2,048	3,072	4,096	5,120	6,144	7,168	8,192
64	4,096	8,192	12,288	16,384	20,480	24,576	28,672	32,768
128	16,384	32,768	49,152	65,536	81,920	98,304	114,688	131,072
256	65,536	131,072	196,608	262,144	327,680	393,216	458,752	524,288
512	262,144	524,288	786,432	1,048,576	1,310,720	1,572,864	1,835,008	2,097,152
1024	1,048,576	2,097,152	3,145,728	4,194,304	5,242,880	6,291,456	7,340,032	8,388,608
2048	4,194,304	8,388,608	12,582,912	16,777,216	20,971,520	25,165,824	29,369,128	33,554,432
4096	16,777,216	33,554,432	50,331,648	67,108,864	83,886,080	100,663,296	117,440,512	134,217,728
8192	67,108,864	134,217,728	201,326,592	268,435,456	335,544,320	402,653,184	469,762,048	536,870,912

2.4.3 Spatial and Gray-Level Resolution

Sampling is the principal factor determining the *spatial resolution* of an image. Basically, spatial resolution is the smallest discernible detail in an image. Suppose that we construct a chart with vertical lines of width W, with the space between the lines also having width W. A *line pair* consists of one such line and its adjacent space. Thus, the width of a line pair is $2W$, and there are $1/2W$ line pairs per unit distance. A widely used definition of resolution is simply the smallest number of discernible line pairs per unit distance; for example, 100 line pairs per millimeter.

Gray-level resolution similarly refers to the smallest discernible change in gray level, but, as noted in Section 2.1.3, measuring discernible changes in gray level is a highly subjective process. We have considerable discretion regarding the number of samples used to generate a digital image, but this is not true for the number of gray levels. Due to hardware considerations, the number of gray levels is usually an integer power of 2, as mentioned in the previous section. The most common number is 8 bits, with 16 bits being used in some applications where enhancement of specific gray-level ranges is necessary. Sometimes we find systems that can digitize the gray levels of an image with 10 or 12 bits of accuracy, but these are the exception rather than the rule.

When an actual measure of physical resolution relating pixels and the level of detail they resolve in the original scene are not necessary, it is not uncommon to refer to an L-level digital image of size $M \times N$ as having a spatial resolution of $M \times N$ pixels and a gray-level resolution of L levels. We will use this terminology from time to time in subsequent discussions, making a reference to actual resolvable detail only when necessary for clarity.

EXAMPLE 2.2: Typical effects of varying the number of samples in a digital image.

■ Figure 2.19 shows an image of size 1024×1024 pixels whose gray levels are represented by 8 bits. The other images shown in Fig. 2.19 are the results of

32
64
128
256
512
1024

FIGURE 2.19 A 1024×1024, 8-bit image subsampled down to size 32×32 pixels. The number of allowable gray levels was kept at 256.

subsampling the 1024 × 1024 image. The subsampling was accomplished by deleting the appropriate number of rows and columns from the original image. For example, the 512 × 512 image was obtained by deleting every other row and column from the 1024 × 1024 image. The 256 × 256 image was generated by deleting every other row and column in the 512 × 512 image, and so on. The number of allowed gray levels was kept at 256.

These images show the dimensional proportions between various sampling densities, but their size differences make it difficult to see the effects resulting from a reduction in the number of samples. The simplest way to compare these effects is to bring all the subsampled images up to size 1024 × 1024 by row and column pixel replication. The results are shown in Figs. 2.20(b) through (f). Figure 2.20(a) is the same 1024 × 1024, 256-level image shown in Fig. 2.19; it is repeated to facilitate comparisons.

Compare Fig. 2.20(a) with the 512 × 512 image in Fig. 2.20(b) and note that it is virtually impossible to tell these two images apart. The level of detail lost is simply too fine to be seen on the printed page at the scale in which these im-

a b c
d e f

FIGURE 2.20 (a) 1024 × 1024, 8-bit image. (b) 512 × 512 image resampled into 1024 × 1024 pixels by row and column duplication. (c) through (f) 256 × 256, 128 × 128, 64 × 64, and 32 × 32 images resampled into 1024 × 1024 pixels.

ages are shown. Next, the 256 × 256 image in Fig. 2.20(c) shows a very slight fine checkerboard pattern in the borders between flower petals and the black background. A slightly more pronounced graininess throughout the image also is beginning to appear. These effects are much more visible in the 128 × 128 image in Fig. 2.20(d), and they become pronounced in the 64 × 64 and 32 × 32 images in Figs. 2.20(e) and (f), respectively.   ■

■ In this example, we keep the number of samples constant and reduce the number of gray levels from 256 to 2, in integer powers of 2. Figure 2.21(a) is a 452 × 374 CAT projection image, displayed with $k = 8$ (256 gray levels). Images such as this are obtained by fixing the X-ray source in one position, thus producing a 2-D image

EXAMPLE 2.3:
Typical effects of varying the number of gray levels in a digital image.

a b
c d

FIGURE 2.21
(a) 452 × 374, 256-level image. (b)–(d) Image displayed in 128, 64, and 32 gray levels, while keeping the spatial resolution constant.

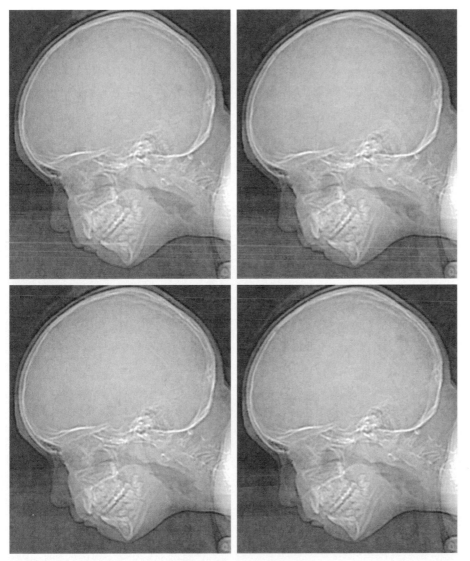

in any desired direction. Projection images are used as guides to set up the parameters for a CAT scanner, including tilt, number of slices, and range.

Figures 2.21(b) through (h) were obtained by reducing the number of bits from $k = 7$ to $k = 1$ while keeping the spatial resolution constant at 452×374 pixels. The 256-, 128-, and 64-level images are visually identical for all practical purposes. The 32-level image shown in Fig. 2.21(d), however, has an almost imperceptible set of very fine ridgelike structures in areas of smooth gray levels (particularly in the skull). This effect, caused by the use of an insufficient number of gray levels in smooth areas of a digital image, is called *false contouring*, so called because the ridges resemble topographic contours in a map. False contouring generally is quite visible in images displayed using 16 or less uniformly spaced gray levels, as the images in Figs. 2.21(e) through (h) show.

e f
g h

FIGURE 2.21
(Continued)
(e)–(h) Image
displayed in 16, 8,
4, and 2 gray
levels. (Original
courtesy of
Dr. David
R. Pickens,
Department of
Radiology &
Radiological
Sciences,
Vanderbilt
University
Medical Center.)

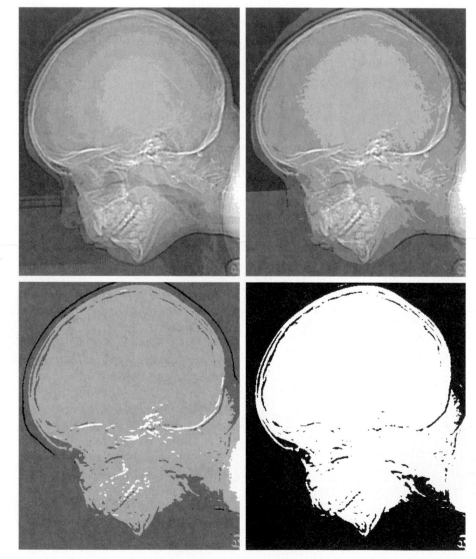

As a very rough rule of thumb, and assuming powers of 2 for convenience, images of size 256 × 256 pixels and 64 gray levels are about the smallest images that can be expected to be reasonably free of objectionable sampling checkerboards and false contouring. ■

The results in Examples 2.2 and 2.3 illustrate the effects produced on image quality by varying N and k independently. However, these results only partially answer the question of how varying N and k affect images because we have not considered yet any relationships that might exist between these two parameters. An early study by Huang [1965] attempted to quantify experimentally the effects on image quality produced by varying N and k simultaneously. The experiment consisted of a set of subjective tests. Images similar to those shown in Fig. 2.22 were used. The woman's face is representative of an image with relatively little detail; the picture of the cameraman contains an intermediate amount of detail; and the crowd picture contains, by comparison, a large amount of detail.

Sets of these three types of images were generated by varying N and k, and observers were then asked to rank them according to their subjective quality. Results were summarized in the form of so-called *isopreference curves* in the Nk-plane (Fig. 2.23 shows average isopreference curves representative of curves corresponding to the images shown in Fig. 2.22). Each point in the Nk-plane represents an image having values of N and k equal to the coordinates of that point. Points lying on an isopreference curve correspond to images of equal subjective quality. It was found in the course of the experiments that the isopreference curves tended to shift right and upward, but their shapes in each of the three image categories were similar to those shown in Fig. 2.23. This is not unexpected, since a shift up and right in the curves simply means larger values for N and k, which implies better picture quality.

The key point of interest in the context of the present discussion is that isopreference curves tend to become more vertical as the detail in the image increases. This result suggests that for images with a large amount of detail only

a b c

FIGURE 2.22 (a) Image with a low level of detail. (b) Image with a medium level of detail. (c) Image with a relatively large amount of detail. (Image (b) courtesy of the Massachusetts Institute of Technology.)

FIGURE 2.23
Representative
isopreference
curves for the
three types of
images in
Fig. 2.22.

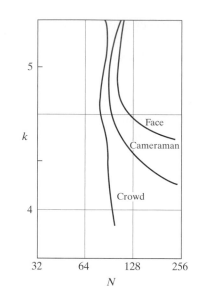

a few gray levels may be needed. For example, the isopreference curve in Fig. 2.23 corresponding to the crowd is nearly vertical. This indicates that, for a fixed value of N, the perceived quality for this type of image is nearly independent of the number of gray levels used (for the range of gray levels shown in Fig. 2.23). It is also of interest to note that perceived quality in the other two image categories remained the same in some intervals in which the spatial resolution was increased, but the number of gray levels actually decreased. The most likely reason for this result is that a decrease in k tends to increase the apparent contrast of an image, a visual effect that humans often perceive as improved quality in an image.

2.4.4 Aliasing and Moiré Patterns

As discussed in more detail in Chapter 4, functions whose area under the curve is finite can be represented in terms of sines and cosines of various frequencies. The sine/cosine component with the highest frequency determines the highest "frequency content" of the function. Suppose that this highest frequency is finite and that the function is of unlimited duration (these functions are called *band-limited functions*). Then, the Shannon sampling theorem [Bracewell (1995)] tells us that, if the function is sampled at a rate equal to or greater than twice its highest frequency, it is possible to recover completely the original function from its samples. If the function is *undersampled*, then a phenomenon called *aliasing* corrupts the sampled image. The corruption is in the form of additional frequency components being introduced into the sampled function. These are called *aliased frequencies*. Note that the *sampling rate* in images is the number of samples taken (in both spatial directions) per unit distance.

As it turns out, except for a special case discussed in the following paragraph, it is impossible to satisfy the sampling theorem in practice. We can only work with sampled data that are finite in duration. We can model the process of convert-

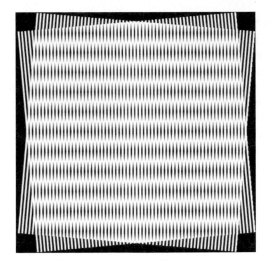

FIGURE 2.24 Illustration of the Moiré pattern effect.

ing a function of unlimited duration into a function of finite duration simply by multiplying the unlimited function by a "gating function" that is valued 1 for some interval and 0 elsewhere. Unfortunately, this function itself has frequency components that extend to infinity. Thus, the very act of limiting the duration of a band-limited function causes it to cease being band limited, which causes it to violate the key condition of the sampling theorem. The principal approach for reducing the aliasing effects on an image is to reduce its high-frequency components by blurring the image (we discuss blurring in detail in Chapter 4) *prior* to sampling. However, aliasing is always present in a sampled image. The effect of aliased frequencies can be seen under the right conditions in the form of so-called *Moiré patterns*[†], as discussed next.

There is one special case of significant importance in which a function of infinite duration can be sampled over a finite interval without violating the sampling theorem. When a function is periodic, it may be sampled at a rate equal to or exceeding twice its highest frequency, and it is possible to recover the function from its samples *provided* that the sampling captures *exactly* an integer number of periods of the function. This special case allows us to illustrate vividly the Moiré effect. Figure 2.24 shows two identical periodic patterns of equally spaced vertical bars, rotated in opposite directions and then superimposed on each other by multiplying the two images. A Moiré pattern, caused by a breakup of the periodicity, is seen in Fig. 2.24 as a 2-D sinusoidal (aliased) waveform (which looks like a corrugated tin roof) running in a vertical direction. A similar pattern can appear when images are digitized (e.g., scanned) from a printed page, which consists of periodic ink dots.

[†] The word *Moiré* appears to have originated with weavers and comes from the word *mohair*, a cloth made from Angora goat hairs.

2.4.5 Zooming and Shrinking Digital Images

We conclude the treatment of sampling and quantization with a brief discussion on how to zoom and shrink a digital image. This topic is related to image sampling and quantization because zooming may be viewed as oversampling, while shrinking may be viewed as undersampling. The key difference between these two operations and sampling and quantizing an original continuous image is that zooming and shrinking are applied to a *digital* image.

Zooming requires two steps: the creation of new pixel locations, and the assignment of gray levels to those new locations. Let us start with a simple example. Suppose that we have an image of size 500×500 pixels and we want to enlarge it 1.5 times to 750×750 pixels. Conceptually, one of the easiest ways to visualize zooming is laying an imaginary 750×750 grid over the original image. Obviously, the spacing in the grid would be less than one pixel because we are fitting it over a smaller image. In order to perform gray-level assignment for any point in the overlay, we look for the closest pixel in the original image and assign its gray level to the new pixel in the grid. When we are done with all points in the overlay grid, we simply expand it to the original specified size to obtain the zoomed image. This method of gray-level assignment is called *nearest neighbor interpolation*. (Pixel neighborhoods are discussed in the next section.)

Pixel replication, the method used to generate Figs. 2.20(b) through (f), is a special case of nearest neighbor interpolation. Pixel replication is applicable when we want to increase the size of an image an integer number of times. For instance, to double the size of an image, we can duplicate each column. This doubles the image size in the horizontal direction. Then, we duplicate each row of the enlarged image to double the size in the vertical direction. The same procedure is used to enlarge the image by any integer number of times (triple, quadruple, and so on). Duplication is just done the required number of times to achieve the desired size. The gray-level assignment of each pixel is predetermined by the fact that new locations are exact duplicates of old locations.

Although nearest neighbor interpolation is fast, it has the undesirable feature that it produces a checkerboard effect that is particularly objectionable at high factors of magnification. Figures 2.20(e) and (f) are good examples of this. A slightly more sophisticated way of accomplishing gray-level assignments is *bilinear interpolation* using the four nearest neighbors of a point. Let (x', y') denote the coordinates of a point in the zoomed image (think of it as a point on the grid described previously), and let $v(x', y')$ denote the gray level assigned to it. For bilinear interpolation, the assigned gray level is given by

$$v(x', y') = ax' + by' + cx'y' + d \tag{2.4-6}$$

where the four coefficients are determined from the four equations in four unknowns that can be written using the four nearest neighbors of point (x', y').

Image shrinking is done in a similar manner as just described for zooming. The equivalent process of pixel replication is row-column deletion. For example, to shrink an image by one-half, we delete every other row and column. We can use the zooming grid analogy to visualize the concept of shrinking by a noninteger factor, except that we now *expand* the grid to fit over the original image, do gray-level nearest neighbor or bilinear interpolation, and then shrink the grid back to its original specified size. To reduce possible aliasing effects, it is a good idea to blur an image slightly before shrinking it. Blurring of digital images is discussed in Chapters 3 and 4.

It is possible to use more neighbors for interpolation. Using more neighbors implies fitting the points with a more complex surface, which generally gives smoother results. This is an exceptionally important consideration in image generation for 3-D graphics [Watt (1993)] and in medical image processing [Lehmann et al. (1999)], but the extra computational burden seldom is justifiable for general-purpose digital image zooming and shrinking, where bilinear interpolation generally is the method of choice.

■ Figures 2.20(d) through (f) are shown again in the top row of Fig. 2.25. As noted earlier, these images were zoomed from 128×128, 64×64, and 32×32 to 1024×1024 pixels using nearest neighbor interpolation. The equivalent results using bilinear interpolation are shown in the second row of Fig. 2.25. The improvements in overall appearance are clear, especially in the 128×128 and

EXAMPLE 2.4:
Image zooming using bilinear interpolation.

a b c
d e f

FIGURE 2.25 Top row: images zoomed from 128×128, 64×64, and 32×32 pixels to 1024×1024 pixels, using nearest neighbor gray-level interpolation. Bottom row: same sequence, but using bilinear interpolation.

64×64 cases. The 32×32 to 1024×1024 image is blurry, but keep in mind that this image was zoomed by a factor of 32. In spite of this, the result of bilinear interpolation shown in Fig. 2.25(f) is a reasonably good rendition of the original image shape, something that is lost in Fig. 2.25(c).   ■

2.5   Some Basic Relationships Between Pixels

In this section, we consider several important relationships between pixels in a digital image. As mentioned before, an image is denoted by $f(x, y)$. When referring in this section to a particular pixel, we use lowercase letters, such as p and q.

2.5.1 Neighbors of a Pixel

A pixel p at coordinates (x, y) has four *horizontal* and *vertical* neighbors whose coordinates are given by

$$(x + 1, y), (x - 1, y), (x, y + 1), (x, y - 1)$$

This set of pixels, called the 4-*neighbors* of p, is denoted by $N_4(p)$. Each pixel is a unit distance from (x, y), and some of the neighbors of p lie outside the digital image if (x, y) is on the border of the image.

The four *diagonal* neighbors of p have coordinates

$$(x + 1, y + 1), (x + 1, y - 1), (x - 1, y + 1), (x - 1, y - 1)$$

and are denoted by $N_D(p)$. These points, together with the 4-neighbors, are called the 8-*neighbors* of p, denoted by $N_8(p)$. As before, some of the points in $N_D(p)$ and $N_8(p)$ fall outside the image if (x, y) is on the border of the image.

2.5.2 Adjacency, Connectivity, Regions, and Boundaries

Connectivity between pixels is a fundamental concept that simplifies the definition of numerous digital image concepts, such as regions and boundaries. To establish if two pixels are connected, it must be determined if they are neighbors and if their gray levels satisfy a specified criterion of similarity (say, if their gray levels are equal). For instance, in a binary image with values 0 and 1, two pixels may be 4-neighbors, but they are said to be connected only if they have the same value.

Let V be the set of gray-level values used to define adjacency. In a binary image, $V = \{1\}$ if we are referring to adjacency of pixels with value 1. In a gray-scale image, the idea is the same, but set V typically contains more elements. For example, in the adjacency of pixels with a range of possible gray-level values 0 to 255, set V could be any subset of these 256 values. We consider three types of adjacency:

(a) 4-*adjacency*. Two pixels p and q with values from V are 4-adjacent if q is in the set $N_4(p)$.
(b) 8-*adjacency*. Two pixels p and q with values from V are 8-adjacent if q is in the set $N_8(p)$.

(c) *m-adjacency* (mixed adjacency). Two pixels p and q with values from V are *m*-adjacent if

    (i) q is in $N_4(p)$, *or*

    (ii) q is in $N_D(p)$ *and* the set $N_4(p) \cap N_4(q)$ has no pixels whose values are from V.

Mixed adjacency is a modification of 8-adjacency. It is introduced to eliminate the ambiguities that often arise when 8-adjacency is used. For example, consider the pixel arrangement shown in Fig. 2.26(a) for $V = \{1\}$. The three pixels at the top of Fig. 2.26(b) show multiple (ambiguous) 8-adjacency, as indicated by the dashed lines. This ambiguity is removed by using *m*-adjacency, as shown in Fig. 2.26(c). Two image subsets S_1 and S_2 are adjacent if some pixel in S_1 is adjacent to some pixel in S_2. It is understood here and in the following definitions that *adjacent* means 4-, 8-, or *m*-adjacent.

A (*digital*) *path* (or *curve*) from pixel p with coordinates (x, y) to pixel q with coordinates (s, t) is a sequence of distinct pixels with coordinates

$$(x_0, y_0), (x_1, y_1), \dots, (x_n, y_n)$$

where $(x_0, y_0) = (x, y)$, $(x_n, y_n) = (s, t)$, and pixels (x_i, y_i) and (x_{i-1}, y_{i-1}) are adjacent for $1 \leq i \leq n$. In this case, n is the *length* of the path. If $(x_0, y_0) = (x_n, y_n)$, the path is a *closed* path. We can define 4-, 8-, or *m*-paths depending on the type of adjacency specified. For example, the paths shown in Fig. 2.26(b) between the northeast and southeast points are 8-paths, and the path in Fig. 2.26(c) is an *m*-path. Note the absence of ambiguity in the *m*-path.

Let S represent a subset of pixels in an image. Two pixels p and q are said to be *connected* in S if there exists a path between them consisting entirely of pixels in S. For any pixel p in S, the *set* of pixels that are connected to it in S is called a *connected component* of S. If it only has one connected component, then set S is called a *connected set*.

Let R be a subset of pixels in an image. We call R a *region* of the image if R is a connected set. The *boundary* (also called *border* or *contour*) of a region R is the set of pixels in the region that have one or more neighbors that are not in R. If R happens to be an entire image (which we recall is a rectangular set of pixels), then its boundary is defined as the set of pixels in the first and last rows and columns of the image. This extra definition is required because an image has no neighbors beyond its border. Normally, when we refer to a region, we are

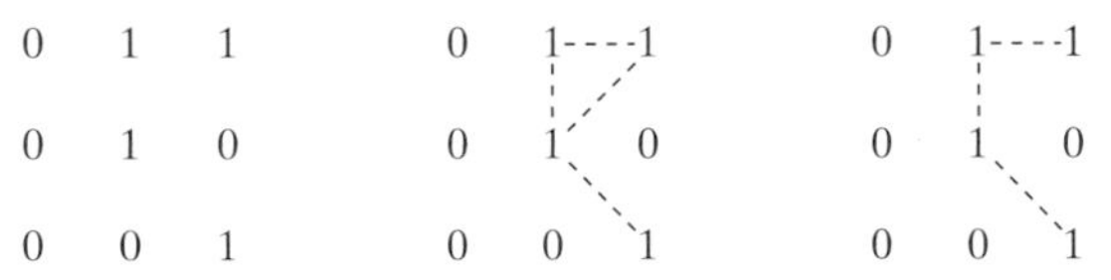

a b c

FIGURE 2.26 (a) Arrangement of pixels; (b) pixels that are 8-adjacent (shown dashed) to the center pixel; (c) *m*-adjacency.

referring to a subset of an image, and any pixels in the boundary of the region that happen to coincide with the border of the image are included implicitly as part of the region boundary.

The concept of an *edge* is found frequently in discussions dealing with regions and boundaries. There is a key difference between these concepts, however. The boundary of a finite region forms a closed path (Problem 2.14) and is thus a "global" concept. As discussed in detail in Chapter 10, edges are formed from pixels with derivative values that exceed a preset threshold. Thus, the idea of an edge is a "local" concept that is based on a measure of gray-level discontinuity at a point. It is possible to link edge points into edge segments, and sometimes these segments are linked in such a way that correspond to boundaries, but this is not always the case. The one exception in which edges and boundaries correspond is in binary images. Depending on the type of connectivity and edge operators used (we discuss these in Chapter 10), the edge extracted from a binary region will be the same as the region boundary. This is intuitive. Conceptually, until we arrive at Chapter 10, it is helpful to think of edges as intensity discontinuities and boundaries as closed paths.

2.5.3 Distance Measures

For pixels p, q, and z, with coordinates (x, y), (s, t), and (v, w), respectively, D is a *distance function* or *metric* if

(a) $D(p, q) \geq 0$ $(D(p, q) = 0$ iff $p = q)$,
(b) $D(p, q) = D(q, p)$, and
(c) $D(p, z) \leq D(p, q) + D(q, z)$.

The *Euclidean distance* between p and q is defined as

$$D_e(p, q) = \left[(x - s)^2 + (y - t)^2\right]^{\frac{1}{2}}. \tag{2.5-1}$$

For this distance measure, the pixels having a distance less than or equal to some value r from (x, y) are the points contained in a disk of radius r centered at (x, y).

The D_4 *distance* (also called *city-block distance*) between p and q is defined as

$$D_4(p, q) = |x - s| + |y - t|. \tag{2.5-2}$$

In this case, the pixels having a D_4 distance from (x, y) less than or equal to some value r form a diamond centered at (x, y). For example, the pixels with D_4 distance ≤ 2 from (x, y) (the center point) form the following contours of constant distance:

```
          2
       2  1  2
    2  1  0  1  2
       2  1  2
          2
```

The pixels with $D_4 = 1$ are the 4-neighbors of (x, y).

The D_8 *distance* (also called *chessboard distance*) between p and q is defined as

$$D_8(p, q) = \max(|x - s|, |y - t|). \qquad (2.5\text{-}3)$$

In this case, the pixels with D_8 distance from (x, y) less than or equal to some value r form a square centered at (x, y). For example, the pixels with D_8 distance ≤ 2 from (x, y) (the center point) form the following contours of constant distance:

$$
\begin{array}{ccccc}
2 & 2 & 2 & 2 & 2 \\
2 & 1 & 1 & 1 & 2 \\
2 & 1 & 0 & 1 & 2 \\
2 & 1 & 1 & 1 & 2 \\
2 & 2 & 2 & 2 & 2 \\
\end{array}
$$

The pixels with $D_8 = 1$ are the 8-neighbors of (x, y).

Note that the D_4 and D_8 distances between p and q are independent of any paths that might exist between the points because these distances involve only the coordinates of the points. If we elect to consider m-adjacency, however, the D_m distance between two points is defined as the shortest m-path between the points. In this case, the distance between two pixels will depend on the values of the pixels along the path, as well as the values of their neighbors. For instance, consider the following arrangement of pixels and assume that p, p_2, and p_4 have value 1 and that p_1 and p_3 can have a value of 0 or 1:

$$
\begin{array}{cc}
p_3 & p_4 \\
p_1 & p_2 \\
p & \\
\end{array}
$$

Suppose that we consider adjacency of pixels valued 1 (i.e., $V = \{1\}$). If p_1 and p_3 are 0, the length of the shortest m-path (the D_m distance) between p and p_4 is 2. If p_1 is 1, then p_2 and p will no longer be m-adjacent (see the definition of m adjacency) and the length of the shortest m-path becomes 3 (the path goes through the points $pp_1p_2p_4$). Similar comments apply if p_3 is 1 (and p_1 is 0); in this case, the length of the shortest m-path also is 3. Finally, if both p_1 and p_3 are 1 the length of the shortest m-path between p and p_4 is 4. In this case, the path goes through the sequence of points $pp_1p_2p_3p_4$.

2.5.4 Image Operations on a Pixel Basis

Numerous references are made in the following chapters to operations between images, such as dividing one image by another. In Eq. (2.4-2), images were represented in the form of matrices. As we know, matrix division is not defined. However, when we refer to an operation like "dividing one image by another," we mean specifically that the division is carried out between *corresponding* pixels in the two images. Thus, for example, if f and g are images, the first element of the image formed by "dividing" f by g is simply the first pixel in f divided by the first pixel in g; of course, the assumption is that none of the pixels in g have value 0. Other arithmetic and logic operations are similarly defined between corresponding pixels in the images involved.

2.6   Linear and Nonlinear Operations

Let H be an operator whose input and output are images. H is said to be a *linear* operator if, for any two images f and g and any two scalars a and b,

$$H(af + bg) = aH(f) + bH(g). \tag{2.6-1}$$

In other words, the result of applying a linear operator to the sum of two images (that have been multiplied by the constants shown) is identical to applying the operator to the images individually, multiplying the results by the appropriate constants, and then adding those results. For example, an operator whose function is to compute the sum of K images is a linear operator. An operator that computes the absolute value of the difference of two images is not. An operator that fails the test of Eq. (2.6-1) is by definition *nonlinear*.

Linear operations are exceptionally important in image processing because they are based on a significant body of well-understood theoretical and practical results. Although nonlinear operations sometimes offer better performance, they are not always predictable, and for the most part are not well understood theoretically.

Summary

The material in this chapter is primarily background information for subsequent discussions. Our treatment of the human visual system, although brief, provides a basic idea of the capabilities of the eye in perceiving pictorial information. The discussion of light and the electromagnetic spectrum is fundamental in understanding the origin of the many images we use in this book. Similarly, the image model developed in Section 2.3.4 is used in the Chapter 4 as the basis for an image enhancement technique called *homomorphic filtering*, and again in Chapter 10 to explain the effect of illumination on the shape of image histograms.

The sampling ideas introduced in Section 2.4 are the foundation for many of the digitizing phenomena likely to be encountered in practice. These ideas can be expanded further once a basic understanding of frequency content is mastered. A detailed discussion of the frequency domain is given in Chapter 4. The concepts of sampling and aliasing effects also are of importance in the context of image acquisition.

The concepts introduced in Section 2.5 are the basic building blocks for processing techniques based on pixel neighborhoods. As shown in the following chapter and in Chapter 5, neighborhood processing methods are at the core of many image enhancement and restoration procedures. When applicable, neighborhood processing is favored in commercial applications of image processing due to their operational speed and simplicity of implementation in hardware and/or firmware. Finally, the concept of a linear operator and the theoretical and conceptual power associated with it will be used extensively in the following three chapters.

References and Further Reading

Additional reading for the material in Section 2.1 regarding the structure of the human eye may be found in Atchison and Smith [2000], and Oyster [1999]. For additional reading on visual perception, see Regan [2000] and Gordon [1997]. The book by Hubel [1988] and the now classic book by Cornsweet [1970] also are of interest. Born and Wolf [1999] is a basic reference that discusses light in terms of electromagnetic theory. Electromagnetic energy propagation is covered in some detail by Felsen and Marcuvitz [1994].

The area of image sensing is quite broad and very fast moving. An excellent source of information on optical and other imaging sensors is the International Society for Optical Engineering (SPIE). The following are representative publications by the SPIE in this area: Blouke et al. [2001], Hoover and Doty [1996], and Freeman [1987].

The image model presented in Section 2.3.4 is from Oppenheim, Schafer, and Stockham [1968]. A reference for the illumination and reflectance values used in that section is the *IES Lighting Handbook* [2000]. For additional reading on image sampling and some of its effects, such as aliasing, see Bracewell [1995]. The early experiments mentioned in Section 2.4.3 on perceived image quality as a function of sampling and quatization were reported by Huang [1965]. The issue of reducing the number of samples and gray levels in an image while minimizing the ensuing degradation is still of current interest, as exemplified by Papamarkos and Atsalakis [2000]. For further reading on image shrinking and zooming, see Sid-Ahmed [1995], Unser et al. [1995], Umbaugh [1998], and Lehmann et al. [1999]. For further reading on the topics covered in Section 2.5, see Rosenfeld and Kak [1982], Marchand-Maillet and Sharaiha [2000], and Ritter and Wilson [2001]. Additional reading on linear systems in the context of image processing may be found in Castleman [1996].

Problems

★ **2.1** Using the background information provided in Section 2.1, and thinking purely in geometric terms, estimate the diameter of the smallest printed dot that the eye can discern if the page on which the dot is printed is 0.2 m away from the eyes. Assume for simplicity that the visual system ceases to detect the dot when the image of the dot on the fovea becomes smaller than the diameter of one receptor (cone) in that area of the retina. Assume further that the fovea can be modeled as a square array of dimensions 1.5 mm × 1.5 mm, and that the cones and spaces between the cones are distributed uniformly throughout this array.

2.2 When you enter a dark theater on a bright day, it takes an appreciable interval of time before you can see well enough to find an empty seat. Which of the visual processes explained in Section 2.1 is at play in this situation?

★ **2.3** Although it is not shown in Fig. 2.10, alternating current certainly is part of the electromagnetic spectrum. Commercial alternating current in the United States has a frequency of 60 Hz. What is the wavelength in kilometers of this component of the spectrum?

2.4 You are hired to design the front end of an imaging system for studying the boundary shapes of cells, bacteria, viruses, and protein. The front end consists, in this case, of the illumination source(s) and corresponding imaging camera(s). The diameters of circles required to enclose individual specimens in each of these categories are 50, 1, 0.1, and 0.01 μm, respectively.

  (a) Can you solve the imaging aspects of this problem with a single sensor and camera? If your answer is yes, specify the illumination wavelength band and the type of camera needed. Identify the camera as being a color camera, far-infrared camera, or whatever appropriate name corresponds to the illumination source.

  (b) If your answer in (a) is no, what type of illumination sources and corresponding imaging sensors would you recommend? Specify the light sources

See inside front cover

Detailed solutions to the problems marked with a star can be found in the book web site. The site also contains suggested projects based on the material in this chapter.

and cameras as requested in part (a). Use the *minimum* number of illumination sources and cameras needed to solve the problem.

2.5 A CCD camera chip of dimensions 7×7 mm, and having 1024×1024 elements, is focused on a square, flat area, located 0.5 m away. How many line pairs per mm will this camera be able to resolve? The camera is equipped with a 35-mm lens. (*Hint:* Model the imaging process as in Fig. 2.3, with the focal length of the camera lens substituting for the focal length of the eye.)

★ 2.6 An automobile manufacturer is automating the placement of certain components on the bumpers of a limited-edition line of sports cars. The components are color coordinated, so the robots need to know the color of each car in order to select the appropriate bumper component. Models come in only four colors: blue, green, red, and white. You are hired to propose a solution based on imaging. How would you solve the problem of automatically determining the color of each car, keeping in mind that *cost* is the most important consideration in your choice of components?

2.7 Suppose that a flat area with center at $(x_0,\ y_0)$ is illuminated by a light source with intensity distribution

$$i(x,\ y)\ =\ Ke^{-[(x-x_0)^2+(y-y_0)^2]}.$$

Assume for simplicity that the reflectance of the area is constant and equal to 1.0, and let $K = 255$. If the resulting image is digitized with k bits of intensity resolution, and the eye can detect an abrupt change of eight shades of intensity between adjacent pixels, what value of k will cause visible false contouring?

2.8 Sketch the image in Problem 2.7 for $k = 2$.

★ 2.9 A common measure of transmission for digital data is the *baud rate*, defined as the number of bits transmitted per second. Generally, transmission is accomplished in packets consisting of a start bit, a byte (8 bits) of information, and a stop bit. Using these facts, answer the following:

   (a) How many minutes would it take to transmit a 1024×1024 image with 256 gray levels using a 56K baud modem?

   (b) What would the time be at 750K baud, a representative speed of a phone DSL (digital subscriber line) connection?

2.10 High-definition television (HDTV) generates images with a resolution of 1125 horizontal TV lines interlaced (where every other line is painted on the tube face in each of two fields, each field being 1/60th of a second in duration). The width-to-height aspect ratio of the images is 16:9. The fact that the horizontal lines are distinct fixes the vertical resolution of the images. A company has designed an image capture system that generates digital images from HDTV images. The resolution of each TV (horizontal) line in their system is in proportion to vertical resolution, with the proportion being the width-to-height ratio of the images. Each pixel in the color image has 24 bits of intensity resolution, 8 bits each of a red, a green, and a blue image. These three "primary" images form a color image. How many bits would it take to store a 2-hour HDTV program?

★ 2.11 Consider the two image subsets, S_1 and S_2, shown in the following figure. For $V = \{1\}$, determine whether these two subsets are (a) 4-adjacent, (b) 8-adjacent, or (c) *m*-adjacent.

	S_1				S_2				
0	0	0	0	0	0	0	1	1	0
1	0	0	1	0	0	1	0	0	1
1	0	0	1	0	1	1	0	0	0
0	0	1	1	1	0	0	0	0	0
0	0	1	1	1	0	0	1	1	1

★ **2.12** Develop an algorithm for converting a one-pixel-thick 8-path to a 4-path.

2.13 Develop an algorithm for converting a one-pixel-thick m-path to a 4-path.

2.14 Show that the boundary of the region, as defined in Section 2.5.2, is a closed path.

★ **2.15** Consider the image segment shown.

    (a) Let $V = \{0, 1\}$ and compute the lengths of the shortest 4-, 8-, and m-path between p and q. If a particular path does not exist between these two points, explain why.

    (b) Repeat for $V = \{1, 2\}$.

$$\begin{array}{cccc} 3 & 1 & 2 & 1\,(q) \\ 2 & 2 & 0 & 2 \\ 1 & 2 & 1 & 1 \\ (p)\,1 & 0 & 1 & 2 \end{array}$$

★ **2.16** **(a)** Give the condition(s) under which the D_4 distance between two points p and q is equal to the shortest 4-path between these points.

    (b) Is this path unique?

2.17 Repeat Problem 2.16 for the D_8 distance.

★ **2.18** In the following chapter, we will deal with operators whose function is to compute the sum of pixel values in a small subimage area, S. Show that these are linear operators.

2.19 The median, ζ, of a set of numbers is such that half the values in the set are below ζ and the other half are above it. For example, the median of the set of values $\{2, 3, 8, 20, 21, 25, 31\}$ is 20. Show that an operator that computes the median of a subimage area, S, is nonlinear.

2.20 A plant produces a line of translucent miniature polymer squares. Stringent quality requirements dictate 100% visual inspection, and the plant manager finds the use of human inspectors increasingly expensive. Inspection is semiautomated. At each inspection station, a robotic mechanism places each polymer square over a light located under an optical system that produces a magnified image of the square. The image completely fills a viewing screen measuring 80 × 80 mm. Defects appear as dark circular blobs, and the inspector's job is to look at the screen and reject any sample that has one or more such dark blobs with a diameter of 0.8 mm or larger, as measured on the scale of the screen. The manager believes that, if she can find a way to automate the process completely, she will increase profits by 50%. She also believes that success in this project will aid her climb up the corporate ladder. After much investigation, the manager decides that the way to solve the problem is to view each inspection screen with a CCD TV camera and feed the output of the camera into an image processing system capable of detecting the blobs, measuring their diameter, and activating the accept/reject buttons previously operated by an inspector. She is able to find a system that can do the job, as long as the smallest defect occupies an area of at least 2×2 pixels in the digital image. The manager hires you to help her specify the camera and lens system, but requires that you use off-the-shelf components. For the lenses, assume that this constraint means any integer multiple of 25 mm or 35 mm, up to 200 mm. For the cameras, it means resolutions of $512 \times 512, 1024 \times 1024,$ or 2048×2048 pixels. The *individual* imaging elements in these cameras are squares measuring 8×8 μm, and the spaces between imaging elements are 2 μm. For this application, the cameras cost much more than the lenses, so the problem should be solved with the lowest-resolution camera possible, based on the choice of lenses. As a consultant, you are to provide a written recommendation, showing in reasonable detail the analysis that led to your conclusion. Use the same imaging geometry suggested in Problem 2.5.

3 *Image Enhancement in the Spatial Domain*

It makes all the difference whether one sees darkness
through the light or brightness through the shadows.

David Lindsay

Preview

The principal objective of enhancement is to process an image so that the result is more suitable than the original image for a specific application. The word *specific* is important, because it establishes at the outset that the techniques discussed in this chapter are very much problem oriented. Thus, for example, a method that is quite useful for enhancing X-ray images may not necessarily be the best approach for enhancing pictures of Mars transmitted by a space probe. Regardless of the method used, however, image enhancement is one of the most interesting and visually appealing areas of image processing.

Image enhancement approaches fall into two broad categories: spatial domain methods and frequency domain methods. The term *spatial domain* refers to the image plane itself, and approaches in this category are based on direct manipulation of pixels in an image. *Frequency domain* processing techniques are based on modifying the Fourier transform of an image. Spatial methods are covered in this chapter, and frequency domain enhancement is discussed in Chapter 4. Enhancement techniques based on various combinations of methods from these two categories are not unusual. We note also that many of the fundamental techniques introduced in this chapter in the context of enhancement are used in subsequent chapters for a variety of other image processing applications.

There is no general theory of image enhancement. When an image is processed for visual interpretation, the viewer is the ultimate judge of how well a particular method works. Visual evaluation of image quality is a highly subjective process, thus making the definition of a "good image" an elusive standard by which to compare algorithm performance. When the problem is one of processing images for machine perception, the evaluation task is somewhat easier. For example, in dealing with a character recognition application, and leaving aside other issues such as computational requirements, the best image processing method would be the one yielding the best machine recognition results. However, even in situations when a clear-cut criterion of performance can be imposed on the problem, a certain amount of trial and error usually is required before a particular image enhancement approach is selected.

3.1 Background

As indicated previously, the term *spatial domain* refers to the aggregate of pixels composing an image. Spatial domain methods are procedures that operate directly on these pixels. Spatial domain processes will be denoted by the expression

$$g(x, y) = T[f(x, y)] \tag{3.1-1}$$

where $f(x, y)$ is the input image, $g(x, y)$ is the processed image, and T is an operator on f, defined over some neighborhood of (x, y). In addition, T can operate on a *set* of input images, such as performing the pixel-by-pixel sum of K images for noise reduction, as discussed in Section 3.4.2.

The principal approach in defining a neighborhood about a point (x, y) is to use a square or rectangular subimage area centered at (x, y), as Fig. 3.1 shows. The center of the subimage is moved from pixel to pixel starting, say, at the top left corner. The operator T is applied at each location (x, y) to yield the output, g, at that location. The process utilizes only the pixels in the area of the image spanned by the neighborhood. Although other neighborhood shapes, such as ap-

FIGURE 3.1 A 3×3 neighborhood about a point (x, y) in an image.

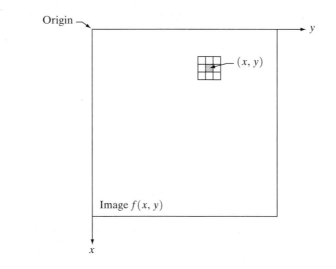

proximations to a circle, sometimes are used, square and rectangular arrays are by far the most predominant because of their ease of implementation.

The simplest form of T is when the neighborhood is of size 1×1 (that is, a single pixel). In this case, g depends only on the value of f at (x, y), and T becomes a *gray-level* (also called an *intensity* or *mapping*) *transformation function* of the form

$$s = T(r) \qquad (3.1\text{-}2)$$

where, for simplicity in notation, r and s are variables denoting, respectively, the gray level of $f(x, y)$ and $g(x, y)$ at any point (x, y). For example, if $T(r)$ has the form shown in Fig. 3.2(a), the effect of this transformation would be to produce an image of higher contrast than the original by darkening the levels below m and brightening the levels above m in the original image. In this technique, known as *contrast stretching*, the values of r below m are compressed by the transformation function into a narrow range of s, toward black. The opposite effect takes place for values of r above m. In the limiting case shown in Fig. 3.2(b), $T(r)$ produces a two-level (binary) image. A mapping of this form is called a *thresholding* function. Some fairly simple, yet powerful, processing approaches can be formulated with gray-level transformations. Because enhancement at any point in an image depends only on the gray level at that point, techniques in this category often are referred to as *point processing*.

Larger neighborhoods allow considerably more flexibility. The general approach is to use a function of the values of f in a predefined neighborhood of (x, y) to determine the value of g at (x, y). One of the principal approaches in this formulation is based on the use of so-called *masks* (also referred to as *filters, kernels, templates,* or *windows*). Basically, a mask is a small (say, 3×3) 2-D array, such as the one shown in Fig. 3.1, in which the values of the mask coefficients determine the nature of the process, such as image sharpening. Enhancement techniques based on this type of approach often are referred to as *mask processing* or *filtering.* These concepts are discussed in Section 3.5.

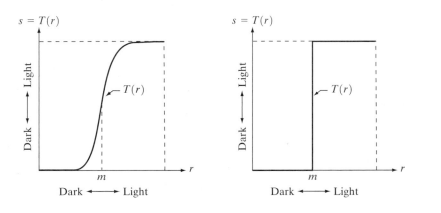

a b

FIGURE 3.2 Gray-level transformation functions for contrast enhancement.

3.2 Some Basic Gray Level Transformations

We begin the study of image enhancement techniques by discussing gray-level transformation functions. These are among the simplest of all image enhancement techniques. The values of pixels, before and after processing, will be denoted by r and s, respectively. As indicated in the previous section, these values are related by an expression of the form $s = T(r)$, where T is a transformation that maps a pixel value r into a pixel value s. Since we are dealing with digital quantities, values of the transformation function typically are stored in a one-dimensional array and the mappings from r to s are implemented via table lookups. For an 8-bit environment, a lookup table containing the values of T will have 256 entries.

As an introduction to gray-level transformations, consider Fig. 3.3, which shows three basic types of functions used frequently for image enhancement: linear (negative and identity transformations), logarithmic (log and inverse-log transformations), and power-law (nth power and nth root transformations). The identity function is the trivial case in which output intensities are identical to input intensities. It is included in the graph only for completeness.

3.2.1 Image Negatives

The negative of an image with gray levels in the range $[0, L-1]$ is obtained by using the negative transformation shown in Fig. 3.3, which is given by the expression

$$s = L - 1 - r. \tag{3.2-1}$$

FIGURE 3.3 Some basic gray-level transformation functions used for image enhancement.

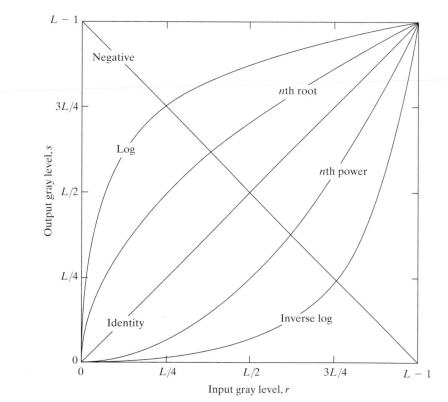

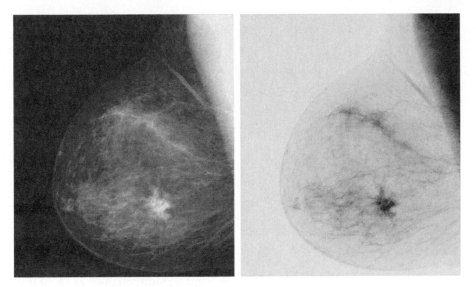

FIGURE 3.4
(a) Original
digital
mammogram.
(b) Negative
image obtained
using the negative
transformation in
Eq. (3.2-1).
(Courtesy of G.E.
Medical Systems.)

Reversing the intensity levels of an image in this manner produces the equivalent of a photographic negative. This type of processing is particularly suited for enhancing white or gray detail embedded in dark regions of an image, especially when the black areas are dominant in size. An example is shown in Fig. 3.4. The original image is a digital mammogram showing a small lesion. In spite of the fact that the visual content is the same in both images, note how much easier it is to analyze the breast tissue in the negative image in this particular case.

3.2.2 Log Transformations

The general form of the log transformation shown in Fig. 3.3 is

$$s = c \log(1 + r) \tag{3.2-2}$$

where c is a constant, and it is assumed that $r \geq 0$. The shape of the log curve in Fig. 3.3 shows that this transformation maps a narrow range of low gray-level values in the input image into a wider range of output levels. The opposite is true of higher values of input levels. We would use a transformation of this type to expand the values of dark pixels in an image while compressing the higher-level values. The opposite is true of the inverse log transformation.

Any curve having the general shape of the log functions shown in Fig. 3.3 would accomplish this spreading/compressing of gray levels in an image. In fact, the power-law transformations discussed in the next section are much more versatile for this purpose than the log transformation. However, the log function has the important characteristic that it compresses the dynamic range of images with large variations in pixel values. A classic illustration of an application in which pixel values have a large dynamic range is the Fourier spectrum, which will be discussed in Chapter 4. At the moment, we are concerned only with the image characteristics of spectra. It is not unusual to encounter spectrum values

FIGURE 3.5
(a) Fourier
spectrum.
(b) Result of
applying the log
transformation
given in
Eq. (3.2-2) with
$c = 1$.

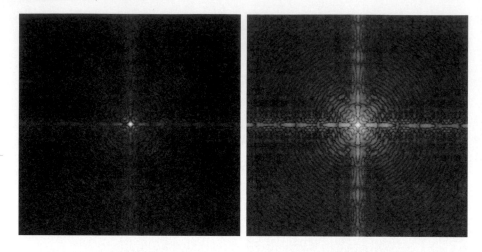

that range from 0 to 10^6 or higher. While processing numbers such as these presents no problems for a computer, image display systems generally will not be able to reproduce faithfully such a wide range of intensity values. The net effect is that a significant degree of detail will be lost in the display of a typical Fourier spectrum.

As an illustration of log transformations, Fig. 3.5(a) shows a Fourier spectrum with values in the range 0 to 1.5×10^6. When these values are scaled linearly for display in an 8-bit system, the brightest pixels will dominate the display, at the expense of lower (and just as important) values of the spectrum. The effect of this dominance is illustrated vividly by the relatively small area of the image in Fig. 3.5(a) that is not perceived as black. If, instead of displaying the values in this manner, we first apply Eq. (3.2-2) (with $c = 1$ in this case) to the spectrum values, then the range of values of the result become 0 to 6.2, a more manageable number. Figure 3.5(b) shows the result of scaling this new range linearly and displaying the spectrum in the same 8-bit display. The wealth of detail visible in this image as compared to a straight display of the spectrum is evident from these pictures. Most of the Fourier spectra seen in image processing publications have been scaled in just this manner.

3.2.3 Power-Law Transformations

Power-law transformations have the basic form

$$s = cr^\gamma \tag{3.2-3}$$

where c and γ are positive constants. Sometimes Eq. (3.2-3) is written as $s = c(r + \varepsilon)^\gamma$ to account for an offset (that is, a measurable output when the input is zero). However, offsets typically are an issue of display calibration and as a result they are normally ignored in Eq. (3.2-3). Plots of s versus r for various values of γ are shown in Fig. 3.6. As in the case of the log transformation, power-law curves with fractional values of γ map a narrow range of dark input values into a wider range of output values, with the opposite being true for high-

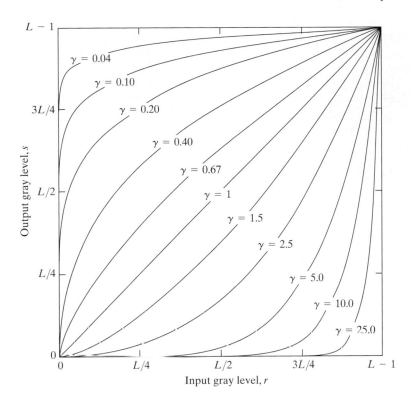

FIGURE 3.6 Plots
of the equation
$s = cr^\gamma$ for
various values of
γ ($c = 1$ in all
cases).

er values of input levels. Unlike the log function, however, we notice here a
family of possible transformation curves obtained simply by varying γ. As expected, we see in Fig. 3.6 that curves generated with values of $\gamma > 1$ have exactly the opposite effect as those generated with values of $\gamma < 1$. Finally, we
note that Eq. (3.2-3) reduces to the identity transformation when $c = \gamma = 1$.

A variety of devices used for image capture, printing, and display respond according to a power law. By convention, the exponent in the power-law equation
is referred to as *gamma* [hence our use of this symbol in Eq. (3.2-3)]. The process
used to correct this power-law response phenomena is called *gamma correction*. For example, cathode ray tube (CRT) devices have an intensity-to-volt-
age response that is a power function, with exponents varying from
approximately 1.8 to 2.5. With reference to the curve for $\gamma = 2.5$ in Fig. 3.6, we
see that such display systems would tend to produce images that are darker
than intended. This effect is illustrated in Fig. 3.7. Figure 3.7(a) shows a simple
gray-scale linear wedge input into a CRT monitor. As expected, the output of
the monitor appears darker than the input, as shown in Fig. 3.7(b). Gamma correction in this case is straightforward. All we need to do is preprocess the input
image before inputting it into the monitor by performing the transformation
$s = r^{1/2.5} = r^{0.4}$. The result is shown in Fig. 3.7(c). When input into the same
monitor, this gamma-corrected input produces an output that is close in appearance to the original image, as shown in Fig. 3.7(d). A similar analysis would

FIGURE 3.7
(a) Linear-wedge gray-scale image.
(b) Response of monitor to linear wedge.
(c) Gamma-corrected wedge.
(d) Output of monitor.

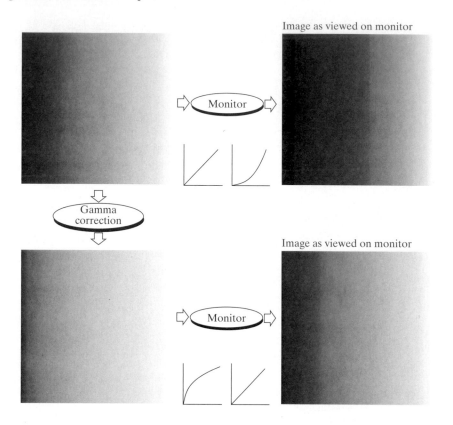

apply to other imaging devices such as scanners and printers. The only difference would be the device-dependent value of gamma (Poynton [1996]).

Gamma correction is important if displaying an image accurately on a computer screen is of concern. Images that are not corrected properly can look either bleached out, or, what is more likely, too dark. Trying to reproduce colors accurately also requires some knowledge of gamma correction because varying the value of gamma correction changes not only the brightness, but also the ratios of red to green to blue. Gamma correction has become increasingly important in the past few years, as use of digital images for commercial purposes over the Internet has increased. It is not unusual that images created for a popular Web site will be viewed by millions of people, the majority of whom will have different monitors and/or monitor settings. Some computer systems even have partial gamma correction built in. Also, current image standards do not contain the value of gamma with which an image was created, thus complicating the issue further. Given these constraints, a reasonable approach when storing images in a Web site is to preprocess the images with a gamma that represents an "average" of the types of monitors and computer systems that one expects in the open market at any given point in time.

EXAMPLE 3.1:
Contrast enhancement using power-law transformations.

■ In addition to gamma correction, power-law transformations are useful for general-purpose contrast manipulation. Figure 3.8(a) shows a magnetic resonance (MR) image of an upper thoracic human spine with a fracture dislocation

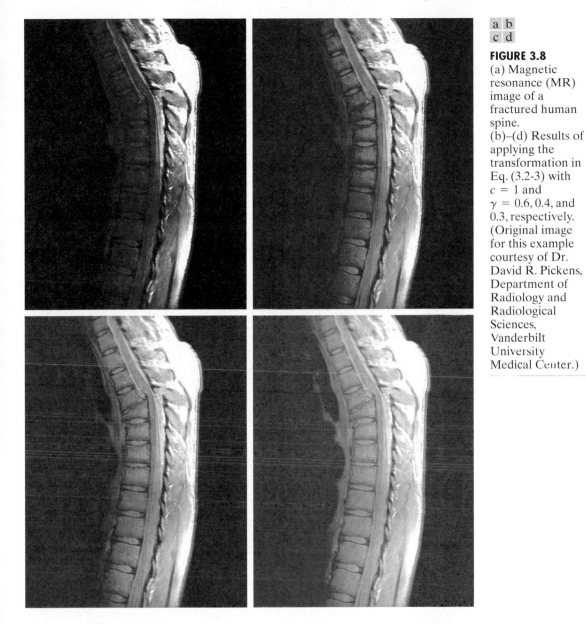

FIGURE 3.8
(a) Magnetic resonance (MR) image of a fractured human spine.
(b)–(d) Results of applying the transformation in Eq. (3.2-3) with $c = 1$ and $\gamma = 0.6, 0.4,$ and 0.3, respectively. (Original image for this example courtesy of Dr. David R. Pickens, Department of Radiology and Radiological Sciences, Vanderbilt University Medical Center.)

and spinal cord impingement. The fracture is visible near the vertical center of the spine, approximately one-fourth of the way down from the top of the picture. Since the given image is predominantly dark, an expansion of gray levels are desirable. This can be accomplished with a power-law transformation with a fractional exponent. The other images shown in the figure were obtained by processing Fig. 3.8(a) with the power-law transformation function of Eq. (3.2-3). The values of gamma corresponding to images (b) through (d) are 0.6, 0.4, and 0.3, respectively (the value of c was 1 in all cases). We note that, as gamma decreased from 0.6 to 0.4, more detail became visible. A further decrease of gamma to 0.3 enhanced a little more detail in the background, but began to reduce contrast to the point where the image started to have a very slight "washed-out" look, especially in the background. By comparing all results, we see that the best enhancement in terms of contrast and discernable detail was obtained with $\gamma = 0.4$. A value of $\gamma = 0.3$ is an approximate limit below which contrast in this particular image would be reduced to an unacceptable level.   ■

EXAMPLE 3.2:
Another
illustration of
power-law
transformations.

■ Figure 3.9(a) shows the opposite problem of Fig. 3.8(a). The image to be enhanced now has a washed-out appearance, indicating that a compression of gray levels is desirable. This can be accomplished with Eq. (3.2-3) using values of γ greater than 1. The results of processing Fig. 3.9(a) with $\gamma = 3.0$, 4.0, and 5.0 are shown in Figs. 3.9(b) through (d). Suitable results were obtained with gamma values of 3.0 and 4.0, the latter having a slightly more appealing appearance because it has higher contrast. The result obtained with $\gamma = 5.0$ has areas that are too dark, in which some detail is lost. The dark region to the left of the main road in the upper left quadrant is an example of such an area.   ■

a b
c d

FIGURE 3.9
(a) Aerial image.
(b)–(d) Results of
applying the
transformation in
Eq. (3.2-3) with
$c = 1$ and
$\gamma = 3.0$, 4.0, and
5.0, respectively.
(Original image
for this example
courtesy of
NASA.)

3.2.4 Piecewise-Linear Transformation Functions

A complementary approach to the methods discussed in the previous three sections is to use piecewise linear functions. The principal advantage of piecewise linear functions over the types of functions we have discussed thus far is that the form of piecewise functions can be arbitrarily complex. In fact, as we will see shortly, a practical implementation of some important transformations can be formulated only as piecewise functions. The principal disadvantage of piecewise functions is that their specification requires considerably more user input.

Contrast stretching

One of the simplest piecewise linear functions is a contrast-stretching transformation. Low-contrast images can result from poor illumination, lack of dynamic range in the imaging sensor, or even wrong setting of a lens aperture during image acquisition. The idea behind contrast stretching is to increase the dynamic range of the gray levels in the image being processed.

Figure 3.10(a) shows a typical transformation used for contrast stretching. The locations of points (r_1, s_1) and (r_2, s_2) control the shape of the transformation

a b
c d

FIGURE 3.10
Contrast stretching.
(a) Form of transformation function. (b) A low-contrast image. (c) Result of contrast stretching. (d) Result of thresholding. (Original image courtesy of Dr. Roger Heady, Research School of Biological Sciences, Australian National University, Canberra, Australia.)

function. If $r_1 = s_1$ and $r_2 = s_2$, the transformation is a linear function that produces no changes in gray levels. If $r_1 = r_2$, $s_1 = 0$ and $s_2 = L - 1$, the transformation becomes a *thresholding function* that creates a binary image, as illustrated in Fig. 3.2(b). Intermediate values of (r_1, s_1) and (r_2, s_2) produce various degrees of spread in the gray levels of the output image, thus affecting its contrast. In general, $r_1 \le r_2$ and $s_1 \le s_2$ is assumed so that the function is single valued and monotonically increasing. This condition preserves the order of gray levels, thus preventing the creation of intensity artifacts in the processed image.

Figure 3.10(b) shows an 8-bit image with low contrast. Fig. 3.10(c) shows the result of contrast stretching, obtained by setting $(r_1, s_1) = (r_{min}, 0)$ and $(r_2, s_2) = (r_{max}, L - 1)$ where r_{min} and r_{max} denote the minimum and maximum gray levels in the image, respectively. Thus, the transformation function stretched the levels linearly from their original range to the full range $[0, L - 1]$. Finally, Fig. 3.10(d) shows the result of using the thresholding function defined previously, with $r_1 = r_2 = m$, the mean gray level in the image. The original image on which these results are based is a scanning electron microscope image of pollen, magnified approximately 700 times.

Gray-level slicing

Highlighting a specific range of gray levels in an image often is desired. Applications include enhancing features such as masses of water in satellite imagery and enhancing flaws in X-ray images. There are several ways of doing level slicing, but most of them are variations of two basic themes. One approach is to display a high value for all gray levels in the range of interest and a low value for all other gray levels. This transformation, shown in Fig. 3.11(a), produces a binary image. The second approach, based on the transformation shown in Fig. 3.11(b), brightens the desired range of gray levels but preserves the background and gray-level tonalities in the image. Figure 3.11(c) shows a gray-scale image, and Fig. 3.11(d) shows the result of using the transformation in Fig. 3.11(a). Variations of the two transformations shown in Fig. 3.11 are easy to formulate.

Bit-plane slicing

Instead of highlighting gray-level ranges, highlighting the contribution made to total image appearance by specific bits might be desired. Suppose that each pixel in an image is represented by 8 bits. Imagine that the image is composed of eight 1-bit planes, ranging from bit-plane 0 for the least significant bit to bit-plane 7 for the most significant bit. In terms of 8-bit bytes, plane 0 contains all the lowest order bits in the bytes comprising the pixels in the image and plane 7 contains all the high-order bits. Figure 3.12 illustrates these ideas, and Fig. 3.14 shows the various bit planes for the image shown in Fig. 3.13. Note that the higher-order bits (especially the top four) contain the majority of the visually significant data. The other bit planes contribute to more subtle details in the image. Separating a digital image into its bit planes is useful for analyzing the relative importance played by each bit of the image, a process that aids in determining the adequacy of the number of bits used to quantize each pixel. Also, this type of decomposition is useful for image compression, as discussed in Chapter 8.

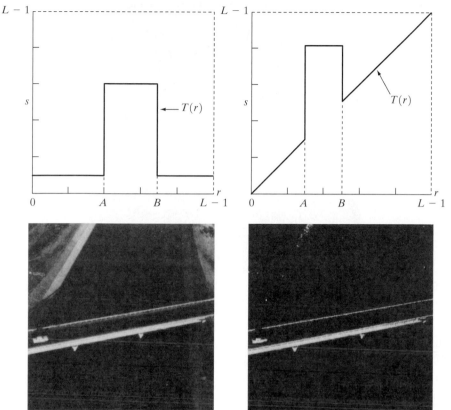

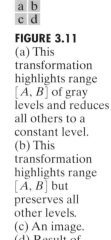

a b
c d

FIGURE 3.11
(a) This transformation highlights range $[A, B]$ of gray levels and reduces all others to a constant level.
(b) This transformation highlights range $[A, B]$ but preserves all other levels.
(c) An image.
(d) Result of using the transformation in (a).

In terms of bit-plane extraction for an 8-bit image, it is not difficult to show that the (binary) image for bit-plane 7 can be obtained by processing the input image with a thresholding gray-level transformation function that (1) maps all levels in the image between 0 and 127 to one level (for example, 0); and (2) maps all levels between 128 and 255 to another (for example, 255). The binary image for bit-plane 7 in Fig. 3.14 was obtained in just this manner. It is left as an exercise (Problem 3.3) to obtain the gray-level transformation functions that would yield the other bit planes.

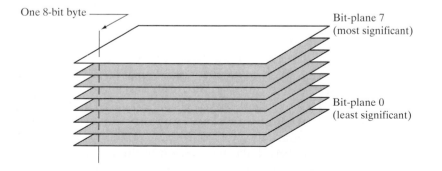

One 8-bit byte

Bit-plane 7
(most significant)

Bit-plane 0
(least significant)

FIGURE 3.12
Bit-plane representation of an 8-bit image.

FIGURE 3.13 An 8-bit fractal image. (A fractal is an image generated from mathematical expressions). (Courtesy of Ms. Melissa D. Binde, Swarthmore College, Swarthmore, PA.)

See inside front cover

Consult the book web site for a review of basic probability theory.

3.3   Histogram Processing

The histogram of a digital image with gray levels in the range $[0, L-1]$ is a discrete function $h(r_k) = n_k$, where r_k is the kth gray level and n_k is the number of pixels in the image having gray level r_k. It is common practice to normalize a histogram by dividing each of its values by the total number of pixels in the image, denoted by n. Thus, a normalized histogram is given by $p(r_k) = n_k/n$, for $k = 0, 1, \ldots, L-1$. Loosely speaking, $p(r_k)$ gives an estimate of the probability of occurrence of gray level r_k. Note that the sum of all components of a normalized histogram is equal to 1.

Histograms are the basis for numerous spatial domain processing techniques. Histogram manipulation can be used effectively for image enhancement, as shown in this section. In addition to providing useful image statistics, we shall see in subsequent chapters that the information inherent in histograms also is quite useful in other image processing applications, such as image compression and segmentation. Histograms are simple to calculate in software and also lend themselves to economic hardware implementations, thus making them a popular tool for real-time image processing.

As an introduction to the role of histogram processing in image enhancement, consider Fig. 3.15, which is the pollen image of Fig. 3.10 shown in four basic gray-level characteristics: dark, light, low contrast, and high contrast. The right side of the figure shows the histograms corresponding to these images. The horizontal axis of each histogram plot corresponds to gray level values, r_k. The vertical axis corresponds to values of $h(r_k) = n_k$ or $p(r_k) = n_k/n$ if the values are normalized. Thus, as indicated previously, these histogram plots are simply plots of $h(r_k) = n_k$ versus r_k or $p(r_k) = n_k/n$ versus r_k.

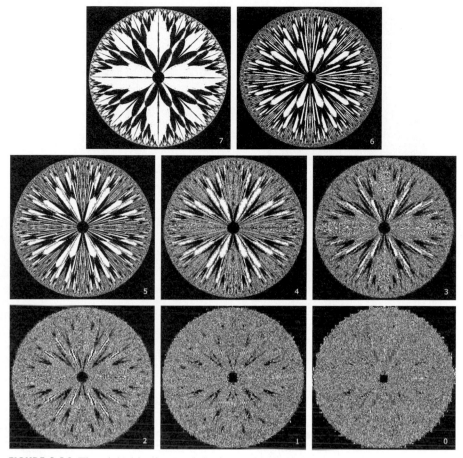

FIGURE 3.14 The eight bit planes of the image in Fig. 3.13. The number at the bottom, right of each image identifies the bit plane.

We note in the dark image that the components of the histogram are concentrated on the low (dark) side of the gray scale. Similarly, the components of the histogram of the bright image are biased toward the high side of the gray scale. An image with low contrast has a histogram that will be narrow and will be centered toward the middle of the gray scale. For a monochrome image this implies a dull, washed-out gray look. Finally, we see that the components of the histogram in the high-contrast image cover a broad range of the gray scale and, further, that the distribution of pixels is not too far from uniform, with very few vertical lines being much higher than the others. Intuitively, it is reasonable to conclude that an image whose pixels tend to occupy the entire range of possible gray levels and, in addition, tend to be distributed uniformly, will have an appearance of high contrast and will exhibit a large variety of gray tones. The net effect will be an image that shows a great deal of gray-level detail and has high dynamic range. It will be shown shortly that it is possible to develop a transformation function that can automatically achieve this effect, based only on information available in the histogram of the input image.

FIGURE 3.15 Four basic image types: dark, light, low contrast, high contrast, and their corresponding histograms. (Original image courtesy of Dr. Roger Heady, Research School of Biological Sciences, Australian National University, Canberra, Australia.)

3.3.1 Histogram Equalization

Consider for a moment continuous functions, and let the variable r represent the gray levels of the image to be enhanced. In the initial part of our discussion we assume that r has been normalized to the interval $[0, 1]$, with $r = 0$ representing black and $r = 1$ representing white. Later, we consider a discrete formulation and allow pixel values to be in the interval $[0, L - 1]$.

For any r satisfying the aforementioned conditions, we focus attention on transformations of the form

$$s = T(r) \qquad 0 \leq r \leq 1 \tag{3.3-1}$$

that produce a level s for every pixel value r in the original image. For reasons that will become obvious shortly, we assume that the transformation function $T(r)$ satisfies the following conditions:

(a) $T(r)$ is single-valued and monotonically increasing in the interval $0 \leq r \leq 1$; and
(b) $0 \leq T(r) \leq 1$ for $0 \leq r \leq 1$.

The requirement in (a) that $T(r)$ be single valued is needed to guarantee that the inverse transformation will exist, and the monotonicity condition preserves the increasing order from black to white in the output image. A transformation function that is not monotonically increasing could result in at least a section of the intensity range being inverted, thus producing some inverted gray levels in the output image. While this may be a desirable effect in some cases, that is not what we are after in the present discussion. Finally, condition (b) guarantees that the output gray levels will be in the same range as the input levels. Figure 3.16 gives an example of a transformation function that satisfies these two conditions. The inverse transformation from s back to r is denoted

$$r = T^{-1}(s) \qquad 0 \leq s \leq 1. \tag{3.3-2}$$

It can be shown by example (Problem 3.8) that even if $T(r)$ satisfies conditions (a) and (b), it is possible that the corresponding inverse $T^{-1}(s)$ may fail to be single valued.

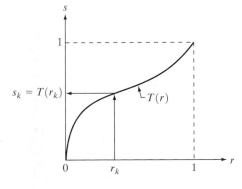

FIGURE 3.16 A gray-level transformation function that is both single valued and monotonically increasing.

The gray levels in an image may be viewed as random variables in the interval $[0, 1]$. One of the most fundamental descriptors of a random variable is its probability density function (PDF). Let $p_r(r)$ and $p_s(s)$ denote the probability density functions of random variables r and s, respectively, where the subscripts on p are used to denote that p_r and p_s are different functions. A basic result from an elementary probability theory is that, if $p_r(r)$ and $T(r)$ are known and $T^{-1}(s)$ satisfies condition (a), then the probability density function $p_s(s)$ of the transformed variable s can be obtained using a rather simple formula:

$$p_s(s) = p_r(r) \left| \frac{dr}{ds} \right|. \tag{3.3-3}$$

Thus, the probability density function of the transformed variable, s, is determined by the gray-level PDF of the input image and by the chosen transformation function.

A transformation function of particular importance in image processing has the form

$$s = T(r) = \int_0^r p_r(w) \, dw \tag{3.3-4}$$

where w is a dummy variable of integration. The right side of Eq. (3.3-4) is recognized as the cumulative distribution function (CDF) of random variable r. Since probability density functions are always positive, and recalling that the integral of a function is the area under the function, it follows that this transformation function is single valued and monotonically increasing, and, therefore, satisfies condition (a). Similarly, the integral of a probability density function for variables in the range $[0, 1]$ also is in the range $[0, 1]$, so condition (b) is satisfied as well.

Given transformation function $T(r)$, we find $p_s(s)$ by applying Eq. (3.3-3). We know from basic calculus (Leibniz's rule) that the derivative of a definite integral with respect to its upper limit is simply the integrand evaluated at that limit. In other words,

$$\frac{ds}{dr} = \frac{dT(r)}{dr}$$

$$= \frac{d}{dr} \left[\int_0^r p_r(w) \, dw \right] \tag{3.3-5}$$

$$= p_r(r).$$

Substituting this result for dr/ds into Eq. (3.3-3), and keeping in mind that all probability values are positive, yields

$$p_s(s) = p_r(r) \left| \frac{dr}{ds} \right|$$

$$= p_r(r) \left| \frac{1}{p_r(r)} \right| \tag{3.3-6}$$

$$= 1 \qquad 0 \le s \le 1.$$

Because $p_s(s)$ is a probability density function, it follows that it must be zero outside the interval $[0, 1]$ in this case because its integral over all values of s must equal 1. We recognize the form of $p_s(s)$ given in Eq. (3.3-6) as a *uniform* probability density function. Simply stated, we have demonstrated that performing the transformation function given in Eq. (3.3-4) yields a random variable s characterized by a uniform probability density function. It is important to note from Eq. (3.3-4) that $T(r)$ depends on $p_r(r)$, but, as indicated by Eq. (3.3-6), the resulting $p_s(s)$ *always* is uniform, *independent* of the form of $p_r(r)$.

For discrete values we deal with probabilities and summations instead of probability density functions and integrals. The probability of occurrence of gray level r_k in an image is approximated by

$$p_r(r_k) = \frac{n_k}{n} \qquad k = 0, 1, 2, \dots, L - 1 \tag{3.3-7}$$

where, as noted at the beginning of this section, n is the total number of pixels in the image, n_k is the number of pixels that have gray level r_k, and L is the total number of possible gray levels in the image. The discrete version of the transformation function given in Eq. (3.3-4) is

$$s_k = T(r_k) = \sum_{j=0}^{k} p_r(r_j) \tag{3.3-8}$$

$$= \sum_{j=0}^{k} \frac{n_j}{n} \qquad k = 0, 1, 2, \dots, L - 1.$$

Thus, a processed (output) image is obtained by mapping each pixel with level r_k in the input image into a corresponding pixel with level s_k in the output image via Eq. (3.3-8). As indicated earlier, a plot of $p_r(r_k)$ versus r_k is called a *histogram*. The transformation (mapping) given in Eq. (3.3-8) is called *histogram equalization* or *histogram linearization*. It is not difficult to show (Problem 3.9) that the transformation in Eq. (3.3-8) satisfies conditions (a) and (b) stated previously in this section.

Unlike its continuous counterpart, it cannot be proved in general that this discrete transformation will produce the discrete equivalent of a uniform probability density function, which would be a uniform histogram. However, as will be seen shortly, use of Eq. (3.3-8) does have the general tendency of spreading the histogram of the input image so that the levels of the histogram-equalized image will span a fuller range of the gray scale.

We discussed earlier in this section the many advantages of having gray-level values that cover the entire gray scale. In addition to producing gray levels that have this tendency, the method just derived has the additional advantage that it is fully "automatic." In other words, given an image, the process of histogram equalization consists simply of implementing Eq. (3.3-8), which is based on information that can be extracted directly from the given image, without the need for further parameter specifications. We note also the simplicity of the computations that would be required to implement the technique.

The inverse transformation from s back to r is denoted by

$$r_k = T^{-1}(s_k) \qquad k = 0, 1, 2, \dots, L - 1 \tag{3.3-9}$$

It can be shown (Problem 3.9) that the inverse transformation in Eq. (3.3-9) satisfies conditions (a) and (b) stated previously in this section only if none of the levels, $r_k, k = 0, 1, 2, \ldots, L - 1$, are missing from the input image. Although the inverse transformation is not used in histogram equalization, it plays a central role in the histogram-matching scheme developed in the next section. We also discuss in that section details of how to implement histogram processing techniques.

EXAMPLE 3.3:
Histogram
equalization.

■ Figure 3.17(a) shows the four images from Fig. 3.15, and Fig. 3.17(b) shows the result of performing histogram equalization on each of these images. The first three results (top to bottom) show significant improvement. As expected, histogram equalization did not produce a significant visual difference in the fourth image because the histogram of this image already spans the full spectrum of the gray scale. The transformation functions used to generate the images in Fig. 3.17(b) are shown in Fig. 3.18. These functions were generated from the histograms of the original images [see Fig. 3.15(b)] using Eq. (3.3-8). Note that transformation (4) has a basic linear shape, again indicating that the gray levels in the fourth input image are nearly uniformly distributed. As was just noted, we would expect histogram equalization in this case to have negligible effect on the appearance of the image.

The histograms of the equalized images are shown in Fig. 3.17(c). It is of interest to note that, while all these histograms are different, the histogram-equalized images themselves are visually very similar. This is not unexpected because the difference between the images in the left column is simply one of contrast, not of content. In other words, since the images have the same content, the increase in contrast resulting from histogram equalization was enough to render any gray-level differences in the resulting images visually indistinguishable. Given the significant contrast differences of the images in the left column, this example illustrates the power of histogram equalization as an adaptive enhancement tool.   ■

3.3.2 Histogram Matching (Specification)

As indicated in the preceding discussion, histogram equalization automatically determines a transformation function that seeks to produce an output image that has a uniform histogram. When automatic enhancement is desired, this is a good approach because the results from this technique are predictable and the method is simple to implement. We show in this section that there are applications in which attempting to base enhancement on a uniform histogram is not the best approach. In particular, it is useful sometimes to be able to specify the shape of the histogram that we wish the processed image to have. The method used to generate a processed image that has a specified histogram is called *histogram matching* or *histogram specification*.

Development of the method

Let us return for a moment to continuous gray levels r and z (considered continuous random variables), and let $p_r(r)$ and $p_z(z)$ denote their corresponding continuous probability density functions. In this notation, r and z denote

FIGURE 3.17 (a) Images from Fig. 3.15. (b) Results of histogram equalization. (c) Corresponding histograms.

FIGURE 3.18
Transformation
functions (1)
through (4) were
obtained from the
histograms of the
images in Fig.
3.17(a), using
Eq. (3.3-8).

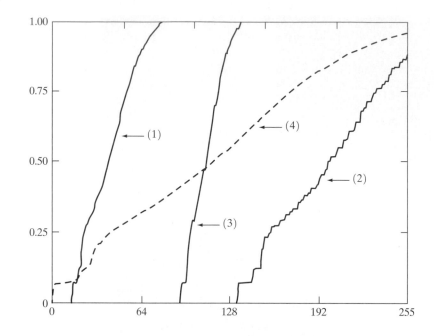

the gray levels of the input and output (processed) images, respectively. We can
estimate $p_r(r)$ from the given input image, while $p_z(z)$ is the *specified* probability
density function that we wish the output image to have.

Let s be a random variable with the property

$$s = T(r) = \int_0^r p_r(w)\, dw \tag{3.3-10}$$

where w is a dummy variable of integration. We recognize this expression as the
continuouos version of histogram equalization given in Eq. (3.3-4). Suppose
next that we define a random variable z with the property

$$G(z) = \int_0^z p_z(t)\, dt = s \tag{3.3-11}$$

where t is a dummy variable of integration. It then follows from these two equations that $G(z) = T(r)$ and, therefore, that z must satisfy the condition

$$z = G^{-1}(s) = G^{-1}[T(r)]. \tag{3.3-12}$$

The transformation $T(r)$ can be obtained from Eq. (3.3-10) once $p_r(r)$ has been
estimated from the input image. Similarly, the transformation function $G(z)$
can be obtained using Eq. (3.3-11) because $p_z(z)$ is given.

Assuming that G^{-1} exists and that it satisfies conditions (a) and (b) in the
previous section, Eqs. (3.3-10) through (3.3-12) show that an image with a specified probability density function can be obtained from an input image by using
the following procedure: (1) Obtain the transformation function $T(r)$ using
Eq. (3.3-10). (2) Use Eq. (3.3-11) to obtain the transformation function $G(z)$.
(3) Obtain the inverse transformation function G^{-1}. (4) Obtain the output image

by applying Eq. (3.3-12) to all the pixels in the input image. The result of this procedure will be an image whose gray levels, z, have the specified probability density function $p_z(z)$.

Although the procedure just described is straightforward in principle, it is seldom possible in practice to obtain analytical expressions for $T(r)$ and for G^{-1}. Fortunately, this problem is simplified considerably in the case of discrete values. The price we pay is the same as in histogram equalization, where only an approximation to the desired histogram is achievable. In spite of this, however, some very useful results can be obtained even with crude approximations.

The discrete formulation of Eq. (3.3-10) is given by Eq. (3.3-8), which we repeat here for convenience:

$$s_k = T(r_k) = \sum_{j=0}^{k} p_r(r_j)$$

$$= \sum_{j=0}^{k} \frac{n_j}{n} \qquad k = 0, 1, 2, \dots, L - 1 \qquad \text{(3.3-13)}$$

where n is the total number of pixels in the image, n_j is the number of pixels with gray level r_j, and L is the number of discrete gray levels. Similarly, the discrete formulation of Eq. (3.3-11) is obtained from the given histogram $p_z(z_i), i = 0, 1, 2, \dots, L - 1$, and has the form

$$v_k = G(z_k) = \sum_{i=0}^{k} p_z(z_i) = s_k \qquad k = 0, 1, 2, \dots, L - 1. \qquad \text{(3.3-14)}$$

As in the continuous case, we are seeking values of z that satisfy this equation. The variable v_k was added here for clarity in the discussion that follows. Finally, the discrete version of Eq. (3.3-12) is given by

$$z_k = G^{-1}[T(r_k)] \qquad k = 0, 1, 2, \dots, L - 1 \qquad \text{(3.3-15)}$$

or, from Eq. (3.3-13),

$$z_k = G^{-1}(s_k) \qquad k = 0, 1, 2, \dots, L - 1. \qquad \text{(3.3-16)}$$

Equations (3.3-13) through (3.3-16) are the foundation for implementing histogram matching for digital images. Equation (3.3-13) is a mapping from the levels in the original image into corresponding levels s_k based on the histogram of the original image, which we compute from the pixels in the image. Equation (3.3-14) computes a transformation function G from the given histogram $p_z(z)$. Finally, Eq. (3.3-15) or its equivalent, Eq. (3.3-16), gives us (an approximation of) the desired levels of the image with that histogram. The first two equations can be implemented easily because all the quantities are known. Implementation of Eq. (3.3-16) is straightforward, but requires additional explanation.

Implementation

We start by noting the following: (1) Each set of gray levels $\{r_j\}, \{s_j\}$, and $\{z_j\}$, $j = 0, 1, 2, \dots, L - 1$, is a one-dimensional array of dimension $L \times 1$. (2) All mappings from r to s and from s to z are simple table lookups between a given pixel value and these arrays. (3) Each of the elements of these arrays, for example, s_k, contains two important pieces of information: The subscript k denotes the location of the element in the array, and s denotes the value at that location. (4) We need to be concerned only with integer pixel values. For example, in the case of an 8-bit image, $L = 256$ and the elements of each of the arrays just mentioned are integers between 0 and 255. This implies that we now work with gray level values in the interval $[0, L - 1]$ instead of the normalized interval $[0, 1]$ that we used before to simplify the development of histogram processing techniques.

In order to see how histogram matching actually can be implemented, consider Fig. 3.19(a), ignoring for a moment the connection shown between this figure and Fig. 3.19(c). Figure 3.19(a) shows a hypothetical discrete transformation function $s = T(r)$ obtained from a given image. The first gray level in the image, r_1, maps to s_1; the second gray level, r_2, maps to s_2; the kth level r_k maps to s_k; and so on (the important point here is the *ordered* correspondence between these values). Each value s_j in the array is precomputed using Eq. (3.3-13), so the process of mapping simply uses the actual value of a pixel as an index in an array to determine the corresponding value of s. This process is particularly easy because we are dealing with integers. For example, the s mapping for an 8-bit pixel with value 127 would be found in the 128th position in array $\{s_j\}$ (recall that we start at 0) out of the possible 256 positions. If we stopped here and mapped the value of each pixel of an input image by the

a b
c

FIGURE 3.19
(a) Graphical interpretation of mapping from r_k to s_k via $T(r)$.
(b) Mapping of z_q to its corresponding value v_q via $G(z)$.
(c) Inverse mapping from s_k to its corresponding value of z_k.

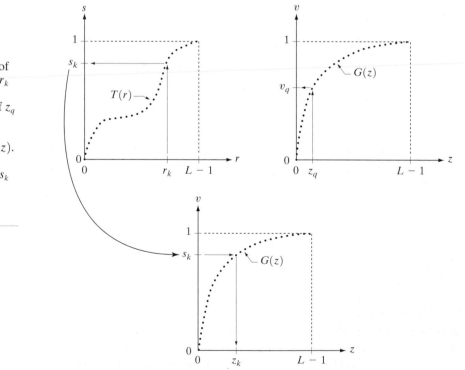

method just described, the output would be a histogram-equalized image, according to Eq. (3.3-8).

In order to implement histogram matching we have to go one step further. Figure 3.19(b) is a hypothetical transformation function G obtained from a given histogram $p_z(z)$ by using Eq. (3.3-14). For any z_q, this transformation function yields a corresponding value v_q. This mapping is shown by the arrows in Fig. 3.19(b). Conversely, given any value v_q, we would find the corresponding value z_q from G^{-1}. In terms of the figure, all this means graphically is that we would reverse the direction of the arrows to map v_q into its corresponding z_q. However, we know from the definition in Eq. (3.3-14) that $v = s$ for corresponding subscripts, so we can use exactly this process to find the z_k corresponding to any value s_k that we computed previously from the equation $s_k = T(r_k)$. This idea is shown in Fig. 3.19(c).

Since we really do not have the z's (recall that finding these values is precisely the objective of histogram matching), we must resort to some sort of iterative scheme to find z from s. The fact that we are dealing with integers makes this a particularly simple process. Basically, because $v_k = s_k$, we have from Eq. (3.3-14) that the z's for which we are looking must satisfy the equation $G(z_k) = s_k$, or $(G(z_k) - s_k) = 0$. Thus, all we have to do to find the value of z_k corresponding to s_k is to iterate on values of z such that this equation is satisfied for $k = 0, 1, 2, \ldots, L - 1$. This is the same thing as Eq. (3.3-16), except that we do not have to find the inverse of G because we are going to iterate on z. Since we are dealing with integers, the closest we can get to satisfying the equation $(G(z_k) - s_k) = 0$ is to let $z_k = \hat{z}$ for each value of k, where $\hat{z}$ is the *smallest* integer in the interval $[0, L - 1]$ such that

$$(G(\hat{z}) - s_k) \geq 0 \qquad k = 0, 1, 2, \ldots, L - 1. \qquad (3.3\text{-}17)$$

Given a value s_k, all this means conceptually in terms of Fig. 3.19(c) is that we would start with $\hat{z} = 0$ and increase it in integer steps until Eq. (3.3-17) is satisfied, at which point we let $z_k = \hat{z}$. Repeating this process for all values of k would yield all the required mappings from s to z, which constitutes the implementation of Eq. (3.3-16). In practice, we would not have to start with $\hat{z} = 0$ each time because the values of s_k are known to increase monotonically. Thus, for $k = k + 1$, we would start with $\hat{z} = z_k$ and increment in integer values from there.

The procedure we have just developed for histogram matching may be summarized as follows:

1. Obtain the histogram of the given image.
2. Use Eq. (3.3-13) to precompute a mapped level s_k for each level r_k.
3. Obtain the transformation function G from the given $p_z(z)$ using Eq. (3.3-14).
4. Precompute z_k for each value of s_k using the iterative scheme defined in connection with Eq. (3.3-17).
5. For each pixel in the original image, if the value of that pixel is r_k, map this value to its corresponding level s_k; then map level s_k into the final level z_k. Use the precomputed values from Steps (2) and (4) for these mappings.

Note that Step (5) implements two mappings for each pixel in the image being processed. The first mapping is nothing more than histogram equalization. If the histogram-equalized image is not required, it obviously would be beneficial to combine both transformations into one in order to save an intermediate step.

Finally, we note that, even in the discrete case, we need to be concerned about G^{-1} satisfying conditions (a) and (b) of the previous section. It is not difficult to show (Problem 3.9) that the only way to guarantee that G^{-1} be single valued and monotonic is to require that G be strictly monotonic (i.e., always increasing), which means simply that none of the values of the specified histogram $p_z(z_i)$ in Eq. (3.3-14) can be zero.

EXAMPLE 3.4:
Comparison between histogram equalization and histogram matching.

■ Figure 3.20(a) shows an image of the Mars moon, Phobos, taken by NASA's *Mars Global Surveyor*. Figure 3.20(b) shows the histogram of Fig. 3.20(a). The image is dominated by large, dark areas, resulting in a histogram characterized by a large concentration of pixels in the dark end of the gray scale. At first glance, one might conclude that histogram equalization would be a good approach to enhance this image, so that details in the dark areas become more visible. It is demonstrated in the following discussion that this is not so.

Figure 3.21(a) shows the histogram equalization transformation [Eq. (3.3-8) or (3.3-13)] obtained from the histogram shown in Fig. 3.20(b). The most relevant characteristic of this transformation function is how fast it rises from gray level 0 to a level near 190. This is caused by the large concentration of pixels in the input histogram having levels very near 0. When this transformation is applied to the levels of the input image to obtain a histogram-equalized result, the net effect is to map a very narrow interval of dark pixels into the upper end of the gray scale of the output image. Because numerous pixels in the input image have levels precisely in this interval, we would expect the result to be an

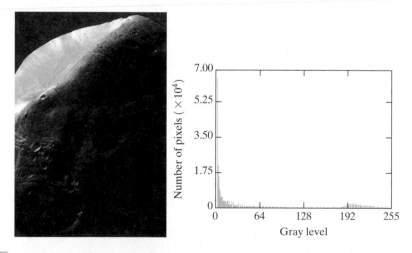

a b

FIGURE 3.20 (a) Image of the Mars moon Phobos taken by NASA's *Mars Global Surveyor.* (b) Histogram. (Original image courtesy of NASA.)

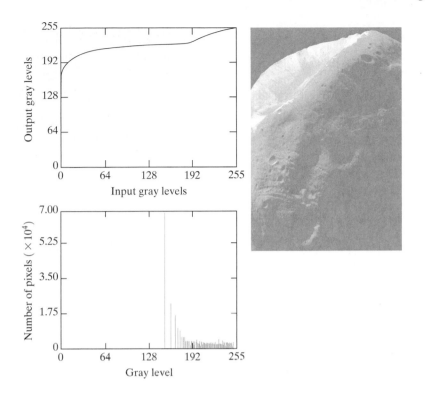

FIGURE 3.21
(a) Transformation function for histogram equalization.
(b) Histogram-equalized image (note the washed-out appearance).
(c) Histogram of (b).

image with a light, washed-out appearance. As shown in Fig. 3.21(b), this is indeed the case. The histogram of this image is shown in Fig. 3.21(c). Note how all the gray levels are biased toward the upper one-half of the gray scale.

Since the problem with the transformation function in Fig. 3.21(a) was caused by a large concentration of pixels in the original image with levels near 0, a reasonable approach is to modify the histogram of that image so that it does not have this property. Figure 3.22(a) shows a *manually specified* function that preserves the general shape of the original histogram, but has a smoother transition of levels in the dark region of the gray scale. Sampling this function into 256 equally spaced discrete values produced the desired specified histogram. The transformation function $G(z)$ obtained from this histogram using Eq. (3.3-14) is labeled transformation (1) in Fig. 3.22(b). Similarly, the inverse transformation $G^{-1}(s)$ from Eq. (3.3-16) [obtained using the iterative technique discussed in connection with Eq. (3.3-17)] is labeled transformation (2) in Fig. 3.22(b). The enhanced image in Fig. 3.22(c) was obtained by applying transformation (2) to the pixels of the histogram-equalized image in Fig. 3.21(b). The improvement of the histogram-specified image over the result obtained by histogram equalization is evident by comparing these two images. It is of interest to note that a rather modest change in the original histogram was all that was required to obtain a significant improvement in enhancement. The histogram of Fig. 3.22(c) is shown in Fig. 3.22(d). The most distinguishing feature of this histogram is how its low end has shifted right toward the lighter region of the gray scale, as desired.   ■

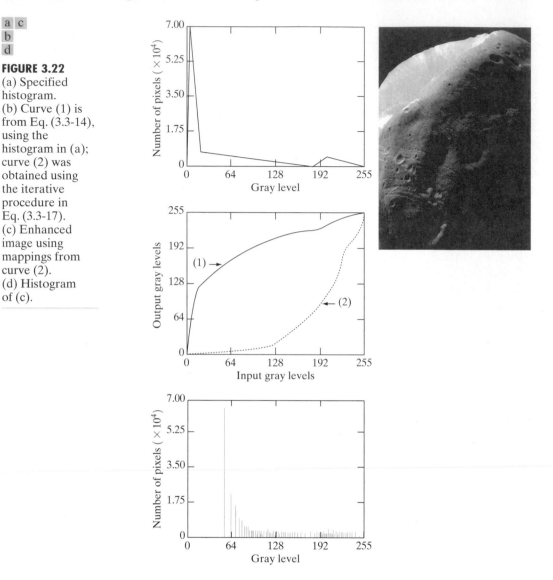

FIGURE 3.22
(a) Specified
histogram.
(b) Curve (1) is
from Eq. (3.3-14),
using the
histogram in (a);
curve (2) was
obtained using
the iterative
procedure in
Eq. (3.3-17).
(c) Enhanced
image using
mappings from
curve (2).
(d) Histogram
of (c).

Although it probably is obvious by now, we emphasize before leaving this section that histogram specification is, for the most part, a trial-and-error process. One can use guidelines learned from the problem at hand, just as we did in the preceding example. At times, there may be cases in which it is possible to formulate what an "average" histogram should look like and use that as the specified histogram. In cases such as these, histogram specification becomes a straightforward process. In general, however, there are no rules for specifying histograms, and one must resort to analysis on a case-by-case basis for any given enhancement task.

3.3.3 Local Enhancement

The histogram processing methods discussed in the previous two sections are *global*, in the sense that pixels are modified by a transformation function based on the gray-level content of an entire image. Although this global approach is suitable for overall enhancement, there are cases in which it is necessary to enhance details over small areas in an image. The number of pixels in these areas may have negligible influence on the computation of a global transformation whose shape does not necessarily guarantee the desired local enhancement. The solution is to devise transformation functions based on the gray-level distribution—or other properties—in the neighborhood of every pixel in the image. Although processing methods based on neighborhoods are the topic of Section 3.5, we discuss local histogram processing here for the sake of clarity and continuity. The reader will have no difficulty in following the discussion.

The histogram processing techniques previously described are easily adaptable to local enhancement. The procedure is to define a square or rectangular neighborhood and move the center of this area from pixel to pixel. At each location, the histogram of the points in the neighborhood is computed and either a histogram equalization or histogram specification transformation function is obtained. This function is finally used to map the gray level of the pixel centered in the neighborhood. The center of the neighborhood region is then moved to an adjacent pixel location and the procedure is repeated. Since only one row or column of the neighborhood changes during a pixel-to-pixel translation of the region, updating the histogram obtained in the previous location with the new data introduced at each motion step is possible (Problem 3.11). This approach has obvious advantages over repeatedly computing the histogram over all pixels in the neighborhood region each time the region is moved one pixel location. Another approach used some times to reduce computation is to utilize nonoverlapping regions, but this method usually produces an undesirable checkerboard effect.

■ Figure 3.23(a) shows an image that has been slightly blurred to reduce its noise content (see Section 3.6.1 regarding blurring). Figure 3.23(b) shows the result of global histogram equalization. As is often the case when this technique is applied to smooth, noisy areas, Fig. 3.23(b) shows considerable enhancement of the noise, with a slight increase in contrast. Note that no new structural details were brought out by this method. However, local histogram equalization using a 7×7 neighborhood revealed the presence of small squares inside the larger dark squares. The small squares were too close in gray level to the larger ones, and their sizes were too small to influence global histogram equalization significantly. Note also the finer noise texture in Fig. 3.23(c), a result of local processing using relatively small neighborhoods. ■

EXAMPLE 3.5: Enhancement using local histograms.

3.3.4 Use of Histogram Statistics for Image Enhancement

Instead of using the image histogram directly for enhancement, we can use instead some statistical parameters obtainable directly from the histogram. Let r denote a discrete random variable representing discrete gray-levels in the range

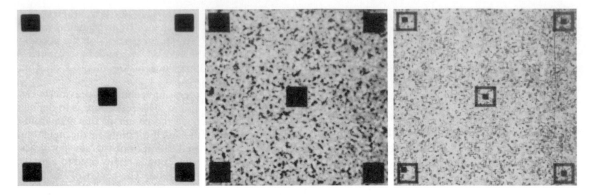

FIGURE 3.23 (a) Original image. (b) Result of global histogram equalization. (c) Result of local histogram equalization using a 7 × 7 neighborhood about each pixel.

$[0, L - 1]$, and let $p(r_i)$ denote the normalized histogram component corresponding to the ith value of r. As indicated previously in this section, we may view $p(r_i)$ as an estimate of the probability of occurrence of gray level r_i. The nth moment of r about its mean is defined as

$$\mu_n(r) = \sum_{i=0}^{L-1} (r_i - m)^n p(r_i) \qquad (3.3\text{-}18)$$

where m is the mean value of r (its average gray level):

$$m = \sum_{i=0}^{L-1} r_i p(r_i). \qquad (3.3\text{-}19)$$

It follows from Eqs. (3.3-18) and (3.3-19) that $\mu_0 = 1$ and $\mu_1 = 0$. The second moment is given by

$$\mu_2(r) = \sum_{i=0}^{L-1} (r_i - m)^2 p(r_i). \qquad (3.3\text{-}20)$$

We recognize this expression as the variance of r, which is denoted conventionally by $\sigma^2(r)$. The standard deviation is defined simply as the square root of the variance. We will revisit moments in Chapter 11 in connection with image description. In terms of enhancement, however, we are interested primarily in the mean, which is a measure of average gray level in an image, and the variance (or standard deviation), which is a measure of average contrast.

We consider two uses of the mean and variance for enhancement purposes. The *global* mean and variance are measured over an entire image and are useful primarily for gross adjustments of overall intensity and contrast. A much more powerful use of these two measures is in local enhancement, where the *local* mean and variance are used as the basis for making changes that depend on image characteristics in a predefined region about each pixel in the image.

Let (x, y) be the coordinates of a pixel in an image, and let S_{xy} denote a neighborhood (subimage) of specified size, centered at (x, y). From Eq. (3.3-19) the mean value $m_{S_{xy}}$ of the pixels in S_{xy} can be computed using the expression

$$m_{S_{xy}} = \sum_{(s,t) \in S_{xy}} r_{s,t}\, p(r_{s,t}) \qquad (3.3\text{-}21)$$

where $r_{s,t}$ is the gray level at coordinates (s, t) in the neighborhood, and $p(r_{s,t})$ is the neighborhood normalized histogram component corresponding to that value of gray level. Similarly, from Eq. (3.3-20), the gray-level variance of the pixels in region S_{xy} is given by

$$\sigma^2_{S_{xy}} = \sum_{(s,t) \in S_{xy}} [r_{s,t} - m_{S_{xy}}]^2 p(r_{s,t}). \qquad (3.3\text{-}22)$$

The local mean is a measure of average gray level in neighborhood S_{xy}, and the variance (or standard deviation) is a measure of contrast in that neighborhood.

An important aspect of image processing using the local mean and variance is the flexibility they afford in developing simple, yet powerful enhancement techniques based on statistical measures that have a close, predictable correspondence with image appearance. We illustrate these characteristics by means of an example.

■ Figure 3.24 shows an SEM (scanning electron microscope) image of a tungsten filament wrapped around a support. The filament in the center of the image and its support are quite clear and easy to study. There is another filament structure on the right side of the image, but it is much darker and its size and other features are not as easily discernable. Local enhancement by contrast manipulation is an ideal approach to try on problems such as this, where part of the image is acceptable, but other parts may contain hidden features of interest.

In this particular case, the problem is to enhance dark areas while leaving the light area as unchanged as possible since it does note require enhancement. We can use the concepts presented in this section to formulate an enhancement method that can tell the difference between dark and light and, at the same time, is capable of enhancing only the dark areas. A measure of whether an area is relatively light or dark at a point (x, y) is to compare the local average gray level $m_{S_{xy}}$ to the average image gray level, called the global mean and denoted M_G. This latter quantity is obtained by letting S encompass the entire image. Thus, we have the first element of our enhancement scheme: We will consider the pixel at a point (x, y) as a candidate for processing if $m_{S_{xy}} \leq k_0 M_G$, where k_0 is a positive constant with value less than 1.0. Since we are interested in enhancing areas that have low contrast, we also need a measure to determine whether the contrast of an area makes it a candidate for enhancement. Thus, we will consider the pixel at a point (x, y) as a candidate for enhancement if $\sigma_{S_{xy}} \leq k_2 D_G$, where D_G is the global standard deviation and k_2 is a positive constant. The value of this constant will be greater than 1.0 if we are interested in enhancing light areas and less than 1.0 for dark areas. Finally, we need to restrict

EXAMPLE 3.6: Enhancement based on local statistics.

the lowest values of contrast we are willing to accept, otherwise the procedure would attempt to enhance even constant areas, whose standard deviation is zero. Thus, we also set a lower limit on the local standard deviation by requiring that $k_1 D_G \leq \sigma_{S_{xy}}$, with $k_1 < k_2$. A pixel at (x, y) that meets all the conditions for local enhancement is processed simply by multiplying it by a specified constant, E, to increase (or decrease) the value of its gray level relative to the rest of the image. The values of pixels that do not meet the enhancement conditions are left unchanged.

A summary of the enhancement method is as follows. Let $f(x, y)$ represent the value of an image pixel at any image coordinates (x, y), and let $g(x, y)$ represent the corresponding enhanced pixel at those coordinates. Then

$$g(x, y) = \begin{cases} E \cdot f(x, y) & \text{if } m_{S_{xy}} \leq k_0 M_G \quad \text{AND} \quad k_1 D_G \leq \sigma_{S_{xy}} \leq k_2 D_G \\ f(x, y) & \text{otherwise} \end{cases}$$

where, as indicated previously, E, k_0, k_1, and k_2 are specified parameters; M_G is the global mean of the input image; and D_G is its global standard deviation.

Normally, making a successful selection of parameters requires a bit of experimentation to gain familiarity with a given image or class of images. In this case, the following values were selected: $E = 4.0$, $k_0 = 0.4$, $k_1 = 0.02$, and $k_2 = 0.4$. The relatively low value of 4.0 for E was chosen so that, when it was multiplied by the levels in the areas being enhanced (which are dark), the result would still tend toward the dark end of the scale, and thus preserve the general visual balance of the image. The value of k_0 was chosen as somewhat less than half the global mean since it is obvious by looking at the image that the areas that require enhancement definitely are dark enough to be below half the global mean. A similar analysis led to the choice of values for k_1 and k_2. Choosing these constants is not a difficult task in general, but their choice

FIGURE 3.24 SEM image of a tungsten filament and support, magnified approximately 130×. (Original image courtesy of Mr. Michael Shaffer, Department of Geological Sciences, University of Oregon, Eugene).

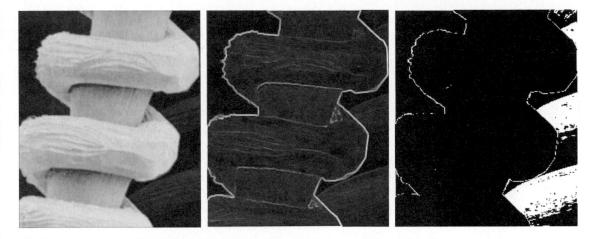

FIGURE 3.25 (a) Image formed from all local means obtained from Fig. 3.24 using Eq. (3.3-21). (b) Image formed from all local standard deviations obtained from Fig. 3.24 using Eq. (3.3-22). (c) Image formed from all multiplication constants used to produce the enhanced image shown in Fig. 3.26.

definitely must be guided by a logical analysis of the enhancement problem at hand. Finally, the choice of size for the local area should be as small as possible in order to preserve detail and keep the computational burden as low as possible. We chose a small (3×3) local region.

Figure 3.25(a) shows the values of $m_{S_{xy}}$ for all values of (x, y). Since the value of $m_{S_{xy}}$ for each (x, y) is the average of the neighboring pixels in a 3×3 area centered at (x, y), we expect the result to be similar to the original image, but

FIGURE 3.26
Enhanced SEM image. Compare with Fig. 3.24. Note in particular the enhanced area on the right side of the image.

slightly blurred. This indeed is the case in Fig. 3.25(a). Figure 3.25(b) shows in image formed using all the values of $\sigma_{S_{xy}}$. Similarly, we can construct an image out the values that multiply $f(x, y)$ at each coordinate pair (x, y) to form $g(x, y)$. Since the values are either 1 or E, the image is binary, as shown in Fig. 3.25(c). The dark areas correspond to 1 and the light areas to E. Thus, any light point in Fig. 3.25(c) signifies a coordinate pair (x, y) at which the enhancement procedure multiplied $f(x, y)$ by E to produce an enhanced pixel. The dark points represent coordinates at which the procedure did not to modify the pixel values.

The enhanced image obtained with the method just described is shown in Fig. 3.26. In comparing this image with the original in Fig. 3.24, we note the obvious detail that has been brought out on the right side of the enhanced image. It is worthwhile to point out that the unenhanced portions of the image (the light areas) were left intact for the most part. We do note the appearance of some small bright dots in the shadow areas where the coil meets the support stem, and around some of the borders between the filament and the background. These are undesirable artifacts created by the enhancement technique. In other words, the points appearing as light dots met the criteria for enhancement and their values were amplified by factor E. Introduction of artifacts is a definite drawback of a method such as the one just described because of the nonlinear way in which they process an image. The key point here, however, is that the image was enhanced in a most satisfactory way as far as bringing out the desired detail. ■

It is not difficult to imagine the numerous ways in which the example just given could be adapted or extended to other situations in which local enhancement is applicable.

3.4 Enhancement Using Arithmetic/Logic Operations

Arithmetic/logic operations involving images are performed on a pixel-by-pixel basis between two or more images (this excludes the logic operation NOT, which is performed on a single image). As an example, subtraction of two images results in a new image whose pixel at coordinates (x, y) is the difference between the pixels in that same location in the two images being subtracted. Depending on the hardware and/or software being used, the actual mechanics of implementing arithmetic/logic operations can be done sequentially, one pixel at a time, or in parallel, where all operations are performed simultaneously.

Logic operations similarly operate on a pixel-by-pixel basis[†]. We need only be concerned with the ability to implement the AND, OR, and NOT logic operators because these three operators are *functionally complete*. In other words, any other logic operator can be implemented by using only these three basic functions. When dealing with logic operations on gray-scale images, pixel values are processed as strings of binary numbers. For example, performing the NOT operation on a black, 8-bit pixel (a string of eight 0's) produces a white pixel

---

[†] Recall that, for two binary variables a and b: aANDb yields 1 only when both a and b are 1; otherwise the result is 0. Similarly, aORb is 0 when both variables are 0; otherwise the result is 1. Finally, if a is 1, NOT (a) is 0, and vice versa.

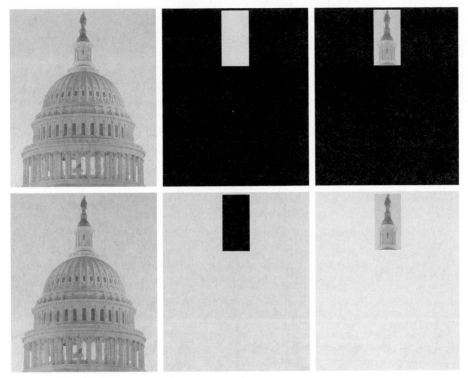

FIGURE 3.27
(a) Original image. (b) AND image mask. (c) Result of the AND operation on images (a) and (b). (d) Original image. (e) OR image mask. (f) Result of operation OR on images (d) and (e).

(a string of eight 1's). Intermediate values are processed the same way, changing all 1's to 0's and vice versa. Thus, the NOT logic operator performs the same function as the negative transformation of Eq. (3.2-1). The AND and OR operations are used for *masking*; that is, for selecting subimages in an image, as illustrated in Fig. 3.27. In the AND and OR image masks, light represents a binary 1 and dark represents a binary 0. Masking sometimes is referred to as *region of interest* (ROI) processing. In terms of enhancement, masking is used primarily to isolate an area for processing. This is done to highlight that area and differentiate it from the rest of the image. Logic operations also are used frequently in conjunction with morphological operations, as discussed in Chapter 9.

Of the four arithmetic operations, subtraction and addition (in that order) are the most useful for image enhancement. We consider division of two images simply as multiplication of one image by the reciprocal of the other. Aside from the obvious operation of multiplying an image by a constant to increase its average gray level, image multiplication finds use in enhancement primarily as a masking operation that is more general than the logical masks discussed in the previous paragraph. In other words, multiplication of one image by another can be used to implement gray-level, rather than binary, masks. We give an example in Section 3.8 of how such a masking operation can be a useful tool. In the remainder of this section, we develop and illustrate methods based on subtraction and addition for image enhancement. Other uses of image multiplication are discussed in Chapter 5, in the context of image restoration.

3.4.1 Image Subtraction

The difference between two images $f(x, y)$ and $h(x, y)$, expressed as

$$g(x, y) = f(x, y) - h(x, y), \qquad (3.4\text{-}1)$$

is obtained by computing the difference between all pairs of corresponding pixels from f and h. The key usefulness of subtraction is the enhancement of *differences* between images. We illustrate this concept by returning briefly to the discussion in Section 3.2.4, where we showed that the higher-order bit planes of an image carry a significant amount of visually relevant detail, while the lower planes contribute more to fine (often imperceptible) detail. Figure 3.28(a) shows the fractal image used earlier to illustrate the concept of bit planes. Figure 3.28(b) shows the result of discarding (setting to zero) the four least significant bit planes of the original image. The images are nearly identical visually, with the exception of a very slight drop in overall contrast due to less variability of the gray-level values in the image of Fig. 3.28(b). The pixel-by-pixel difference between these two images is shown in Fig. 3.28(c). The differences in pixel values are so small that the difference image appears nearly black when displayed on an 8-bit

a b
c d

FIGURE 3.28
(a) Original fractal image.
(b) Result of setting the four lower-order bit planes to zero.
(c) Difference between (a) and (b).
(d) Histogram-equalized difference image. (Original image courtesy of Ms. Melissa D. Binde, Swarthmore College, Swarthmore, PA).

display. In order to bring out more detail, we can perform a contrast stretching transformation, such as those discussed in Sections 3.2 or 3.3. We chose histogram equalization, but an appropriate power-law transformation would have done the job also. The result is shown in Fig. 3.28(d). This is a very useful image for evaluating the effect of setting to zero the lower-order planes.

■ One of the most commercially successful and beneficial uses of image subtraction is in the area of medical imaging called *mask mode radiography*. In this case $h(x, y)$, the *mask*, is an X-ray image of a region of a patient's body captured by an intensified TV camera (instead of traditional X-ray film) located opposite an X-ray source. The procedure consists of injecting a contrast medium into the patient's bloodstream, taking a series of images of the same anatomical region as $h(x, y)$, and subtracting this mask from the series of incoming images after injection of the contrast medium. The net effect of subtracting the mask from each sample in the incoming stream of TV images is that the areas that are different between $f(x, y)$ and $h(x, y)$ appear in the output image as enhanced detail. Because images can be captured at TV rates, this procedure in essence gives a movie showing how the contrast medium propagates through the various arteries in the area being observed.

Figure 3.29(a) shows an X-ray image of the top of a patient's head prior to injection of an iodine medium into the bloodstream. The camera yielding this image was positioned above the patient's head, looking down. As a reference point, the bright spot in the lower one-third of the image is the core of the spinal column. Figure 3.29(b) shows the difference between the mask (Fig. 3.29a) and an image taken some time after the medium was introduced into the bloodstream. The bright arterial paths carrying the medium are unmistakably enhanced in Fig. 3.29(b). These arteries appear quite bright because they are not subtracted out (that is, they are not part of the mask image). The overall background is much darker than that in Fig. 3.29(a) because differences between areas of little change yield low values, which in turn appear as dark shades of gray in the difference image. Note, for instance, that the spinal cord, which is bright in Fig. 3.29(a), appears quite dark in Fig. 3.29(b) as a result of subtraction. ■

EXAMPLE 3.7:
Use of image subtraction in mask mode radiography.

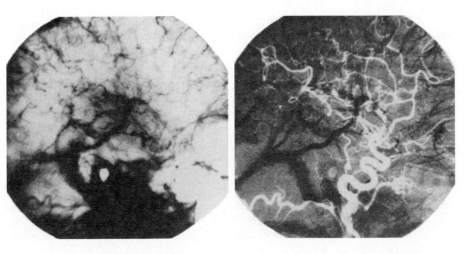

a b

FIGURE 3.29
Enhancement by image subtraction. (a) Mask image. (b) An image (taken after injection of a contrast medium into the bloodstream) with mask subtracted out.

A few comments on implementation are an order before we leave this section. In practice, most images are displayed using 8 bits (even 24-bit color images consists of three separate 8-bit channels). Thus, we expect image values not to be outside the range from 0 to 255. The values in a difference image can range from a minimum of -255 to a maximum of 255, so some sort of scaling is required to display the results. There are two principal ways to scale a difference image. One method is to add 255 to every pixel and then divide by 2. It is not guaranteed that the values will cover the entire 8-bit range from 0 to 255, but all pixel values definitely will be within this range. This method is fast and simple to implement, but it has the limitations that the full range of the display may not be utilized and, potentially more serious, the truncation inherent in the division by 2 will generally cause loss in accuracy.

If more accuracy and full coverage of the 8-bit range are desired, then we can resort to another approach. First, the value of the minimum difference is obtained and its negative added to all the pixels in the difference image (this will create a modified difference image whose minimum values is 0). Then, all the pixels in the image are scaled to the interval $[0, 255]$ by multiplying each pixel by the quantity 255/Max, where Max is the maximum pixel value in the modified difference image. It is evident that this approach is considerably more complex and difficult to implement.

Before leaving this section we note also that change detection via image subtraction finds another major application in the area of segmentation, which is the topic of Chapter 10. Basically, segmentation techniques attempt to subdivide an image into regions based on a specified criterion. Image subtraction for segmentation is used when the criterion is "changes." For instance, in tracking (segmenting) moving vehicles in a sequence of images, subtraction is used to remove all stationary components in an image. What is left should be the moving elements in the image, plus noise.

3.4.2 Image Averaging

Consider a noisy image $g(x, y)$ formed by the addition of noise $\eta(x, y)$ to an original image $f(x, y)$; that is,

$$g(x, y) = f(x, y) + \eta(x, y) \tag{3.4-2}$$

where the assumption is that at every pair of coordinates (x, y) the noise is uncorrelated[†] and has zero average value. The objective of the following procedure is to reduce the noise content by adding a set of noisy images, $\{g_i(x, y)\}$.

If the noise satisfies the constraints just stated, it can be shown (Problem 3.15) that if an image $\bar{g}(x, y)$ is formed by averaging K different noisy images,

$$\bar{g}(x, y) = \frac{1}{K} \sum_{i=1}^{K} g_i(x, y) \tag{3.4-3}$$

[†] Recall that the variance of a random variable x with mean m is defined as $E[(x - m)^2]$, where $E\{\cdot\}$ is the expected value of the argument. The covariance of two random variables x_i and x_j is defined as $E[(x_i - m_i)(x_j - m_j)]$. If the variables are *uncorrelated*, their covariance is 0.

then it follows that

$$E\{\bar{g}(x, y)\} = f(x, y) \tag{3.4-4}$$

and

$$\sigma^2_{\bar{g}(x, y)} = \frac{1}{K} \sigma^2_{\eta(x, y)} \tag{3.4-5}$$

where $E\{\bar{g}(x, y)\}$ is the expected value of $\bar{g}$, and $\sigma^2_{\bar{g}(x, y)}$ and $\sigma^2_{\eta(x, y)}$ are the variances of $\bar{g}$ and η, all at coordinates (x, y). The standard deviation at any point in the average image is

$$\sigma_{\bar{g}(x, y)} = \frac{1}{\sqrt{K}} \sigma_{\eta(x, y)}. \tag{3.4-6}$$

As K increases, Eqs. (3.4-5) and (3.4-6) indicate that the variability (noise) of the pixel values at each location (x, y) decreases. Because $E\{\bar{g}(x, y)\} = f(x, y)$, this means that $\bar{g}(x, y)$ approaches $f(x, y)$ as the number of noisy images used in the averaging process increases. In practice, the images $g_i(x, y)$ must be registered (aligned) in order to avoid the introduction of blurring and other artifacts in the output image.

■ An important application of image averaging is in the field of astronomy, where imaging with very low light levels is routine, causing sensor noise frequently to render single images virtually useless for analysis. Figure 3.30(a) shows an image of a galaxy pair called NGC 3314, taken by NASA's Hubble Space Telescope with a wide field planetary camera. NGC 3314 lies about 140 million light-years from Earth, in the direction of the southern-hemisphere constellation Hydra. The bright stars forming a pinwheel shape near the center of the front galaxy have formed recently from interstellar gas and dust. Figure 3.30(b) shows the same image, but corrupted by uncorrelated Gaussian noise with zero mean and a standard deviation of 64 gray levels. This image is useless for all practical purposes. Figures 3.30(c) through (f) show the results of averaging 8, 16, 64, and 128 images, respectively. We see that the result obtained with $K = 128$ is reasonably close to the original in visual appearance.

EXAMPLE 3.8:
Noise reduction by image averaging.

We can get a better appreciation from Fig. 3.31 for how reduction in the visual appearance of noise takes place as a function of increasing K. This figure shows the difference images between the original [Fig. 3.30(a)] and each of the averaged images in Figs. 3.30(c) through (f). The histograms corresponding to the difference images are also shown in the figure. As usual, the vertical scale in the histograms represents number of pixels and is in the range $[0, 2.6 \times 10^4]$. The horizontal scale represents gray level and is in the range $[0, 255]$. Notice in the histograms that the mean and standard deviation of the difference images decrease as K increases. This is as expected because, according to Eqs. (3.4-3) and (3.4-4), the average image should approach the original as K increases. We can also see the effect of a decreasing mean in the difference images on the left column of Fig. 3.31, which become darker as the K increases.

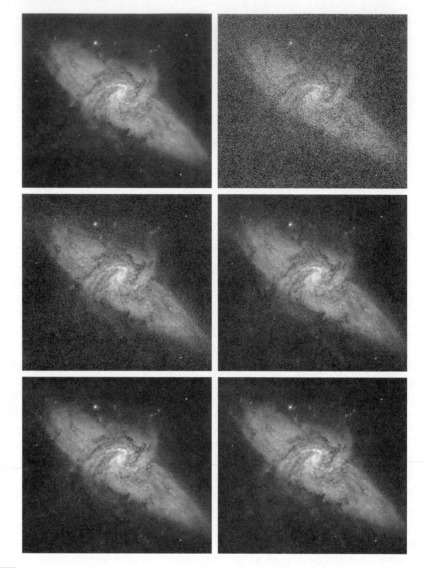

a b
c d
e f

FIGURE 3.30 (a) Image of Galaxy Pair NGC 3314. (b) Image corrupted by additive Gaussian noise with zero mean and a standard deviation of 64 gray levels. (c)–(f) Results of averaging $K = 8, 16, 64,$ and 128 noisy images. (Original image courtesy of NASA.)

Addition is the discrete formulation of continuous integration. In astronomical observations, a process equivalent to the method just described is to use the integrating capabilities of CCD or similar sensors for noise reduction by observing the same scene over long periods of time. The net effect, however, is analogous to the procedure just discussed. Cooling the sensor further reduces its noise level.   ■

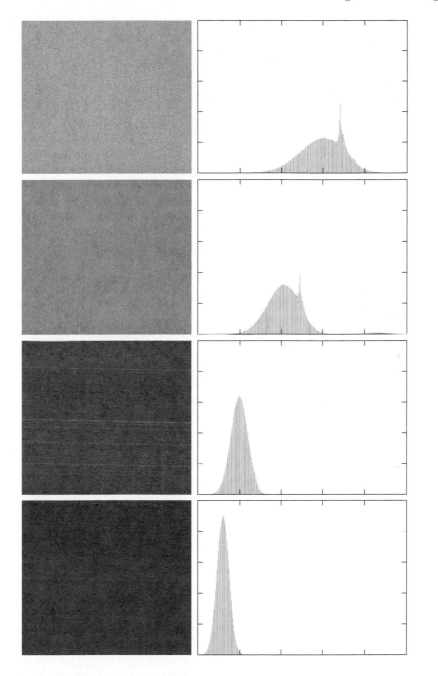

FIGURE 3.31
(a) From top to
bottom:
Difference images
between
Fig. 3.30(a) and
the four images in
Figs. 3.30(c)
through (f),
respectively.
(b) Corresponding
histograms.

As in the case of image subtraction, adding two or more 8-bit images requires special care when it comes to displaying the result on an 8-bit display. The values in the sum of K, 8-bit images can range from 0 to $255 \times K$. Scaling back to 8 bits in this case consists simply of dividing the result by K. Naturally, some accuracy will be lost in the process, but this is unavoidable if the display has to be limited to 8 bits.

It is possible in some implementations of image averaging to have negative values when noise is added to an image. In fact, in the example just given, this was precisely the case because Gaussian random variables with zero mean and nonzero variance have negative as well as positive values. The images in the example were scaled using the second scaling method discussed at the end of the previous section. That is, the minimum value in a given average image was obtained and its negative was added to the image. Then all the pixels in the modified image were scaled to the range $[0, 255]$ by multiplying each pixel in the modified image by the quantity $255/\text{Max}$, where Max was the maximum pixel value in that image.

3.5 Basics of Spatial Filtering

As mentioned in Section 3.1, some neighborhood operations work with the values of the image pixels in the neighborhood *and* the corresponding values of a subimage that has the same dimensions as the neighborhood. The subimage is called a *filter, mask, kernel, template,* or *window,* with the first three terms being the most prevalent terminology. The values in a filter subimage are referred to as *coefficients,* rather than pixels.

The concept of filtering has its roots in the use of the Fourier transform for signal processing in the so-called *frequency domain.* This topic is discussed in more detail in Chapter 4. In the present chapter, we are interested in filtering operations that are performed directly on the pixels of an image. We use the term *spatial filtering* to differentiate this type of process from the more traditional frequency domain filtering.

The mechanics of spatial filtering are illustrated in Fig. 3.32. The process consists simply of moving the filter mask from point to point in an image. At each point (x, y), the *response* of the filter at that point is calculated using a predefined relationship. For *linear* spatial filtering (see Section 2.6 regarding linearity), the response is given by a sum of products of the filter coefficients and the corresponding image pixels in the area spanned by the filter mask. For the 3×3 mask shown in Fig. 3.32, the result (or response), R, of linear filtering with the filter mask at a point (x, y) in the image is

$$R = w(-1, -1)f(x - 1, y - 1) + w(-1, 0)f(x - 1, y) + \cdots$$
$$+ w(0, 0)f(x, y) + \cdots + w(1, 0)f(x + 1, y) + w(1, 1)f(x + 1, y + 1),$$

which we see is the sum of products of the mask coefficients with the corresponding pixels directly under the mask. Note in particular that the coefficient $w(0, 0)$ coincides with image value $f(x, y)$, indicating that the mask is centered at (x, y) when the computation of the sum of products takes place. For a mask of size $m \times n$, we assume that $m = 2a + 1$ and $n = 2b + 1$, where a and b are nonnegative integers. All this says is that our focus in the following discussion will be on masks of *odd* sizes, with the smallest meaningful size being 3×3 (we exclude from our discussion the trivial case of a 1×1 mask).

Image origin

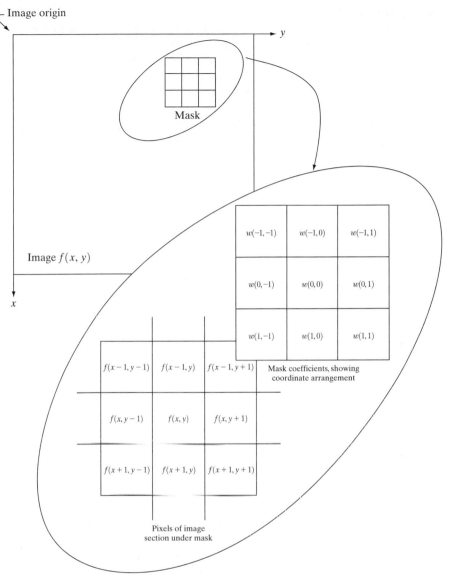

Mask

Image $f(x, y)$

x

y

$w(-1,-1)$	$w(-1,0)$	$w(-1,1)$
$w(0,-1)$	$w(0,0)$	$w(0,1)$
$w(1,-1)$	$w(1,0)$	$w(1,1)$

Mask coefficients, showing
coordinate arrangement

$f(x-1,y-1)$	$f(x-1,y)$	$f(x-1,y+1)$
$f(x,y-1)$	$f(x,y)$	$f(x,y+1)$
$f(x+1,y-1)$	$f(x+1,y)$	$f(x+1,y+1)$

Pixels of image
section under mask

FIGURE 3.32 The mechanics of spatial filtering. The magnified drawing shows a 3 × 3 mask and the image section directly under it; the image section is shown displaced out from under the mask for ease of readability.

In general, linear filtering of an image f of size $M \times N$ with a filter mask of size $m \times n$ is given by the expression:

$$g(x, y) = \sum_{s=-a}^{a} \sum_{t=-b}^{b} w(s, t) f(x + s, y + t) \qquad (3.5\text{-}1)$$

where, from the previous paragraph, $a = (m - 1)/2$ and $b = (n - 1)/2$. To generate a complete filtered image this equation must be applied for $x = 0, 1, 2, \ldots, M - 1$ and $y = 0, 1, 2, \ldots, N - 1$. In this way, we are assured that the mask processes all pixels in the image. It is easily verified when $m = n = 3$ that this expression reduces to the example given in the previous paragraph.

As discussed in Chapter 4, the process of linear filtering given in Eq. (3.5-1) is similar to a frequency domain concept called *convolution*. For this reason, linear spatial filtering often is referred to as "convolving a mask with an image." Similarly, filter masks are sometimes called *convolution masks*. The term *convolution kernel* also is in common use.

When interest lies on the response, R, of an $m \times n$ mask at any point (x, y), and not on the mechanics of implementing mask convolution, it is common practice to simplify the notation by using the following expression:

$$R = w_1 z_1 + w_2 z_2 + \ldots + w_{mn} z_{mn} \qquad (3.5\text{-}2)$$

$$= \sum_{i=1}^{mn} w_i z_i$$

where the w's are mask coefficients, the z's are the values of the image gray levels corresponding to those coefficients, and mn is the total number of coefficients in the mask. For the 3×3 general mask shown in Fig. 3.33 the response at any point (x, y) in the image is given by

$$R = w_1 z_1 + w_2 z_2 + \ldots + w_9 z_9 \qquad (3.5\text{-}3)$$

$$= \sum_{i=1}^{9} w_i z_i.$$

We make special mention of this simple formula because it is seen frequently in the published literature on image processing.

Nonlinear spatial filters also operate on neighborhoods, and the mechanics of sliding a mask past an image are the same as was just outlined. In general, however, the filtering operation is based conditionally on the values of the pixels in the neighborhood under consideration, and they do not explicitly use coefficients in the sum-of-products manner described in Eqs. (3.5-1) and (3.5-2). As shown in Section 3.6.2, for example, noise reduction can be achieved effectively with a nonlinear filter whose basic function is to compute the median gray-level value in the neighborhood in which the filter is located. Computation of the median is a nonlinear operation, as is computation of the variance, which we used in Section 3.3.4.

FIGURE 3.33
Another representation of a general 3×3 spatial filter mask.

An important consideration in implementing neighborhood operations for spatial filtering is the issue of what happens when the center of the filter approaches the border of the image. Consider for simplicity a square mask of size $n \times n$. At least one edge of such a mask will coincide with the border of the image when the center of the mask is at a distance of $(n - 1)/2$ pixels away from the border of the image. If the center of the mask moves any closer to the border, one or more rows or columns of the mask will be located outside the image plane. There are several ways to handle this situation. The simplest is to limit the excursions of the center of the mask to be at a distance no less than $(n - 1)/2$ pixels from the border. The resulting filtered image will be smaller than the original, but all the pixels in the filtered imaged will have been processed with the full mask. If the result is required to be the same size as the original, then the approach typically employed is to filter all pixels only with the section of the mask that is fully contained in the image. With this approach, there will be bands of pixels near the border that will have been processed with a partial filter mask. Other approaches include "padding" the image by adding rows and columns of 0's (or other constant gray level), or padding by replicating rows or columns. The padding is then stripped off at the end of the process. This keeps the size of the filtered image the same as the original, but the values of the padding will have an effect near the edges that becomes more prevalent as the size of the mask increases. The only way to obtain a perfectly filtered result is to accept a somewhat smaller filtered image by limiting the excursions of the center of the filter mask to a distance no less than $(n - 1)/2$ pixels from the border of the original image.

3.6 Smoothing Spatial Filters

Smoothing filters are used for blurring and for noise reduction. Blurring is used in preprocessing steps, such as removal of small details from an image prior to (large) object extraction, and bridging of small gaps in lines or curves. Noise reduction can be accomplished by blurring with a linear filter and also by nonlinear filtering.

3.6.1 Smoothing Linear Filters

The output (response) of a smoothing, linear spatial filter is simply the average of the pixels contained in the neighborhood of the filter mask. These filters sometimes are called *averaging filters*. For reasons explained in Chapter 4, they also are referred to a *lowpass filters*.

The idea behind smoothing filters is straightforward. By replacing the value of every pixel in an image by the average of the gray levels in the neighborhood defined by the filter mask, this process results in an image with reduced "sharp" transitions in gray levels. Because random noise typically consists of sharp transitions in gray levels, the most obvious application of smoothing is noise reduction. However, edges (which almost always are desirable features of an image) also are characterized by sharp transitions in gray levels, so averaging filters have the undesirable side effect that they blur edges. Another application of this type of process includes the smoothing of false contours that result

FIGURE 3.34 Two 3 × 3 smoothing (averaging) filter masks. The constant multiplier in front of each mask is equal to the sum of the values of its coefficients, as is required to compute an average.

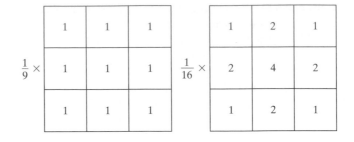

from using an insufficient number of gray levels, as discussed in Section 2.4.3. A major use of averaging filters is in the reduction of "irrelevant" detail in an image. By "irrelevant" we mean pixel regions that are small with respect to the size of the filter mask. This latter application is illustrated later in this section.

Figure 3.34 shows two 3 × 3 smoothing filters. Use of the first filter yields the standard average of the pixels under the mask. This can best be seen by substituting the coefficients of the mask into Eq. (3.5-3):

$$R = \frac{1}{9} \sum_{i=1}^{9} z_i,$$

which is the average of the gray levels of the pixels in the 3 × 3 neighborhood defined by the mask. Note that, instead of being 1/9, the coefficients of the filter are all 1's. The idea here is that it is computationally more efficient to have coefficients valued 1. At the end of the filtering process the entire image is divided by 9. An $m \times n$ mask would have a normalizing constant equal to $1/mn$. A spatial averaging filter in which all coefficients are equal is sometimes called a *box filter*.

The second mask shown in Fig. 3.34 is a little more interesting. This mask yields a so-called *weighted average*, terminology used to indicate that pixels are multiplied by different coefficients, thus giving more importance (weight) to some pixels at the expense of others. In the mask shown in Fig. 3.34(b) the pixel at the center of the mask is multiplied by a higher value than any other, thus giving this pixel more importance in the calculation of the average. The other pixels are inversely weighted as a function of their distance from the center of the mask. The diagonal terms are further away from the center than the orthogonal neighbors (by a factor of $\sqrt{2}$) and, thus, are weighed less than these immediate neighbors of the center pixel. The basic strategy behind weighing the center point the highest and then reducing the value of the coefficients as a function of increasing distance from the origin is simply an attempt to reduce blurring in the smoothing process. We could have picked other weights to accomplish the same general objective. However, the sum of all the coefficients in the mask of Fig. 3.34(b) is equal to 16, an attractive feature for computer implementation because it has an integer power of 2. In practice, it is difficult in general to see differences between images smoothed by using either of the masks in Fig. 3.34, or similar arrangements, because the area these masks span at any one location in an image is so small.

With reference to Eq. (3.5-1), the general implementation for filtering an $M \times N$ image with a weighted averaging filter of size $m \times n$ (m and n odd) is given by the expression

$$g(x, y) = \frac{\displaystyle\sum_{s=-a}^{a} \sum_{t=-b}^{b} w(s, t) f(x + s, y + t)}{\displaystyle\sum_{s=-a}^{a} \sum_{t=-b}^{b} w(s, t)} \tag{3.6-1}$$

The parameters in this equation are as defined in Eq. (3.5-1). As before, it is understood that the complete filtered image is obtained by applying Eq. (3.6-1) for $x = 0, 1, 2, \ldots, M - 1$ and $y = 0, 1, 2, \ldots, N - 1$. The denominator in Eq. (3.6-1) is simply the sum of the mask coefficients and, therefore, it is a constant that needs to be computed only once. Typically, this scale factor is applied to all the pixels of the output image after the filtering process is completed.

■ The effects of smoothing as a function of filter size are illustrated in Fig. 3.35, which shows an original image and the corresponding smoothed results obtained using square averaging filters of sizes $n = 3, 5, 9, 15$, and 35 pixels, respectively. The principal features of these results are as follows: For $n = 3$, we note a general slight blurring throughout the entire image but, as expected, details that are of approximately the same size as the filter mask are affected considerably more. For example, the 3×3 and 5×5 squares, the small letter "a," and the fine grain noise show significant blurring when compared to the rest of the image. A positive result is that the noise is less pronounced. Note that the jagged borders of the characters and gray circles have been pleasingly smoothed.

EXAMPLE 3.9:
Image smoothing with masks of various sizes.

The result for $n = 5$ is somewhat similar, with a slight further increase in blurring. For $n = 9$ we see considerably more blurring, and the 20% black circle is not nearly as distinct from the background as in the previous three images, illustrating the blending effect that blurring has on objects whose gray level content is close to that of its neighboring pixels. Note the significant further smoothing of the noisy rectangles. The results for $n = 15$ and 35 are extreme with respect to the sizes of the objects in the image. This type of excessive blurring is generally used to eliminate small objects from an image. For instance, the three small squares, two of the circles, and most of the noisy rectangle areas have been blended into the background of the image in Fig. 3.35(f). Note also in this figure the pronounced black border. This is a result of padding the border of the original image with 0's (black) and then trimming off the padded area. Some of the black was blended into all filtered images, but became truly objectionable for the images smoothed with the larger filters.   ■

As mentioned earlier, an important application of spatial averaging is to blur an image for the purpose getting a gross representation of objects of interest, such that the intensity of smaller objects blends with the background and larger objects become "bloblike" and easy to detect. The size of the mask establishes the relative size of the objects that will be blended with the background. As an illustration, consider Fig. 3.36(a), which is an image from the Hubble telescope in orbit around the Earth. Figure 3.36(b) shows the result of applying a

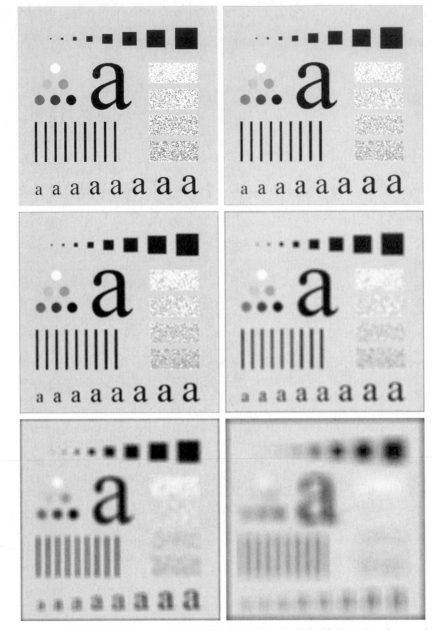

FIGURE 3.35 (a) Original image, of size 500×500 pixels. (b)–(f) Results of smoothing with square averaging filter masks of sizes $n = 3, 5, 9, 15,$ and $35,$ respectively. The black squares at the top are of sizes 3, 5, 9, 15, 25, 35, 45, and 55 pixels, respectively; their borders are 25 pixels apart. The letters at the bottom range in size from 10 to 24 points, in increments of 2 points; the large letter at the top is 60 points. The vertical bars are 5 pixels wide and 100 pixels high; their separation is 20 pixels. The diameter of the circles is 25 pixels, and their borders are 15 pixels apart; their gray levels range from 0% to 100% black in increments of 20%. The background of the image is 10% black. The noisy rectangles are of size 50×120 pixels.

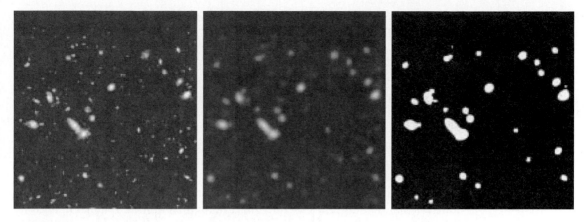

a b c

FIGURE 3.36 (a) Image from the Hubble Space Telescope. (b) Image processed by a 15 × 15 averaging mask. (c) Result of thresholding (b). (Original image courtesy of NASA.)

15 × 15 averaging mask to this image. We see that a number of objects have either blended with the background or their intensity has diminished considerably. It is typical to follow an operation like this with thresholding to eliminate objects based on their intensity. The result of using the thresholding function of Fig. 3.2(b) with a threshold value equal to 25% of the highest intensity in the blurred image is shown in Fig. 3.36(c). Comparing this result with the original image, we see that it is a reasonable representation of what we would consider to be the largest, brightest objects in that image.

3.6.2 Order-Statistic Filters

Order-statistic filters are nonlinear spatial filters whose response is based on ordering (ranking) the pixels contained in the image area encompassed by the filter, and then replacing the value of the center pixel with the value determined by the ranking result. The best-known example in this category is the *median filter*, which, as its name implies, replaces the value of a pixel by the median of the gray levels in the neighborhood of that pixel (the original value of the pixel is included in the computation of the median). Median filters are quite popular because, for certain types of random noise, they provide excellent noise-reduction capabilities, with considerably less blurring than linear smoothing filters of similar size. Median filters are particularly effective in the presence of *impulse noise*, also called *salt-and-pepper noise* because of its appearance as white and black dots superimposed on an image.

The median, ξ, of a set of values is such that half the values in the set are less than or equal to ξ, and half are greater than or equal to ξ. In order to perform median filtering at a point in an image, we first sort the values of the pixel in question and its neighbors, determine their median, and assign this value to that pixel. For example, in a 3 × 3 neighborhood the median is the 5th largest value, in a 5 × 5 neighborhood the 13th largest value, and so on. When several values in a neighborhood are the same, all equal values are grouped. For example, suppose that a 3×3 neighborhood has values $(10, 20, 20, 20, 15, 20, 20, 25, 100)$. These values are sorted as $(10, 15, 20, 20, 20, 20, 20, 25, 100)$, which results in a median of 20. Thus, the principal function of median filters is to force points with distinct gray levels to be more like their neighbors. In fact, isolated clusters of pixels that are light or dark with respect to their neighbors, and whose area is less than $n^2/2$ (one-half the filter area), are eliminated by an $n \times n$ median filter. In this case "eliminated" means forced to the median intensity of the neighbors. Larger clusters are affected considerably less.

Although the median filter is by far the most useful order-statistic filter in image processing, it is by no means the only one. The median represents the 50th percentile of a ranked set of numbers, but the reader will recall from basic statistics that ranking lends itself to many other possibilities. For example, using the 100th percentile results in the so-called *max filter*, which is useful in finding the brightest points in an image. The response of a 3×3 max filter is given by $R = \max\{z_k \mid k = 1, 2, \ldots, 9\}$. The 0th percentile filter is the *min filter*, used for the opposite purpose. Median, max, and mean filters are considered in more detail in Chapter 5.

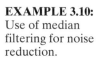

EXAMPLE 3.10:
Use of median filtering for noise reduction.

■ Figure 3.37(a) shows an X-ray image of a circuit board heavily corrupted by salt-and-pepper noise. To illustrate the point about the superiority of median filtering over average filtering in situations such as this, we show in Fig. 3.37(b) the result of processing the noisy image with a 3×3 neighborhood averaging mask, and in Fig. 3.37(c) the result of using a 3×3 median filter. The image processed with the averaging filter has less visible noise, but the price paid is significant blurring. The superiority in all respects of median over average filtering in this case is quite evident. In general, median filtering is much better suited than averaging for the removal of salt-and-pepper noise. ■

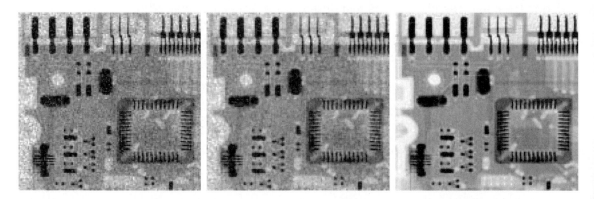

a b c

FIGURE 3.37 (a) X-ray image of circuit board corrupted by salt-and-pepper noise. (b) Noise reduction with a 3×3 averaging mask. (c) Noise reduction with a 3×3 median filter. (Original image courtesy of Mr. Joseph E. Pascente, Lixi, Inc.)

3.7 Sharpening Spatial Filters

The principal objective of sharpening is to highlight fine detail in an image or to enhance detail that has been blurred, either in error or as a natural effect of a particular method of image acquisition. Uses of image sharpening vary and include applications ranging from electronic printing and medical imaging to industrial inspection and autonomous guidance in military systems.

In the last section, we saw that image blurring could be accomplished in the spatial domain by pixel averaging in a neighborhood. Since averaging is analogous to integration, it is logical to conclude that sharpening could be accomplished by spatial differentiation. This, in fact, is the case, and the discussion in this section deals with various ways of defining and implementing operators for sharpening by digital differentiation. Fundamentally, the strength of the response of a derivative operator is proportional to the degree of discontinuity of the image at the point at which the operator is applied. Thus, image differentiation enhances edges and other discontinuities (such as noise) and deemphasizes areas with slowly varying gray-level values.

3.7.1 Foundation

In the two sections that follow, we consider in some detail sharpening filters that are based on first- and second-order derivatives, respectively. Before proceeding with that discussion, however, we stop to look at some of the fundamental properties of these derivatives in a digital context. To simplify the explanation, we focus attention on one-dimensional derivatives. In particular, we are interested in the behavior of these derivatives in areas of constant gray level (flat segments), at the onset and end of discontinuities (step and ramp discontinuities), and along gray-level ramps. These types of discontinuities can be used to model noise points, lines, and edges in an image. The behavior of derivatives during transitions into and out of these image features also is of interest.

The derivatives of a digital function are defined in terms of differences. There are various ways to define these differences. However, we require that any definition we use for a first derivative (1) must be zero in flat segments (areas of constant gray-level values); (2) must be nonzero at the onset of a gray-level step or ramp; and (3) must be nonzero along ramps. Similarly, any definition of a second derivative (1) must be zero in flat areas; (2) must be nonzero at the onset and end of a gray-level step or ramp; and (3) must be zero along ramps of constant slope. Since we are dealing with digital quantities whose values are finite, the maximum possible gray-level change also is finite, and the shortest distance over which that change can occur is between adjacent pixels.

A basic definition of the first-order derivative of a one-dimensional function $f(x)$ is the difference

$$\frac{\partial f}{\partial x} = f(x + 1) - f(x).$$

We used a partial derivative here in order to keep the notation the same as when we consider an image function of two variables, $f(x, y)$, at which time we will be dealing with partial derivatives along the two spatial axes. Use of a partial derivative in the present discussion does not affect in any way the nature of what we are trying to accomplish.

Similarly, we define a second-order derivative as the difference

$$\frac{\partial^2 f}{\partial x^2} = f(x + 1) + f(x - 1) - 2f(x).$$

It is easily verified that these two definitions satisfy the conditions stated previously regarding derivatives of the first and second order. To see this, and also to highlight the fundamental similarities and differences between first- and second-order derivatives in the context of image processing, consider the example shown in Fig. 3.38.

Figure 3.38(a) shows a simple image that contains various solid objects, a line, and a single noise point. Figure 3.38(b) shows a horizontal gray-level profile (scan line) of the image along the center and including the noise point. This profile is the one-dimensional function we will use for illustrations regarding this figure. Figure 3.38(c) shows a simplification of the profile, with just enough num-

a b
c

FIGURE 3.38
(a) A simple image. (b) 1-D horizontal gray-level profile along the center of the image and including the isolated noise point.
(c) Simplified profile (the points are joined by dashed lines to simplify interpretation).

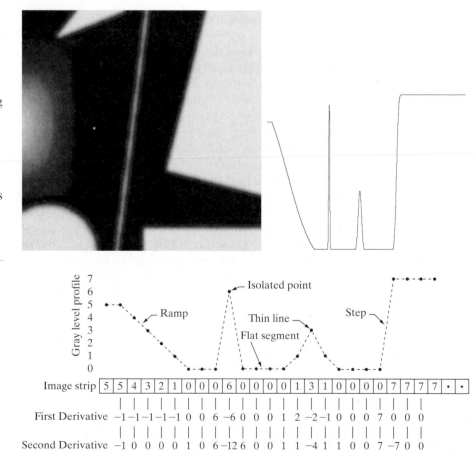

bers to make it possible for us to analyze how the first- and second-order derivatives behave as they encounter a noise point, a line, and then the edge of an object. In our simplified diagram the transition in the ramp spans four pixels, the noise point is a single pixel, the line is three pixels thick, and the transition into the gray-level step takes place between adjacent pixels. The number of gray levels was simplified to only eight levels.

Let us consider the properties of the first and second derivatives as we traverse the profile from left to right. First, we note that the first-order derivative is nonzero along the entire ramp, while the second-order derivative is nonzero only at the onset and end of the ramp. Because edges in an image resemble this type of transition, we conclude that first-order derivatives produce "thick" edges and second-order derivatives, much finer ones. Next we encounter the isolated noise point. Here, the response at and around the point is much stronger for the second- than for the first-order derivative. Of course, this is not unexpected. A second-order derivative is much more aggressive than a first-order derivative in enhancing sharp changes. Thus, we can expect a second-order derivative to enhance fine detail (including noise) much more than a first-order derivative. The thin line is a fine detail, and we see essentially the same difference between the two derivatives. If the maximum gray level of the line had been the same as the isolated point, the response of the second derivative would have been stronger for the latter. Finally, in this case, the response of the two derivatives is the same at the gray-level step (in most cases when the transition into a step is not from zero, the second derivative will be weaker). We also note that the second derivative has a transition from positive back to negative. In an image, this shows as a thin double line. This "double-edge" effect is an issue that will be important in Chapter 10, where we use derivatives for edge detection. It is of interest also to note that if the gray level of the thin line had been the same as the step, the response of the second derivative would have been stronger for the line than for the step.

In summary, comparing the response between first- and second-order derivatives, we arrive at the following conclusions. (1) First-order derivatives generally produce thicker edges in an image. (2) Second-order derivatives have a stronger response to fine detail, such as thin lines and isolated points. (3) First-order derivatives generally have a stronger response to a gray-level step. (4) Second-order derivatives produce a double response at step changes in gray level. We also note of second-order derivatives that, for similar changes in gray-level values in an image, their response is stronger to a line than to a step, and to a point than to a line.

In most applications, the second derivative is better suited than the first derivative for image enhancement because of the ability of the former to enhance fine detail. For this, and for reasons of simpler implementation and extensions, we will focus attention initially on uses of the second derivative for enhancement. First-order derivatives are discussed in Section 3.7.3. Although the principle of use of first derivatives in image processing is for edge extraction, they do have important uses in image enhancement. In fact, we show in Section 3.8 that they can be used in conjunction with the second derivative to obtain some impressive enhancement results.

3.7.2 Use of Second Derivatives for Enhancement–The Laplacian

In this section we consider in some detail the use of two-dimensional, second-order derivatives for image enhancement. The approach basically consists of defining a discrete formulation of the second-order derivative and then constructing a filter mask based on that formulation. We are interested in *isotropic* filters, whose response is independent of the direction of the discontinuities in the image to which the filter is applied. In other words, isotropic filters are *rotation invariant*, in the sense that rotating the image and then applying the filter gives the same result as applying the filter to the image first and then rotating the result.

Development of the method

It can be shown (Rosenfeld and Kak [1982]) that the simplest isotropic derivative operator is the *Laplacian*, which, for a function (image) $f(x, y)$ of two variables, is defined as

$$\nabla^2 f = \frac{\partial^2 f}{\partial x^2} + \frac{\partial^2 f}{\partial y^2}. \tag{3.7-1}$$

Because derivatives of any order are linear operations, the Laplacian is a linear operator.

In order to be useful for digital image processing, this equation needs to be expressed in discrete form. There are several ways to define a digital Laplacian using neighborhoods. Whatever the definition, however, it has to satisfy the properties of a second derivative outlined in Section 3.7.1. The definition of the digital second derivative given in that section is one of the most used. Taking into account that we now have two variables, we use the following notation for the partial second-order derivative in the *x*-direction:

$$\frac{\partial^2 f}{\partial x^2} = f(x + 1, y) + f(x - 1, y) - 2f(x, y) \tag{3.7-2}$$

and, similarly in the *y*-direction, as

$$\frac{\partial^2 f}{\partial y^2} = f(x, y + 1) + f(x, y - 1) - 2f(x, y) \tag{3.7-3}$$

The digital implementation of the two-dimensional Laplacian in Eq. (3.7-1) is obtained by summing these two components:

$$\nabla^2 f = \left[f(x + 1, y) + f(x - 1, y) + f(x, y + 1) + f(x, y - 1) \right] \\ - 4f(x, y). \tag{3.7-4}$$

This equation can be implemented using the mask shown in Fig. 3.39(a), which gives an isotropic result for rotations in increments of 90°. The mechanics of implementation are given in Eq. (3.5-1) and are illustrated in Section 3.6.1 for the linear smoothing filters. We simply are using different coefficients here.

The diagonal directions can be incorporated in the definition of the digital Laplacian by adding two more terms to Eq. (3.7-4), one for each of the two diagonal directions. The form of each new term is the same as either Eq. (3.7-2)

0	1	0	1	1	1
1	−4	1	1	−8	1
0	1	0	1	1	1
0	−1	0	−1	−1	−1
−1	4	−1	−1	8	−1
0	−1	0	−1	−1	−1

FIGURE 3.39
(a) Filter mask used to implement the digital Laplacian, as defined in Eq. (3.7-4). (b) Mask used to implement an extension of this equation that includes the diagonal neighbors. (c) and (d) Two other implementations of the Laplacian.

or (3.7-3), but the coordinates are along the diagonals. Since each diagonal term also contains a $-2f(x, y)$ term, the total subtracted from the difference terms now would be $-8f(x, y)$. The mask used to implement this new definition is shown in Fig. 3.39(b). This mask yields isotropic results for increments of 45°. The other two masks shown in Fig. 3.39 also are used frequently in practice. They are based on a definition of the Laplacian that is the negative of the one we used here. As such, they yield equivalent results, but the difference in sign must be kept in mind when combining (by addition or subtraction) a Laplacian-filtered image with another image.

Because the Laplacian is a derivative operator, its use highlights gray-level discontinuities in an image and deemphasizes regions with slowly varying gray levels. This will tend to produce images that have grayish edge lines and other discontinuities, all superimposed on a dark, featureless background. Background features can be "recovered" while still preserving the sharpening effect of the Laplacian operation simply by adding the original and Laplacian images. As noted in the previous paragraph, it is important to keep in mind which definition of the Laplacian is used. If the definition used has a negative center coefficient, then we *subtract*, rather than add, the Laplacian image to obtain a sharpened result. Thus, the basic way in which we use the Laplacian for image enhancement is as follows:

$$g(x, y) = \begin{cases} f(x, y) - \nabla^2 f(x, y) & \text{if the center coefficient of the} \\ & \text{Laplacian mask is negative} \\ f(x, y) + \nabla^2 f(x, y) & \text{if the center coefficient of the} \\ & \text{Laplacian mask is positive.} \end{cases} \quad (3.7-5)$$

Use of this equation is illustrated next.

EXAMPLE 3.11:
Imaging
sharpening with
the Laplacian.

■ Figure 3.40(a) shows an image of the North Pole of the moon. Figure 3.40(b) shows the result of filtering this image with the Laplacian mask in Fig. 3.39(b). Since the Laplacian image contains both positive and negative values, a typical way to scale it is to use the approach discussed at the end of Section 3.4.1. Sometimes one encounters the absolute value being used for this purpose, but this really is not correct because it produces double lines of nearly equal magnitude, which can be confusing.

The image shown in Fig. 3.40(c) was scaled in the manner just described for display purposes. Note that the dominant features of the image are edges and sharp gray-level discontinuities of various gray-level values. The background, previously near black, is now gray due to the scaling. This grayish appearance is typical of Laplacian images that have been scaled properly. Finally, Fig. 3.40(d)

a b
c d

FIGURE 3.40
(a) Image of the
North Pole of the
moon.
(b) Laplacian-
filtered image.
(c) Laplacian
image scaled for
display purposes.
(d) Image
enhanced by
using Eq. (3.7-5).
(Original image
courtesy of
NASA.)

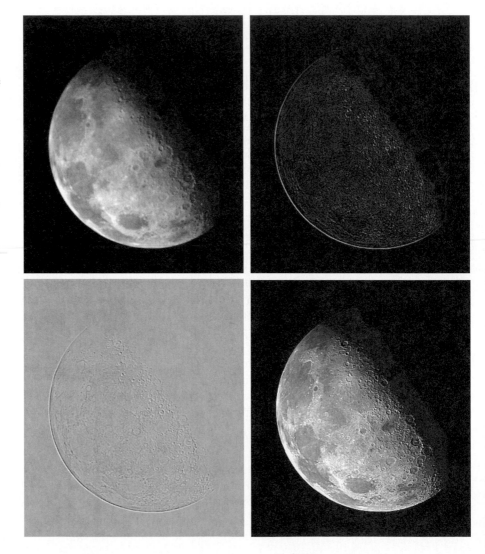

shows the result obtained using Eq. (3.7-5). The detail in this image is unmistakably clearer and sharper than in the original image. Adding the original image to
the Laplacian restored the overall gray level variations in the image, with the
Laplacian increasing the contrast at the locations of gray-level discontinuities.
The net result is an image in which small details were enhanced and the background tonality was reasonably preserved. Results like these have made
Laplacian-based enhancement a fundamental tool used frequently for sharpening digital images.   ■

Simplifications

In the previous example, we implemented Eq. (3.7-5) by first computing the
Laplacian-filtered image and then subtracting it from the original image. This
was done for instructional purposes to illustrate each step in the procedure. In
practice, Eq. (3.7-5) is usually implemented with one pass of a single mask. The
coefficients of the single mask are easily obtained by substituting Eq. (3.7-4)
for $\nabla^2 f(x, y)$ in the first line of Eq. (3.7-5):

$$
\begin{aligned}
g(x, y) = f(x, y) &- \big[f(x + 1, y) + f(x - 1, y) \\
&+ f(x, y + 1) + f(x, y - 1) \big] + 4f(x, y) \\
&- 5f(x, y) - \big[f(x + 1, y) + f(x - 1, y) \\
&+ f(x, y + 1) + f(x, y - 1) \big].
\end{aligned} \tag{3.7-6}
$$

This equation can be implemented using the mask shown in Fig. 3.41(a). The
mask shown in Fig. 3.41(b) would be used if the diagonal neighbors also were
included in the calculation of the Laplacian. Identical masks would have resulted if we had substituted the negative of Eq. (3.7-4) into the second line of
Eq. (3.7-5).

EXAMPLE 3.12:
Image
enhancement
using a composite
Laplacian mask.

■ The results obtainable with the mask containing the diagonal terms usually
are a little sharper than those obtained with the more basic mask of Fig. 3.41(a).
This property is illustrated by the Laplacian-filtered images shown in
Figs. 3.41(d) and (e), which were obtained by using the masks in Figs. 3.41(a) and
(b), respectively. By comparing the filtered images with the original image shown
in Fig. 3.41(c), we note that both masks produced effective enhancement, but the
result using the mask in Fig. 3.41(b) is visibly sharper. Figure 3.41(c) is a scanning electron microscope (SEM) image of a tungsten filament following thermal failure; the magnification is approximately 250×.)   ■

   Because the Laplacian is a linear operator, we could have arrived at the same
composite masks in Figs. 3.41(a) and (b) by noting that Eq. (3.7-5) is the difference between (sum of) two linear processes. That is, $f(x, y)$ be may viewed
as itself processed with a mask that has a unit coefficient in the center and zeros
elsewhere. The second term in the equation is the same image processed with
one of the Laplacian masks of Fig. 3.39. Due to linearity, the result obtained in
Eq. (3.7-5) with the unit-center mask and one of those Laplacian masks would
be the same as the result obtained with a single mask formed by subtracting
(adding) the Laplacian mask from (to) the unity-center mask.

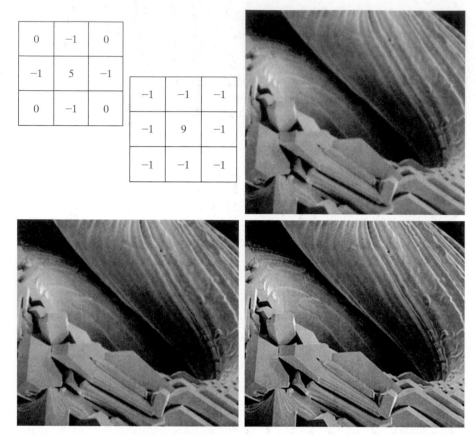

FIGURE 3.41 (a) Composite Laplacian mask. (b) A second composite mask. (c) Scanning electron microscope image. (d) and (e) Results of filtering with the masks in (a) and (b), respectively. Note how much sharper (e) is than (d). (Original image courtesy of Mr. Michael Shaffer, Department of Geological Sciences, University of Oregon, Eugene.)

Unsharp masking and high-boost filtering

A process used for many years in the publishing industry to sharpen images consists of subtracting a blurred version of an image from the image itself. This process, called *unsharp masking*, is expressed as

$$f_s(x, y) = f(x, y) - \overline{f}(x, y) \tag{3.7-7}$$

where $f_s(x, y)$ denotes the sharpened image obtained by unsharp masking, and $\overline{f}(x, y)$ is a blurred version of $f(x, y)$. The origin of unsharp masking is in darkroom photography, where it consists of clamping together a blurred negative to a corresponding positive film and then developing this combination to produce a sharper image.

A slight further generalization of unsharp masking is called *high-boost filtering*. A high-boost filtered image, f_{hb}, is defined at any point (x, y) as

$$f_{hb}(x, y) = Af(x, y) - \overline{f}(x, y) \tag{3.7-8}$$

0	−1	0
−1	$A + 4$	−1
0	−1	0

−1	−1	−1
−1	$A + 8$	−1
−1	−1	−1

FIGURE 3.42 The high-boost filtering technique can be implemented with either one of these masks, with $A \geq 1$.

where $A \geq 1$ and, as before, $\bar{f}$ is a blurred version of f. This equation may be written as

$$f_{hb}(x, y) = (A - 1)f(x, y) + f(x, y) - \bar{f}(x, y). \qquad (3.7\text{-}9)$$

By using Eq. (3.7-7), we obtain

$$f_{hb}(x, y) = (A - 1)f(x, y) + f_s(x, y) \qquad (3.7\text{-}10)$$

as the expression for computing a high boost filtered image.

Equation (3.7-10) is applicable in general and does not state explicitly how the sharp image is obtained. If we elect to use the Laplacian, then we know that $f_s(x, y)$ can be obtained using Eq. (3.7-5). In this case, Eq. (3.7-10) becomes

$$f_{hb} = \begin{cases} Af(x, y) - \nabla^2 f(x, y) & \text{if the center coefficient of the} \\ & \text{Laplacian mask is negative} \\ Af(x, y) + \nabla^2 f(x, y) & \text{if the center coefficient of the} \\ & \text{Laplacian mask is positive.} \end{cases} \qquad (3.7\text{-}11)$$

High-boost filtering can be implemented with one pass using either of the two masks shown in Fig. 3.42. Note that, when $A = 1$, high boost filtering becomes "standard" Laplacian sharpening. As the value of A increases past 1, the contribution of the sharpening process becomes less and less important. Eventually, if A is large enough, the high-boost image will be approximately equal to the original image multiplied by a constant.

■ One of the principal applications of boost filtering is when the input image is darker than desired. By varying the boost coefficient, it generally is possible to obtain an overall increase in average gray level of the image, thus helping to brighten the final result. Figure 3.43 shows such an application. Part (a) of this figure is a darker version of the image in Fig. 3.41(c). Figure 3.43(b) shows the Laplacian computed using the mask in Fig. 3.42(b), with $A = 0$. Figure 3.43(c) was obtained using the mask in Fig. 3.42(b) with $A = 1$. As expected, the image has been sharpened, but it is still as dark as the original. Finally, Fig. 3.43(d) shows the result of using $A = 1.7$. This is a much more acceptable result, in which the average gray level has increased, thus making the image lighter and more natural. ■

EXAMPLE 3.13: Image enhancement with a high-boost filter.

a b
c d

FIGURE 3.43
(a) Same as
Fig. 3.41(c), but
darker.
(b) Laplacian of
(a) computed with
the mask in
Fig. 3.42(b) using
$A = 0$.
(c) Laplacian
enhanced image
using the mask in
Fig. 3.42(b) with
$A = 1$. (d) Same
as (c), but using
$A = 1.7$.

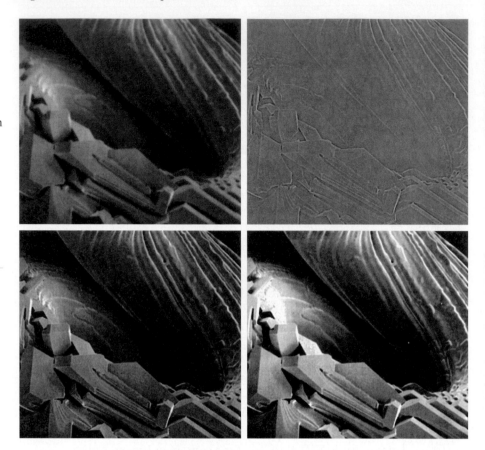

3.7.3 Use of First Derivatives for Enhancement—The Gradient

First derivatives in image processing are implemented using the magnitude of
the gradient. For a function $f(x, y)$, the gradient of f at coordinates (x, y) is defined as the two-dimensional column *vector*

$$\nabla \mathbf{f} = \begin{bmatrix} G_x \\ G_y \end{bmatrix} = \begin{bmatrix} \dfrac{\partial f}{\partial x} \\ \dfrac{\partial f}{\partial y} \end{bmatrix}. \tag{3.7-12}$$

The magnitude of this vector is given by

$$\begin{aligned} \nabla f &= \mathrm{mag}(\nabla \mathbf{f}) \\ &= [G_x^2 + G_y^2]^{1/2} \\ &= \left[\left(\frac{\partial f}{\partial x} \right)^2 + \left(\frac{\partial f}{\partial y} \right)^2 \right]^{1/2}. \end{aligned} \tag{3.7-13}$$

The components of the gradient vector itself are linear operators, but the magnitude of this vector obviously is not because of the squaring and square root

operations. On the other hand, the partial derivatives in Eq. (3.7-12) are not rotation invariant (isotropic), but the magnitude of the gradient vector is. Although it is not strictly correct, the magnitude of the gradient vector often is referred to as the *gradient*. In keeping with tradition, we will use this term in the following discussions, explicitly referring to the vector or its magnitude only in cases where confusion is likely.

The computational burden of implementing Eq. (3.7-13) over an entire image is not trivial, and it is common practice to approximate the magnitude of the gradient by using absolute values instead of squares and square roots:

$$\nabla f \approx |G_x| + |G_y|. \qquad (3.7\text{-}14)$$

This equation is simpler to compute and it still preserves relative changes in gray levels, but the isotropic feature property is lost in general. However, as in the case of the Laplacian, the isotropic properties of the digital gradient defined in the following paragraph are preserved only for a limited number of rotational increments that depend on the masks used to approximate the derivatives. As it turns out, the most popular masks used to approximate the gradient give the same result only for vertical and horizontal edges and thus the isotropic properties of the gradient are preserved only for multiples of 90°. These results are independent of whether Eq. (3.7-13) or (3.7-14) is used, so nothing of significance is lost in using the simpler of the two equations.

As in the case of the Laplacian, we now define digital approximations to the preceding equations, and from there formulate the appropriate filter masks. In order to simplify the discussion that follows, we will use the notation in Fig. 3.44(a) to denote image points in a 3×3 region. For example, the center point, z_5, denotes $f(x, y)$, z_1 denotes $f(x - 1, y - 1)$, and so on. As indicated in Section 3.7.1, the simplest approximations to a first-order derivative that satisfy the conditions stated in that section are $G_x = (z_8 - z_5)$ and $G_y = (z_6 - z_5)$. Two other definitions proposed by Roberts [1965] in the early development of digital image processing use cross differences:

$$G_x = (z_9 - z_5) \quad \text{and} \quad G_y = (z_8 - z_6). \qquad (3.7\text{-}15)$$

If we elect to use Eq. (3.7-13), then we compute the gradient as

$$\nabla f = \left[(z_9 - z_5)^2 + (z_8 - z_6)^2 \right]^{1/2} \qquad (3.7\text{-}16)$$

If we use absolute values, then substituting the quantities in Eq. (3.7-15) into Eq. (3.7-14) gives us the following approximation to the gradient:

$$\nabla f \approx |z_9 - z_5| + |z_8 - z_6|. \qquad (3.7\text{-}17)$$

This equation can be implemented with the two masks shown in Figs. 3.44(b) and (c). These masks are referred to as the *Roberts cross-gradient operators*.

Masks of even size are awkward to implement. The smallest filter mask in which we are interested is of size 3×3. An approximation using absolute values, still at point z_5, but using a 3×3 mask, is

$$\nabla f \approx |(z_7 + 2z_8 + z_9) - (z_1 + 2z_2 + z_3)|$$
$$+ |(z_3 + 2z_6 + z_9) - (z_1 + 2z_4 + z_7)|. \qquad (3.7\text{-}18)$$

a
b c
d e

FIGURE 3.44
A 3×3 region of an image (the z's are gray-level values) and masks used to compute the gradient at point labeled z_5. All masks coefficients sum to zero, as expected of a derivative operator.

z_1	z_2	z_3
z_4	z_5	z_6
z_7	z_8	z_9

-1	0
0	1

0	-1
1	0

-1	-2	-1
0	0	0
1	2	1

-1	0	1
-2	0	2
-1	0	1

The difference between the third and first rows of the 3×3 image region approximates the derivative in the x-direction, and the difference between the third and first columns approximates the derivative in the y-direction. The masks shown in Figs. 3.44(d) and (e), called the *Sobel operators*, can be used to implement Eq. (3.7-18) via the mechanics given in Eq. (3.5-1). The idea behind using a weight value of 2 is to achieve some smoothing by giving more importance to the center point (we discuss this in more detail in Chapter 10). Note that the coefficients in all the masks shown in Fig. 3.44 sum to 0, indicating that they would give a response of 0 in an area of constant gray level, as expected of a derivative operator.

EXAMPLE 3.14:
Use of the gradient for edge enhancement.

■ The gradient is used frequently in industrial inspection, either to aid humans in the detection of defects or, what is more common, as a preprocessing step in automated inspection. We will have more to say about this in Chapters 10 and 11. However, it will be instructive at this point to consider a simple example to show how the gradient can be used to enhance defects and eliminate slowly changing background features. In this particular example, the enhancement is used as a preprocessing step for automated inspection, rather than for human analysis.

Figure 3.45(a) shows an optical image of a contact lens, illuminated by a lighting arrangement designed to highlight imperfections, such as the two edge

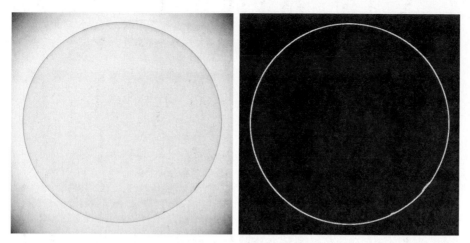

FIGURE 3.45
(a) Optical image of contact lens (note defects on the boundary at 4 and 5 o'clock).
(b) Sobel gradient.
(Original image courtesy of Mr. Pete Sites, Perceptics Corporation.)

defects in the lens boundary seen at 4 and 5 o'clock. Figure 3.45(b) shows the gradient obtained using Eq. (3.7-14) with the two Sobel masks in Figs. 3.44(d) and (e). The edge defects also are quite visible in this image, but with the added advantage that constant or slowly varying shades of gray have been eliminated, thus simplifying considerably the computational task required for automated inspection. Note also that the gradient process highlighted small specs that are not readily visible in the gray scale image (specs like these can be foreign matter, air pockets in a supporting solution, or miniscule imperfections in the lens). The ability to enhance small discontinuities in an otherwise flat gray field is another important feature of the gradient. ■

3.8 Combining Spatial Enhancement Methods

With a few exceptions, like combining blurring with thresholding in Section 3.6.1, we have focused attention thus far on individual enhancement approaches. Frequently, a given enhancement task will require application of several complementary enhancement techniques in order to achieve an acceptable result. In this section we illustrate by means of an example how to combine several of the approaches developed in this chapter to address a difficult enhancement task.

The image shown in Fig. 3.46(a) is a nuclear whole body bone scan, used to detect diseases such as bone infection and tumors. Our objective is to enhance this image by sharpening it and by bringing out more of the skeletal detail. The narrow dynamic range of the gray levels and high noise content make this image difficult to enhance. The strategy we will follow is to utilize the Laplacian to highlight fine detail, and the gradient to enhance prominent edges. For reasons that will be explained shortly, a smoothed version of the gradient image will be used to mask the Laplacian image (see Section 3.4 regarding masking). Finally, we will attempt to increase the dynamic range of the gray levels by using a gray-level transformation.

Figure 3.46 (b) shows the Laplacian of the original image, obtained using the mask in Fig. 3.39(d). This image was scaled (for display only) using the same technique as in Fig. 3.40. We can obtain a sharpened image at this point

a b
c d

FIGURE 3.46
(a) Image of
whole body bone
scan.
(b) Laplacian of
(a). (c) Sharpened
image obtained
by adding (a) and
(b). (d) Sobel of
(a).

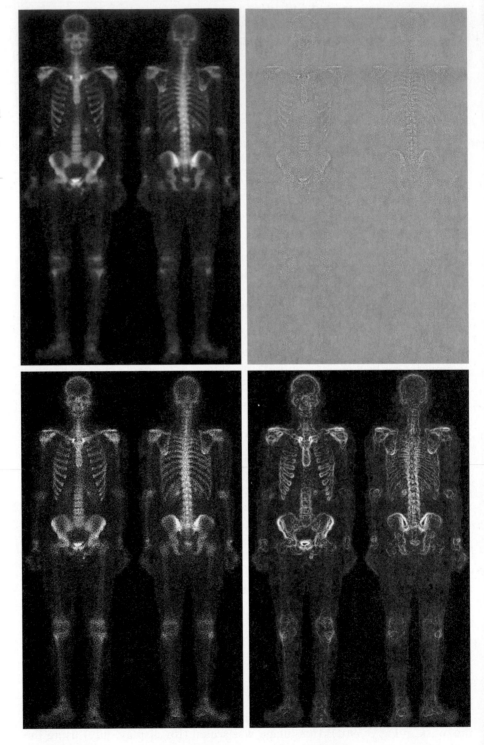

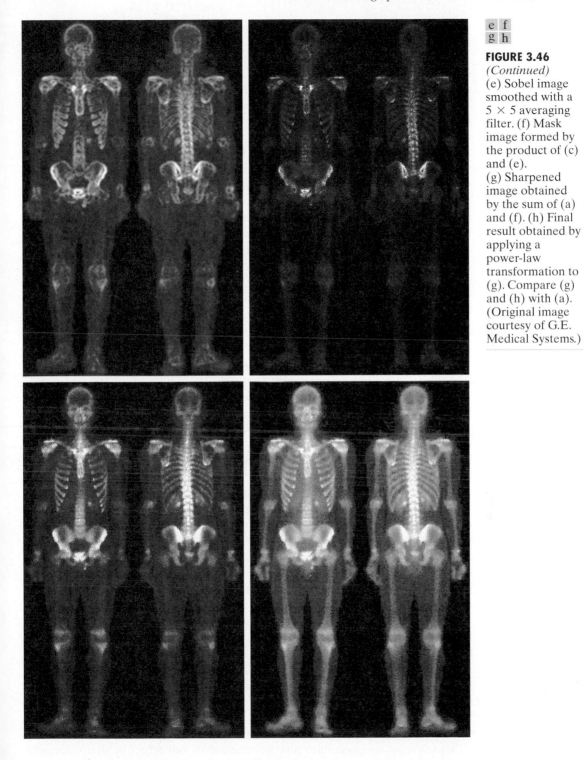

e f
g h

FIGURE 3.46
(Continued)
(e) Sobel image smoothed with a 5×5 averaging filter. (f) Mask image formed by the product of (c) and (e). (g) Sharpened image obtained by the sum of (a) and (f). (h) Final result obtained by applying a power-law transformation to (g). Compare (g) and (h) with (a). (Original image courtesy of G.E. Medical Systems.)

simply by adding Figs. 3.46(a) and (b), which are an implementation of the second line in Eq. (3.7-5) (we used a mask with a positive center coefficient). Just by looking at the noise level in (b), we would expect a rather noisy sharpened image if we added Figs. 3.46(a) and (b), a fact that is confirmed by the result shown in Fig. 3.46(c). One way that comes immediately to mind to reduce the noise is to use a median filter. However, median filtering is a nonlinear process capable of removing image features. This is unacceptable in medical image processing.

An alternate approach is to use a mask formed from a smoothed version of the gradient of the original image. The motivation behind this is straightforward and is based on the properties of first- and second-order derivatives explained in Section 3.7.1. The Laplacian, being a second-order derivative operator, has the definite advantage that it is superior in enhancing fine detail. However, this causes it to produce noisier results than the gradient. This noise is most objectionable in smooth areas, where it tends to be more visible. The gradient has a stronger response in areas of significant gray-level transitions (gray-level ramps and steps) than does the Laplacian. The response of the gradient to noise and fine detail is lower than the Laplacian's and can be lowered further by smoothing the gradient with an averaging filter. The idea, then, is to smooth the gradient and multiply it by the Laplacian image. In this context, we may view the smoothed gradient as a mask image. The product will preserve details in the strong areas while reducing noise in the relatively flat areas. This process can be viewed roughly as combining the best features of the Laplacian and the gradient. The result is added to the original to obtain a final sharpened image, and could even be used in boost filtering.

Figure 3.46(d) shows the Sobel gradient of the original image, computed using Eq. (3.7-14). Components G_x and G_y were obtained using the masks in Figs. 3.44(d) and (e), respectively. As expected from our discussion in Section 3.7.1, edges are much more dominant in this image than in the Laplacian image. The smoothed gradient image shown in Fig. 3.46(e) was obtained by using an averaging filter of size 5×5. The two gradient images were scaled for display in the same manner as the two Laplacian images. Because the smallest possible value of a gradient image is 0, the background is black in the scaled gradient images, rather than gray as in the scaled Laplacian. The fact that Figs. 3.46(d) and (e) are much brighter than Fig. 3.46(b) is again evidence that the gradient of an image with significant edge content has values that are higher in general than in a Laplacian image.

The product of the Laplacian and smoothed-gradient image is shown in Fig. 3.46(f). Note the dominance of the strong edges and the relative lack of visible noise, which is the key objective behind masking the Laplacian with a smoothed gradient image. Adding the product image to the original resulted in the sharpened image shown in Fig. 3.46(g). The significant increase in sharpness of detail in this image over the original is evident in most parts of the image, including the ribs, spinal chord, pelvis, and skull. This type of improvement would not have been possible by using the Laplacian or gradient alone.

The sharpening procedure just discussed does not affect in an appreciable way the dynamic range of the gray levels in an image. Thus, the final step in our enhancement task is to increase the dynamic range of the sharpened image. As we discussed in some detail in Sections 3.2 and 3.3, there are a number of gray-level transformation functions that can accomplish this objective. We do know from the results in Section 3.3.2 that histogram equalization is not likely to work well on images that have dark gray-level distributions like our images have here. Histogram specification could be a solution, but the dark characteristics of the images with which we are dealing lend themselves much better to a power-law transformation. Since we wish to spread the gray levels, the value of γ in Eq. (3.2-3) has to be less than 1. After a few trials with this equation we arrived at the result shown in Fig. 3.46(h), obtained with $\gamma = 0.5$ and $c = 1$. Comparing this image with Fig. 3.46(g), we see that significant new detail is visible in Fig. 3.46(h). The areas around the wrists, hands, ankles, and feet are good examples of this. The skeletal bone structure also is much more pronounced, including the arm and leg bones. Note also the faint definition of the outline of the body, and of body tissue. Bringing out detail of this nature by expanding the dynamic range of the gray levels also enhanced noise, but Fig. 3.46(h) represents a significant visual improvement over the original image.

The approach just discussed is representative of the types of processes that can be linked in order to achieve results that are not possible with a single technique. The way in which the results are used depends on the application. The final user of the type of images shown in this section is likely to be a radiologist. For a number of reasons that are beyond the scope of our discussion, physicians are unlikely to rely on enhanced results to arrive at a diagnosis. However, enhanced images are quite useful in highlighting details that can serve as clues for further analysis in the original image or sequence of images. In other areas, the enhanced result may indeed be the final product. Examples are found in the printing industry, in image-based product inspection, in forensics, in microscopy, in surveillance, and in a host of other areas where the principal objective of enhancement is to obtain an image with a higher content of visual detail.

Summary

The material presented in this chapter is representative of spatial domain techniques commonly used in practice for image enhancement. This area of image processing is a dynamic field, and new techniques and applications are reported routinely in professional literature and in new product announcements. For this reason, the topics included in this chapter were selected for their value as fundamental material that would serve as a foundation for understanding the state of the art in enhancement techniques, as well as for further study in this field. In addition to enhancement, this chapter served the purpose of introducing a number of concepts, such as filtering with spatial masks, that will be used in numerous occasions throughout the remainder of the book. In the following chapter, we deal with enhancement from a complementary viewpoint in the frequency domain. Between these two chapters, the reader will have developed a solid foundation for the terminology and some of the most fundamental tools used in image processing. The fact that these tools were introduced in the context of image enhancement is likely to aid in the understanding of how they operate on digital images.

References and Further Reading

The material in Section 3.1 is from Gonzalez [1986]. Additional reading for the material in Section 3.2 may be found in Schowengerdt [1983], Poyton [1996], and Russ [1999]. See also the paper by Tsujii et al. [1998] regarding the optimization of image displays. Early references on histogram processing are Hummel [1974], Gonzalez and Fittes [1977], and Woods and Gonzalez [1981]. Stark [2000] gives some interesting generalizations of histogram equalization for adaptive contrast enhancement. Other approaches for contrast enhancement are exemplified by Centeno and Haertel [1997] and Cheng and Xu [2000]. For enhancement based on an ideal image model, see Highnam and Brady [1997]. For extensions of the local histogram equalization method, see Caselles et al. [1999], and Zhu et al. [1999]. See Narendra and Fitch [1981] on the use and implementation of local statistics for image enhancement. Kim et al. [1997] present an interesting approach combining the gradient with local statistics for image enhancement.

Image subtraction (Section 3.4.1) is a generic image processing tool widely used for change detection. As noted in that section, one of the principal applications of digital image subtraction is in mask mode radiography, where patient motion is a problem because motion smears the results. The problem of motion during image subtraction has received significant attention over the years, as exemplified in the survey article by Meijering et al. [1999]. The method of noise reduction by image averaging (Section 3.4.2) was first proposed by Kohler and Howell [1963]. See Peebles [1993] regarding the expected value of the mean and variance of a sum of random variables.

For additional reading on linear spatial filters and their implementation, see Umbaugh [1998], Jain [1989], and Rosenfeld and Kak [1982]. Rank-order filters are discussed in these references as well. Wilburn [1998] discusses generalizations of rank-order filters. The book by Pitas and Venetsanopoulos [1990] also deals with median and other nonlinear spatial filters. A special issue of *IEEE Transactions in Image Processing* [1996] is dedicated to the topic of nonlinear image processing. The material on high-boost filtering is from Schowengerdt [1983]. We will encounter again many of the spatial filters introduced in this chapter in discussions dealing with image restoration (Chapter 5) and edge detection (Chapter 10).

Problems

See inside front cover

Detailed solutions to the problems marked with a star can be found in the book web site. The site also contains suggested projects based on the material in this chapter.

3.1   Exponentials of the form $e^{-\alpha r^2}$, with α a positive constant, are useful for constructing smooth gray-level transformation functions. Start with this basic function and construct transformation functions having the general shapes shown in the following figures. The constants shown are *input* parameters, and your proposed transformations must include them in their specification. (For simplicity in your answers, L_0 is not a required parameter in the third curve.)

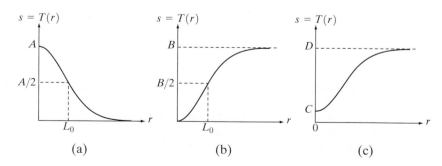

(a)   (b)   (c)

3.2 ★ **(a)** Give a continuous function for implementing the contrast stretching transformation shown in Fig. 3.2(a). In addition to m, your function must include a parameter, E, for controlling the slope of the function as it transitions from low to high gray-level values. Your function should be normalized so that its minimum and maximum values are 0 and 1, respectively.

   (b) Sketch a family of transformations as a function of parameter E, for a fixed value $m = L/2$, where L is the number of gray levels in the image.

   (c) What is the smallest value of E that will make your function *effectively* perform as the function in Fig. 3.2(b)? In other words, your function does not have to be identical to Fig. 3.2(b). It just has to yield the same result of producing a binary image. Assume that you are working with 8-bit images, and let $m = 128$. Also, let C be the smallest positive number representable in the computer you are using.

3.3 Propose a set of gray-level-slicing transformations capable of producing all the individual bit planes of an 8-bit monochrome image. (For example, a transformation function with the property $T(r) = 0$ for r in the range $[0, 127]$, and $T(r) = 255$ for r in the range $[128, 255]$ produces an image of the 7th bit plane in an 8-bit image.)

3.4 ★ **(a)** What effect would setting to zero the lower-order bit planes have on the histogram of an image in general?

   (b) What would be the effect on the histogram if we set to zero the higher-order bit planes instead?

★ **3.5** Explain why the discrete histogram equalization technique does not, in general, yield a flat histogram.

3.6 Suppose that a digital image is subjected to histogram equalization. Show that a second pass of histogram equalization will produce exactly the same result as the first pass.

3.7 In some applications it is useful to model the histogram of input images as Gaussian probability density functions of the form

$$p_r(r) = \frac{1}{\sqrt{2\pi}\sigma} e^{-\frac{(r-m)^2}{2\sigma^2}}$$

where m and σ are the mean and standard deviation of the Gaussian PDF. The approach is to let m and σ be measures of average gray level and contrast of a given image. What is the transformation function you would use for histogram equalization?

★ **3.8** Assuming continuous values, show by example that it is possible to have a case in which the transformation function given in Eq. (3.3-4) satisfies Conditions (a) and (b) in Section 3.3.1, but its inverse may fail to be single valued.

3.9 **(a)** Show that the discrete transformation function given in Eq. (3.3-8) for histogram equalization satisfies conditions (a) and (b) in Section 3.3.1.

   (b) Show by example that this does not hold in general for the inverse discrete transformation function given in Eq. (3.3-9).

   ★ **(c)** Show that the inverse discrete transformation in Eq. (3.3-9) satisfies Conditions (a) and (b) in Section 3.3.1 if none of the gray levels r_k, $k = 0, 1, \dots, L - 1$, are missing.

3.10 An image has the gray level PDF $p_r(r)$ shown in the following diagram. It is desired to transform the gray levels of this image so that they will have the specified $p_z(z)$ shown. Assume continuous quantities and find the transformation (in terms of r and z) that will accomplish this.

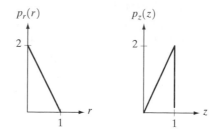

★3.11 Propose a method for updating the local histogram for use in the local enhancement technique discussed in Section 3.3.3.

3.12 Two images, $f(x, y)$ and $g(x, y)$, have histograms h_f and h_g. Give the conditions under which you can determine the histograms of

(a) $f(x, y) + g(x, y)$

(b) $f(x, y) - g(x, y)$

(c) $f(x, y) \times g(x, y)$

(d) $f(x, y) \div g(x, y)$

in terms of h_f and h_g. Explain how to obtain the histogram in each case.

3.13 Consider two 8-bit images whose gray levels span the full range from 0 to 255.

(a) Discuss the limiting effect of repeatedly subtracting image (b) from image (a).

(b) Would reversing the order of the images yield a different result?

★3.14 Image subtraction is used often in industrial applications for detecting missing components in product assembly. The approach is to store a "golden" image that corresponds to a correct assembly; this image is then subtracted from incoming images of the same product. Ideally, the differences would be zero if the new products are assembled correctly. Difference images for products with missing components would be nonzero in the area where they differ from the golden image. What conditions do you think have to be met in practice for this method to work?

3.15 Prove the validity of Eqs. (3.4-4) and (3.4-5).

3.16 In an industrial application, X-ray imaging is to be used to inspect the inside of certain composite castings. The objective is to look for voids in the castings, which typically appear as small blobs in the image. However, due to properties in of the casting material and X-ray energy used, high noise content often makes inspection difficult, so the decision is made to use image averaging to reduce the noise and thus improve visible contrast. In computing the average, it is important to keep the number of images as small as possible to reduce the time the parts have to remain stationary during imaging. After numerous experiments, it is concluded that decreasing the noise variance by a factor of 10 is sufficient. If the imaging device can produce 30 frames/s, how long would the castings have to remain stationary during imaging to achieve the desired decrease in variance? Assume that the noise is uncorrelated and has zero mean.

3.17   The implementation of linear spatial filters requires moving the center of a mask throughout an image and, at each location, computing the sum of products of the mask coefficients with the corresponding pixels at that location (see Section 3.5). In the case of lowpass filtering, all coefficients are 1, allowing use of a so-called *box-filter* or *moving-average* algorithm, which consists of updating only the part of the computation that changes from one location to the next.

★ **(a)** Formulate such an algorithm for an $n \times n$ filter, showing the nature of the computations involved and the scanning sequence used for moving the mask around the image.

(b) The ratio of the number of computations performed by a brute-force implementation to the number of computations performed by the box-filter algorithm is called the *computational advantage*. Obtain the computational advantage in this case and plot it as a function of n for $n > 1$. The $1/n^2$ scaling factor is common to both approaches, so you need not consider it in obtaining the computational advantage. Assume that the image has an outer border of zeros that is thick enough to allow you to ignore border effects in your analysis.

3.18   Discuss the limiting effect of repeatedly applying a 3×3 lowpass spatial filter to a digital image. You may ignore border effects.

3.19 ★ **(a)** It was stated in Section 3.6.2 that isolated clusters of dark or light (with respect to the background) pixels whose area is less than one-half the area of a median filter are eliminated (forced to the median value of the neighbors) by the filter. Assume a filter of size $n \times n$, with n odd, and explain why this is so.

(b) Consider an image having various sets of pixel clusters. Assume that all points in a cluster are lighter or darker than the background (but not both simultaneously in the same cluster), and that the area of each cluster is less than or equal to $n^2/2$. In terms of n, under what condition would one or more of these clusters cease to be isolated in the sense described in part (a)?

★ **3.20**   **(a)** Develop a procedure for computing the median of an $n \times n$ neighborhood.

(b) Propose a technique for updating the median as the center of the neighborhood is moved from pixel to pixel.

3.21   **(a)** In a character recognition application, text pages are reduced to binary form using a thresholding transformation function of the form shown in Fig. 3.2(b). This is followed by a procedure that thins the characters until they become strings of binary 1's on a background of 0's. Due to noise, the binarization and thinning processes result in broken strings of characters with gaps ranging from 1 to 3 pixels. One way to "repair" the gaps is to run an averaging mask over the binary image to blur it, and thus create bridges of nonzero pixels between gaps. Give the (odd) size of the smallest averaging mask capable of performing this task.

(b) After bridging the gaps, it is desired to threshold the image in order to convert it back to binary form. For your answer in (a), what is the minimum value of the threshold required to accomplish this, without causing the segments to break up again?

★ **3.22**   The three images shown were blurred using square averaging masks of sizes $n = 23, 25$, and 45, respectively. The vertical bars on the left lower part of (a) and (c) are blurred, but a clear separation exists between them. However, the bars have merged in image (b), in spite of the fact that the mask that produced this image is significantly smaller than the mask that produced image (c). Explain this.

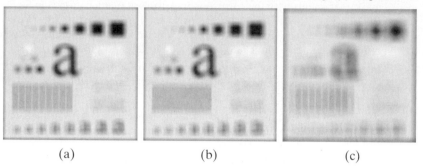

(a)   (b)   (c)

3.23   Consider an application such as the one shown in Fig. 3.36, in which it is desired to eliminate objects smaller than those enclosed in a square of size $q \times q$ pixels. Suppose that we want to reduce the average gray level of those objects to one-tenth of their original average gray level. In this way, those objects will be closer to the gray level of the background and they can then be eliminated by thresholding. Give the (odd) size of the smallest averaging mask that will accomplish the desired reduction in average gray level in only one pass of the mask over the image.

3.24   In a given application an averaging mask is applied to input images to reduce noise, and then a Laplacian mask is applied to enhance small details. Would the result be the same if the order of these operations were reversed?

★ **3.25**   Show that the Laplacian operation defined in Eq. (3.7-1) is isotropic (invariant to rotation). You will need the following equations relating coordinates after axis rotation by an angle θ:

$$x = x' \cos\theta - y' \sin\theta$$
$$y = x' \sin\theta + y' \cos\theta$$

where (x, y) are the unrotated and (x', y') are the rotated coordinates.

3.26   Give a 3×3 mask for performing unsharp masking in a single pass through an image.

★ **3.27**   Show that subtracting the Laplacian from an image is proportional to unsharp masking. Use the definition for the Laplacian given in Eq. (3.7-4).

3.28   **(a)** Show that the magnitude of the gradient given in Eq. (3.7-13) is an isotropic operation. (See Problem 3.25.)

(b) Show that the isotropic property is lost in general if the gradient is computed using Eq. (3.7-14).

3.29   A CCD TV camera is used to perform a long-term study by observing the same area 24 hours a day, for 30 days. Digital images are captured and transmitted to a central location every 5 minutes. The illumination of the scene changes from natural daylight to artificial lighting. At no time is the scene without illumination, so it is always possible to obtain an image. Because the range of illumination is such that it is always in the linear operating range of the camera, it is decided not to employ any compensating mechanisms on the camera itself. Rather, it is decided to use digital techniques to postprocess, and thus normalize, the images to the equivalent of constant illumination. Propose a method to do this. You are at liberty to use any method you wish, but state clearly all the assumptions you made in arriving at your design.

Image Enhancement in the Frequency Domain

Enhance: To make greater (as in value, desirability, or attractiveness).

Frequency: The number of times that a periodic function repeats the same sequence of values during a unit variation of the independent variable.

Webster's New Collegiate Dictionary

Preview

Although significant effort was devoted in the previous chapter to spatial techniques for image enhancement, a thorough understanding of this area is impossible without having at least a working knowledge of how the Fourier transform and the frequency domain can be used for image processing. A solid understanding of these topics can be developed without having to become a signal processing expert. The key lies in focusing on the fundamentals and their relevance to digital image processing. The notation, usually a source of trouble for the beginner, is clarified significantly in this chapter by emphasizing the connection between image characteristics and the mathematical tools used to represent them. This chapter is concerned primarily with helping the reader develop a basic understanding of the Fourier transform and the frequency domain, and how they apply to image enhancement. Later, in Chapters 5, 8, 10, and 11, we discuss other applications of the Fourier transform.

We begin the discussion with a brief outline of the origins of the Fourier transform and its impact on countless branches of mathematics, science, and engineering. Next, we give an introduction to the Fourier transform and the frequency domain, and establish the notation and reasons why these tools are so useful for image enhancement. This is followed by sections that parallel the spatial smoothing and sharpening filtering techniques discussed in Chapter 3, except that all filtering is done in the frequency domain. After discussing other uses of the Fourier transform for image enhancement, we conclude the chapter with a discussion of issues related to implementing the Fourier transform in the context of image processing.

4.1 Background

The French mathematician Jean Baptiste Joseph Fourier was born in 1768 in the town of Auxerre, about midway between Paris and Dijon. The contribution for which he is most remembered was outlined in a memoir in 1807 and published in 1822 in his book, *La Théorie Analitique de la Chaleur* (The Analytic Theory of Heat). This book was translated into English 55 years later by Freeman (see Freeman [1878]). Basically, Fourier's contribution in this particular field states that any function that periodically repeats itself can be expressed as the sum of sines and/or cosines of different frequencies, each multiplied by a different coefficient (we now call this sum a *Fourier series*). It does not matter how complicated the function is; as long as it is periodic and meets some mild mathematical conditions, it can be represented by such a sum. This is now taken for granted, but at the time it first appeared it was a revolutionary concept to which it took mathematicians all over the world over a century to "adjust." At that time, regularity in functions was a mainstay of mathematical thinking. With this type of cultural mindset, the concept that complicated functions could be represented as a sum of simple sines and cosines was not at all intuitive (Fig. 4.1), so it is not surprising that Fourier's ideas in this regard were met with skepticism.

Even functions that are not periodic (but whose area under the curve is finite) can be expressed as the integral of sines and/or cosines multiplied by a weighing function. The formulation in this case is the *Fourier transform*, and its utility is even greater than the Fourier series in most practical problems. Both representations share the important characteristic that a function, expressed in either a Fourier series or transform, can be reconstructed (recovered) completely via an inverse process, with no loss of information. This is one of the most important characteristics of these representations because they allow us to work in the "Fourier domain" and then return to the original domain of the function without losing any information.

Ultimately, it was the utility of the Fourier series and transform in solving practical problems that made them widely used and studied as fundamental tools. The application of Fourier initial ideas was in the field of heat diffusion, where they allowed the formulation of differential equations representing heat flow in such a way that solutions could be obtained for the first time. During the past century, and especially in the past 50 years, entire industries and academic disciplines have flourished as a result of Fourier's ideas. The advent of digital computation and the "discovery" of a fast Fourier transform (FFT) algorithm in the late 1950s (more about this later) revolutionized the field of signal processing. These two core technologies allowed for the first time practical processing and meaningful interpretation of a host of signals of exceptional human and industrial importance, from medical monitors and scanners to modern electronic communications.

We will be dealing only with functions (images) of finite duration, so the Fourier transform is the tool in which we are interested. The material in the fol-

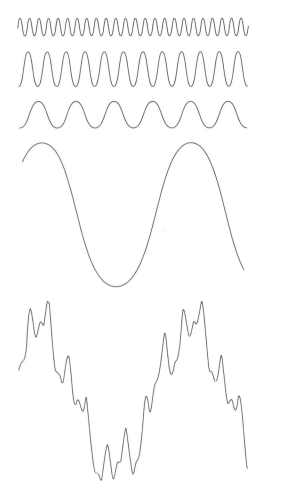

FIGURE 4.1 The function at the bottom is the sum of the four functions above it. Fourier's idea in 1807 that periodic functions could be represented as a weighted sum of sines and cosines was met with skepticism.

lowing section introduces the Fourier transform and the frequency domain. It is shown that Fourier techniques provide a meaningful and practical way to study and implement a host of image enhancement approaches. In some cases, these approaches are similar to the ones we developed in Chapter 3. In others, they are complementary.

4.2 Introduction to the Fourier Transform and the Frequency Domain

This section introduces the Fourier transform in one and two dimensions. The focus is mostly on a discrete formulation of the continuous transform and some of its properties.

4.2.1 The One-Dimensional Fourier Transform and its Inverse

The Fourier transform, $F(u)$, of a single variable, continuous function, $f(x)$, is defined by the equation

$$F(u) = \int_{-\infty}^{\infty} f(x)e^{-j2\pi ux}\, dx \qquad (4.2\text{-}1)$$

where $j = \sqrt{-1}$. Conversely, given $F(u)$, we can obtain $f(x)$ by means of the *inverse* Fourier transform

$$f(x) = \int_{-\infty}^{\infty} F(u)e^{j2\pi ux}\, du. \qquad (4.2\text{-}2)$$

These two equations comprise the *Fourier transform pair*. They indicate the important fact mentioned in the previous section that a function can be recovered from its transform. These equations are easily extended to two variables, u and v:

$$F(u, v) = \int_{-\infty}^{\infty}\int_{-\infty}^{\infty} f(x, y)e^{-j2\pi(ux+vy)}\, dx\, dy \qquad (4.2\text{-}3)$$

and, similarly for the inverse transform,

$$f(x, y) = \int_{-\infty}^{\infty}\int_{-\infty}^{\infty} F(u, v)e^{j2\pi(ux+vy)}\, du\, dv. \qquad (4.2\text{-}4)$$

Our interest is in discrete functions, so we will not dwell on these equations here. However, in some cases, the reader may find them easier to manipulate than their discrete equivalents in proving the validity of properties of the 2-D Fourier transform.

The Fourier transform of a discrete function of one variable, $f(x), x = 0, 1, 2, \ldots, M - 1$, is given by the equation

$$F(u) = \frac{1}{M}\sum_{x=0}^{M-1} f(x)e^{-j2\pi ux/M} \qquad \text{for } u = 0, 1, 2, \ldots, M - 1. \quad (4.2\text{-}5)$$

This *discrete Fourier transform* (DFT) is the foundation for most of the work in this chapter. Similarly, given $F(u)$, we can obtain the original function back using the inverse DFT:

$$f(x) = \sum_{u=0}^{M-1} F(u)e^{j2\pi ux/M} \qquad \text{for } x = 0, 1, 2, \ldots, M - 1. \quad (4.2\text{-}6)$$

The $1/M$ multiplier in front of the Fourier transform sometimes is placed in front of the inverse instead. Other times (not as often) both equations are multiplied by $1/\sqrt{M}$. The location of the multiplier does not matter. If two multipliers are used, the only requirement is that their product be equal to $1/M$. Considering their importance, these equations really are very simple.

In order to compute $F(u)$ in Eq. (4.2-5) we start by substituting $u = 0$ in the exponential term and then summing for *all* values of x. We then substitute $u = 1$ in the exponential and repeat the summation over all values of x. We repeat this process for all M values of u in order to obtain the complete Fourier transform. It takes approximately M^2 summations and multiplications to compute the discrete Fourier transform (reduction of this number is an important topic of discussion in Section 4.6). Like $f(x)$, the transform is a discrete quantity, and it has the same number of components as $f(x)$. Similar comments apply to the computation of the inverse Fourier transform.

An important property of the discrete transform pair is that, unlike the continuos case, we need not be concerned about the existence of the DFT or its inverse. The discrete Fourier transform and its inverse always exist. This can be shown by substituting either of Eqs. (4.2-5) or (4.2-6) into the other and making use of the orthogonality of the exponential terms (Problem 4.1). We will get an identity, indicating the existence of the two functions. Of course, there is always the question of what happens if $f(x)$ has infinite values, but we deal strictly with finite quantities in this book. These comments are directly applicable to two-dimensional (and higher) functions. Thus, for digital image processing, existence of either the discrete transform or its inverse is not an issue.

The concept of the frequency domain, mentioned numerous times in this chapter and in Chapter 3, follows directly from Euler's formula:

$$e^{j\theta} = \cos\theta + j\sin\theta. \qquad (4.2\text{-}7)$$

Substituting this expression into Eq. (4.2-5), and using the fact that $\cos(-\theta) = \cos\theta$, gives us

$$F(u) = \frac{1}{M}\sum_{x=0}^{M-1} f(x)[\cos 2\pi ux/M \quad j\sin 2\pi ux/M] \qquad (4.2\text{-}8)$$

for $u = 0, 1, 2, \ldots, M - 1$. Thus, we see that *each* term of the Fourier transform [that is, the value of $F(u)$ for each value of u] is composed of the sum of *all* values of the function $f(x)$. The values of $f(x)$, in turn, are multiplied by sines and cosines of various frequencies. The domain (values of u) over which the values of $F(u)$ range is appropriately called the *frequency domain*, because u determines the frequency of the components of the transform. (The x's also affect the frequencies, but they are summed out and they all make the same contributions for each value of u.) Each of the M terms of $F(u)$ is called a *frequency component* of the transform. Use of the terms *frequency domain* and *frequency components* is really no different from the terms *time domain* and *time components*, which we would use to express the domain and values of $f(x)$ if x were a time variable.

A useful analogy is to compare the Fourier transform to a glass prism. The prism is a physical device that separates light into various color components, each depending on its wavelength (or frequency) content. The Fourier transform may be viewed as a "mathematical prism" that separates a function into various components, also based on frequency content. When we consider light, we talk about its spectral or frequency content. Similarly, the Fourier transform lets us characterize a function by its frequency content. This is a powerful concept that lies at the heart of linear filtering.

In general, we see from Eqs. (4.2-5) or (4.2-8) that the components of the Fourier transform are complex quantities. As in the analysis of complex numbers, we find it convenient sometimes to express $F(u)$ in polar coordinates:

$$F(u) = |F(u)|e^{j\phi(u)} \tag{4.2-9}$$

where

$$|F(u)| = \left[R^2(u) + I^2(u)\right]^{1/2} \tag{4.2-10}$$

is called the *magnitude* or *spectrum* of the Fourier transform, and

$$\phi(u) = \tan^{-1}\left[\frac{I(u)}{R(u)}\right] \tag{4.2-11}$$

is called the *phase angle* or *phase spectrum* of the transform. In Eqs. (4.2-10) and (4.2-11), $R(u)$ and $I(u)$ are the real and imaginary parts of $F(u)$, respectively. In terms of image enhancement we are concerned primarily with properties of the spectrum. Another quantity that is used later in this chapter is the *power spectrum*, defined as the square of the Fourier spectrum:

$$\begin{aligned} P(u) &= |F(u)|^2 \\ &= R^2(u) + I^2(u). \end{aligned} \tag{4.2-12}$$

The term *spectral density* also is used to refer to the power spectrum.

EXAMPLE 4.1:
Fourier spectra of two simple 1-D functions.

■ Before proceeding, it will be helpful to consider a simple one-dimensional example of the DFT. Figure 4.2(a) shows a function and Fig. 4.2(b) shows its Fourier spectrum. Both $f(x)$ and $F(u)$ are discrete quantities, but the points in the plots are linked to make them easier to follow visually. In this example, $M = 1024$, $A = 1$, and K is only 8 points. Also note that the spectrum is centered at $u = 0$. As shown in the following section, this is accomplished by multiplying $f(x)$ by $(-1)^x$ before taking the transform. The next two figures depict basically the same thing, but with $K = 16$ points. The important features to note are that (1) the height of the spectrum doubled as the area under the curve in the x-domain doubled, and (2) the number of zeros in the spectrum in the same interval doubled as the length of the function doubled. This "reciprocal" nature of the Fourier transform pair is most useful in interpreting results of image processing in the frequency domain. ■

In the discrete transform of Eq. (4.2-5), the function $f(x)$ for $x = 0, 1, 2, \ldots, M - 1$, represents M samples from its continuous counterpart. It is important to keep in mind that these samples are *not* necessarily always taken at integer values of x in the interval $[0, M - 1]$. They are taken at equally spaced, but otherwise arbitrary, points. This is usually represented by letting x_0 denote the first (arbitrarily located) point in the sequence. The first value of the sampled function is then $f(x_0)$. The next sample has taken a fixed interval Δx units away to give $f(x_0 + \Delta x)$. The kth sample gives us $f(x_0 + k\Delta x)$, and the final sample is $f(x_0 + [M - 1]\Delta x)$. Thus, in the discrete case, when we write $f(k)$, it is understood that we are utilizing shorthand notation that really means $f(x_0 + k\Delta x)$. In terms of the notation we have used thus far, $f(x)$ is then understood to mean

$$f(x) \triangleq f(x_0 + x\Delta x) \tag{4.2-13}$$

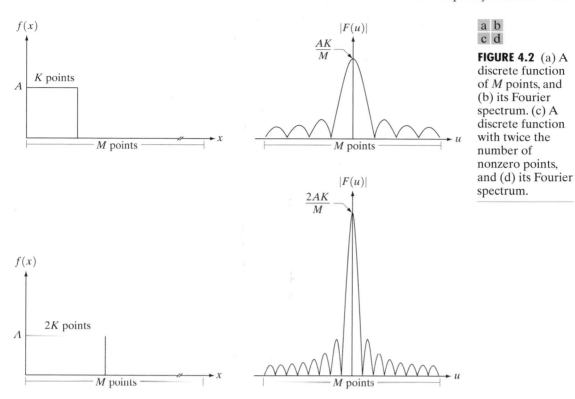

FIGURE 4.2 (a) A discrete function of M points, and (b) its Fourier spectrum. (c) A discrete function with twice the number of nonzero points, and (d) its Fourier spectrum.

when dealing with discrete variables. The variable u has a similar interpretation, but the sequence always starts at true zero frequency. Thus, the sequence for the values of u is $0, \Delta u, 2\Delta u, \dots, [M - 1]\Delta u$. Then, $F(u)$ is understood to mean

$$F(u) \triangleq F(u\Delta u) \qquad (4.2\text{-}14)$$

for $u = 0, 1, 2, \dots, M - 1$. This type of shorthand notation simplifies equations considerably and is much easier to follow.

Given the inverse relationship between a function and its transform illustrated in Fig. 4.2, it is not surprising that Δx and Δu are inversely related by the expression

$$\Delta u = \frac{1}{M\Delta x}. \qquad (4.2\text{-}15)$$

This relationship is useful when measurements are an issue in the images being processed. For instance, in an application of electron microscopy the image samples may be spaced 1 micron apart, and certain characteristics in the frequency domain (like periodicity components) may have implications regarding the structure of the physical sample. For the most part in subsequent discussions in this book we use the variables x and u without making reference to specific sampling or other measurement considerations.

4.2.2 The Two-Dimensional DFT and Its Inverse

Extension of the one-dimensional discrete Fourier transform and its inverse to two dimensions is straightforward. The discrete Fourier transform of a function (image) $f(x, y)$ of size $M \times N$ is given by the equation

$$F(u, v) = \frac{1}{MN} \sum_{x=0}^{M-1} \sum_{y=0}^{N-1} f(x, y) e^{-j2\pi(ux/M + vy/N)}. \qquad (4.2\text{-}16)$$

As in the 1-D case, this expression must be computed for values of $u = 0, 1, 2, \ldots, M - 1$, and also for $v = 0, 1, 2, \ldots, N - 1$. Similarly, given $F(u, v)$, we obtain $f(x, y)$ via the *inverse* Fourier transform, given by the expression

$$f(x, y) = \sum_{u=0}^{M-1} \sum_{v=0}^{N-1} F(u, v) e^{j2\pi(ux/M + vy/N)} \qquad (4.2\text{-}17)$$

for $x = 0, 1, 2, \ldots, M - 1$ and $y = 0, 1, 2, \ldots, N - 1$. Equations (4.2-16) and (4.2-17) comprise the *two-dimensional, discrete Fourier transform (DFT) pair*. The variables u and v are the *transform* or *frequency variables*, and x and y are the *spatial* or *image variables*. As in the one-dimensional case, the location of the $1/MN$ constant is not important. Sometimes it is located in front of the inverse transform. Other times it is found split into two equal terms of $1/\sqrt{MN}$ multiplying the transform and its inverse.

We define the Fourier spectrum, phase angle, and power spectrum as in the previous section:

$$|F(u, v)| = \left[R^2(u, v) + I^2(u, v) \right]^{1/2} \qquad (4.2\text{-}18)$$

$$\phi(u, v) = \tan^{-1} \left[\frac{I(u, v)}{R(u, v)} \right] \qquad (4.2\text{-}19)$$

and

$$P(u, v) = |F(u, v)|^2 \qquad (4.2\text{-}20)$$
$$= R^2(u, v) + I^2(u, v)$$

where $R(u, v)$ and $I(u, v)$ are the real and imaginary parts of $F(u, v)$, respectively.

It is common practice to multiply the input image function by $(-1)^{x+y}$ prior to computing the Fourier transform. Due to the properties of exponentials, it is not difficult to show (see Section 4.6) that

$$\Im[f(x, y)(-1)^{x+y}] = F(u - M/2, v - N/2) \qquad (4.2\text{-}21)$$

where $\Im[\,\cdot\,]$ denotes the Fourier transform of the argument. This equation states that the origin of the Fourier transform of $f(x, y)(-1)^{x+y}$ [that is, $F(0, 0)$] is located at $u = M/2$ and $v = N/2$. In other words, multiplying $f(x, y)$ by $(-1)^{x+y}$ shifts the origin of $F(u, v)$ to frequency coordinates $(M/2, N/2)$, which is the center of the $M \times N$ area occupied by the 2-D DFT. We refer to this area of the frequency domain as the *frequency rectangle*. It extends from $u = 0$ to $u = M - 1$, and from $v = 0$ to $v = N - 1$ (keep in mind that u and v are integers). In order to guarantee that these shifted coordinates are integers, we require that M and N be even numbers. When implementing the Fourier transform in a computer, the limits of summations are from $u = 1$ to M and $v = 1$ to N. The actual center of the transform will then be at $u = (M/2) + 1$ and $v = (N/2) + 1$.

The value of the transform at $(u, v) = (0, 0)$ is, from Eq. (4.2-16),

$$F(0, 0) = \frac{1}{MN} \sum_{x=0}^{M-1} \sum_{y=0}^{N-1} f(x, y), \qquad (4.2-22)$$

which we see is the average of $f(x, y)$. In other words, if $f(x, y)$ is an image, the value of the Fourier transform at the origin is equal to the average gray level of the image. Because both frequencies are zero at the origin, $F(0, 0)$ sometimes is called the *dc* component of the spectrum. This terminology is from electrical engineering, where "dc" signifies direct current (i.e., current of zero frequency).

If $f(x, y)$ is real, its Fourier transform is conjugate symmetric; that is,

$$F(u, v) = F^*(-u, -v) \qquad (4.2-23)$$

where "*" indicates the standard conjugate operation on a complex number. From this, it follows that

$$|F(u, v)| = |F(-u, -v)|, \qquad (4.2-24)$$

which says that the spectrum of the Fourier transform is symmetric. Conjugate symmetry and the centering property discussed previously truly simplify the specification of circularly symmetric filters in the frequency domain, as shown in the following section.

Finally, as in the 1-D case, we have the following relationships between samples in the spatial and frequency domains:

$$\Delta u = \frac{1}{M \Delta x} \qquad (4.2-25)$$

and

$$\Delta v = \frac{1}{N \Delta y}. \qquad (4.2-26)$$

The significance of these variables is identical in meaning to the explanation given in Section 4.2.1 for 1-D variables.

■ Figure 4.3(a) shows a white rectangle of size 20×40 pixels superimposed on a black background of size 512×512 pixels. This image was multiplied by $(-1)^{x+y}$ prior to computing the Fourier transform in order to center the spectrum, which is shown in Fig. 4.3(b). (Note the location, labels, and origin of the axes in both figures. We follow this convention throughout all discussions of images and their corresponding Fourier spectra.) In Fig. 4.3(b), the separation of spectrum zeros in the u-direction is exactly twice the separation of zeros in the v direction. This corresponds inversely to the 1-to-2 size ratio of the rectangle in the image. The spectrum was processed prior to displaying by using the log transformation in Eq. (3.2-2) to enhance gray-level detail. A value of $c = 0.5$ was used in the transformation in order to decrease overall intensity. Most Fourier spectra shown in this chapter are similarly processed by a log transformation. ■

EXAMPLE 4.2:
Centered spectrum of a simple 2-D function.

a b

FIGURE 4.3
(a) Image of a
20×40 white
rectangle on a
black background
of size 512×512
pixels.
(b) Centered
Fourier spectrum
shown after
application
of the log
transformation
given in
Eq. (3.2-2).
Compare with
Fig. 4.2.

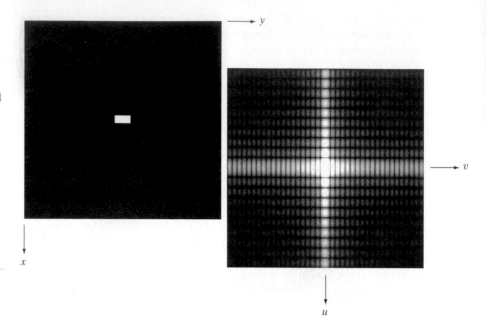

4.2.3 Filtering in the Frequency Domain

As noted in the past two sections, the frequency domain is nothing more than the space defined by values of the Fourier transform and its frequency variables (u, v). In this section, we attach "meaning" to the frequency domain, as it relates to image processing.

Some basic properties of the frequency domain

We start by observing in Eq. (4.2-16) that *each* term of $F(u, v)$ contains *all* values of $f(x, y)$, modified by the values of the exponential terms. Thus, with the exception of trivial cases, it usually is impossible to make direct associations between specific components of an image and its transform. However, some general statements can be made about the relationship between the frequency components of the Fourier transform and spatial characteristics of an image. For instance, since frequency is directly related to rate of change, it is not difficult intuitively to associate frequencies in the Fourier transform with patterns of intensity variations in an image. We showed in the previous section that the slowest varying frequency component $(u = v = 0)$ corresponds to the average gray level of an image. As we move away from the origin of the transform, the low frequencies correspond to the slowly varying components of an image. In an image of a room, for example, these might correspond to smooth gray-level variations on the walls and floor. As we move further away from the origin, the higher frequencies begin to correspond to faster and faster gray level changes in the image. These are the edges of objects and other components of an image characterized by abrupt changes in gray level, such as noise.

■ An illustration will help fix these ideas. The image shown in Fig. 4.4(a) is a scanning electron microscope image of an integrated circuit, magnified approximately 2500 times. Aside from the interesting construction of the device itself, we note two principal features: strong edges that run approximately at ±45°, and the two white oxide protrusions resulting from thermally induced failure. The Fourier spectrum in Fig. 4.4(b) shows prominent components along the ±45° directions that correspond to the edges just mentioned. Looking carefully along the vertical axis, we see a vertical component that is off-axis slightly to the left. This component was caused by the edges of the oxide protrusions. Note how the off-axis angle of the frequency component corresponds to the inclination off horizontal of the long white element, and note also the zeros in the vertical frequency component, corresponding to the narrow vertical span of the oxide protrusions.

This example is typical of the types of associations that can be made in general between the frequency and spatial domains. As we show throughout this chapter, even these types of gross associations, along with the relationships mentioned previously between frequency content and rate of change of gray levels in an image, can lead to some very useful enhancement results.   ■

EXAMPLE 4.3:
An image and its Fourier spectrum, showing some important features.

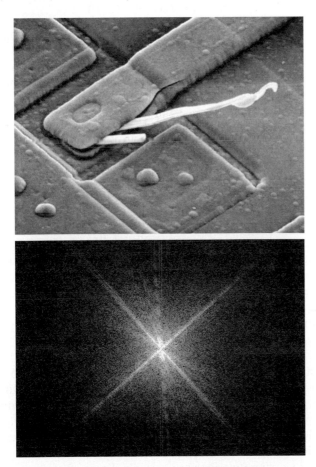

a
b

FIGURE 4.4
(a) SEM image of a damaged integrated circuit. (b) Fourier spectrum of (a). (Original image courtesy of Dr. J. M. Hudak, Brockhouse Institute for Materials Research, McMaster University, Hamilton, Ontario, Canada.)

Basics of filtering in the frequency domain

Filtering in the frequency domain is straightforward. It consists of the following steps:

1. Multiply the input image by $(-1)^{x+y}$ to center the transform, as indicated in Eq. (4.2-21).
2. Compute $F(u, v)$, the DFT of the image from (1).
3. Multiply $F(u, v)$ by a *filter* function $H(u, v)$.
4. Compute the inverse DFT of the result in (3).
5. Obtain the real part of the result in (4).
6. Multiply the result in (5) by $(-1)^{x+y}$.

The reason that $H(u, v)$ is called a *filter* (the term *filter transfer function* also is used commonly) is because it suppresses certain frequencies in the transform while leaving others unchanged. The analogy from everyday life is a screen filter that passes certain objects and suppresses others, based strictly on their size.

In equation form, let $f(x, y)$ represent the input image in Step 1 and $F(u, v)$ its Fourier transform. Then the Fourier transform of the output image is given by

$$G(u, v) = H(u, v)F(u, v). \tag{4.2-27}$$

The multiplication of H and F involves two-dimensional functions and is defined on an element-by-element basis. That is, the first element of H multiplies the first element of F, the second element of H multiplies the second element of F, and so on. In general, the components of F are complex quantities, but the filters with which we deal in this book typically are real. In this case, each component of H multiplies *both* the real and imaginary parts of the corresponding component in F. Such filters are called *zero-phase-shift* filters. As their name implies, these filters do not change the phase of the transform, a fact that can be seen in Eq. (4.2-19) by noting that the multiplier of the real and imaginary parts would cancel out because they have the same value.

The filtered image is obtained simply by taking the inverse Fourier transform of $G(u, v)$:

$$\text{Filtered Image} = \Im^{-1}\left[G(u, v)\right]. \tag{4.2-28}$$

The final image is obtained by taking the real part of this result and multiplying it by $(-1)^{x+y}$ to cancel the multiplication of the input image by this quantity. The inverse Fourier transform is, in general, complex. However, when the input image and the filter function are real, the imaginary components of the inverse transform should all be zero. In practice, the inverse DFT generally has parasitic imaginary components due to computational round-off errors. These components are ignored.

The filtering procedure just outlined is summarized in Fig. 4.5 in a slightly more general form that includes pre- and postprocessing stages. In addition to the $(-1)^{x+y}$ process, examples of other processes might include cropping of the input image to its closest even dimensions (required for proper transform centering), gray-level scaling, conversion to floating point on input, and conversion to an 8-bit integer format on the output. Multiple filtering stages and other pre- and postprocessing functions are possible. There are numerous variations

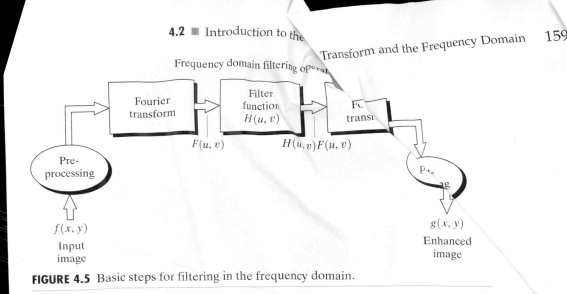

Frequency domain filtering operat...

Fourier
transform

$F(u, v)$

Filter
function
$H(u, v)$

$H(u, v)F(u, v)$

F...
transi...

Pre-
processing

$f(x, y)$

Input
image

P...
...g

$g(x, y)$

Enhanced
image

FIGURE 4.5 Basic steps for filtering in the frequency domain.

of this basic theme. The important point to keep in mind is that the filtering process is based on modifying the transform of an image in some way via a filter function, and then taking the inverse of the result to obtain the processed output image.

Some basic filters and their properties

At this point we have established the foundation for filtering in the frequency domain. The next logical step is to look at some specific filters and see how they affect images. The earlier discussion of Eq. (4.2-22) gives us a perfect lead into an introductory example of filtering. Suppose that we wish to force the average value of an image to zero. According to Eq. (4.2-22), the average value of an image is given by $F(0, 0)$. If we set this term to zero in the frequency domain and take the inverse transform, then the average value of the resulting image will be zero. Assuming that the transform has been centered as discussed in Eq. (4.2-21), we can do this operation by multiplying all values of $F(u, v)$ by the filter function:

$$H(u, v) = \begin{cases} 0 & \text{if } (u, v) = (M/2, N/2) \\ 1 & \text{otherwise.} \end{cases} \qquad (4.2\text{-}29)$$

All this filter would do is set $F(0, 0)$ to zero and leave all other frequency components of the Fourier transform untouched, as desired. The processed image (with zero average value) can then be obtained by taking the inverse Fourier transform of $H(u, v)F(u, v)$, as indicated in Eq. (4.2-28). As stated earlier, both the real and imaginary parts of $F(u, v)$ are multiplied by the filter function $H(u, v)$.

The filter just discussed is called a *notch filter* because it is a constant function with a hole (notch) at the origin. The result of processing the image in Fig. 4.4(a) with this filter is shown in Fig. 4.6. Note the drop in overall average gray level resulting from forcing the average value to zero; note also the "byproduct" result of making prominent edges stand out. (In reality the average of the *displayed* image cannot be zero because the image has to have

FIGURE 4.6
Result of filtering
the image in
Fig. 4.4(a) with a
notch filter that
set to 0 the
$F(0, 0)$ term in
the Fourier
transform.

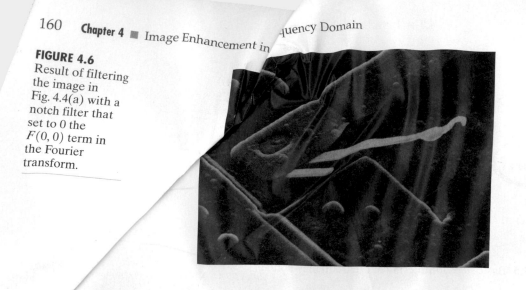

negative values for its average gray level to be zero and displays cannot handle
negative quantities. Figure 4.6 was displayed in the "standard" way, which is to
display the most negative value as 0, or black, with all other values scaled up
from that.) As shown in Section 5.4.3, notch filters are exceptionally useful tools
when it is possible to identify spatial image effects caused by specific, localized
frequency domain components.

Low frequencies in the Fourier transform are responsible for the general
gray-level appearance of an image over smooth areas, while high frequencies are
responsible for detail, such as edges and noise. These ideas are discussed in more
detail in the sections that follow, but it will be instructive to complement our illustration of the notch filter with an example of filters in these other two categories. A filter that attenuates high frequencies while "passing" low frequencies
is called a *lowpass filter*. A filter that has the opposite characteristic is appropriately called a *highpass filter*. We would expect a lowpass-filtered image to
have less sharp detail than the original because the high frequencies have been
attenuated. Similarly, a highpass-filtered image would have less gray level variations in smooth areas and emphasized transitional (e.g., edge) gray-level detail. Such an image will appear sharper.

Figure 4.7 illustrates the effects of lowpass and highpass filtering the image
in Fig. 4.4(a). The left part of the figure shows the filters and the right part shows
the results of filtering using the procedure summarized in Fig. 4.5. The filters,
$H(u, v)$, shown are both circularly symmetric. After shifting their origin to the
center of the frequency rectangle occupied by $F(u, v)$, they were multiplied by
the centered transform, as outlined in our discussion of Eqs. (4.2-27), (4.2-28),
and Fig. 4.5. Taking the real part of each result and multiplying it by $(-1)^{x+y}$
yielded the images on the right. As expected, the image in Fig. 4.7(b) is blurred,
and the image in Fig. 4.7(d) is sharp, with little smooth gray-level detail because
the $F(0, 0)$ term has been set to zero. This is typical of highpassed results, and
a procedure often followed is to add a constant to the filter so that it will not
completely eliminate $F(0, 0)$. The result of using this procedure is shown in
Fig. 4.8. The improvement over Fig. 4.7(d) is evident.

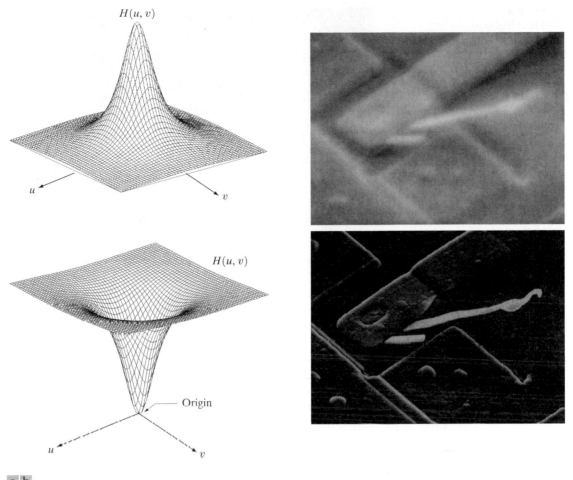

a b
c d

FIGURE 4.7 (a) A two-dimensional lowpass filter function. (b) Result of lowpass filtering the image in Fig. 4.4(a). (c) A two-dimensional highpass filter function. (d) Result of highpass filtering the image in Fig. 4.4(a).

4.2.4 Correspondence between Filtering in the Spatial and Frequency Domains

In the previous chapter we arrived at forms for various spatial filters using intuition and/or a mathematical formulation, such as the Laplacian. In this section we establish a direct link between some of those spatial filters and their frequency domain counterparts.

The most fundamental relationship between the spatial and frequency domains is established by a well-known result called the *convolution theorem*. The reader is already familiar with the basic concepts and mechanics of convolution in the spatial domain, which were introduced and illustrated in Section 3.5.

FIGURE 4.8
Result of highpass filtering the image in Fig. 4.4(a) with the filter in Fig. 4.7(c), modified by adding a constant of one-half the filter height to the filter function. Compare with Fig. 4.4(a).

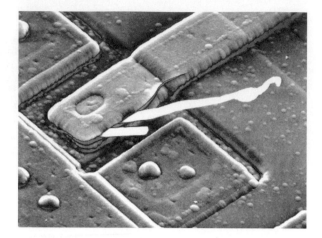

The process by which we move a mask from pixel to pixel in an image, and compute a predefined quantity at each pixel, is the foundation of the convolution process. Formally, the discrete convolution of two functions $f(x, y)$ and $h(x, y)$ of size $M \times N$ is denoted by $f(x, y) * h(x, y)$ and is defined by the expression

$$f(x, y) * h(x, y) = \frac{1}{MN} \sum_{m=0}^{M-1} \sum_{n=0}^{N-1} f(m, n)h(x - m, y - n). \quad (4.2\text{-}30)$$

With the exception of the leading constant, the minus signs, and the limits of the summation, this expression is similar in *form* to Eq. (3.5-1). The minus signs, in particular, simply mean that function h is mirrored about the origin. This is inherent in the definition of convolution. Equation (4.2-30) is really nothing more than an implementation for (1) flipping one function about the origin; (2) shifting that function with respect to the the other by changing the values of (x, y); and (3) computing a sum of products over all values of m and n, for *each* displacement (x, y). The displacements (x, y) are integer increments that stop when the functions no longer overlap.

Letting $F(u, v)$ and $H(u, v)$ denote the Fourier transforms of $f(x, y)$ and $h(x, y)$, respectively, one-half of the convolution theorem simply states that $f(x, y) * h(x, y)$ and $F(u, v)H(u, v)$ constitute a Fourier transform pair. This result is formally stated as

$$f(x, y) * h(x, y) \Leftrightarrow F(u, v)H(u, v). \quad (4.2\text{-}31)$$

The double arrow is used to indicate that the expression on the left (spatial convolution) can be obtained by taking the *inverse* Fourier transform of the expression on the right [the product $F(u, v)H(u, v)$ in the frequency domain]. Conversely, the expression on the right can be obtained by taking the *forward* Fourier transform of the expression on the left. An analogous result is that convolution in the frequency domain reduces to multiplication in the spatial domain, and vice versa; that is,

$$f(x, y)h(x, y) \Leftrightarrow F(u, v) * H(u, v). \quad (4.2\text{-}32)$$

These two results comprise the convolution theorem. It is important to keep in mind that there is nothing complicated about what has just been stated. We already know what the convolution operation is all about. The other part of the process, multiplication, is simply the element-by-element product of the two functions.

We need one more concept before completing the tie between the spatial and frequency domains. An *impulse function* of *strength A*, located at coordinates (x_0, y_0), is denoted by $A\delta(x - x_0, y - y_0)$ and is *defined* by the expression

$$\sum_{x=0}^{M-1} \sum_{y=0}^{N-1} s(x, y) A\delta(x - x_0, y - y_0) = As(x_0, y_0). \qquad (4.2\text{-}33)$$

In words, this equation states that the summation of a function $s(x, y)$ multiplied by an impulse is simply the value of the function at the location of the impulse, multiplied by the strength of the impulse. It is understood that the limits of the summation are the same as the limits spanned by the function. We point out that $A\delta(x - x_0, y - y_0)$ also is an image of size $M \times N$. It is composed of all zeros, except at coordinates (x_0, y_0), where the value of the image is A.

By letting either f or h in Eq. (4.2-30) be an impulse function, and using the definition in Eq. (4.2-33), we would conclude after a little manipulation that convolution of a function with an impulse "copies" the value of that function at the location of the impulse. This characteristic is called the *sifting property* of the impulse function. Of particular importance at the moment is the case of a unit impulse located at the origin, which is denoted as $\delta(x, y)$. In this case,

$$\sum_{x=0}^{M-1} \sum_{y=0}^{N-1} s(x, y)\delta(x, y) = s(0, 0). \qquad (4.2\text{-}34)$$

Armed with these simple tools, we are now in a position to establish a most interesting and useful tie between filtering in the spatial and frequency domains. From Eq. (4.2-16), we can compute the Fourier transform of a unit impulse at the origin,

$$F(u, v) = \frac{1}{MN} \sum_{x=0}^{M-1} \sum_{y=0}^{N-1} \delta(x, y)e^{-j2\pi(ux/M + vy/N)}$$
$$= \frac{1}{MN} \qquad (4.2\text{-}35)$$

where the second step follows from Eq. (4.2-34). Thus, we see that the Fourier transform of an impulse at the origin of the spatial domain is a *real* constant (this means that the phase angle is zero). If the impulse were located elsewhere, the transform would have complex components. The magnitude would be the same, with the translation of the impulse being reflected in a nonzero phase angle in the transform.

Now suppose that we let $f(x, y) = \delta(x, y)$ and carry out the convolution defined in Eq. (4.2-30). Using Eq. (4.2-34) again gives us

$$f(x, y) * h(x, y) = \frac{1}{MN} \sum_{m=0}^{M-1} \sum_{n=0}^{N-1} \delta(m, n)h(x - m, y - n)$$
$$= \frac{1}{MN} h(x, y) \qquad (4.2\text{-}36)$$

where the last step follows from Eq. (4.2-34) by noting that the variables in the summation are m and n. By combining the results of Eqs. (4.2-35) and (4.2-36) with Eq. (4.2-31), we obtain

$$f(x, y) * h(x, y) \Leftrightarrow F(u, v)H(u, v)$$
$$\delta(x, y) * h(x, y) \Leftrightarrow \Im[\delta(x, y)]H(u, v) \qquad (4.2\text{-}37)$$
$$h(x, y) \Leftrightarrow H(u, v).$$

Using only the properties of the impulse function and the convolution theorem, we have established that filters in the spatial and frequency domains constitute a Fourier transform pair. Thus, given a filter in the frequency domain, we can obtain the corresponding filter in the spatial domain by taking the inverse Fourier transform of the former. The reverse also is true.

Note that *all* functions in the preceding development are of the same size, $M \times N$. Therefore, in practice, specifying a filter in the frequency domain and then taking the inverse transform to compute an equivalent spatial domain filter of the same size does not really help matters from a computational point of view. As discussed in Section 4.6, if both filters are of the same size, it generally is more efficient computationally to do the filtering in the frequency domain. But we use much smaller filters in the spatial domain. This is precisely the connection in which we are interested. Filtering often is more intuitive in the frequency domain. However, whenever possible, it makes more sense to filter in the spatial domain using small filter masks. Equation (4.2-37) tells us that we can specify filters in the frequency domain, take their inverse transform, and then use the resulting filter in the spatial domain as a guide for constructing smaller spatial filter masks (more formal approaches are discussed in Section 4.6.7). This is illustrated next. Keep in mind during the following discussion that the Fourier transform and its inverse are linear processes (Problem 4.2), so the discussion is by definition limited to linear filtering.

Filters based on Gaussian functions are of particular importance because their shapes are easily specified and because both the forward and inverse Fourier transforms of a Gaussian function are real Gaussian functions. We will limit the discussion here to one variable to simplify the notation. Two-dimensional functions are discussed later in this chapter.

Let $H(u)$ denote a frequency domain, Gaussian filter function given by the equation

$$H(u) = Ae^{-u^2/2\sigma^2} \qquad (4.2\text{-}38)$$

where σ is the standard deviation of the Gaussian curve. It can be shown (Problem 4.4) that the corresponding filter in the spatial domain is

$$h(x) = \sqrt{2\pi}\sigma A e^{-2\pi^2\sigma^2 x^2}. \qquad (4.2\text{-}39)$$

These two equations represent an important result for two reasons: (1) They constitute a Fourier transform pair, both components of which are Gaussian *and* real. This facilitates analysis considerably because we do not have to be concerned with complex numbers. In addition, Gaussian curves are intuitive and easy to manipulate. (2) These functions behave reciprocally with respect to one another. In other words, when $H(u)$ has a broad profile (large value of σ), $h(x)$ has a narrow profile, and vice versa. In fact, when σ approaches infinity, $H(u)$ tends toward a constant function and $h(x)$ tends toward an impulse. This is exactly the type of reciprocal behavior we saw in Section 4.2, in connection with

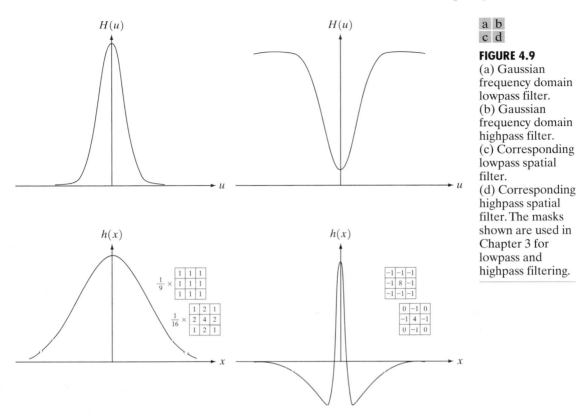

FIGURE 4.9
(a) Gaussian frequency domain lowpass filter.
(b) Gaussian frequency domain highpass filter.
(c) Corresponding lowpass spatial filter.
(d) Corresponding highpass spatial filter. The masks shown are used in Chapter 3 for lowpass and highpass filtering.

Figs. 4.2 and 4.3. These two properties help considerably in developing a solid understanding of the properties of filtering in both the spatial and frequency domains because they lend themselves to familiar analytical interpretations.

A plot of a Gaussian filter in the frequency domain is shown in Fig. 4.9(a). The reader will recognize this shape of $H(u)$ as a lowpass filter. The corresponding lowpass filter in the spatial domain is shown in Fig. 4.9(c). Our interest is in the general shape of $h(x)$, which we generally want to use as a guide to specify the coefficients of a smaller filter in the spatial domain. A glaring similarity between the two filters is that all the values are positive in both domains. Thus, we arrive at the conclusion that we can implement lowpass filtering in the spatial domain by using a mask with all positive coefficients, just as we did in Section 3.6.1. Two of the masks from that section are shown in Fig. 4.9(c) for reference. Another important characteristic is the reciprocal relationship discussed in the previous paragraph. The narrower the frequency domain filter, the more it will attenuate the low frequencies, resulting in increased blurring. In the spatial domain this means a wider filter, which in turn implies a larger mask, as illustrated in Example 3.9.

More complex filters can be constructed from the basic Gaussian function of Eq. (4.2-38). For instance, we can construct a highpass filter as a difference of Gaussians, as follows:

$$H(u) = Ae^{-u^2/2\sigma_1^2} - Be^{-u^2/2\sigma_2^2} \tag{4.2-40}$$

with $A \geq B$ and $\sigma_1 > \sigma_2$. The corresponding filter in the spatial domain is

$$h(x) = \sqrt{2\pi}\sigma_1 A e^{-2\pi^2 \sigma_1^2 x^2} - \sqrt{2\pi}\sigma_2 B e^{-2\pi^2 \sigma_2^2 x^2}. \qquad (4.2\text{-}41)$$

Plots of these two functions are shown in Figs. 4.9(b) and (d), respectively. We note again the reciprocity in width, but the most important feature here is that the spatial filter has both negative and positive values. In fact, it is interesting to note that once the values turn negative, they never turn positive again. Two of the masks we used in Chapter 3 for highpass filtering are shown in Fig. 4.9(d). The similarity in form between the spatial curve and the filters is unmistakable.

In Chapter 3, we specified the shapes of lowpass and highpass filters based strictly on spatial domain considerations. It is important to note that we could have arrived at the basic shapes of all the small spatial filter masks shown in Fig. 4.9 by following the alternate path provided by the frequency domain analysis we have just completed. Although we have gone through significant effort to get here, the reader is assured that it is impossible to truly understand filtering in the frequency domain without the foundation we have just established.

A question that often arises at this point in the development of frequency domain techniques is the issue of computational complexity. Why do in the frequency domain what could be done (at least partially) in the spatial domain using small spatial masks? The basic answer is twofold. First, as we have seen, the frequency domain carries with it a significant degree of intuitiveness regarding how to specify filters. The second part of the answer depends on the size of the spatial masks and is usually answered with respect to comparable implementations.

A benchmark used frequently for this purpose is implementation of convolution in the spatial and frequency domains. Spatial convolution is given in Eq. (4.2-30), and we know from the convolution theorem that we can obtain the same result via the frequency domain by taking the inverse transform of the product of the transforms of the two functions. Suppose that we implemented both approaches in software on the same machine [using the fast Fourier transform (FFT) algorithm discussed in Section 4.6.6 for frequency domain computations]. We would find that the frequency domain implementation runs faster for surprisingly small values of M and N. For instance, a comparison by Brigham [1988] showed that, for the 1-D case, the FFT approach is faster if the number of points is greater than 32. Although this number is somewhat dependent on other factors, such as the machine and algorithms used, it certainly is well below the values that we encounter in image processing.

The frequency domain may be viewed as a "laboratory" in which we take advantage of the correspondence between frequency content and image appearance. As is demonstrated numerous times later in this chapter, some enhancement tasks that would be exceptionally difficult or impossible to formulate directly in the spatial domain become almost trivial in the frequency domain. Once we have selected a specific filter via experimentation in the frequency domain, the actual implementation of the method usually is done in the spatial domain. One approach is to specify small spatial masks that attempt to capture the "essence" of the full filter function in the spatial domain, as was explained in Fig. 4.9. A more formal approach is to design a 2-D digital filter by using approximations based on mathematical or statistical criteria. We touch on this point again in Section 4.6.7.

4.3 Smoothing Frequency-Domain Filters

As indicated in Section 4.2.3, edges and other sharp transitions (such as noise) in the gray levels of an image contribute significantly to the high-frequency content of its Fourier transform. Hence smoothing (blurring) is achieved in the frequency domain by attenuating a specified range of high-frequency components in the transform of a given image.

Our basic "model" for filtering in the frequency domain is given by Eq. (4.2-27), which we repeat here for convenience:

$$G(u, v) = H(u, v)F(u, v) \qquad (4.3-1)$$

where $F(u, v)$ is the Fourier transform of the image to be smoothed. The objective is to select a filter transfer function $H(u, v)$ that yields $G(u, v)$ by attenuating the high-frequency components of $F(u, v)$. All filtering done in this section is based on the procedure outlined in Section 4.2.3, including the use of zero-phase-shift filters.

We consider three types of lowpass filters: ideal, Butterworth, and Gaussian filters. These three filters cover the range from very sharp (ideal) to very smooth (Gaussian) filter functions. The Butterworth filter has a parameter, called the filter *order*. For high values of this parameter the Butterworth filter approaches the form of the ideal filter. For lower-order values, the Butterworth filter has a smooth form similar to the Gaussian filter. Thus, the Butterworth filter may be viewed as a transition between two "extremes."

4.3.1 Ideal Lowpass Filters

The simplest lowpass filter we can envision is a filter that "cuts off" all high-frequency components of the Fourier transform that are at a distance greater than a specified distance D_0 from the origin of the (centered) transform. Such a filter is called a two-dimensional (2-D) *ideal lowpass filter* (ILPF) and has the transfer function

$$H(u, v) = \begin{cases} 1 & \text{if } D(u, v) \leq D_0 \\ 0 & \text{if } D(u, v) > D_0 \end{cases} \qquad (4.3-2)$$

where D_0 is a specified nonnegative quantity, and $D(u, v)$ is the distance from point (u, v) to the center of the frequency rectangle. If the image in question is of size $M \times N$, we know that its transform also is of this size, so the center of the frequency rectangle is at $(u, v) = (M/2, N/2)$ due to the fact that the transform has been centered, as discussed in connection with Eq. (4.2-21). In this case, the distance from any point (u, v) to the center (origin) of the Fourier transform is given by

$$D(u, v) = \left[(u - M/2)^2 + (v - N/2)^2\right]^{1/2}. \qquad (4.3-3)$$

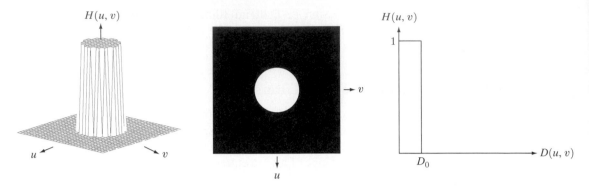

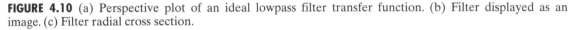

a b c

FIGURE 4.10 (a) Perspective plot of an ideal lowpass filter transfer function. (b) Filter displayed as an image. (c) Filter radial cross section.

Figure 4.10(a) shows a 3-D perspective plot of $H(u, v)$ as a function of u and v, and Fig. 4.10(b) shows $H(u, v)$ displayed as an image. The name *ideal filter* indicates that *all* frequencies inside a circle of radius D_0 are passed with no attenuation, whereas *all* frequencies outside this circle are completely attenuated. The lowpass filters considered in this chapter are radially symmetric about the origin. This means that a cross section extending as a function of distance from the origin along a radial line is sufficient to specify the filter, as Fig. 4.10(c) shows. The complete filter transfer function can be visualized by rotating the cross section 360° about the origin.

For an ideal lowpass filter cross section, the point of transition between $H(u, v) = 1$ and $H(u, v) = 0$ is called the *cutoff frequency*. In the case of Fig. 4.10, for example, the cutoff frequency is D_0. The sharp cutoff frequencies of an ideal lowpass filter cannot be realized with electronic components, although they can certainly be implemented in a computer. The effects of using these "nonphysical" filters on a digital image are discussed later in this section.

The lowpass filters introduced in this section are compared by studying their behavior as a function of the same cutoff frequencies. One way to establish a set of standard cutoff frequency loci is to compute circles that enclose specified amounts of total image power P_T. This quantity is obtained by summing the components of the power spectrum at each point (u, v), for $u = 0, 1, 2, \ldots, M - 1$ and $v = 0, 1, 2, \ldots, N - 1$; that is,

$$P_T = \sum_{u=0}^{M-1} \sum_{v=0}^{N-1} P(u, v) \tag{4.3-4}$$

where $P(u, v)$ is given in Eq. (4.2-20). If the transform has been centered, a circle of radius r with origin at the center of the frequency rectangle encloses α percent of the power, where

$$\alpha = 100 \left[\sum_u \sum_v P(u, v)/P_T \right] \tag{4.3-5}$$

and the summation is taken over the values of (u, v) that lie inside the circle or on its boundary.

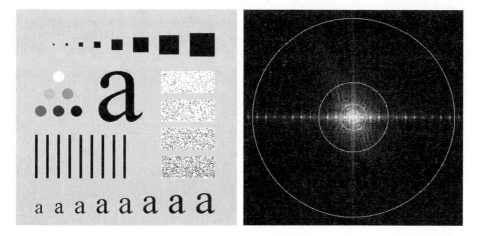

a b

FIGURE 4.11 (a) An image of size 500×500 pixels and (b) its Fourier spectrum. The superimposed circles have radii values of 5, 15, 30, 80, and 230, which enclose 92.0, 94.6, 96.4, 98.0, and 99.5% of the image power, respectively.

■ Figure 4.11(a) shows the test pattern we used in Fig. 3.35 to illustrate spatial blurring. The Fourier spectrum of this image is shown in Fig. 4.11(b). The circles superimposed on the spectrum have radii of 5, 15, 30, 80, and 230 pixels (the circle of radius 5 is not easily visible). These circles enclose α percent of the image power, for $\alpha = 92.0, 94.6, 96.4, 98$, and 99.5%, respectively. The spectrum falls off rapidly, with 92% of the total power being enclosed by a relatively small circle of radius 5.

Figure 4.12 shows the results of applying ideal lowpass filters with cutoff frequencies at the radii shown in Fig. 4.11(b). Figure 4.12(b) is useless for all practical purposes, unless the objective of blurring in this case is to eliminate all detail in the image, except the "blobs" representing the largest objects. The severe blurring in this image is a clear indication that most of the sharp detail information in the picture is contained in the 8% power removed by the filter. As the filter radius increases, less and less power is removed, resulting in less severe blurring. Note that the images in Figs. 4.12(c) through (e) are characterized by "ringing," which becomes finer in texture as the amount of high-frequency content removed decreases. Ringing is evident even in the image in which only 2% of the total power was removed. This ringing behavior is a characteristic of ideal filters, as will be explained shortly. Finally, close observation of the result for $\alpha = 99.5$ shows very slight blurring in the noisy squares but, for the most part, this image is quite close to the original. This indicates that little edge information is contained in the upper 0.5% of the spectrum power in this particular case.

It is clear from this example that ideal lowpass filtering is not very practical. However, because ideal filters can be implemented in a computer, it is useful to study their behavior as part of our development of filtering concepts. Also, as shown in the discussion that follows, some interesting insight is gained by attempting to explain the ringing property of ILPFs in the spatial domain. ■

EXAMPLE 4.4:
Image power as a function of distance from the origin of the DFT.

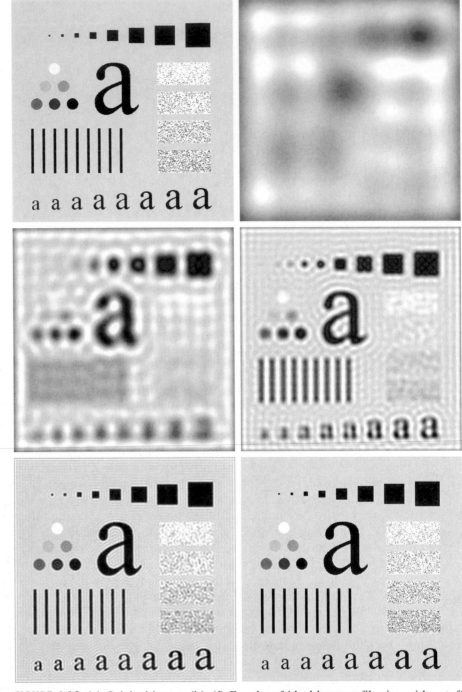

FIGURE 4.12 (a) Original image. (b)–(f) Results of ideal lowpass filtering with cutoff frequencies set at radii values of 5, 15, 30, 80, and 230, as shown in Fig. 4.11(b). The power removed by these filters was 8, 5.4, 3.6, 2, and 0.5% of the total, respectively.

The blurring and ringing properties of the ILPF can be explained by reference to the convolution theorem discussed in Section 4.2.4. The Fourier transforms of the original image $f(x, y)$ and the blurred image $g(x, y)$ are related in the frequency domain by the equation

$$G(u, v) = H(u, v)F(u, v)$$

where, as before, $H(u, v)$ is the filter function and F and G are the Fourier transforms of the two images just mentioned. The convolution theorem tells us that the corresponding process in the spatial domain is

$$g(x, y) = h(x, y) * f(x, y)$$

where $h(x, y)$ is the inverse Fourier transform of the filter transfer function $H(u, v)$.

The key to understanding blurring as a convolution process in the spatial domain lies in the nature of $h(x, y)$. For instance, the ILPF of radius 5 that caused so much blurring in the preceding example is shown in Fig. 4.13(a). This is the function $H(u, v)$ in the *frequency* domain. The spatial filter function $h(x, y)$ was obtained in the standard way: (1) $H(u, v)$ was multiplied by $(-1)^{u+v}$ for centering; (2) this was followed by the inverse DFT; and (3) the real part of the inverse DFT was multiplied by $(-1)^{x+y}$. Figure 4.13(b) shows the result of this process.

We see that the filter $h(x, y)$ has two major distinctive characteristics: a dominant component at the origin, and concentric, circular components about the center component. The center component is primarily responsible for blurring. The concentric components are responsible primarily for the ringing characteristic of ideal filters. Both the radius of the center component and the number of circles per unit distance from the origin are inversely proportional to the value of the cutoff frequency of the ideal filter. The insert at the top is a gray-level profile of a horizontal scan line through the center of the spatial filter. The axis shown indicates zero amplitude, so we see that the spatial filter has negative values. This normally is not a serious problem because the larger center component dominates the convolution result. However, the filtered image can have negative values, so scaling normally is required.

Suppose next that $f(x, y)$ is a simple image composed of five bright pixels on a black background, as Fig. 4.13(c) shows. These bright points may be viewed as approximations to impulses, whose strength depends on the intensity of the points. Then the convolution of $h(x, y)$ and $f(x, y)$ is simply a process of "copying" $h(x, y)$ at the location of each impulse, as noted in Section 4.2.4. The result of this operation, shown in Fig. 4.13(d), explains how the original points are blurred as a consequence of convolving $f(x, y)$ with the blurring filter function $h(x, y)$. Note also that ringing was introduced during the same process. In fact, the ringing is so severe in this case that distortion is caused by their interference with one another. These concepts are extended conceptually to more complex images by considering each pixel as an impulse whose strength is proportional to the gray level of the pixel. The insert at the bottom of Fig. 4.13 shows the gray-level profile of a diagonal scan line through the center of the filtered image.

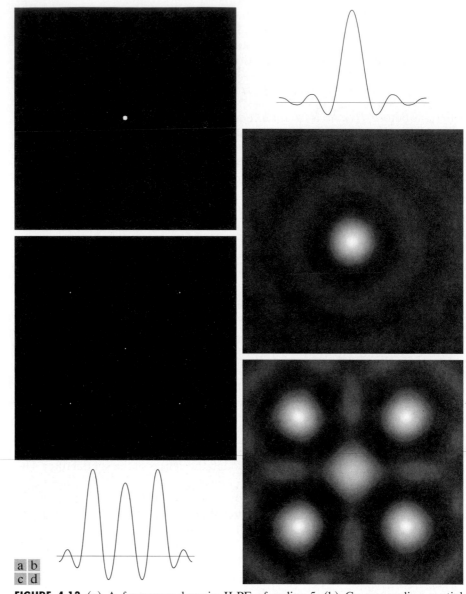

a b
c d

FIGURE 4.13 (a) A frequency-domain ILPF of radius 5. (b) Corresponding spatial filter (note the ringing). (c) Five impulses in the spatial domain, simulating the values of five pixels. (d) Convolution of (b) and (c) in the spatial domain.

The reciprocal nature between $H(u, v)$ and $h(x, y)$, along with the convolution process just discussed, explains mathematically why the blurring and ringing are more severe the narrower the filter in the frequency domain. This type of reciprocal behavior should be routine to the reader by now. In the next two sections we show that it is possible to achieve blurring with little or no ringing, which is our main objective.

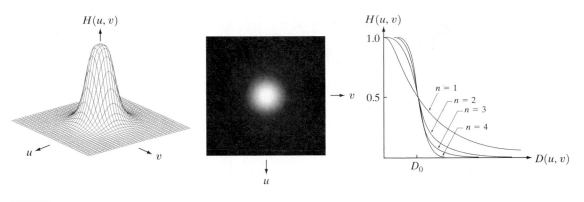

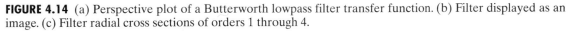

FIGURE 4.14 (a) Perspective plot of a Butterworth lowpass filter transfer function. (b) Filter displayed as an image. (c) Filter radial cross sections of orders 1 through 4.

4.3.2 Butterworth Lowpass Filters

The transfer function of a Butterworth lowpass filter (BLPF) of order n, and with cutoff frequency at a distance D_0 from the origin, is defined as

$$H(u, v) = \frac{1}{1 + \left[D(u, v)/D_0 \right]^{2n}} \tag{4.3-6}$$

where $D(u, v)$ is given by Eq. (4.3-3). A perspective plot, image display, and radial cross sections of the BLPF function are shown in Fig. 4.14.

Unlike the ILPF, the BLPF transfer function does not have a sharp discontinuity that establishes a clear cutoff between passed and filtered frequencies. For filters with smooth transfer functions, defining a cutoff frequency locus at points for which $H(u, v)$ is down to a certain fraction of its maximum value is customary. In the case of Eq. (4.3-6), $H(u, v) = 0.5$ (down 50% from its maximum value of 1) when $D(u, v) = D_0$.

■ Figure 4.15 shows the results of applying the BLPF of Eq. (4.3-6) to Fig. 4.15(a), with $n = 2$ and D_0 equal to the five radii shown in Fig. 4.11(b). Unlike the results shown in Fig. 4.12 for the ILPF, we note here a smooth transition in blurring as a function of increasing cutoff frequency. Moreover, no ringing is visible in any of the images processed with this particular BLPF, a fact attributed to the filter's smooth transition between low and high frequencies. ■

EXAMPLE 4.5: Butterworth lowpass filtering.

A Butterworth filter of order 1 has no ringing. Ringing generally is imperceptible in filters of order 2, but can become a significant factor in filters of higher order. Figure 4.16 shows an interesting comparison between the *spatial* representation of BLPFs of various orders (with cutoff frequency of 5 pixels). Shown also is the gray-level profile along a horizontal scan line through the center of each filter. These filters were obtained and displayed by using the same procedure we used to generate Fig. 4.13(b). In order to facilitate comparisons,

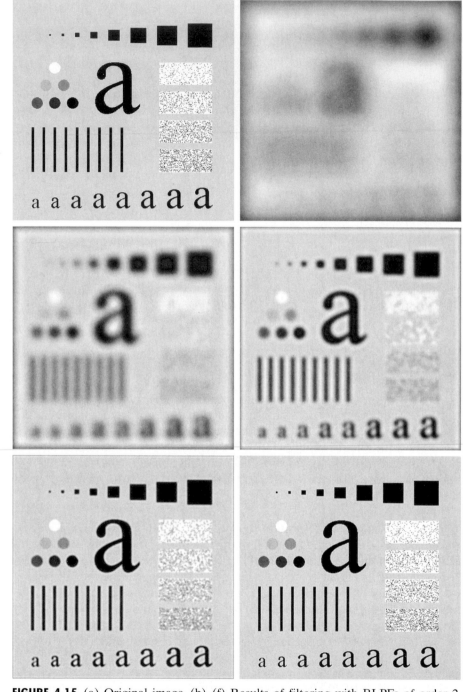

a b
c d
e f   **FIGURE 4.15** (a) Original image. (b)–(f) Results of filtering with BLPFs of order 2, with cutoff frequencies at radii of 5, 15, 30, 80, and 230, as shown in Fig. 4.11(b). Compare with Fig. 4.12.

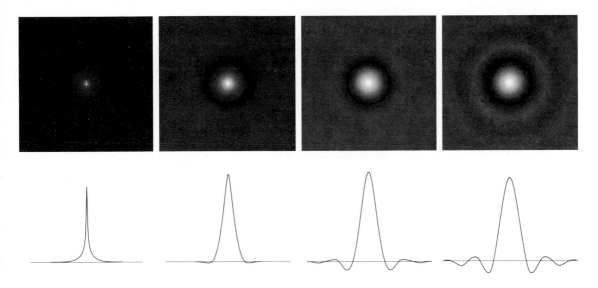

a b c d

FIGURE 4.16 (a)–(d) Spatial representation of BLPFs of order 1, 2, 5, and 20, and corresponding gray-level profiles through the center of the filters (all filters have a cutoff frequency of 5). Note that ringing increases as a function of filter order.

additional enhancing with a gamma transformation [see Eq. (3.2-3)] was applied to the images of Fig. 4.16 to accentuate even more the components further away from the origin. The BLPF of order 1 [Fig. 4.16(a)] has neither ringing nor negative values. The filter of order 2 does show mild ringing and small negative values, but they certainly are less pronounced than in the ILPF. As the remaining images show, ringing in the BLPF becomes significant for higher-order filters. A Butterworth filter of order 20 already exhibits the characteristics of the ILPF, which can be seen by comparing Figs. 4.16(d) and 4.13(b). In the limit, both filters are identical. In general, BLPFs of order 2 are a good compromise between effective lowpass filtering and acceptable ringing characteristics.

4.3.3 Gaussian Lowpass Filters

Gaussian lowpass filters (GLPFs) of one dimension were introduced in Section 4.2.4 as an aid in exploring some important relationships between the spatial and frequency domains. The form of these filters in two dimensions is given by

$$H(u, v) = e^{-D^2(u,v)/2\sigma^2} \tag{4.3-7}$$

where, as in Eq. (4.3-3), $D(u, v)$ is the distance from the origin of the Fourier transform, which we assume has been shifted to the center of the frequency rectangle using the procedure outlined in Section 4.2.3. We did not use a constant in front of the filter as in Section 4.2.4 to be consistent with all the other filters discussed in the present section, which have a value of 1 at the origin. As before,

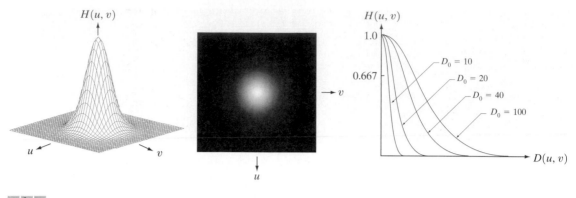

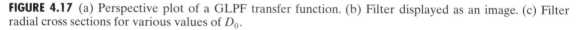

a b c

FIGURE 4.17 (a) Perspective plot of a GLPF transfer function. (b) Filter displayed as an image. (c) Filter radial cross sections for various values of D_0.

σ is a measure of the spread of the Gaussian curve. By letting $\sigma = D_0$, we can express the filter in a more familiar form in terms of the notation in this section:

$$H(u, v) = e^{-D^2(u,v)/2D_0^2} \qquad (4.3\text{-}8)$$

where D_0 is the cutoff frequency. When $D(u, v) = D_0$, the filter is down to 0.607 of its maximum value.

As discussed in Section 4.2.4, the inverse Fourier transform of the Gaussian lowpass filter also is Gaussian. We already saw in that section the advantages of this property as an analysis tool. In terms of our current interest, this also means that a *spatial* Gaussian filter, obtained by computing the inverse Fourier transform of Eq. (4.3-7) or (4.3-8), will have no ringing. A perspective plot, image display, and radial cross sections of a GLPF function are shown in Fig. 4.17.

EXAMPLE 4.6:
Gaussian lowpass filtering.

■ Figure 4.18 shows the results of applying the GLPF of Eq. (4.3-8) to Fig. 4.18(a), with D_0 equal to the five radii shown in Fig. 4.11(b). As in the case of the BLPF (Fig. 4.15), we note a smooth transition in blurring as a function of increasing cutoff frequency. The GLPF did not achieve as much smoothing as the BLPF of order 2 for the same value of cutoff frequency, as can be seen, for example, by comparing Figs. 4.15(c) and 4.18(c). This is expected, because the profile of the GLPF is not as "tight" as the profile of the BLPF of order 2. However, the results are quite comparable in general, and we are assured of no ringing in the case of the GLPF. This is an important characteristic in practice, especially in situations where any type of artifact (e.g., in medical imaging) is not acceptable. In cases where tight control of the transition between low and high frequencies about the cutoff frequency are needed, then the BLPF presents a more suitable choice. The price of this additional control over the filter profile is the possibility of ringing. ■

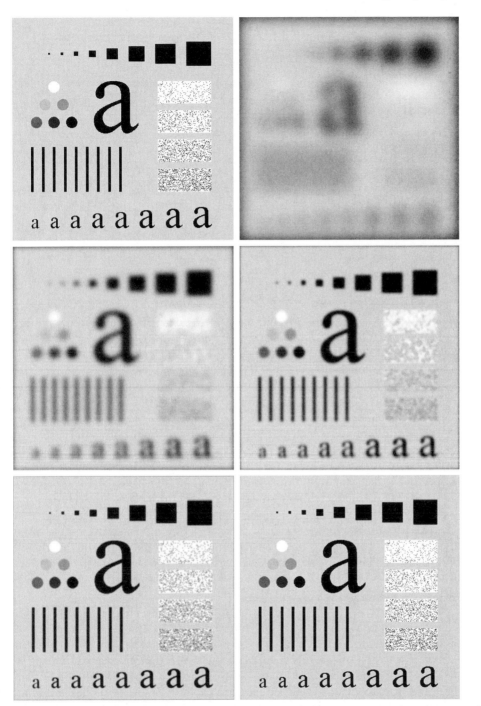

FIGURE 4.18 (a) Original image. (b)–(f) Results of filtering with Gaussian lowpass filters with cutoff frequencies set at radii values of 5, 15, 30, 80, and 230, as shown in Fig. 4.11(b). Compare with Figs. 4.12 and 4.15.

a b

FIGURE 4.19
(a) Sample text of
poor resolution
(note broken
characters in
magnified view).
(b) Result of
filtering with a
GLPF (broken
character
segments were
joined).

Historically, certain computer
programs were written using
only two digits rather than
four to define the applicable
year. Accordingly, the
company's software may
recognize a date using "00"
as 1900 rather than the year
2000.

Historically, certain computer
programs were written using
only two digits rather than
four to define the applicable
year. Accordingly, the
company's software may
recognize a date using "00"
as 1900 rather than the year
2000.

4.3.4 Additional Examples of Lowpass Filtering

The lowpass filtering results given thus far have been with images of good quality in order to illustrate and compare filter effects. In the following discussion
we show a few practical applications of lowpass filtering. The first example is
from the field of machine perception, with application to character recognition;
the second is from the printing and publishing industry; and the third is related
to processing satellite and aerial images.

Figure 4.19 shows a sample of text of poor resolution. One encounters text
like this, for example, in fax transmissions, duplicated material, and historical
records. As poor text goes, this particular sample is free of additional difficulties like smudges, creases, and torn sections. The magnified section in Fig. 4.19(a)
shows that the characters in this particular document have distorted shapes
due to lack of resolution, and many of the characters are broken. Although humans fill these gaps visually without difficulty, a machine recognition system
has real difficulties reading broken characters. The approach used most often
to handle this problem is to bridge small gaps in the input image by blurring it.
Figure 4.19(b) shows how well characters can be "repaired" by this simple
process using a Gaussian lowpass filter with $D_0 = 80$. The images are of size
444×508 pixels.

Lowpass filtering is a staple in the printing and publishing industry, where it
is used for numerous preprocessing functions, including unsharp masking, as
discussed in Section 3.7.2. "Cosmetic" processing is another use of lowpass filtering prior to printing. Figure 4.20 shows an application of lowpass filtering to
produce a smoother, softer-looking result from a sharp original. For human
faces, the typical objective is to reduce the sharpness of fine skin lines and small
blemishes. The magnified sections in Figs. 4.20(b) and (c) clearly show a significant reduction in fine skin lines around the eyes in this particular case. In fact,
the smoothed images look quite soft and pleasing.

a b c

FIGURE 4.20 (a) Original image (1028 × 732 pixels). (b) Result of filtering with a GLPF with $D_0 = 100$. (c) Result of filtering with a GLPF with $D_0 = 80$. Note reduction in skin fine lines in the magnified sections of (b) and (c).

Figure 4.21 shows two applications of lowpass filtering on the same image, but with totally different objectives. Figure 4.21(a) is a 588 × 600 very high resolution radiometer (VHRR) image showing part of the Gulf of Mexico (dark) and Florida (light), taken from a NOAA satellite (note horizontal sensor scan lines). The boundaries between water bodies were caused by loop currents. This image is illustrative of remotely sensed images in which sensors (for a number of reasons beyond the present discussion) have the tendency to produce pronounced scan lines along the direction in which the scene is being scanned. Lowpass filtering is a crude but simple way to reduce the effect of these lines, as Fig. 4.21(b) shows (we consider more effective approaches in Chapter 5). This image was obtained using a Gaussian lowpass filter with $D_0 = 30$. The resulting reduction in the effect of the scan lines can simplify the detection of features like the interface boundaries between ocean currents. Figure 4.21(c) shows the result of considerably more aggressive Gaussian lowpass filtering $(D_0 = 10)$. Here the objective is to blur out as much detail as possible while leaving large features recognizable. For instance, this type of filtering would be part of a preprocessing stage for an image analysis system that was

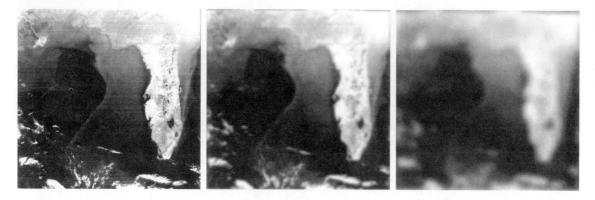

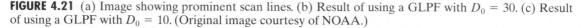

FIGURE 4.21 (a) Image showing prominent scan lines. (b) Result of using a GLPF with $D_0 = 30$. (c) Result of using a GLPF with $D_0 = 10$. (Original image courtesy of NOAA.)

looking for features in an image bank. An example of such features could be lakes of a given size [such as Lake Okeechobee in the lower eastern region of Florida, shown as a nearly round dark region in Fig. 4.21(c)]. Lowpass filtering helps simplify the analysis by averaging out features smaller than the ones of interest.

4.4 Sharpening Frequency Domain Filters

In the previous section we showed that an image can be blurred by attenuating the high-frequency components of its Fourier transform. Because edges and other abrupt changes in gray levels are associated with high-frequency components, image sharpening can be achieved in the frequency domain by a *high-pass filtering* process, which attenuates the low-frequency components without disturbing high-frequency information in the Fourier transform. As in Section 4.3, we consider only zero-phase-shift filters that are radially symmetric. All filtering in this section is based on the procedure outlined in Section 4.2.3.

Because the intended function of the filters in this section is to perform precisely the reverse operation of the ideal lowpass filters discussed in the previous section, the transfer function of the highpass filters discussed in this section can be obtained using the relation

$$H_{\text{hp}}(u, v) = 1 - H_{\text{lp}}(u, v) \tag{4.4-1}$$

where $H_{\text{lp}}(u, v)$ is the transfer function of the corresponding lowpass filter. That is, when the lowpass filter attenuates frequencies, the highpass filter passes them, and vice versa.

In this section we consider ideal, Butterworth, and Gaussian highpass filters. As in the previous section, we illustrate the characteristics of these filters in both the frequency and spatial domains. Figure 4.22 shows typical 3-D plots, image representations, and cross sections for these filters. As before, we see that

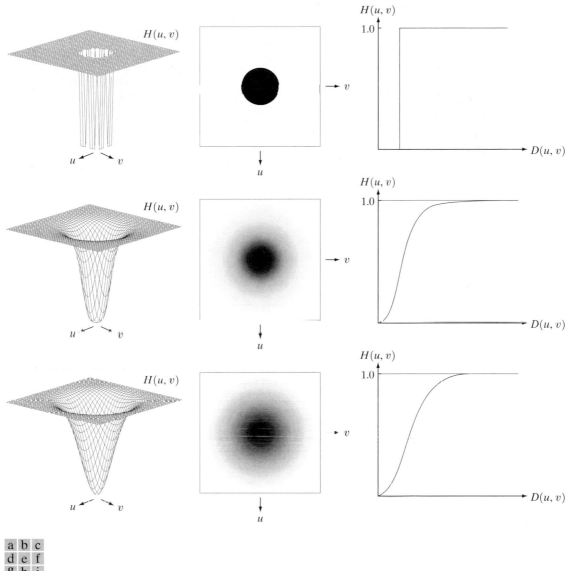

FIGURE 4.22 Top row: Perspective plot, image representation, and cross section of a typical ideal highpass filter. Middle and bottom rows: The same sequence for typical Butterworth and Gaussian highpass filters.

the Butterworth filter represents a transition between the sharpness of the ideal filter and the total smoothness of the Gaussian filter. Figure 4.23 illustrates what these filters look like in the spatial domain. Recall that a spatial representation of a frequency domain filter is obtained by (1) multiplying $H(u, v)$ by $(-1)^{u+v}$ for centering; (2) computing the inverse DFT; and (3) multiplying the real part of the inverse DFT by $(-1)^{x+y}$. Important characteristics of these figures are discussed in the sections that follow.

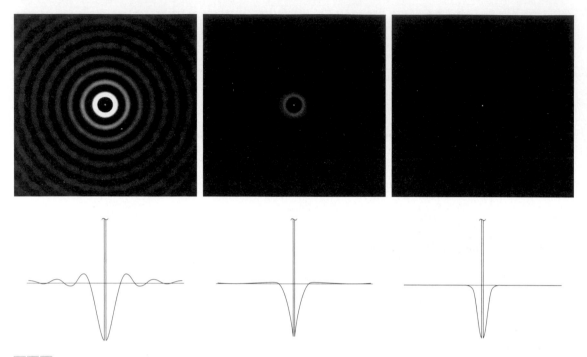

a b c

FIGURE 4.23 Spatial representations of typical (a) ideal, (b) Butterworth, and (c) Gaussian frequency domain highpass filters, and corresponding gray-level profiles.

4.4.1 Ideal Highpass Filters

A 2-D ideal highpass filter (IHPF) is defined as

$$H(u, v) = \begin{cases} 0 & \text{if } D(u, v) \leq D_0 \\ 1 & \text{if } D(u, v) > D_0 \end{cases} \tag{4.4-2}$$

where D_0 is the cutoff distance measured from the origin of the frequency rectangle, and $D(u, v)$ is given in Eq. (4.3-3). This expression follows directly from Eqs. (4.4-1) and (4.4-2). As intended, this filter is the opposite of the ideal lowpass filter in the sense that it sets to zero all frequencies inside a circle of radius D_0 while passing, without attenuation, all frequencies outside the circle. As in the case of the ideal lowpass filter, the IHPF is not physically realizable with electronic components. However, since it can be implemented in a computer, we consider it for completeness. The discussion will be brief.

Because of the way they are related [Eq. (4.4-1)], we can expect IHPFs to have the same ringing properties as ILPFs [see Fig. 4.23(a)]. This is demonstrated clearly in Fig. 4.24, which consists of various IHPF results using the original image in Fig. 4.11(a) with D_0 set to 15, 30, and 80 pixels, respectively. The ringing in Fig. 4.24(a) is so severe that it produced distorted, thickened object boundaries (e.g., look at the large letter "a"). Edges of the top three circles do not show well because they are not as strong as the other edges in the image (the gray level of

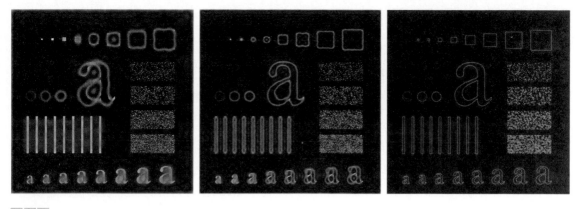

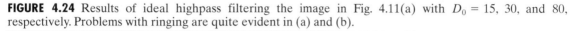

a b c

FIGURE 4.24 Results of ideal highpass filtering the image in Fig. 4.11(a) with $D_0 = 15$, 30, and 80, respectively. Problems with ringing are quite evident in (a) and (b).

these three objects is much closer to the background gray level, giving discontinuities of smaller magnitude). Looking at the "spot" size of the spatial representation of the IHPF in Fig. 4.23(a) and keeping in mind that filtering in the spatial domain is convolution of the filter with the image helps explain why the smaller objects and lines appear almost solid white. Look in particular at the three small squares in the top row and the thin, vertical bars. The situation improved somewhat with $D_0 = 30$. Edge distortion still is quite evident, but now we begin to see filtering on the smaller objects. Due to the now familiar inverse relationship between the frequency and spatial domains, we know that the spot size of this filter is smaller than the spot of the filter with $D_0 = 5$. The result for $D_0 = 80$ is more of what a highpass-filtered image should look like. Here, the edges are much cleaner and less distorted, and the smaller objects have been filtered properly.

4.4.2 Butterworth Highpass Filters

The transfer function of the Butterworth highpass filter (BHPF) of order n and with cutoff frequency locus at a distance D_0 from the origin is given by

$$H(u, v) = \frac{1}{1 + \left[D_0/D(u, v)\right]^{2n}} \tag{4.4-3}$$

where $D(u, v)$ is given in Eq. (4.3-3). Equation (4.4-3) follows directly from Eqs. (4.4-1) and (4.3-6). The middle row of Fig. 4.22 shows an image and cross section of a BHPF function.

As in the case of lowpass filters, we can expect Butterworth highpass filters to behave smoother than IHPFs. The performance of a BHPF, of order 2 and with D_0 set to the same values as in Fig. 4.24, is shown in Fig. 4.25. The boundaries are much less distorted than in Fig. 4.24, even for the smallest value of cutoff frequency. Since the center spot sizes of the IHPF and the BHPF are similar [see Figs. 4.23(a) and (b)], the performance of the two filters in terms of filtering the smaller objects is comparable. The transition into higher values of cutoff frequencies is much smoother with the BHPF.

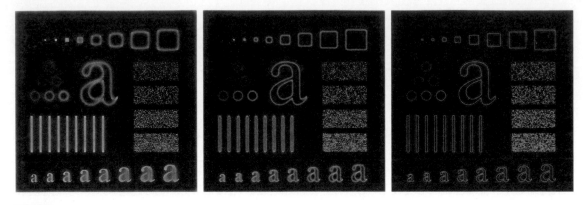

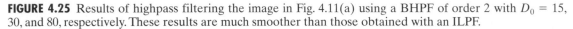

a b c

FIGURE 4.25 Results of highpass filtering the image in Fig. 4.11(a) using a BHPF of order 2 with $D_0 = 15$, 30, and 80, respectively. These results are much smoother than those obtained with an ILPF.

4.4.3 Gaussian Highpass Filters

The transfer function of the Gaussian highpass filter (GHPF) with cutoff frequency locus at a distance D_0 from the origin is given by

$$H(u, v) = 1 - e^{-D^2(u,v)/2D_0^2} \qquad (4.4\text{-}4)$$

where $D(u, v)$ is given in Eq. (4.3-3). This equation follows directly from Eqs. (4.4-1) and (4.3-8). The third row in Fig. 4.22 shows a perspective plot, image, and cross section of the GHPF function. Following the same format as for the BHPF, we show in Fig. 4.26 comparable results using GHPFs. As expected, the results obtained are smoother than with the previous two filters. Even the filtering of the smaller objects and thin bars is cleaner with the Gaussian filter.

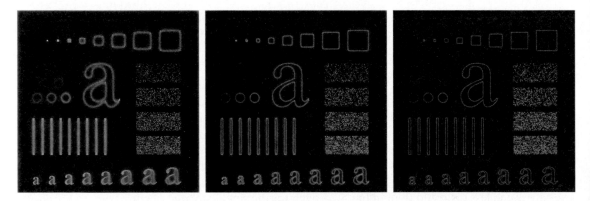

a b c

FIGURE 4.26 Results of highpass filtering the image of Fig. 4.11(a) using a GHPF with $D_0 = 15$, 30, and 80, respectively. Compare with Figs. 4.24 and 4.25.

As discussed in Section 4.2.4, it is possible to construct highpass filters as the difference of Gaussian lowpass filters. These difference filters have more parameters and, therefore, allow more control over the filter shape. However, the simple filter of Eq. (4.4-4) usually is quite adequate in practice, and it is an easier formulation for experimenting.

4.4.4 The Laplacian in the Frequency Domain

It can be shown that

$$\Im\left[\frac{d^n f(x)}{dx^n}\right] = (ju)^n F(u). \qquad (4.4\text{-}5)$$

From this simple expression, it follows that

$$\Im\left[\frac{\partial^2 f(x, y)}{\partial x^2} + \frac{\partial^2 f(x, y)}{\partial y^2}\right] = (ju)^2 F(u, v) + (jv)^2 F(u, v)$$
$$= -(u^2 + v^2)F(u, v). \qquad (4.4\text{-}6)$$

The expression inside the brackets on the left side of Eq. (4.4-6) is recognized as the Laplacian of $f(x, y)$, defined in Eq. (3.7-1). Thus, we have the important result

$$\Im\left[\nabla^2 f(x, y)\right] = -(u^2 + v^2)F(u, v), \qquad (4.4\text{-}7)$$

which simply says that the Laplacian can be implemented in the frequency domain by using the filter

$$H(u, v) = -(u^2 + v^2). \qquad (4.4\text{-}8)$$

As in all filtering operations in this chapter, the assumption is that the origin of $F(u, v)$ has been centered by performing the operation $f(x, y)(-1)^{x+y}$ prior to taking the transform of the image. As discussed earlier, if f (and F) are of size $M \times N$, this operation shifts the center transform so that $(u, v) = (0, 0)$ is at point $(M/2, N/2)$ in the frequency rectangle. As before, the center of the filter function also needs to be shifted:

$$H(u, v) = -[(u - M/2)^2 + (v - N/2)^2]. \qquad (4.4\text{-}9)$$

The Laplacian-filtered image in the spatial domain is obtained by computing the inverse Fourier transform of $H(u, v)F(u, v)$:

$$\nabla^2 f(x, y) = \Im^{-1}\{-[(u - M/2)^2 + (v - N/2)^2]F(u, v)\}. \quad (4.4\text{-}10)$$

Conversely, computing the Laplacian in the spatial domain using Eq. (3.7-1) and computing the Fourier transform of the result is equivalent to multiplying $F(u, v)$ by $H(u, v)$. We express this dual relationship in the familiar Fourier-transform-pair notation

$$\nabla^2 f(x, y) \Leftrightarrow -[(u - M/2)^2 + (v - N/2)^2]F(u, v). \qquad (4.4\text{-}11)$$

The spatial domain Laplacian filter function obtained by taking the inverse Fourier transform of Eq. (4.4-9) has some interesting properties, as Fig. 4.27 shows. Figure 4.27(a) is a 3-D perspective plot of Eq. (4.4-9). The function is

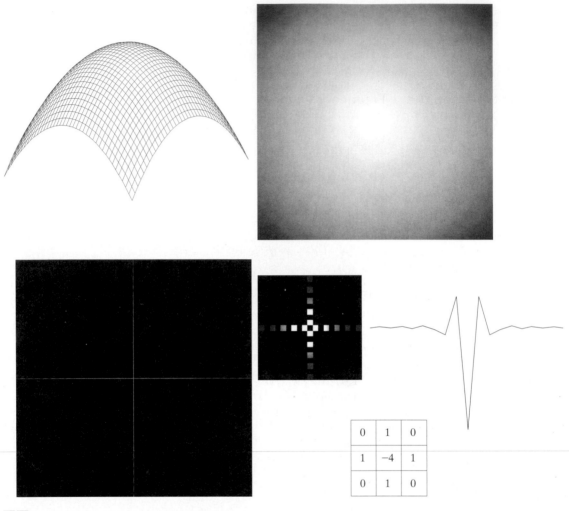

FIGURE 4.27 (a) 3-D plot of Laplacian in the frequency domain. (b) Image representation of (a). (c) Laplacian in the spatial domain obtained from the inverse DFT of (b). (d) Zoomed section of the origin of (c). (e) Gray-level profile through the center of (d). (f) Laplacian mask used in Section 3.7.

centered at $(M/2, N/2)$, and its value at the top of the dome is zero. All other values are negative. Figure 4.27(b) shows $H(u, v)$ as an image, also centered. Figure 4.27(c) is the Laplacian in the *spatial* domain, obtained by multiplying by $H(u, v)$ by $(-1)^{u+v}$, taking the inverse Fourier transform, and multiplying the real part of the result by $(-1)^{x+y}$. Figure 4.27(d) is a zoomed section at about the origin of Fig. 4.27(c). Figure 4.27(e) is a horizontal gray-level profile passing through the center of the zoomed section. Finally, Fig. 4.27(f) shows the mask we used in Section 3.7 to implement the definition of the discrete

Laplacian in the spatial domain, given in Eq. (3.7-4). A horizontal profile through the center of this mask has the same basic shape as the profile in Fig. 4.27(e) (that is, a negative value between two smaller positive values). It is of interest to note that, had we started in the frequency domain, we would have been led to a spatial mask similar in form to Fig. 3.39(a).

As in Eq. (3.7-5), we form an enhanced image $g(x, y)$ by subtracting the Laplacian from the original image:

$$g(x, y) = f(x, y) - \nabla^2 f(x, y). \qquad (4.4\text{-}12)$$

The Laplacian is subtracted from (rather than added to) the original because of the negative sign in Eq. (4.4-8). This is consistent with the formulation in the spatial domain given in Eq. (3.7-5). We would have arrived at the same conclusion by noting that the center spike in Fig. 4.27(e) is negative, with its immediate neighbors being positive.

As in the spatial domain, where we obtained the enhanced image with a single mask, it is possible to perform the entire operation in the frequency domain with only one filter, given by $H(u, v) = \left[1 + \left[(u - M/2)^2 + (v - N/2)^2\right]\right]$. In this case the enhanced image is obtained with a single inverse transform operation:

$$g(x, y) = \Im^{-1}\left\{\left[1 + ((u - M/2)^2 + (v - N/2)^2)\right]F(u, v)\right\}. \qquad (4.4\text{-}13)$$

In order to obtain the proper result with this implementation, care must be exercised in scaling the filter because the squared filter variables can be several orders of magnitude greater than 1.

■ Figure 4.28(a) is the same as Fig. 3.40(a). Figure 4.28(b) shows the result of filtering this image in the frequency domain using Eq. (4.4-10). As is typical of Laplacian-filtered images, which contain both positive and negative values of comparable magnitudes, scaling is necessary here. Figure 4.28(c) shows the image scaled (for display purposes only) so that its most negative value is scaled to zero and the maximum positive value is scaled to the maximum displayable gray level (255 in this case). Finally, Fig. 4.28(d) shows the enhanced result obtained by using Eq. (4.4-12). The increase in sharpness of small feature detail is unmistakable, as expected of results obtained by using the Laplacian. The sequence of images just presented should be compared with Fig. 3.40, which shows exactly the same sequence of steps, but computed using only spatial domain techniques. The results are identical for all practical purposes.   ■

EXAMPLE 4.7:
Illustration of the Laplacian in the frequency domain.

4.4.5 Unsharp Masking, High-Boost Filtering, and High-Frequency Emphasis Filtering

All the filtered images in Sections 4.4.1 through 4.4.3 have one thing in common: Their average background intensity has been reduced to near black. This is due to the fact that the highpass filters we applied to those images eliminate the zero-frequency component of their Fourier transforms (see the discussion in Section 4.2.3 regarding this phenomenon). As discussed in Section 3.7.2, the solution to this problem consists of adding a portion of the image back to the filtered result. In fact, enhancement using the Laplacian does precisely this, by adding back the entire image to the filtered result. Sometimes it is advantageous

a	b
c	d

FIGURE 4.28
(a) Image of the North Pole of the moon.
(b) Laplacian filtered image.
(c) Laplacian image scaled.
(d) Image enhanced by using Eq. (4.4-12). (Original image courtesy of NASA.)

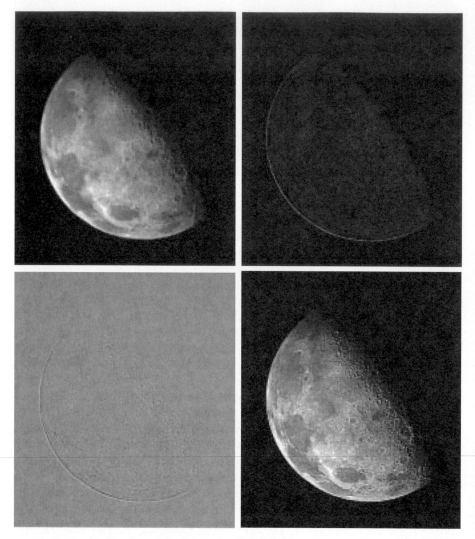

to increase the contribution made by the original image to the overall filtered result. This approach, called *high-boost filtering*, is a generalization of *unsharp masking*. These concepts were introduced in Section 3.7.2. We repeat them here using frequency domain concepts and notation.

Unsharp masking consists simply of generating a sharp image by subtracting from an image a blurred version of itself. Using frequency domain terminology, this means obtaining a highpass-filtered image by subtracting from the image a lowpass-filtered version of itself. That is,

$$f_{hp}(x,y) = f(x,y) - f_{lp}(x,y). \qquad (4.4\text{-}14)$$

High-boost filtering generalizes this by multiplying $f(x,y)$ by a constant $A \geq 1$:

$$f_{hb} = Af(x,y) - f_{lp}(x,y). \qquad (4.4\text{-}15)$$

Thus, high-boost filtering gives us the flexibility to increase the contribution made by the image to the overall enhanced result. This equation may be written as

$$f_{hb}(x, y) = (A - 1)f(x, y) + f(x, y) - f_{lp}(x, y). \qquad (4.4\text{-}16)$$

Then, using Eq. (4.4-14), we obtain

$$f_{hb}(x, y) = (A - 1)f(x, y) + f_{hp}(x, y). \qquad (4.4\text{-}17)$$

This result is based on a highpass rather than a lowpass image. When $A = 1$, high-boost filtering reduces to regular highpass filtering. As A increases past 1, the contribution made by the image itself becomes more dominant.

From Eq. (4.4-14), $F_{hp}(u, v) = F(u, v) - F_{lp}(u, v)$. But, $F_{lp}(u, v) = H_{lp}(u, v)F(u, v)$, where H_{lp} is the transfer function of a lowpass filter. Therefore, unsharp masking can be implemented directly in the frequency domain by using the composite filter

$$H_{hp}(u, v) = 1 - H_{lp}(u, v). \qquad (4.4\text{-}18)$$

Note that this result agrees with Eq. (4.4-1). Similarly, high-boost filtering can be implemented with the composite filter

$$H_{hb}(u, v) = (A - 1) + H_{hp}(u, v) \qquad (4.4\text{-}19)$$

with $A \geq 1$. The process consists of multiplying this filter by the (centered) transform of the input image and then taking the inverse transform of the product. Multiplication of the real part of this result by $(-1)^{x+y}$ gives us the high-boost filtered image $f_{hb}(x, y)$ in the spatial domain.

EXAMPLE 4.8:
High boost filtering in the frequency domain.

■ Figure 4.29 shows the same sequence as in Fig. 3.43, but using frequency domain computations. Figure 4.29(a) is the input image, and Fig. 4.29(b) is the highpass-filtered image. In order to compare the results in this example with those in Fig. 3.43, we used the Laplacian as the highpass filter, computed using Eq. (4.4-10). We elected not to use a composite filter to make scaling of the Laplacian easier.

The image in Fig. 4.29(c) was obtained using Eq. (4.4-17) with $A = 2$. As in Fig. 3.43(c), this image is sharper, but it still is too dark. Figure 4.29(d) was obtained with $A = 2.7$, which in effect means that the input image was multiplied by 1.7 before the Laplacian was subtracted from it. As in Fig. 3.43, this is an improved result. However, Fig. 4.29(d) is not as sharp as Fig. 3.43(d). The reason for this is that a frequency domain representation of the Laplacian is closer to the mask that excludes the diagonal neighbors [see Fig. 4.27(f)]. We know from the example in Fig. 3.41 that a mask that includes the diagonal neighbors produces slightly sharper results. The differences are not generally noticeable when the details are small (as in the moon example), but they do become evident for images with larger features. ■

Sometimes it is advantageous to accentuate the contribution to enhancement made by the high-frequency components of an image. In this case, we simply multiply a highpass filter function by a constant and add an offset so that the

a b
c d

FIGURE 4.29
Same as Fig. 3.43,
but using
frequency domain
filtering. (a) Input
image.
(b) Laplacian of
(a). (c) Image
obtained using
Eq. (4.4-17) with
$A = 2$. (d) Same
as (c), but with
$A = 2.7$. (Original
image courtesy of
Mr. Michael
Shaffer,
Department of
Geological
Sciences,
University of
Oregon, Eugene.)

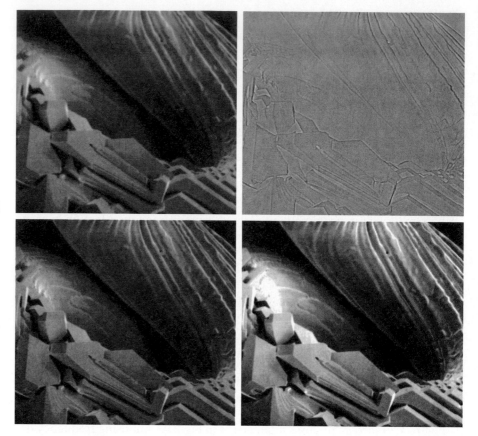

zero frequency term is not eliminated by the filter. This process, called *high-frequency emphasis*, has a filter transfer function given by

$$H_{\text{hfe}}(u, v) = a + bH_{\text{hp}}(u, v) \qquad (4.4\text{-}20)$$

where $a \geq 0$ and $b > a$. Typical values of a are in the range 0.25 to 0.5 and typical values of b are in the range 1.5 to 2.0. With reference to Eq. (4.4-17), we see that high-frequency emphasis reduces to high-boost filtering when $a = (A - 1)$ and $b = 1$. When $b > 1$, the high frequencies are emphasized, thus giving this procedure its name.

EXAMPLE 4.9:
High-frequency
emphasis filtering.

■ Figure 4.30(a) shows a chest X-ray with a narrow range of gray levels. Our main objective is to sharpen the image. X-rays cannot be focused in the same manner that lenses are focused, and the resulting images generally tend to be slightly blurred. Since the gray levels in this particular image are biased toward the dark end of the gray scale, we also take the opportunity here to give an example of how spatial domain processing can be used to complement frequency domain filtering.

Figure 4.30(b) shows the result of highpass filtering using a Butterworth filter of order 2 and a value of D_0 equal to 5% of the image vertical dimension.

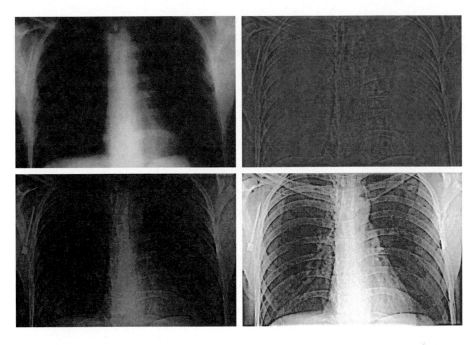

FIGURE 4.30
(a) A chest X-ray image. (b) Result of Butterworth highpass filtering. (c) Result of high-frequency emphasis filtering. (d) Result of performing histogram equalization on (c). (Original image courtesy Dr. Thomas R. Gest, Division of Anatomical Sciences, University of Michigan Medical School.)

Highpass filtering is not overly sensitive to this parameter, as long as the radius of the filter is not so small that frequencies near the origin of the transform are passed. As expected, the filtered result is rather featureless, but it shows faintly the principal edges in the image. The advantage of high-emphasis filtering (with $a = 0.5$ and $b = 2.0$ in this case) is shown in the image of Fig. 4.30(c). Although the image is still dark, the gray-level tonality due to the low frequency components was not lost.

As indicated in Section 3.3, an image characterized by gray levels in a narrow range of the gray scale is an ideal candidate for histogram equalization. As shown in Fig. 4.30(d), this indeed was an appropriate method to further enhance the image. Note the clarity of the bone structure and other details that simply are not visible in any of the other three images. The final enhanced image is a little noisy, but this is typical of X-ray images when their gray scale is expanded. The result obtained using a combination of high-frequency emphasis and histogram equalization is superior to the result that would be obtained by using either method alone. ■

4.5 Homomorphic Filtering

The illumination-reflectance model introduced in Section 2.3.4 can be used to develop a frequency domain procedure for improving the appearance of an image by simultaneous gray-level range compression and contrast enhancement. From the discussion in Section 2.3.4, an image $f(x, y)$ can be expressed as the product of illumination and reflectance components:

$$f(x, y) = i(x, y)r(x, y). \tag{4.5-1}$$

Equation (4.5-1) cannot be used directly to operate separately on the frequency components of illumination and reflectance because the Fourier transform of the product of two functions is not separable; in other words,

$$\Im\{f(x, y)\} \neq \Im\{i(x, y)\}\Im\{r(x, y)\}.$$

Suppose, however, that we define

$$
\begin{aligned}
z(x, y) &= \ln f(x, y) \\
&= \ln i(x, y) + \ln r(x, y).
\end{aligned}
\tag{4.5-2}
$$

Then

$$
\begin{aligned}
\Im\{z(x, y)\} &= \Im\{\ln f(x, y)\} \\
&= \Im\{\ln i(x, y)\} + \Im\{\ln r(x, y)\}
\end{aligned}
\tag{4.5-3}
$$

or

$$Z(u, v) = F_i(u, v) + F_r(u, v) \tag{4.5-4}$$

where $F_i(u, v)$ and $F_r(u, v)$ are the Fourier transforms of $\ln i(x, y)$ and $\ln r(x, y)$, respectively.

If we process $Z(u, v)$ by means of a filter function $H(u, v)$ then, from Eq. (4.2-27),

$$
\begin{aligned}
S(u, v) &= H(u, v)Z(u, v) \\
&= H(u, v)F_i(u, v) + H(u, v)F_r(u, v)
\end{aligned}
\tag{4.5-5}
$$

where $S(u, v)$ is the Fourier transform of the result. In the spatial domain,

$$
\begin{aligned}
s(x, y) &= \Im^{-1}\{S(u, v)\} \\
&= \Im^{-1}\{H(u, v)F_i(u, v)\} + \Im^{-1}\{H(u, v)F_r(u, v)\}.
\end{aligned}
\tag{4.5-6}
$$

By letting

$$i'(x, y) = \Im^{-1}\{H(u, v)F_i(u, v)\} \tag{4.5-7}$$

and

$$r'(x, y) = \Im^{-1}\{H(u, v)F_r(u, v)\}, \tag{4.5-8}$$

Eq. (4.5-6) can be expressed in the form

$$s(x, y) = i'(x, y) + r'(x, y). \tag{4.5-9}$$

Finally, as $z(x, y)$ was formed by taking the logarithm of the original image $f(x, y)$, the inverse (exponential) operation yields the desired enhanced image, denoted by $g(x, y)$; that is,

$$
\begin{aligned}
g(x, y) &= e^{s(x, y)} \\
&= e^{i'(x, y)} \cdot e^{r'(x, y)} \\
&= i_0(x, y)r_0(x, y)
\end{aligned}
\tag{4.5-10}
$$

where

$$i_0(x, y) = e^{i'(x, y)} \tag{4.5-11}$$

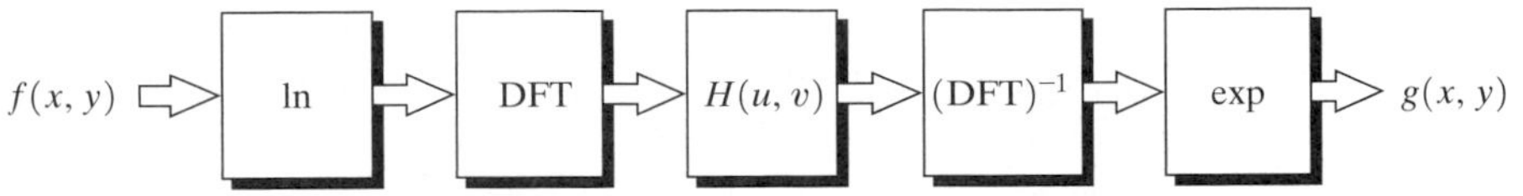

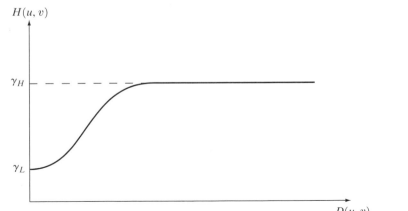

$f(x, y)$ ⇨ | ln | ⇨ | DFT | ⇨ | $H(u, v)$ | ⇨ | $(DFT)^{-1}$ | ⇨ | exp | ⇨ $g(x, y)$

FIGURE 4.31
Homomorphic filtering approach for image enhancement.

and

$$r_0(x, y) = e^{r'(x, y)} \qquad (4.5\text{-}12)$$

are the illumination and reflectance components of the output image.

The enhancement approach using the foregoing concepts is summarized in Fig. 4.31. This method is based on a special case of a class of systems known as *homomorphic systems*. In this particular application, the key to the approach is the separation of the illumination and reflectance components achieved in the form shown in Eq. (4.5-4). The *homomorphic filter function* $H(u, v)$ can then operate on these components separately, as indicated in Eq. (4.5-5).

The illumination component of an image generally is characterized by slow spatial variations, while the reflectance component tends to vary abruptly, particularly at the junctions of dissimilar objects. These characteristics lead to associating the low frequencies of the Fourier transform of the logarithm of an image with illumination and the high frequencies with reflectance. Although these associations are rough approximations, they can be used to advantage in image enhancement.

A good deal of control can be gained over the illumination and reflectance components with a homomorphic filter. This control requires specification of a filter function $H(u, v)$ that affects the low- and high-frequency components of the Fourier transform in different ways. Figure 4.32 shows a cross section of such a filter. If the parameters γ_L and γ_H are chosen so that $\gamma_L < 1$ and $\gamma_H > 1$, the filter function shown in Fig. 4.32 tends to decrease the contribution made by the low frequencies (illumination) and amplify the contribution made by high frequencies (reflectance). The net result is simultaneous dynamic range compression and contrast enhancement.

$H(u, v)$

γ_H

γ_L

$D(u, v)$

FIGURE 4.32
Cross section of a circularly symmetric filter function. $D(u, v)$ is the distance from the origin of the centered transform.

FIGURE 4.33
(a) Original
image. (b) Image
processed by
homomorphic
filtering (note
details inside
shelter).
(Stockham.)

The curve shape shown in Fig. 4.32 can be approximated using the basic form of any of the ideal highpass filters discussed in the previous section. For example, using a slightly modified form of the Gaussian highpass filter gives us

$$H(u, v) = (\gamma_H - \gamma_L)\left[1 - e^{-c\left(D^2(u, v)/D_0^2\right)}\right] + \gamma_L \qquad (4.5\text{-}13)$$

where $D^2(u, v)$ is given in Eq. (4.3-3) and the constant c has been introduced to control the sharpness of the slope of the filter function as it transitions between γ_L and γ_H. This type of filter is similar to the high-frequency emphasis filter we discussed at the end of Section 4.4.

EXAMPLE 4.10:
Enhancement by
homomorphic
filtering.

■ Figure 4.33 is typical of the results that can be obtained with the homomorphic filtering function in Fig. 4.32. In the original image shown in Fig. 4.33(a) the details inside the shelter are obscured by the glare from the outside walls. Figure 4.33(b) shows the result of processing this image by homomorphic filtering, with $\gamma_L = 0.5$ and $\gamma_H = 2.0$ in the filter function of Fig. 4.32. A reduction of dynamic range in the brightness, together with an increase in contrast, brought out the details of objects inside the shelter and balanced the gray levels of the outside wall. The enhanced image also is sharper. ■

4.6 Implementation

In this section we discuss details basic to implementing the Fourier transform. We begin with a summary of some additional properties of the 2-D Fourier transform and conclude with a brief review of the fast Fourier transform (FFT).

4.6.1 Some Additional Properties of the 2-D Fourier Transform

Translation

The Fourier transform pair has the following translation properties:

$$f(x, y)e^{j2\pi(u_0 x/M + v_0 y/N)} \Leftrightarrow F(u - u_0, v - v_0) \qquad (4.6\text{-}1)$$

and

$$f(x - x_0, y - y_0) \Leftrightarrow F(u, v)e^{-j2\pi(ux_0/M + vy_0/N)}, \qquad (4.6\text{-}2)$$

where, as introduced in Eq. (4.2-31), the double arrow is used to designate a Fourier transform pair. When $u_0 = M/2$ and $v_0 = N/2$, it follows that

$$e^{j2\pi(u_0x/M + v_0y/N)} = e^{j\pi(x+y)}$$
$$= (-1)^{x+y}.$$

In this case, Eq. (4.6-1) becomes

$$f(x, y)(-1)^{x+y} \Leftrightarrow F(u - M/2, v - N/2) \qquad (4.6\text{-}3)$$

and, similarly,

$$f(x - M/2, y - N/2) \Leftrightarrow F(u, v)(-1)^{(u+v)}. \qquad (4.6\text{-}4)$$

We see that Eq. (4.6-3) is the same as Eq. (4.2-21), which we used for centering the transform. These results are based on the variables u and v having values in the range $[0, M - 1]$ and $[0, N - 1]$, respectively. In a computer implementation these variables will run from $u = 1$ to M and $v = 1$ to N, in which case the actual center of the transform will be at $u = (M/2) + 1$ and $v = (N/2) + 1$.

Distributivity and scaling

From the definition of the Fourier transform it follows that

$$\Im[f_1(x, y) + f_2(x, y)] = \Im[f_1(x, y)] + \Im[f_2(x, y)] \qquad (4.6\text{-}5)$$

and, in general, that

$$\Im[f_1(x, y) \cdot f_2(x, y)] \neq \Im[f_1(x, y)] \cdot \Im[f_2(x, y)]. \qquad (4.6\text{-}6)$$

In other words, the Fourier transform is distributive over addition, but not over multiplication. Identical comments apply to the inverse Fourier transform. Similarly, for two scalars a and b,

$$af(x, y) \Leftrightarrow aF(u, v) \qquad (4.6\text{-}7)$$

and

$$f(ax, by) \Leftrightarrow \frac{1}{|ab|} F(u/a, v/b). \qquad (4.6\text{-}8)$$

Rotation

If we introduce the polar coordinates

$$x = r\cos\theta \qquad y = r\sin\theta \qquad u = \omega\cos\varphi \qquad v = \omega\sin\varphi$$

then $f(x, y)$ and $F(u, v)$ become $f(r, \theta)$ and $F(\omega, \varphi)$, respectively. Direct substitution into the definition of the Fourier transform yields

$$f(r, \theta + \theta_0) \Leftrightarrow F(\omega, \varphi + \theta_0). \qquad (4.6\text{-}9)$$

This expression indicates that rotating $f(x, y)$ by an angle θ_0 rotates $F(u, v)$ by the same angle. Similarly, rotating $F(u, v)$ rotates $f(x, y)$ by the same angle.

Periodicity and conjugate symmetry

The discrete Fourier transform has the following periodicity properties:

$$F(u, v) = F(u + M, v) = F(u, v + N) = F(u + M, v + N). \quad (4.6\text{-}10)$$

The inverse transform also is periodic:

$$f(x, y) = f(x + M, y) = f(x, y + N) = f(x + M, y + N). \quad (4.6\text{-}11)$$

The idea of conjugate symmetry was introduced in Section 4.2, and is repeated here for convenience:

$$F(u, v) = F^*(-u, -v) \quad (4.6\text{-}12)$$

from which it follows that the spectrum also is symmetric about the origin:

$$|F(u, v)| = |F(-u, -v)|. \quad (4.6\text{-}13)$$

The validity of these equations is easily established from Eqs. (4.2-16) and (4.2-17).

The importance of the periodicity property is illustrated in Fig. 4.34(a), which shows the spectrum for a one-dimensional transform $F(u)$ [see Eqs. (4.2-5) and (4.2-10)]. From Eq. (4.6-10), $F(u) = F(u + M)$, from which it follows that $|F(u)| = |F(u + M)|$. Also, because of Eq. (4.6-13), $|F(u)| = |F(-u)|$. The periodicity property indicates that $F(u)$ has a period of length M, and the symmetry property indicates that the spectrum is centered on the origin, as Fig. 4.34(a) shows. This figure and the preceding comments demonstrate that the magnitudes of the transform values from $(M/2) + 1$ to $M - 1$ are reflections of the values in the half period to the left of the origin. Because the discrete Fourier transform has been formulated for values of u in the interval $[0, M - 1]$, the result of this formulation yields two back-to-back half periods in this interval. To

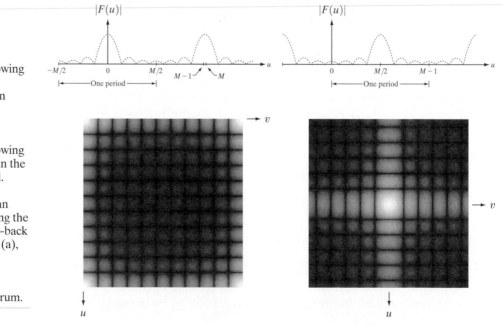

a b
c d

FIGURE 4.34
(a) Fourier spectrum showing back-to-back half periods in the interval $[0, M - 1]$.
(b) Shifted spectrum showing a full period in the same interval.
(c) Fourier spectrum of an image, showing the same back-to-back properties as (a), but in two dimensions.
(d) Centered Fourier spectrum.

display one full period, all that is necessary is to move the origin of the transform to the point $u = M/2$, as Fig. 4.34(b) shows. To do so, we simply multiply $f(x)$ by $(-1)^x$ prior to taking the transform, as indicated in Eq. (4.6-3).

Two-dimensional spectra are analyzed in a similar manner. Figures 4.34(c) and (d) show what a typical spectrum looks like before and after centering (only one period is shown). The center of the transform in this case is at the top, left corner of the picture, and four 2-D periodic components extend "back to back" from the four corners of the image toward its center, which contains the highest-frequency component. By contrast, the centered transform, obtained using Eq. (4.6-3), has the origin (zero frequency components) at the center of Fig. 4.34(d), and its components extend outward from there with conjugate symmetry. Figure 4.34(d) clearly shows that centering the transform not only helps with visualization but, as mentioned several times already in this chapter, it also simplifies filtering.

Separability

The discrete Fourier transform in Eq. (4.2-16) can be expressed in the separable form

$$
\begin{aligned}
F(u, v) &= \frac{1}{M} \sum_{x-0}^{M-1} e^{j2\pi ux/M} \frac{1}{N} \sum_{y-0}^{N-1} f(x, y) e^{-j2\pi vy/N} \\
&- \frac{1}{M} \sum_{x=0}^{M-1} F(x, v) e^{-j2\pi ux/M}
\end{aligned}
\tag{4.6-14}
$$

where

$$
F(x, v) - \frac{1}{N} \sum_{y=0}^{N-1} f(x, y) e^{-j2\pi vy/N}.
\tag{4.6-15}
$$

For *each* value of x, and for values of $v = 0, 1, 2, \ldots, N - 1$, this equation is a complete 1-D Fourier transform. In other words, $F(x, v)$ is the Fourier transform along one row of $f(x, y)$. By varying x from 0 to $M - 1$, we compute the Fourier transform along all rows of $f(x, y)$. Thus far, the frequency variable u has remained constant. To complete the 2-D transform we have to vary u from 0 to $M - 1$ in Eq. (4.6-14). After a little thought, it becomes evident that this involves computing the 1-D transform along each *column* of $F(x, v)$. This is an important result. It tells us that we can compute the 2-D transform by first computing a 1-D transform along each row of the input image, and then computing a 1-D transform along each column of this intermediate result. The same comments hold if we reverse the order of computation: columns first, followed by rows. The procedure is summarized in Fig. 4.35.

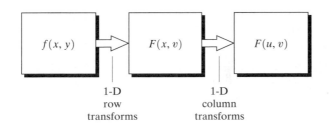

1-D
row
transforms

1-D
column
transforms

FIGURE 4.35
Computation of the 2-D Fourier transform as a series of 1-D transforms.

A similar development applies to computing the 2-D inverse Fourier transform. We first compute 1-D inverse transforms along each row of $F(u, v)$ and then compute inverse 1-D transforms along each column of the intermediate result. As shown in the following section, it is possible to implement the inverse transform using a 1-D *forward* Fourier transform algorithm.

4.6.2 Computing the Inverse Fourier Transform Using a Forward Transform Algorithm

As noted in the previous section, 2-D Fourier transforms can be computed via the application of 1-D transforms. The 1-D Fourier transform pair was defined in Section 4.2.1. We repeat the two equations here for convenience:

$$F(u) = \frac{1}{M} \sum_{x=0}^{M-1} f(x) e^{-j2\pi ux/M} \tag{4.6-16}$$

for $u = 0, 1, 2, \ldots, M - 1$, and

$$f(x) = \sum_{u=0}^{M-1} F(u) e^{j2\pi ux/M} \tag{4.6-17}$$

for $x = 0, 1, 2, \ldots, M - 1$. Taking the complex conjugate of Eq. (4.6-17) and dividing both sides by M yields

$$\frac{1}{M} f^*(x) = \frac{1}{M} \sum_{u=0}^{M-1} F^*(u) e^{-j2\pi ux/M}. \tag{4.6-18}$$

Comparing this result with Eq. (4.6-16) shows that the right side of Eq. (4.6-18) is in the form of the forward Fourier transform. Therefore, inputting $F^*(u)$ into an algorithm designed to compute the forward transform gives the quantity $f^*(x)/M$. Taking the complex conjugate and multiplying by M yields the desired inverse $f(x)$. A similar analysis for two variables yields:

$$\frac{1}{MN} f^*(x, y) = \frac{1}{MN} \sum_{u=0}^{M-1} \sum_{v=0}^{N-1} F^*(u, v) e^{-j2\pi(ux/M + vy/N)}, \tag{4.6-19}$$

which is in the form of a 2-D forward Fourier transform. If $f(x)$ or $f(x, y)$ are real functions (e.g., an image), then the complex conjugate on the left of Eq. (4.6-19) is unnecessary; we simply take the real part of the result, ignoring the parasitic complex terms that are typical in most Fourier transform computations.

Computation of the 2-D transform by successive passes of the 1-D transform is a frequent source of confusion when the technique we have just developed is used to obtain the inverse. Keep in mind the procedure outlined in the previous section, and avoid being misled by Eq. (4.6-18). In other words, when a 1-D algorithm is used to compute the 2-D inverse, we do not compute the complex conjugate after processing each row or column. Instead, the function $F^*(u, v)$ is treated as if it were $f(x, y)$ in the forward, 2-D transform procedure summarized in Fig. 4.35. The complex conjugate (or real part, if applicable) of the result, multiplied by MN, yields the proper inverse $f(x, y)$. We emphasize that the preceding comments regarding the constants M and N are based on the definition of the discrete Fourier transform that has all constants associated with the forward transform. As indicated in Sections 4.2.1 and 4.2.2, it is not un-usual to encounter the constants distributed in a different way between the for-ward and inverse transforms. Therefore, in order to avoid being off by a scale factor, care must be exercised in the placement of the constants when comput-ing the inverse transform if these constants are distributed differently from the way that we do in this book.

4.6.3 More on Periodicity: the Need for Padding

It was explained in Section 4.2.4 that, based on the convolution theorem, mul-tiplication in the frequency domain is equivalent to convolution in the spatial domain, and vice versa. When working with discrete variables and the Fourier transform, we need to keep in mind the periodicity of the various functions in-volved (Section 4.6.1). Although it may not be intuitive, this periodicity is a mathematical byproduct of the way in which the discrete Fourier transform pair is defined. Periodicity is part of the process, and it cannot be ignored.

Figure 4.36 illustrates the significance of periodicity. The left column of this figure shows convolution computed using the 1-D version of Eq. (4.2-30):

$$f(x) * h(x) = \frac{1}{M} \sum_{m=0}^{M-1} f(m)h(x - m). \qquad (4.6\text{-}20)$$

We also take the opportunity here to explain convolution in a little more detail. To simplify the notation, simple numbers instead of general symbols are used for the height and length of the functions. Figures 4.36(a) and (b) show the two functions we wish to convolve. Each function consists of 400 points. The first step in convolution is to mirror (flip) one of the functions about the origin. In this case this was done to the second function, which is shown as $h(-m)$ in Fig. 4.36(c). The next step is to "slide" $h(-m)$ past $f(m)$. This is done by adding a constant, x, to $h(-m)$; that is, we form $h(x - m)$, as shown in Fig. 4.36(d). Note that this is only *one* displacement value. This simple step is a frequent source of confusion when first encountered. It helps to remember that this is precisely what convolution is all about. In other words, to perform convolution we flip one of the functions and slide it past the other. At each displacement (each value of x) the entire summation in Eq. (4.6-20) is carried out. This summation is noth-ing more than the sum of products of f and h at a given displacement. The dis-placement x ranges over all values required to completely slide h past f. Figure 4.36(e) shows the result of completely sliding h past f and computing Eq. (4.6-20) at each value of x. In this case x had to range for 0 to 799 for $h(x - m)$ to slide completely past f. This figure is the convolution of the two functions. Keep clear-ly in mind that the variable in convolution is x.

We know from the convolution theorem introduced in Section 4.2 [see Eq. (4.2-31)] that we can obtain exactly the same result given in Eq. (4.6-20) by tak-ing the inverse Fourier transform of the product $F(u)H(u)$. However, we also know from the discussion on periodicity earlier in this section that the discrete Fourier transform automatically takes the input functions as periodic. In other words, using the DFT allows us to perform convolution in the frequency domain, but the func-tions are treated as periodic, with a period equal to the length of the functions.

FIGURE 4.36 Left: convolution of two discrete functions. Right: convolution of the same functions, taking into account the implied periodicity of the DFT. Note in (j) how data from adjacent periods corrupt the result of convolution.

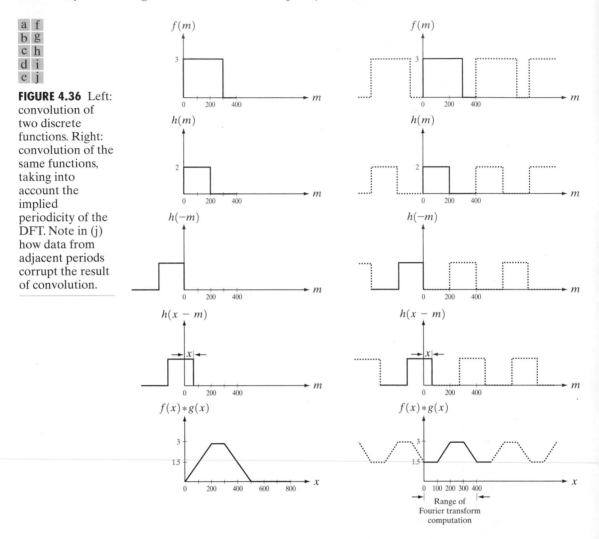

We can examine the implications of this periodicity with the aid of the right column of Fig. 4.36. Figure 4.36(f) is the same as Fig. 4.36(a), but with periods of the same function extending infinitely in both directions (extended sections are shown dashed). The same applies to Figs. 4.36(g) through (i). We now perform convolution by sliding $h(x - m)$ past $f(m)$. The sliding is accomplished by varying x, as before. However, now the periodic extensions of $h(x - m)$ introduce values that were not there in our computations on the left side of Fig. 4.36. For example, when $x = 0$ in Fig. 4.36(i), we see that part of the first extended period to the right of $h(x - m)$ lies inside the part of $f(m)$ in Fig. 4.36(f) that starts at the origin (shown solid). As $h(x - m)$ slides to the right, the section that was inside $f(m)$ starts to move out to the right, but it is replaced by an identical section from the left side of $h(x - m)$. This causes the convolution to have a constant value, as shown in the segment $[0, 100]$ in Fig. 4.36(j). The segment from 100 through 400 is correct, but periodicity starts again, thus causing part of the tail of the convolution function to be lost, as can be seen by comparing the solid lines in Figs. 4.36(j) and 4.36(e).

In the frequency domain the procedure would be to compute the Fourier transforms of the functions in Figs. 4.36(a) and (b). According to the convolution theorem, the two transforms would then be multiplied and the inverse Fourier transform taken. The result would be the 400 points comprising the convolution shown in solid in Fig. 4.36(j). This simple illustration shows that failure to handle the periodicity issue properly will give incorrect results if the convolution function is obtained using the Fourier transform. The result will have erroneous data at the beginning and have missing data at the end.

The solution to this problem is straightforward. Assume that f and h consist of A and B points, respectively. We append zeros to both functions so that they have identical periods, denoted by P. This procedure yields *extended*, or *padded*, functions given by

$$f_e(x) = \begin{cases} f(x) & 0 \le x \le A - 1 \\ 0 & A \le x \le P \end{cases} \tag{4.6-21}$$

and

$$g_e(x) = \begin{cases} g(x) & 0 \le x \le B - 1 \\ 0 & B \le x \le P. \end{cases} \tag{4.6-22}$$

It can be shown (Brigham [1988]) that, unless we choose $P \ge A + B - 1$, the individual periods of the convolution will overlap. We already saw in Fig. 4.36 the result of this phenomenon, which is commonly referred to as *wraparound error*. If $P = A + B - 1$, the periods will be adjacent. If $P > A + B - 1$, the periods will be separated, with the degree of separation being equal to the difference between P and $A + B - 1$.

The results obtained after extending the functions in Fig. 4.36(a) and (b) are shown in Figs. 4.37(a) and (b). In this case we chose $P = A + B - 1$ (799), so we know that the periods of the convolution will be adjacent. Following a procedure identical to the previous explanation, we arrive at the convolution function shown in Fig. 4.37(e). One period of this result is identical to Fig. 4.36(e), which we know to be correct. Thus, if we wanted to compute the convolution in the frequency domain, we would (1) obtain the Fourier transform of the two extended sequences (each of which is 800 points long); (2) multiply the two transforms; and (3) compute the inverse Fourier transform. The result would be the correct 800-point convolution function shown in the period highlighted in Fig. 4.37(e).

Extension of these concepts to 2-D functions follows the same line of reasoning. Suppose that we have two images $f(x, y)$ and $h(x, y)$ of sizes $A \times B$ and $C \times D$, respectively. As in the 1-D case, these arrays must be assumed periodic with some period P in the x-direction and Q in the y-direction. Wraparound error in 2-D convolution is avoided by choosing

$$P \ge A + C - 1 \tag{4.6-23}$$

and

$$Q \ge B + D - 1. \tag{4.6-24}$$

FIGURE 4.37
Result of
performing
convolution with
extended
functions.
Compare
Figs. 4.37(e) and
4.36(e).

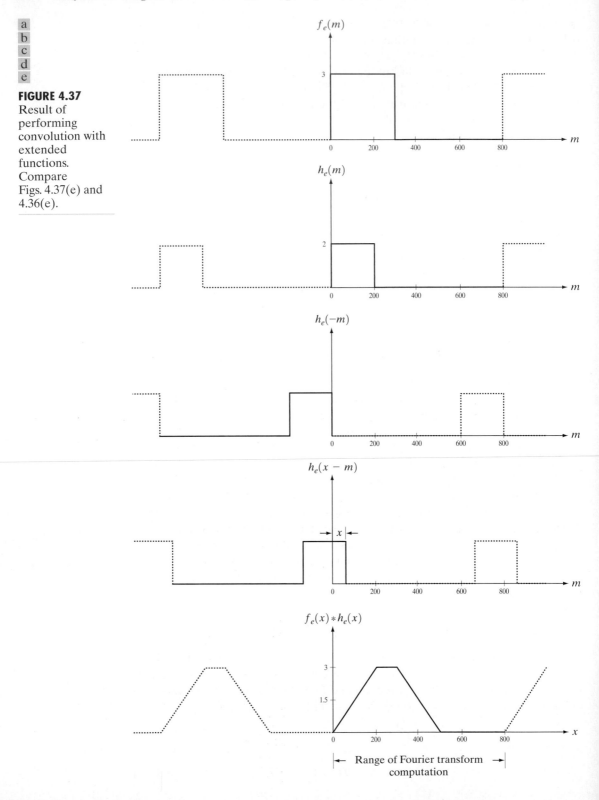

Range of Fourier transform
computation

The periodic sequences are formed by extending $f(x, y)$ and $h(x, y)$ as follows:

$$f_e(x, y) = \begin{cases} f(x, y) & 0 \le x \le A - 1 \quad \text{and} \quad 0 \le y \le B - 1 \\ 0 & A \le x \le P \quad \text{or} \quad B \le y \le Q \end{cases} \quad (4.6\text{-}25)$$

and

$$h_e(x, y) = \begin{cases} h(x, y) & 0 \le x \le C - 1 \quad \text{and} \quad 0 \le y \le D - 1 \\ 0 & C \le x \le P \quad \text{or} \quad D \le y \le Q \end{cases} \quad (4.6\text{-}26)$$

The issue of padding is central to filtering. When we implement any of the frequency domain filters discussed in this chapter, we do it by multiplying the filter transfer function by the transform of the image we wish to process. By the convolution theorem, we know that this is the same as convolving the spatial representation of the filter with the image. Thus, unless proper padding is implemented, the results will be erroneous. This is illustrated in Fig. 4.38. For the sake of simplicity in the figure, we assume that f and h are square and that they are

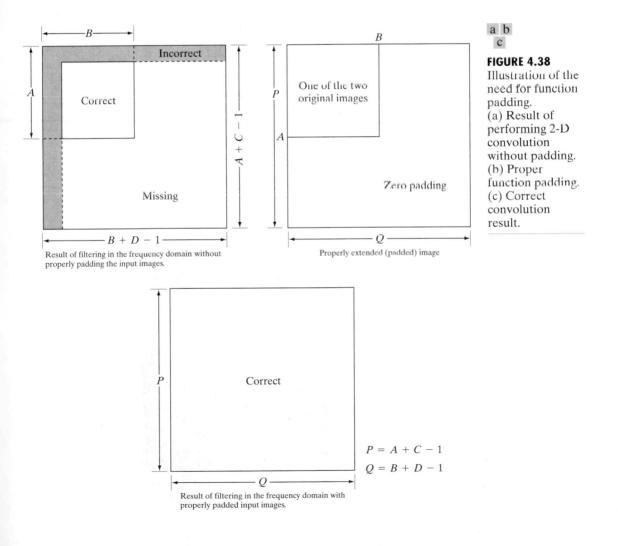

a b
c

FIGURE 4.38
Illustration of the need for function padding.
(a) Result of performing 2-D convolution without padding.
(b) Proper function padding.
(c) Correct convolution result.

both of the same size, where h is the inverse DFT of the filter $H(u, v)$. Figure 4.38(a) shows what the result of filtering would be if the images were not padded. This is the generic result we would obtain if we computed the Fourier transform of an input image that was not padded, multiplied it by a filter function of the same size (also not padded) and computed the inverse transform. The result would be of size $A \times B$, the same as the input image, as shown in the top, left quadrant in Fig. 4.38(a). As in the 1-D case, the leading edges of the image would contain erroneous data induced by periodicity (shown shaded), and there would be missing data at the trailing edges. By properly padding the input image *and* the filter function as shown in Fig. 4.38(b), the result would be a correctly filtered image of size $P \times Q$, as shown in Fig. 4.38(c). This image is twice the size of the original in both directions, and thus has four times as many pixels. However, as will be shown shortly, the area of interest typically is cropped out of this larger image.

It is important to note that the approach just described calls for the frequency domain filter function to be multiplied by $(-1)^{u+v}$, inverse transformed, padded with 0's, and then forward transformed. All other aspects of filtering are as described in Section 4.2.3. Note also that the inverse Fourier transform of the filter has both real and imaginary parts. Although the imaginary components of the filters with which we deal in this book typically are many orders of magnitude smaller than the real components, it not good practice in general to ignore imaginary components in intermediate Fourier computations. Thus, both real and imaginary components are padded prior to generating the padded frequency domain filter via computation of the forward transform.

Figure 4.39 shows the padded spatial representation (only the real part is shown) of the ideal lowpass filter used to generate Fig. 4.12(c). The padded area of 0's is shown in black. An ideal lowpass filter was selected for illustration because it has the most visible "structure" in the spatial domain. The padding used

FIGURE 4.39 Padded lowpass filter in the spatial domain (only the real part is shown).

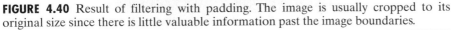

FIGURE 4.40 Result of filtering with padding. The image is usually cropped to its original size since there is little valuable information past the image boundaries.

was the minimum size required, which, when images and filters are squares of the same size, simply doubles the size in both dimensions.

Figure 4.40 shows the result of filtering with padded functions using the approach just discussed. It is easy to visualize how convolving the filter in Fig. 4.39 with a padded version of Fig. 4.12(a) would generate Fig. 4.40. It also is evident in this case that three-quarters of the result contain no valuable information, so cropping back to the original image gives the desired filtered result. We are assured by using padding that the cropped image is free of wraparound error.

4.6.4 The Convolution and Correlation Theorems

Convolution was introduced in Section 4.2.4 and its implementation was discussed in additional detail in Section 4.6.3. We repeat it briefly here to facilitate comparison with a similar process called *correlation*. The discrete convolution of two functions $f(x, y)$ and $h(x, y)$ of size $M \times N$ is denoted by $f(x, y) * h(x, y)$ and is defined by the expression

$$f(x, y) * h(x, y) = \frac{1}{MN} \sum_{m=0}^{M-1} \sum_{n=0}^{N-1} f(m, n)h(x - m, y - n). \quad (4.6-27)$$

From the discussion in Section 4.2.4, we know that the convolution theorem consists of the following relationships between the two functions and their Fourier transforms:

$$f(x, y) * h(x, y) \Leftrightarrow F(u, v)H(u, v) \quad (4.6-28)$$

and

$$f(x, y)h(x, y) \Leftrightarrow F(u, v) * H(u, v). \quad (4.6-29)$$

The correlation of two functions $f(x, y)$ and $h(x, y)$ is defined as

$$f(x, y) \circ h(x, y) = \frac{1}{MN} \sum_{m=0}^{M-1} \sum_{n=0}^{N-1} f^*(m, n) h(x + m, y + n) \quad (4.6\text{-}30)$$

where f^* denotes the complex conjugate of f. We normally deal with real functions (images), in which case $f^* = f$. The correlation function has exactly the same *form* as the convolution function given in Eq. (4.6-27), with the exception of the complex conjugate and the fact that the second term in the summation has positive instead of negative signs. This means that h is not mirrored about the origin. Everything else in the implementation of correlation is identical to convolution, including the need for padding.

Given the similarity of convolution and correlation, it is not surprising that there is a *correlation theorem*, analogous to the convolution theorem. Let $F(u, v)$ and $H(u, v)$ denote the Fourier transforms of $f(x, y)$ and $h(x, y)$, respectively. One-half of the correlation theorem states that spatial correlation, $f(x, y) \circ h(x, y)$, and the frequency domain product, $F^*(u, v)H(u, v)$, constitute a Fourier transform pair. This result, formally stated as

$$f(x, y) \circ h(x, y) \Leftrightarrow F^*(u, v)H(u, v), \quad (4.6\text{-}31)$$

indicates that correlation in the spatial domain can be obtained by taking the inverse Fourier transform of the product $F^*(u, v)H(u, v)$, where F^* is the complex conjugate of F. An analogous result is that correlation in the frequency domain reduces to multiplication in the spatial domain; that is,

$$f^*(x, y)h(x, y) \Leftrightarrow F(u, v) \circ H(u, v). \quad (4.6\text{-}32)$$

These two results comprise the correlation theorem. It is assumed that all functions have been properly extended by padding.

As we know by now, convolution is the tie between filtering in the spatial and frequency domains. The principal use of correlation is for matching. In *matching*, $f(x, y)$ is an image containing objects or regions. If we want to determine whether f contains a particular object or region in which we are interested, we let $h(x, y)$ be that object or region (we normally call this image a *template*). Then, if there is a match, the correlation of the two functions will be maximum at the location where h finds a correspondence in f. Preprocessing, like scaling and alignment, is necessary in most practical applications, but the bulk of the process is performing the correlation.

Finally, we point out that the term *cross correlation* often is used in place of the term *correlation* to clarify that the images being correlated are different. This is as opposed to *autocorrelation*, in which both images are identical. In the latter case, we have the *autocorrelation theorem*, which follows directly from Eq. (4.6-31):

$$f(x, y) \circ f(x, y) \Leftrightarrow |F(u, v)|^2. \quad (4.6\text{-}33)$$

On the right side, we used the fact that the product of a complex quantity and its complex conjugate is the magnitude of the complex quantity squared. In words, this result states that the Fourier transform of the spatial autocorrelation is the power spectrum defined in Eq. (4.2-20). Similarly,

$$|f(x, y)|^2 \Leftrightarrow F(u, v) \circ F(u, v). \quad (4.6\text{-}34)$$

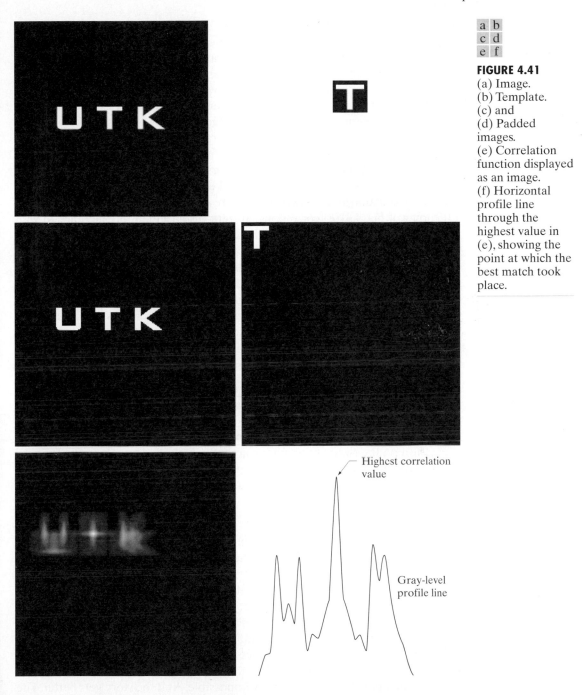

a b
c d
e f

FIGURE 4.41
(a) Image.
(b) Template.
(c) and
(d) Padded
images.
(e) Correlation
function displayed
as an image.
(f) Horizontal
profile line
through the
highest value in
(e), showing the
point at which the
best match took
place.

Highest correlation
value

Gray-level
profile line

■ Figure 4.41 shows a simple illustration of image padding and correlation. Figure 4.41(a) is the image and Fig. 4.41(b) is the template. The image and template are of size 256×256 and 38×42 pixels, respectively. In this case, $A = B = 256, C = 38$, and $D = 42$. This gives the minimum values for the extended functions: $P = A + C - 1 = 293$ and $Q = B + D - 1 = 297$. We

EXAMPLE 4.11:
Image correlation.

chose equal padding dimensions of 298×298. The padded images are shown in Figs. 4.41(c) and (d). The spatial correlation of the two padded images is displayed as an image in Fig. 4.41(e). As indicated in Eq. (4.6-31), the correlation function was obtained by computing the transforms of the padded images, taking the complex conjugate of one of them (we chose the template), multiplying the two transforms, and computing the inverse DFT. It is left as an exercise for the reader (Problem 4.23) to discuss what Fig. 4.41(e) would look like if we had taken the complex conjugate of the other transform instead.

As expected, we see in Fig. 4.41(e) that the highest value of the correlation function occurs at the point where the template is exactly on top of the "T" in the image. As in convolution, it is important to keep in mind that the variables in the correlation function in the spatial domain are *displacements*. For example, the top, left corner of Fig. 4.41(e) corresponds to zero displacement of one function with respect to the other. The value of each pixel in Fig. 4.41(e) is the value of the correlation function at *one* location of displacement; that is, for one specific value of the pair (x, y) in Eq. (4.6-30). Also, we note that the correlation function has the same dimensions as the padded images. Finally, Fig. 4.41(f) shows a horizontal gray-level profile passing through the highest value in Fig. 4.41(e). This figure simply confirms that the highest peak in the correlation function is located at the point where the best match of the template and the image occurs. ■

4.6.5 Summary of Properties of the 2-D Fourier Transform

All the properties of the Fourier transform discussed in this chapter are summarized in Table 4.1. A footnote identifies the items requiring that functions be padded in order to avoid incorrect results. As before, the double arrows are used to denote that the expressions form a Fourier transform pair. That is, the expression on the right of the double arrows is obtained by taking the forward Fourier transform of the expression on the left; the expression on the left is obtained by taking the inverse Fourier transform of the expression on the right.

4.6.6 The Fast Fourier Transform

As indicated in Section 4.1, one of the main reasons that the DFT has become an essential tool in signal processing was the development of the fast Fourier transform (FFT). Computing the 1-D Fourier transform of M points using Eq. (4.2-5) directly requires on the order of M^2 multiplication/addition operations. The FFT accomplishes the same task on the order of $M \log_2 M$ operations. If, for example, $M = 1024$, the brute-force method will require approximately 10^6 operations, while the FFT will require approximately 10^4 operations. This is a computational advantage of 100 to 1. If this advantage does not seem significant, imagine being able to complete a given project in one year as opposed to 100. This is the difference between possible and practically impossible. And the story gets better. The bigger the problem, the greater the computational advantage. If, for instance, $M = 8192 \left(2^{13}\right)$, the computational advantage grows to 600 to 1. These types of numbers are great motivators for wanting to learn more about how an FFT algorithm works. In this section we take a look at the derivation of a fundamental decomposition of the DFT that leads to the FFT. The focus is on the FFT of one variable. As indicated in Section 4.6.1, the 2-D Fourier transform can be obtained by successive passes of a 1-D transform algorithm.

The FFT algorithm developed in this section is based on the so-called *successive doubling method*. For notational convenience we express Eq. (4.2-5) in the form

$$F(u) = \frac{1}{M} \sum_{x=0}^{M-1} f(x) W_M^{ux} \tag{4.6-35}$$

where

$$W_M = e^{-j2\pi/M} \tag{4.6-36}$$

and M is assumed to be of the form

$$M = 2^n \tag{4.6-37}$$

with n being a positive integer. Hence, M can be expressed as

$$M = 2K \tag{4.6-38}$$

with K also being a positive integer. Substitution of Eq. (4.6-38) into Eq. (4.6-35) yields

$$F(u) = \frac{1}{2K} \sum_{x=0}^{2K-1} f(x) W_{2K}^{ux}$$

$$= \frac{1}{2} \left[\frac{1}{K} \sum_{x=0}^{K-1} f(2x) W_{2K}^{u(2x)} + \frac{1}{K} \sum_{x=0}^{K-1} f(2x+1) W_{2K}^{u(2x+1)} \right]. \tag{4.6-39}$$

However, it can be shown using Eq. (4.6-36) that $W_{2K}^{2ux} = W_K^{ux}$, so Eq. (4.6-39) can be expressed as

$$F(u) = \frac{1}{2} \left[\frac{1}{K} \sum_{x=0}^{K-1} f(2x) W_K^{ux} + \frac{1}{K} \sum_{x=0}^{K-1} f(2x+1) W_K^{ux} W_{2K}^{u} \right]. \tag{4.6-40}$$

Defining

$$F_{\text{even}}(u) = \frac{1}{K} \sum_{x=0}^{K-1} f(2x) W_K^{ux} \tag{4.6-41}$$

for $u = 0, 1, 2, \ldots, K - 1$, and

$$F_{\text{odd}}(u) = \frac{1}{K} \sum_{x=0}^{K-1} f(2x+1) W_K^{ux} \tag{4.6-42}$$

for $u = 0, 1, 2, \ldots, K - 1$, reduces Eq. (4.6-40) to

$$F(u) = \frac{1}{2} \left[F_{\text{even}}(u) + F_{\text{odd}}(u) W_{2K}^{u} \right]. \tag{4.6-43}$$

Also, because $W_M^{u+M} = W_M^{u}$ and $W_{2M}^{u+M} = -W_{2M}^{u}$, Eqs. (4.6-41) through (4.6-43) give

$$F(u + K) = \frac{1}{2} \left[F_{\text{even}}(u) - F_{\text{odd}}(u) W_{2K}^{u} \right]. \tag{4.6-44}$$

TABLE 4.1
Summary of some important properties of the 2-D Fourier transform.

Property	Expression(s)
Fourier transform	$F(u, v) = \dfrac{1}{MN} \displaystyle\sum_{x=0}^{M-1} \sum_{y=0}^{N-1} f(x, y) e^{-j2\pi(ux/M + vy/N)}$
Inverse Fourier transform	$f(x, y) = \displaystyle\sum_{u=0}^{M-1} \sum_{v=0}^{N-1} F(u, v) e^{j2\pi(ux/M + vy/N)}$
Polar representation	$F(u, v) = \|F(u, v)\| e^{-j\phi(u, v)}$
Spectrum	$\|F(u, v)\| = \left[R^2(u, v) + I^2(u, v) \right]^{1/2}, \quad R = \text{Real}(F)$ and $I = \text{Imag}(F)$
Phase angle	$\phi(u, v) = \tan^{-1}\left[\dfrac{I(u, v)}{R(u, v)} \right]$
Power spectrum	$P(u, v) = \|F(u, v)\|^2$
Average value	$\bar{f}(x, y) = F(0, 0) = \dfrac{1}{MN} \displaystyle\sum_{x=0}^{M-1} \sum_{y=0}^{N-1} f(x, y)$
Translation	$f(x, y) e^{j2\pi(u_0 x/M + v_0 y/N)} \Leftrightarrow F(u - u_0, v - v_0)$ $f(x - x_0, y - y_0) \Leftrightarrow F(u, v) e^{-j2\pi(ux_0/M + vy_0/N)}$ When $x_0 = u_0 = M/2$ and $y_0 = v_0 = N/2$, then $f(x, y)(-1)^{x+y} \Leftrightarrow F(u - M/2, v - N/2)$ $f(x - M/2, y - N/2) \Leftrightarrow F(u, v)(-1)^{u+v}$
Conjugate symmetry	$F(u, v) = F^*(-u, -v)$ $\|F(u, v)\| = \|F(-u, -v)\|$
Differentiation	$\dfrac{\partial^n f(x, y)}{\partial x^n} \Leftrightarrow (ju)^n F(u, v)$ $(-jx)^n f(x, y) \Leftrightarrow \dfrac{\partial^n F(u, v)}{\partial u^n}$
Laplacian	$\nabla^2 f(x, y) \Leftrightarrow -(u^2 + v^2) F(u, v)$
Distributivity	$\Im[f_1(x, y) + f_2(x, y)] = \Im[f_1(x, y)] + \Im[f_2(x, y)]$ $\Im[f_1(x, y) \cdot f_2(x, y)] \neq \Im[f_1(x, y)] \cdot \Im[f_2(x, y)]$
Scaling	$af(x, y) \Leftrightarrow aF(u, v), f(ax, by) \Leftrightarrow \dfrac{1}{\|ab\|} F(u/a, v/b)$
Rotation	$x = r\cos\theta \quad y = r\sin\theta \quad u = \omega\cos\varphi \quad v = \omega\sin\varphi$ $f(r, \theta + \theta_0) \Leftrightarrow F(\omega, \varphi + \theta_0)$
Periodicity	$F(u, v) = F(u + M, v) = F(u, v + N) = F(u + M, v + N)$ $f(x, y) = f(x + M, y) = f(x, y + N) = f(x + M, y + N)$
Separability	See Eqs. (4.6-14) and (4.6-15). Separability implies that we can compute the 2-D transform of an image by first computing 1-D transforms along each row of the image, and then computing a 1-D transform along each column of this intermediate result. The reverse, columns and then rows, yields the same result.

TABLE 4.1
(continued)
Summary of some
important
properties of the
2-D Fourier
transform.

Property	Expression(s)
Computation of the inverse Fourier transform using a forward transform algorithm	$$\frac{1}{MN} f^*(x, y) = \frac{1}{MN} \sum_{u=0}^{M-1} \sum_{v=0}^{N-1} F^*(u, v) e^{-j2\pi(ux/M+vy/N)}$$ This equation indicates that inputting the function $F^*(u, v)$ into an algorithm designed to compute the forward transform (right side of the preceding equation) yields $f^*(x, y)/MN$. Taking the complex conjugate and multiplying this result by MN gives the desired inverse.
Convolution[†]	$$f(x, y) * h(x, y) = \frac{1}{MN} \sum_{m=0}^{M-1} \sum_{n=0}^{N-1} f(m, n) h(x - m, y - n)$$
Correlation[†]	$$f(x, y) \circ h(x, y) = \frac{1}{MN} \sum_{m=0}^{M-1} \sum_{n=0}^{N-1} f^*(m, n) h(x + m, y + n)$$
Convolution theorem[†]	$f(x, y) * h(x, y) \Leftrightarrow F(u, v) H(u, v);$ $f(x, y) h(x, y) \Leftrightarrow F(u, v) * H(u, v)$
Correlation theorem[†]	$f(x, y) \circ h(x, y) \Leftrightarrow F^*(u, v) H(u, v);$ $f^*(x, y) h(x, y) \Leftrightarrow F(u, v) \circ H(u, v)$

Some useful FT pairs:

Impulse	$\delta(x, y) \Leftrightarrow 1$
Gaussian	$A2\pi\sigma^2 e^{\,2\pi^2\sigma^2(x^2+y^2)} \Leftrightarrow A e^{-(u^2+v^2)/2\sigma^2}$
Rectangle	$\text{rect}[a, b] \Leftrightarrow ab \dfrac{\sin(\pi ua)}{(\pi ua)} \dfrac{\sin(\pi vb)}{(\pi vb)} e^{-j\pi(ua+vb)}$
Cosine	$\cos(2\pi u_0 x + 2\pi v_0 y) \Leftrightarrow$ $\frac{1}{2}\left[\delta(u + Mu_0, v + Nv_0) + \delta(u - Mu_0, v \quad Nv_0)\right]$
Sine	$\sin(2\pi u_0 x + 2\pi v_0 y) \Leftrightarrow$ $j\frac{1}{2}\left[\delta(u + Mu_0, v + Nv_0) - \delta(u - Mu_0, v - Nv_0)\right]$

[†] Assumes that functions have been extended by zero padding.

Careful analysis of Eqs. (4.6-41) through (4.6-44) reveals some interesting properties of these expressions. An M-point transform can be computed by dividing the original expression into two parts, as indicated in Eqs. (4.6-43) and (4.6-44). Computing the first half of $F(u)$ requires evaluation of the two $(M/2)$-point transforms given in Eqs. (4.6-41) and (4.6-42). The resulting values of $F_{\text{even}}(u)$ and $F_{\text{odd}}(u)$ are then substituted into Eq. (4.6-43) to obtain $F(u)$ for $u = 0, 1, 2, \ldots, (M/2 - 1)$. The other half then follows directly from Eq. (4.6-44) *without* additional transform evaluations.

In order to examine the computational implications of this procedure, let $m(n)$ and $a(n)$ represent the number of complex multiplications and additions, respectively, required to implement it. As before, the number of samples is 2^n, where n is a positive integer. Suppose first that $n = 1$. A two-point transform requires the evaluation of $F(0)$; then $F(1)$ follows from Eq. (4.6-44). To obtain $F(0)$ first requires computing and $F_{even}(0)$ and $F_{odd}(0)$. In this case $K = 1$ and Eqs. (4.6-41) and (4.6-42) are one-point transforms. Because the Fourier transform of a single point is the sample itself, however, no multiplications or additions are required to obtain $F_{even}(0)$ and $F_{odd}(0)$. One multiplication of $F_{odd}(0)$ by W_2^0 and one addition yield $F(0)$ from Eq. (4.6-43). Then $F(1)$ follows from (4.6-44) with one more addition (subtraction is considered to be the same as addition). As $F_{odd}(0)W_2^0$ had already been computed, the total number of operations required for a two-point transform consists of $m(1) = 1$ multiplication and $a(1) = 2$ additions.

The next allowed value for n is 2. According to the preceding development, a four-point transform can be divided into two parts. The first half of $F(u)$ requires evaluation of two two-point transforms, as given in Eqs. (4.6-41) and (4.6-42) for $K = 2$. A two-point transform requires $m(1)$ multiplications and $a(1)$ additions, so evaluation of these two equations requires a total of $2m(1)$ multiplications and $2a(1)$ additions. Two further multiplications and additions are necessary to obtain $F(0)$ and $F(1)$ from Eq. (4.6-43). Because $F_{odd}(u)W_{2K}^u$ already had been computed for $u = \{0, 1\}$, two more additions give $F(2)$ and $F(3)$. The total is then $m(2) = 2m(1) + 2$ and $a(2) = 2a(1) + 4$.

When n is equal to 3, two four-point transforms are considered in the evaluation of $F_{even}(u)$ and $F_{odd}(u)$. They require $2m(2)$ multiplications and $2a(2)$ additions. Four more multiplications and eight more additions yield the complete transform. The total then is $m(3) = 2m(2) + 4$ and $a(3) = 2a(2) + 8$.

Continuing this argument for any positive integer value of n leads to recursive expressions for the number of multiplications and additions required to implement the FFT:

$$m(n) = 2m(n-1) + 2^{n-1} \quad n \geq 1 \tag{4.6-45}$$

and

$$a(n) = 2a(n-1) + 2^n \quad n \geq 1 \tag{4.6-46}$$

where $m(0) = 0$ and $a(0) = 0$ because the transform of a single point does not require any additions or multiplications.

Implementation of Eqs. (4.6-41) through (4.6-44) constitutes the successive doubling FFT algorithm. This name comes from the method of computing a two-point transform from two one-point transforms, a four-point transform from two two-point transforms, and so on, for any M equal to an integer power of 2. It is left as an exercise (Problem 4.25) to show that

$$m(n) = \frac{1}{2} M \log_2 M \tag{4.6-47}$$

and

$$a(n) = M \log_2 M. \tag{4.6-48}$$

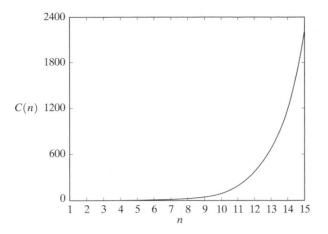

FIGURE 4.42
Computational advantage of the FFT over a direct implementation of the 1-D DFT. Note that the advantage increases rapidly as a function of n.

The computational advantage of the FFT over a direct implementation of the 1-D DFT is defined as

$$C(M) = \frac{M^2}{M \log_2 M}$$

$$= \frac{M}{\log_2 M}. \tag{4.6-49}$$

Because it is assumed that $M = 2^n$, we can express Eq. (4.6-49) in terms of n:

$$C(n) = \frac{2^n}{n}. \tag{4.6-50}$$

A plot of this function is shown in Fig. 4.42. As noted earlier, the advantage increases rapidly as a function of n. For instance, when $n = 15$ (32,768 points), the FFT has nearly a 2200 to 1 advantage over the DFT. Thus, we would expect that the FFT can be computed nearly 2200 times faster than the DFT on the same machine.

There are so many excellent sources that cover details of the FFT that we will not dwell on this topic further (see, for example, Brigham [1988]). Virtually all comprehensive signal and image processing software packages have generalized implementations of the FFT that also handle cases in which the number of points is not an integer power of 2 (at the expense of computational efficiency). Free FFT programs also are readily available, principally over the Internet.

4.6.7 Some Comments on Filter Design

All the filters discussed in this chapter are specified in equation form. In order to use the filters, we simply sample the equations for the desired values of (u, v). This process results in the filter function $H(u, v)$. In all our examples, this function was multiplied by the (centered) DFT of the input image, and the inverse DFT was computed. All forward and inverse Fourier transforms in this chapter were computed with an FFT algorithm, using the procedure summarized in Fig. 4.35 and Section 4.6.2.

The approach to filtering discussed in this chapter is focused strictly on fundamentals, the focus being specifically to explain the effects of filtering in the frequency domain as clearly as possible. We know of no better way to do that than to treat filtering the way we did here. One can view this development as the basis for "prototyping" a filter. In other words, given a problem for which we want to find a filter, the frequency domain approach is an ideal tool for experimenting, quickly and with full control over filter parameters.

Once a filter for a specific application has been found, it often is of interest to implement the filter directly in the spatial domain, using firmware and/or hardware. Petrou and Bosdogianni [1999] present a nice tie between two-dimensional frequency domain filters and the corresponding digital filters. On the design of 2-D digital filters see Lu and Antoniou [1992].

Summary

The material presented in this chapter, combined with the development in Chapter 3, is a comprehensive foundation for image enhancement. Undoubtedly, it is clear to the reader by now that the area of image enhancement really is a collection of tools that have been found in practice to produce acceptable results in given applications. Most of the tools themselves are well grounded in mathematical and statistical concepts, but their use is strictly problem oriented. In other words, image enhancement is more an art than a science, and the definition of a "properly enhanced" image is highly subjective. In Chapter 5 we extend some of the mathematical concepts developed in this and the last chapter to the area of image restoration. Unlike enhancement, restoration techniques for "improving" images tend to be based on objective, rather than subjective, criteria. As such, restoration techniques are considerably more structured than the methods we have covered thus far.

Another major objective of this chapter was the development of the Fourier transform. Although this was done in the context of image enhancement, the methods developed are perfectly general, as will be seen by the various applications of the DFT in subsequent chapters.

References

For additional reading on the material in Section 4.1 see Hubbard [1998]. The books by Bracewell [2000] and Bracewell [1995] are a good introduction to the continuous Fourier transform and its extensions to two dimensions for image processing. These two books, in addition to Lim [1990], Castleman [1996], Petrou and Bosdogianni [1999], and Brigham [1988], provide comprehensive background for most of the discussion in Section 4.2.

For additional reading on the material in Sections 4.3 and 4.4, see Castleman [1996], Pratt [1991], and Hall [1979]. Effective handling of issues on filter implementation (like ringing) still is a topic of interest, as exemplified by Bakir and Reeves [2000]. For unsharp masking and high-boost filtering, see Schowengerdt [1983]. The material on homomorphic filtering (Section 4.5) is based on a paper by Stockham [1972]; see also the books by Oppenheim and Schafer [1975] and Pitas and Venetsanopoulos [1990]. Brinkman et al. [1998] combine unsharp masking and homomorphic filtering for the enhancement of magnetic resonance images. For the generation of digital filters (Section 4.6.7) based on the frequency domain formulations discussed in this chapter, see Lu and Antoniou [1992] and Petrou and Bosdogianni [1999].

As noted in Section 4.1, "discovery" of the fast Fourier transform (Section 4.6.6) was a major milestone in the popularization of the DFT as a fundamental signal processing tool. Our presentation of the FFT in Section 4.6.6 is based on a paper by Cooley and Tuckey [1965] and on the book by Brigham [1988], who also discusses several implementations of the FFT, including bases other than 2. Formulation of the fast Fourier transform is often credited to Cooley and Tukey [1965]. The FFT, however, has an interesting history worth sketching here. In response to the Cooley-Tukey paper, Rudnick [1966] reported that he was using a similar technique, whose number of operations also was proportional to $M \log_2 M$ and which was based on a method published by Danielson and Lanczos [1942]. These authors, in turn, referenced Runge [1903, 1905] as the source of their technique. The latter two papers, together with the lecture notes of Runge and König [1924], contain the essential computational advantages of present FFT algorithms. Similar techniques also were published by Yates [1937], Stumpff [1939], Good [1958], and Thomas [1963]. A paper by Cooley, Lewis, and Welch [1967a] presents a historical summary and an interesting comparison of results prior to the 1965 Cooley-Tukey paper.

Problems

★**4.1**   Show that $F(u)$ and $f(x)$ in Eqs. (4.2-5) and (4.2-6) are a Fourier transform pair. You can do this by substituting Eq. (4.2-6) for $f(x)$ into Eq. (4.2-5) and showing that the left and right sides are equal. Repeat the process by substituting Eq. (4.2-5) for $F(u)$ into Eq. (4.2-6). You will need the following orthogonality property of exponentials:

$$\sum_{x=0}^{M-1} e^{j2\pi rx/M} e^{-j2\pi ux/M} = \begin{cases} M & \text{if } r = u \\ 0 & \text{otherwise.} \end{cases}$$

See inside front cover

Detailed solutions to the problems marked with a star can be found in the book web site. The site also contains suggested projects based on the material in this chapter.

4.2   Show that the Fourier transform and its inverse are linear processes (see Section 2.6 regarding linearity).

4.3   Let $F(u, v)$ denote the DFT of an image $f(x, y)$. We know from the discussion in Section 4.2.3 that multiplying $F(u, v)$ by a filter function $H(u, v)$ and taking the inverse Fourier transform will alter the appearance of the image, depending on the nature of the filter. Suppose that $H(u, v) = A$, a positive constant. The net effect of filtering will be to multiply the image by the same constant. Using the convolution theorem, explain mathematically why the pixels in the spatial domain representation are multiplied by the same constant.

★**4.4**   A Gaussian lowpass filter in the frequency domain has the transfer function

$$H(u, v) = Ae^{-(u^2+v^2)/2\sigma^2}.$$

Show that the corresponding filter in the *spatial* domain has the form

$$h(x, y) = A2\pi\sigma^2 e^{-2\pi^2\sigma^2(x^2+y^2)}.$$

(*Hint:* Treat the variables as continuous to simplify manipulation.)

4.5   As shown in Eq. (4.4-1), a highpass filter has the transfer function

$$H_{hp}(u, v) = 1 - H_{lp}(u, v)$$

where $H_{lp}(u, v)$ is the transfer function of the corresponding lowpass filter. Using the result in Problem 4.4, what is the form of the *spatial domain* Gaussian highpass filter?

4.6   ★**(a)** Prove the validity of Eq. (4.2-21).

(b) Prove the validity of Eqs. (4.6-1) and (4.6-2).

4.7   What is the source of nearly periodic bright points in the horizontal axis of the spectrum in Fig. 4.11(b)?

★**4.8**   Each of the spatial filters shown in Fig. 4.23 has a strong spike at the origin. Explain the source of these spikes.

4.9   Consider the images shown. The image on the right was obtained by (a) multiplying the image on the left by $(-1)^{x+y}$; (b) computing the DFT; (c) taking the complex conjugate of the transform; (d) computing the inverse DFT; and (e) multiplying the real part of the result by $(-1)^{x+y}$. Explain (mathematically) why the image on the right appears as it does.

4.10   Show that if a filter transfer function $H(u, v)$ is real and symmetric, then the corresponding spatial domain filter $h(x, y)$ also is real and symmetric.

★**4.11**   Prove the validity of the convolution theorem. For simplicity, limit your development to continuos functions of one variable.

4.12   Consider the images shown. The image on the right was obtained by lowpass filtering the image on the left with a Gaussian lowpass filter, and then highpass filtering the result with a Gaussian highpass filter. The dimension of the images is 420×344, and $D_0 = 25$ was used for both filters.

(a)   Consider the figure on the right. Explain why the center part of the ring appears so bright and solid, when the dominant characteristic of the filtered image consists of edges on the outer boundary of objects (e.g., fingers, wrist bones) with a darker area in between. In other words, would you not expect the highpass filter to render the constant area inside the ring dark, since a highpass filter eliminates the dc term?

(b)   Do you think the result would have been different if the order of the filtering process had been reversed?

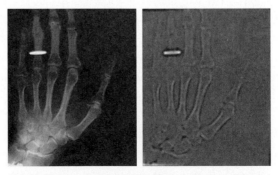

Original image courtesy of Dr. Thomas R. Gest, Division of Anatomical Sciences, University of Michigan Medical School.

4.13 Given an image of size $M \times N$, you are asked to perform an experiment that consists of repeatedly lowpass filtering the image using a Gaussian lowpass filter with a given cutoff D_0. You may ignore computational round-off errors. Let k_{min} denote the smallest positive number representable in the machine in which the proposed experiment will be conducted.

★ **(a)** Let K denote the number of applications of the filter. Can you predict, without doing the experiment what the result (image) will be for a sufficiently large value of K? If so, what is that result?

(b) Derive an expression for the *minimum* value of K that will guarantee the result that you predicted.

4.14 Suppose that you form a lowpass *spatial* filter that averages the four immediate neighbors of a point (x, y) but excludes the point itself.

(a) Find the equivalent filter $H(u, v)$ in the frequency domain.

(b) Show that your result is a lowpass filter.

★ **4.15** The basic approach used to approximate a discrete derivative (Section 3.7) involves taking differences of the form $f(x + 1, y) - f(x, y)$.

(a) Obtain the filter transfer function, $H(u, v)$, for performing the equivalent process in the frequency domain.

(b) Show that $H(u, v)$ is a highpass filter.

4.16 Consider the sequence of images shown. The image on the left is a segment of an X-ray image of a commercial printed circuit board. The images following it are, respectively, the results of subjecting the image to 1, 10, and 100 passes of a Gaussian highpass filter with $D_0 = 30$. The images are of size 330×334 pixels, with each pixel being represented by 8 bits of gray. They have been scaled for display, but this has no effect on the problem statement.

(a) There is a hint in these results that indicates that the image will cease to change after some finite number of passes. Show whether or not this in fact is the case. You may ignore computational round-off errors. Let k_{min} denote the smallest positive number representable in the machine in which the proposed experiment will be conducted.

(b) If you determined in (a) that changes would cease after a finite number of iterations, determine the minimum value of that number.

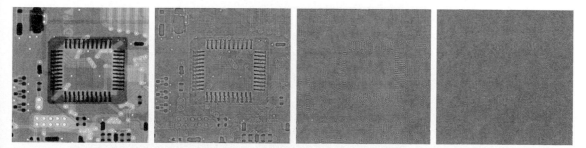

Original image courtesy of Mr. Joseph E. Pascente, Lixi, Inc.

4.17 As shown in Fig. 4.30, combining high-frequency emphasis and histogram equalization is an effective method for achieving edge sharpening and contrast enhancement.

(a) Show whether or not it matters which process is applied first.

(b) If the order does matter, give a rationale for using one or the other method first.

★ **4.18** Can you think of a way to use the Fourier transform to compute (or partially compute) the magnitude of the gradient for use in image differentiation (see Section 3.7.3)? If your answer is yes, give a method to do it. If your answers is no, explain why.

4.19 In Section 4.4.4 we started with the definition of the Laplacian and derived a filter for performing the equivalent operation via the frequency domain. Start with the mask approximation shown in Fig. 4.27(f) and find the frequency domain filter that would implement that approximation.

4.20 Use the transfer function of a Butterworth highpass filter of order n to construct a homomorphic filter. Your filter must exhibit the characteristic shape shown in Fig. 4.32 and must include the parameters shown in that figure.

★ **4.21** The need for image padding when filtering in the frequency domain was discussed in some detail in Section 4.6.3. We showed in that section that images needed to be padded by appending zeros to the ends of rows and columns in the image (see the following image on the left). Do you think it would make a difference if we centered the image and surrounded it by a border of zeros instead (see image on the right), but without changing the total number of zeros used? Explain.

Original image courtesy of NASA.

4.22 The two Fourier spectra shown are of the same image. The spectrum on the left corresponds to the original image, and the spectrum on the right was obtained after the image was padded with zeros.

(a) Explain the difference in overall contrast.

(b) Explain the significant increase in signal strength along the vertical and horizontal axes of the spectrum shown on the right.

4.23 The correlation function shown in Fig. 4.41(e) was computed via the frequency domain, according to Eq. (4.6-31). The complex conjugate of the transform of the template in Fig. 4.41(d) was used in the computation. Sketch what Fig. 4.41(e) would look like if, instead, we had used the conjugate of the transform of the image.

4.24 With reference to Fig. 4.41, sketch what the correlation function [Fig. 4.41(e)] would look like if the letters UTK were only one pixel away from

★ **(a)** The left border of the image

★ **(b)** The right border of the image

(c) The top of the image

(d) The bottom of the image

In (a) and (b), assume that the letters are centered with respect to an imaginary horizontal line passing through the center of all the letters and also through the center of the image. In (c) and (d) they are centered with respect to a vertical line passing through the center of the letter T and also through the center of the image.

★ **4.25** Show the validity of Eqs. (4.6-47) and (4.6-48). (*Hint:* Use proof by induction.)

4.26 Suppose that you are given a set of images generated by an experiment dealing with the analysis of stellar events. Each image contains a set of bright, widely scattered dots corresponding to stars in a sparsely occupied section of the universe. The problem is that the stars are barely visible, due to superimposed illumination resulting from atmospheric dispersion. If these images are modeled as the product of a constant illumination component with a set of impulses, give an enhancement procedure based on homomorphic filtering designed to bring out the image components due to the stars themselves.

4.27 A skilled medical technician is charged with the job of inspecting a certain class of images generated by an electron microscope. In order to simplify the inspection task, the technician decides to use digital image enhancement and, to this end, examines a set of representative images and finds the following problems: (1) bright, isolated dots that are of no interest; (2) lack of sharpness; (3) not enough contrast in some images; and (4) shifts in the average gray-level value, when this value should be V to perform correctly certain intensity measurements. The technician wants to correct these problems and then display in white all gray levels in the band between I_1 and I_2, while keeping normal tonality in the remaining gray levels. Propose a sequence of processing steps that the technician can follow to achieve the desired goal. You may use techniques from both Chapters 3 and 4.

5 *Image Restoration*

Things which we see are not by themselves what we see. . . .
It remains completely unknown to us what the objects may be by
themselves and apart from the receptivity of our senses. We know
nothing but our manner of perceiving them.

Immanuel Kant

Preview

As in image enhancement, the ultimate goal of restoration techniques is to improve an image in some predefined sense. Although there are areas of overlap, image enhancement is largely a subjective process, while image restoration is for the most part an objective process. Restoration attempts to reconstruct or recover an image that has been degraded by using a priori knowledge of the degradation phenomenon. Thus restoration techniques are oriented toward modeling the degradation and applying the inverse process in order to recover the original image.

This approach usually involves formulating a criterion of goodness that will yield an optimal estimate of the desired result. By contrast, enhancement techniques basically are heuristic procedures designed to manipulate an image in order to take advantage of the psychophysical aspects of the human visual system. For example, contrast stretching is considered an enhancement technique because it is based primarily on the pleasing aspects it might present to the viewer, whereas removal of image blur by applying a deblurring function is considered a restoration technique.

The material developed in this chapter is strictly introductory. We consider the restoration problem only from the point where a degraded, *digital* image is given; thus we consider topics dealing with sensor, digitizer, and display degradations only superficially. These subjects, although of importance in the overall treatment of image restoration applications, are beyond the scope of the present discussion.

As in Chapters 3 and 4, some restoration techniques are best formulated in the spatial domain, while others are better suited for the frequency domain. For example, spatial processing is applicable when the only degradation is additive noise. On the other hand, degradations such as image blur are difficult to approach in the spatial domain using small masks. In this case, frequency domain filters based on various criteria of optimality are the approaches of choice. These filters also take into account the presence of noise. As in Chapter 4 (see comments in Section 4.6.7), a restoration filter that solves a given application in the frequency domain often is used as the basis for generating a digital filter that will be more suitable for routine operation using a hardware/firmware implementation.

5.1 A Model of the Image Degradation/Restoration Process

As Fig. 5.1 shows, the degradation process is modeled in this chapter as a degradation function that, together with an additive noise term, operates on an input image $f(x, y)$ to produce a degraded image $g(x, y)$. Given $g(x, y)$, some knowledge about the degradation function H, and some knowledge about the additive noise term $\eta(x, y)$, the objective of restoration is to obtain an estimate $\hat{f}(x, y)$ of the original image. We want the estimate to be as close as possible to the original input image and, in general, the more we know about H and η, the closer $\hat{f}(x, y)$ will be to $f(x, y)$. The approach used throughout most of this chapter is based on various types of image restoration filters.

It is shown in Section 5.5 that if H is a linear, position-invariant process, then the degraded image is given in the *spatial domain* by

$$g(x, y) = h(x, y) * f(x, y) + \eta(x, y) \tag{5.1-1}$$

where $h(x, y)$ is the spatial representation of the degradation function and, as in Chapter 4, the symbol "$*$" indicates convolution. We know from the discussion in Sections 4.2.4 and 4.6.4 that convolution in the spatial domain is equal to multiplication in the frequency domain, so we may write the model in Eq. (5.1-1) in an equivalent *frequency domain* representation:

$$G(u, v) = H(u, v)F(u, v) + N(u, v) \tag{5.1-2}$$

where the terms in capital letters are the Fourier transforms of the corresponding terms in Eq. (5.1-1). These two equations are the bases for most of the material in this chapter.

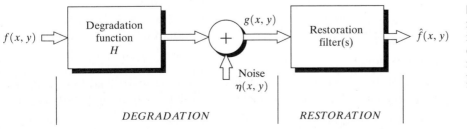

FIGURE 5.1 A model of the image degradation/restoration process.

In the following three sections, we assume that H is the identity operator, and we deal only with degradations due to noise. Beginning in Section 5.6 we consider a number of important image degradations functions and look at several methods for image restoration in the presence of both H and η.

5.2   Noise Models

The principal sources of noise in digital images arise during image acquisition (digitization) and/or transmission. The performance of imaging sensors is affected by a variety of factors, such as environmental conditions during image acquisition, and by the quality of the sensing elements themselves. For instance, in acquiring images with a CCD camera, light levels and sensor temperature are major factors affecting the amount of noise in the resulting image. Images are corrupted during transmission principally due to interference in the channel used for transmission. For example, an image transmitted using a wireless network might be corrupted as a result of lightning or other atmospheric disturbance.

5.2.1   Spatial and Frequency Properties of Noise

Relevant to our discussion are parameters that define the spatial characteristics of noise, and whether the noise is correlated with the image. Frequency properties refer to the frequency content of noise in the Fourier sense (i.e., as opposed to the electromagnetic spectrum). For example, when the Fourier spectrum of noise is constant, the noise usually is called *white noise*. This terminology is a carry over from the physical properties of white light, which contains nearly all frequencies in the visible spectrum in equal proportions. From the discussion in Chapter 4, it is not difficult to show that the Fourier spectrum of a function containing all frequencies in equal proportions is a constant.

With the exception of spatially periodic noise (Section 5.2.3), we assume in this chapter that noise is independent of spatial coordinates, and that it is uncorrelated with respect to the image itself (that is, there is no correlation between pixel values and the values of noise components). Although these assumptions are at least partially invalid in some applications (quantum-limited imaging, such as in X-ray and nuclear-medicine imaging, is a good example), the complexities of dealing with spatially dependent and correlated noise are beyond the scope of our discussion.

5.2.2   Some Important Noise Probability Density Functions

Based on the assumptions in the previous section, the *spatial* noise descriptor with which we shall be concerned is the statistical behavior of the gray-level values in the noise component of the model in Fig. 5.1. These may be considered random variables, characterized by a probability density function (PDF). The following are among the most common PDFs found in image processing applications.

See inside front cover

Consult the book web site for a brief review of probability theory.

Gaussian noise

Because of its mathematical tractability in both the spatial and frequency domains, Gaussian (also called *normal*) noise models are used frequently in practice. In fact, this tractability is so convenient that it often results in Gaussian models being used in situations in which they are marginally applicable at best.

The PDF of a Gaussian random variable, z, is given by

$$p(z) = \frac{1}{\sqrt{2\pi}\sigma} e^{-(z-\mu)^2/2\sigma^2} \tag{5.2-1}$$

where z represents gray level, μ is the mean of average value of z, and σ is its standard deviation. The standard deviation squared, σ^2, is called the *variance* of z. A plot of this function is shown in Fig. 5.2(a). When z is described by Eq. (5.2-1), approximately 70% of its values will be in the range $[(\mu - \sigma), (\mu + \sigma)]$, and about 95% will be in the range $[(\mu - 2\sigma), (\mu + 2\sigma)]$.

Rayleigh noise

The PDF of Rayleigh noise is given by

$$p(z) = \begin{cases} \dfrac{2}{b}(z - a)e^{-(z-a)^2/b} & \text{for } z \geq a \\ 0 & \text{for } z < a. \end{cases} \tag{5.2-2}$$

The mean and variance of this density are given by

$$\mu = a + \sqrt{\pi b/4} \tag{5.2-3}$$

and

$$\sigma^2 = \frac{b(4 - \pi)}{4}. \tag{5.2-4}$$

Figure 5.2(b) shows a plot of the Rayleigh density. Note the displacement from the origin and the fact that the basic shape of this density is skewed to the right. The Rayleigh density can be quite useful for approximating skewed histograms.

Erlang (Gamma) noise

The PDF of Erlang noise is given by

$$p(z) = \begin{cases} \dfrac{a^b z^{b-1}}{(b-1)!} e^{-az} & \text{for } z \geq 0 \\ 0 & \text{for } z < 0 \end{cases} \tag{5.2-5}$$

where the parameters are such that $a > 0$, b is a positive integer, and "!" indicates factorial. The mean and variance of this density are given by

$$\mu = \frac{b}{a} \tag{5.2-6}$$

and

$$\sigma^2 = \frac{b}{a^2}. \tag{5.2-7}$$

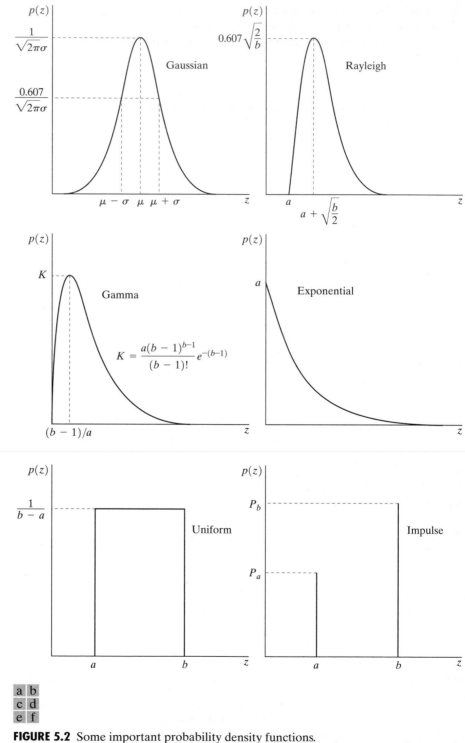

FIGURE 5.2 Some important probability density functions.

Figure 5.2(c) shows a plot of this density. Although Eq. (5.2-5) often is referred to as the *gamma density*, strictly speaking this is correct only when the denominator is the gamma function, $\Gamma(b)$. When the denominator is as shown, the density is more appropriately called the *Erlang density*.

Exponential noise

The PDF of *exponential* noise is given by

$$p(z) = \begin{cases} ae^{-az} & \text{for } z \geq 0 \\ 0 & \text{for } z < 0 \end{cases} \qquad (5.2\text{-}8)$$

where $a > 0$. The mean and variance of this density function are

$$\mu = \frac{1}{a} \qquad (5.2\text{-}9)$$

and

$$\sigma^2 = \frac{1}{a^2}. \qquad (5.2\text{-}10)$$

Note that this PDF is a special case of the Erlang PDF, with $b = 1$. Figure 5.2(d) shows a plot of this density function.

Uniform noise

The PDF of *uniform* noise is given by

$$p(z) = \begin{cases} \dfrac{1}{b-a} & \text{if } a \leq z \leq b \\ 0 & \text{otherwise.} \end{cases} \qquad (5.2\text{-}11)$$

The mean of this density function is given by

$$\mu = \frac{a+b}{2} \qquad (5.2\text{-}12)$$

and its variance by

$$\sigma^2 = \frac{(b-a)^2}{12}. \qquad (5.2\text{-}13)$$

Figure 5.2(e) shows a plot of the uniform density.

Impulse (salt-and-pepper) noise

The PDF of (*bipolar*) *impulse* noise is given by

$$p(z) = \begin{cases} P_a & \text{for } z = a \\ P_b & \text{for } z = b \\ 0 & \text{otherwise} \end{cases} \qquad (5.2\text{-}14)$$

If $b > a$, gray-level b will appear as a light dot in the image. Conversely, level a will appear like a dark dot. If either P_a or P_b is zero, the impulse noise is called *unipolar.* If neither probability is zero, and especially if they are approximately equal, impulse noise values will resemble salt-and-pepper granules randomly distributed over the image. For this reason, bipolar impulse noise also is called *salt-and-pepper* noise. *Shot* and *spike* noise also are terms used to refer to this type of noise. In our discussion we will use the terms *impulse* or *salt-and-pepper* noise interchangeably.

Noise impulses can be negative or positive. Scaling usually is part of the image digitizing process. Because impulse corruption usually is large compared with the strength of the image signal, impulse noise generally is digitized as extreme (pure black or white) values in an image. Thus, the assumption usually is that a and b are "saturated" values, in the sense that they are equal to the minimum and maximum allowed values in the digitized image. As a result, negative impulses appear as black (pepper) points in an image. For the same reason, positive impulses appear white (salt) noise. For an 8-bit image this means that $a = 0$ (black) and $b = 255$ (white). Figure 5.2(f) shows the PDF of impulse noise.

As a group, the preceding PDFs provide useful tools for modeling a broad range of noise corruption situations found in practice. For example, Gaussian noise arises in an image due to factors such as electronic circuit noise and sensor noise due to poor illumination and/or high temperature. The Rayleigh density is helpful in characterizing noise phenomena in range imaging. The exponential and gamma densities find application in laser imaging. Impulse noise is found in situations where quick transients, such as faulty switching, take place during imaging, as mentioned in the previous paragraph. The uniform density is perhaps the least descriptive of practical situations. However, the uniform density is quite useful as the basis for numerous random number generators that are used in simulations (Peebles [1993]).

EXAMPLE 5.1:
Sample noisy images and their histograms.

■ Figure 5.3 shows a test pattern well-suited for illustrating the noise models just discussed. This is a suitable pattern to use because it is composed of simple, constant areas that span the gray scale from black to near white in only three increments. This facilitates visual analysis of the characteristics of the various noise components added to the image.

Figure 5.4 shows the test pattern after addition of the six types of noise discussed thus far in this section. Shown below each image is the histogram computed directly from that image. The parameters of the noise were chosen in each case so that the histogram corresponding to the three gray levels in the test pattern would start to merge. This made the noise quite visible, without obscuring the basic structure of the underlying image.

We see a close correspondence in comparing the histograms in Fig. 5.4 with the PDFs in Fig. 5.2. The histogram for the salt-and-pepper example has an extra peak at the white end of the spectrum because the noise components were pure black and white, and the lightest component of the test pattern (the circle) is light gray. With the exception of slightly different overall intensity, it is difficult to differentiate visually between the first five images in Fig. 5.4, even though their histograms are significantly different. The salt-and-pepper appearance of

FIGURE 5.3 Test pattern used to illustrate the characteristics of the noise PDFs shown in Fig. 5.2.

the image corrupted by impulse noise is the only one that is visually indicative of the type of noise causing the degradation.   ■

5.2.3 Periodic Noise

Periodic noise in an image arises typically from electrical or electromechanical interference during image acquisition. This is the only type of spatially dependent noise that will be considered in this chapter. As discussed in Section 5.4, periodic noise can be reduced significantly via frequency domain filtering. For example, consider the image shown in Fig. 5.5(a). This image is severely corrupted by (spatial) sinusoidal noise of various frequencies. The Fourier transform of a pure sinusoid is a pair of conjugate impulses located at the conjugate frequencies of the sine wave (Table 4.1). Thus, if the amplitude of a sine wave in the spatial domain is strong enough, we would expect to see in the spectrum of the image a pair of impulses for each sine wave in the image. As shown in Fig. 5.5(b), this is indeed the case, with the impulses appearing in an approximate circle because the frequency values in this particular case are so arranged. We will have much more to say in Section 5.4 about this and other examples of periodic noise.

5.2.4 Estimation of Noise Parameters

The parameters of periodic noise typically are estimated by inspection of the Fourier spectrum of the image. As noted in the previous section, periodic noise tends to produce frequency spikes that often can be detected even by visual analysis. Another approach is to attempt to infer the periodicity of noise components directly from the image, but this is possible only in simplistic cases. Automated analysis is possible in situations in which the noise spikes are either exceptionally pronounced, or when some knowledge is available about the general location of the frequency components of the interference.

The parameters of noise PDFs may be known partially from sensor specifications, but it is often necessary to estimate them for a particular imaging

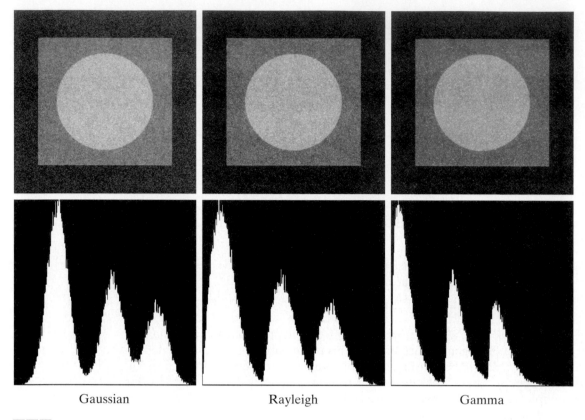

Gaussian	Rayleigh	Gamma

a b c
d e f

FIGURE 5.4 Images and histograms resulting from adding Gaussian, Rayleigh, and gamma noise to the image in Fig. 5.3.

arrangement. If the imaging system is available, one simple way to study the characteristics of system noise is to capture a set of images of "flat" environments. For example, in the case of an optical sensor, this is as simple as imaging a solid gray board that is illuminated uniformly. The resulting images typically are good indicators of system noise.

When only images already generated by the sensor are available, frequently it is possible to estimate the parameters of the PDF from small patches of reasonably constant gray level. For example, the vertical strips (of 150×20 pixels) shown in Fig. 5.6 were cropped from the Gaussian, Rayleigh, and uniform images in Fig. 5.4. The histograms shown were calculated using image data from these small strips. The histograms in Fig. 5.4 that correspond to the histograms in Fig. 5.6 are the ones in the middle of the group of three in Figs. 5.4(d), (e), and (k). We see that the shapes of these histograms correspond quite closely to the shapes of the histograms in Fig. 5.6. Their heights are different due to scaling, but the shapes are unmistakably similar.

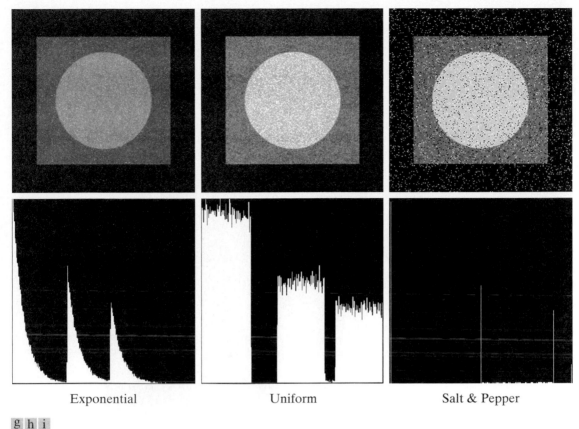

Exponential   Uniform   Salt & Pepper

g h i
j k l

FIGURE 5.4 *(Continued)*   Images and histograms resulting from adding exponential, uniform, and impulse noise to the image in Fig. 5.3.

The simplest way to use the data from the image strips is for calculating the mean and variance of the gray levels. Consider a strip (subimage) denoted by S. We can use the following sample approximations from basic statistics:

$$\mu = \sum_{z_i \in S} z_i p(z_i) \qquad (5.2\text{-}15)$$

and

$$\sigma^2 = \sum_{z_i \in S} (z_i - \mu)^2 p(z_i) \qquad (5.2\text{-}16)$$

where the z_i's are the gray-level values of the pixels in S, and $p(z_i)$ are the corresponding normalized histogram values.

The shape of the histogram identifies the closest PDF match. If the shape is approximately Gaussian, then the mean and variance is all we need because the Gaussian PDF is completely specified by these two parameters. For the

a
b

FIGURE 5.5
(a) Image corrupted by sinusoidal noise.
(b) Spectrum (each pair of conjugate impulses corresponds to one sine wave). (Original image courtesy of NASA.)

other shapes discussed in Section 5.2.2, we use the mean and variance to solve for the parameters a and b. Impulse noise is handled differently because the estimate needed is of the actual probability of occurrence of white and black pixels. Obtaining this estimate requires that both black and white pixels be visible, so a midgray, relatively constant area is needed in the image in order to be able to compute a histogram. The heights of the peaks corresponding to black and white pixels are the estimates of P_a and P_b in Eq. (5.2-14).

5.3 Restoration in the Presence of Noise Only–Spatial Filtering

When the only degradation present in an image is noise, Eqs. (5.1-1) and (5.1-2) become

$$g(x, y) = f(x, y) + \eta(x, y) \tag{5.3-1}$$

and

$$G(u, v) = F(u, v) + N(u, v). \tag{5.3-2}$$

The noise terms are unknown, so subtracting them from $g(x, y)$ or $G(u, v)$ is not a realistic option. In the case of periodic noise, it usually is possible to

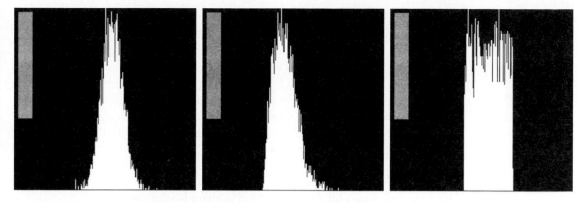

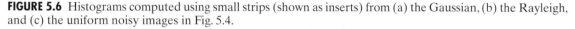

FIGURE 5.6 Histograms computed using small strips (shown as inserts) from (a) the Gaussian, (b) the Rayleigh, and (c) the uniform noisy images in Fig. 5.4.

estimate $N(u, v)$ from the spectrum of $G(u, v)$, as noted in Section 5.2.3. In this case $N(u, v)$ can be subtracted from $G(u, v)$ to obtain an estimate of the original image. In general, however, this type of knowledge is the exception, rather than the rule.

Spatial filtering is the method of choice in situations when only additive noise is present. This topic was discussed in detail in Sections 3.5 and 3.6 in connection with image enhancement. In fact, enhancement and restoration become almost indistinguishable disciplines in this particular case. With the exception of the nature of the computation performed by a specific filter, the mechanics for implementing all the filters that follow are exactly as discussed in Section 3.5.

5.3.1 Mean Filters

In this section we discuss briefly the noise-reduction spatial filters introduced in Section 3.6 and develop several other filters whose performance is in many cases superior to the filters discussed in that section.

Arithmetic mean filter

This is the simplest of the mean filters. Let S_{xy} represent the set of coordinates in a rectangular subimage window of size $m \times n$, centered at point (x, y). The arithmetic mean filtering process computes the average value of the corrupted image $g(x, y)$ in the area defined by S_{xy}. The value of the restored image $\hat{f}$ at any point (x, y) is simply the arithmetic mean computed using the pixels in the region defined by S_{xy}. In other words,

$$\hat{f}(x, y) = \frac{1}{mn} \sum_{(s,t) \in S_{xy}} g(s, t). \tag{5.3-3}$$

This operation can be implemented using a convolution mask in which all coefficients have value $1/mn$. As discussed in Section 3.6.1, a mean filter simply smoothes local variations in an image. Noise is reduced as a result of blurring.

Geometric mean filter

An image restored using a *geometric mean* filter is given by the expression

$$\hat{f}(x, y) = \left[\prod_{(s,t) \in S_{xy}} g(s, t) \right]^{\frac{1}{mn}}. \tag{5.3-4}$$

Here, each restored pixel is given by the product of the pixels in the subimage window, raised to the power $1/mn$. As shown in Example 5.2, a geometric mean filter achieves smoothing comparable to the arithmetic mean filter, but it tends to lose less image detail in the process.

Harmonic mean filter

The *harmonic mean* filtering operation is given by the expression

$$\hat{f}(x, y) = \frac{mn}{\displaystyle\sum_{(s,t) \in S_{xy}} \frac{1}{g(s, t)}}. \tag{5.3-5}$$

The harmonic mean filter works well for salt noise, but fails for pepper noise. It does well also with other types of noise like Gaussian noise.

Contraharmonic mean filter

The *contraharmonic* mean filtering operation yields a restored image based on the expression

$$\hat{f}(x, y) = \frac{\displaystyle\sum_{(s,t) \in S_{xy}} g(s, t)^{Q+1}}{\displaystyle\sum_{(s,t) \in S_{xy}} g(s, t)^{Q}} \tag{5.3-6}$$

where Q is called the *order* of the filter. This filter is well suited for reducing or virtually eliminating the effects of salt-and-pepper noise. For positive values of Q, the filter eliminates pepper noise. For negative values of Q it eliminates salt noise. It cannot do both simultaneously. Note that the contraharmonic filter reduces to the arithmetic mean filter if $Q = 0$, and to the harmonic mean filter if $Q = -1$.

EXAMPLE 5.2:
Illustration of
mean filters.

■ Figure 5.7(a) shows an X-ray image of a circuit board, and Fig. 5.7(b) shows the same image, but corrupted with additive Gaussian noise of zero mean and variance of 400. For this type of image this is a significant level of noise. Figures 5.7(c) and (d) show, respectively, the result of filtering the noisy image with an arithmetic mean filter of size 3 × 3 and a geometric mean filter of the same size. Although both filters did a reasonable job of attenuating the contribution due to noise, the geometric mean filter did not blur the image as much as the arithmetic filter. For instance, the connector fingers at the top of the image are much sharper in Fig. 5.7(d) than in (c). The same is true in other parts of the image (see Problem 5.1).

Figure 5.8(a) shows the same circuit image, but corrupted now by pepper noise with probability of 0.1. Similarly, Fig. 5.8(b) shows the image corrupted by salt noise with the same probability. Figure 5.8(c) shows the result of filtering Fig. 5.8(a) using a contraharmonic mean filter with $Q = 1.5$, and Fig. 5.8(d)

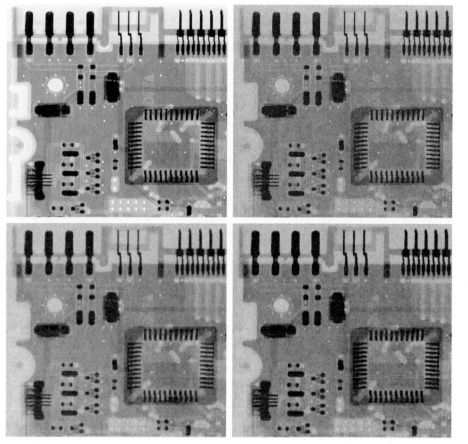

FIGURE 5.7 (a) X-ray image. (b) Image corrupted by additive Gaussian noise. (c) Result of filtering with an arithmetic mean filter of size 3 × 3. (d) Result of filtering with a geometric mean filter of the same size. (Original image courtesy of Mr. Joseph E. Pascente, Lixi, Inc.)

shows the result of filtering Fig. 5.8(b) with $Q = -1.5$. Both filters did a good job in reducing the effect of the noise. The positive-order filter did a better job of cleaning the background, at the expense of blurring the dark areas. The opposite was true of the negative-order filter.

In general, the arithmetic and geometric mean filters (particularly the latter) are well suited for random noise like Gaussian or uniform noise. The contraharmonic filter is well suited for impulse noise, but it has the disadvantage that it must be known whether the noise is dark or light in order to select the proper sign for Q. The results of choosing the wrong sign for Q can be disastrous, as Fig. 5.9 shows. Some of the filters discussed in the following section eliminate this shortcoming. ■

5.3.2 Order-Statistics Filters

Order-statistics filters were introduced in Section 3.6.2. We now expand the discussion in that section and introduce some additional order-statistics filters. As noted in Section 3.6.2, order-statistics filters are spatial filters whose response is based on ordering (ranking) the pixels contained in the image area encompassed by the filter. The response of the filter at any point is determined by the ranking result.

a b
c d

FIGURE 5.8
(a) Image
corrupted by
pepper noise with
a probability of
0.1. (b) Image
corrupted by salt
noise with the
same probability.
(c) Result of
filtering (a) with a
3×3
contraharmonic
filter of order 1.5.
(d) Result of
filtering (b) with
$Q = -1.5$.

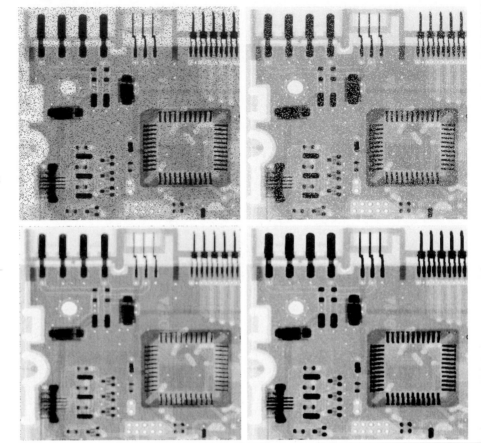

Median filter

The best-known order-statistics filter is the *median filter*, which, as its name implies, replaces the value of a pixel by the median of the gray levels in the neighborhood of that pixel:

$$\hat{f}(x, y) = \underset{(s,t)\in S_{xy}}{\text{median}}\{g(s, t)\}. \tag{5.3-7}$$

The original value of the pixel is included in the computation of the median. Median filters are quite popular because, for certain types of random noise, they provide excellent noise-reduction capabilities, with considerably less blurring than linear smoothing filters of similar size. Median filters are particularly effective in the presence of both bipolar and unipolar impulse noise. In fact, as Example 5.3 shows, the median filter yields excellent results for images corrupted by this type of noise. Computation of the median and implementation of this filter are discussed in detail in Section 3.6.2.

Max and min filters

Although the median filter is by far the order-statistics filter most used in image processing, it is by no means the only one. The median represents the 50th per-

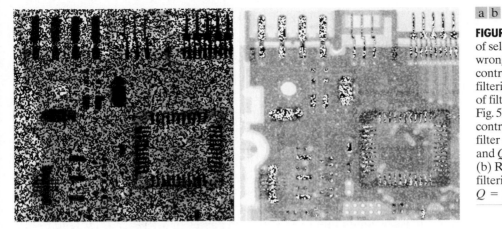

FIGURE 5.9 Results of selecting the wrong sign in contraharmonic filtering. (a) Result of filtering Fig. 5.8(a) with a contraharmonic filter of size 3×3 and $Q = -1.5$. (b) Result of filtering 5.8(b) with $Q = 1.5$.

centile of a ranked set of numbers, but the reader will recall from basic statistics that ranking lends itself to many other possibilities. For example, using the 100th percentile results in the so-called *max filter*, given by

$$\hat{f}(x, y) = \max_{(s,t) \in S_{xy}} \{g(s, t)\}. \tag{5.3-8}$$

This filter is useful for finding the brightest points in an image. Also, because pepper noise has very low values, it is reduced by this filter as a result of the max selection process in the subimage area S_{xy}.

The 0th percentile filter is the *min filter*:

$$\hat{f}(x, y) = \min_{(s,t) \in S_{xy}} \{g(s, t)\}. \tag{5.3-9}$$

This filter is useful for finding the darkest points in an image. Also, it reduces salt noise as a result of the min operation.

Midpoint filter

The midpoint filter simply computes the midpoint between the maximum and minimum values in the area encompassed by the filter:

$$\hat{f}(x, y) = \frac{1}{2} \left[\max_{(s,t) \in S_{xy}} \{g(s, t)\} + \min_{(s,t) \in S_{xy}} \{g(s, t)\} \right]. \tag{5.3-10}$$

Note that this filter combines order statistics and averaging. This filter works best for randomly distributed noise, like Gaussian or uniform noise.

Alpha-trimmed mean filter

Suppose that we delete the $d/2$ lowest and the $d/2$ highest gray-level values of $g(s, t)$ in the neighborhood S_{xy}. Let $g_r(s, t)$ represent the remaining $mn - d$ pixels. A filter formed by averaging these remaining pixels is called an *alpha-trimmed mean* filter:

$$\hat{f}(x, y) = \frac{1}{mn - d} \sum_{(s,t) \in S_{xy}} g_r(s, t) \tag{5.3-11}$$

where the value of d can range from 0 to $mn - 1$. When $d = 0$, the alpha-trimmed filter reduces to the arithmetic mean filter discussed in the previous section. If we choose $d = mn - 1$, the filter becomes a median filter. For other values of d, the alpha-trimmed filter is useful in situations involving multiple types of noise, such as a combination of salt-and-pepper and Gaussian noise.

EXAMPLE 5.3:
Illustration of order-statistics filters.

■ Figure 5.10(a) shows the circuit image corrupted by impulse noise with probabilities $P_a = P_b = 0.1$. Figure 5.10(b) shows the result of median filtering with a filter of size 3×3. The improvement over Fig. 5.10(a) is significant, but several noise points are still visible. A second pass [of the image in Fig. 5.10(b)] with the median filter removed most of these points, leaving only few, barely visible noise points. These are removed with a third pass of the filter. These results are good examples of the power of median filtering in handling impulselike additive noise. Keep in mind that repeated passes of a median filter tend to blur the image, so it is desirable to keep the number of passes as low as possible.

Figure 5.11(a) shows the result of applying the max filter to the pepper noise image of Fig. 5.8(a). The filter did a reasonable job of removing the pepper

a b
c d

FIGURE 5.10
(a) Image corrupted by salt-and-pepper noise with probabilities $P_a = P_b = 0.1$.
(b) Result of one pass with a median filter of size 3×3.
(c) Result of processing (b) with this filter.
(d) Result of processing (c) with the same filter.

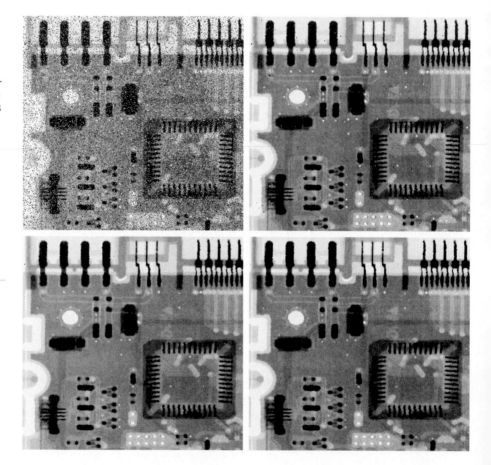

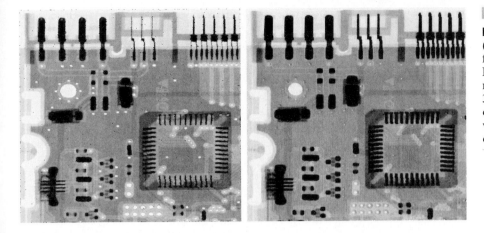

FIGURE 5.11
(a) Result of filtering Fig. 5.8(a) with a max filter of size 3×3. (b) Result of filtering 5.8(b) with a min filter of the same size.

noise, but we note that it also removed (set to a light gray level) some dark pixels from the borders of the dark objects. Figure 5.11(b) shows the result of applying the min filter to the image in Fig. 5.8(b). In this particular case, the min filter did a better job than the max filter on noise removal, but it removed some white points around the border of light objects. These made the light objects smaller and some of the dark objects larger (like the connector fingers in the top of the image) because white points around these objects were set to a dark level.

The alpha-trimmed filter is illustrated next. Figure 5.12(a) shows the circuit image corrupted this time by additive, uniform noise of variance 800 and zero mean. This is a high level of noise corruption that is made worse by further addition of salt-and-pepper noise with $P_a = P_b = 0.1$, as Fig. 5.12(b) shows. The high level of noise in this image warrants use of larger filters. Figures 5.12(c) through (f) show the results obtained using arithmetic mean, geometric mean, median, and alpha-trimmed mean (with $d = 5$) filters of size 5×5. As expected, the arithmetic and geometric mean filters (especially the latter) do not do well because of the presence of impulse noise. The median and alpha-trimmed filters performed much better, with the alpha-trimmed filter giving slightly better noise reduction. Note, for example, that the fourth connector finger from the top, left, is slightly smoother in the alpha-trimmed result. This is not unexpected because, for a high value of d, the alpha-trimmed filter approaches the performance of the median filter, but still retains some smoothing capabilities. ■

5.3.3 Adaptive Filters

Once selected, the filters discussed thus far are applied to an image without regard for how image characteristics vary from one point to another. In this section we take a look at two simple *adaptive* filters whose behavior changes based on statistical characteristics of the image inside the filter region defined by the $m \times n$ rectangular window S_{xy}. As shown in the following discussion, adaptive filters are capable of performance superior to that of the filters discussed thus far. The price paid for improved filtering power is an increase in filter complexity. Keep in mind that we still are dealing with the case in which the degraded

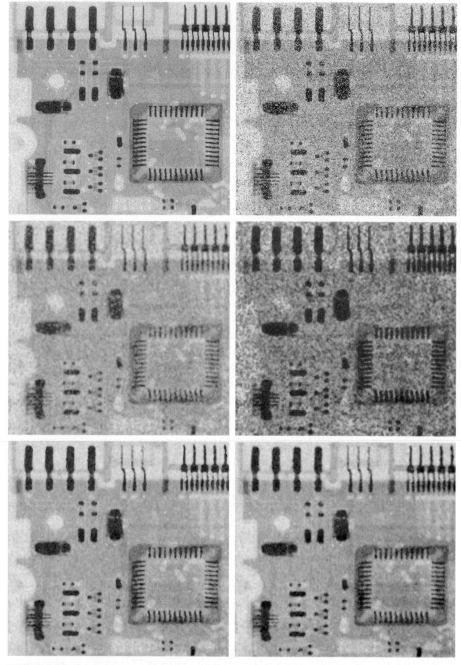

FIGURE 5.12 (a) Image corrupted by additive uniform noise. (b) Image additionally corrupted by additive salt-and-pepper noise. Image in (b) filtered with a 5×5: (c) arithmetic mean filter; (d) geometric mean filter; (e) median filter; and (f) alpha-trimmed mean filter with $d = 5$.

image is equal to the original image plus noise. No other types of degradations are being considered yet.

Adaptive, local noise reduction filter

The simplest statistical measures of a random variable are its mean and variance. These are reasonable parameters on which to base an adaptive filter because they are quantities closely related to the appearance of an image. The mean gives a measure of average gray level in the region over which the mean is computed, and the variance gives a measure of average contrast in that region.

Our filter is to operate on a local region, S_{xy}. The response of the filter at any point (x, y) on which the region is centered is to be based on four quantities: (a) $g(x, y)$, the value of the noisy image at (x, y); (b) σ_η^2, the variance of the noise corrupting $f(x, y)$ to form $g(x, y)$; (c) m_L, the local mean of the pixels in S_{xy}; and (d) σ_L^2, the local variance of the pixels in S_{xy}. We want the behavior of the filter to be as follows:

1. If σ_η^2 is zero, the filter should return simply the value of $g(x, y)$. This is the trivial, zero-noise case in which $g(x, y)$ is equal to $f(x, y)$.
2. If the local variance is high relative to σ_η^2, the filter should return a value close to $g(x, y)$. A high local variance typically is associated with edges, and these should be preserved.
3. If the two variances are equal, we want the filter to return the arithmetic mean value of the pixels in S_{xy}. This condition occurs when the local area has the same properties as the overall image, and local noise is to be reduced simply by averaging.

An adaptive expression for obtaining $\hat{f}(x, y)$ based on these assumptions may be written as

$$\hat{f}(x, y) = g(x, y) - \frac{\sigma_\eta^2}{\sigma_L^2}\left[g(x, y) - m_L\right]. \tag{5.3-12}$$

The only quantity that needs to be known or estimated is the variance of the overall noise, σ_η^2. The other parameters are computed from the pixels in S_{xy} at each location (x, y) on which the filter window is centered. A tacit assumption in Eq. (5.3-12) is that $\sigma_\eta^2 \leq \sigma_L^2$. The noise in our model is additive and position independent, so this is a reasonable assumption to make because S_{xy} is a subset of $g(x, y)$. However, we seldom have exact knowledge of σ_η^2. Therefore, it is possible for this condition to be violated in practice. For that reason, a test should be built into an implementation of Eq. (5.3-12) so that the ratio is set to 1 if the condition $\sigma_\eta^2 > \sigma_L^2$ occurs. This makes this filter nonlinear. However, it prevents nonsensical results (i.e., negative gray levels, depending on the value of m_L) due to a potential lack of knowledge about the variance of the image noise. Another approach is to allow the negative values to occur, and then rescale the gray level values at the end. The result then would be a loss of dynamic range in the image.

EXAMPLE 5.4:
Illustration of
adaptive, local
noise-reduction
filtering.

■ Figure 5.13(a) shows the circuit image, corrupted this time by additive Gaussian noise of zero mean and a variance of 1000. This is a significant level of noise corruption, but it makes an ideal test bed on which to compare relative filter performance. Figure 5.13(b) is the result of processing the noisy image with an arithmetic mean filter of size 7×7. The noise is smoothed out, but at the cost of significant blurring in the image. Similar comments are applicable to Fig. 5.13(c), which shows the result of processing the noisy image with a geometric mean filter, also of size 7×7. The differences between these two filtered images are analogous to those we discussed in Example 5.2; only the degree of blurring is different.

Figure 5.13(d) shows the result of using the adaptive filter of Eq. (5.3-12) with $\sigma_\eta^2 = 1000$. The improvements in this result compared to the two previous filters are significant. In terms of overall noise reduction, the adaptive filter achieved results similar to the arithmetic and geometric mean filters. However, the image filtered with the adaptive filter is much sharper. For example, the connector fingers at the top of the image are sharper in Fig. 5.13(d). Other features, such as holes and the eight legs of the dark component on the

a	b
c	d

FIGURE 5.13
(a) Image
corrupted by
additive Gaussian
noise of zero
mean and
variance 1000.
(b) Result of
arithmetic mean
filtering.
(c) Result of
geometric mean
filtering.
(d) Result of
adaptive noise
reduction
filtering. All filters
were of size
7×7.

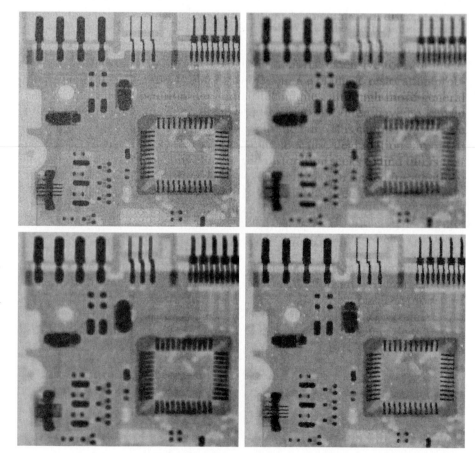

lower left-hand side of the image, are significantly clearer in Fig. 5.13(d). These results are typical of what can be achieved with an adaptive filter. As mentioned at the beginning of this section, the price paid for the improved performance is additional filter complexity.

The preceding results used a value for σ_η^2 that matched the variance of the noise exactly. If this quantity is not known and an estimate is used that is too low, the algorithm will return an image that closely resembles the original because the corrections will be smaller than they should be. Estimates that are too high will cause the ratio of the variances to be clipped at 1.0, and the algorithm will subtract the mean from the image more frequently than it would normally do so. If negative values are allowed and the image is rescaled at the end, the result will be a loss of dynamic range, as mentioned previously.   ■

Adaptive median filter

The median filter discussed in Section 5.3.2 performs well as long as the spatial density of the impulse noise is not large (as a rule of thumb, P_a and P_b less than 0.2). It is shown in this section that adaptive median filtering can handle impulse noise with probabilities even larger than these. An additional benefit of the adaptive median filter is that it seeks to preserve detail while smoothing nonimpulse noise, something that the "traditional" median filter does not do. As in all the filters discussed in the preceding sections, the adaptive median filter also works in a rectangular window area S_{xy}. Unlike those filters, however, the adaptive median filter changes (increases) the size of S_{xy} during filter operation, depending on certain conditions listed in this section. Keep in mind that the output of the filter is a single value used to replace the value of the pixel at (x, y), the particular point on which the window S_{xy} is centered at a given time.

Consider the following notation:

$$z_{min} = \text{minimum gray level value in } S_{xy}$$
$$z_{max} = \text{maximum gray level value in } S_{xy}$$
$$z_{med} = \text{median of gray levels in } S_{xy}$$
$$z_{xy} = \text{gray level at coordinates } (x, y)$$
$$S_{max} = \text{maximum allowed size of } S_{xy}.$$

The adaptive median filtering algorithm works in two levels, denoted level A and level B, as follows:

Level A: $A1 = z_{med} - z_{min}$
$A2 = z_{med} - z_{max}$
If $A1 > 0$ AND $A2 < 0$, Go to level B
Else increase the window size
If window size $\leq S_{max}$ repeat level A
Else output z_{med}.

Level B: $B1 = z_{xy} - z_{min}$
$B2 = z_{xy} - z_{max}$
If $B1 > 0$ AND $B2 < 0$, output z_{xy}
Else output z_{med}.

The key to understanding the mechanics of this algorithm is to keep in mind that it has three main purposes: to remove salt-and-pepper (impulse) noise, to provide smoothing of other noise that may not be impulsive, and to reduce distortion, such as excessive thinning or thickening of object boundaries. The values z_{min} and z_{max} are considered statistically by the algorithm to be "impulselike" noise components, even if these are not the lowest and highest possible pixel values in the image.

With these observations in mind, we see that the purpose of level A is to determine if the median filter output, z_{med}, is an impulse (black *or* white) or not. If the condition $z_{min} < z_{med} < z_{max}$ holds, then z_{med} cannot be an impulse for the reason mentioned in the previous paragraph. In this case, we go to level B and test to see if the point in the center of the window, z_{xy}, is itself an impulse (recall that z_{xy} is the point being processed). If the condition $B1 > 0$ AND $B2 < 0$ is true, then $z_{min} < z_{xy} < z_{max}$, and z_{xy} cannot be an impulse for the same reason that z_{med} was not. In this case, the algorithm outputs the unchanged pixel value, z_{xy}. By not changing these "intermediate-level" points, distortion is reduced in the image. If the condition $B1 > 0$ AND $B2 < 0$ is false, then either $z_{xy} = z_{min}$ or $z_{xy} = z_{max}$. In either case, the value of the pixel is an extreme value and the algorithm outputs the median value z_{med}, which we know from level A is not a noise impulse. The last step is what the standard median filter does. The problem is that the standard median filter replaces every point in the image by the median of the corresponding neighborhood. This causes unnecessary loss of detail.

Continuing with the explanation, suppose that level A *does* find an impulse (i.e., it fails the test that would cause it to branch to level B). The algorithm then increases the size of the window and repeats level A. This looping continues until the algorithm either finds a median value that is not an impulse (and branches to level B), or the maximum window size is reached. If the maximum window size is reached, the algorithm returns the value of z_{med}. Note that there is no guarantee that this value is not an impulse. The smaller the noise probabilities P_a and/or P_b are, or the larger S_{max} is allowed to be, the less likely it is that a premature exit condition will occur. This is plausible. As the density of the impulses increases, it stands to reason that we would need a larger window to "clean up" the noise spikes.

Every time the algorithm outputs a value, the window S_{xy} is moved to the next location in the image. The algorithm then is reinitialized and applied to the pixels in the new location. As indicated in Problem 3.20, the median value can be updated iteratively using only the new pixels, thus reducing computational overhead.

EXAMPLE 5.5:
Illustration of adaptive median filtering.

■ Figure 5.14(a) shows the circuit image corrupted by salt-and-pepper noise with probabilities $P_a = P_b = 0.25$, which is 2.5 times the noise level used in Fig. 5.10(a). Here the noise level is high enough to obscure most of the detail in the image. As a basis for comparison, the image was filtered first using the smallest median filter required to remove most visible traces of impulse noise. A 7×7 median filter was required to do this, and the result is shown in

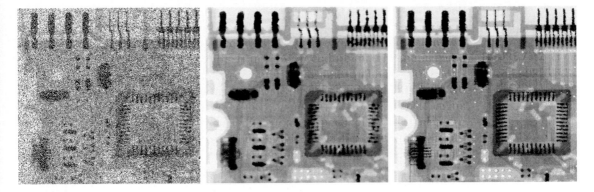

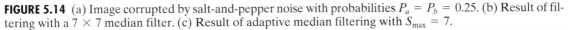

a b c

FIGURE 5.14 (a) Image corrupted by salt-and-pepper noise with probabilities $P_a = P_b = 0.25$. (b) Result of filtering with a 7×7 median filter. (c) Result of adaptive median filtering with $S_{max} = 7$.

Fig. 5.14(b). Although the noise was effectively removed, the filter caused significant loss of detail in the image. For instance, some of the connector fingers at the top of the image appear distorted or broken. Other image details are similarly distorted.

Figure 5.14(c) shows the result of using the adaptive median filter with $S_{max} - 7$. The level of noise removal was similar to the median filter. However, the adaptive filter preserved sharpness and detail to the point where improvements over Fig. 5.14(b) are quite significant. The connector fingers are less distorted, and some other features that were either obscured or distorted beyond recognition by the median filter appear sharper and better defined in Fig. 5.14(c). Two notable examples are the feed-through small white holes throughout the board, and the dark component with eight legs in the bottom, left quadrant of the image.

Considering the high level of noise in Fig. 5.14(a), the adaptive algorithm performed quite well. The choice of maximum allowed window size depends on the application, but a reasonable starting value can be estimated by experimenting with various sizes of the standard median filter first. This will establish a visual baseline regarding expectations on the performance of the adaptive algorithm. ■

5.4 Periodic Noise Reduction by Frequency Domain Filtering

In Chapter 4 we discussed lowpass and highpass frequency domain filters as fundamental tools for image enhancement. In this section we discuss the more specialized bandreject, bandpass, and notch filters as tools for periodic noise reduction or removal.

a b c

FIGURE 5.15 From left to right, perspective plots of ideal, Butterworth (of order 1), and Gaussian bandreject filters.

5.4.1 Bandreject Filters

Bandreject filters remove or attenuate a band of frequencies about the origin of the Fourier transform. An ideal bandreject filter is given by the expression

$$H(u, v) = \begin{cases} 1 & \text{if } D(u, v) < D_0 - \dfrac{W}{2} \\ 0 & \text{if } D_0 - \dfrac{W}{2} \le D(u, v) \le D_0 + \dfrac{W}{2} \\ 1 & \text{if } D(u, v) > D_0 + \dfrac{W}{2} \end{cases} \qquad (5.4\text{-}1)$$

where $D(u, v)$ is the distance from the origin of the centered frequency rectangle, as given in Eq. (4.3-3), W is the width of the band, and D_0 is its radial center. Similarly, a Butterworth bandreject filter of order n is given by the expression

$$H(u, v) = \dfrac{1}{1 + \left[\dfrac{D(u, v)W}{D^2(u, v) - D_0^2} \right]^{2n}} \qquad (5.4\text{-}2)$$

and a Gaussian bandreject filter is given by

$$H(u, v) = 1 - e^{-\frac{1}{2}\left[\frac{D^2(u, v) - D_0^2}{D(u, v)W} \right]^2}. \qquad (5.4\text{-}3)$$

Figure 5.15 shows perspective plots of these three filters.

EXAMPLE 5.6:
Use of bandreject filtering for periodic noise removal.

■ One of the principal applications of bandreject filtering is for noise removal in applications where the general location of the noise component(s) in the frequency domain is approximately known. A good example is an image corrupted by additive periodic noise that can be approximated as two-dimensional sinusoidal functions. It is not difficult to show that the Fourier transform of a sine consists of two impulses that are mirror images of each other about the origin of the transform. Their locations are given in Table 4.1. The impulses are both imaginary (the real part of the Fourier transform of a sine is zero) and are complex conjugates of each other. We will have more to say about this topic in Sections 5.4.3 and 5.4.4. Our purpose at the moment is to illustrate bandreject filtering.

FIGURE 5.16
(a) Image corrupted by sinusoidal noise. (b) Spectrum of (a). (c) Butterworth bandreject filter (white represents 1). (d) Result of filtering. (Original image courtesy of NASA.)

Figure 5.16(a), which is the same as Fig. 5.5(a), shows an image heavily corrupted by sinusoidal noise of various frequencies. The noise components are easily seen as symmetric pairs of bright dots in the Fourier spectrum shown in Fig. 5.16(b). In this example, the components lie on an approximate circle about origin of the transform, so a circularly symmetric bandreject filter is a good choice. Figure 5.16(c) shows a Butterworth bandreject filter of order 4, with the appropriate radius and width to enclose completely the noise impulses. Since it is desirable in general to remove as little as possible from the transform, sharp, narrow filters are common in bandreject filtering. The result of filtering Fig. 5.16(a) with this filter is shown in Fig. 5.16(d). The improvement is quite evident. Even small details and textures were restored effectively by this simple filtering approach. It is worth noting also that it would not be possible to get equivalent results by a direct spatial domain filtering approach using small convolution masks.   ■

5.4.2 Bandpass Filters

A *bandpass* filter performs the opposite operation of a bandreject filter. In Section 4.4 we showed how a highpass filter can be obtained from a corresponding lowpass filter by using Eq. (4.4-1). Similarly, the transfer function $H_{bp}(u, v)$ of a bandpass filter is obtained from a corresponding bandreject filter with transfer function $H_{br}(u, v)$ by using the equation

$$H_{bp}(u, v) = 1 - H_{br}(u, v). \tag{5.4-4}$$

FIGURE 5.17
Noise pattern of
the image in
Fig. 5.16(a)
obtained by
bandpass filtering.

It is left as an exercise for the reader (Problem 5.12) to derive expressions for the bandpass filters corresponding to Eqs. (5.4-1) through (5.4-3).

EXAMPLE 5.7:
Bandpass filtering
for extracting
noise patterns.

■ Performing straight bandpass filtering on an image is not a common procedure because it generally removes too much image detail. However, bandpass filtering is quite useful in isolating the effect on an image of selected frequency bands. This is illustrated in Fig. 5.17. This image was generated by (1) using Eq. (5.4-4) to obtain the bandpass filter corresponding to the bandreject filter used in the previous example; and (2) taking the inverse transform of the bandpass-filtered transform. Most image detail was lost, but the information that remains is most useful, as it is clear that the noise pattern recovered using this method is quite close to the noise that corrupted the image in Fig. 5.16(a). In other words, bandpass filtering helped isolate the noise pattern. This is a useful result because it simplifies analysis of the noise, reasonably independently of image content.   ■

5.4.3 Notch Filters

A *notch* filter rejects (or passes) frequencies in predefined neighborhoods about a center frequency. Figure 5.18 shows 3-D plots of ideal, Butterworth, and Gaussian notch (reject) filters. Due to the symmetry of the Fourier transform, notch filters must appear in symmetric pairs about the origin in order to obtain meaningful results. The one exception to this rule is if the notch filter is located at the origin, in which case it appears by itself. Although we show only one pair for illustrative purposes, the number of pairs of notch filters that can be implemented is arbitrary. The shape of the notch areas also can be arbitrary (e.g., rectangular).

The transfer function of an ideal notch reject filter of radius D_0, with centers at (u_0, v_0) and, by symmetry, at $(-u_0, -v_0)$, is

$$H(u, v) = \begin{cases} 0 & \text{if } D_1(u, v) \le D_0 \quad \text{or} \quad D_2(u, v) \le D_0 \\ 1 & \text{otherwise} \end{cases} \qquad (5.4\text{-}5)$$

where

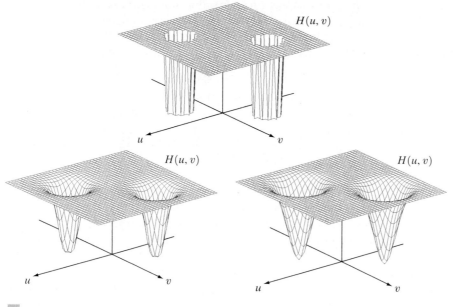

a
b c

FIGURE 5.18 Perspective plots of (a) ideal, (b) Butterworth (of order 2), and (c) Gaussian notch (reject) filters.

$$D_1(u, v) = \left[(u - M/2 - u_0)^2 + (v - N/2 - v_0)^2\right]^{1/2} \qquad (5.4\text{-}6)$$

and

$$D_2(u, v) = \left[(u - M/2 + u_0)^2 + (v - N/2 + v_0)^2\right]^{1/2}. \qquad (5.4\text{-}7)$$

As usual, the assumption is that the center of the frequency rectangle has been shifted to the point $(M/2, N/2)$, according to the filtering procedure outlined in Section 4.2.3. Therefore, the values of (u_0, v_0) are with respect to the shifted center.

The transfer function of a Butterworth notch reject filter of order n is given by

$$H(u, v) = \frac{1}{1 + \left[\dfrac{D_0^2}{D_1(u, v) D_2(u, v)}\right]^n} \qquad (5.4\text{-}8)$$

where $D_1(u, v)$ and $D_2(u, v)$ are given in Eqs. (5.4-6) and (5.4-7), respectively. A Gaussian notch reject filter has the form

$$H(u, v) = 1 - e^{-\frac{1}{2}\left[\frac{D_1(u, v) D_2(u, v)}{D_0^2}\right]} \qquad (5.4\text{-}9)$$

It is interesting to note that these three filters become highpass filters if $u_0 = v_0 = 0$.

As shown in the previous section for bandpass filters, we can obtain notch filters that *pass*, rather than suppress, the frequencies contained in the notch areas. Since these filters perform exactly the opposite function as the notch reject filters given in Eqs. (5.4-5), (5.4-8), and (5.4-9), their transfer functions are given by

$$H_{np}(u, v) = 1 - H_{nr}(u, v) \qquad (5.4\text{-}10)$$

where $H_{np}(u, v)$ is the transfer function of the notch pass filter corresponding to the notch reject filter with transfer function $H_{nr}(u, v)$. It is left as an exercise for the reader (Problem 5.13) to derive equations for the notch pass filters corresponding to the reject filters just discussed, and to show that they become lowpass filters when $u_0 = v_0 = 0$.

EXAMPLE 5.8:
Removal of periodic noise by notch filtering.

■ Figure 5.19(a) shows the same image as Fig. 4.21(a). When we discussed lowpass filtering of that image in Section 4.3.4, we indicated that there were better ways to reduce the effect of the scan lines. The notch filtering approach that follows reduces the noise in this image, without introducing appreciable blurring. Unless blurring is desirable for reasons we discussed in Section 4.3, notch filtering is preferable if a suitable filter can be found.

Just by looking at the nearly horizontal lines of the noise pattern in Fig. 5.19(a), we expect its contribution in the frequency domain to be concentrated along the vertical axis. However, the noise is not dominant enough to have a clear pattern along this axis, as is evident from the spectrum shown in Fig. 5.19(b). We can get an idea of what the noise contribution looks like by constructing a simple ideal notch pass filter along the vertical axis of the Fourier transform, as shown in Fig. 5.19(c). The spatial representation of the noise pattern (inverse transform of the notch pass–filtered result) is shown in Fig. 5.19(d). This noise pattern corresponds closely to the pattern in Fig. 5.19(a). Having thus constructed a suitable notch pass filter that isolates the noise to a reasonable degree, we can obtain the corresponding notch reject filter from Eq. (5.4-10). The result of processing the image with the notch reject filter is shown in Fig. 5.19(e). This image contains significantly fewer visible noise scan lines than Fig. 5.19(a).   ■

5.4.4 Optimum Notch Filtering

Clearly defined interference patterns are not common. Images derived from electro-optical scanners, such as those used in space and aerial imaging, sometimes are corrupted by coupling and amplification of low-level signals in the scanners' electronic circuitry. The resulting images tend to contain pronounced, 2-D periodic structures superimposed on the scene data with patterns more complex than those we have studied thus far.

Figure 5.20(a), an example of this type of periodic image degradation, shows a digital image of the Martian terrain taken by the *Mariner 6* spacecraft. The interference pattern is quite similar to the one shown in Fig. 5.16(a), but the former pattern is considerably more subtle and, consequently, harder to detect in the frequency plane. Figure 5.20(b) shows the Fourier spectrum of the image in question. The starlike components were caused by the interference, and several pairs of components are present, indicating that the pattern contained more than just one sinusoidal component.

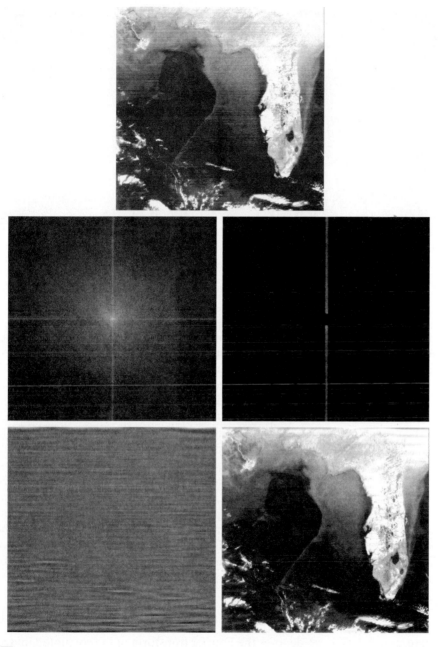

a
b c
d e

FIGURE 5.19 (a) Satellite image of Florida and the Gulf of Mexico (note horizontal sensor scan lines). (b) Spectrum of (a). (c) Notch pass filter shown superimposed on (b). (d) Inverse Fourier transform of filtered image, showing noise pattern in the spatial domain. (e) Result of notch reject filtering. (Original image courtesy of NOAA.)

FIGURE 5.20
(a) Image of the
Martian terrain
taken by
Mariner 6.
(b) Fourier
spectrum showing
periodic
interference.
(Courtesy of
NASA.)

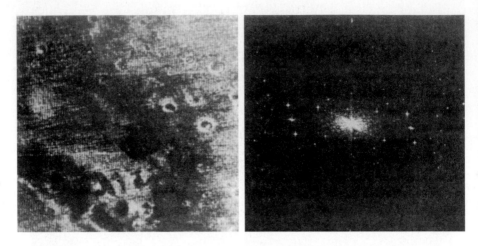

When several interference components are present, the methods discussed in the preceding sections are not always acceptable because they may remove too much image information in the filtering process (a highly undesirable feature when images are unique and/or expensive to acquire). In addition, the interference components generally are not single-frequency bursts. Instead, they tend to have broad skirts that carry information about the interference pattern. These skirts are not always easily detectable from the normal transform background. Alternative filtering methods that reduce the effect of these drawbacks are quite useful in many applications. The method discussed here is optimum, in the sense that it minimizes local variances of the restored estimate $\hat{f}(x, y)$.

The procedure consists of first isolating the principal contributions of the interference pattern and then subtracting a variable, weighted portion of the pattern from the corrupted image. Although we develop the procedure in the context of a specific application, the basic approach is quite general and can be applied to other restoration tasks in which multiple periodic interference is a problem.

The first step is to extract the principal frequency components of the interference pattern. This can be done by placing a notch pass filter, $H(u, v)$, at the location of each spike. If $H(u, v)$ is constructed to pass only components associated with the interference pattern, then, from the discussion in Sections 5.4.2 and 5.4.3, it follows that the Fourier transform of the interference noise pattern is given by the expression

$$N(u, v) = H(u, v)G(u, v) \qquad (5.4-11)$$

where, as usual, $G(u, v)$, denotes the Fourier transform of the corrupted image.

Formation of $H(u, v)$ requires considerable judgment about what is or is not an interference spike. For this reason, the notch pass filter generally is constructed interactively by observing the spectrum of $G(u, v)$ on a display. After a particular filter has been selected, the corresponding pattern in the spatial domain is obtained from the expression

$$\eta(x, y) = \Im^{-1}\{H(u, v)G(u, v)\}. \qquad (5.4-12)$$

Because the corrupted image is assumed to be formed by the addition of the un-corrupted image $f(x, y)$ and the interference, if $\eta(x, y)$ were known complete-ly, subtracting the pattern from $g(x, y)$ to obtain $f(x, y)$ would be a simple matter, as discussed earlier in this chapter. The problem, of course, is that this filtering procedure usually yields only an approximation of the true pattern. The effect of components not present in the estimate of $\eta(x, y)$ can be mini-mized instead by subtracting from $g(x, y)$ a *weighted* portion of $\eta(x, y)$ to ob-tain an estimate of $f(x, y)$:

$$\hat{f}(x, y) = g(x, y) - w(x, y)\eta(x, y) \tag{5.4-13}$$

where, as before, $\hat{f}(x, y)$ is the estimate of $f(x, y)$ and $w(x, y)$ is to be determined. The function $w(x, y)$ is called a *weighting* or *modulation* function, and the objec-tive of the procedure is to select this function so that the result is optimized in some meaningful way. One approach is to select $w(x, y)$ so that the variance of the estimate $\hat{f}(x, y)$ is minimized over a specified neighborhood of every point (x, y).

Consider a neighborhood of size $(2a + 1)$ by $(2b + 1)$ about a point (x, y). The "local" variance of $\hat{f}(x, y)$ at coordinates (x, y) can be estimated from the samples as follows:

$$\sigma^2(x, y) = \frac{1}{(2a + 1)(2b + 1)} \sum_{s=-a}^{a} \sum_{t=-b}^{b} [\hat{f}(x + s, y + t) - \overline{\hat{f}}(x, y)]^2 \tag{5.4-14}$$

where $\overline{\hat{f}}(x, y)$ is the average value of $\hat{f}$ in the neighborhood; that is,

$$\overline{\hat{f}}(x, y) = \frac{1}{(2a + 1)(2b + 1)} \sum_{s=-a}^{a} \sum_{t=-b}^{b} \hat{f}(x + s, y + t). \tag{5.4-15}$$

Points on or near the edge of the image can be treated by considering partial neighborhoods.

Substituting Eq. (5.4-13) into Eq. (5.4-14) yields

$$\sigma^2(x, y) = \frac{1}{(2a + 1)(2b + 1)} \sum_{s=-a}^{a} \sum_{t=-b}^{b} \{[g(x + s, y + t) \\ - w(x + s, y + t)\eta(x + s, y + t)] \\ - [\overline{g}(x, y) - \overline{w(x, y)\eta(x, y)}]\}^2. \tag{5.4-16}$$

Assuming that $w(x, y)$ remains essentially constant over the neighborhood gives the approximation

$$w(x + s, y + t) = w(x, y) \tag{5.4-17}$$

for $-a \le s \le a$ and $-b \le t \le b$. This assumption also results in the expression

$$\overline{w(x, y)\eta(x, y)} = w(x, y)\overline{\eta}(x, y) \tag{5.4-18}$$

in the neighborhood. With these approximations, Eq. (5.4-16) becomes

$$\sigma^2(x, y) = \frac{1}{(2a + 1)(2b + 1)} \sum_{s=-a}^{a} \sum_{t=-b}^{b} \{[g(x + s, y + t) \\ - w(x, y)\eta(x + s, y + t)] \\ - [\overline{g}(x, y) - w(x, y)\overline{\eta}(x, y)]\}^2. \tag{5.4-19}$$

To minimize $\sigma^2(x, y)$, we solve

$$\frac{\partial \sigma^2(x, y)}{\partial w(x, y)} = 0 \qquad (5.4\text{-}20)$$

for $w(x, y)$. The result is

$$w(x, y) = \frac{\overline{g(x, y)\eta(x, y)} - \bar{g}(x, y)\bar{\eta}(x, y)}{\overline{\eta^2}(x, y) - \bar{\eta}^2(x, y)}. \qquad (5.4\text{-}21)$$

To obtain the restored image $\hat{f}(x, y)$, we compute $w(x, y)$ from Eq. (5.4-21) and then make use of Eq. (5.4-13). As $w(x, y)$ is assumed to be constant in a neighborhood, computing this function for every value of x and y in the image is unnecessary. Instead, $w(x, y)$ is computed for *one* point in each nonoverlapping neighborhood (preferably the center point) and then used to process all the image points contained in that neighborhood.

EXAMPLE 5.9:
Illustration of
optimum notch
filtering.

■ Figures 5.21 through 5.23 show the result of applying the preceding technique to the image shown in Fig. 5.20(a). This image is of size 512×512 pixels, and a neighborhood with $a = b = 15$ was selected. Figure 5.21 shows the Fourier spectrum of the corrupted image. The origin was not shifted to the center of the frequency plane in this particular case, so $u = v = 0$ is at the top, left corner of the transform image shown in Fig. 5.21. Figure 5.22(a) shows the spectrum of $N(u, v)$, where only the noise spikes are present. Figure 5.22(b) shows the

FIGURE 5.21 Fourier spectrum (without shifting) of the image shown in Fig. 5.20(a). (Courtesy of NASA.)

a b

FIGURE 5.22 (a) Fourier spectrum of $N(u, v)$, and (b) corresponding noise interference pattern $\eta(x, y)$. (Courtesy of NASA.)

interference pattern $\eta(x, y)$ obtained by taking the inverse Fourier transform of $N(u, v)$. Note the similarity between this pattern and the structure of the noise present in Fig. 5.20(a). Finally, Fig. 5.23 shows the processed image obtained by using Eq. (5.4-13). The periodic interference was removed for all practical purposes.   ■

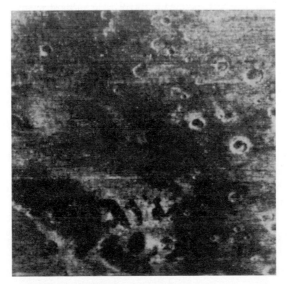

FIGURE 5.23 Processed image. (Courtesy of NASA.)

5.5   Linear, Position-Invariant Degradations

The input-output relationship in Fig. 5.1 before the restoration stage is expressed as

$$g(x, y) = H[f(x, y)] + \eta(x, y). \qquad (5.5\text{-}1)$$

See inside front cover
Consult the book web site
for a brief review of linear
system theory.

For the moment, let us assume that $\eta(x, y) = 0$ so that $g(x, y) = H[f(x, y)]$. Based on the discussion in Section 2.6, H is *linear* if

$$H[af_1(x, y) + bf_2(x, y)] = aH[f_1(x, y)] + bH[f_2(x, y)] \qquad (5.5\text{-}2)$$

where a and b are scalars and $f_1(x, y)$ and $f_2(x, y)$ are any two input images. If $a = b = 1$, Eq. (5.5-2) becomes

$$H[f_1(x, y) + f_2(x, y)] = H[f_1(x, y)] + H[f_2(x, y)] \qquad (5.5\text{-}3)$$

which is called the property of *additivity*. This property simply says that, if H is a linear operator, the response to a sum of two inputs is equal to the sum of the two responses.

With $f_2(x, y) = 0$, Eq. (5.5-2) becomes

$$H[af_1(x, y)] = aH[f_1(x, y)] \qquad (5.5\text{-}4)$$

which is called the property of *homogeneity*. It says that the response to a constant multiple of any input is equal to the response to that input multiplied by the same constant. Thus a linear operator possesses both the property of additivity and the property of homogeneity.

An operator having the input-output relationship $g(x, y) = H[f(x, y)]$ is said to be *position* (or *space*) *invariant* if

$$H[f(x - \alpha, y - \beta)] = g(x - \alpha, y - \beta) \qquad (5.5\text{-}5)$$

for any $f(x, y)$ and any α and β. This definition indicates that the response at any point in the image depends only on the *value* of the input at that point, not on its *position*.

With a slight (but equivalent) change in notation in the definition of the discrete impulse function in Eq. (4.2-33), $f(x, y)$ can be expressed in terms of a continuous impulse function:

$$f(x, y) = \int_{-\infty}^{\infty} \int_{-\infty}^{\infty} f(\alpha, \beta)\delta(x - \alpha, y - \beta)\, d\alpha\, d\beta. \qquad (5.5\text{-}6)$$

This, in fact, is the *definition* using continuous variables of a unit impulse located at coordinates (x, y).

Assume again for a moment that $\eta(x, y) = 0$. Then, substitution of Eq. (5.5-6) into Eq. (5.5-1) results in the expression

$$g(x, y) = H[f(x, y)] = H\left[\int_{-\infty}^{\infty} \int_{-\infty}^{\infty} f(\alpha, \beta)\delta(x - \alpha, y - \beta)\, d\alpha\, d\beta \right]. \qquad (5.5\text{-}7)$$

If H is a linear operator and we extend the additivity property to integrals, then

$$g(x, y) = \int_{-\infty}^{\infty} \int_{-\infty}^{\infty} H[f(\alpha, \beta)\delta(x - \alpha, y - \beta)]\, d\alpha\, d\beta. \qquad (5.5\text{-}8)$$

Because $f(\alpha, \beta)$ is independent of x and y, and using the homogeneity property, it follows that

$$g(x, y) = \int_{-\infty}^{\infty} \int_{-\infty}^{\infty} f(\alpha, \beta) H[\delta(x - \alpha, y - \beta)] \, d\alpha \, d\beta. \qquad (5.5\text{-}9)$$

The term

$$h(x, \alpha, y, \beta) = H[\delta(x - \alpha, y - \beta)] \qquad (5.5\text{-}10)$$

is called the *impulse response* of H. In other words, if $\eta(x, y) = 0$ in Eq. (5.5-1), then $h(x, \alpha, y, \beta)$ is the response of H to an impulse of strength 1 at coordinates (x, y). In optics, the impulse becomes a point of light and $h(x, \alpha, y, \beta)$ is commonly referred to as the *point spread function* (PSF). This name arises from the fact that all physical optical systems blur (spread) a point of light to some degree, with the amount of blurring being determined by the quality of the optical components.

Substituting Eq. (5.5-10) into Eq. (5.5-9) yields the expression

$$g(x, y) = \int_{-\infty}^{\infty} \int_{-\infty}^{\infty} f(\alpha, \beta) h(x, \alpha, y, \beta) \, d\alpha \, d\beta \qquad (5.5\text{-}11)$$

which is called the *superposition* (or *Fredholm*) *integral of the first kind*. This expression is a fundamental result that is at the core of linear system theory. It states that if the response of H to an impulse is known, the response to *any* input $f(\alpha, \beta)$ can be calculated by means of Eq. (5.5-11). In other words, a linear system H is completely characterized by its impulse response.

If H is position invariant, then, from Eq. (5.5-5),

$$H[\delta(x - \alpha, y - \beta)] = h(x - \alpha, y - \beta). \qquad (5.5\text{-}12)$$

Equation (5.5-11) reduces in this case to

$$g(x, y) = \int_{-\infty}^{\infty} \int_{-\infty}^{\infty} f(\alpha, \beta) h(x - \alpha, y - \beta) \, d\alpha \, d\beta. \qquad (5.5\text{-}13)$$

This expression is called the *convolution integral*; it is the continuous-variable equivalent of the discrete convolution expression in Eq. (4.2-30). This integral tells us that knowing the impulse response of a linear system allows us to compute its response, g, to any input f. The result is simply the convolution of the impulse response and the input function.

In the presence of additive noise, the expression of the linear degradation model [Eq. (5.5-11)] becomes

$$g(x, y) = \int_{-\infty}^{\infty} \int_{-\infty}^{\infty} f(\alpha, \beta) h(x, \alpha, y, \beta) \, d\alpha \, d\beta + \eta(x, y). \qquad (5.5\text{-}14)$$

If H is position invariant, Eq. (5.5-14) becomes

$$g(x, y) = \int_{-\infty}^{\infty} \int_{-\infty}^{\infty} f(\alpha, \beta) h(x - \alpha, y - \beta) \, d\alpha \, d\beta + \eta(x, y). \qquad (5.5\text{-}15)$$

The values of the noise term $\eta(x, y)$ are random, and are assumed to be independent of position. Using the familiar notation for convolution, we can write Eq. (5.5-15) as

$$g(x, y) = h(x, y) * f(x, y) + \eta(x, y) \qquad (5.5\text{-}16)$$

or, based on the convolution theorem, we can express it in the frequency domain, as

$$G(u, v) = H(u, v)F(u, v) + N(u, v). \qquad (5.5\text{-}17)$$

These two expressions agree with Eqs. (5.1-1) and (5.1-2). Keep in mind that, for discrete quantities, all products are term by term. For example, term ij of $H(u, v)F(u, v)$ is the product of term ij of $H(u, v)$ and term ij of $F(u, v)$.

In summary, the preceding discussion indicates that a linear, spatially-invariant degradation system with additive noise can be modeled in the spatial domain as the convolution of the degradation (point spread) function with an image, followed by the addition of noise. Based on the convolution theorem (Sections 4.2.4 and 4.6.4), the same process can be expressed in the frequency domain as the product of the transforms of the image and degradation, followed by the addition of the transform of the noise. When working in the frequency domain, we make use of an FFT algorithm, as discussed in Section 4.6. Keep in mind also the need for function padding in the implementation of discrete Fourier transforms, as outlined in Section 4.6.3.

Many types of degradations can be approximated by linear, position-invariant processes. The advantage of this approach is that the extensive tools of linear system theory then become available for the solution of image restoration problems. Nonlinear and position-dependent techniques, although more general (and usually more accurate), introduce difficulties that often have no known solution or are very difficult to solve computationally. This chapter focuses on linear, space-invariant restoration techniques. Because degradations are modeled as being the result of convolution, and restoration seeks to find filters that apply the process in reverse, the term *image deconvolution* is used frequently to signify linear image restoration. Similarly, the filters used in the restoration process often are called *deconvolution filters*.

5.6 Estimating the Degradation Function

There are three principal ways to estimate the degradation function for use in image restoration: (1) observation, (2) experimentation, and (3) mathematical modeling. These methods are discussed in the following sections. The process of restoring an image by using a degradation function that has been estimated in some way sometimes is called *blind deconvolution,* due to the fact that the true degradation function is seldom known completely.

5.6.1 Estimation by Image Observation

Suppose that we are given a degraded image without any knowledge about the degradation function H. One way to estimate this function is to gather information from the image itself. For example, if the image is blurred, we can look at a small section of the image containing simple structures, like part of an object and the background. In order to reduce the effect of noise in our observation, we would look for areas of strong signal content. Using sample gray levels of the object and background, we can construct an unblurred image of the same size and characteristics as the observed subimage. Let the observed subimage be denoted by $g_s(x, y)$, and let the constructed subimage (which in reality is our estimate of the original image in that area) be denoted by $\hat{f}_s(x, y)$. Then, assuming that the effect of noise is negligible because of our choice of a strong-signal area, it follows from Eq. (5.5-17) that

$$H_s(u, v) = \frac{G_s(u, v)}{\hat{F}_s(u, v)}. \tag{5.6-1}$$

From the characteristics of this function we then deduce the complete function $H(u, v)$ by making use of the fact that we are assuming position invariance. For example, suppose that a radial plot of $H_s(u, v)$ turns out to have the shape of Butterworth lowpass filter. We can use that information to construct a function $H(u, v)$ on a larger scale, but having the same shape.

5.6.2 Estimation by Experimentation

If equipment similar to the equipment used to acquire the degraded image is available, it is possible in principle to obtain an accurate estimate of the degradation. Images similar to the degraded image can be acquired with various system settings until they are degraded as closely as possible to the image we wish to restore. Then the idea is to obtain the impulse response of the degradation by imaging an impulse (small dot of light) using the same system settings. As noted in Section 5.5, a linear, space-invariant system is described completely by its impulse response.

An impulse is simulated by a bright dot of light, as bright as possible to reduce the effect of noise. Then, recalling that the Fourier transform of an impulse is a constant, it follows from Eq. (5.5-17) that

$$H(u, v) = \frac{G(u, v)}{A} \tag{5.6-2}$$

where, as before, $G(u, v)$ is the Fourier transform of the observed image and A is a constant describing the strength of the impulse. Figure 5.24 shows an example.

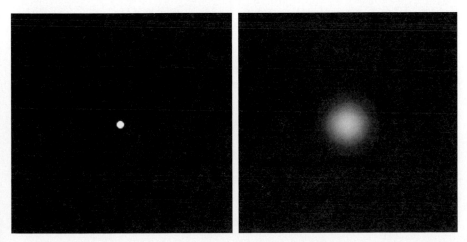

a b

FIGURE 5.24
Degradation estimation by impulse characterization. (a) An impulse of light (shown magnified). (b) Imaged (degraded) impulse.

5.6.3 Estimation by Modeling

Degradation modeling has been used for many years because of the insight it affords into the image restoration problem. In some cases, the model can even take into account environmental conditions that cause degradations. For example, a degradation model proposed by Hufnagel and Stanley [1964] is based on the physical characteristics of atmospheric turbulence. This model has a familiar form:

$$H(u, v) = e^{-k(u^2 + v^2)^{5/6}} \qquad (5.6\text{-}3)$$

where k is a constant that depends on the nature of the turbulence. With the exception of the 5/6 power on the exponent, this equation has the same form as the Gaussian lowpass filter discussed in Section 4.3.3. In fact, the Gaussian LPF is used sometimes to model mild, uniform blurring. Figure 5.25 shows examples obtained by simulating blurring an image using Eq. (5.6-3) with values $k = 0.0025$ (severe turbulence in this case), $k = 0.001$ (mild turbu-

a b
c d

FIGURE 5.25
Illustration of the atmospheric turbulence model.
(a) Negligible turbulence.
(b) Severe turbulence, $k = 0.0025$.
(c) Mild turbulence, $k = 0.001$.
(d) Low turbulence, $k = 0.00025$.
(Original image courtesy of NASA.)

lence), and $k = 0.00025$ (low turbulence). All images are of size 480×480 pixels.

Another major approach in modeling is to derive a mathematical model starting from basic principles. We illustrate this procedure by treating in some detail the case in which an image has been blurred by uniform linear motion between the image and the sensor during image acquisition. Suppose that an image $f(x, y)$ undergoes planar motion and that $x_0(t)$ and $y_0(t)$ are the time varying components of motion in the x- and y-directions, respectively. The total exposure at any point of the recording medium (say, film or digital memory) is obtained by integrating the instantaneous exposure over the time interval during which the imaging system shutter is open.

Assuming that shutter opening and closing takes place instantaneously, and that the optical imaging process is perfect, isolates the effect of image motion. Then, if T is the duration of the exposure, it follows that

$$g(x, y) = \int_0^T f[x - x_0(t), y - y_0(t)] \, dt \tag{5.6-4}$$

where $g(x, y)$ is the blurred image.

From Eq. (4.2-3), the Fourier transform of Eq. (5.6-4) is

$$G(u, v) = \int_{-\infty}^{\infty} \int_{-\infty}^{\infty} g(x, y) e^{-j2\pi(ux+vy)} \, dx \, dy$$

$$= \int_{-\infty}^{\infty} \int_{-\infty}^{\infty} \left[\int_0^T f[x - x_0(t), y - y_0(t)] \, dt \right] e^{-j2\pi(ux+vy)} \, dx \, dy. \tag{5.6-5}$$

Reversing the order of integration allows Eq. (5.6-5) to be expressed in the form

$$G(u, v) = \int_0^T \left[\int_{-\infty}^{\infty} \int_{-\infty}^{\infty} f[x - x_0(t), y - y_0(t)] e^{-j2\pi(ux+vy)} \, dx \, dy \right] dt. \tag{5.6-6}$$

The term inside the outer brackets is the Fourier transform of the displaced function $f[x - x_0(t), y - y_0(t)]$. Using Eq. (4.6-2) then yields the expression

$$G(u, v) = \int_0^T F(u, v) e^{-j2\pi[ux_0(t)+vy_0(t)]} \, dt$$

$$= F(u, v) \int_0^T e^{-j2\pi[ux_0(t)+vy_0(t)]} \, dt \tag{5.6-7}$$

where the last step follows from the fact that $F(u, v)$ is independent of t.

By defining

$$H(u, v) = \int_0^T e^{-j2\pi[ux_0(t)+vy_0(t)]} \, dt \tag{5.6-8}$$

Eq. (5.6-7) may be expressed in the familiar form

$$G(u, v) = H(u, v) F(u, v). \tag{5.6-9}$$

If the motion variables $x_0(t)$ and $y_0(t)$ are known, the transfer function $H(u, v)$ can be obtained directly from Eq. (5.6-8). As an illustration, suppose that the image in question undergoes uniform linear motion in the x-direction only, at a rate given by $x_0(t) = at/T$. When $t = T$, the image has been displaced by a total distance a. With $y_0(t) = 0$, Eq. (5.6-8) yields

$$
\begin{aligned}
H(u, v) &= \int_0^T e^{-j2\pi u x_0(t)} \, dt \\
&= \int_0^T e^{-j2\pi u a t/T} \, dt \\
&= \frac{T}{\pi u a} \sin(\pi u a) e^{-j\pi u a}.
\end{aligned}
\tag{5.6-10}
$$

It is noted that H vanishes at values of u given by $u = n/a$, where n is an integer. If we allow the y-component to vary as well, with the motion given by $y_0 = bt/T$, then the degradation function becomes

$$
H(u, v) = \frac{T}{\pi(ua + vb)} \sin[\pi(ua + vb)] e^{-j\pi(ua + vb)}.
\tag{5.6-11}
$$

EXAMPLE 5.10:
Image blurring
due to motion.

■ The blurring characteristics of Eq. (5.6-11) are illustrated next. Figure 5.26(b) is an image blurred by computing the Fourier transform of the image in Fig. 5.26(a), multiplying the transform by $H(u, v)$ from Eq. (5.6-11), and taking the inverse transform. The images are of size 688×688 pixels, and the parameters used in Eq. (5.6-11) were $a = b = 0.1$ and $T = 1$. As discussed in Sections 5.8 and 5.9, recovery of the original image from its blurred counterpart presents some interesting challenges, particularly when noise is present in the degraded image.   ■

FIGURE 5.26 (a) Original image. (b) Result of blurring using the function in Eq. (5.6-11) with $a = b = 0.1$ and $T = 1$.

5.7 Inverse Filtering

The material in this section is our first step in studying restoration of images degraded by a degradation function H, which is given or obtained by a method such as those discussed in the previous section. The simplest approach to restoration is direct inverse filtering, where we compute an estimate, $\hat{F}(u, v)$, of the transform of the original image simply by dividing the transform of the degraded image, $G(u, v)$, by the degradation function:

$$\hat{F}(u, v) = \frac{G(u, v)}{H(u, v)}. \tag{5.7-1}$$

The divisions are between individual elements of the functions, as explained in connection with Eq. (5.5-17). Substituting the right side of Eq. (5.5-17) for $G(u, v)$ in Eq. (5.7-1) yields

$$\hat{F}(u, v) = F(u, v) + \frac{N(u, v)}{H(u, v)}. \tag{5.7-2}$$

This is an interesting expression. It tells us that even if we know the degradation function we cannot recover the undegraded image [the inverse Fourier transform of $F(u, v)$] exactly because $N(u, v)$ is a random function whose Fourier transform is not known. There is more bad news. If the degradation has zero or very small values, then the ratio $N(u, v)/H(u, v)$ could easily dominate the estimate $\hat{F}(u, v)$. This, in fact, is frequently the case, as will be demonstrated shortly.

One approach to get around the zero or small-value problem is to limit the filter frequencies to values near the origin. From Eq. (4.2-22) we know that $H(0, 0)$ is equal to the average value of $h(x, y)$ and that this is usually the highest value of $H(u, v)$ in the frequency domain. Thus, by limiting the analysis to frequencies near the origin, we reduce the probability of encountering zero values. This approach is illustrated in the following example.

■ The image shown in Fig. 5.25(b) was inverse filtered with Eq. (5.7-1) using the exact inverse of the degradation function that generated that image. That is, the degradation function used was

EXAMPLE 5.11:
Inverse filtering.

$$H(u, v) = e^{-k[(u-M/2)^2 + (v-N/2)^2]^{5/6}}$$

with $k = 0.0025$. The $M/2$ and $N/2$ constants are offset values; they center the function so that it will correspond with the centered Fourier transform, as discussed on numerous occasions in the previous chapter. In this case, $M = N = 480$. We know that a Gaussian-shape function has no zeros, so that will not be a concern here. However, in spite of this, the degradation values became so small that the result of full inverse filtering [Fig. 5.27(a)] is useless. The reasons for this poor result are as discussed in connection with Eq. (5.7-2).

Figures 5.27(b) through (d) show the results of cutting off values of the ratio $G(u, v)/H(u, v)$ outside a radius of 40, 70, and 85, respectively. The cut off was implemented by applying to the ratio a Butterworth lowpass function of order 10. This provided a sharp (but smooth) transition at the desired radius. Radii near 70 yielded the best visual results [Fig. 5.27(c)]. Radius values below that

a b
c d

FIGURE 5.27
Restoring
Fig. 5.25(b) with
Eq. (5.7-1).
(a) Result of
using the full
filter. (b) Result
with H cut off
outside a radius of
40; (c) outside a
radius of 70; and
(d) outside a
radius of 85.

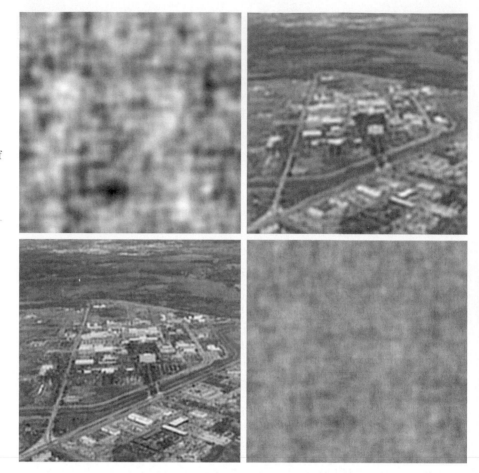

tended toward blurred images, as illustrated in Fig. 5.27(b), which was obtained using a radius of 40. Values above 70 started to produce degraded images, as illustrated in Fig. 5.27(d), which was obtained using a radius of 85. The image content is almost visible behind a "curtain" of noise, but the noise definitely dominates the result. Further increases in radius values produced images that looked more and more like Fig. 5.27(a). ■

The results in the preceding example are illustrative of the poor performance of direct inverse filtering in general. The basic theme of the sections that follow is how to improve on direct inverse filtering.

5.8 Minimum Mean Square Error (Wiener) Filtering

The inverse filtering approach discussed in the previous section makes no explicit provision for handling noise. In this section we discuss an approach that incorporates both the degradation function and statistical characteristics of noise into the restoration process. The method is founded on considering images and noise as random processes, and the objective is to find an estimate $\hat{f}$ of the uncorrupted image f such that the mean square error between them is minimized. This error measure is given by

$$e^2 = E\{(f - \hat{f})^2\} \tag{5.8-1}$$

where $E\{\cdot\}$ is the expected value of the argument. It is assumed that the noise and the image are uncorrelated; that one or the other has zero mean; and that the gray levels in the estimate are a linear function of the levels in the degraded image. Based on these conditions, the minimum of the error function in Eq. (5.8-1) is given in the frequency domain by the expression

$$
\begin{aligned}
\hat{F}(u, v) &= \left[\frac{H^*(u, v) S_f(u, v)}{S_f(u, v)|H(u, v)|^2 + S_\eta(u, v)} \right] G(u, v) \\
&= \left[\frac{H^*(u, v)}{|H(u, v)|^2 + S_\eta(u, v)/S_f(u, v)} \right] G(u, v) \\
&= \left[\frac{1}{H(u, v)} \frac{|H(u, v)|^2}{|H(u, v)|^2 + S_\eta(u, v)/S_f(u, v)} \right] G(u, v)
\end{aligned}
\tag{5.8-2}
$$

where we used the fact that the product of a complex quantity with its conjugate is equal to the magnitude of the complex quantity squared. This result is known as the *Wiener filter*, after N. Wiener [1942], who first proposed the concept in the year shown. The filter, which consists of the terms inside the brackets, also is commonly referred to as the *minimum mean square error filter* or the *least square error filter*. We include references at the end of the chapter to sources containing detailed derivations of the Wiener filter. Note from the first line in Eq. (5.8-2) that the Wiener filter does not have the same problem as the inverse filter with zeros in the degradation function, unless both $H(u, v)$ and $S_\eta(u, v)$ are zero for the same value(s) of u and v.

The terms in Eq. (5.8-2) are as follows:

$H(u, v) =$ degradation function

$H^*(u, v) =$ complex conjugate of $H(u, v)$

$|H(u, v)|^2 = H^*(u, v)H(u, v)$

$S_\eta(u, v) = |N(u, v)|^2 =$ power spectrum of the noise [see Eq. (4.2-20)]

$S_f(u, v) = |F(u, v)|^2 =$ power spectrum of the undegraded image.

As before, $H(u, v)$ is the transform of the degradation function and $G(u, v)$ is the transform of the degraded image. The restored image in the spatial domain is given by the inverse Fourier transform of the frequency-domain estimate $\hat{F}(u, v)$. Note that if the noise is zero, then the noise power spectrum vanishes and the Wiener filter reduces to the inverse filter.

When we are dealing with spectrally white noise, the spectrum $|N(u, v)|^2$ is a constant, which simplifies things considerably. However, the power spectrum of the undegraded image seldom is known. An approach used frequently when these quantities are not known or cannot be estimated is to approximate Eq. (5.8-2) by the expression

$$\hat{F}(u, v) = \left[\frac{1}{H(u, v)} \frac{|H(u, v)|^2}{|H(u, v)|^2 + K} \right] G(u, v) \tag{5.8-3}$$

a b c

FIGURE 5.28 Comparison of inverse- and Wiener filtering. (a) Result of full inverse filtering of Fig. 5.25(b). (b) Radially limited inverse filter result. (c) Wiener filter result.

where K is a specified constant. The following examples illustrate the use of this expression.

EXAMPLE 5.12:
Comparison of inverse and Wiener filtering.

■ Figure 5.28 illustrates the power of Wiener filtering over direct inverse filtering. Figure 5.28(a) is the full inverse-filtered result shown in Fig. 5.27(a). Similarly, Fig. 5.28(b) is the radially limited inverse filter result of Fig, 5.27(c). These images are duplicated here for convenience in making comparisons. Figure 5.28(c) shows the result obtained using Eq. (5.8-3) with the degradation function used in Example 5.11. The value of K was chosen interactively to yield the best visual results. The power of Wiener filtering over the direct inverse approach is evident in this example. By comparing Figs. 5.25(a) and 5.28(c), we see that the Wiener filter yielded a result very close in appearance to the original image. ■

EXAMPLE 5.13:
Further comparisons of Wiener filtering.

■ The first row of Fig. 5.29 shows, from left to right, the blurred image of Fig. 5.26(b) heavily corrupted by additive Gaussian noise of zero mean and variance of 650; the result of direct inverse filtering; and the result of Wiener filtering. The Wiener filter of Eq. (5.8-3) was used, with $H(u, v)$ from Example 5.10, and with K chosen interactively to give the best possible visual result. As expected, the inverse filter produced an unusable image. Note that the noise in the inverse filter image is so strong that its structure is in the direction of the *deblurring* filter. The Wiener filter result is by no means perfect, but it does give us a hint as to image content. With some difficulty, the text is readable.

The second row of Fig. 5.29 shows the same sequence, but with the level of noise variance reduced by one order of magnitude. This reduction had little effect on the inverse filter, but the Wiener results are considerably improved. The text now is much easier to read. In the third row of Fig. 5.29, the noise variance has decreased more than five orders of magnitude from the first row. In fact, image 5.29(g) has no visible noise. The inverse filter result is interesting in this case. The noise is still quite visible, but the text can be seen through a "curtain" of noise. This is a good example of the comments made regarding Eq. (5.7-2). In other words, as is evident in Fig. 5.29(h), the inverse filter was

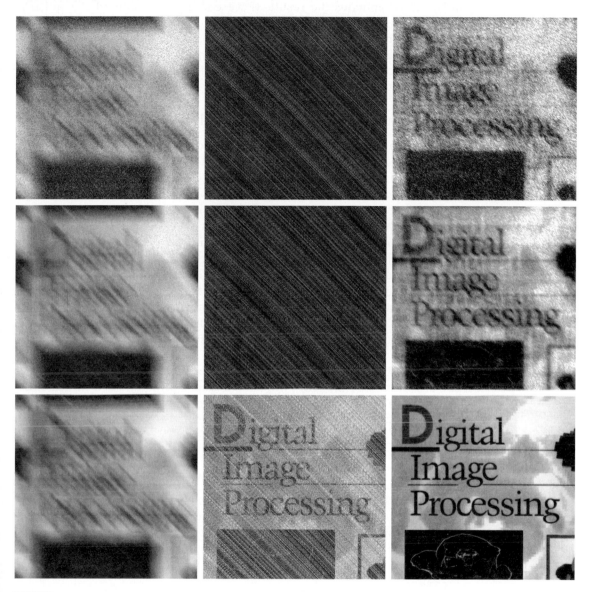

a	b	c
d	e	f
g	h	i

FIGURE 5.29 (a) Image corrupted by motion blur and additive noise. (b) Result of inverse filtering. (c) Result of Wiener filtering. (d)–(f) Same sequence, but with noise variance one order of magnitude less. (g)–(i) Same sequence, but noise variance reduced by five orders of magnitude from (a). Note in (h) how the deblurred image is quite visible through a "curtain" of noise.

quite capable of essentially eliminating the blur in the image. However, the noise still dominates the result. If we could "look" behind the noise in Figs. 5.29(b) and (e), the characters also would show with little blurring. The Wiener filter result in Fig. 5.29(i) is excellent, being quite close visually to the original image shown in Fig. 5.26(a). These types of results are representative of what is possible with Wiener filtering, as long as a reasonable estimate of the degradation function is available.   ■

5.9 Constrained Least Squares Filtering

The problem of having to know something about the degradation function H is common to all methods discussed in this chapter. However, the Wiener filter presents an additional difficulty: The power spectra of the undegraded image and noise must be known. We showed in the previous section that it is possible to achieve excellent results using the approximation given in Eq. (5.8-3). However, a constant estimate of the ratio of the power spectra is not always a suitable solution.

The method discussed in this section requires knowledge of only the mean and variance of the noise. As discussed in Section 5.2.4, these parameters usually can be calculated from a given degraded image, so this is an important advantage. Another difference is that the Wiener filter is based on minimizing a statistical criterion and, as such, it is optimal in an average sense. The algorithm presented in this section has the notable feature that it yields an optimal result for *each* image to which it is applied. Of course, it is important to keep in mind that these optimality criteria, while satisfying from a theoretical point of view, are not related to the dynamics of visual perception. As a result, the choice of one algorithm over the other will almost always be determined (at least partially) by the perceived visual quality of the resulting images.

By using the definition of convolution given in Eq. (4.2-30), we can express Eq. (5.5-16) in vector-matrix form, as follows:

$$\mathbf{g} = \mathbf{Hf} + \boldsymbol{\eta}. \tag{5.9-1}$$

See inside front cover
Consult the book web site for a brief review of vectors and matrices.

For example, suppose that $g(x, y)$ is of size $M \times N$. Then we can form the first N elements of the vector $\mathbf{g}$ by using the image elements in first row of $g(x, y)$, the next N elements from the second row, and so on. The resulting vector will have dimensions $MN \times 1$. These are also the dimensions of $\mathbf{f}$ and $\boldsymbol{\eta}$, as these vectors are formed in the same manner. The matrix $\mathbf{H}$ then has dimensions $MN \times MN$. Its elements are given by the elements of the convolution given in Eq. (4.2-30).

It would be reasonable to come to the conclusion that the restoration problem can now be reduced to simple matrix manipulations. Unfortunately, this is not the case. For instance, suppose that we are working with images of medium size; say $M = N = 512$. Then the vectors in Eq. (5.9-1) would be of dimension $262,144 \times 1$, and matrix $\mathbf{H}$ would be of dimensions $262,144 \times 262,144$. Manipulating vectors and matrices of these sizes is not a trivial task. The problem is complicated further by the fact $\mathbf{H}$ is highly sensitive to noise (after the experiences we had with the effect of noise in the previous two sections, this should not be a surprise). However, formulating the restoration problem in matrix form does facilitate derivation of restoration techniques.

Although we do not fully derive the method of constrained least squares that we are about to present, this method has its roots in a matrix formulation. We give references at the end of the chapter to sources where such derivations are covered in detail. Central to the method is the issue of the sensitivity of **H** to noise. One way to alleviate the noise sensitivity problem is to base optimality of restoration on a measure of smoothness, such as the second derivative of an image (our old friend the Laplacian). To be meaningful, the restoration must be constrained by the parameters of the problems at hand. Thus, what is desired is to find the minimum of a criterion function, C, defined as

$$C = \sum_{x=0}^{M-1} \sum_{y=0}^{N-1} \left[\nabla^2 f(x, y) \right]^2 \tag{5.9-2}$$

subject to the constraint

$$\|\mathbf{g} - \mathbf{H}\hat{\mathbf{f}}\|^2 = \|\boldsymbol{\eta}\|^2 \tag{5.9-3}$$

where $\|\mathbf{w}\|^2 \triangleq \mathbf{w}^T\mathbf{w}$ is the Euclidean vector norm,[†] and $\hat{\mathbf{f}}$ is the estimate of the undegraded image. The Laplacian operator ∇^2 is defined in Eq. (3.7-1).

The frequency domain solution to this optimization problem is given by the expression

$$\hat{F}(u, v) = \left[\frac{H^*(u, v)}{|H(u, v)|^2 + \gamma |P(u, v)|^2} \right] G(u, v) \tag{5.9-4}$$

where γ is a parameter that must be adjusted so that the constraint in Eq. (5.9-3) is satisfied, and $P(u, v)$ is the Fourier transform of the function

$$p(x, y) = \begin{bmatrix} 0 & -1 & 0 \\ -1 & 4 & -1 \\ 0 & -1 & 0 \end{bmatrix}. \tag{5.9-5}$$

We recognize this function as the Laplacian operator introduced in Section 3.7.2. As noted earlier, it is important to keep in mind that $p(x, y)$, as well as all other relevant spatial domain functions, must be properly padded with zeros prior to computing their Fourier transforms for use in Eq. (5.9-4), as discussed in Section 4.6.3. Note that Eq. (5.9-4) reduces to inverse filtering if γ is zero.

■ Figure 5.30 shows the result of processing Figs. 5.29(a), (d), and (g) with constrained least squares filters, in which the values of γ were selected manually to yield the best visual results. This is the same procedure we used to generate the Wiener filtered results in Fig. 5.29(c), (f), and (i). By comparing the constrained least squares and Wiener results, it is noted that the former yielded slightly better results for the high- and medium-noise cases, with both filters generating essentially equal results for the low-noise case. It is not unexpected that the constrained least squares filter would outperform the Wiener filter when selecting the parameters manually for better visual results. The parameter γ in Eq. (5.9-4) is a scalar, while the value of K in Eq. (5.8-3) is an approximation to

EXAMPLE 5.14:
Comparison of Wiener and constrained least squares filtering.

[†] Recall that, for a vector $\mathbf{w}$ with n components, $\mathbf{w}^T\mathbf{w} = \sum_{k=1}^{n} w_k^2$, where w_k is the kth component of $\mathbf{w}$.

a b c

FIGURE 5.30 Results of constrained least squares filtering. Compare (a), (b), and (c) with the Wiener filtering results in Figs. 5.29(c), (f), and (i), respectively.

the ratio of two unknown frequency domain functions, whose ratio seldom is constant. Thus, it stands to reason that a result based on manually selecting γ would be a more accurate estimate of the undegraded image. ■

As shown in the preceding example, it is possible to adjust the parameter γ interactively until acceptable results are achieved. If we are interested in optimality, however, then the parameter γ must be adjusted so that the constraint in Eq. (5.9-3) is satisfied. A procedure for computing γ by iteration is as follows.

Define a "residual" vector $\mathbf{r}$ as

$$\mathbf{r} = \mathbf{g} - \mathbf{H}\hat{\mathbf{f}}. \tag{5.9-6}$$

Since, from the solution in Eq. (5.9-4), $\hat{F}(u, v)$ (and by implication $\hat{\mathbf{f}}$) is a function of γ, then $\mathbf{r}$ also is a function of this parameter. It can be shown (Hunt [1973]) that

$$\phi(\gamma) = \mathbf{r}^T\mathbf{r}$$
$$= \|\mathbf{r}\|^2 \tag{5.9-7}$$

is a monotonically increasing function of γ. What we want to do is adjust gamma so that

$$\|\mathbf{r}\|^2 = \|\mathbf{\eta}\|^2 \pm a \tag{5.9-8}$$

where a is an accuracy factor. In view of Eq. (5.9-6), if $\|\mathbf{r}\|^2 = \|\mathbf{\eta}\|^2$, the constraint in Eq. (5.9-3) will be strictly satisfied.

Because $\phi(\gamma)$ is monotonic, finding the desired value of γ is not difficult. One approach is to

1. Specify an initial value of γ.
2. Compute $\|\mathbf{r}\|^2$.
3. Stop if Eq. (5.9-8) is satisfied; otherwise return to Step 2 after increasing γ if $\|\mathbf{r}\|^2 < \|\mathbf{\eta}\|^2 - a$ or decreasing γ if $\|\mathbf{r}\|^2 > \|\mathbf{\eta}\|^2 + a$. Use the new value of γ in Eq. (5.9-4) to recompute the optimum estimate $\hat{F}(u, v)$.

Other procedures, such as a Newton-Raphson algorithm, can be used to improve the speed of convergence.

In order to use this algorithm, we need the quantities $\|\mathbf{r}\|^2$ and $\|\boldsymbol{\eta}\|^2$. To compute $\|\mathbf{r}\|^2$, we note from Eq. (5.9-6) that

$$R(u, v) = G(u, v) - H(u, v)\hat{F}(u, v) \qquad (5.9\text{-}9)$$

from which we obtain $r(x, y)$ by computing the inverse transform of $R(u, v)$. Then

$$\|\mathbf{r}\|^2 = \sum_{x=0}^{M-1} \sum_{y=0}^{N-1} r^2(x, y). \qquad (5.9\text{-}10)$$

Computation of $\|\boldsymbol{\eta}\|^2$ leads to an interesting result. First, consider the variance of the noise over the entire image, which we estimate by the sample-average method:

$$\sigma_\eta^2 = \frac{1}{MN} \sum_{x=0}^{M-1} \sum_{y=0}^{N-1} [\eta(x, y) - m_\eta]^2 \qquad (5.9\text{-}11)$$

where

$$m_\eta = \frac{1}{MN} \sum_{x=0}^{M-1} \sum_{y=0}^{N-1} \eta(x, y) \qquad (5.9\text{-}12)$$

is the sample mean. With reference to the *form* of Eq. (5.9-10), we note that the double summation in Eq. (5.9-11) is equal to $\|\boldsymbol{\eta}\|^2$. This gives us the expression

$$\|\boldsymbol{\eta}\|^2 = MN[\sigma_\eta^2 + m_\eta^2]. \qquad (5.9\text{-}13)$$

This is a most useful result. It tells us that we can implement an optimum restoration algorithm by having knowledge of only the mean and variance of the noise. These quantities are not difficult to estimate (Section 5.2.4), assuming that the noise and image gray-level values are not correlated. This is a basic assumption of all the methods discussed in this chapter.

■ Figure 5.31(a) shows the result obtained by using the algorithm just described to estimate the optimum filter for restoring Fig. 5.25(b). The initial value used for γ was 10^{-5}, the correction factor for adjusting γ was 10^{-6}, and the value for a was 0.25. The noise parameters specified were the same used to generate Fig. 5.25(a): a noise variance of 10^{-5}, and zero mean. The restored result is almost as good as Fig. 5.28(c), which was obtained by Wiener filtering with K manually specified for best visual results. Figure 5.31(b) shows what can happen if the wrong estimate of noise parameters are used. In this case, the noise variance specified was 10^{-2} and the mean was left at a value of 0. The result in this case is considerably more blurred. ■

EXAMPLE 5.15:
Iterative estimation of the optimum constrained least squares filter.

As stated a the beginning of this section, it is important to keep in mind that optimum restoration in the sense of constrained least squares does not necessarily imply "best" in the visual sense. Depending on the nature and magnitude of the degradation and noise, the other parameters in the algorithm for iteratively determining the optimum estimate also play a role in the final result. In general, automatically determined restoration filters yield inferior results to manual adjustment of filter parameters. This is particularly true of the constrained least squares filter, which is completely specified by a single, scalar parameter.

a | b

FIGURE 5.31
(a) Iteratively
determined
constrained least
squares
restoration of
Fig. 5.16(b), using
correct noise
parameters.
(b) Result
obtained with
wrong noise
parameters.

5.10   Geometric Mean Filter

It is possible to generalize slightly the Wiener filter discussed in the Section 5.8.
The generalization is in the form of the so-called *geometric mean filter*:

$$\hat{F}(u, v) = \left[\frac{H^*(u, v)}{|H(u, v)|^2} \right]^\alpha \left[\frac{H^*(u, v)}{|H(u, v)|^2 + \beta \left[\dfrac{S_\eta(u, v)}{S_f(u, v)} \right]} \right]^{1-\alpha} G(u, v) \qquad (5.10\text{-}1)$$

with α and β being positive, real constants. The geometric mean filter consists
of the two expressions in brackets raised to the powers α and $1 - \alpha$, respectively.

When $\alpha = 1$ this filter reduces to the inverse filter. With $\alpha = 0$ the filter becomes the so-called *parametric Wiener filter*, which reduces to the standard Wiener
filter when $\beta = 1$. If $\alpha = 1/2$, the filter becomes a product of the two quantities
raised to the same power, which is the definition of the geometric mean, thus giving the filter its name. With $\beta = 1$, as α decreases below $1/2$, the filter performance will tend more toward the inverse filter. Similarly, when α increases above
$1/2$, the filter will behave more like the Wiener filter. When $\alpha = 1/2$ and $\beta = 1$,
the filter also is commonly referred to as the *spectrum equalization filter*. Equation (5.10-1) is quite useful when implementing restoration filters because it really represents a family of filters combined into a single expression.

5.11   Geometric Transformations

We conclude this chapter with an introductory discussion on the use of geometric transformations for image restoration. Unlike the techniques discussed
so far, geometric transformations modify the spatial relationships between pixels in an image. Geometric transformations often are called *rubber-sheet
transformations*, because they may be viewed as the process of "printing" an

image on a sheet of rubber and then stretching this sheet according to some predefined set of rules.

In terms of digital image processing, a geometric transformation consists of two basic operations: (1) a *spatial transformation*, which defines the "rearrangement" of pixels on the image plane; and (2) *gray-level interpolation*, which deals with the assignment of gray levels to pixels in the spatially transformed image. We discuss in the following sections the fundamental ideas underlying these concepts, and their use in the context of image restoration.

5.11.1 Spatial Transformations

Suppose that an image f with pixel coordinates (x, y) undergoes geometric distortion to produce an image g with coordinates (x', y'). This transformation may be expressed as

$$x' = r(x, y) \tag{5.11-1}$$

and

$$y' = s(x, y) \tag{5.11-2}$$

where $r(x, y)$ and $s(x, y)$ are the spatial transformations that produced the geometrically distorted image $g(x', y')$. For example, if $r(x, y) = x/2$ and $s(x, y) = y/2$, the "distortion" is simply a shrinking of the size of $f(x, y)$ by one-half in both spatial directions.

If $r(x, y)$ and $s(x, y)$ were known analytically, recovering $f(x, y)$ from the distorted image $g(x', y')$ by applying the transformations in reverse might be possible theoretically. In practice, however, formulating a single set of analytical functions $r(x, y)$ and $s(x, y)$ that describe the geometric distortion process over the entire image plane generally is not possible. The method used most frequently to overcome this difficulty is to formulate the spatial relocation of pixels by the use of *tiepoints*, which are a subset of pixels whose location in the input (distorted) and output (corrected) images is known precisely.

Figure 5.32 shows quadrilateral regions in a distorted and corresponding corrected image. The vertices of the quadrilaterals are corresponding tiepoints.

FIGURE 5.32
Corresponding tiepoints in two image segments.

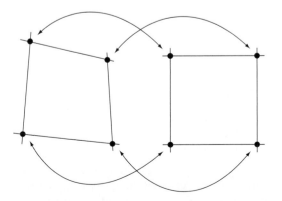

Suppose that the geometric distortion process within the quadrilateral regions is modeled by a pair of bilinear equations so that

$$r(x, y) = c_1 x + c_2 y + c_3 xy + c_4 \tag{5.11-3}$$

and

$$s(x, y) = c_5 x + c_6 y + c_7 xy + c_8. \tag{5.11-4}$$

Then, from Eqs. (5.11-1) and (5.11-2),

$$x' = c_1 x + c_2 y + c_3 xy + c_4 \tag{5.11-5}$$

and

$$y' = c_5 x + c_6 y + c_7 xy + c_8. \tag{5.11-6}$$

Since there are a total of eight known tiepoints, these equations can be solved for the eight coefficients c_i, $i = 1, 2, \ldots, 8$. The coefficients constitute the geometric distortion model used to transform *all* pixels within the quadrilateral region defined by the tiepoints used to obtain the coefficients. In general, enough tiepoints are needed to generate a set of quadrilaterals that cover the entire image, with each quadrilateral having its own set of coefficients.

Once we have the coefficients, the procedure used to generate the corrected (i.e., restored) image is not difficult. If we want to find the value of the undistorted image at any point (x_0, y_0), we simply need to know where in the distorted image $f(x_0, y_0)$ was mapped. This we find out by substituting (x_0, y_0) into Eqs. (5.11-5) and (5.11-6) to obtain the geometrically distorted coordinates (x_0', y_0'). The value of the point in the undistorted image that was mapped to (x_0', y_0') is $g(x_0', y_0')$. So we obtain the restored image value simply by letting $\hat{f}(x_0, y_0) = g(x_0', y_0')$. For example, to generate $\hat{f}(0, 0)$, we substitute $(x, y) = (0, 0)$ into Eqs. (5.11-5) and (5.11-6) to obtain a pair of coordinates (x', y') from those equations. Then we let $\hat{f}(0, 0) = g(x', y')$, where x' and y' are the coordinate values just obtained. Next, we substitute $(x, y) = (0, 1)$ into Eqs. (5.11-5) and (5.11-6), obtain another pair of values (x', y'), and let $f(0, 1) = g(x', y')$ for those coordinate values. The procedure continues pixel by pixel and row by row until an array whose size does not exceed the size of image g is obtained. A column (rather than a row) scan would yield identical results. Also, a bookkeeping procedure is needed to keep track of which quadrilaterals apply at a given pixel location in order to use the proper coefficients.

Tiepoints are established by a number of different techniques, depending on the application. For instance, some image generation systems having physical artifacts (such as metallic points) embedded on the imaging sensor itself. These produce a *known* set of points (called *reseau marks*) directly on the image as it is acquired. If the image is distorted later by some other process (such as an image display or image reconstruction process), then the image can be geometrically corrected using the technique just described.

5.11.2 Gray-Level Interpolation

The method discussed in the preceding section steps through integer values of the coordinates (x, y) to yield the restored image $\hat{f}(x, y)$. However, depending on the values of the coefficients c_i, Eqs. (5.11-5) and (5.11-6) can yield noninteger val-

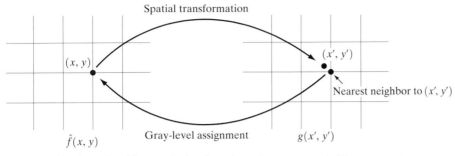

Spatial transformation

(x, y)

(x', y')

Nearest neighbor to (x', y')

$\hat{f}(x, y)$ Gray-level assignment $g(x', y')$

FIGURE 5.33 Gray-level interpolation based on the nearest neighbor concept.

ues for x' and y'. Because the distorted image g is digital, its pixel values are defined only at integer coordinates. Thus using noninteger values for x' and y' causes a mapping into locations of g for which no gray levels are defined. Inferring what the gray-level values at those locations should be, based only on the pixel values at integer coordinate locations, then becomes necessary. The technique used to accomplish this is called *gray-level interpolation*.

The simplest scheme for gray-level interpolation is based on a nearest neighbor approach. This method, also called *zero-order interpolation*, is illustrated in Fig. 5.33. This figure shows (1) the mapping of integer (x, y) coordinates into fractional coordinates (x', y') by means of Eqs. (5.11-5) and (5.11-6); (2) the selection of the closest integer coordinate neighbor to (x', y'); and (3) the assignment of the gray level of this nearest neighbor to the pixel located at (x, y).

Although nearest neighbor interpolation is simple to implement, this method often has the drawback of producing undesirable artifacts, such as distortion of straight edges in images of high resolution. Smoother results can be obtained by using more sophisticated techniques, such as *cubic convolution interpolation*, which fits a surface of the $\sin(z)/z$ type through a much larger number of neighbors (say, 16) in order to obtain a smooth estimate of the gray level at any desired point. Typical areas in which smoother approximations generally are required include 3-D graphics (Watt [1993]) and medical imaging (Lehman et al. [1999]). The price paid for smoother approximations is additional computational burden. For general-purpose image processing a *bilinear interpolation* approach that uses the gray levels of the four nearest neighbors usually is adequate. This approach is straightforward. Because the gray level of each of the four integral nearest neighbors of a nonintegral pair of coordinates (x', y') is known, the gray-level value at these coordinates, denoted $v(x', y')$, can be interpolated from the values of its neighbors by using the relationship

$$v(x', y') = ax' + by' + cx'y' + d \qquad (5.11-7)$$

where the four coefficients are easily determined from the four equations in four unknowns that can be written using the four known neighbors of (x', y'). When these coefficients have been determined, $v(x', y')$ is computed and this value is assigned to the location in $f(x, y)$ that yielded the spatial mapping into location (x', y'). It is easy to visualize this procedure with the aid of Fig. 5.33. The exception is that, instead of using the gray-level value of the nearest neighbor to (x', y'), we actually interpolate a value at location (x', y') and use this value for the gray-level assignment at (x, y).

EXAMPLE 5.16:
Illustration of geometric transformations.

■ Figure 5.34(a) shows an image with 25 regularly spaced tiepoints (highlighted to enhance visibility of the points in the picture). Figure 5.34(b) shows a simple rearrangement of the tiepoints to create geometric distortion. With reference to the procedure discussed in connection with Eqs. (5.11-5) and (5.11-6), the coefficients of these equations are a result of the mapping from the undistorted to the distorted coordinates. Once the coefficients are known, we have the model, and we can either distort an image (for demonstration purposes) or we can recover an image that was geometrically distorted under the set of conditions defined by the coefficients.

Suppose that we want to distort the image in Fig. 5.34(a). We simply substitute the value of any pixel (x_0, y_0) from that image into Eqs. (5.11-5) and (5.11-6) and generate the corresponding coordinates (x'_0, y'_0), which we round off to the closest integer values. The value of the distorted image at that point is given by letting $g(x'_0, y'_0) = f(x_0, y_0)$, or we can use gray-level interpolation on the values of f in the neighborhood of (x_0, y_0). This is the same process described in connection with Eqs. (5.11-5) and (5.11-6). We are simply applying it in reverse.

The result of distorting Fig. 5.34(a) by the method just discussed is shown in Fig. 5.34(c), where the nearest neighbor gray-level assignment scheme was used. Note that this is fairly severe distortion. If this were the given image, we would use the method discussed in connection with Eqs. (5.11-5) and (5.11-6), and one of the gray-level assignment techniques discussed in Section 5.11.2. The result of this procedure is shown in Fig. 5.34(d). The nearest neighbor gray-level assignment method was employed again. Note that the geometric correction was reasonable, but there is a significant number of errors in gray-level assignments, especially along the boundaries between the gray and black regions. Figures 5.34(e) and (f) show the same sequence of experiments, but using bilinear gray-level interpolation instead. The improvements are particularly visible in the boundaries between the gray and black regions.

The images just discussed are so regular and have such few gray levels in the sharp boundaries that almost any type of geometric distortion will cause significant degradation. When images have more texture, geometric correction errors tend to be less noticeable. For example, consider Fig. 5.35. Figure 5.35(b) is the result of geometrically distorting Fig. 5.35(a) in the same manner as Fig. 5.34(e). This distortion in Fig. 5.35(b) is not nearly as noticeable. The differences between Figs. 5.35(a) and (b) are not insignificant, as the difference image in Fig. 5.35(c) shows. They simply are not as visible because of the variety of texture in this image. Finally, Fig. 5.35(d) shows the geometrically corrected image. For all practical purposes, this image is of the same quality as the original.   ■

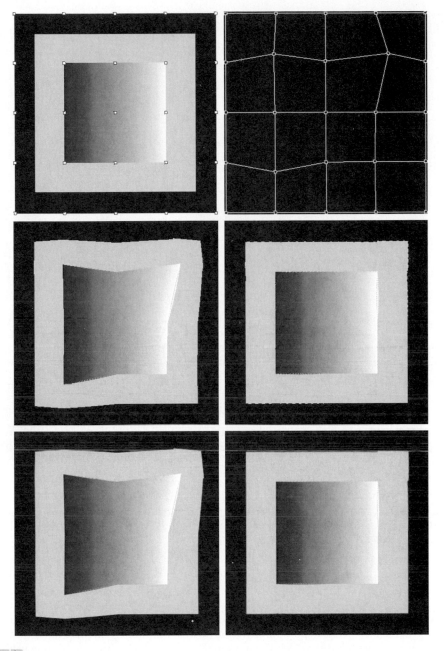

FIGURE 5.34 (a) Image showing tiepoints. (b) Tiepoints after geometric distortion. (c) Geometrically distorted image, using nearest neighbor interpolation. (d) Restored result. (e) Image distorted using bilinear interpolation. (f) Restored image.

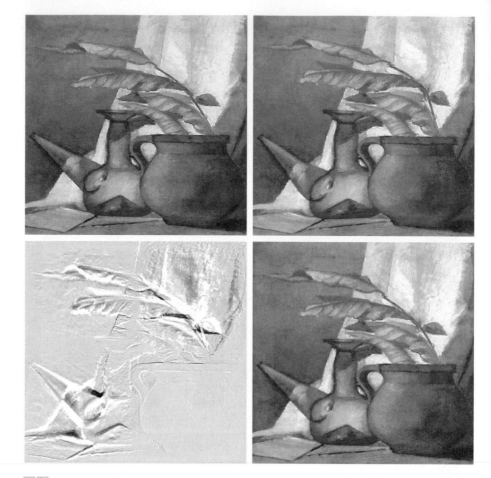

a	b
c	d

FIGURE 5.35 (a) An image before geometric distortion. (b) Image geometrically distorted using the same parameters as in Fig. 5.34(e). (c) Difference between (a) and (b). (d) Geometrically restored image.

Summary

The principal results in this chapter are based on the assumption that image degradation can be modeled as a linear, position-invariant process followed by additive noise that is not correlated with image values. Even when these assumptions are not entirely valid, it often is possible to obtain useful results by using the methods developed in the preceding sections.

Some of the restoration techniques derived in this chapter are based on various criteria of optimality. The use of the word *optimal* in this context refers strictly to a mathematical concept, not to optimal response of the human visual system. In fact, the present lack of knowledge about visual perception precludes a general formulation of the image restoration problem that takes into account observer preferences and capabilities. In view of these limitations, the advantage of the concepts introduced in this chapter is the development of fundamental approaches that have reasonably predictable behavior and are supported by a solid body of knowledge.

As in Chapters 3 and 4, certain restoration tasks, such as random noise reduction, are carried out in the spatial domain using convolution masks. The frequency domain was found ideal for reducing periodic noise and for modeling some important degradations, such as blur caused by motion during image acquisition. We also found the frequency domain to be a useful tool for formulating restoration filters, such as the Wiener and constrained least squares filters.

As mentioned in Chapter 4, the frequency domain offers an intuitive, solid base for experimentation. Once an approach (filter) has been found to perform satisfactorily for a given application, implementation usually is carried out via the design of a digital filter that approximates the frequency-domain solution, but runs much faster in a computer or in a dedicated hardware/firmware system. Digital filter design is beyond the scope of this book, but references relevant to this topic are included in the section that follows.

References and Further Reading

For additional reading on the linear model of degradation presented in Section 5.1, see Castleman [1996] and Pratt [1991]. The book by Peebles [1993] provides an intermediate-level coverage of noise probability density functions and their properties (Section 5.2). The book by Papoulis [1991] is more advanced and covers these concepts in more detail. References for Section 5.3 are Umbaugh [1998], Boie and Cox [1992], Hwang and Haddad [1995], Wilburn [1998], and Eng and Ma [2001]. The general area of adaptive filter design is good background for the adaptive filters discussed in Section 5.3. The book by Haykin [1996] is a good introduction to this topic. The filters in Section 5.4 are direct extensions of the material in Chapter 4. For additional reading on the material of Section 5.5, see Rosenfeld and Kak [1982] and Pratt [1991].

The topic of estimating the degradation function (Section 5.6) is an area of considerable current interest. Some of the early techniques for estimating the degradation function are given in Andrews and Hunt [1977], Rosenfeld and Kak [1982], Bates and McDonnell [1986], and Stark [1987]. Since the degradation function seldom is known exactly, a number of techniques have been proposed over the years, in which specific aspects of restoration are emphasized. For example, Geman and Reynolds [1992], and Hurn and Jennison [1996], deal with issues of preserving sharp transitions in gray levels in an attempt to emphasize sharpness, while Boyd and Meloche [1998] are concerned with restoring thin objects in degraded images. Examples of techniques that deal with image blur are Yitzhaky et al. [1998], Harikumar and Bresler [1999], Mesarović [2000], and Giannakis and Heath [2000]. Restoration of sequences of images also is of considerable interest. The book by Kokaram [1998] provides a good foundation in this area.

The filtering approaches discussed in Sections 5.7 through 5.10 have been explained in various way over the years in numerous books and articles on image processing. There are two major approaches underpinning the development of these filters. One is based on a general formulation using matrix theory, as introduced by Andrews and Hunt [1977]. This approach is elegant and general, but it is difficult for newcomers to the field because it lacks intuitiveness. Approaches based directly on frequency domain filtering (the approach we followed in this chapter) usually are easier to follow by those who first encounter restoration, but lack the unifying mathematical rigor of the matrix approach. Both approaches arrive at the same results, but our experience in teaching this material in a variety of settings indicates that students first entering this field favor the latter approach by a significant margin. Complementary readings for our coverage of the filtering concepts presented in Sections 5.7 through 5.10 are Castleman [1996], Umbaugh [1998], and Petrou and Bosdogianni [1999]. This last reference also presents a nice tie between two-dimensional frequency domain filters and the corresponding digital filters. On the design of 2-D digital filters, see Lu and Antoniou [1992]. Although we do not cover it in this chapter, the area of computerized axial tomography sometimes is included as a topic in restoration. A good introduction to this field is given by Kak and Slaney [2001]. For further basic reading on the material of Section 5.11, see Sonka et al. [1999]. The papers by Unser et al. [1995] and by Carey et al. [1999] also are of interest.

Problems

★ **5.1**   The white bars in the test pattern shown are 7 pixels wide and 210 pixels high. The separation between bars is 17 pixels. What would this image look like after application of

   (a) A 3×3 arithmetic mean filter?

   (b) A 7×7 arithmetic mean filter?

   (c) A 9×9 arithmetic mean filter?

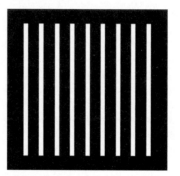

Note: This problem and the ones that follow it, related to filtering this image, may seem a bit tedious. However, they are worth the effort, as they help develop a real understanding of how these filters work. After you understand how a particular filter affects the image, your answer can be a brief verbal description of the result. For example, "the resulting image will consist of vertical bars 3 pixels wide and 206 pixels high." Be sure to describe any deformation of the bars, such as rounded corners. You may ignore image border effects, in which the masks only partially contain image pixels.

5.2   Repeat Problem 5.1 using a geometric mean filter.

★ **5.3**   Repeat Problem 5.1 using a harmonic mean filter.

5.4   Repeat Problem 5.1 using a contraharmonic mean filter with $Q = 1$.

★ **5.5**   Repeat Problem 5.1 using a contraharmonic mean filter with $Q = -1$.

5.6   Repeat Problem 5.1 using a median filter.

★ **5.7**   Repeat Problem 5.1 using a max filter.

5.8   Repeat Problem 5.1 using a min filter.

★ **5.9**   Repeat Problem 5.1 using a midpoint filter.

5.10   The two subimages shown were extracted from the top, right corners of Figs. 5.7(c) and (d), respectively. Thus, the subimage on the left is the result of using an arithmetic mean filter of size 3 × 3; the other subimage is the result of using a geometric mean filter of the same size.

★ **(a)** Explain why the subimage obtained with geometric mean filtering is less blurred. *Hint:* Start your analysis by examining a 1-D step edge profile (see Fig. 3.38 for an example of a step edge).

(b) Explain why the black components in the right image are thicker.

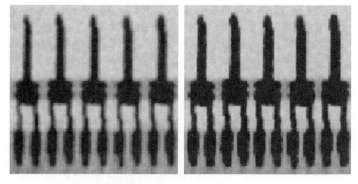

5.11 Refer to the contraharmonic filter given in Eq. (5.3-6).

(a) Explain why the filter is effective in elimination pepper noise when Q is positive.

(b) Explain why the filter is effective in eliminating salt noise when Q is negative.

(c) Explain why the filter gives poor results (such as the results shown in Fig. 5.9) when the wrong polarity is chosen for Q.

(d) Discuss the behavior of the filter when $Q = -1$.

(e) Discuss (for positive and negative Q) the behavior of the filter in areas of constant gray levels.

★ **5.12** Obtain equations for the bandpass filters corresponding to the bandreject filters in Eqs. (5.4-1) through (5.4-3).

5.13 Obtain equations for the notch pass filters corresponding to the notch reject filters discussed in Section 5.4.3. Show that they become lowpass filters when $u_0 = v_0 = 0$.

★ **5.14** Show that the Fourier transform of the 2-D continuous sine function

$$f(x, y) = A \sin(u_0 x + v_0 y)$$

is the pair of conjugate impulses

$$F(u, v) = -j\frac{A}{2}\left[\delta\left(u - \frac{u_0}{2\pi}, v - \frac{v_0}{2\pi}\right) - \delta\left(u + \frac{u_0}{2\pi}, v + \frac{v_0}{2\pi}\right)\right].$$

Hint: Use the continuous version of the Fourier transform in Eq. (4.2-3), and express the sine in terms of exponentials.

5.15 Start with Eq. (5.4-19) and derive Eq. (5.4-21).

★ **5.16** Consider a linear, position-invariant image degradation system with impulse response

$$h(x - \alpha, y - \beta) = e^{-[(x-\alpha)^2 + (y-\beta)^2]}.$$

Suppose that the input to the system is an image consisting of a line of infinitesimal width located at $x = a$, and modeled by $f(x, y) = \delta(x - a)$, where δ is the impulse function. Assuming no noise, what is the output image $g(x, y)$?

5.17   During acquisition, an image undergoes uniform linear motion in the vertical direction for a time T_1. The direction of motion then switches to the horizontal direction for a time interval T_2. Assuming that the time it takes the image to change directions is negligible, and that shutter opening and closing times are negligible also, give an expression for the blurring function, $H(u, v)$.

★**5.18**   Consider the problem of image blurring caused by uniform acceleration in the x-direction. If the image is at rest at time $t = 0$ and accelerates with a uniform acceleration $x_0(t) = at^2/2$ for a time T, find the blurring function $H(u, v)$. You may assume that shutter opening and closing times are negligible.

5.19   A space probe is designed to transmit images from a planet as it approaches it for landing. During the last stages of landing, one of the control thrusters fails, resulting in rapid rotation of the craft about its vertical axis. The images sent during the last two seconds prior to landing are blurred as a consequence of this circular motion. The camera is located in the bottom of the probe, along its vertical axis, and pointing down. Fortunately, the rotation of the craft is also about its vertical axis, so the images are blurred by uniform rotational motion. During the acquisition time of each image the craft rotation was limited to $\pi/8$ radians. The image acquisition process can be modeled as an ideal shutter that is open only during the time the craft rotated the $\pi/8$ radians. You may assume that vertical motion was negligible during image acquisition. Formulate a solution for restoring the images.

★**5.20**   The image shown is a blurred, 2-D projection of a volumetric rendition of a heart. It is known that each of the cross hairs on the right, bottom part of the image was 3 pixels wide, 30 pixels long, and had gray-level values of 255 before blurring. Provide a step-by-step procedure indicating how you would use the information just given to you obtain the blurring function $H(u, v)$.

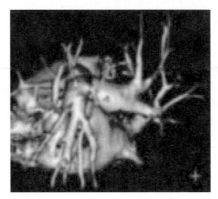

Original image courtesy of G.E. Medical Systems.

5.21   A certain X-ray imaging geometry produces a blurring degradation that can be modeled as the convolution of the sensed image with the spatial, circularly symmetric function

$$h(r) = \left[(r^2 - \sigma^2)/\sigma^4\right]e^{-r^2/2\sigma^2}$$

where $r^2 = x^2 + y^2$. Show that the degradation in the frequency domain is given by the expression

$$H(u, v) = -2\pi\sigma^2(u^2 + v^2)e^{-2\pi^2\sigma^2(u^2 + v^2)}.$$

★ **5.22** Using the transfer function in Problem 5.21, give the expression for a Wiener filter, assuming that the ratio of power spectra of the noise and undegraded signal is a constant.

5.23 Using the transfer function in Problem 5.21, give the resulting expression for the constrained least squares filter.

5.24 Assume that the model in Fig. 5.1 is linear and position invariant and that the noise and image are uncorrelated. Show that the power spectrum of the output is

$$|G(u,v)|^2 = |H(u,v)|^2|F(u,v)|^2 + |N(u,v)|^2.$$

Refer to Eqs. (5.5-17) and (4.2-20).

5.25 Cannon [1974] suggested a restoration filter $R(u, v)$ satisfying the condition

$$|F(u,v)|^2 = |R(u,v)|^2|G(u,v)|^2$$

and based on the premise of forcing the power spectrum of the restored image, $|\hat{F}(u,v)|^2$, to equal the power spectrum of the original image, $|F(u,v)|^2$.

★ **(a)** Find $R(u, v)$ in terms of $|F(u,v)|^2$, $|H(u,v)|^2$, and $|N(u,v)|^2$. *Hint:* Refer to Fig. 5.1, Eq. (5.5-17), and Problem 5.24.

(b) Use your result in (a) to state a result in the form of Eq. (5.8-2).

5.26 An astronomer working with a large-scale telescope observes that her images are a little blurry. The manufacturer tells the astronomer that the unit is operating within specifications. The telescope lenses focus images onto a high-resolution, CCD imaging array, and the images are then converted by the telescope electronics into digital images. Trying to improve the situation by conducting controlled lab experiments with the lenses and imaging sensors is not possible due to the size and weight of the telescope components. The astronomer, having heard about your success as an image processing expert, calls you to help her formulate a digital image processing solution for sharpening the images a little more. How would you go about solving this problem, given that the only images you can obtain are images of stellar bodies?

★ **5.27** A professor of archeology doing research on currency exchange practices during the Roman Empire recently became aware that four Roman coins crucial to his research are listed in the holdings of the British Museum in London. Unfortunately, he was told after arriving there that the coins recently had been stolen. Further research on his part revealed that the museum keeps photographs of every item for which it is responsible. Unfortunately, the photos of the coins in question are blurred to the point where the date and other small markings are not readable. The cause of the blurring was the camera being out of focus when the pictures were taken. As an image processing expert and friend of the professor, you are asked as a favor to determine whether computer processing can be utilized to restore the images to the point where the professor can read the markings. You are told that the original camera used to take the photos is still available, as are other representative coins of the same era. Propose a step-by-step solution to this problem.

5.28 Suppose that, instead of using quadrilaterals, you use triangular regions in Section 5.11 to establish a spatial transformation and gray-level interpolation. What would be the equations analogous to Eqs. (5.11-5), (5.11-6), and (5.11-7) for triangular regions?

6 *Color Image Processing*

It is only after years of preparation that the young artist should touch color—not color used descriptively, that is, but as a means of personal expression.

Henri Matisse

For a long time I limited myself to one color—as a form of discipline.

Pablo Picasso

Preview

The use of color in image processing is motivated by two principal factors. First, color is a powerful descriptor that often simplifies object identification and extraction from a scene. Second, humans can discern thousands of color shades and intensities, compared to about only two dozen shades of gray. This second factor is particularly important in manual (i.e., when performed by humans) image analysis.

Color image processing is divided into two major areas: *full-color* and *pseudo-color* processing. In the first category, the images in question typically are acquired with a full-color sensor, such as a color TV camera or color scanner. In the second category, the problem is one of assigning a color to a particular monochrome intensity or range of intensities. Until recently, most digital color image processing was done at the pseudocolor level. However, in the past decade, color sensors and hardware for processing color images have become available at reasonable prices. The result is that full-color image processing techniques are now used in a broad range of applications, including publishing, visualization, and the Internet.

It will become evident in the discussions that follow that some of the gray-scale methods covered in previous chapters are directly applicable to color images. Others require reformulation to be consistent with the properties of the color spaces developed in this chapter. The techniques described here are far from exhaustive; they illustrate the range of methods available for color image processing.

6.1 Color Fundamentals

Although the process followed by the human brain in perceiving and interpreting color is a physiopsychological phenomenon that is not yet fully understood, the physical nature of color can be expressed on a formal basis supported by experimental and theoretical results.

In 1666, Sir Isaac Newton discovered that when a beam of sunlight passes through a glass prism, the emerging beam of light is not white but consists instead of a continuous spectrum of colors ranging from violet at one end to red at the other. As Fig. 6.1 shows, the color spectrum may be divided into six broad regions: violet, blue, green, yellow, orange, and red. When viewed in full color (Fig. 6.2), no color in the spectrum ends abruptly, but rather each color blends smoothly into the next.

Basically, the colors that humans and some other animals perceive in an object are determined by the nature of the light reflected from the object. As illustrated in Fig. 6.2, visible light is composed of a relatively narrow band of frequencies in the electromagnetic spectrum. A body that reflects light that is balanced in all visible wavelengths appears white to the observer. However, a body that favors reflectance in a limited range of the visible spectrum exhibits some shades of color. For example, green objects reflect light with wavelengths primarily in the 500 to 570 nm range while absorbing most of the energy at other wavelengths.

Characterization of light is central to the science of color. If the light is achromatic (void of color), its only attribute is its *intensity*, or amount. Achromatic light is what viewers see on a black and white television set, and it has been an implicit component of our discussion of image processing thus far. As defined in Chapter 2, and used numerous times since, the term *gray level* refers to a scalar measure of intensity that ranges from black, to grays, and finally to white.

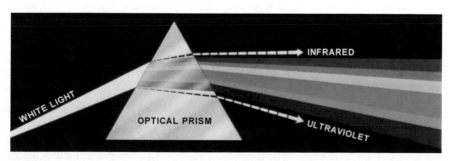

FIGURE 6.1 Color spectrum seen by passing white light through a prism. (Courtesy of the General Electric Co., Lamp Business Division.)

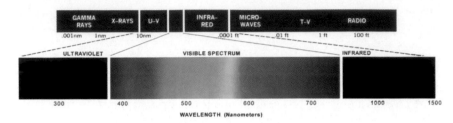

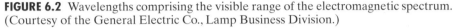

FIGURE 6.2 Wavelengths comprising the visible range of the electromagnetic spectrum. (Courtesy of the General Electric Co., Lamp Business Division.)

Chromatic light spans the electromagnetic spectrum from approximately 400 to 700 nm. Three basic quantities are used to describe the quality of a chromatic light source: radiance, luminance, and brightness. *Radiance* is the total amount of energy that flows from the light source, and it is usually measured in watts (W). *Luminance*, measured in lumens (lm), gives a measure of the amount of energy an observer *perceives* from a light source. For example, light emitted from a source operating in the far infrared region of the spectrum could have significant energy (radiance), but an observer would hardly perceive it; its luminance would be almost zero. Finally, *brightness* is a subjective descriptor that is practically impossible to measure. It embodies the achromatic notion of intensity and is one of the key factors in describing color sensation.

As noted in Section 2.1.1, cones are the sensors in the eye responsible for color vision. Detailed experimental evidence has established that the 6 to 7 million cones in the human eye can be divided into three principal sensing categories, corresponding roughly to red, green, and blue. Approximately 65% of all cones are sensitive to red light, 33% are sensitive to green light, and only about 2% are sensitive to blue (but the blue cones are the most sensitive). Figure 6.3 shows average experimental curves detailing the absorption of light by the red, green, and blue cones in the eye. Due to these absorption characteristics of the human eye, colors are seen as variable combinations of the so-called *primary colors* red (R), green (G), and blue (B). For the purpose of standardization, the CIE (Commission Internationale de l'Eclairage—the International Commission on Illumination) designated in 1931 the following specific wavelength values to the three primary colors: blue = 435.8 nm, green = 546.1 nm, and red = 700 nm. This standard was set before the detailed experimental curves shown Fig. 6.3 became available in 1965. Thus, the CIE standards correspond only approximately with experimental data. We note from Figs. 6.2 and 6.3 that no single color may be called red, green, or blue. Also, it is important to keep in mind that having three specific primary color wavelengths for the purpose of standardization does not mean that these three fixed RGB components acting alone can generate all spectrum colors. Use of the word *primary* has been widely misinterpreted to mean that the three standard primaries, when mixed in various intensity proportions, can produce *all* visible colors. As we will see shortly, this interpretation is not

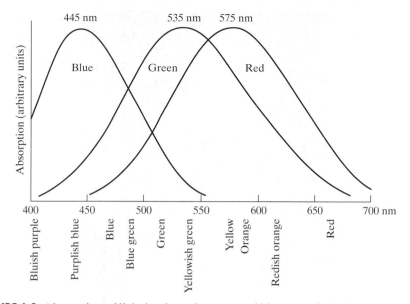
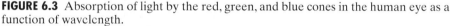

FIGURE 6.3 Absorption of light by the red, green, and blue cones in the human eye as a function of wavelength.

correct unless the wavelength also is allowed to vary, in which case we would no longer have three fixed, standard primary colors.

The primary colors can be added to produce the *secondary* colors of light— magenta (red plus blue), cyan (green plus blue), and yellow (red plus green). Mixing the three primaries, or a secondary with its opposite primary color, in the right intensities produces white light. This result is shown in Fig. 6.4(a), which also illustrates the three primary colors and their combinations to produce the secondary colors.

Differentiating between the primary colors of light and the primary colors of pigments or colorants is important. In the latter, a primary color is defined as one that subtracts or absorbs a primary color of light and reflects or transmits the other two. Therefore, the primary colors of pigments are magenta, cyan, and yellow, and the secondary colors are red, green, and blue. These colors are shown in Fig. 6.4(b). A proper combination of the three pigment primaries, or a secondary with its opposite primary, produces black.

Color television reception is an example of the additive nature of light colors. The interior of many color TV tubes is composed of a large array of triangular dot patterns of electron-sensitive phosphor. When excited, each dot in a triad is capable of producing light in one of the primary colors. The intensity of the red-emitting phosphor dots is modulated by an electron gun inside the tube, which generates pulses corresponding to the "red energy" seen by the TV camera. The green and blue phosphor dots in each triad are modulated in the same manner. The effect, viewed on the television receiver, is that the three primary colors from each phosphor triad are "added" together and received by

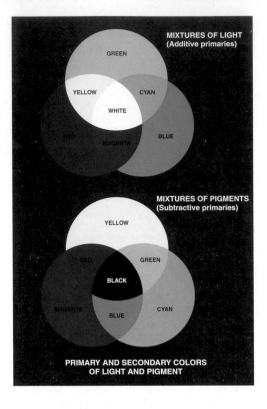

a
b

FIGURE 6.4 Primary and secondary colors of light and pigments. (Courtesy of the General Electric Co., Lamp Business Division.)

the color-sensitive cones in the eye as a full-color image. Thirty successive image changes per second in all three colors complete the illusion of a continuous image display on the screen.

The characteristics generally used to distinguish one color from another are *brightness, hue*, and *saturation*. As indicated earlier in this section, brightness embodies the achromatic notion of intensity. Hue is an attribute associated with the dominant wavelength in a mixture of light waves. Hue represents dominant color as perceived by an observer. Thus, when we call an object red, orange, or yellow, we are specifying its hue. Saturation refers to the relative purity or the amount of white light mixed with a hue. The pure spectrum colors are fully saturated. Colors such as pink (red and white) and lavender (violet and white) are less saturated, with the degree of saturation being inversely proportional to the amount of white light added.

Hue and saturation taken together are called *chromaticity*, and, therefore, a color may be characterized by its brightness and chromaticity. The amounts of red, green, and blue needed to form any particular color are called the *tristimulus* values and are denoted, X, Y, and Z, respectively. A color is then specified by its *trichromatic coefficients*, defined as

$$x = \frac{X}{X + Y + Z} \tag{6.1-1}$$

$$y = \frac{Y}{X + Y + Z} \tag{6.1-2}$$

and

$$z = \frac{Z}{X + Y + Z}. \tag{6.1-3}$$

It is noted from these equations that[†]

$$x + y + z = 1. \tag{6.1-4}$$

For any wavelength of light in the visible spectrum, the tristimulus values needed to produce the color corresponding to that wavelength can be obtained directly from curves or tables that have been compiled from extensive experimental results (Poynton [1996]. See also the early references by Walsh [1958] and by Kiver [1965]).

Another approach for specifying colors is to use the *CIE chromaticity diagram* (Fig. 6.5), which shows color composition as a function of x (red) and y (green). For any value of x and y, the corresponding value of z (blue) is obtained from Eq. (6.1-4) by noting that $z = 1 - (x + y)$. The point marked green in Fig. 6.5, for example, has approximately 62% green and 25% red content. From Eq. (6.1-4), the composition of blue is approximately 13%.

The positions of the various spectrum colors—from violet at 380 nm to red at 780 nm—are indicated around the boundary of the tongue-shaped chromaticity diagram. These are the pure colors shown in the spectrum of Fig. 6.2. Any point not actually on the boundary but within the diagram represents some mixture of spectrum colors. The point of equal energy shown in Fig. 6.5 corresponds to equal fractions of the three primary colors; it represents the CIE standard for white light. Any point located on the boundary of the chromaticity chart is fully saturated. As a point leaves the boundary and approaches the point of equal energy, more white light is added to the color and it becomes less saturated. The saturation at the point of equal energy is zero.

The chromaticity diagram is useful for color mixing because a straight-line segment joining any two points in the diagram defines all the different color variations that can be obtained by combining these two colors additively. Consider, for example, a straight line drawn from the red to the green points shown in Fig. 6.5. If there is more red light than green light, the exact point representing the new color will be on the line segment, but it will be closer to the red point than to the green point. Similarly, a line drawn from the point of equal energy to any point on the boundary of the chart will define all the shades of that particular spectrum color.

[†] The use x, y, z in this context follows notational convention. These should not be confused with the use of (x, y) to denote spatial coordinates in other sections of the book.

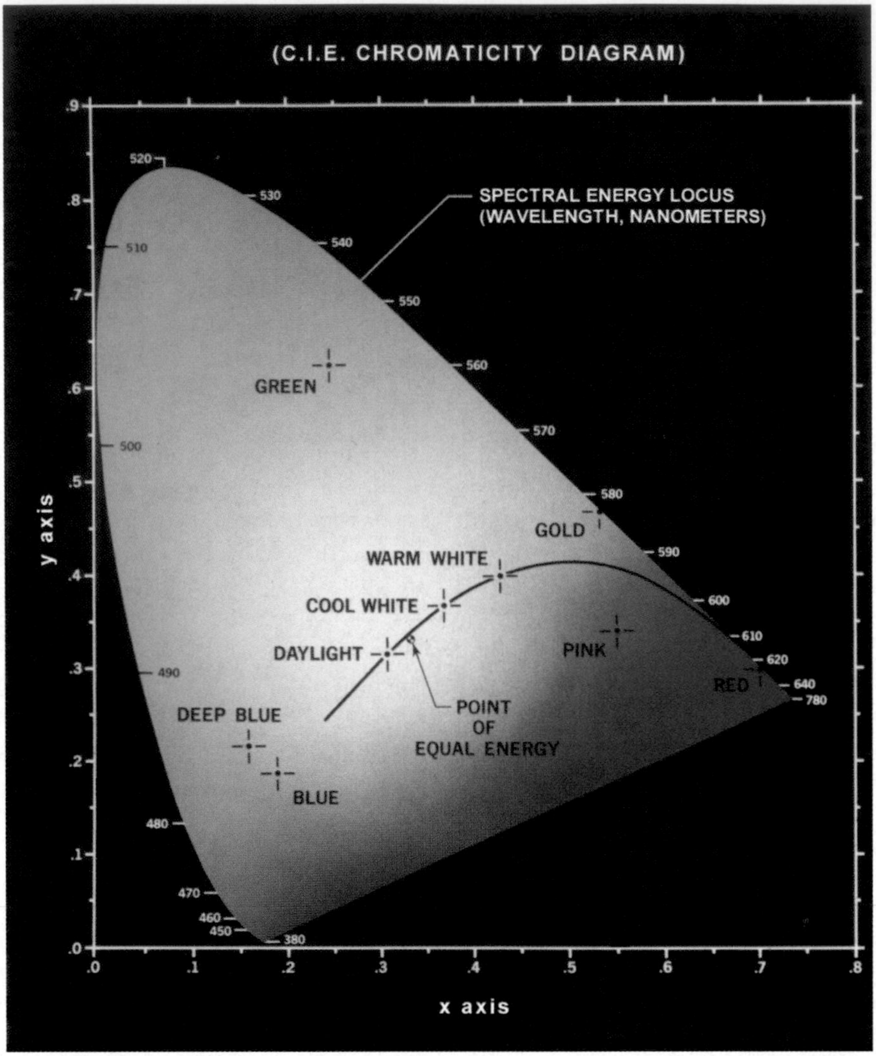

FIGURE 6.5
Chromaticity
diagram.
(Courtesy of the
General Electric
Co., Lamp
Business
Division.)

Extension of this procedure to three colors is straightforward. To determine the range of colors that can be obtained from any three given colors in the chromaticity diagram, we simply draw connecting lines to each of the three color points. The result is a triangle, and any color inside the triangle can be produced by various combinations of the three initial colors. A triangle with vertices at any three *fixed* colors cannot enclose the entire color region in Fig. 6.5. This observation supports graphically the remark made earlier that not all colors can be obtained with three single, fixed primaries.

The triangle in Figure 6.6 shows a typical range of colors (called the *color gamut*) produced by RGB monitors. The irregular region inside the triangle is representative of the color gamut of today's high-quality color printing devices.

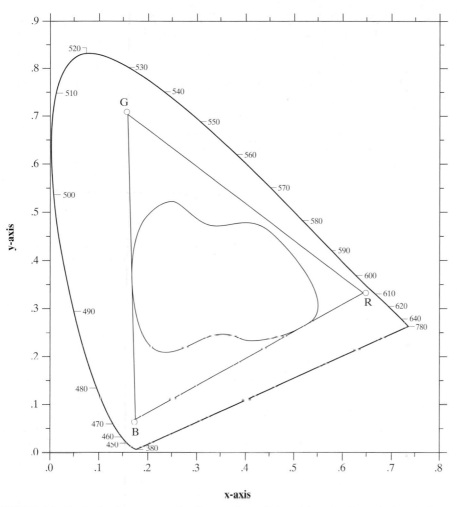

FIGURE 6.6 Typical color gamut of color monitors (triangle) and color printing devices (irregular region).

The boundary of the color printing gamut is irregular because color printing is a combination of additive and subtractive color mixing, a process that is much more difficult to control than that of displaying colors on a monitor, which is based on the addition of three highly controllable light primaries.

6.2 Color Models

The purpose of a color model (also called *color space* or *color system*) is to facilitate the specification of colors in some standard, generally accepted way. In essence, a color model is a specification of a coordinate system and a subspace within that system where each color is represented by a single point.

Most color models in use today are oriented either toward hardware (such as for color monitors and printers) or toward applications where color manipulation is a goal (such as in the creation of color graphics for animation). In terms of digital image processing, the hardware-oriented models most commonly used in practice are the RGB (red, green, blue) model for color monitors and a broad class of color video cameras; the CMY (cyan, magenta, yellow) and CMYK (cyan, magenta, yellow, black) models for color printing; and the HSI (hue, saturation, intensity) model, which corresponds closely with the way humans describe and interpret color. The HSI model also has the advantage that it decouples the color and gray-scale information in an image, making it suitable for many of the gray-scale techniques developed in this book. There are numerous color models in use today due to the fact that color science is a broad field that encompasses many areas of application. It is tempting to dwell on some of these models here simply because they are interesting and informative. However, keeping to the task at hand, the models discussed in this chapter are leading models for image processing. Having mastered the material in this chapter, the reader will have no difficulty in understanding additional color models in use today.

6.2.1 The RGB Color Model

In the RGB model, each color appears in its primary spectral components of red, green, and blue. This model is based on a Cartesian coordinate system. The color subspace of interest is the cube shown in Fig. 6.7, in which RGB values are at three corners; cyan, magenta, and yellow are at three other corners; black is at the origin; and white is at the corner farthest from the origin. In this model, the gray scale (points of equal RGB values) extends from black to white along the line joining these two points. The different colors in this model are points on or inside the cube, and are defined by vectors extending from the origin. For

FIGURE 6.7
Schematic of the RGB color cube. Points along the main diagonal have gray values, from black at the origin to white at point $(1, 1, 1)$.

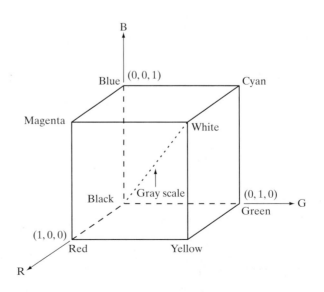

convenience, the assumption is that all color values have been normalized so that the cube shown in Fig. 6.7 is the unit cube. That is, all values of R, G, and B are assumed to be in the range $[0, 1]$.

Images represented in the RGB color model consist of three component images, one for each primary color. When fed into an RGB monitor, these three images combine on the phosphor screen to produce a composite color image. The number of bits used to represent each pixel in RGB space is called the *pixel depth*. Consider an RGB image in which each of the red, green, and blue images is an 8-bit image. Under these conditions each RGB *color* pixel [that is, a triplet of values (R, G, B)] is said to have a depth of 24 bits (3 image planes times the number of bits per plane). The term *full-color* image is used often to denote a 24-bit RGB color image. The total number of colors in a 24-bit RGB image is $(2^8)^3 = 16,777,216$. Figure 6.8 shows the 24-bit RGB color cube corresponding to the diagram in Fig. 6.7.

■ The cube shown in Fig. 6.8 is a solid, composed of the $(2^8)^3 = 16,777,216$ colors mentioned in the preceding paragraph. A convenient way to view these colors is to generate color planes (faces or cross sections of the cube). This is accomplished simply by fixing one of the three colors and allowing the other two to vary. For instance, a cross-sectional plane through the center of the cube and parallel to the GB-plane in Figs. 6.8 and 6.7 is the plane $(127, G, B)$ for $G, B = 0, 1, 2, \ldots, 255$. Here we used the actual pixel values rather than the mathematically convenient normalized values in the range $[0, 1]$ because the former values are the ones actually used in a computer to generate colors. Figure 6.9(a) shows that an image of the cross-sectional plane is viewed simply by feeding the three individual component images into a color monitor. In the component images, 0 represents black and 255 represents white (note that these are gray-scale images). Finally, Fig. 6.9(b) shows the three hidden surface planes of the cube in Fig. 6.8, generated in the same manner.

It is of interest to note that *acquiring* a color image is basically the process shown in Fig. 6.9 in reverse. A color image can be acquired by using three filters, sensitive to red, green, and blue, respectively. When we view a color scene with a monochrome camera equipped with one of these filters, the result is a monochrome image whose intensity is proportional to the response of that filter.

EXAMPLE 6.1:
Generating the hidden face planes and a cross section of the RGB color cube.

FIGURE 6.8 RGB 24-bit color cube.

FIGURE 6.9
(a) Generating
the RGB image of
the cross-sectional
color plane
$(127, G, B)$.
(b) The three
hidden surface
planes in the color
cube of Fig. 6.8.

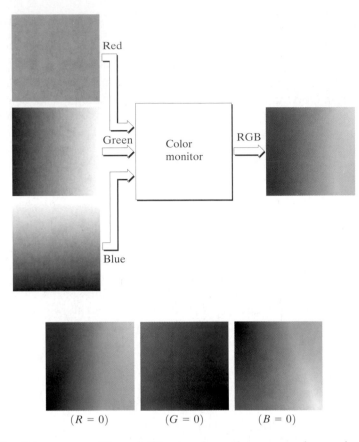

$(R = 0)$ $(G = 0)$ $(B = 0)$

Repeating this process with each filter produces three monochrome images that are the RGB component images of the color scene. (In practice, RGB color image sensors usually integrate this process into a single device.) Clearly, displaying these three RGB component images in the form shown in Fig. 6.9(a) would yield an RGB color rendition of the original color scene. ■

While high-end display cards and monitors provide a reasonable rendition of the colors in a 24-bit RGB image, many systems in use today are limited to 256 colors. Also, there are numerous applications in which it simply makes no sense to use more than a few hundred, and sometimes fewer, colors. A good example of this is provided by the pseudocolor image processing techniques discussed in Section 6.3. Given the variety of systems in current use, it is of considerable interest to have a subset of colors that are likely to be reproduced faithfully, reasonably independently of viewer hardware capabilities. This subset of colors is called the set of *safe RGB colors*, or the set of *all-systems-safe colors*. In Internet applications, they are called *safe Web colors* or *safe browser colors*.

On the assumption that 256 colors is the minimum number of colors that can be reproduced faithfully by any system in which a desired result is likely to be displayed, it is useful to have an accepted standard notation to refer to these colors. Forty of these 256 colors are known to be processed differently by various operating systems, leaving only 216 colors that are common to most systems.

Number System		Color Equivalents				
Hex	00	33	66	99	CC	FF
Decimal	0	51	102	153	204	255

TABLE 6.1
Valid values of each RGB component in a safe color.

These 216 colors have become the de facto standard for safe colors, especially in Internet applications. They are used whenever it is desired that the colors viewed by most people appear the same.

Each of the 216 safe colors is formed from three RGB values as before, but each value can only be 0, 51, 102, 153, 204, or 255. Thus, RGB triplets of these values give us $(6)^3 = 216$ possible values (note that all values are divisible by 3). It is customary to express these values in the hexagonal number system, as shown in Table 6.1. Recall that hex numbers 0, 1, 2, ..., 9, A, B, C, D, E, F correspond to decimal numbers 0, 1, 2, ..., 9, 10, 11, 12, 13, 14, 15. Recall also that $(0)_{16} = (0000)_2$ and $(F)_{16} = (1111)_2$. Thus, for example, $(FF)_{16} = (255)_{10} = (11111111)_2$ and we see that a grouping of two hex numbers forms an 8-bit byte.

Since it takes three numbers to form an RGB color, each safe color is formed from three of the two digit hex numbers in Table 6.1. For example, the purest red is FF0000. The values 000000 and FFFFFF represent black and white, respectively. Keep in mind that the same result is obtained by using the more familiar decimal notation. For instance, the brightest red in decimal notation has $R = 255$ (FF) and $G = B = 0$.

Figure 6.10(a) shows the 216 safe colors, organized in descending RGB values. The square in the top left array has value FFFFFF (white), the second

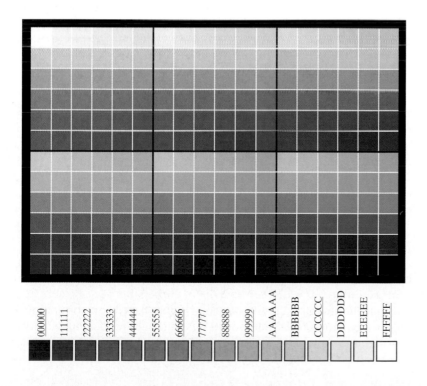

a
b

FIGURE 6.10
(a) The 216 safe RGB colors.
(b) All the grays in the 256-color RGB system (grays that are part of the safe color group are shown underlined).

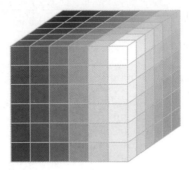

FIGURE 6.11 The RGB safe-color cube.

square to its right has value FFFFCC, the third square has value FFFF99, and so on for the first row. The second row of that same array has values FFCCFF, FFCCCC, FFCC99, and so on. The final square of that array has value FF0000 (the brightest possible red). The second array to the right of the one just examined starts with value CCFFFF and proceeds in the same manner, as do the other remaining four arrays. The final (bottom right) square of the last array has value 000000 (black). It is important to note that not all possible 8-bit gray colors are included in the 216 safe colors. Figure 6.10(b) shows the hex codes for *all* the possible gray colors in a 256-color RGB system. Some of these values are outside of the safe color set but are represented properly (in terms on their relative intensities) by most display systems. The grays from the safe color group, $(KKKKKK)_{16}$, for $K = 0, 3, 6, 9, C, F$, are shown underlined in Fig. 6.10(b).

Figure 6.11 shows the RGB safe-color cube. Unlike the full-color cube in Fig. 6.8, which is solid, the cube in Fig. 6.11 has valid colors only on the surface planes. As shown in Fig. 6.10(a), each plane has a total of 36 colors, so the entire surface of the safe-color cube is covered by 216 different colors, as expected.

6.2.2 The CMY and CMYK Color Models

As indicated in Section 6.1, cyan, magenta, and yellow are the secondary colors of light or, alternatively, the primary colors of pigments. For example, when a surface coated with cyan pigment is illuminated with white light, no red light is reflected from the surface. That is, cyan subtracts red light from reflected white light, which itself is composed of equal amounts of red, green, and blue light.

Most devices that deposit colored pigments on paper, such as color printers and copiers, require CMY data input or perform an RGB to CMY conversion internally. This conversion is performed using the simple operation

$$\begin{bmatrix} C \\ M \\ Y \end{bmatrix} = \begin{bmatrix} 1 \\ 1 \\ 1 \end{bmatrix} - \begin{bmatrix} R \\ G \\ B \end{bmatrix} \tag{6.2-1}$$

where, again, the assumption is that all color values have been normalized to the range $[0, 1]$. Equation (6.2-1) demonstrates that light reflected from a surface coated with pure cyan does not contain red (that is, $C = 1 - R$ in the equation).

Similarly, pure magenta does not reflect green, and pure yellow does not reflect blue. Equation (6.2-1) also reveals that RGB values can be obtained easily from a set of CMY values by subtracting the individual CMY values from 1. As indicated earlier, in image processing this color model is used in connection with generating hardcopy output, so the inverse operation from CMY to RGB generally is of little practical interest.

According to Fig. 6.4, equal amounts of the pigment primaries, cyan, magenta, and yellow should produce black. In practice, combining these colors for printing produces a muddy-looking black. So, in order to produce true black (which is the predominant color in printing), a fourth color, *black*, is added, giving rise to the CMYK color model. Thus, when publishers talk about "four-color printing," they are referring to the three colors of the CMY color model plus black.

6.2.3 The HSI Color Model

As we have seen, creating colors in the RGB and CMY models and changing from one model to the other is a straightforward process. As noted earlier, these color systems are ideally suited for hardware implementations. In addition, the RGB system matches nicely with the fact that the human eye is strongly perceptive to red, green, and blue primaries. Unfortunately, the RGB, CMY, and other similar color models are not well suited for *describing* colors in terms that are practical for human interpretation. For example, one does not refer to the color of an automobile by giving the percentage of each of the primaries composing its color. Furthermore, we do not think of color images as being composed of three primary images that combine to form that single image.

When humans view a color object, we describe it by its hue, saturation, and brightness. Recall from the discussion in Section 6.1 that hue is a color attribute that describes a pure color (pure yellow, orange, or red), whereas saturation gives a measure of the degree to which a pure color is diluted by white light. Brightness is a subjective descriptor that is practically impossible to measure. It embodies the achromatic notion of *intensity* and is one of the key factors in describing color sensation. We do know that intensity (gray level) is a most useful descriptor of monochromatic images. This quantity definitely is measurable and easily interpretable. The model we are about to present, called the *HSI* (hue, saturation, intensity) *color model*, decouples the intensity component from the color-carrying information (hue and saturation) in a color image. As a result, the HSI model is an ideal tool for developing image processing algorithms based on color descriptions that are natural and intuitive to humans, who, after all, are the developers and users of these algorithms. We can summarize by saying that RGB is ideal for image color generation (as in image capture by a color camera or image display in a monitor screen), but its use for color *description* is much more limited. The material that follows provides a very effective way to do this.

As discussed in Example 6.1, an RGB color image can be viewed as three monochrome intensity images (representing red, green, and blue), so it should come as no surprise that we should be able to extract intensity from an RGB image. This becomes rather clear if we take the color cube from Fig. 6.7 and stand it on the black $(0, 0, 0)$ vertex, with the white vertex $(1, 1, 1)$ directly

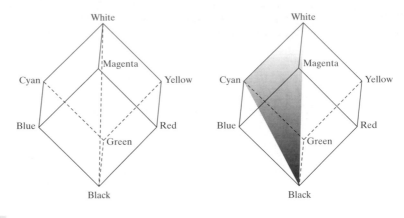

a   b

FIGURE 6.12 Conceptual relationships between the RGB and HSI color models.

above it, as shown in Fig. 6.12(a). As noted in connection with Fig. 6.7, the intensity (gray scale) is along the line joining these two vertices. In the arrangement shown in Fig. 6.12, the line (intensity axis) joining the black and white vertices is vertical. Thus, if we wanted to determine the intensity component of any color point in Fig. 6.12, we would simply pass a plane *perpendicular* to the intensity axis and containing the color point. The intersection of the plane with the intensity axis would give us a point with intensity value in the range $[0, 1]$. We also note with a little thought that the saturation (purity) of a color increases as a function of distance from the intensity axis. In fact, the saturation of points on the intensity axis is zero, as evidenced by the fact that all points along this axis are gray.

In order to see how hue can be determined also from a given RGB point, consider Fig. 6.12(b), which shows a plane defined by three points (black, white, and cyan). The fact that the black and white points are contained in the plane tells us that the intensity axis also is contained in the plane. Furthermore, we see that *all* points contained in the plane segment defined by the intensity axis and the boundaries of the cube have the *same* hue (cyan in this case). We would arrive at the same conclusion by recalling from Section 6.1 that all colors generated by three colors lie in the triangle defined by those colors. If two of those points are black and white and the third is a color point, all points on the triangle would have the same hue because the black and white components cannot change the hue (of course, the intensity and saturation of points in this triangle would be different). By rotating the shaded plane about the vertical intensity axis, we would obtain different hues. From these concepts we arrive at the conclusion that the hue, saturation, and intensity values required to form the HSI space can be obtained from the RGB color cube. That is, we can convert any RGB point to a corresponding point in the HSI color model by working out the geometrical formulas describing the reasoning outlined in the preceding discussion.

The key point to keep in mind regarding the cube arrangement in Fig. 6.12 and its corresponding HSI color space is that the HSI space is represented by

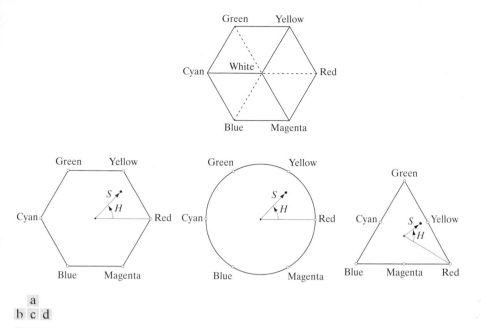

FIGURE 6.13 Hue and saturation in the HSI color model. The dot is an arbitrary color point. The angle from the red axis gives the hue, and the length of the vector is the saturation. The intensity of all colors in any of these planes is given by the position of the plane on the vertical intensity axis.

a vertical intensity axis and the locus of color points that lie on planes *perpendicular* to this axis. As the planes move up and down the intensity axis, the boundaries defined by the intersection of each plane with the faces of the cube have either a triangular or hexagonal shape. This can be visualized much more readily by looking at the cube down its gray-scale axis, as shown in Fig. 6.13(a). In this plane we see that the primary colors are separated by 120°. The secondary colors are 60° from the primaries, which means that the angle between secondaries also is 120°. Figure 6.13(b) shows the same hexagonal shape and an arbitrary color point (shown as a dot). The hue of the point is determined by an angle from some reference point. Usually (but not always) an angle of 0° from the red axis designates 0 hue, and the hue increases counterclockwise from there. The saturation (distance from the vertical axis) is the length of the vector from the origin to the point. Note that the origin is defined by the intersection of the color plane with the vertical intensity axis. The important components of the HSI color space are the vertical intensity axis, the length of the vector to a color point, and the angle this vector makes with the red axis. Therefore, it is not unusual to see the HSI planes defined is terms of the hexagon just discussed, a triangle, or even a circle, as Figs. 6.13(c) and (d) show. The shape chosen really does not matter, since any one of these shapes can be warped into one of the other two by a geometric transformation. Figure 6.14 shows the HSI model based on color triangles and also on circles.

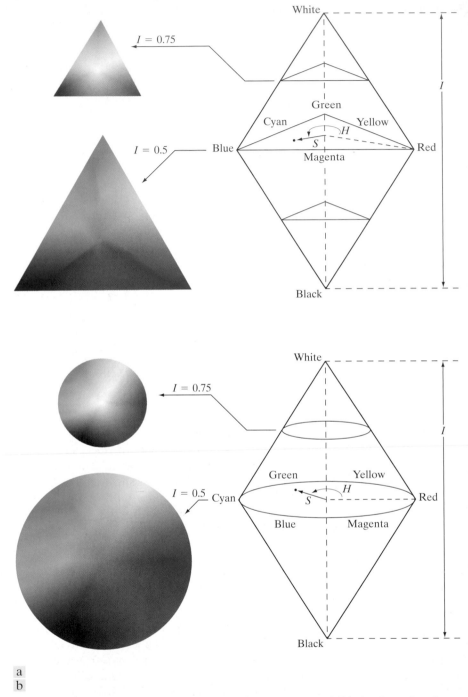

a
b

FIGURE 6.14 The HSI color model based on (a) triangular and (b) circular color planes. The triangles and circles are perpendicular to the vertical intensity axis.

Converting colors from RGB to HSI

Given an image in RGB color format, the H component of each RGB pixel is obtained using the equation

$$H = \begin{cases} \theta & \text{if } B \le G \\ 360 - \theta & \text{if } B > G \end{cases}$$

(6.2-2)

See inside front cover

Consult the book web site for a detailed derivation of the conversion equations between RGB and HSI, and vice versa.

with

$$\theta = \cos^{-1}\left\{ \frac{\frac{1}{2}[(R - G) + (R - B)]}{[(R - G)^2 + (R - B)(G - B)]^{1/2}} \right\}.$$

The saturation component is given by

$$S = 1 - \frac{3}{(R + G + B)}\left[\min(R, G, B)\right].$$

(6.2-3)

Finally, the intensity component is given by

$$I = \frac{1}{3}(R + G + B).$$

(6.2-4)

It is assumed that the RGB values have been normalized to the range $[0, 1]$ and that angle θ is measured with respect to the red axis of the HSI space, as indicated in Fig. 6.13. Hue can be normalized to the range $[0, 1]$ by dividing by $360°$ all values resulting from Eq. (6.2-2). The other two HSI components already are in this range if the given RGB values are in the interval $[0, 1]$.

The results in Eqs. (6.2-2) through (6.2-4) can be derived from the geometry shown in Figs. 6.12 and 6.13. The derivation is tedious and would not add significantly to the present discussion. The interested reader can consult the book's references or web site for a proof of these equations, as well as for the following HSI to RGB conversion results.

Converting colors from HSI to RGB

Given values of HSI in the interval $[0, 1]$, we now want to find the corresponding RGB values in the same range. The applicable equations depend on the values of H. There are three sectors of interest, corresponding to the $120°$ intervals in the separation of primaries (see Fig. 6.13). We begin by multiplying H by $360°$, which returns the hue to its original range of $[0°, 360°]$.

RG sector $(0° \le H < 120°)$: When H is in this sector, the RGB components are given by the equations

$$B = I(1 - S)$$

(6.2-5)

$$R = I\left[1 + \frac{S \cos H}{\cos(60° - H)}\right]$$

(6.2-6)

and

$$G = 3I - (R + B).$$

(6.2-7)

GB sector $(120° \leq H < 240°)$: If the given value of H is in this sector, we first subtract 120° from it:

$$H = H - 120°. \tag{6.2-8}$$

Then the RGB components are

$$R = I(1 - S) \tag{6.2-9}$$

$$G = I\left[1 + \frac{S \cos H}{\cos(60° - H)}\right] \tag{6.2-10}$$

and

$$B = 3I - (R + G). \tag{6.2-11}$$

BR sector $(240° \leq H \leq 360°)$: Finally, if H is in this range, we subtract 240° from it:

$$H = H - 240°. \tag{6.2-12}$$

Then the RGB components are

$$G = I(1 - S) \tag{6.2-13}$$

$$B = I\left[1 + \frac{S \cos H}{\cos(60° - H)}\right] \tag{6.2-14}$$

and

$$R = 3I - (G + B). \tag{6.2-15}$$

Uses of these equations for image processing are discussed in several of the following sections.

EXAMPLE 6.2:
The HSI values corresponding to the image of the RGB color cube.

■ Figure 6.15 shows the hue, saturation, and intensity images for the RGB values shown in Fig. 6.8. Figure 6.15(a) is the hue image. Its most distinguishing feature is the discontinuity in value along a 45° line in the front (red) plane of the cube. To understand the reason for this discontinuity, refer to Fig. 6.8, draw a line from the red to the white vertices of the cube, and select a point in the middle of this line. Starting at that point, draw a path to the right, following the cube around until you return to the starting point. The major colors encountered in this path are yellow, green, cyan, blue, magenta, and back to red. According to Fig. 6.13, the values of hue along this path should increase from 0° to 360° (i.e., from the lowest to highest possible values of hue). This is precisely what Fig. 6.15(a) shows because the lowest value is represented as black and the highest value as white in the gray scale. In fact, the hue image was originally normalized to the range $[0, 1]$ and then scaled to 8 bits; that is, it was converted to the range $[0, 255]$, for display.

The saturation image in Fig. 6.15(b) shows progressively darker values toward the white vertex of the RGB cube, indicating that colors become less and less

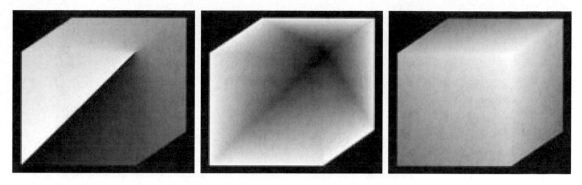

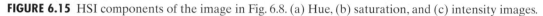

a b c

FIGURE 6.15 HSI components of the image in Fig. 6.8. (a) Hue, (b) saturation, and (c) intensity images.

saturated as they approach white. Finally, every pixel in the intensity image shown in Fig. 6.15(c) is the average of the RGB values at the corresponding pixel in Fig. 6.8. ■

Manipulating HSI component images

In the following discussion, we take a look at some simple techniques for manipulating HSI component images. This will help develop familiarity with these components and also help deepen our understanding of the HSI color model. Figure 6.16(a) shows an image composed of the primary and secondary RGB colors. Figures 6.16(b) through (d) show the H, S, and I components of this image. These images were generated using Eqs. (6.2-2) through (6.2-4). Recall from the discussion earlier in this section that the gray-level values in Fig. 6.16(b) correspond to angles; thus, for example, because red corresponds to $0°$, the red region in Fig. 6.16(a) mapped to a black region in the hue image. Similarly, the gray levels in Fig. 6.16(c) correspond to saturation (they were scaled to $[0, 255]$ for display), and the gray levels in Fig. 6.16(d) are average intensities.

To change the individual color of any region in the RGB image, we change the values of the corresponding region in the hue image of Fig. 6.16(b). Then we convert the new H image, along with the unchanged S and I images, back to RGB using the procedure explained in connection with Eqs. (6.2-5) through (6.2-15). To change the saturation (purity) of the color in any region, we follow the same procedure, except that we make the changes in the saturation image in HSI space. Similar comments apply to changing the average intensity of any region. Of course, these changes can be made simultaneously. For example, the image in Fig. 6.17(a) was obtained by changing to 0 the pixels corresponding to the blue and green regions in Fig. 6.16(b). In Fig. 6.17(b) we reduced by half the saturation of the cyan region in component image S from Fig. 6.16(c). In Fig. 6.17(c) we reduced by half the intensity of the central white region in the intensity image of Fig. 6.16(d). The result of converting this modified HSI image back to RGB is shown in Fig. 6.17(d). As expected, we see in this figure that

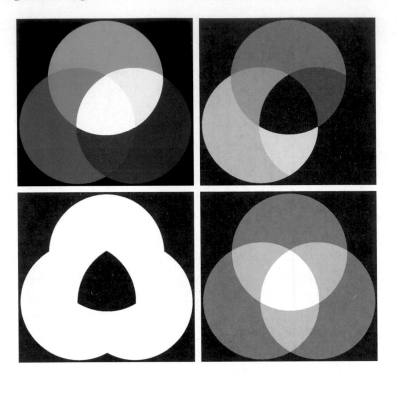

a b
c d

FIGURE 6.16 (a) RGB image and the components of its corresponding HSI image: (b) hue, (c) saturation, and (d) intensity.

the outer portions of all circles are now red; the purity of the cyan region was diminished, and the central region became gray rather than white. Although these results are simple, they illustrate clearly the power of the HSI color model in allowing *independent* control over hue, saturation, and intensity, quantities with which we are quite familiar when describing colors.

6.3 Pseudocolor Image Processing

Pseudocolor (also called *false color*) image processing consists of assigning colors to gray values based on a specified criterion. The term *pseudo* or *false* color is used to differentiate the process of assigning colors to monochrome images from the processes associated with true color images, a topic discussed starting in Section 6.4. The principal use of pseudocolor is for human visualization and interpretation of gray-scale events in an image or sequence of images. As noted

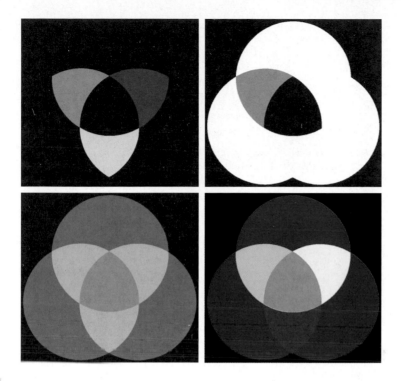

a b
c d

FIGURE 6.17 (a)–(c) Modified HSI component images. (d) Resulting RGB image. (See Fig. 6.16 for the original HSI images.)

at the beginning of this chapter, one of the principal motivations for using color is the fact that humans can discern thousands of color shades and intensities, compared to only two dozen or so shades of gray.

6.3.1 Intensity Slicing

The technique of *intensity* (sometimes called *density*) *slicing* and color coding is one of the simplest examples of pseudocolor image processing. If an image is interpreted as a 3-D function (intensity versus spatial coordinates), the method can be viewed as one of placing planes parallel to the coordinate plane of the image; each plane then "slices" the function in the area of intersection. Figure 6.18 shows an example of using a plane at $f(x, y) = l_i$ to slice the image function into two levels.

If a different color is assigned to each side of the plane shown in Fig. 6.18, any pixel whose gray level is above the plane will be coded with one color, and any pixel below the plane will be coded with the other. Levels that lie on the plane

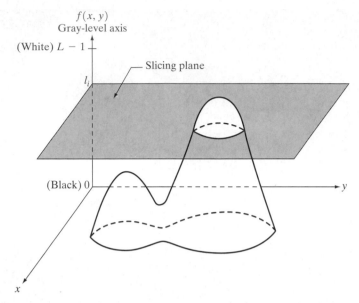

FIGURE 6.18 Geometric interpretation of the intensity-slicing technique.

itself may be arbitrarily assigned one of the two colors. The result is a two-color image whose relative appearance can be controlled by moving the slicing plane up and down the gray-level axis.

In general, the technique may be summarized as follows. Let $[0, L-1]$ represent the gray scale (see Section 2.3.4), let level l_0 represent black $[f(x, y) = 0]$, and level l_{L-1} represent white $[f(x, y) = L-1]$. Suppose that P planes perpendicular to the intensity axis are defined at levels $l_1, l_2, \ldots, l_P$. Then, assuming that $0 < P < L-1$, the P planes partition the gray scale into $P+1$ intervals, V_1, $V_2, \ldots, V_{P+1}$. Gray-level to color assignments are made according to the relation

$$f(x, y) = c_k \qquad \text{if } f(x, y) \in V_k \tag{6.3-1}$$

where c_k is the color associated with the kth intensity interval V_k defined by the partitioning planes at $l = k-1$ and $l = k$.

The idea of planes is useful primarily for a geometric interpretation of the intensity-slicing technique. Figure 6.19 shows an alternative representation that defines the same mapping as in Fig. 6.18. According to the mapping function shown in Fig. 6.19, any input gray level is assigned one of two colors, depending on whether it is above or below the value of l_i. When more levels are used, the mapping function takes on a staircase form.

EXAMPLE 6.3:
Intensity slicing.

■ A simple, but practical, use of intensity slicing is shown in Fig. 6.20. Figure 6.20(a) is a monochrome image of the Picker Thyroid Phantom (a radiation test pattern), and Fig. 6.20(b) is the result of intensity slicing this image into eight color regions. Regions that appear of constant intensity in the monochrome image

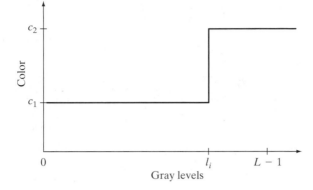

FIGURE 6.19 An alternative representation of the intensity-slicing technique.

are really quite variable, as shown by the various colors in the sliced image. The left lobe, for instance, is a dull gray in the monochrome image, and picking out variations in intensity is difficult. By contrast, the color image clearly shows eight different regions of constant intensity, one for each of the colors used. ■

In the preceding simple example, the gray scale was divided into intervals and a different color was assigned to each region, without regard for the meaning of the gray levels in the image. Interest in that case was simply to view the different gray levels constituting the image. Intensity slicing assumes a much more meaningful and useful role when subdivision of the gray scale is based on physical characteristics of the image. For instance, Fig. 6.21(a) shows an X-ray image of a weld

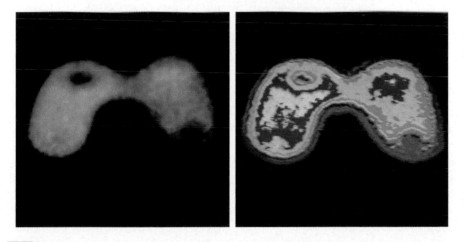

a b

FIGURE 6.20 (a) Monochrome image of the Picker Thyroid Phantom. (b) Result of density slicing into eight colors. (Courtesy of Dr. J. L. Blankenship, Instrumentation and Controls Division, Oak Ridge National Laboratory.)

a
b

FIGURE 6.21
(a) Monochrome
X-ray image of a
weld. (b) Result
of color coding.
(Original image
courtesy of
X-TEK Systems,
Ltd.)

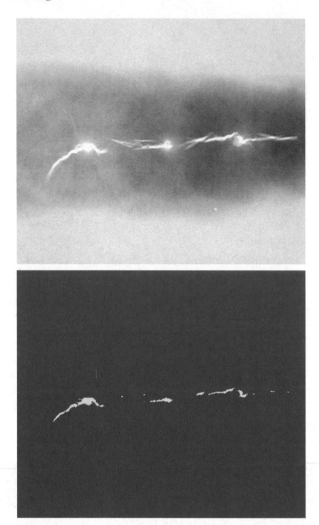

(the horizontal dark region) containing several cracks and porosities (the bright, white streaks running horizontally through the middle of the image). It is known that when there is a porosity or crack in a weld, the full strength of the X-rays going through the object saturates the imaging sensor on the other side of the object (see Section 2.3). Thus, gray levels of value 255 in an 8-bit image coming from such a system automatically imply a problem with the weld. If a human were to be the ultimate judge of the analysis and manual processes were employed to inspect welds (still a common procedure today), a simple color coding that assigns one color to level 255 and another to all other gray levels would simplify the inspector's job considerably. Figure 6.21(b) shows the result. No explanation is required to arrive at the conclusion that human error rates would be lower if images were displayed in the form of Fig. 6.21(b), instead of the form shown in Fig. 6.21(a). In other words, if the exact values of gray levels one is looking for are known, intensity slicing is a simple but powerful aid in visualization, especially if numerous images are involved. The following is a significantly more complex example.

■ Measurement of rainfall levels, especially in the tropical regions of the Earth, is of interest in diverse applications dealing with the environment. Accurate measurements using ground-based sensors are difficult and expensive to acquire, and total rainfall figures are even more difficult to obtain because a significant portion of precipitation occurs over the ocean. One approach for obtaining rainfall figures is to use a satellite. The TRMM (Tropical Rainfall Measuring Mission) satellite utilizes, among others, three sensors specially designed to detect rain: a precipitation radar, a microwave imager, and a visible and infrared scanner (see Sections 1.3 and 2.3 regarding image sensing modalities).

The results from the various rain sensors are processed, resulting in estimates of average rainfall over a given time period in the area monitored by the sensors. From these estimates, it is not difficult to generate gray-scale images whose intensity values correspond directly to rainfall, with each pixel representing a physical land area whose size depends on the resolution of the sensors. Such an intensity image is shown in Fig. 6.22(a), where the area monitored by the satellite

EXAMPLE 6.4:
Use of color to highlight rainfall levels.

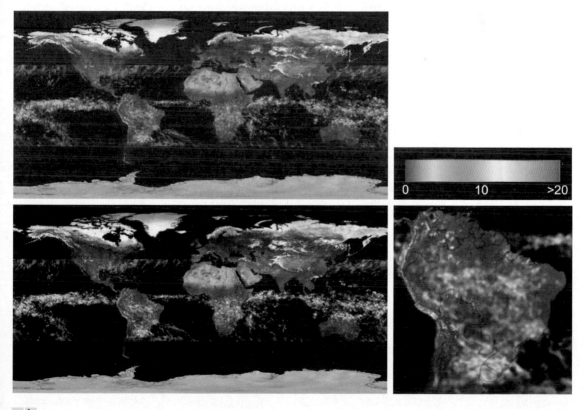

a b
c d

FIGURE 6.22 (a) Gray-scale image in which intensity (in the lighter horizontal band shown) corresponds to average monthly rainfall. (b) Colors assigned to intensity values. (c) Color-coded image. (d) Zoom of the South America region. (Courtesy of NASA.)

is the slightly lighter horizontal band in the middle one-third of the picture (these are the tropical regions). In this particular example, the rainfall values are average monthly values (in inches) over a three-year period.

Visual examination of this picture for rainfall patterns is quite difficult, if not impossible. However, suppose that we code gray levels from 0 to 255 using the colors shown in Fig. 6.22(b). Values toward the blues signify low values of rainfall, with the opposite being true for red. Note that the scale tops out at pure red for values of rainfall greater than 20 inches. Figure 6.22(c) shows the result of color coding the gray image with the color map just discussed. The results are much easier to interpret, as shown in this figure and in the zoomed area of Fig. 6.22(d). In addition to providing global coverage, this type of data allows meteorologists to calibrate ground-based rain monitoring systems with greater precision than ever before.   ■

6.3.2 Gray Level to Color Transformations

Other types of transformations are more general and thus are capable of achieving a wider range of pseudocolor enhancement results than the simple slicing technique discussed in the preceding section. An approach that is particularly attractive is shown in Fig. 6.23. Basically, the idea underlying this approach is to perform three independent transformations on the gray level of any input pixel. The three results are then fed separately into the red, green, and blue channels of a color television monitor. This method produces a composite image whose color content is modulated by the nature of the transformation functions. Note that these are transformations on the gray-level values of an image and are not functions of position.

The method discussed in the previous section is a special case of the technique just described. There, piecewise linear functions of the gray levels (Fig. 6.19) are used to generate colors. The method discussed in this section, on the other hand, can be based on smooth, nonlinear functions, which, as might be expected, gives the technique considerable flexibility.

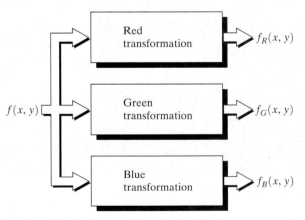

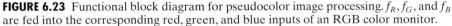

FIGURE 6.23 Functional block diagram for pseudocolor image processing. f_R, f_G, and f_B are fed into the corresponding red, green, and blue inputs of an RGB color monitor.

■ Figure 6.24(a) shows two monochrome images of luggage obtained from an airport X-ray scanning system. The image on the left contains ordinary articles. The image on the right contains the same articles, as well as a block of simulated plastic explosives. The purpose of this example is to illustrate the use of gray level to color transformations to obtain various degrees of enhancement.

EXAMPLE 6.5:
Use of
pseudocolor for
highlighting
explosives
contained in
luggage.

Figure 6.25 shows the transformation functions used. These sinusoidal functions contain regions of relatively constant value around the peaks as well as regions that change rapidly near the valleys. Changing the phase and frequency of each sinusoid can emphasize (in color) ranges in the gray scale. For instance, if all three transformations have the same phase and frequency, the output image will be monochrome. A small change in the phase between the three transformations produces little change in pixels whose gray levels correspond to peaks in the sinusoids, especially if the sinusoids have broad profiles (low frequencies). Pixels with gray-level values in the steep section of the sinusoids are assigned a much stronger color content as a result of significant differences between the amplitudes of the three sinusoids caused by the phase displacement between them.

The image shown in Fig. 6.24(b) was obtained with the transformation functions in Fig. 6.25(a), which shows the gray-level bands corresponding to the explosive, garment bag, and background, respectively. Note that the explosive and background have quite different gray levels, but they were both coded with

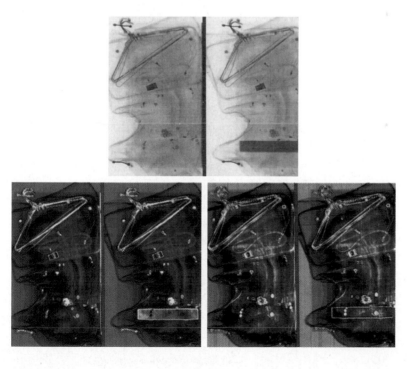

a
b c

FIGURE 6.24 Pseudocolor enhancement by using the gray-level to color transformations in Fig. 6.25. (Original image courtesy of Dr. Mike Hurwitz, Westinghouse.)

approximately the same color as a result of the periodicity of the sine waves. The image shown in Fig. 6.24(c) was obtained with the transformation functions in Fig. 6.25(b). In this case the explosives and garment bag intensity bands were mapped by similar transformations and thus received essentially the same color assignments. Note that this mapping allows an observer to "see" through the explosives. The background mappings were about the same as those used for Fig. 6.24(b), producing almost identical color assignments.   ■

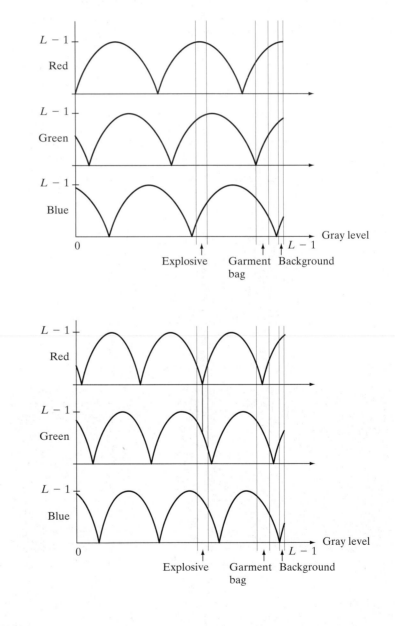

a
b

FIGURE 6.25 Transformation functions used to obtain the images in Fig. 6.24.

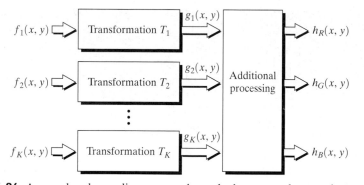

FIGURE 6.26 A pseudocolor coding approach used when several monochrome images are available.

The approach shown in Fig. 6.23 is based on a single monochrome image. Often, it is of interest to combine several monochrome images into a single color composite, as shown in Fig. 6.26. A frequent use of this approach (illustrated in Example 6.6) is in multispectral image processing, where different sensors produce individual monochrome images, each in a different spectral band. The types of additional processes shown in Fig. 6.26 can be techniques such as color balancing (see Section 6.5.4), combining images, and selecting the three images for display based on knowledge about response characteristics of the sensors used to generate the images.

■ Figures 6.27(a) through (d) show four spectral satellite images of Washington, D.C., including part of the Potomac River. The first three images are in the visible red, green, and blue, and the fourth is in the near infrared (see Table 1.1 and Fig. 1.10). Figure 6.27(e) is the full-color image obtained by combining the first three images into an RGB image. Full-color images of dense areas are difficult to interpret, but one notable feature of this image is the difference in color in various parts of the Potomac River. Figure 6.27(f) is a little more interesting. This image was formed by replacing the red component of Fig. 6.27(e) with the near-infrared image. From Table 1.1, we know that this band is strongly responsive to the biomass components of a scene. Figure 6.27(f) shows quite clearly the difference between biomass (in red) and the human-made features in the scene, composed primarily of concrete and asphalt, which appear bluish in the image.

The type of processing just illustrated is quite powerful in helping visualize events of interest in complex images, especially when those events are beyond our normal sensing capabilities. Figure 6.28 is an excellent illustration of this. These are images of the Jupiter moon Io, shown in pseudocolor by combining several of the sensor images from the *Galileo* spacecraft, some of which are in spectral regions not visible to the eye. However, by understanding the physical and chemical processes likely to affect sensor response, it is possible to combine the sensed images into a meaningful pseudocolor map. One way to combine the sensed image data is by how they show either differences in surface chemical composition or changes in the way the surface reflects sunlight. For example, in the pseudocolor image in Fig. 6.28(b), bright red depicts material newly ejected

EXAMPLE 6.6:
Color coding of multispectral images.

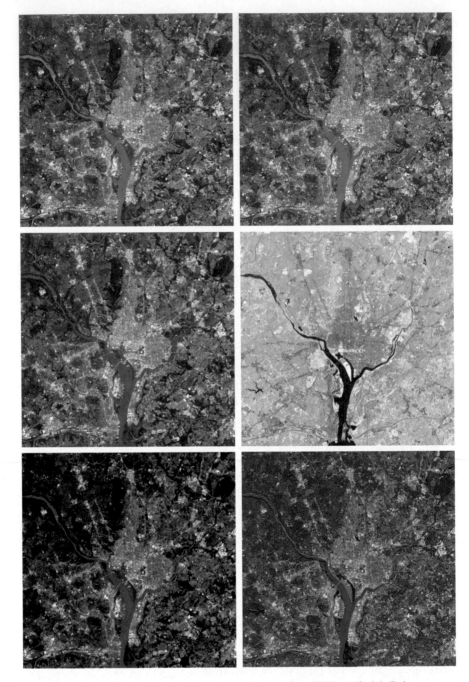

a b
c d
e f

FIGURE 6.27 (a)–(d) Images in bands 1–4 in Fig. 1.10 (see Table 1.1). (e) Color composite image obtained by treating (a), (b), and (c) as the red, green, blue components of an RGB image. (f) Image obtained in the same manner, but using in the red channel the near-infrared image in (d). (Original multispectral images courtesy of NASA.)

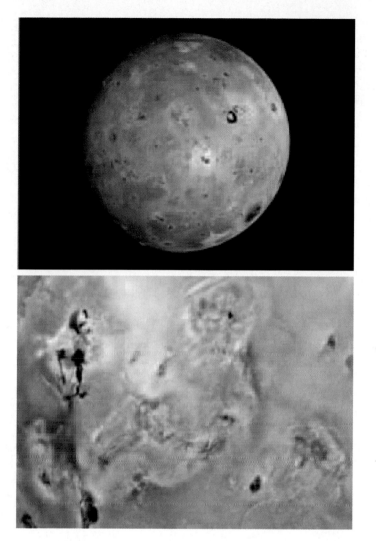

FIGURE 6.28
(a) Pseudocolor
rendition of
Jupiter Moon Io.
(b) A close-up.
(Courtesy of
NASA.)

from an active volcano on Io, and the surrounding yellow materials are older sulfur deposits. This image conveys these characteristics much more readily than would be possible by analyzing the component images individually.   ■

6.4 Basics of Full-Color Image Processing

In this section we begin the study of processing techniques applicable to full-color images. Although they are far from being exhaustive, the techniques developed in the sections that follow are illustrative of how full-color images are handled for a variety of image processing tasks. Full-color image processing approaches fall into two major categories. In the first category, we process each component image individually and then form a composite processed color image from the individually processed components. In the second category, we work with color pixels directly. Because full-color images have at least three components, color pixels really are vectors. For example, in the RGB system, each color point can be interpreted as a vector extending from the origin to that point in the RGB coordinate system (see Fig. 6.7).

Let **c** represent an arbitrary vector in RGB color space:

$$\mathbf{c} = \begin{bmatrix} c_R \\ c_G \\ c_B \end{bmatrix} = \begin{bmatrix} R \\ G \\ B \end{bmatrix}. \tag{6.4-1}$$

This equation indicates that the components of **c** are simply the RGB components of a color image at a point. We take into account the fact that the color components are a function of coordinates (x, y) by using the notation

$$\mathbf{c}(x, y) = \begin{bmatrix} c_R(x, y) \\ c_G(x, y) \\ c_B(x, y) \end{bmatrix} = \begin{bmatrix} R(x, y) \\ G(x, y) \\ B(x, y) \end{bmatrix}. \tag{6.4-2}$$

For an image of size $M \times N$, there are MN such vectors, $\mathbf{c}(x, y)$, for $x = 0, 1, 2, \ldots, M - 1; y = 0, 1, 2, \ldots, N - 1$.

It is important to keep clearly in mind that Eq. (6.4-2) depicts a vector whose components are *spatial* variables in x and y. This is a frequent source of confusion that can be avoided by focusing on the fact that our interest lies on spatial processes. That is, we are interested in image processing techniques formulated in x and y. The fact that the pixels are now color pixels introduces a factor that, in its easiest formulation, allows us to process a color image by processing each of its component images separately, using standard gray-scale image processing methods. However, the results of individual color component processing are not always equivalent to direct processing in color vector space, in which case we must formulate new approaches.

In order for per-color-component and vector-based processing to be equivalent, two conditions have to be satisfied: First, the process has to be applicable to both vectors and scalars. Second, the operation on each component of a vector must be independent of the other components. As an illustration, Fig. 6.29

a b

FIGURE 6.29
Spatial masks for gray-scale and RGB color images.

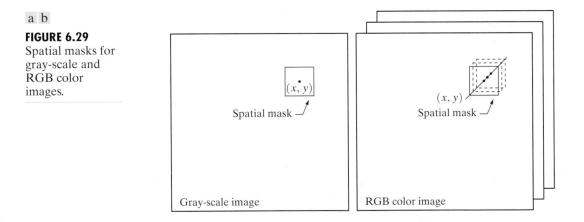

Gray-scale image   RGB color image

shows neighborhood spatial processing of gray-scale and full-color images. Suppose that the process is neighborhood averaging. In Fig. 6.29(a), averaging would be accomplished by summing the gray levels of all the pixels in the neighborhood and dividing by the total number of pixels in the neighborhood. In Fig. 6.29(b), averaging would be done by summing all the vectors in the neighborhood and dividing each component by the total number of vectors in the neighborhood. But each component of the average vector is the sum of the pixels in the image corresponding to that component, which is the same as the result that would be obtained if the averaging were done on a per-color-component basis and then the vector was formed. We show this in more detail in the following sections. We also show methods in which the results of the two approaches are not the same.

6.5 Color Transformations

The techniques described in this section, collectively called *color transformations*, deal with processing the components of a color image within the context of a *single* color model, as opposed to the conversion of those components between models (like the RGB-to-HSI and HSI-to-RGB conversion transformations of Section 6.2.3).

6.5.1 Formulation

As with the gray-level transformation techniques of Chapter 3, we model color transformations using the expression

$$g(x, y) = T[f(x, y)] \tag{6.5-1}$$

where $f(x, y)$ is a color input image, $g(x, y)$ is the transformed or processed color output image, and T is an operator on f over a spatial neighborhood of (x, y). The principal difference between this equation and Eq. (3.1-1) is in its interpretation. The pixel values here are triplets or quartets (i.e., groups of three or four values) from the color space chosen to represent the images, as illustrated in Fig. 6.29(b).

Analogous to the approach we used to introduce the basic gray-level transformations in Section 3.2, we will restrict attention in this section to color transformations of the form

$$s_i = T_i(r_1, r_2, \ldots, r_n), \qquad i = 1, 2, \ldots, n \tag{6.5-2}$$

where, for notational simplicity, r_i and s_i are variables denoting the color components of $f(x, y)$ and $g(x, y)$ at any point (x, y), n is the number of color components, and $\{T_1, T_2, \ldots, T_n\}$ is a set of *transformation* or *color mapping functions* that operate on r_i to produce s_i. Note that n transformations, T_i, combine to implement the single transformation function, T, in Eq. (6.5-1). The color space chosen to describe the pixels of f and g determines the value of n. If the RGB color space is selected, for example, $n = 3$ and $r_1, r_2,$ and r_3 denote the red, green, and blue components of the input image, respectively. If the CMYK or HSI color spaces are chosen, $n = 4$ or $n = 3$.

Figure 6.30(a) shows a high-resolution color image of a bowl of strawberries and cup of coffee that was digitized from a large format ($4'' \times 5''$) color negative. The second row of the figure contains the components of the initial CMYK scan. In these images, black represents 0 and white represents 1 in each CMYK color component. Thus, we see that the strawberries are composed of large amounts of magenta and yellow because the images corresponding to these two CMYK components are the brightest. Black is used sparingly and is generally confined to the coffee and shadows within the bowl of strawberries. When the CMYK image is converted to RGB, as shown in the third row of the figure, the strawberries are seen to contain a large amount of red and very little (although some) green and blue. The last row of Fig. 6.30 shows the HSI components of Fig. 6.30(a)—computed using Eqs. (6.2-2) through (6.2-4). As expected, the intensity component is a monochrome rendition of the full-color original. In addition, the strawberries are relatively pure in color; they possess the highest saturation or least dilution by white light of any of the hues in the image. Finally, we note some difficulty in interpreting the hue component. The problem is compounded by the fact that (1) there is a discontinuity in the HSI model where 0° and 360° meet, and (2) hue is undefined for a saturation of 0 (i.e., for white, black, and pure grays). The discontinuity of the model is most apparent around the strawberries, which are depicted in gray level values near both black (0) and white (1). The result is an unexpected mixture of highly contrasting gray levels to represent a single color—red.

Any of the color space components in Fig. 6.30 can be used in conjunction with Eq. (6.5-2). In theory, any transformation can be performed in any color model. In practice, however, some operations are better suited to specific models. For a given transformation, the cost of converting between representations must be factored into the decision regarding the color space in which to implement it. Suppose, for example, that we wish to modify the intensity of the image in Fig. 6.30(a) using

$$g(x, y) = kf(x, y) \tag{6.5-3}$$

where $0 < k < 1$. In the HSI color space, this can be done with the simple transformation

$$s_3 = kr_3 \tag{6.5-4}$$

where $s_1 = r_1$ and $s_2 = r_2$. Only HSI intensity component r_3 is modified. In the RGB color space, three components must be transformed:

$$s_i = kr_i \qquad i = 1, 2, 3. \tag{6.5-5}$$

The CMY space requires a similar set of linear transformations:

$$s_i = kr_i + (1 - k) \qquad i = 1, 2, 3. \tag{6.5-6}$$

Although the HSI transformation involves the fewest number of operations, the computations required to convert an RGB or CMY(K) image to the HSI space more than offsets (in this case) the advantages of the simpler transformation—

Full color

Cyan	Magenta	Yellow	Black

Red	Green	Blue

Hue	Saturation	Intensity

FIGURE 6.30 A full-color image and its various color-space components. (Original image courtesy of Med Data Interactive.)

the conversion calculations are more computationally intense than the intensity transformation itself. Regardless of the color space selected, however, the output is the same. Figure 6.31(b) shows the result of applying any of the transformations in Eqs. (6.5-4) through (6.5-6) to the image of Fig. 6.30(a) using $k = 0.7$. The mapping functions themselves are depicted graphically in Figs. 6.31(c) through (e).

a b
c d e

FIGURE 6.31
Adjusting the intensity of an image using color transformations.
(a) Original image. (b) Result of decreasing its intensity by 30% (i.e., letting $k = 0.7$).
(c)–(e) The required RGB, CMY, and HSI transformation functions.
(Original image courtesy of MedData Interactive.)

It is important to note that each transformation defined in Eqs. (6.5-4) through (6.5-6) depends only on one component within its color space. For example, the red output component, s_1, in Eq. (6.5-5) is independent of the green (r_2) and blue (r_3) inputs; it depends only on the red (r_1) input. Transformations of this type are among the simplest and most used color processing tools and can be carried out on a per-color-component basis, as mentioned at the beginning of our discussion. In the remainder of this section we examine several such transformations and discuss a case in which the component transformation functions are dependent on all the color components of the input image and, therefore, cannot be done on an individual color component basis.

6.5.2 Color Complements

The hues directly opposite one another on the *color circle*† of Fig. 6.32 are called *complements*. Our interest in complements stems from the fact that they are analogous to the gray-scale negatives of Section 3.2.1. As in the gray-scale case, color complements are useful for enhancing detail that is embedded in dark regions of a color image—particularly when the regions are dominant in size.

EXAMPLE 6.7:
Computing color image complements.

■ Figures 6.33(a) and (c) show the image from Fig. 6.30(a) and its color complement. The RGB transformations used to compute the complement are plotted in Fig. 6.33(b). They are identical to the gray-scale negative

---

†The color circle originated with Sir Isaac Newton, who in the seventeenth century joined the ends of the color spectrum to form the first color circle.

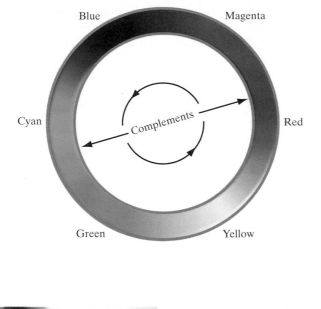

FIGURE 6.32
Complements on
the color circle.

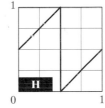

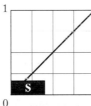
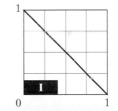

a b
c d

FIGURE 6.33
Color
complement
transformations.
(a) Original
image.
(b) Complement
transformation
functions.
(c) Complement
of (a) based on
the RGB mapping
functions. (d) An
approximation of
the RGB
complement using
HSI
transformations.

transformation defined in Section 3.2.1. Note that the computed complement is reminiscent of conventional photographic color film negatives. Reds of the original image are replaced by cyans in the complement. When the original image is black, the complement is white, and so on. Each of the hues in the complement image can be predicted from the original image using the color circle of Fig. 6.32. And each of the RGB component transforms involved in the computation of the complement is a function of *only* the corresponding input color component.

Unlike the intensity transformations of Fig. 6.31, the RGB complement transformation functions used in this example do not have a straightforward HSI space equivalent. It is left as an exercise for the reader (see Problem 6.18) to show that the saturation component of the complement cannot be computed from the saturation component of the input image alone. Figure 6.33(d) provides an approximation of the complement using the hue, saturation, and intensity transformations given in Fig. 6.33(b). Note that the saturation component of the input image is unaltered; it is responsible for the visual differences between Figs. 6.33(c) and (d).   ■

6.5.3 Color Slicing

Highlighting a specific range of colors in an image is useful for separating objects from their surroundings. The basic idea is either to (1) display the colors of interest so that they stand out from the background or (2) use the region defined by the colors as a mask for further processing. The most straightforward approach is to extend the gray-level slicing techniques of Section 3.2.4. Because a color pixel is an *n*-dimensional quantity, however, the resulting color transformation functions are more complicated than their gray-scale counterparts in Fig. 3.11. In fact, the required transformations are more complex than the color component transforms considered thus far. This is because all practical color slicing approaches require each pixel's transformed color components to be a function of all *n* original pixel's color components.

One of the simplest ways to "slice" a color image is to map the colors outside some range of interest to a nonprominent neutral color. If the colors of interest are enclosed by a cube (or *hypercube* for $n > 3$) of width W and centered at a prototypical (e.g., average) color with components $(a_1, a_2, \ldots, a_n)$, the necessary set of transformations is

$$s_i = \begin{cases} 0.5 & \text{if } \left[|r_j - a_j| > \dfrac{W}{2} \right]_{\text{any } 1 \le j \le n}, \\ r_i & \text{otherwise} \end{cases} \qquad i = 1, 2, \ldots, n. \quad (6.5\text{-}7)$$

These transformations highlight the colors around the prototype by forcing all other colors to the midpoint of the reference color space (an arbitrarily chosen neutral point). For the RGB color space, for example, a suitable neutral point is middle gray or color $(0.5, 0.5, 0.5)$.

If a sphere is used to specify the colors of interest, Eq. (6.5-7) becomes

$$s_i = \begin{cases} 0.5 & \text{if } \sum_{j=1}^{n} (r_j - a_j)^2 > R_0^2 \\ r_i & \text{otherwise} \end{cases} \quad , \quad i = 1, 2, \dots, n. \quad (6.5\text{-}8)$$

Here, R_0 is the radius of the enclosing sphere (or hypersphere for $n > 3$) and $(a_1, a_2, \dots, a_n)$ are the components of its center (i.e., the prototypical color). Other useful variations of Eqs. (6.5-7) and (6.5-8) include implementing multiple color prototypes and reducing the intensity of the colors outside the region of interest—rather than setting them to a neutral constant.

■ Equations (6.5-7) and (6.5-8) can be used to separate the edible part of the strawberries in Fig. 6.30(a) from the background cups, bowl, coffee, and table. Figures 6.34(a) and (b) show the results of applying both transformations. In each case, a prototype red with RGB color coordinate $(0.6863, 0.1608, 0.1922)$ was selected from the most prominent strawberry; W and R_0 were chosen so that the highlighted region would not expand to undesirable portions of the image. The actual values, $W = 0.2549$ and $R_0 = 0.1765$, were determined interactively. Note that the sphere-based transformation of Eq. (6.5-8) is slightly better, in the sense that it includes more of the strawberries' red areas. A sphere of radius 0.1765 does not completely enclose a cube of width 0.2549 but is itself not completely enclosed by the cube. ■

EXAMPLE 6.8:
An illustration of color slicing.

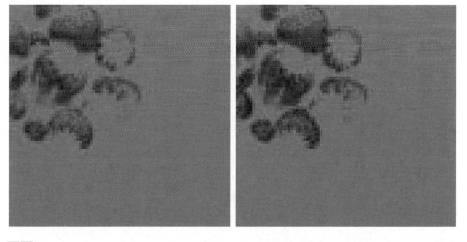

a b

FIGURE 6.34 Color slicing transformations that detect (a) reds within an RGB cube of width $W = 0.2549$ centered at $(0.6863, 0.1608, 0.1922)$, and (b) reds within an RGB sphere of radius 0.1765 centered at the same point. Pixels outside the cube and sphere were replaced by color $(0.5, 0.5, 0.5)$.

6.5.4 Tone and Color Corrections

Color transformations can be performed on most desktop computers. In conjunction with digital cameras, flatbed scanners, and inkjet printers, they turn a personal computer into a *digital darkroom*—allowing tonal adjustments and color corrections, the mainstays of high-end color reproduction systems, to be performed without the need for traditionally outfitted wet processing (i.e., darkroom) facilities. Although tone and color corrections are useful in other areas of imaging, the focus of the current discussion is on the most common uses—photo enhancement and color reproduction.

The effectiveness of the transformations examined in this section is judged ultimately in print. Since these transformations are developed, refined, and evaluated on monitors, it is necessary to maintain a high degree of color consistency between the monitors used and the eventual output devices. In fact, the colors of the monitor should represent accurately any digitally scanned source images, as well as the final printed output. This is best accomplished with a *device-independent color model* that relates the color gamuts (see Section 6.1) of the monitors and output devices, as well as any other devices being used, to one another. The success of this approach is a function of the quality of the *color profiles* used to map each device to the model and the model itself. The model of choice for many *color management systems* (CMS) is the CIE $L^*a^*b^*$ model, also called CIELAB (CIE [1978], Robertson [1977]). The $L^*a^*b^*$ color components are given by the following equations:

$$L^* = 116 \cdot h\left(\frac{Y}{Y_W}\right) - 16 \tag{6.5-9}$$

$$a^* = 500\left[h\left(\frac{X}{X_W}\right) - h\left(\frac{Y}{Y_W}\right)\right] \tag{6.5-10}$$

$$b^* = 200\left[h\left(\frac{Y}{Y_W}\right) - h\left(\frac{Z}{Z_W}\right)\right] \tag{6.5-11}$$

where

$$h(q) = \begin{cases} \sqrt[3]{q} & q > 0.008856 \\ 7.787q + 16/116 & q \leq 0.008856 \end{cases} \tag{6.5-12}$$

and X_W, Y_W, and Z_W are reference white tristimulus values—typically the white of a perfectly reflecting diffuser under CIE standard D65 illumination (defined by $x = 0.3127$ and $y = 0.3290$ in the CIE chromaticity diagram of Fig. 6.5). The $L^*a^*b^*$ color space is *colorimetric* (i.e., colors perceived as matching are encoded identically), *perceptually uniform* (i.e., color differences among various hues are perceived uniformly—see the classic paper by MacAdams [1942]), and *device independent*. While not a directly displayable format (conversion to another color space is required), its gamut encompasses the entire visible spectrum and can represent accurately the colors of any display, print, or input device. Like the HSI system, the $L^*a^*b^*$ system is an excellent decoupler of intensity (represented by lightness L^*) and color (represented by a^* for red minus green and $b*$ for green minus blue), making it useful in both image manipulation (tone and contrast editing) and image compression applications.[†]

The principal benefit of calibrated imaging systems is that they allow tonal and color imbalances to be corrected interactively and independently—that is, in two sequential operations. Before color irregularities, like over- and under-saturated colors, are resolved, problems involving the image's tonal range are corrected. The *tonal range* of an image, also called its *key type*, refers to its general distribution of color intensities. Most of the information in *high-key* images is concentrated at high (or light) intensities; the colors of *low-key* images are located predominantly at low intensities; *middle-key* images lie in between. As in the monochrome case, it is often desirable to distribute the intensities of a color image equally between the highlights and the shadows. The following examples demonstrate a variety of color transformations for the correction of tonal and color imbalances.

■ Transformations for modifying image tones normally are selected interactively. The idea is to adjust experimentally the image's brightness and contrast to provide maximum detail over a suitable range of intensities. The colors themselves are not changed. In the RGB and CMY(K) spaces, this means mapping all three (or four) color components with the same transformation function; in the HSI color space, only the intensity component is modified.

Figure 6.35 shows typical transformations used for correcting three common tonal imbalances—flat, light, and dark images. The S-shaped curve in the first row of the figure is ideal for boosting contrast [see Fig. 3.2(a)]. Its midpoint is anchored so that highlight and shadow areas can be lightened and darkened, respectively. (The inverse of this curve can be used to correct excessive contrast.) The transformations in the second and third rows of the figure correct light and dark images and are reminiscent of the power-law transformations in Fig. 3.6. Although the color components are discrete, as are the actual transformation functions, the transformation functions themselves are displayed and manipulated as continuous quantities—typically constructed from piecewise linear or higher order (for smoother mappings) polynomials. Note that the keys of the images in Fig. 6.35 are directly observable; they could also be determined using the histograms of the images' color components.   ■

EXAMPLE 6.9:
Tonal
transformations.

■ After the tonal characteristics of an image have been properly established, any color imbalances can be addressed. Although color imbalances can be determined objectively by analyzing—with a color spectrometer—a known color in an image, accurate visual assessments are possible when white areas, where the RGB or CMY(K) components should be equal, are present. As can be seen in Fig. 6.36, skin tones also are excellent subjects for visual color assessments because humans are highly perceptive of proper skin color. Vivid colors, such as bright red objects, are of little value when it comes to visual color assessment.

EXAMPLE 6.10:
Color balancing.

[†] Studies indicate that the degree to which the luminance (lightness) information is separated from the color information in $L*a*b*$ is greater than in other color models—such as CIELUV, YIQ, YUV, YCC, and XYZ (Kasson and Plouffe [1992]).

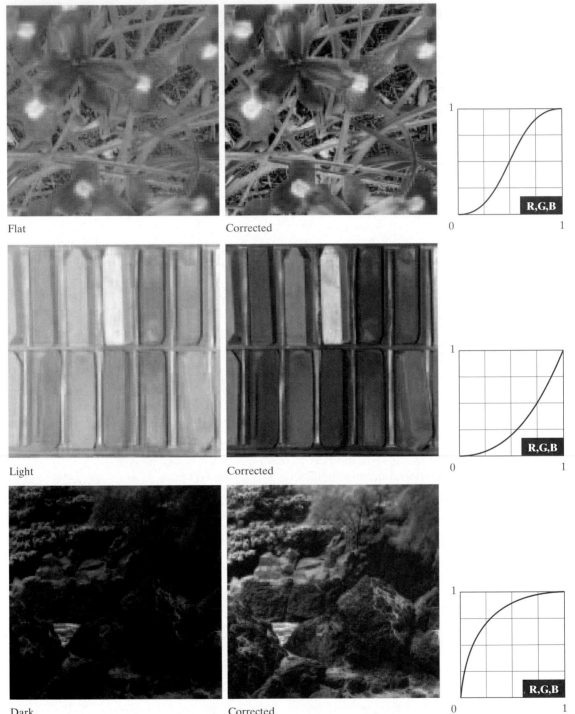

Flat   Corrected

Light   Corrected

Dark   Corrected

FIGURE 6.35 Tonal corrections for flat, light (high key), and dark (low key) color images. Adjusting the red, green, and blue components equally does not always alter the image hues significantly.

Original/Corrected

FIGURE 6.36 Color balancing corrections for CMYK color images.

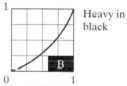

Heavy in black

Weak in black

Heavy in cyan

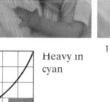

Weak in cyan

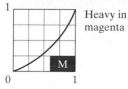

Heavy in magenta

Weak in magenta

Heavy in yellow

Weak in yellow

When a color imbalance is noted, there are a variety of ways to correct it. When adjusting the color components of an image, it is important to realize that every action affects the overall color balance of the image. That is, the perception of one color is affected by its surrounding colors. Nevertheless, the color wheel of Fig. 6.32 can be used to predict how one color component will affect others. Based on the color wheel, for example, the proportion of any color can be increased by decreasing the amount of the opposite (or complementary) color in the image. Similarly, it can be increased by raising the proportion of the two immediately adjacent colors or decreasing the percentage of the two colors adjacent to the complement. Suppose, for instance, that there is an abundance of magenta in an RGB image. It can be decreased by (1) removing both red and blue or (2) adding green.

Figure 6.36 shows the transformations used to correct simple CMYK output imbalances. Note that the transformations depicted are the functions required for correcting the images; the inverses of these functions were used to generate the associated color imbalances. Together, the images are analogous to a color ring-around print of a darkroom environment and are useful as a reference tool for identifying color printing problems. Note, for example, that too much red can be due to excessive magenta (per the bottom left image) or too little cyan (as shown in the rightmost image of the second row).   ■

6.5.5 Histogram Processing

Unlike the interactive enhancement approaches of the previous section, the gray-level histogram processing transformations of Section 3.3 can be applied to color images in an automated way. Recall that histogram equalization automatically determines a transformation that seeks to produce an image with a uniform histogram of intensity values. In the case of monochrome images, it was shown (see Fig. 3.17) to be reasonably successful at handling low-, high-, and middle-key images. Since color images are composed of multiple components, however, consideration must be given to adapting the gray-scale technique to more than one component and/or histogram. As might be expected, it is generally unwise to histogram equalize the components of a color image independently. This results in erroneous color. A more logical approach is to spread the color intensities uniformly, leaving the colors themselves (e.g., hues) unchanged. The following example shows that the HSI color space is ideally suited to this type of approach.

EXAMPLE 6.11:
Histogram equalization in the HSI color space.

■ Figure 6.37(a) shows a color image of a caster stand containing cruets and shakers whose intensity component spans the entire (normalized) range of possible values, $[0, 1]$. As can be seen in the histogram of its intensity component prior to processing [Fig. 6.37(b)], the image contains a large number of dark colors that reduce the median intensity to 0.36. Histogram equalizing the intensity component, without altering the hue and saturation, resulted in the image shown in Fig. 6.37(c). Note that the overall image is significantly brighter and that several moldings and the grain of the wooden table on which the caster is sitting are now visible. Figure 6.37(b) shows the intensity histogram of the new image, as well as the intensity transformation used to equalize the intensity component [see Eq. (3.3-8)].

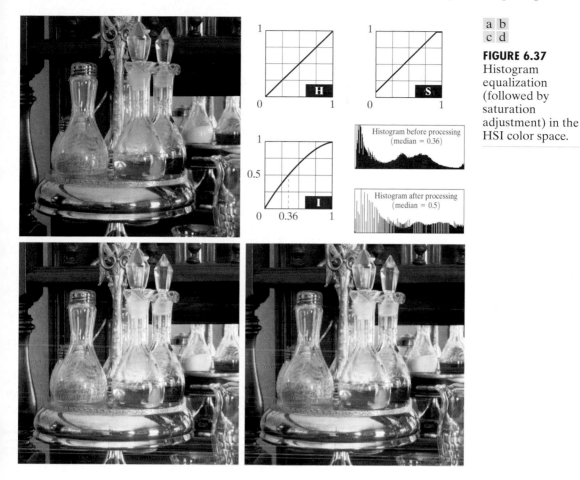

FIGURE 6.37
Histogram
equalization
(followed by
saturation
adjustment) in the
HSI color space.

Although the intensity equalization process did not alter the values of hue
and saturation of the image, it did impact the overall color perception. Note, in
particular, the loss of vibrancy in the oil and vinegar in the cruets. Figure 6.37(d)
shows the result of correcting this partially by increasing the image's saturation
component, subsequent to histogram equalization, using the transformation in
Fig. 6.37(b). This type of adjustment is common when working with the intensity component in HSI space because changes in intensity usually affect the relative appearance of colors in an image.   ■

6.6 Smoothing and Sharpening

The next step beyond transforming each pixel of a color image without regard
to its neighbors (as in the previous section) is to modify its value based on the
characteristics of the surrounding pixels. In this section, the basics of this type
of neighborhood processing are illustrated within the context of color image
smoothing and sharpening.

6.6.1 Color Image Smoothing

With reference to Fig. 6.29(a) and the discussion in Section 3.6, gray-scale image smoothing can be viewed as a spatial filtering operation in which the coefficients of the filtering mask are all 1's. As the mask is slid across the image to be smoothed, each pixel is replaced by the average of the pixels in the neighborhood defined by the mask. As can be seen in Fig. 6.29(b), this concept is easily extended to the processing of full-color images. The principal difference is that instead of scalar gray-level values we must deal with component vectors of the form given in Eq. (6.4-2).

Let S_{xy} denote the set of coordinates defining a neighborhood centered at (x, y) in an RGB color image. The average of the RGB component vectors in this neighborhood is

$$\bar{\mathbf{c}}(x, y) = \frac{1}{K} \sum_{(s,t) \in S_{xy}} \mathbf{c}(s, t). \tag{6.6-1}$$

It follows from Eq. (6.4-2) and the properties of vector addition that

See inside front cover
Consult the book web site for a brief review of vectors and matrices.

$$\bar{\mathbf{c}}(x, y) = \begin{bmatrix} \dfrac{1}{K} \sum\limits_{(s,t) \in S_{xy}} R(s, t) \\[2mm] \dfrac{1}{K} \sum\limits_{(s,t) \in S_{xy}} G(s, t) \\[2mm] \dfrac{1}{K} \sum\limits_{(s,t) \in S_{xy}} B(s, t) \end{bmatrix}. \tag{6.6-2}$$

We recognize the components of this vector as the scalar images that would be obtained by independently smoothing each plane of the starting RGB image using conventional gray-scale neighborhood processing. Thus, we conclude that smoothing by neighborhood averaging can be carried out on a per-color-plane basis. The result is the same as when the averaging is performed using RGB color vectors.

EXAMPLE 6.12:
Color image smoothing by neighborhood averaging.

■ Consider the color image shown in Fig. 6.38(a). The red, green, and blue planes of this image are depicted in Figs. 6.38(b) through (d). Figures 6.39(a) through (c) show the image's HSI components. In accordance with the discussion in the preceding paragraph, we can smooth the RGB image in Fig. 6.38 using the 5×5 gray-level averaging mask of Section 3.6. We simply smooth independently each of the RGB color planes and then combine the processed planes to form a smoothed full-color result. The image so computed is shown in Fig. 6.40(a). Note that it appears as we would expect from the discussion and examples in Section 3.6.

In Section 6.2 it was noted that an important advantage of the HSI color model is that it decouples intensity (closely related to gray scale) and color information. This makes it suitable for many gray-scale processing techniques and suggests that it might be more efficient to smooth only the intensity component of the HSI representation in Fig. 6.39. To illustrate the merits and/or consequences of this approach, we next smooth only the intensity component

a b
c d

FIGURE 6.38
(a) RGB image.
(b) Red
component image.
(c) Green
component.
(d) Blue
component.

a b c

FIGURE 6.39 HSI components of the RGB color image in Fig. 6.38(a). (a) Hue. (b) Saturation. (c) Intensity.

a b c

FIGURE 6.40 Image smoothing with a 5 × 5 averaging mask. (a) Result of processing each RGB component image. (b) Result of processing the intensity component of the HSI image and converting to RGB. (c) Difference between the two results.

(leaving the hue and saturation components unmodified) and convert the processed result to an RGB image for display. The smoothed color image is shown in Fig. 6.40(b). Note that it is similar to Fig. 6.40(a), but, as can be seen from the difference image in Fig. 640(c), is not identical. This is due to the fact that the average of two pixels of differing color is a mixture of the two colors, not either of the original colors. By smoothing only the intensity image, the pixels in Fig. 6.40(b) maintain their original hue and saturation—and thus their original color. Finally, we note that the difference (between the smoothed results in this example) would increase as the size of the smoothing mask increases.   ■

6.6.2 Color Image Sharpening

In this section we consider image sharpening using the Laplacian (see Section 3.7.2). From vector analysis, we know that the Laplacian of a vector is defined as a vector whose components are equal to the Laplacian of the individual scalar components of the input vector. In the RGB color system, the Laplacian of vector **c** in Eq. (6.4-2) is

$$\nabla^2\big[\mathbf{c}(x, y)\big] = \begin{bmatrix} \nabla^2 R(x, y) \\ \nabla^2 G(x, y) \\ \nabla^2 B(x, y) \end{bmatrix}, \tag{6.6-3}$$

which, as in the previous section, tells us that we can compute the Laplacian of a full-color image by computing the Laplacian of each component image separately.

EXAMPLE 6.13:
Sharpening with the Laplacian.

■ Figure 6.41(a) was obtained using Eq. (3.7-6) to compute the Laplacians of the RGB component images in Fig. 6.38 and combining them to produce the sharpened full-color result. Figure 6.41(b) shows a similarly sharpened image based on the HSI components in Fig. 6.39. This result was generated by combining the Laplacian of the intensity component with the unchanged hue and saturation

a b c

FIGURE 6.41 Image sharpening with the Laplacian. (a) Result of processing each RGB channel. (b) Result of processing the intensity component and converting to RGB. (c) Difference between the two results.

components. The difference between the RGB and HSI-based results is shown in Fig. 6.41(c); it results from the same factors explained in Example 6.12.   ■

6.7 Color Segmentation

Segmentation is a process that partitions an image into regions. Although segmentation is the topic of Chapter 10, we consider color segmentation briefly here for the sake of continuity. The reader will have no difficulty in following the discussion.

6.7.1 Segmentation in HSI Color Space

If we wish to segment an image based on color, and, in addition, we want to carry out the process on individual planes, it is natural to think first of the HSI space because color is conveniently represented in the hue image. Typically, saturation is used as a masking image in order to isolate further regions of interest in the hue image. The intensity image is used less frequently for segmentation of color images because it carries no color information. The following example is typical of how segmentation is performed in the HSI system.

■ Suppose that it is of interest to segment the reddish region in the lower left of the image in Fig. 6.42(a). Although it was generated by pseudocolor methods, this image can be processed (segmented) as a full-color image without loss of generality. Figures 6.42(b) through (d) are its HSI component images. Note by comparing Fig. 6.42(a) and (b) that the region in which we are interested has relatively high values of hue, indicating that the colors are on the blue-magenta side of red (see Fig. 6.13). Figure 6.42(e) shows a binary mask generated by thresholding the saturation image with a threshold equal to 10% of the maximum value in the saturation image. Any pixel value greater than the threshold was set to 1 (white). All others were set to 0 (black).

Figure 6.42(f) is the product of the mask with the hue image, and Fig. 6.42(g) is the histogram of the product image (note that the gray scale is in the range [0, 1]).

EXAMPLE 6.14: Segmentation in HSI space.

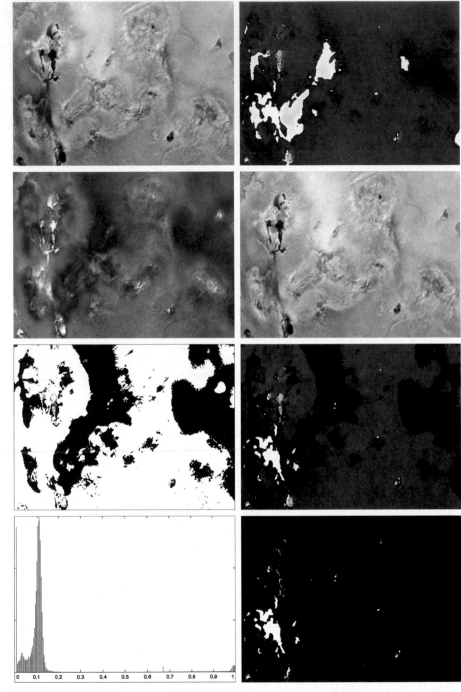

FIGURE 6.42 Image segmentation in HSI space. (a) Original. (b) Hue. (c) Saturation. (d) Intensity. (e) Binary saturation mask (black = 0). (f) Product of (b) and (e). (g) Histogram of (f). (h) Segmentation of red components in (a).

We see in the histogram that high values (which are the values of interest) are grouped at the very high end of the gray scale, near 1.0. The result of thresholding the product image with threshold value of 0.9 resulted in the binary image shown in Fig. 6.42(h). The spatial location of the white points in this image identifies the points in the original image that have the reddish hue of interest. This was far from a perfect segmentation because there are points in the original image that we certainly would say have a reddish hue, but that were not identified by this segmentation method. However, it can be shown by experimentation that the regions shown in white in Fig. 6.42(h) are about the best this method can do in identifying the reddish components of the original image. The segmentation method discussed in the following section is capable of yielding considerably better results.   ■

6.7.2 Segmentation in RGB Vector Space

Although, as mentioned numerous times in this chapter, working in HSI space is more intuitive, segmentation is one area in which better results generally are obtained by using RGB color vectors. The approach is straightforward. Suppose that the objective is to segment objects of a specified color range in an RGB image. Given a set of sample color points representative of the colors of interest, we obtain an estimate of the "average" color that we wish to segment. Let this average color be denoted by the RGB vector **a**. The objective of segmentation is to classify each RGB pixel in a given image as having a color in the specified range or not. In order to perform this comparison, it is necessary to have a measure of similarity. One of the simplest measures is the Euclidean distance. Let **z** denote an arbitrary point in RGB space. We say that **z** is similar to **a** if the distance between them is less than a specified threshold, D_0. The Euclidean distance between **z** and **a** is given by

$$D(\mathbf{z}, \mathbf{a}) = \|\mathbf{z} - \mathbf{a}\|$$

$$= \left[(\mathbf{z} - \mathbf{a})^T (\mathbf{z} - \mathbf{a}) \right]^{\frac{1}{2}} \qquad (6.7\text{-}1)$$

$$= \left[(z_R - a_R)^2 + (z_G - a_G)^2 + (z_B - a_B)^2 \right]^{\frac{1}{2}}$$

where the subscripts R, G, and B, denote the RGB components of vectors **a** and **z**. The locus of points such that $D(\mathbf{z}, \mathbf{a}) \leq D_0$ is a solid sphere of radius D_0, as illustrated in Fig. 6.43(a). Points contained within or on the surface of the

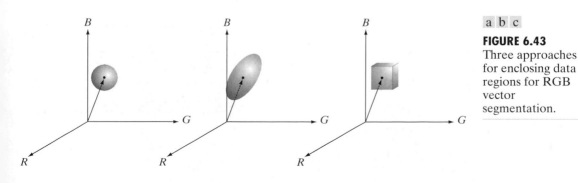

a b c

FIGURE 6.43
Three approaches for enclosing data regions for RGB vector segmentation.

sphere satisfy the specified color criterion; points outside the sphere do not. Coding these two sets of points in the image with, say, black and white, produces a binary segmented image.

A useful generalization of Eq. (6.7-1) is a distance measure of the form

$$D(\mathbf{z}, \mathbf{a}) = \left[(\mathbf{z} - \mathbf{a})^T \mathbf{C}^{-1} (\mathbf{z} - \mathbf{a}) \right]^{\frac{1}{2}} \qquad (6.7\text{-}2)$$

where $\mathbf{C}$ is the covariance matrix[†] of the samples representative of the color we wish to segment. The locus of points such that $D(\mathbf{z}, \mathbf{a}) \leq D_0$ describes a solid 3-D elliptical body [Fig. 6.43(b)] with the important property that its principal axes are oriented in the direction of maximum data spread. When $\mathbf{C} = \mathbf{I}$, the 3×3 identity matrix, Eq. (6.7-2) reduces to Eq. (6.7-1). Segmentation is as described in the preceding paragraph.

Because distances are positive and monotonic, we can work with the distance squared instead, thus avoiding root computations. However, implementing Eq. (6.7-1) or (6.7-2) is computationally expensive for images of practical size, even if the square roots are not computed. A compromise is to use a bounding box, as illustrated in Fig. 6.43(c). In this approach, the box is centered on $\mathbf{a}$, and its dimensions along each of the color axes is chosen proportional to the standard deviation of the samples along each of the axis. Computation of the standard deviations is done only once using sample color data.

Given an arbitrary color point, we segment it by determining whether or not it is on the surface or inside the box, as with the distance formulations. However, determining whether a color point is inside or outside the box is much simpler computationally when compared to a spherical or elliptical enclosure. Note that the preceding discussion is a generalization of the method introduced in Section 6.5.3 in connection with color slicing.

EXAMPLE 6.15:
Color image segmentation in RGB space.

▪ The rectangular region shown Fig. 6.44(a) contains samples of reddish colors we wish to segment out of the color image. This is the same problem we considered in Example 6.14 using hue, but here we approach the problem using RGB color vectors. The approach followed was to compute the mean vector $\mathbf{a}$ using the color points contained within the rectangle in Fig. 6.44(a), and then to compute the standard deviation of the red, green, and blue values of those samples. A box was centered at $\mathbf{a}$, and its dimensions along each of the RGB axes were selected as 1.25 times the standard deviation of the data along the corresponding axis. For example, let σ_R denote the standard deviation of the red components of the sample points. Then the dimensions of the box along the R-axis extended from $(a_R - 1.25\sigma_R)$ to $(a_R + 1.25\sigma_R)$, where a_R denotes the red component of average vector $\mathbf{a}$. The result of coding each point in the entire color image as white if it was on the surface or inside the box, and as black otherwise, is shown in Fig. 6.44(b). Note how the segmented region was generalized from the color samples enclosed by the rectangle. In fact, by comparing

---

[†] Computation of the covariance matrix of a set of vector samples is discussed in detail in Section 11.4.

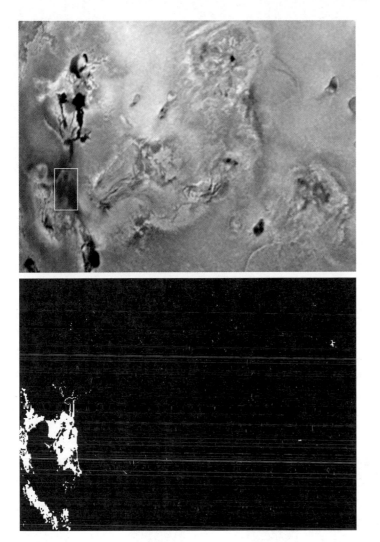

a
b

FIGURE 6.44
Segmentation in RGB space.
(a) Original image with colors of interest shown enclosed by a rectangle.
(b) Result of segmentation in RGB vector space. Compare with Fig. 6.42(h).

Figs. 6.44(b) and. 6.42(h), we see that segmentation in the RGB vector space yielded results that are much more accurate, in the sense that they correspond much more closely with what we would define as "reddish" points in the original color image.   ■

6.7.3 Color Edge Detection

As discussed in Chapter 10, edge detection is an important tool for image segmentation. In this section, we are interested in the issue of computing edges on an individual-image basis versus computing edges directly in color vector space. The details of edge-based segmentation are given in Section 10.1.3.

Edge detection by gradient operators was introduced in Section 3.7.3 in connection with edge enhancement. Unfortunately, the gradient discussed in Section 3.7.3 is not defined for vector quantities. Thus, we know immediately that computing the gradient on individual images and then using the results to form a color image will lead to erroneous results. A simple example will help illustrate the reason why.

Consider the two $M \times M$ color images (M odd) in Fig. 6.45(d) and (h), composed of the three component images in Figs. 6.45(a) through (c) and 6.45(e) through (g), respectively. If, for example, we compute the gradient image of each of the component images [see Eq. (3.7-13)] and add the results to form the two corresponding RGB gradient images, the value of the gradient at point $[(M + 1)/2, (M + 1)/2]$ would be the same in both cases. Intuitively, we would expect the gradient at that point to be stronger for the image in Fig. 6.45(d) because the edges of the R, G, and B images are in the same direction in that image, as opposed to the image in Fig. 6.45(h), in which only two of the edges are in the same direction. Thus we see from this simple example that processing the three individual planes to form a composite gradient image can yield erroneous results. If the problem is one of just detecting edges, then the individual-component approach usually yields acceptable results. If accuracy is an issue, however, then obviously we need a new definition of the gradient applicable to vector quantities. We discuss next a method proposed by Di Zenzo [1986] for doing this.

The problem at hand is to define the gradient (magnitude and direction) of the vector $\mathbf{c}$ in Eq. (6.4-2) at any point (x, y). As was just mentioned, the gradient we studied in Section 3.7.3 is applicable to a *scalar* function $f(x, y)$; it is

a b c d
e f g h

FIGURE 6.45 (a)–(c) R, G, and B component images and (d) resulting RGB color image. (f)–(g) R, G, and B component images and (h) resulting RGB color image.

not applicable to vector functions. The following is one of the various ways in which we can extend the concept of a gradient to vector functions. Recall that for a scalar function $f(x, y)$, the gradient is a vector pointing in the direction of maximum rate of change of f at coordinates (x, y).

Let $\mathbf{r}, \mathbf{g}$, and $\mathbf{b}$ be unit vectors along the R, G, and B axis of RGB color space (Fig. 6.7), and define the vectors

$$\mathbf{u} = \frac{\partial R}{\partial x}\mathbf{r} + \frac{\partial G}{\partial x}\mathbf{g} + \frac{\partial B}{\partial x}\mathbf{b} \qquad (6.7\text{-}3)$$

and

$$\mathbf{v} = \frac{\partial R}{\partial y}\mathbf{r} + \frac{\partial G}{\partial y}\mathbf{g} + \frac{\partial B}{\partial y}\mathbf{b}. \qquad (6.7\text{-}4)$$

Let the quantities g_{xx}, g_{yy}, and g_{xy} be defined in terms of the dot product of these vectors, as follows:

$$g_{xx} = \mathbf{u} \cdot \mathbf{u} = \mathbf{u}^T\mathbf{u} = \left|\frac{\partial R}{\partial x}\right|^2 + \left|\frac{\partial G}{\partial x}\right|^2 + \left|\frac{\partial B}{\partial x}\right|^2 \qquad (6.7\text{-}5)$$

$$g_{yy} = \mathbf{v} \cdot \mathbf{v} = \mathbf{v}^T\mathbf{v} = \left|\frac{\partial R}{\partial y}\right|^2 + \left|\frac{\partial G}{\partial y}\right|^2 + \left|\frac{\partial B}{\partial y}\right|^2 \qquad (6.7\text{-}6)$$

and

$$g_{xy} = \mathbf{u} \cdot \mathbf{v} = \mathbf{u}^T\mathbf{v} = \frac{\partial R}{\partial x}\frac{\partial R}{\partial y} + \frac{\partial G}{\partial x}\frac{\partial G}{\partial y} + \frac{\partial B}{\partial x}\frac{\partial B}{\partial y}. \qquad (6.7\text{-}7)$$

Keep in mind that R, G, and B, and consequently the g's, are functions of x and y. Using this notation, it can be shown (Di Zenzo [1986]) that the direction of maximum rate of change of $\mathbf{c}(x, y)$ is given by the angle

$$\theta = \frac{1}{2}\tan^{-1}\left[\frac{2g_{xy}}{(g_{xx} - g_{yy})}\right] \qquad (6.7\text{-}8)$$

and that the value of the rate of change at (x, y), in the direction of θ, is given by

$$F(\theta) = \left\{\frac{1}{2}\left[(g_{xx} + g_{yy}) + (g_{xx} - g_{yy})\cos 2\theta + 2g_{xy}\sin 2\theta\right]\right\}^{\frac{1}{2}}. \qquad (6.7\text{-}9)$$

Because $\tan(\alpha) = \tan(\alpha \pm \pi)$, if θ_0 is a solution to Eq. (6.7-8), so is $\theta_0 \pm \pi/2$. Furthermore, $F(\theta) = F(\theta + \pi)$, so F needs to be computed only for values of θ in the half-open interval $[0, \pi)$. The fact that Eq. (6.7-8) provides two values 90° apart means that this equation associates with each point (x, y) a pair of orthogonal directions. Along one of those directions F is maximum, and it is minimum along the other. The derivation of these results is rather lengthy, and we would gain little in terms of the fundamental objective of our current discussion by detailing it here. The interested reader should consult the paper by Di Zenzo [1986] for details. The partial derivatives required for implementing Eqs. (6.7-5) through (6.7-7) can be computed using, for example, the Sobel operators discussed in Section 3.7.3.

EXAMPLE 6.16:
Edge detection in vector space.

■ Figure 6.46(b) is the gradient of the image in Fig. 6.46(a), obtained using the vector method just discussed. Figure 6.46(c) shows the image obtained by computing the gradient of each RGB component image and forming a composite gradient image by adding the corresponding values of the three component images at each coordinate (x, y). The edge detail of the vector gradient image is more complete than the detail in the individual-plane gradient image in Fig. 6.46(c); for example, see the detail around the subject's right eye. The image in Fig. 6.46(d) shows the difference between the two gradient images at each point (x, y). It is important to note that both approaches yielded reasonable results. Whether the extra detail in Fig. 6.46(b) is worth the added computational burden (as opposed to implementation of the Sobel operators, which

a b
c d

FIGURE 6.46
(a) RGB image.
(b) Gradient computed in RGB color vector space.
(c) Gradients computed on a per-image basis and then added.
(d) Difference between (b) and (c).

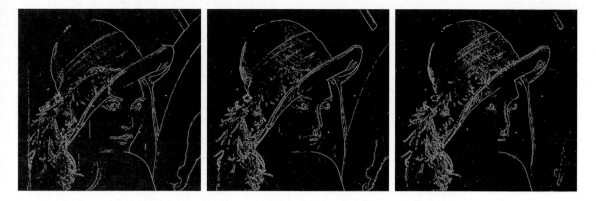

a b c

FIGURE 6.47 Component gradient images of the color image in Fig. 6.46. (a) Red component, (b) green component, and (c) blue component. These three images were added and scaled to produce the image in Fig. 6.46(c).

were used to generate the gradient of the individual planes) can only be determined by the requirements of a given problem. Figure 6.47 shows the three component gradient images, which, when added and scaled, were used to obtain Fig. 6.46(c). ■

6.8 Noise in Color Images

The noise models discussed in Section 5.2 are applicable to color images. Usually, the noise content of a color image has the same characteristics in each color channel, but it is possible for color channels to be affected differently by noise. One possibility is for the electronics of a particular channel to malfunction. However, different noise levels are more likely to be caused by differences in the relative strength of illumination available to each of the color channels. For example, use of a red (reject) filter in a CCD camera will reduce the strength of illumination available to the red sensor. CCD sensors are noisier at lower levels of illumination, so the resulting red component of an RGB image would tend to be noisier than the other two component images in this situation.

■ In this example we take a brief look at noise in color images and how noise carries over when converting from one color model to another. Figures 6.48(a) through (c) show the three color planes of an RGB image corrupted by Gaussian noise, and Fig. 6.48(d) is the composite RGB image. Note that fine grain noise such as this tends to be less visually noticeable in a color image than it is in a monochrome image. Figures 6.49(a) through (c) show the result of converting the RGB image in Fig. 6.48(d) to HSI. Compare these results with the HSI components of the original image (Fig. 6.39) and note how significantly

EXAMPLE 6.17:
Illustration of the effects of converting noisy RGB images to HSI.

a b
c d

FIGURE 6.48
(a)–(c) Red,
green, and blue
component
images corrupted
by additive
Gaussian noise of
mean 0 and
variance 800.
(d) Resulting
RGB image.
[Compare (d)
with Fig. 6.46(a).]

a b c

FIGURE 6.49 HSI components of the noisy color image in Fig. 6.48(d). (a) Hue. (b) Saturation. (c) Intensity.

FIGURE 6.50
(a) RGB image
with green plane
corrupted by salt-
and-pepper noise.
(b) Hue
component of
HSI image.
(c) Saturation
component.
(d) Intensity
component.

degraded the hue and saturation components of the noisy image are. This is due to the nonlinearity of the cos and min operations in Eqs. (6.2-2) and (6.2-3), respectively. On the other hand, the intensity component in Fig. 6.49(c) is slightly smoother than any of the three noisy RGB component images. This is due to the fact that the intensity image is the average of the RGB images, as indicated in Eq. (6.2-4). (Recall the discussion in Section 3.4.2 regarding the fact that image averaging reduces random noise.)

In cases when, say, only one RGB channel is affected by noise, conversion to HSI spreads the noise to all HSI component images. Figure 6.50 shows an example. Figure 6.50(a) shows an RGB image whose green image is corrupted by salt-and-pepper noise, in which the probability of either salt or pepper is 0.05. The HSI component images in Figs. 6.50(b) through (d) show clearly how the noise spread from the green RGB channel to all the HSI images. Of course, this is not unexpected because computation of the HSI components makes use of all RGB components, as shown in Section 6.2.3. ■

As is true of the processes we have discussed thus far, filtering of full-color images can be carried out on a per-image basis or directly in color vector space, depending on the process. For example, noise reduction by using an averaging filter is the process discussed in Section 6.6.1, which we know gives the same result in vector space as it does if the component images are processed independently. Other filters, however, cannot be formulated in this manner. Examples include the class of order statistics filters discussed in Section 5.3.2. For instance, to implement a median filter in color vector space it is necessary to find a scheme for ordering vectors in a way that the median makes sense. While this was a simple process when dealing with scalars, the process is considerably more complex when dealing with vectors. A discussion of vector ordering is beyond the scope of our discussion here, but the book by Plataniotis and Venetsanopoulos [2000] is a good reference on vector ordering and some of the filters based on the ordering concept.

6.9   Color Image Compression

Since the number of bits required to represent color is typically three to four times greater than the number employed in the representation of gray levels, *data compression* plays a central role in the storage and transmission of color images. With respect to the RGB, CMY(K), and HSI images of the previous sections, the *data* that are the object of any compression are the components of each color pixel (e.g., the red, green, and blue components of the pixels in an RGB image); they are the means by which the color information is conveyed. *Compression* is the process of reducing or eliminating redundant and/or irrelevant data. Although compression is the topic of Chapter 8, we illustrate the concept briefly in the following example using a color image.

EXAMPLE 6.18:
A color image compression example.

■ Figure 6.51(a) shows a 24-bit RGB full-color image of an iris in which 8 bits each are used to represent the red, green, and blue components. Figure 6.51(b) was reconstructed from a compressed version of the image in (a) and is, in fact, a compressed and subsequently decompressed approximation of it. Although the compressed image is not directly displayable—it must be decompressed before input to a color monitor—the compressed image contains only 1 data bit (and thus 1 storage bit) for every 230 bits of data in the original image. Assuming that the compressed image could be transmitted over, say, the Internet, in 1 minute, transmission of the original image would require almost 4 hours. Of course, the transmitted data would have to be decompressed for viewing, but the decompression could be done in a matter of seconds. The JPEG 2000 compression algorithm used to generate Fig. 6.51(b) is a recently introduced standard that is described in detail in Section 8.6.2. Note that the reconstructed approximation image is slightly blurred. This is a characteristic of many *lossy* compression techniques; it can be reduced or eliminated by altering the level of compression.   ■

a
b

FIGURE 6.51
Color image
compression.
(a) Original RGB
image. (b) Result
of compressing
and
decompressing
the image in (a).

Summary

The material in this chapter is an introduction to color image processing and covers topics selected to give the reader a solid background in the techniques used in this branch of image processing. Our treatment of color fundamentals and color models was prepared as foundation material for a field that is in its own right wide in technical scope and areas of application. In particular, we focused on color models that we felt are not only useful in digital image processing but would also provide the tools necessary for further study in this area of color image processing. The discussion of pseudocolor and full-color processing on an individual image basis provides a tie to techniques that were covered in some detail in Chapters 3 through 5.

The material on color vector spaces is a departure from methods that we had studied before and highlights some important differences between gray-scale and full-color processing. In terms of techniques, the areas of direct color vector processing are numerous and include processes such as median and other order filters, adaptive and morphological filters, image restoration, image compression, and many others. These processes are not equivalent to color processing carried out on the individual component images of a color image. The references in the following section provide a pointer to further results in this field.

Our treatment of noise in color images also points out that the vector nature of the problem, along with the fact that color images are routinely transformed from one working space to another, has implications on the issue of how to reduce noise in these images. In some cases, noise filtering can be done on a per-image basis, but others, such as median filtering, require special treatment to reflect the fact that color pixels are vector quantities, as mentioned in the previous paragraph.

Although segmentation is the topic of Chapter 10 and image data compression is the topic of Chapter 8, we gained the advantage of continuity by introducing them here in the context of color image processing. As will become evident in subsequent discussions, many of the techniques developed in those chapters are applicable to the discussion in this chapter.

References and Further Reading

For a comprehensive reference on the science of color, see Malacara [2001]. Regarding the physiology of color, see Gegenfurtner and Sharpe [1999]. These two references, along with the early books by Walsh [1958] and by Kiver [1965], provide ample supplementary material for the discussion in Section 6.1. For further reading on color models (Section 6.2), see Fortner and Meyer [1997], Poynton [1996], and Fairchild [1998]. For a detailed derivation of the HSI model equations in Section 6.2.3 see the paper by Smith [1978] or consult the book web site. The topic of pseudocolor (Section 6.3) is closely tied to the general area of data visualization. Wolff and Yaeger [1993] is a good basic reference on the use of pseudocolor. The book by Thorell and Smith [1990] also is of interest. For a discussion on the vector representation of color signals (Section 6.4), see Plataniotis and Venetsanopoulos [2000].

References for Section 6.5 are Benson [1985], Robertson [1977], and CIE [1978]. See also the classic paper by MacAdam [1942]. The material on color image filtering (Section 6.6) is based on the vector formulation introduced in Section 6.4 and on our discussion of spatial filtering in Chapter 3. Segmentation of color images (Section 6.7) has been a topic of much attention during the past ten years. The papers by Liu and Yang [1994] and by Shafarenko et al. [1998] are representative of work in this field. A special issue of the *IEEE Transactions on Image Processing* [1997] also is of interest. The discussion on color edge detection (Section 6.7.3) is from Di Zenzo [1986]. The book by Plataniotis and Venetsanopoulos [2000] does a good job of summarizing a variety of approaches to the segmentation of color images. The discussion in Section 6.8 is based on the noise models introduced in Section 5.2. References on image compression (Section 6.9) are listed at the end of Chapter 8.

See inside front cover

Detailed solutions to the problems marked with a star can be found in the book web site. The site also contains suggested projects based on the material in this chapter.

Problems

6.1 Give the percentages of red (X), green (Y), and blue (Z) light required to generate the point labeled "warm white" in Fig. 6.5.

★ **6.2** Consider any two valid colors c_1 and c_2 with coordinates (x_1, y_1) and (x_2, y_2) in the chromaticity diagram of Fig. 6.5. Derive the necessary general expression(s) for computing the relative percentages of colors c_1 and c_2 composing a given color that is known to lie on the straight line joining these two colors.

6.3   Consider any three valid colors $c_1, c_2,$ and c_3 with coordinates $(x_1, y_1), (x_2, y_2),$ and (x_3, y_3) in the chromaticity diagram of Fig. 6.5. Derive the necessary general expression(s) for computing the relative percentages of $c_1, c_2,$ and c_3 composing a given color that is known to lie within the triangle whose vertices are at the coordinates of $c_1, c_2,$ and c_3.

★6.4   In an automated assembly application, three classes of parts are to be color coded in order to simplify detection. However, only a monochrome TV camera is available to acquire digital images. Propose a technique for using this camera to detect the three different colors.

6.5   In a simple RGB image, the $R, G,$ and B component images have the horizontal intensity profiles shown in the following diagram. What color would a person see in the middle column of this image?

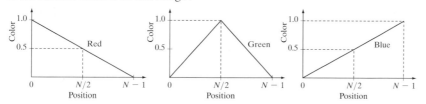

★6.6   Sketch the RGB components of the following image as they would appear on a monochrome monitor. All colors are at maximum intensity and saturation. In working this problem, consider the middle gray border as part of the image.

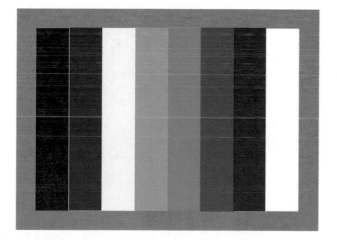

6.7   How many different shades of gray are there in a color RGB system in which each RGB image is an 8-bit image?

6.8   Consider the RGB color cube shown in Fig 6.8, and answer each of the following:
   ★ (a) Describe how the gray levels vary in the $R, G,$ and B primary images that make up the front face of the color cube.
   (b) Suppose that we replace every color in the RGB cube by its CMY color. This new cube is displayed on an RGB monitor. Label with a color name the eight vertices of the new cube that you would see on the screen.

(c) What can you say about the colors on the edges of the RGB color cube regarding saturation?

6.9 (a) Sketch the CMY components of the image in Problem 6.6 as they would appear on a monochrome monitor.

(b) If the CMY components sketched in (a) are fed into the red, green, and blue inputs of a color monitor, respectively, describe the resulting image.

★**6.10** Derive the CMY intensity mapping function of Eq. (6.5-6) from its RGB counterpart in Eq. (6.5-5).

6.11 Consider the entire 216 safe-color array shown in Fig. 6.10(a). Label each cell by its (row, column) designation, so that the top left cell is (1, 1) and the rightmost bottom cell is (12, 18). At which cells will you find

(a) The purest green?

(b) The purest blue?

★**6.12** Sketch the HSI components of the image in Problem 6.6 as they would appear on a monochrome monitor.

6.13 Propose a method for generating a color band similar to the one shown in the zoomed section entitled *Visible Spectrum* in Fig. 6.2. Note that the band starts at a dark purple on the left and proceeds toward pure red on the right. (*Hint:* Use the HSI color model.)

★**6.14** Propose a method for generating a color version of the image shown diagrammatically in Fig. 6.13(c). Give your answer in the form of a flow chart. Assume that the intensity value is fixed and given. (*Hint:* Use the HSI color model.)

6.15 Consider the following image composed of solid color squares. For discussing your answer, choose a gray scale consisting of eight shades of gray, 0 through 7, where 0 is black and 7 is white. Suppose that the image is converted to HSI color space. In answering the following questions, use specific numbers for the grade shades if they make sense. Otherwise, the relationships "same as," "lighter than," or "darker than" are sufficient. If you cannot assign a specific gray level or one of these relationships to the image you are discussing, give the reason.

(a) Sketch the hue image.

(b) Sketch the saturation image.

(c) Sketch the intensity image.

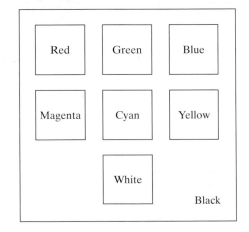

6.16 The following 8-bit images are (left to right) the *H*, *S*, and *I* component images from Fig. 6.16. The numbers indicate gray-level values. Answer the following questions, explaining the basis for your answer in each. If it is not possible to answer a question based on the given information, state why you cannot do so.

   ★ **(a)** Give the gray-level values of all regions in the hue image.

   (b) Give the gray-level value of all regions in the saturation image.

   (c) Give the gray-level values of all regions in the intensity image.

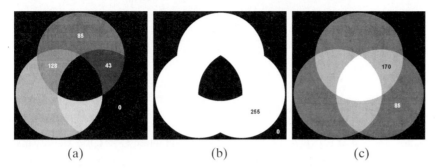

       (a)   (b)   (c)

6.17 Refer to Fig. 6.27 in answering the following:

   (a) Why does the image in Fig. 6.27(f) exhibit predominantly red tones?

   (b) Suggest an automated procedure for coding the water in Fig. 6.27 in a bright-blue color.

   (c) Suggest an automated procedure for coding the predominantly man-made components in a bright yellow color. [*Hint*: Work with Fig. 6.27(f).]

★ **6.18** Show that the saturation component of the complement of a color image cannot be computed from the saturation component of the input image alone.

6.19 Explain the shape of the hue transformation function for the complement approximation in Fig. 6.33(b) using the HSI color model.

★ **6.20** Derive the CMY transformations to generate the complement of a color image.

6.21 Draw the general shape of the transformation functions used to correct excessive contrast in the RGB color space.

★ **6.22** Assume that the monitor and printer of an imaging system are imperfectly calibrated. An image that looks balanced on the monitor appears yellowish in print. Describe general transformations that might correct the imbalance.

6.23 Compute the *L*a*b** components of the image in Problem 6.6 assuming

$$
\begin{bmatrix} X \\ Y \\ Z \end{bmatrix} = \begin{bmatrix} 0.588 & 0.179 & 0.183 \\ 0.29 & 0.606 & 0.105 \\ 0 & 0.068 & 1.021 \end{bmatrix} \begin{bmatrix} R \\ G \\ B \end{bmatrix}.
$$

This matrix equation defines the tristimulus values of the colors generated by standard National Television System Committee (NTSC) color TV phosphors viewed under *D*65 standard illumination (Benson [1985]).

★ **6.24** How would you implement the color equivalent of gray scale histogram matching (specification) from Section 3.3.2?

6.25 Consider the following 500 × 500 RGB color image, where the squares are pure red, green, and blue.

(a) Suppose that we convert this image to HSI, blur the H component image with a 25×25 averaging mask, and convert back to RGB. What would the result look like?

(b) Repeat, blurring only the saturation component this time.

Green	Red
Blue	Green

6.26   Show that Eq. (6.7-2) reduces to Eq. (6.7-1) when $\mathbf{C} = \mathbf{I}$, the identity matrix.

6.27 ★ **(a)**   With reference to the discussion in Section 6.7.2, give a procedure (in flow chart form) for determining whether a color vector (point) $\mathbf{z}$ is inside a cube with sides W, centered at an average color vector $\mathbf{a}$. Distance computations are not allowed.

(b)   This process also can be implemented on an image-by-image basis if the box is lined up with the axes. Show how you would do it.

6.28   Sketch the surface in RGB space for the points that satisfy the equation

$$D(\mathbf{z}, \mathbf{a}) = \left[(\mathbf{z} - \mathbf{a})^T \mathbf{C}^{-1} (\mathbf{z} - \mathbf{a}) \right]^{\frac{1}{2}} = D_0$$

where D_0 is a specified nonzero constant. Assume that $\mathbf{a} = \mathbf{0}$ and that

$$\mathbf{C} = \begin{bmatrix} 8 & 0 & 0 \\ 0 & 1 & 0 \\ 0 & 0 & 1 \end{bmatrix}.$$

6.29   Refer to Section 6.7.3. One might think that a logical approach for defining the gradient of an RGB image at any point (x, y) would be to compute the gradient *vector* (see Section 3.7.3) of each component image and then form a gradient vector for the color image by summing the three individual gradient vectors. Unfortunately, this method can at times yield erroneous results. Specifically, it is possible for a color image with clearly defined edges to have a zero gradient if this method were used. Give an example of such an image. (*Hint:* Set one of the color planes to a constant value to simplify your analysis.)

7 Wavelets and Multiresolution Processing

All this time, the guard was looking at her, first
through a telescope, then through a microscope, and
then through an opera glass.

Lewis Carrol, *Through the Looking Glass*

Preview

Although the Fourier transform has been the mainstay of transform-based image processing since the late 1950s, a more recent transformation, called the *wavelet transform*, is now making it even easier to compress, transmit, and analyze many images. Unlike the Fourier transform, whose basis functions are sinusoids, wavelet transforms are based on small waves, called *wavelets*, of varying frequency *and limited duration*. This allows them to provide the equivalent of a musical score for an image, revealing not only what notes (or frequencies) to play but also when to play them. Conventional Fourier transforms, on the other hand, provide only the notes or frequency information; temporal information is lost in the transformation process.

In 1987, wavelets were first shown to be the foundation of a powerful new approach to signal processing and analysis called *multiresolution* theory (Mallat [1987]). Multiresolution theory incorporates and unifies techniques from a variety of disciplines, including subband coding from signal processing, quadrature mirror filtering from digital speech recognition, and pyramidal image processing. As its name implies, multiresolution theory is concerned with the representation and analysis of signals (or images) at more than one resolution. The appeal of such an approach is obvious—features that might go undetected at one resolution may be easy to spot at another. Although the imaging community's interest in multiresolution analysis was limited until the late 1980s, it is now difficult to keep up with the number of papers, theses, and books devoted to the subject.

In this chapter, we examine wavelet-based transformations from a multiresolution point of view. Although they can be presented in other ways, this approach simplifies both the mathematics and physical interpretations. We begin with an overview of imaging techniques that influenced the formulation of multiresolution theory. Our objective is to introduce the theory's fundamental concepts within the context of image processing and simultaneously provide a brief historical perspective of the method and its application. The bulk of the chapter is focused on the development of a multiresolution-based toolkit for the representation and processing of images. To demonstrate the usefulness of the toolkit, examples ranging from image coding to noise removal and edge detection are provided. In the next chapter, wavelets will be used for image compression, an application in which they have received considerable attention.

7.1 Background

When we look at images, generally we see connected regions of similar texture and gray level that combine to form objects. If the objects are small in size or low in contrast, we normally examine them at high resolutions; if they are large in size or high in contrast, a coarse view is all that is required. If both small and large objects—or low and high contrast objects—are present simultaneously, it can be advantageous to study them at several resolutions. This, of course, is the fundamental motivation for multiresolution processing.

From a mathematical viewpoint, images are two-dimensional arrays of intensity values with locally varying statistics that result from different combinations of abrupt features like edges and contrasting homogeneous regions. As illustrated in Fig. 7.1—an image that will be examined repeatedly in the remainder of the

FIGURE 7.1 A natural image and its local histogram variations.

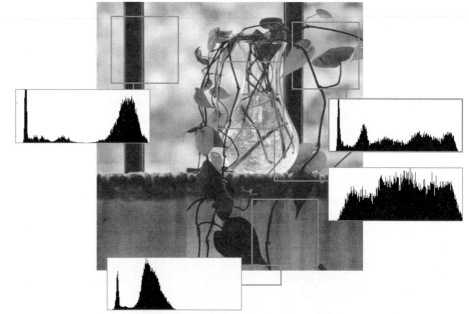

section—even first-order statistics vary significantly from one part of an image to another. They defy simple statistical modeling over the entire image.

7.1.1 Image Pyramids

A powerful, but conceptually simple structure for representing images at more than one resolution is the *image pyramid* (Burt and Adelson [1983]). Originally devised for machine vision and image compression applications, an image pyramid is a collection of decreasing resolution images arranged in the shape of a pyramid. As can be seen in Fig. 7.2(a), the base of the pyramid contains a high-resolution representation of the image being processed; the apex contains a low-resolution approximation. As you move up the pyramid, both size and resolution decrease. Since base level J is size $2^J \times 2^J$ or $N \times N$, where $J = \log_2 N$, intermediate level j is size $2^j \times 2^j$, where $0 \le j \le J$. Fully populated pyramids are composed of $J + 1$ resolution levels from $2^J \times 2^J$ to $2^0 \times 2^0$, but most pyramids are truncated to $P + 1$ levels, where $j = J - P, \dots, J - 2, J - 1, J$ and $1 \le P \le J$. That is, we normally limit ourselves to P reduced resolution approximations of the original image; a 1×1 or single pixel approximation of a 512×512 image, for example, is of little value. The total number of elements in a $P + 1$ level pyramid for $P > 0$ is

$$N^2\left(1 + \frac{1}{(4)^1} + \frac{1}{(4)^2} + \cdots + \frac{1}{(4)^P}\right) \le \frac{4}{3}N^2.$$

Figure 7.2(b) shows a simple system for constructing image pyramids. The *level j − 1 approximation* output is used to create *approximation pyramids*,

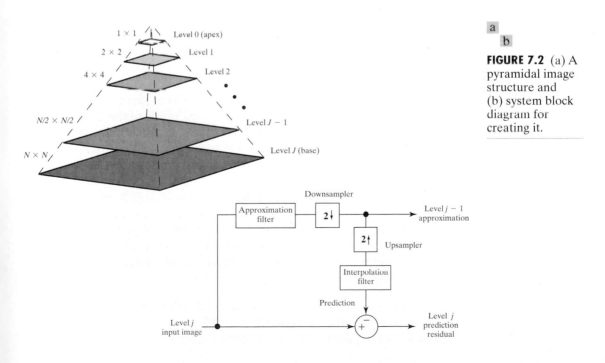

a
b
FIGURE 7.2 (a) A pyramidal image structure and (b) system block diagram for creating it.

which contain one or more approximations of the original image. Both the original image, which is at the base of the pyramid, and its P reduced resolution approximations can be accessed and manipulated directly. The *level j prediction residual* output of Fig. 7.2(b) is used to build *prediction residual pyramids*. These pyramids contain a low-resolution approximation of the original image at level $J - P$ and information for the construction of P higher-resolution approximations at the other levels. The information at level j is the difference between the level j approximation of the corresponding approximation pyramid and an estimate of that approximation based on the level $j - 1$ prediction residual. This difference can be coded—and therefore stored and transmitted—more efficiently than the approximation.

As the block diagram of Fig. 7.2(b) suggests, approximation and prediction residual pyramids are computed in an iterative fashion. A $P + 1$ level pyramid is built by executing the operations in the block diagram P times. During the first iteration or pass, $j = J$ and the original $2^J \times 2^J$ image is applied as the *level J input image*. This produces the *level J − 1 approximation* and *level J prediction residual* results. For passes $j = J - 1, J - 2, \ldots, J - P + 1$ (in that order), the previous iteration's *level j − 1 approximation* output is used as the input. Each pass is composed of three sequential steps:

1. Compute a reduced-resolution approximation of the input image. This is done by filtering the input and downsampling (i.e., subsampling) the filtered result by a factor of 2. A variety of filtering operations can be used, including neighborhood averaging, which produces a *mean pyramid*, lowpass Gaussian filtering (see Section 4.2.4), which produces a *Gaussian pyramid*, or no filtering, which results in a *subsampling pyramid*. The quality of the generated approximation, labeled the *level j − 1 approximation* in Fig. 7.2(b), is a function of the filter selected. Without the filter, aliasing can become pronounced in the upper levels of the pyramid, as the subsampled points may not be good representatives of the areas from which they were taken.

2. Upsample the output of the previous step—again by a factor of 2—and filter the result. This creates a *prediction* image with the same resolution as the input. By interpolating intensities between the pixels of the *Step 1* output, the *interpolation filter* determines how accurately the *prediction* approximates the input to *Step 1*. If the *interpolation filter* is omitted, the prediction is an upsampled version of the *Step 1* output and the blocking effects of pixel replication may become visible.

3. Compute the difference between the *prediction* of *Step 2* and the input to *Step 1*. This difference, labeled the *level j prediction residual*, can be later used to reconstruct progressively the original image (see Example 7.1). In the absence of quantization error, a prediction residual pyramid can be used to generate the corresponding approximation pyramid, including the original image, without error.

Executing this procedure P times produces two intimately related $P + 1$ level approximation and prediction residual pyramids. The *level j − 1 approximation* outputs are used to populate the approximation pyramid; the *level j prediction residual* outputs are placed in the prediction residual pyramid. If a prediction residual pyramid is not needed, *Steps 2* and *3*, together with the upsampler, *interpolation filter*, and summer of Fig. 7.2(b), can be omitted.

■ Figure 7.3 shows one possible approximation and prediction residual pyramid for the vase of Fig. 7.1. The approximation pyramid in Fig. 7.3(a) is a Gaussian pyramid. Filtering was performed in the spatial domain using a 5 × 5 lowpass Gaussian convolution kernel of the type depicted in Fig. 4.9(c) of Section 4.2.4. As can be seen, the resulting pyramid contains the original 512 × 512 resolution image (at its base) and three low-resolution approximations (of resolution 256 × 256, 128 × 128, and 64 × 64). That is, $P = 3$ and the pyramid has been truncated to four levels—levels 9, 8, 7, and 6 out of a possible $\log_2(512) + 1$ or 10 levels. Note the reduction in detail that accompanies the lower resolutions of the pyramid. The level 6 (i.e., 64 × 64) approximation is suitable for locating the window stiles, for example, but not for finding the philodendron's stems. In general, a pyramid's lower-resolution levels can be used for the analysis of large

EXAMPLE 7.1:
Gaussian and Laplacian pyramids.

a
b

FIGURE 7.3 Two image pyramids and their statistics: (a) a Gaussian (approximation) pyramid and (b) a Laplacian (prediction residual) pyramid.

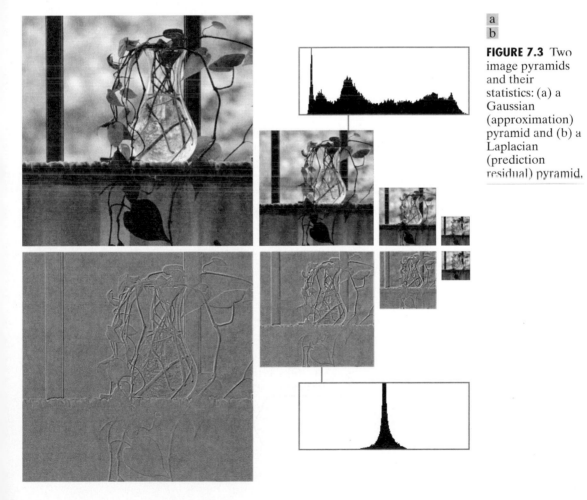

structures or overall image context; its high-resolution images are appropriate for analyzing individual object characteristics. Such a coarse to fine analysis strategy is particularly useful in pattern recognition.

The Laplacian pyramid in Fig. 7.3(b) contains the prediction residuals needed to compute its Gaussian counterpart in 7.3(a). To build the Gaussian pyramid, we begin with the Laplacian pyramid's level 6 64×64 approximation image, predict the Gaussian pyramid's level 7 128×128 resolution approximation (by upsampling and filtering), and add the Laplacian's level 7 prediction residual. This process is repeated using successively computed approximation images until the original 512×512 image is generated. Note that the first-order statistics of the prediction residual images in the Laplacian pyramid are highly peaked around zero. Unlike their Gaussian counterparts, these images can be highly compressed by assigning fewer bits to the more probable values (see the variable length codes of Section 8.1.1). Finally, we note that the prediction residuals in Fig. 7.3(b) are scaled to make the smaller prediction errors more visible; the prediction residual histogram, however, is based on prescaled residuals, with level 128 representing zero error. ■

7.1.2 Subband Coding

Another important imaging technique with ties to multiresolution analysis is *subband coding*. In subband coding, an image is decomposed into a set of band-limited components, called *subbands*, which can be reassembled to reconstruct the original image without error. Originally developed for speech and image compression, each subband is generated by bandpass filtering the input. Since the bandwidth of the resulting subbands is smaller than that of the original image, the subbands can be downsampled without loss of information. Reconstruction of the original image is accomplished by upsampling, filtering, and summing the individual subbands.

Figure 7.4(a) shows the principal components of a two-band subband coding and decoding system. The input of the system is a one-dimensional, band-limited

a
b

FIGURE 7.4 (a) A two-band filter bank for one-dimensional subband coding and decoding, and (b) its spectrum splitting properties.

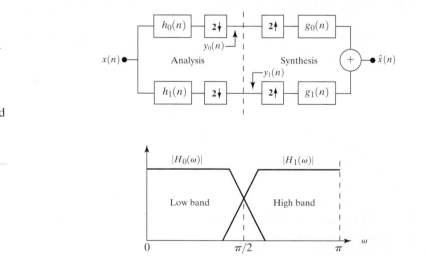

discrete-time signal $x(n)$ for $n = 0, 1, 2,\ldots$; the output sequence, $\hat{x}(n)$, is formed through the decomposition of $x(n)$ into $y_0(n)$ and $y_1(n)$ via *analysis filters* $h_0(n)$ and $h_1(n)$, and subsequent recombination via *synthesis filters* $g_0(n)$ and $g_1(n)$. Note that filters $h_0(n)$ and $h_1(n)$ are half-band digital filters whose idealized transfer characteristics, H_0 and H_1, are shown in Fig. 7.4(b). Filter H_0 is a low-pass filter whose output is an *approximation* of $x(n)$; filter H_1 is a highpass filter whose output is the high frequency or *detail* part of $x(n)$. All filtering is performed in the time domain by convolving each filter's input with its impulse response—its response to a unit amplitude impulse function, $\delta(n)$. We wish to select $h_0(n), h_1(n), g_0(n)$, and $g_1(n)$ (or, alternately H_0, H_1, G_0, and G_1) so that the input can be reconstructed perfectly. That is, so that $\hat{x}(n) = x(n)$.

The *Z-transform*, a generalization of the discrete Fourier transform, is the ideal tool for studying discrete-time, sampled-data systems like the one in Fig. 7.4(a). The Z-transform of sequence $x(n)$ for $n = 0, 1, 2,\ldots$ is

$$X(z) = \sum_{-\infty}^{\infty} x(n)z^{-n} \tag{7.1-1}$$

where z is a complex variable. [If $e^{j\omega}$ is substituted for z, Eq. (7.1-1) becomes the discrete Fourier transform of $x(n)$]. Our interest in the Z-transform stems from the ease with which it handles sampling rate changes. Downsampling by a factor of 2 in the time domain corresponds to the simple Z-domain operation

$$x_{\text{down}}(n) = x(2n) \Leftrightarrow X_{\text{down}}(z) = \frac{1}{2}\left[X(z^{1/2}) + X(-z^{1/2})\right] \tag{7.1-2}$$

where the double arrow indicates that the expressions on the left and right form a Z-transform pair. In a similar manner, upsampling—again by a factor of 2—is defined by the transform pair

$$x^{\text{up}}(n) = \begin{cases} x(n/2) & n = 0, 2, 4, \ldots \\ 0 & \text{otherwise} \end{cases} \Leftrightarrow X^{\text{up}}(z) = X(z^2). \tag{7.1-3}$$

If sequence $x(n)$ is downsampled and subsequently upsampled to yield $\hat{x}(n)$, Eqs. (7.1-2) and (7.1-3) combine to yield

$$\hat{X}(z) = \frac{1}{2}\left[X(z) + X(-z)\right] \tag{7.1-4}$$

where $\hat{x}(n) = Z^{-1}[\hat{X}(z)]$ is the resulting downsampled-upsampled sequence. The $X(-z)$ term in this equation is the Z-transform of an *aliased* or *modulated* version of sequence $x(n)$. Its inverse Z-transform is

$$Z^{-1}[X(-z)] = (-1)^n x(n). \tag{7.1-5}$$

With this brief introduction to the Z-transform, consider again the subband coding and decoding system of Fig. 7.4(a). In accordance with Eq. (7.1-4), we can express the system's output as

$$\hat{X}(z) = \frac{1}{2} G_0(z) \big[H_0(z)X(z) + H_0(-z)X(-z) \big] \tag{7.1-6}$$

$$+ \frac{1}{2} G_1(z) \big[H_1(z)X(z) + H_1(-z)X(-z) \big]$$

where, for example, the output of filter $h_0(n)$ in Fig. 7.4(a) is defined by the transform pair

$$h_0(n) * x(n) = \sum_k h_0(n-k)x(k) \Leftrightarrow H_0(z)X(z). \tag{7.1-7}$$

As with Fourier transforms, convolution in the time (or spatial) domain is equivalent to multiplication in the Z-domain. Rearranging terms in Eq. (7.1-6), we then get

$$\hat{X}(z) = \frac{1}{2} \big[H_0(z)G_0(z) + H_1(z)G_1(z) \big] X(z) \tag{7.1-8}$$

$$+ \frac{1}{2} \big[H_0(-z)G_0(z) + H_1(-z)G_1(z) \big] X(-z)$$

where the second component—by virtue of the fact that it contains the $-z$ dependence—represents the aliasing that is introduced by the downsampling-upsampling process.

For error-free reconstruction of the input, $\hat{x}(n) = x(n)$ and $\hat{X}(z) = X(z)$. Thus, we impose the following conditions:

$$H_0(-z)G_0(z) + H_1(-z)G_1(z) = 0 \tag{7.1-9}$$
$$H_0(z)G_0(z) + H_1(z)G_1(z) = 2. \tag{7.1-10}$$

Equation (7.1-9) eliminates aliasing by forcing the second term of Eq. (7.1-8) to zero; Eq. (7.1-10) eliminates amplitude distortion by reducing the first term to $X(z)$. Both can be incorporated into the single matrix expression

$$\big[G_0(z) \quad G_1(z) \big] \mathbf{H}_m(z) = \big[2 \quad 0 \big] \tag{7.1-11}$$

where *analysis modulation matrix* $\mathbf{H}_m(z)$ is

$$\mathbf{H}_m(z) = \begin{bmatrix} H_0(z) & H_0(-z) \\ H_1(z) & H_1(-z) \end{bmatrix}. \tag{7.1-12}$$

Assuming $\mathbf{H}_m(z)$ is nonsingular (i.e., it has a common left and right inverse), we can transpose Eq. (7.1-11) and left multiply by inverse $(\mathbf{H}_m^T(z))^{-1}$ to get

$$\begin{bmatrix} G_0(z) \\ G_1(z) \end{bmatrix} = \frac{2}{\det(\mathbf{H}_m(z))} \begin{bmatrix} H_1(-z) \\ -H_0(-z) \end{bmatrix} \tag{7.1-13}$$

where $\det(\mathbf{H}_m(z))$ denotes the determinant of $\mathbf{H}_m(z)$.

Equations (7.1-9) through (7.1-13) reveal several important characteristics of perfect reconstruction filter banks. Matrix Eq. (7.1-13), for instance, tells us that $G_1(z)$ is a function of $H_0(-z)$, while $G_0(z)$ is a function of $H_1(-z)$. The analysis and synthesis filters are *cross-modulated*. That is, diagonally opposed filters in the block diagram of Fig. 7.4(a) are functionally related by $-z$ in the Z-domain. For finite impulse response (FIR) filters, the determinant of the modulation matrix is a pure delay, i.e., $\det(\mathbf{H}_m(z)) = \alpha z^{-(2k+1)}$ (see, for example, Vetterli and Kovacevic [1995]). Thus, the exact form of the cross-modulation is a function of α. The $z^{-(2k+1)}$ term can be considered arbitrary since it is a shift that only changes the overall delay of the filter. Ignoring the delay, letting $\alpha = 2$, and taking the inverse Z-transform of Eq. (7.1-13), for instance, we get

$$g_0(n) = (-1)^n h_1(n) \tag{7.1-14}$$
$$g_1(n) = (-1)^{n+1} h_0(n).$$

If $\alpha = -2$, the resulting expressions are sign reversed:

$$g_0(n) = (-1)^{n+1} h_1(n) \tag{7.1-15}$$
$$g_1(n) = (-1)^n h_0(n).$$

Thus, FIR synthesis filters are cross-modulated copies of the analysis filters — with one (and only one) being sign reversed.

Equations (7.1-9) through (7.1-13) can also be used to demonstrate the biorthogonality (defined below) of the analysis and synthesis filters. To do this, let $P(z)$ be the product of the lowpass analysis and synthesis filter transfer functions. Substituting for G_0 from Eq. (7.1-13), we get

$$P(z) = G_0(z)H_0(z) = \frac{2}{\det(\mathbf{H}_m(z))} H_0(z)H_1(-z). \tag{7.1-16}$$

Since $\det(\mathbf{H}_m(z)) = -\det(\mathbf{H}_m(-z))$, product $G_1(z)H_1(z)$ can be similarly defined as

$$G_1(z)H_1(z) = \frac{-2}{\det(\mathbf{H}_m(z))} H_0(-z)H_1(z) = P(-z). \tag{7.1-17}$$

Thus, $G_1(z)H_1(z) = P(-z) = G_0(-z)H_0(-z)$ and Eq. (7.1-10) becomes

$$G_0(z)H_0(z) + G_0(-z)H_0(-z) = 2. \tag{7.1-18}$$

Taking the inverse Z-transform, we see that

$$\sum_k g_0(k)h_0(n-k) + (-1)^n \sum_k g_0(k)h_0(n-k) = 2\delta(n)$$

where, as usual, impulse function $\delta(n)$ is 1 if $n = 0$ and 0 otherwise. Since the odd indexed terms cancel, additional simplification yields[†]

$$\sum_k g_0(k)h_0(2n-k) = \langle g_0(k), h_0(2n-k)\rangle = \delta(n). \tag{7.1-19}$$

[†] The vector inner product of sequences $x(n)$ and $y(n)$ is $\langle x, y\rangle = \sum_n x^*(n)y(n)$, where the * denotes the complex conjugate operation. If $x(n)$ and $y(n)$ are real, $\langle x, y\rangle = \langle y, x\rangle$.

By starting over from Eqs. (7.1-9) and (7.1-10) and expressing G_0 and H_0 as a function of G_1 and H_1, we can also show that

$$\langle g_1(k), h_1(2n - k)\rangle = \delta(n) \qquad (7.1\text{-}20)$$
$$\langle g_0(k), h_1(2n - k)\rangle = 0$$

and

$$\langle g_1(k), h_0(2n - k)\rangle = 0,$$

which, in combination with Eq. (7.1-19), establishes the more general expression

$$\langle h_i(2n - k), g_j(k)\rangle = \delta(i - j)\delta(n), \quad i, j = \{0, 1\}. \qquad (7.1\text{-}21)$$

Filter banks satisfying this condition are called *biorthogonal*. Moreover, the analysis and synthesis filter impulse responses of all two-band, real-coefficient, perfect reconstruction filter banks are subject to the biorthogonality constraint. Example biorthogonal, FIR filters include the *biorthogonal spline* family (Cohen, Daubechies, and Feauveau [1992]) and the *biorthogonal coiflet* family (Tian and Wells [1995]).

Table 7.1 gives three general solutions to Eqs. (7.1-9) and (7.1-10). While each satisfies the biorthogonality requirement of Eq. (7.1-21), they are generated in different ways and define unique classes of perfect reconstruction filters. For each class, a "prototype" filter is designed to a particular specification and the remaining filters are computed from the prototype. Columns 1 and 2 of Table 7.1 are classic results from the filter bank literature—namely, *quadrature mirror filters* (QMFs) (Croisier, Estaban, and Galand [1976]) and *conjugate quadrature filters* (CQFs) (Smith and Barnwell [1986]). The filters in column three, which are later used in the development of the fast wavelet transform (see Section 7.4), are called *orthonormal*. They move a step beyond biorthogonality and require

$$\langle g_i(n), g_j(n + 2m)\rangle = \delta(i - j)\delta(m), \quad i, j = \{0, 1\}, \qquad (7.1\text{-}22)$$

which defines orthonormality for perfect reconstruction filter banks. Note that in the expression for $G_1(z)$ in row three of the table, $2K$ denotes the length or number of coefficients (i.e., filter taps) in each filter. As can be seen, G_1 is related to lowpass synthesis filter G_0 by modulation (see Eq. (7.1-5)), time reversal, and odd shift[†]. In addition, both H_0 and H_1 are time-reversed versions of the corresponding synthesis filters, G_0 and G_1, respectively. Taking the inverse Z-transform of the appropriate entries from column 3 of Table 7.1, we get that

$$g_1(n) = (-1)^n g_0(2K - 1 - n) \qquad (7.1\text{-}23)$$
$$h_i(n) = g_i(2K - 1 - n), \quad i = \{0, 1\}$$

where h_0, h_1, g_0, and g_1 are the impulse responses of the defined orthonormal filters. Examples include the Smith and Barnwell filter (Smith and Barnwell

---

[†]The Z-transform pairs for time reversal and shift are $x(-n) \Leftrightarrow X(z^{-1})$ and $x(n - k) \Leftrightarrow z^{-k}X(z)$, respectively.

Filter	QMF	CQF	Orthonormal
$H_0(z)$	$H_0^2(z) - H_0^2(-z) = 2$	$\begin{array}{l}H_0(z)H_0(z^{-1}) +\\ H_0(-z)H_0(-z^{-1}) = 2\end{array}$	$G_0(z^{-1})$
$H_1(z)$	$H_0(-z)$	$z^{-1}H_0(-z^{-1})$	$G_1(z^{-1})$
$G_0(z)$	$H_0(z)$	$H_0(z^{-1})$	$\begin{array}{l}G_0(z)G_0(z^{-1}) +\\ G_0(-z)G_0(-z^{-1}) = 2\end{array}$
$G_1(z)$	$-H_0(-z)$	$zH_0(-z)$	$-z^{-2K+1}G_0(-z^{-1})$

TABLE 7.1
Perfect reconstruction filter families.

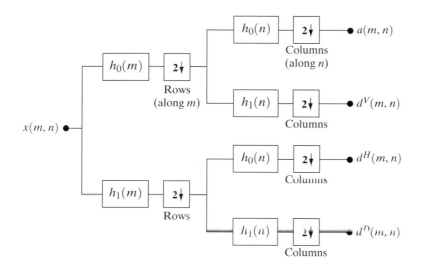

FIGURE 7.5 A two-dimensional, four-band filter bank for subband image coding.

[1984]), Daubechies filters (Daubechies [1988]), and the Vaidyanathan and Hoang filter (Vaidyanathan and Hoang [1988]).

The one-dimensional filters in Table 7.1 can be used as two-dimensional separable filters for the processing of images. As can be seen in Fig. 7.5, separable filters are first applied in one dimension (e.g., vertically) and then in the other (e.g., horizontally). Moreover, downsampling is performed in two stages—once before the second filtering operation to reduce the overall number of computations. The resulting filtered outputs, denoted $a(m, n)$, $d^V(m, n)$, $d^H(m, n)$, and $d^D(m, n)$ in Fig. 7.5, are called the approximation, vertical detail, horizontal detail, and diagonal detail subbands of the image, respectively. One or more of these subbands can be split into four smaller subbands, which can be split again, and so on.

■ Figure 7.6 shows the impulse responses of four 8-tap orthonormal filters. The coefficients of lowpass filter $h_0(n)$ for $0 \le n \le 7$ are -0.01059740, $0.03288301, 0.03084138, -0.18703481, -0.02798376, 0.63088076, 0.71484657$, and 0.23037781 (Daubechies [1992]); the coefficients of the remaining orthonormal filters can be computed using Eq. (7.1-23). Note (by visual inspection) the cross-modulation of the analysis and synthesis filters in Fig. 7.6. It is relatively easy to

EXAMPLE 7.2:
A four-band subband coding of the vase in Fig. 7.1.

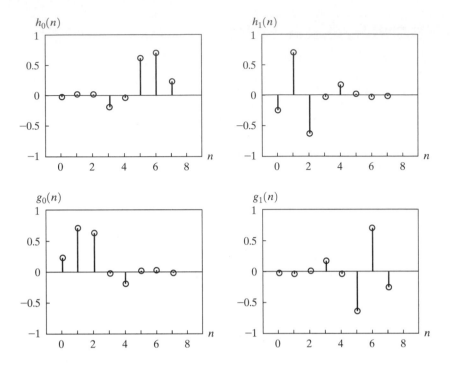

show numerically that the filters are both biorthogonal (satisfy Eq. 7.1-21) and orthogonal (satisfy Eq. 7.1-22). Moreover, they fulfill Eqs. (7.1-9) and (7.1-10) and support error-free reconstruction of the decomposed input.

A four-band split of the 512 × 512 image of a vase in Fig. 7.1, based on the filters in Fig. 7.6, is shown in Fig. 7.7. Each quadrant of this image is a subband of size 256 × 256. Beginning with the upper left corner and proceeding in a clockwise manner, the four quadrants contain approximation subband a, horizontal detail subband d^H, diagonal detail subband d^D, and vertical detail subband d^V, respectively. All subbands, except the approximation subband in the upper left quadrant, have been scaled to make their underlying structure more visible. Note the aliasing that is present in d^H and d^V, which is due to downsampling the barely discernable window screen in Fig. 7.1. Reconstruction of the original image from the subbands via synthesis filters $g_0(n)$ and $g_1(n)$ will cancel this aliasing in accordance with Eq. (7.1-9). To perform the reconstruction, a filter bank that roughly mirrors the system in Fig. 7.5 is required. In the new filter bank, the $h_i(n)$ for $i = \{0, 1\}$ filters are replaced by their $g_i(n)$ counterparts, and upsamplers and summers are added. ∎

7.1.3 The Haar Transform

The third and final imaging-related operation with ties to multiresolution analysis that we will look at is the Haar transform (Haar [1910]). Within the context of this chapter, its importance stems from the fact that its basis functions are the oldest and simplest known orthonormal wavelets. They will be used in a number of examples in the sections that follow.

FIGURE 7.7 A
four-band split of
the vase in Fig. 7.1
using the subband
coding system of
Fig. 7.5.

The Haar transform can be expressed in matrix form

$$\mathbf{T} = \mathbf{HFH}^T \tag{7.1-24}$$

where $\mathbf{F}$ is an $N \times N$ image matrix, $\mathbf{H}$ is an $N \times N$ transformation matrix, and $\mathbf{T}$ is the resulting $N \times N$ transform. For the Haar transform, transformation matrix $\mathbf{H}$ contains the Haar basis functions, $h_k(z)$. They are defined over the continuous, closed interval $z \in [0, 1]$ for $k = 0, 1, 2, \ldots, N - 1$, where $N = 2^n$. To generate $\mathbf{H}$, we define the integer k such that $k = 2^p + q - 1$, where $0 \le p \le n - 1$, $q = 0$ or 1 for $p = 0$, and $1 \le q \le 2^p$ for $p \ne 0$. Then the Haar basis functions are

$$h_0(z) = h_{00}(z) = \frac{1}{\sqrt{N}}, \quad z \in [0, 1] \tag{7.1-25}$$

and

$$h_k(z) = h_{pq}(z) = \frac{1}{\sqrt{N}} \begin{cases} 2^{p/2} & (q-1)/2^p \le z < (q-0.5)/2^p \\ -2^{p/2} & (q-0.5)/2^p \le z < q/2^p \\ 0 & \text{otherwise}, z \in [0, 1]. \end{cases} \tag{7.1-26}$$

The ith row of an $N \times N$ Haar transformation matrix contains the elements of $h_i(z)$ for $z = 0/N, 1/N, 2/N, \ldots, (N-1)/N$. If $N = 4$, for example, k, q, and p assume the values

k	p	q
0	0	0
1	0	1
2	1	1
3	1	2

and the 4×4 transformation matrix, $\mathbf{H}_4$, is

$$\mathbf{H}_4 = \frac{1}{\sqrt{4}} \begin{bmatrix} 1 & 1 & 1 & 1 \\ 2 & 1 & -1 & -1 \\ \sqrt{2} & -\sqrt{2} & 0 & 0 \\ 0 & 0 & \sqrt{2} & -\sqrt{2} \end{bmatrix}. \tag{7.1-27}$$

In a similar manner, the 2×2 transformation matrix, $\mathbf{H}_2$, is

$$\mathbf{H}_2 = \frac{1}{\sqrt{2}} \begin{bmatrix} 1 & 1 \\ 1 & -1 \end{bmatrix}. \tag{7.1-28}$$

Its basis functions define the only two-tap FIR filter bank that will satisfy the QMF prototype filter specification in row 1 and column 1 of Table 7.1. The coefficients of the corresponding QMF analysis filters, $h_0(n)$ and $h_1(n)$, are the elements of the first and second rows of matrix $\mathbf{H}_2$, respectively.

EXAMPLE 7.3:
Haar functions in a discrete wavelet transform.

■ Figure 7.8(a) shows a multiresolution-based decomposition of Fig. 7.1 using the Haar basis functions of Eqs. (7.1-25) and (7.1-26). Unlike the pyramidal structure in Fig. 7.3, this representation, called the *discrete wavelet transform* and developed later in the chapter, contains the same number of pixels as the original image. In addition,

1. Its local statistics are relatively constant and easily modeled. See Fig. 7.8(a).
2. Many of its values are close to zero. This makes it an excellent candidate for image compression.
3. Both coarse and fine resolution approximations of the original image can be extracted from it. Figures 7.8(b)–(d) were reconstructed from the subimages of Fig. 7.8(a).

In database applications, these properties make it easy for users to access low-quality versions of images during searches and later retrieve additional data to refine them.

Finally, we note that Fig. 7.8(a) is not the Haar transform of Eqs. (7.1-24) through (7.1-26), but it bears a close resemblance to both the subband coding result of Fig. 7.7 and the Laplacian pyramid of Fig. 7.3(b). As in the previous results, the subimages in Fig. 7.8(a) are scaled to make their underlying structure more visible. The approximation images in Fig. 7.8(b)–(d) are of size

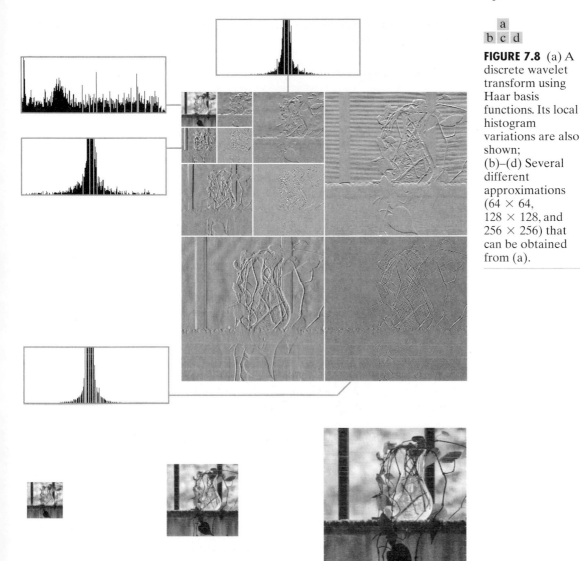

FIGURE 7.8 (a) A discrete wavelet transform using Haar basis functions. Its local histogram variations are also shown; (b)–(d) Several different approximations (64 × 64, 128 × 128, and 256 × 256) that can be obtained from (a).

64 × 64, 128 × 128, and 256 × 256, respectively. A perfect 512 × 512 reconstruction of the original image is also possible.   ▩

7.2 Multiresolution Expansions

The previous section introduced three well-known imaging techniques that played an important role in the development of a unique mathematical theory called multiresolution analysis (MRA). In MRA, a *scaling function* is used to create a series of approximations of a function or image, each differing by a factor of 2 from its nearest neighboring approximations. Additional functions, called *wavelets*, are then used to encode the difference in information between adjacent approximations.

7.2.1 Series Expansions

A signal or function $f(x)$ can often be better analyzed as a linear combination of expansion functions

$$f(x) = \sum_k \alpha_k \varphi_k(x) \tag{7.2-1}$$

where k is an integer index of the finite or infinite sum, the α_k are real-valued *expansion coefficients*, and the $\varphi_k(x)$ are real-valued *expansion functions*. If the expansion is unique—that is, there is only one set of α_k for any given $f(x)$—the $\varphi_k(x)$ are called *basis functions*, and the *expansion set*, $\{\varphi_k(x)\}$, is called a *basis* for the class of functions that can be so expressed. The expressible functions form a *function space* that is referred to as the *closed span* of the expansion set, denoted

$$V = \overline{\underset{k}{\text{Span}} \{\varphi_k(x)\}}. \tag{7.2-2}$$

To say that $f(x) \in V$ means that $f(x)$ is in the closed span of $\{\varphi_k(x)\}$ and can be written in the form of Eq. (7.2-1).

For any function space V and corresponding expansion set $\{\varphi_k(x)\}$, there is a set of *dual* functions, denoted $\{\tilde{\varphi}_k(x)\}$, that can be used to compute the α_k coefficients of Eq. (7.2-1) for any $f(x) \in V$. These coefficients are computed by taking the *integral inner products*[†] of the dual $\tilde{\varphi}_k(x)$'s and function $f(x)$. That is,

$$\alpha_k = \langle \tilde{\varphi}_k(x), f(x) \rangle = \int \tilde{\varphi}_k^*(x) f(x)\, dx \tag{7.2-3}$$

where the * denotes the complex conjugate operation. Depending on the orthogonality of the expansion set, this computation assumes one of three possible forms. Problem 7.10 at the end of the chapter illustrates the three cases using vectors in two-dimensional Euclidean space.

Case 1: If the expansion functions form an orthonormal basis for V, meaning that

$$\langle \varphi_j(x), \varphi_k(x) \rangle = \delta_{jk} = \begin{cases} 0 & j \neq k \\ 1 & j = k \end{cases} \tag{7.2-4}$$

the basis and its dual are equivalent. That is, $\varphi_k(x) = \tilde{\varphi}_k(x)$ and Eq. (7.2-3) becomes

$$\alpha_k = \langle \varphi_k(x), f(x) \rangle. \tag{7.2-5}$$

The α_k are computed as the inner products of the basis functions and $f(x)$.

[†]The integral inner product of two real or complex-valued functions $f(x)$ and $g(x)$ is $\langle f(x), g(x) \rangle = \int f^*(x) g(x)\, dx$. If $f(x)$ is real, $f^*(x) = f(x)$ and $\langle f(x), g(x) \rangle = \int f(x) g(x)\, dx$.

Case 2: If the expansion functions are not orthonormal, but are an orthogonal basis for V, then

$$\langle \varphi_j(x), \varphi_k(x) \rangle = 0 \quad j \neq k \tag{7.2-6}$$

and the basis functions and their duals are called *biorthogonal*. The α_k are computed using Eq. (7.2-3), and the biorthogonal basis and its dual are such that

$$\langle \varphi_j(x), \tilde{\varphi}_k(x) \rangle = \delta_{jk} = \begin{cases} 0 & j \neq k \\ 1 & j = k. \end{cases} \tag{7.2-7}$$

Case 3: If the expansion set is not a basis for V, but supports the expansion defined in Eq. (7.2-1), it is a spanning set in which there is more than one set of α_k for any $f(x) \in V$. The expansion functions and their duals are said to be *overcomplete* or redundant. They form a *frame* in which[†]

$$A\|f(x)\|^2 \leq \sum_k |\langle \varphi_k(x), f(x) \rangle|^2 \leq B\|f(x)\|^2 \tag{7.2-8}$$

for some $A > 0, B < \infty$, and all $f(x) \in V$. Dividing this equation by the norm squared of $f(x)$, we see that A and B "frame" the normalized inner products of the expansion coefficients and the function. Equations similar to (7.2-3) and (7.2-5) can be used to find the expansion coefficients for frames. If $A = B$, the expansion set is called a *tight frame* and it can be shown that (Daubechies [1992])

$$f(x) - \frac{1}{A} \sum_k \langle \varphi_k(x), f(x) \rangle \varphi_k(x). \tag{7.2-9}$$

Except for the A^{-1} term, which is a measure of the frame's redundancy, this is identical to the expression obtained by substituting Eq. (7.2-5) (for orthonormal bases) into Eqs. (7.2-1).

7.2.2 Scaling Functions

Now consider the set of expansion functions composed of integer translations and binary scalings of the real, square-integrable function $\varphi(x)$; that is, the set $\{\varphi_{j,k}(x)\}$ where

$$\varphi_{j,k}(x) = 2^{j/2}\varphi(2^j x - k) \tag{7.2-10}$$

for all $j, k \in \mathbf{Z}$ and $\varphi(x) \in L^2(\mathbf{R})$.[‡] Here, k determines the position of $\varphi_{j,k}(x)$ along the x-axis, j determines $\varphi_{j,k}(x)$'s width—how broad or narrow it is along the x-axis—and $2^{j/2}$ controls its height or amplitude. Because the shape of $\varphi_{j,k}(x)$ changes with j, $\varphi(x)$ is called a *scaling function*. By choosing $\varphi(x)$ wisely, $\{\varphi_{j,k}(x)\}$ can be made to span $L^2(\mathbf{R})$, the set of all measurable, square-integrable functions.

[†] The norm of $f(x)$, denoted $\|f(x)\|$, is defined as the square root of the inner product of $f(x)$ with itself.
[‡] The notation $L^2(\mathbf{R})$, where $\mathbf{R}$ is the set of real numbers, denotes the set of measurable, square-integrable one-dimensional functions; $\mathbf{Z}$ is the set of integers.

If we restrict j in Eq. (7.2-10) to a specific value, say $j = j_0$, the resulting expansion set, $\{\varphi_{j_0,k}(x)\}$, is a subset of $\{\varphi_{j,k}(x)\}$. It will not span $L^2(\mathbf{R})$, but a subspace within it. Using the notation of the previous section, we can define that subspace as

$$V_{j_0} = \overline{\underset{k}{\mathrm{Span}}\{\varphi_{j_0,k}(x)\}}. \qquad (7.2\text{-}11)$$

That is, V_{j_0} is the span of $\varphi_{j_0,k}(x)$ over k. If $f(x) \in V_{j_0}$, it can be written

$$f(x) = \sum_k \alpha_k \varphi_{j_0,k}(x). \qquad (7.2\text{-}12)$$

More generally, we will denote the subspace spanned over k for any j as

$$V_j = \overline{\underset{k}{\mathrm{Span}}\{\varphi_{j,k}(x)\}}. \qquad (7.2\text{-}13)$$

As will be seen in the following example, increasing j increases the size of V_j, allowing functions with smaller variations or finer detail to be included in the subspace. This is a consequence of the fact that, as j increases, the $\varphi_{j,k}(x)$ that are used to represent the subspace functions become narrower and separated by smaller changes in x.

EXAMPLE 7.4:
The Haar scaling function.

■ Consider the unit-height, unit-width scaling function (Haar [1910])

$$\varphi(x) = \begin{cases} 1 & 0 \le x < 1 \\ 0 & \text{otherwise.} \end{cases} \qquad (7.2\text{-}14)$$

Figures 7.9(a)–(d) show four of the many expansion functions that can be generated by substituting this pulse-shaped scaling function into Eq. (7.2-10). Note that when $j = 1$, as opposed to $j = 0$, the resulting expansion functions are narrower and closer together.

Figure 7.9(e) shows a member of subspace V_1. This function does not belong to V_0 because the V_0 expansion functions in 7.9(a) and (b) are too coarse to represent it. Higher-resolution functions like those in 7.9(c) and (d) are required. They can be used, as shown in (e), to represent the function by the three-term expansion

$$f(x) = 0.5\varphi_{1,0}(x) + \varphi_{1,1}(x) - 0.25\varphi_{1,4}(x).$$

To conclude the example, Fig. 7.9(f) illustrates the decomposition of $\varphi_{0,0}(x)$ as a sum of V_1 expansion functions. In a similar manner, any V_0 expansion function can be decomposed using

$$\varphi_{0,k}(x) = \frac{1}{\sqrt{2}} \varphi_{1,2k}(x) + \frac{1}{\sqrt{2}} \varphi_{1,2k+1}(x).$$

Thus, if $f(x)$ is an element of V_0, it is also an element of V_1. This is because all V_0 expansion functions are a part of V_1. Mathematically, we write that V_0 is a subspace of V_1, denoted $V_0 \subset V_1$. ■

The simple scaling function in the preceding example obeys the four fundamental requirements of multiresolution analysis (Mallat [1989a]):

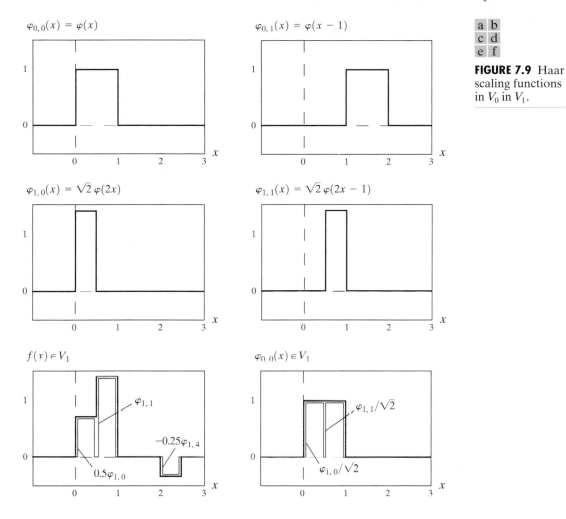

a b
c d
e f

FIGURE 7.9 Haar
scaling functions
in V_0 in V_1.

MRA Requirement 1: The scaling function is orthogonal to its integer translates.

This is easy to see in the case of the Haar function, since whenever it has a value of 1, its integer translates are 0, so that the product of the two is 0. The Haar scaling function is said to have *compact support*, which means that is 0 everywhere outside a finite interval called the support. In fact, its support is 1; it is 0 outside the half open interval $[0, 1)$. It should be noted that the requirement for orthogonal integer translates becomes harder to satisfy as the support of the scaling function becomes larger than 1.

MRA Requirement 2: The subspaces spanned by the scaling function at low scales are nested within those spanned at higher scales.

As can be seen in Fig. 7.10, subspaces containing high-resolution functions must also contain all lower resolution functions. That is,

$$V_{-\infty} \subset \cdots \subset V_{-1} \subset V_0 \subset V_1 \subset V_2 \subset \cdots \subset V_\infty. \qquad (7.2\text{-}15)$$

FIGURE 7.10 The nested function spaces spanned by a scaling function.

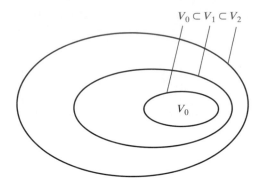

$$V_0 \subset V_1 \subset V_2$$

V_0

Moreover the subspaces satisfy the intuitive condition that if $f(x) \in V_j$, then $f(2x) \in V_{j+1}$. The fact that the Haar scaling function meets this requirement should not be taken to indicate that any function with a support of 1 automatically satisfies the condition. It is left as an exercise for the reader to show that the equally simple function

$$\varphi(x) = \begin{cases} 1 & 0.25 \le x < 0.75 \\ 0 & \text{elsewhere} \end{cases}$$

is not a valid scaling function for a multiresolution analysis (see Problem 7.11).

MRA Requirement 3: The only function that is common to all V_j is $f(x) = 0$.

If we consider the coarsest possible expansion functions (i.e., $j = -\infty$), the only representable function is the function of no information. That is,

$$V_{-\infty} = \{0\}. \tag{7.2-16}$$

MRA Requirement 4: Any function can be represented with arbitrary precision.

Though it may not be possible to expand a particular $f(x)$ at an arbitrarily coarse resolution, as was the case for the function in Fig. 7.9(e), all measurable, square-integrable functions can be represented in the limit as $j \to \infty$. That is,

$$V_{\infty} = \{L^2(\mathbf{R})\}. \tag{7.2-17}$$

Under these conditions, the expansion functions of subspace V_j can be expressed as a weighted sum of the expansion functions of subspace V_{j+1}. Using Eq. (7.2-12), we let

$$\varphi_{j,k}(x) = \sum_n \alpha_n \varphi_{j+1,n}(x)$$

where the index of summation has been changed to n for clarity. Substituting for $\varphi_{j+1,n}(x)$ from Eq. (7.2-10) and changing variable α_n to $h_\varphi(n)$, this becomes

$$\varphi_{j,k}(x) = \sum_n h_\varphi(n) 2^{(j+1)/2} \varphi(2^{j+1}x - n).$$

Since $\varphi(x) = \varphi_{0,0}(x)$, both j and k can be set to 0 to obtain the simpler non-subscripted expression

$$\varphi(x) = \sum_n h_\varphi(n) \sqrt{2} \varphi(2x - n). \qquad (7.2\text{-}18)$$

The $h_\varphi(n)$ coefficients in this recursive equation are called *scaling function coefficients*; h_φ is referred to as a *scaling vector*. The alternate notations $h(n)$ and $h_0(n)$ are often used in the literature, but we wish to avoid confusion with our earlier discussion of subband analysis filters. Equation (7.2-18) is fundamental to multiresolution analysis and is called the *refinement equation*, the *MRA equation*, or the *dilation equation*. It states that the expansion functions of any subspace can be built from double-resolution copies of themselves—that is, from expansion functions of the next higher resolution space. The choice of a reference subspace, V_0, is arbitrary.

EXAMPLE 7.5:
Haar scaling function coefficients.

■ The scaling function coefficients for the Haar function of Eq. (7.2-14) are $h_\varphi(0) = h_\varphi(1) = 1/\sqrt{2}$, the first row of matrix $\mathbf{H}_2$ in Eq. (7.1-28). Thus, Eq. (7.2-18) yields

$$\varphi(x) = \frac{1}{\sqrt{2}}\left[\sqrt{2}\varphi(2x)\right] + \frac{1}{\sqrt{2}}\left[\sqrt{2}\varphi(2x - 1)\right].$$

This decomposition was illustrated graphically for $\varphi_{0,0}(x)$ in Fig. 7.9(f), where the bracketed terms of the preceding expression are seen to be $\varphi_{1,0}(x)$ and $\varphi_{1,1}(x)$. Additional simplification yields $\varphi(x) = \varphi(2x) + \varphi(2x - 1)$. ■

7.2.3 Wavelet Functions

Given a scaling function that meets the MRA requirements of the previous section, we can define a *wavelet function* $\psi(x)$ that, together with its integer translates and binary scalings, spans the difference between any two adjacent scaling subspaces, V_j and V_{j+1}. The situation is illustrated graphically in Fig. 7.11. We define the set $\{\psi_{j,k}(x)\}$ of wavelets

$$\psi_{j,k}(x) = 2^{j/2}\psi(2^j x - k) \qquad (7.2\text{-}19)$$

$$V_2 = V_1 \oplus W_1 = V_0 \oplus W_0 \oplus W_1$$

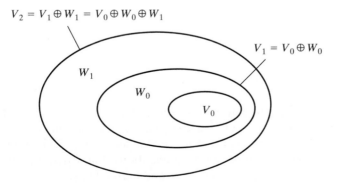

$V_1 = V_0 \oplus W_0$

FIGURE 7.11 The relationship between scaling and wavelet function spaces.

for all $k \in \mathbf{Z}$ that spans the W_j spaces in the figure. As with scaling functions, we write

$$W_j = \overline{\underset{k}{\text{Span}} \{\psi_{j,k}(x)\}} \tag{7.2-20}$$

and note that if $f(x) \in W_j$,

$$f(x) = \sum_k \alpha_k \psi_{j,k}(x). \tag{7.2-21}$$

The scaling and wavelet function subspaces in Fig. 7.11 are related by

$$V_{j+1} = V_j \oplus W_j \tag{7.2-22}$$

where $\oplus$ denotes the union of spaces (like the union of sets). The orthogonal complement of V_j in V_{j+1} is W_j, and all members of V_j are orthogonal to the members of W_j. Thus,

$$\langle \varphi_{j,k}(x), \psi_{j,l}(x) \rangle = 0 \tag{7.2-23}$$

for all appropriate $j, k, l \in \mathbf{Z}$.

We can now express the space of all measurable, square-integrable functions as

$$L^2(\mathbf{R}) = V_0 \oplus W_0 \oplus W_1 \oplus \ldots \tag{7.2-24}$$

or

$$L^2(\mathbf{R}) = V_1 \oplus W_1 \oplus W_2 \oplus \ldots \tag{7.2-25}$$

or even

$$L^2(\mathbf{R}) = \cdots \oplus W_{-2} \oplus W_{-1} \oplus W_0 \oplus W_1 \oplus W_2 \oplus \ldots, \tag{7.2-26}$$

which eliminates the scaling function and represents a function in terms of wavelets alone. Note that if $f(x)$ is an element of V_1, but not V_0, an expansion using Eq. (7.2-24) contains an *approximation* of $f(x)$ using V_0 scaling functions; wavelets from W_0 would encode the *difference* between this approximation and the actual function. Equations (7.2-24) through (7.2-26) can be generalized to yield

$$L^2(\mathbf{R}) = V_{j_0} \oplus W_{j_0} \oplus W_{j_0+1} \oplus \ldots \tag{7.2-27}$$

where j_0 is an arbitrary starting scale.

Since wavelet spaces reside within the spaces spanned by the next higher resolution scaling functions (see Fig. 7.11), any wavelet function—like its scaling function counterpart of Eq. (7.2-18)—can be expressed as a weighted sum of shifted, double-resolution scaling functions. That is, we can write

$$\psi(x) = \sum_n h_\psi(n) \sqrt{2} \varphi(2x - n) \tag{7.2-28}$$

where the $h_\psi(n)$ are called the *wavelet function coefficients* and h_ψ is the *wavelet vector* [the notation $h_1(n)$ is often used in the literature]. Using the condition that wavelets span the orthogonal complement spaces in Fig. 7.11, and that integer wavelet translates are orthogonal, it can be shown that $h_\psi(n)$ is related to $h_\varphi(n)$ by (see, for example, Burrus, Gopinath, and Guo [1998])

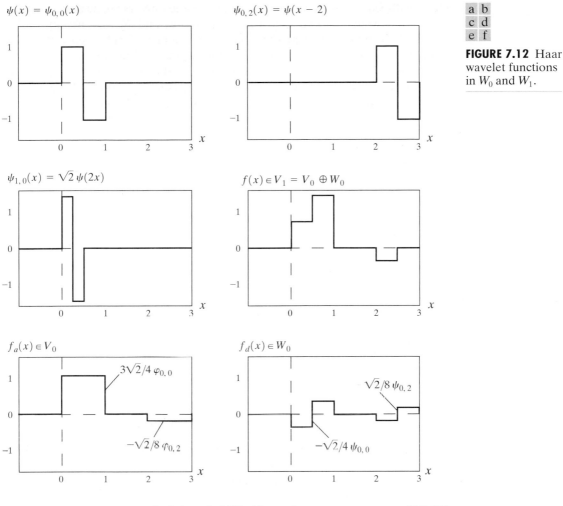

FIGURE 7.12 Haar wavelet functions in W_0 and W_1.

$$h_\psi(n) = (-1)^n h_\varphi(1 - n). \tag{7.2-29}$$

Note the similarity of this result and Eq. (7.1-23), the relationship governing the impulse responses of orthonormal subband coding and decoding filters.

■ In the previous example, the Haar scaling vector was defined as $h_\varphi(0) = h_\varphi(1) = 1/\sqrt{2}$. Using Eq. (7.2-29), the corresponding wavelet vector is $h_\psi(0) = (-1)^0 h_\varphi(1 - 0) = 1/\sqrt{2}$ and $h_\psi(1) = (-1)^1 h_\varphi(1 - 1) = -1/\sqrt{2}$. Note that these coefficients correspond to the second row of matrix $\mathbf{H}_2$ in Eq. (7.1-28). Substituting these values into Eq. (7.2-28), we get $\psi(x) = \varphi(2x) - \varphi(2x - 1)$, which is plotted in Fig. 7.12(a). Thus, the Haar wavelet function is

EXAMPLE 7.6: The Haar wavelet function coefficients.

$$\psi(x) = \begin{cases} 1 & 0 \le x < 0.5 \\ -1 & 0.5 \le x < 1 \\ 0 & \text{elsewhere.} \end{cases} \tag{7.2-30}$$

Using with Eq. (7.2-19), we can now generate the universe of scaled and translated Haar wavelets. Two such wavelets, $\psi_{0,2}(x)$ and $\psi_{1,0}(x)$, are plotted in Figs. 7.12(b) and (c), respectively. Note that wavelet $\psi_{1,0}(x)$ for space W_1 is narrower than $\psi_{0,2}(x)$ for W_0; it can be used to represent finer detail.

Figure 7.12(d) shows a function of subspace V_1 that is not in subspace V_0. This function was considered in an earlier example [see Fig. 7.9(e)]. Although the function cannot be represented accurately in V_0, Eq. (7.2-22) indicates that it can be expanded using V_0 and W_0 expansion functions. The resulting expansion is

$$f(x) = f_a(x) + f_d(x)$$

where

$$f_a(x) = \frac{3\sqrt{2}}{4}\varphi_{0,0}(x) - \frac{\sqrt{2}}{8}\varphi_{0,2}(x)$$

and

$$f_d(x) = \frac{-\sqrt{2}}{4}\psi_{0,0}(x) - \frac{\sqrt{2}}{8}\psi_{0,2}(x).$$

Here, $f_a(x)$ is an approximation of $f(x)$ using V_0 scaling functions, while $f_d(x)$ is the difference $f(x) - f_a(x)$ as a sum of W_0 wavelets. The two expansions, which are shown in Figs. 7.12(e) and (f), divide $f(x)$ in a manner similar to a lowpass and highpass filter. The low frequencies of $f(x)$ are captured in $f_a(x)$—it assumes the average value of $f(x)$ in each integer interval—while the high-frequency details are encoded in $f_d(x)$.   ■

7.3   Wavelet Transforms in One Dimension

We can now formally define several closely related wavelet transformations: the generalized *wavelet series expansion*, the *discrete wavelet transform*, and the *continuous wavelet transform*. Their counterparts in the Fourier domain are the Fourier series expansion, the discrete Fourier transform, and the integral Fourier transform, respectively. In Section 7.4, we will define a computationally efficient implementation of the discrete wavelet transform called the *fast wavelet transform*.

7.3.1 The Wavelet Series Expansions

We begin by defining the *wavelet series expansion* of function $f(x) \in L^2(\mathbf{R})$ relative to wavelet $\psi(x)$ and scaling function $\varphi(x)$. In accordance with Eq. (7.2-27), we can write

$$f(x) = \sum_k c_{j_0}(k)\varphi_{j_0,k}(x) + \sum_{j=j_0}^{\infty}\sum_k d_j(k)\psi_{j,k}(x) \qquad (7.3\text{-}1)$$

where j_0 is an arbitrary starting scale and the $c_{j_0}(k)$'s and $d_j(k)$'s are relabeled α_k's from Eqs. (7.2-12) and (7.2-21), respectively. The $c_{j_0}(k)$'s are normally called the *approximation* or *scaling coefficients*; the $d_j(k)$'s are referred to as the *detail* or *wavelet coefficients*. This is because the first sum in Eq. (7.3-1) uses scaling functions to provide an approximation of $f(x)$ at scale j_0 [unless $f(x) \in V_{j_0}$ and it is exact]. For each higher scale $j \geq j_0$ in the second sum, a finer resolution function—a sum of wavelets—is added to the approximation to provide increasing detail. If the expansion functions form an orthonormal basis or tight frame, which is often the case, the expansion coefficients are calculated—based on Eqs. (7.2-5) and (7.2-9)—as

$$c_{j_0}(k) = \langle f(x), \varphi_{j_0,k}(x) \rangle = \int f(x) \varphi_{j_0,k}(x)\, dx \qquad (7.3\text{-}2)$$

and

$$d_j(k) = \langle f(x), \psi_{j,k}(x) \rangle = \int f(x) \psi_{j,k}(x)\, dx. \qquad (7.3\text{-}3)$$

If the expansion functions are part of a biorthogonal basis, the φ and ψ terms in these equations must be replaced by their dual functions, $\tilde{\varphi}$ and $\tilde{\psi}$, respectively.

■ Consider the simple function

$$y = \begin{cases} x^2 & 0 \leq x < 1 \\ 0 & \text{otherwise} \end{cases}$$

EXAMPLE 7.7:
The Haar wavelet series expansion of $y = x^2$.

shown in Fig. 7.13(a). Using Haar wavelets—see Eqs. (7.2-14) and (7.2-30)—and a starting scale $j_0 = 0$, Eqs. (7.3-2) and (7.3-3) can be used to compute the following expansion coefficients:

$$c_0(0) = \int_0^1 x^2 \varphi_{0,0}(x)\, dx = \int_0^1 x^2\, dx = \frac{x^3}{3}\Big|_0^1 = \frac{1}{3}$$

$$d_0(0) = \int_0^1 x^2 \psi_{0,0}(x)\, dx = \int_0^{0.5} x^2\, dx - \int_{0.5}^1 x^2\, dx = -\frac{1}{4}$$

$$d_1(0) = \int_0^1 x^2 \psi_{1,0}(x)\, dx = \int_0^{0.25} x^2\sqrt{2}\, dx - \int_{0.25}^{0.5} x^2\sqrt{2}\, dx = -\frac{\sqrt{2}}{32}$$

$$d_1(1) = \int_0^1 x^2 \psi_{1,1}(x)\, dx = \int_{0.5}^{0.75} x^2\sqrt{2}\, dx - \int_{0.75}^1 x^2\sqrt{2}\, dx = -\frac{3\sqrt{2}}{32}$$

Substituting these values into Eq. (7.3-1), we get the wavelet series expansion

$$y = \underbrace{\frac{1}{3}\varphi_{0,0}(x)}_{V_0} + \underbrace{\left[-\frac{1}{4}\psi_{0,0}(x)\right]}_{W_0} + \underbrace{\left[-\frac{\sqrt{2}}{32}\psi_{1,0}(x) - \frac{3\sqrt{2}}{32}\psi_{1,1}(x)\right]}_{W_1} + \cdots$$

$$\underbrace{}_{V_1 = V_0 \oplus W_0}$$

$$\underbrace{}_{V_2 = V_1 \oplus W_1 = V_0 \oplus W_0 \oplus W_1}$$

The first term in this expansion uses $c_0(0)$ to generate a subspace V_0 approximation of the function being expanded. This approximation is shown in Fig. 7.13(b) and is the average value of the original function. The second term

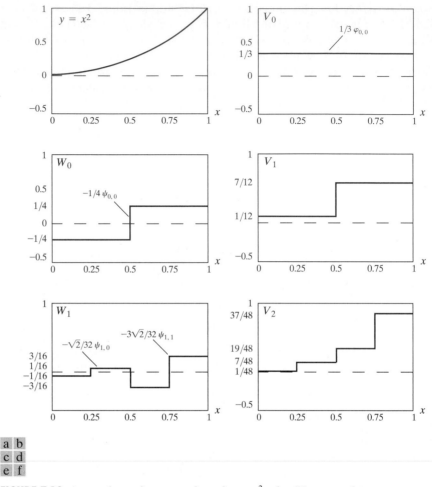

a b
c d
e f

FIGURE 7.13 A wavelet series expansion of $y = x^2$ using Haar wavelets.

uses $d_0(0)$ to refine the approximation by adding a level of detail from subspace W_0. The added detail and resulting V_1 approximation are shown in Figs. 7.13(c) and (d), respectively. Another level of detail is added by the subspace W_1 coefficients $d_1(0)$ and $d_1(1)$. This additional detail is shown in Fig. 7.13(e), and the resulting V_2 approximation is depicted in 7.13(f). Note that the expansion is now beginning to resemble the original function. As higher scales (or greater levels of detail) are added, the approximation becomes a more precise representation of the function, realizing it in the limit as $j \to \infty$. ■

7.3.2 The Discrete Wavelet Transform

Like the Fourier series expansion, the wavelet series expansion of the previous section maps a function of a continuous variable into a sequence of coefficients. If the function being expanded is a sequence of numbers, like samples of a continuous function $f(x)$, the resulting coefficients are called the *discrete wavelet transform* (DWT) of $f(x)$. For this case, the series expansion defined in Eqs. (7.3-1) through (7.3-3) becomes the DWT transform pair

$$W_\varphi(j_0, k) = \frac{1}{\sqrt{M}} \sum_x f(x)\varphi_{j_0,k}(x) \qquad (7.3\text{-}5)$$

$$W_\psi(j, k) = \frac{1}{\sqrt{M}} \sum_x f(x)\psi_{j,k}(x) \qquad (7.3\text{-}6)$$

for $j \geq j_0$ and

$$f(x) = \frac{1}{\sqrt{M}} \sum_k W_\varphi(j_0, k)\varphi_{j_0,k}(x) + \frac{1}{\sqrt{M}} \sum_{j=j_0}^{\infty} \sum_k W_\psi(j, k)\psi_{j,k}(x). \qquad (7.3\text{-}7)$$

Here, $f(x)$, $\varphi_{j_0,k}(x)$, and $\psi_{j,k}(x)$ are functions of the discrete variable $x = 0$, $1, 2, \ldots, M - 1$. For example, $f(x) = f(x_0 + x\Delta x)$ for some x_0, Δx, and $x = 0, 1, 2, \ldots, M - 1$. Normally, we let $j_0 = 0$ and select M to be a power of 2 (i.e., $M = 2^J$) so that the summations are performed over $x = 0, 1$, $2, \ldots, M - 1, j = 0, 1, 2, \ldots, J - 1$, and $k = 0, 1, 2, \ldots, 2^j - 1$. For Haar wavelets, the discretized scaling and wavelet functions employed in the transform (i.e., the basis functions) correspond to the rows of the $M \times M$ Haar transformation matrix of Section 7.1.3. The transform itself is composed of M coefficients, the minimum scale is 0, and the maximum scale is $J - 1$. For reasons noted in Section 7.3.1 and illustrated in Example 7.6, the coefficients defined in Eqs. (7.3-5) and (7.3-6) are usually called *approximation* and *detail coefficients*, respectively.

The $W_\varphi(j_0, k)$'s and $W_\psi(j, k)$'s in Eqs. (7.3-5) to (7.3-7) correspond to the $c_{j_0}(k)$'s and $d_j(k)$'s of the wavelet series expansion in the previous section. (This change of variables is not necessary but paves the way for the standard notation used for the continuous wavelet transform of the next section.) Note that the integrations in the series expansion have been replaced by summations, and a $1/\sqrt{M}$ normalizing factor, reminiscent of the DFT in Section 4.2.1, has been added to both the forward and inverse expressions. This factor could alternately be incorporated into the forward or inverse alone as $1/M$. Finally, it should be remembered that Eqs. (7.3-5) through (7.3-7) are valid for orthonormal bases and tight frames alone. For biorthogonal bases, the φ and ψ terms in Eqs. (7.3-5) and (7.3-6) must be replaced by their duals, $\tilde{\varphi}$ and $\tilde{\psi}$, respectively.

■ To illustrate the use of Eqs. (7.3-5) through (7.3-7), consider the discrete function of four points: $f(0) = 1$, $f(1) = 4$, $f(2) = -3$, and $f(3) = 0$. Since $M = 4$, $J = 2$ and, with $j_0 = 0$, the summations are performed over $x = 0, 1$, $2, 3, j = 0, 1$, and $k = 0$ for $j = 0$ or $k = 0, 1$ for $j = 1$. We will use the Haar scaling and wavelet functions and assume that the four samples of $f(x)$ are distributed over the support of the basis functions, which is 1. Substituting the four samples into Eq. (7.3-5), we find that

$$W_\varphi(0, 0) = \frac{1}{2} \sum_{x=0}^{3} f(x)\varphi_{0,0}(x)$$

$$= \frac{1}{2}[1 \cdot 1 + 4 \cdot 1 - 3 \cdot 1 + 0 \cdot 1] = 1$$

EXAMPLE 7.8:
Computing a one-dimensional discrete wavelet transform.

because $\varphi_{0,0}(x) = 1$ for $x = 0, 1, 2, 3$. Note that we have employed uniformly spaced samples of the Haar scaling function for $j = 0$ and $k = 0$. The values correspond to the first row of Haar transformation matrix $\mathbf{H}_4$ of Section 7.1.3. Continuing with Eq. (7.3-6) and similarly spaced samples of $\psi_{j,k}(x)$, which correspond to rows 2, 3, and 4 of $\mathbf{H}_4$, we get

$$W_\psi(0, 0) = \frac{1}{2}\left[1 \cdot 1 + 4 \cdot 1 - 3 \cdot (-1) + 0 \cdot (-1)\right] = 4$$

$$W_\psi(1, 0) = \frac{1}{2}\left[1 \cdot \sqrt{2} + 4 \cdot (-\sqrt{2}) - 3 \cdot 0 + 0 \cdot 0\right] = -1.5\sqrt{2}$$

$$W_\psi(1, 1) = \frac{1}{2}\left[1 \cdot 0 + 4 \cdot 0 - 3 \cdot \sqrt{2} + 0 \cdot (-\sqrt{2})\right] = -1.5\sqrt{2}.$$

Thus, the discrete wavelet transform of our simple four-sample function relative to the Haar wavelet and scaling function is $\{1, 4, -1.5\sqrt{2}, -1.5\sqrt{2}\}$, where the transform coefficients have been arranged in the order in which they were computed.

Equation (7.3-7) lets us reconstruct the original function from its transform. Iterating through its summation indices, we get

$$f(x) = \frac{1}{2}\Big[W_\varphi(0, 0)\varphi_{0,0}(x) + W_\psi(0, 0)\psi_{0,0}(x) +$$

$$W_\psi(1, 0)\psi_{1,0}(x) + W_\psi(1, 1)\psi_{1,1}(x)\Big]$$

for $x = 0, 1, 2, 3$. If $x = 0$, for instance,

$$f(0) = \frac{1}{2}\left[1 \cdot 1 + 4 \cdot 1 - 1.5\sqrt{2} \cdot (\sqrt{2}) - 1.5\sqrt{2} \cdot 0\right] = 1.$$

As in the forward case, uniformly spaced samples of the scaling and wavelet functions are used in the computation of the inverse.   ■

The four-point DWT in the preceding example is an illustration of a two-scale decomposition of $f(x)$—that is, $j = \{0, 1\}$. The underlying assumption was that starting scale j_0 was zero, but other starting scales are possible. It is left as an exercise for the reader (see Problem 7.16) to compute the single-scale transform $\{2.5\sqrt{2}, -1.5\sqrt{2}, -1.5\sqrt{2}, -1.5\sqrt{2}\}$, which results when the starting scale is one. Thus, Eqs. (7.3-5) and (7.3-6) define a "family" of transforms that differ in starting scale j_0.

7.3.3 The Continuous Wavelet Transform

The natural extension of the discrete wavelet transform is the *continuous wavelet transform* (CWT), which transforms a continuous function into a highly redundant function of two continuous variables—translation and scale. The resulting transform is easy to interpret and valuable for time-frequency analysis. Although our interest is in discrete images, it is covered here for completeness.

The continuous wavelet transform of a continuous, square-integrable function, $f(x)$, relative to a real-valued wavelet, $\psi(x)$, is

$$W_\psi(s, \tau) = \int_{-\infty}^{\infty} f(x)\psi_{s,\tau}(x)\, dx \qquad (7.3\text{-}8)$$

where

$$\psi_{s,\tau}(x) = \frac{1}{\sqrt{s}}\,\psi\!\left(\frac{x - \tau}{s}\right) \qquad (7.3\text{-}9)$$

and s and τ are called *scale* and *translation* parameters, respectively. Given $W_\psi(s, \tau)$, $f(x)$ can be obtained using the *inverse continuous wavelet transform*

$$f(x) = \frac{1}{C_\psi} \int_0^{\infty} \int_{-\infty}^{\infty} W_\psi(s, \tau)\, \frac{\psi_{s,\tau}(x)}{s^2}\, d\tau\, ds \qquad (7.3\text{-}10)$$

where

$$C_\psi = \int_{-\infty}^{\infty} \frac{|\Psi(u)|^2}{|u|}\, du \qquad (7.3\text{-}11)$$

and $\Psi(u)$ is the Fourier transform of $\psi(x)$. Equations (7.3-8) through (7.3-11) define a reversible transformation as long as the so-called *admissibility criterion*, $C_\psi < \infty$, is satisfied (Grossman and Morlet [1984]). In most cases, this simply means that $\Psi(0) = 0$ and $\Psi(u) \to 0$ as $u \to \infty$ fast enough to make $C_\psi < \infty$.

The preceding equations are reminiscent of their discrete counterparts—Eqs. (7.2-19), (7.3-1), (7.3-3), (7.3-6), and (7.3-7). The following similarities should be noted:

1. The continuous translation parameter, τ, takes the place of the integer translation parameter, k.
2. The continuous scale parameter, s, is inversely related to the binary scale parameter, 2^j. This is because s appears in the denominator of $\psi((x - \tau)/s)$ in Eq. (7.3-9). Thus, wavelets used in continuous transforms are compressed or reduced in width when $0 < s < 1$ and dilated or expanded when $s > 1$. Wavelet scale and our traditional notion of frequency are inversely related.
3. The continuous transform is similar to a series expansion [see Eq. (7.3-1)] or discrete transform [see Eq. (7.3-6)] in which the starting scale $j_0 = -\infty$. This—in accordance with Eq. (7.2-26)—eliminates explicit scaling function dependence, so that the function is represented in terms of wavelets alone.
4. Like the discrete transform, the continuous transform can be viewed as a set of transform coefficients, $\{W_\psi(s, \tau)\}$, that measure the similarity of $f(x)$ with a set of basis functions, $\{\psi_{s,\tau}(x)\}$. In the continuous case, however, both sets are infinite. Because $\psi_{s,\tau}(x)$ is real valued and $\psi_{s,\tau}(x) = \psi_{s,\tau}^*(x)$, each coefficient from Eq. (7.3-8) is the integral inner product, $\langle f(x), \psi_{s,\tau}(x) \rangle$, of $f(x)$ and $\psi_{s,\tau}(x)$.

■ The *Mexican hat* wavelet,

$$\psi(x) = \left(\frac{2}{\sqrt{3}}\,\pi^{-1/4}\right)(1 - x^2)e^{-x^2/2} \qquad (7.3\text{-}12)$$

EXAMPLE 7.9:
A one-dimensional continuous wavelet transform.

a b
c d

FIGURE 7.14 The continuous wavelet transform (c and d) and Fourier spectrum (b) of a continuous one-dimensional function (a).

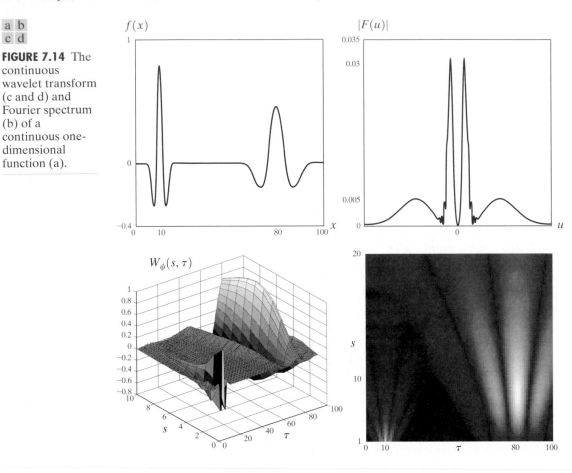

gets its name from its distinctive shape [see Fig. 7.14(a)]. It is proportional to the second derivative of the Gaussian probability function, has an average value of zero, and is compactly supported (i.e., dies out rapidly as $|x| \to \infty$). Although it satisfies the admissibility requirement for the existence of continuous, reversible transforms, there is not an associated scaling function, and the computed transform does not result in an orthogonal analysis. Its most distinguishing features are its symmetry and the existence of the explicit expression of Eq. (7.3-12).

The continuous, one-dimensional function in Fig. 7.14(a) is the sum of two Mexican hat wavelets:

$$f(x) = \psi_{1,10}(x) + \psi_{6,80}(x)$$

Its Fourier spectrum, shown in Fig. 7.14(b), reveals the close connection between scaled wavelets and Fourier frequency bands. The spectrum contains two broad frequency bands (or peaks) that correspond to the function's two Gaussian-like perturbations.

Figure 7.14(c) shows a portion ($1 \le s \le 10$ and $\tau \le 100$) of the CWT of the function in Fig. 7.14(a) relative to the Mexican hat wavelet. Unlike the Fourier spectrum in Fig. 7.14(b), it provides both spatial and frequency information. Note, for example, that when $s = 1$, the transform achieves a maximum at $\tau = 10$, which corresponds to the location of the $\psi_{1,10}(x)$ component of $f(x)$. Because the transform provides an objective measure of the similarity between $f(x)$ and the wavelets for which it is computed, it is easy to see how it can be used for feature detection. We simply need wavelets that match the features of interest. Similar observations can be drawn from the intensity plot in Fig. 7.14(d), where the absolute value of the transform, $|W_\psi(s, \tau)|$, is displayed as gray levels between black and white. Note that the continuous wavelet transform turns a one-dimensional function into a two-dimensional result. ■

7.4 The Fast Wavelet Transform

The *fast wavelet transform* (FWT) is a computationally efficient implementation of the discrete wavelet transform (DWT) that exploits a surprising but fortunate relationship between the coefficients of the DWT at adjacent scales. Also called Mallat's *herringbone algorithm* (Mallat [1989a, b]), the FWT resembles the two-band subband coding scheme of Section 7.1.2.

Consider again the multiresolution refinement equation

$$\varphi(x) = \sum_n h_\varphi(n)\sqrt{2}\varphi(2x - n). \tag{7.4-1}$$

Scaling x by 2^j, translating it by k, and letting $m = 2k + n$ gives

$$\varphi(2^j x - k) = \sum_n h_\varphi(n)\sqrt{2}\varphi(2(2^j x - k) - n)$$

$$= \sum_m h_\varphi(m - 2k)\sqrt{2}\varphi(2^{j+1}x - m). \tag{7.4-2}$$

Note that scaling vector h_φ can be thought of as the "weights" used to expand $\varphi(2^j x - k)$ as a sum of scale $j + 1$ scaling functions. A similar sequence of operations—beginning with Eq. (7.2-28)—provides an analogous result for $\psi(2^j x - k)$. That is,

$$\psi(2^j x - k) = \sum_m h_\psi(m - 2k)\sqrt{2}\varphi(2^{j+1}x - m) \tag{7.4-3}$$

where scaling vector $h_\varphi(n)$ in Eq. (7.4-2) is replaced by wavelet vector $h_\psi(n)$ in Eq. (7.4-3).

Now consider Eqs. (7.3-5) and (7.3-6) of Section 7.2.2. They define the discrete wavelet transform. Substituting Eq. (7.2-19)—the wavelet defining equation—into Eq. (7.3-6), we get

$$W_\psi(j, k) = \frac{1}{\sqrt{M}} \sum_x f(x) 2^{j/2} \psi(2^j x - k), \tag{7.4-4}$$

which, upon replacing $\psi(2^j x - k)$ with the right side of Eq. (7.4-3), becomes

$$W_\psi(j, k) = \frac{1}{\sqrt{M}} \sum_x f(x) 2^{j/2} \left[\sum_m h_\psi(m - 2k)\sqrt{2}\varphi(2^{j+1}x - m) \right]. \tag{7.4-5}$$

Interchanging the sum and integral and rearranging terms then gives

$$W_\psi(j, k) = \sum_m h_\psi(m - 2k) \left[\frac{1}{\sqrt{M}} \sum_x f(x) 2^{(j+1)/2} \varphi(2^{j+1}x - m) \right] \quad (7.4\text{-}6)$$

where the bracketed quantity is identical to Eq. (7.3-5) with $j_0 = j + 1$. To see this, substitute Eq. (7.2-10) into Eq. (7.3-5) and let j_0 be $j + 1$. We can therefore write

$$W_\psi(j, k) = \sum_m h_\psi(m - 2k) W_\varphi(j + 1, m) \quad (7.4\text{-}7)$$

and note that the DWT detail coefficients at scale j are a function of the DWT approximation coefficients at scale $j + 1$. Recognizing Eqs. (7.4-2) and (7.3-5) as the starting point of a similar derivation involving the DWT approximation coefficients, we find similarly that

$$W_\varphi(j, k) = \sum_m h_\varphi(m - 2k) W_\varphi(j + 1, m). \quad (7.4\text{-}8)$$

Equations (7.4-7) and (7.4-8) reveal a remarkable relationship between the DWT coefficients of adjacent scales. Comparing these results to Eq. (7.1-7), we see that both $W_\varphi(j, k)$ and $W_\psi(j, k)$, the scale j approximation and the detail coefficients, can be computed by convolving $W_\varphi(j + 1, k)$, the scale $j + 1$ approximation coefficients, with the time-reversed scaling and wavelet vectors, $h_\varphi(-n)$ and $h_\psi(-n)$, and subsampling the results. Figure 7.15 reduces these operations to block diagram form. Note that it is identical to the analysis portion of the two-band subband coding and decoding system of Fig. 7.4, with $h_0(n) = h_\varphi(-n)$ and $h_1(n) = h_\psi(-n)$. We can therefore write

$$W_\psi(j, k) = h_\psi(-n) * W_\varphi(j + 1, n) \Big|_{n=2k,\, k \ge 0} \quad (7.4\text{-}9)$$

and

$$W_\varphi(j, k) = h_\varphi(-n) * W_\varphi(j + 1, n) \Big|_{n=2k,\, k \ge 0} \quad (7.4\text{-}10)$$

where the convolutions are evaluated at instants $n = 2k$ for $k \ge 0$. Evaluating convolutions at nonnegative, even indices is equivalent to filtering and downsampling by 2.

To conclude the development of the FWT, we simply note that the filter bank in Fig. 7.15 can be "iterated" to create multistage structures for computing DWT

FIGURE 7.15 An FWT analysis bank.

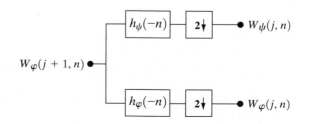

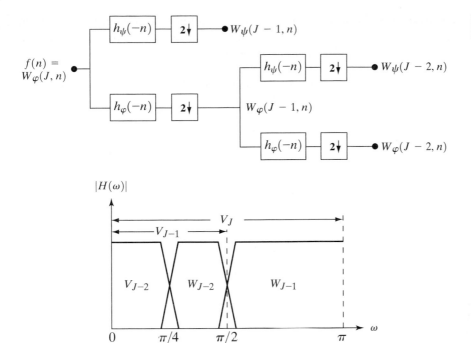

FIGURE 7.16
(a) A two-stage or two-scale FWT analysis bank and (b) its frequency splitting characteristics.

coefficients at two or more successive scales. For example, Fig. 7.16(a) shows a two-stage filter bank for generating the coefficients at the two highest scales of the transform. Note that the highest scale coefficients are assumed to be samples of the function itself. That is, $W_\varphi(J, n) = f(n)$, where J is the highest scale. [In accordance with Section 7.2.2, $f(x) \in V_J$, where V_J is the scaling space in which $f(x)$ resides.] The first filter bank in Fig. 7.16(a) splits the original function into a lowpass, approximation component, which corresponds to scaling coefficients $W_\varphi(J - 1, n)$, and a highpass, detail component, corresponding to coefficients $W_\psi(J - 1, n)$. This is graphically illustrated in Fig. 7.16(b), where scaling space V_J is split into wavelet subspace W_{J-1} and scaling subspace V_{J-1}. The spectrum of the original function is split into two half-band components. The second filter bank of Fig. 7.16(a) splits the spectrum and subspace V_{J-1}, the lower half-band, into quarter-band subspaces W_{J-2} and V_{J-2}—with corresponding DWT coefficients $W_\psi(J - 1, n)$ and $W_\varphi(J - 1, n)$, respectively.

The two-stage filter bank of Fig. 7.16 is easily extended to any number of scales. A third filter bank, for example, would operate on the $W_\varphi(J - 2, n)$ coefficients, splitting scaling space V_{J-2} into two eighth-band subspaces W_{J-3} and V_{J-3}. Normally, we choose 2^J samples of $f(x)$ and employ P filter banks (à la Fig. 7.15) to generate a P-scale FWT at scales $J - 1, J - 2, \ldots, J - P$. The highest scale (i.e., $J - 1$) coefficients are computed first; the lowest scale (i.e., $J - P$) last. If function $f(x)$ is sampled above the Nyquist rate, as is usually the case, its samples are good approximations of the scaling coefficients at the sampling resolution and can be used as the starting high-resolution scaling coefficient inputs. In other words, no wavelet or detail coefficients are needed at the sampling scale. The highest-resolution scaling functions act as "delta functions" in Eqs. (7.3-5) and (7.3-6), allowing $f(n)$ to be used as the scale J approximation or scaling coefficient input to the first two-band filter bank (Odegard, Gopinath, and Burrus [1992]).

EXAMPLE 7.10:
Computing a 1-D fast wavelet transform.

■ To illustrate these concepts, consider the discrete function $f(n) = \{1, 4, -3, 0\}$ from Example 7.8. As in that example, we will recompute the transform based on Haar scaling and wavelet functions. Here, however, we will not use the basis functions directly, as was done in the DWT of Example 7.8, but the corresponding scaling and wavelet vectors from Examples 7.5 and 7.6:

$$h_\varphi(n) = \begin{cases} 1/\sqrt{2} & n = 0, 1 \\ 0 & \text{otherwise} \end{cases} \quad (7.4\text{-}11)$$

and

$$h_\psi(n) = \begin{cases} 1/\sqrt{2} & n = 0 \\ -1/\sqrt{2} & n = 1 \\ 0 & \text{otherwise.} \end{cases} \quad (7.4\text{-}12)$$

These are the functions used to build the FWT filter banks; they provide the filter coefficients.

Since the DWT computed in Example 7.8 was composed of elements $\{W_\varphi(0,0), W_\psi(0,0), W_\psi(1,0), W_\psi(1,1)\}$, we will compute the corresponding two-scale FWT for scales $j = \{0, 1\}$. That is, $J = 2$ (there are $2^J = 2^2$ samples) and $P = 2$ (we are working with scales $J - 1 = 2 - 1 = 1$ and $J - P = 2 - 2 = 0$ in that order). The transform will be computed using the two-stage filter bank of Fig. 7.16(a). Figure 7.17 shows the sequences that result from the required FWT convolutions and downsamplings. Note that function $f(n)$ itself is the scaling or

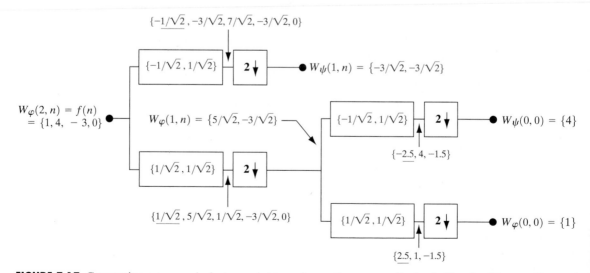

FIGURE 7.17 Computing a two-scale fast wavelet transform of sequence $\{1, 4, -3, 0\}$ using Haar scaling and wavelet vectors.

approximation input to the leftmost filter bank. To compute the $W_\psi(1, k)$ coefficients that appear at the end of the upper branch of Fig. 7.17, for example, we first convolve $f(n)$ with $h_\psi(-n)$. As explained in Section 4.6.3, this requires flipping one of the functions about the origin, sliding it past the other, and computing the sum of the point-to-point product of the two functions. For sequences $\{1, 4, -3, 0\}$ and $\{-1/\sqrt{2}, 1/\sqrt{2}\}$, this produces $\{-1/\sqrt{2}, -3/\sqrt{2}, 7/\sqrt{2}, -3/\sqrt{2}, 0\}$, where the second term corresponds to index $k = 2n = 0$. (In Fig. 7.17, underlined values represent negative indices, i.e., $n < 0$.) When downsampled by taking the even-indexed points, we get $W_\psi(1, k) = \{-3/\sqrt{2}, -3/\sqrt{2}\}$ for $k = \{0, 1\}$. We can alternately use Eq. (7.4-9) to compute

$$
W_\psi(1, k) = h_\psi(-n) * W_\varphi(2, n)\Big|_{n=2k, k \geq 0} = h_\psi(-n) * f(n)\Big|_{n=2k, k \geq 0}
$$

$$
= \sum_l h_\psi(2k - l)x(l)\Big|_{k=0, 1}
$$

$$
= \frac{1}{\sqrt{2}} x(2k) - \frac{1}{\sqrt{2}} x(2k + 1)\Big|_{k=0, 1}.
$$

Here, we have substituted $2k$ for n in the convolution and employed l as a dummy variable of convolution (i.e., for displacing the two sequences relative to one another). There are only two terms in the expanded sum because there are only two nonzero values in the time-reversed wavelet vector $h_\psi(-n)$. Substituting $k = 0$, we find that $W_\psi(1, 0) = -3/\sqrt{2}$; for $k = 1$, we get $W_\psi(1, 1) = -3/\sqrt{2}$. Thus, the filtered and downsampled sequence is $\{-3/\sqrt{2}, -3/\sqrt{2}\}$, which matches the earlier result. The remaining convolutions and downsamplings are performed in a similar manner.   ■

As one might expect, an equally efficient inverse transform for the reconstruction of $f(x)$ from DWT/FWT approximation and detail coefficients, $W_\varphi(j, k)$ and $W_\psi(j, k)$, can also be formulated. Called the *inverse fast wavelet transform* (FWT^{-1}), it uses the scaling and wavelet vectors employed in the forward transform, together with the level j approximation and detail coefficients, to generate the level $j + 1$ approximation coefficients. Noting the similarity between the FWT analysis bank in Fig. 7.15 and the two-band subband analysis portion of Fig. 7.4(a), we can immediately postulate the required FWT^{-1} *synthesis filter bank*. Figure 7.18 details its structure, which is identical to the synthesis portion of the two-band subband coding and decoding system in Fig. 7.4(a). Equation (7.1-23) of Section 7.1.2 defines the

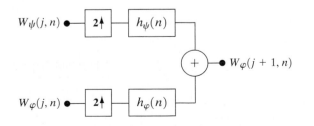

FIGURE 7.18 The FWT^{-1} synthesis filter bank.

$W_\psi(j, n)$ ●—— 2↑ —— $h_\psi(n)$

＋ —● $W_\varphi(j + 1, n)$

$W_\varphi(j, n)$ ●—— 2↑ —— $h_\varphi(n)$

relevant synthesis filters. As noted there, perfect reconstruction (for two-band orthonormal filters) requires $g_i(n) = h_i(-n)$ for $i = \{0, 1\}$. That is, the synthesis and analysis filters must be time-reversed versions of one another. Since the FWT analysis filters (see Fig. 7.15) are $h_0(n) = h_\varphi(-n)$ and $h_1(n) = h_\psi(-n)$, the required FWT^{-1} synthesis filters are $g_0(n) = h_0(-n) = h_\varphi(n)$ and $g_1(n) = h_1(-n) = h_\psi(n)$. It should be remembered, however, that it is also possible to use biorthogonal analysis and synthesis filters, which are not time-reversed versions of one another. Biorthogonal analysis and synthesis filters are cross-modulated per Eqs. (7.1-14) and (7.1-15).

The FWT^{-1} filter bank in Figure 7.18 implements the computation

$$W_\varphi(j + 1, k) = h_\varphi(k) * W_\varphi^{\text{up}}(j, k) + h_\psi(k) * W_\psi^{\text{up}}(j, k)\Big|_{k \geq 0} \quad (7.4\text{-}13)$$

where W^{up} signifies upsampling by 2 (i.e., inserting zeros between the elements of W so that it is twice its original length). The upsampled coefficients are filtered, by convolution with $h_\varphi(n)$ and $h_\psi(n)$, and added to generate a higher scale approximation. In essence, a better approximation of $f(x)$, with greater detail and resolution, is created. As with the forward FWT, the inverse filter bank can be iterated as shown in Fig. 7.19, where a two-scale structure for computing the final two scales of a FWT^{-1} reconstruction is depicted. This coefficient combining process can be extended to any number of scales and guarantees perfect reconstruction of function $f(x)$.

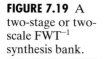

EXAMPLE 7.11:
Computing a 1-D inverse fast wavelet transform.

■ Computation of the inverse fast wavelet transform mirrors its forward counterpart. Figure 7.20 illustrates the process for the sequence considered in Example 7.10. To begin the calculation, the level 0 approximation and detail coefficients are upsampled to yield $\{1, 0\}$ and $\{4, 0\}$, respectively. Convolution with filters $g_0(n) = h_\varphi(n) = \{1/\sqrt{2}, 1/\sqrt{2}\}$ and $g_1(n) = h_\psi(n) = \{1/\sqrt{2}, -1/\sqrt{2}\}$ produces $\{1/\sqrt{2}, 1/\sqrt{2}, 0\}$ and $\{4/\sqrt{2}, -4/\sqrt{2}, 0\}$, which when added give $W_\varphi(1, n) = \{5/\sqrt{2}, -3/\sqrt{2}\}$. Thus, the level 1 approximation of Fig. 7.20, which matches the computed approximation in Fig. 7.17, is reconstructed. Continuing in this manner, $f(n)$ is formed at the right of the second synthesis filter bank. ■

We conclude our discussion of the fast wavelet transform by noting several differences between the FWT and the FFT—the first being their numerical complexity. The number of mathematical operations involved in the computa-

FIGURE 7.19 A two-stage or two-scale FWT^{-1} synthesis bank.

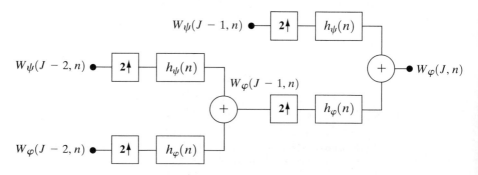

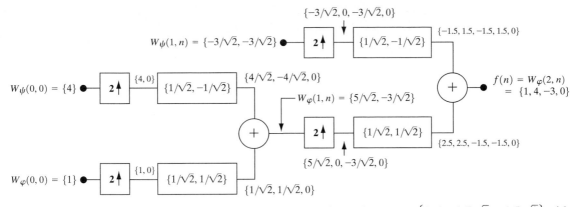

FIGURE 7.20 Computing a two-scale inverse fast wavelet transform of sequence $\{1, 4, -1.5\sqrt{2}, -1.5\sqrt{2}\}$ with Haar scaling and wavelet vectors.

tion of the FWT of a length $M = 2^J$ sequence is on the order of $O(M)$. That is, the number of floating-point multiplications and additions (using filter banks) is linear with respect to the length of the sequence. This compares favorably with the FFT algorithm, which requires $O(M \log M)$.

The second difference relates to the transforms' basis functions. While the Fourier basis functions (i.e., sinusoids) guarantee the existence of the FFT, the existence of the FWT depends upon the availability of a scaling function for the wavelets being used, as well as the orthogonality (or biorthogonality) of the scaling function and corresponding wavelets. Thus, the Mexican hat wavelet of Eq. (7.3-12), which does not have a companion scaling function, cannot be used in the computation of the FWT. In other words, we cannot construct a filter bank like that of Fig. 7.15 for the Mexican hat wavelet; it does not satisfy the underlying assumptions of the FWT approach.

Finally, we note that while time and frequency are usually viewed as different domains when representing functions, they are inextricably linked. When you try to analyze a function simultaneously in time and frequency, you run into the following problem: If you want precise information about time, you must put up with some vagueness about frequency, and vice versa. This is the *Heisenberg uncertainty principle* applied to information processing. To illustrate graphically the principle, each basis function used in the representation of a function is viewed schematically as a *tile* in a *time-frequency plane*. The tile, also called a *Heisenberg cell* or *Heisenberg box*, shows where the basis function's energy is concentrated. Basis functions that are orthonormal are characterized by nonoverlapping tiles.

Figure 7.21 shows the time-frequency tiles for (a) a delta function (i.e., conventional time domain) basis, (b) a sinusoidal (FFT) basis, and (c) an FWT basis. Note that the standard time domain basis pinpoints the instants when events occur but provides no frequency information. A sinusoidal basis, on the other hand, pinpoints the frequencies that are present in events that occur over long periods but provides no time resolution. The time and frequency resolution of

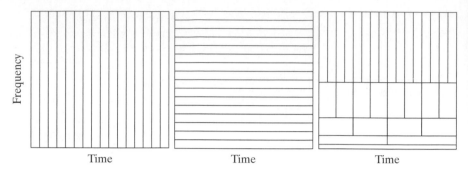

FIGURE 7.21 Time-frequency tilings for (a) sampled data, (b) FFT, and (c) FWT basis functions.

the FWT tiles vary, but the area of each tile is the same. That is, each tile represents an equal portion of the time-frequency plane. At low frequencies, the tiles are shorter (i.e., have better frequency resolution or less ambiguity regarding frequency) but are wider (which corresponds to poorer time resolution or more ambiguity regarding time). At high frequencies, tile width is smaller (so the time resolution is improved) and tile height is greater (which means the frequency resolution is poorer). This fundamental difference between the FFT and FWT was noted in the introduction to the chapter and is important in the analysis of nonstationary functions whose frequencies vary in time.

7.5   Wavelet Transforms in Two Dimensions

The one-dimensional transforms of the previous sections are easily extended to two-dimensional functions like images. In two dimensions, a two-dimensional scaling function, $\varphi(x, y)$, and three two-dimensional wavelets, $\psi^H(x, y), \psi^V(x, y)$, and $\psi^D(x, y)$, are required. Each is the product of a one-dimensional scaling function φ and corresponding wavelet ψ. Excluding products that produce one-dimensional results, like $\varphi(x)\psi(x)$, the four remaining products produce the *separable* scaling function

$$\varphi(x, y) = \varphi(x)\varphi(y) \tag{7.5-1}$$

and separable, "directionally sensitive" wavelets

$$\psi^H(x, y) = \psi(x)\varphi(y) \tag{7.5-2}$$
$$\psi^V(x, y) = \varphi(x)\psi(y) \tag{7.5-3}$$
$$\psi^D(x, y) = \psi(x)\psi(y). \tag{7.5-4}$$

These wavelets measure functional variations—intensity or gray-level variations for images—along different directions: ψ^H measures variations along columns (for example, horizontal edges), ψ^V responds to variations along rows (like vertical edges), and ψ^D corresponds to variations along diagonals. The directional sensitivity is a natural consequence of the separability imposed by

Eqs. (7.5-2) to (7.5-4); it does not increase the computational complexity of the two-dimensional transform discussed in this section.

Given separable two-dimensional scaling and wavelet functions, extension of the one-dimensional DWT to two dimensions is straightforward. We first define the scaled and translated basis functions:

$$\varphi_{j,m,n}(x, y) = 2^{j/2}\varphi(2^j x - m, 2^j y - n), \tag{7.5-5}$$

$$\psi^i_{j,m,n}(x, y) = 2^{j/2}\psi^i(2^j x - m, 2^j y - n), \quad i = \{H, V, D\} \tag{7.5-6}$$

where index i identifies the directional wavelets in Eqs. (7.5-2) to (7.5-4). Rather than an exponent, i is a superscript that assumes the values H, V, and D. The discrete wavelet transform of function $f(x, y)$ of size $M \times N$ is then

$$W_\varphi(j_0, m, n) = \frac{1}{\sqrt{MN}} \sum_{x=0}^{M-1}\sum_{y=0}^{N-1} f(x, y)\varphi_{j_0,m,n}(x, y) \tag{7.5-7}$$

$$W^i_\psi(j, m, n) = \frac{1}{\sqrt{MN}} \sum_{x=0}^{M-1}\sum_{y=0}^{N-1} f(x, y)\psi^i_{j,m,n}(x, y) \quad i = \{H, V, D\}. \tag{7.5-8}$$

As in the one-dimensional case, j_0 is an arbitrary starting scale and the $W_\varphi(j_0, m, n)$ coefficients define an approximation of $f(x, y)$ at scale j_0. The $W^i_\psi(j, m, n)$ coefficients add horizontal, vertical, and diagonal details for scales $j \geq j_0$. We normally let $j_0 = 0$ and select $N = M = 2^J$ so that $j = 0, 1, 2, \ldots, J - 1$ and $m, n = 0, 1, 2, \ldots, 2^j - 1$. Given the W_φ and W^i_ψ of Eqs. (7.5-7) and (7.5-8), $f(x, y)$ is obtained via the inverse discrete wavelet transform

$$f(x, y) = \frac{1}{\sqrt{MN}} \sum_m \sum_n W_\varphi(j_0, m, n)\varphi_{j_0,m,n}(x, y) \tag{7.5-9}$$

$$+ \frac{1}{\sqrt{MN}} \sum_{i=H,V,D} \sum_{j=j_0}^{\infty} \sum_m \sum_n W^i_\psi(j, m, n)\psi^i_{j,m,n}(x, y).$$

Like the one-dimensional discrete wavelet transform, the two dimensional DWT can be implemented using digital filters and downsamplers. With separable two-dimensional scaling and wavelet functions, we simply take the one-dimensional FWT of the rows of $f(x, y)$, followed by the one-dimensional FWT of the resulting columns. Figure 7.22(a) shows the process in block diagram form. Note that, like its one-dimensional counterpart in Fig. 7.15, the two-dimensional FWT "filters" the scale $j + 1$ approximation coefficients to construct the scale j approximation and detail coefficients. In the two-dimensional case, however, we get three sets of detail coefficients—the horizontal, vertical, and diagonal details.

The single-scale filter bank of Fig. 7.22(a) can be "iterated" (by tying the approximation output to the input of another filter bank) to produce a P scale transform in which scale $j = J - 1, J - 2, \ldots, J - P$. As in the one-dimensional case, image $f(x, y)$ is used as the $W_\varphi(J, m, n)$ input. Convolving its rows with $h_\varphi(-n)$ and $h_\psi(-n)$ and downsampling its columns, we get two subimages whose horizontal resolutions are reduced by a factor of 2. The highpass or detail component characterizes the image's high-frequency information with vertical orientation; the lowpass, approximation component contains its low-frequency, vertical information. Both subimages are then filtered columnwise and downsampled to yield four quarter-size output subimages— $W_\varphi, W^H_\psi, W^V_\psi$, and W^D_ψ.

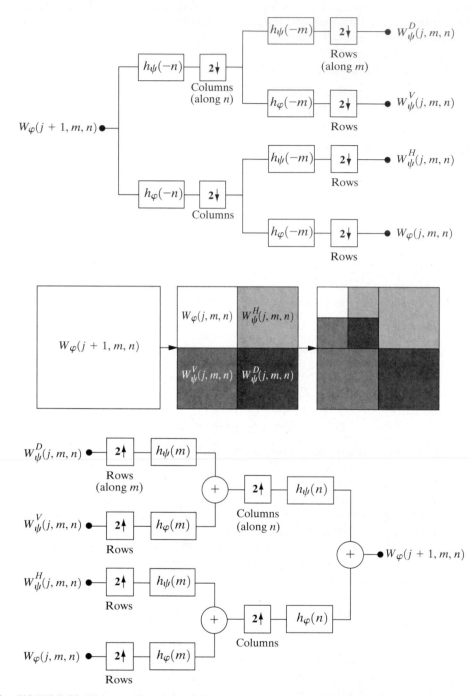

FIGURE 7.22 The two-dimensional fast wavelet transform: (a) the analysis filter bank; (b) the resulting decomposition; and (c) the synthesis filter bank.

These subimages, which are shown in the middle of Fig. 7.22(b), are the inner products of $f(x, y)$ and the two-dimensional scaling and wavelet functions in Eqs. (7.5-1) through (7.5-4), followed by downsampling by two in each dimension. Two iterations of the filtering process produces the two-scale decomposition at the far right of Fig. 7.22(b).

Figure 7.22(c) shows the synthesis filter bank that reverses the process described above. As would be expected, the reconstruction algorithm is similar to the one-dimensional case. At each iteration, four scale j approximation and detail subimages are upsampled and convolved with two one-dimensional filters— one operating on the subimages' columns and the other on its rows. Addition of the results yields the scale $j + 1$ approximation, and the process is repeated until the original image is reconstructed.

■ Consider the two-dimensional FWT shown in Fig. 7.23. Here, we see a sequence of filter-based decompositions of the 128×128 computer-generated image in Fig. 7.23(a). To generate the results, the two-dimensional filter bank of Fig. 7.22(a) and the *symlet* reconstruction filters of Figs. 7.24(a) and (b) were used. Figures 7.23(b), (c), and (d) are the resulting decompositions. In the first decomposition [Fig. 7.23(b)], the original image—a series of sinusoidal pulses on a black background—is the filter bank input; in all subsequent decompositions, approximation image W_φ—a subimage from the upper-left-hand corner of

EXAMPLE 7.12:
Computing a 2-D fast wavelet transform.

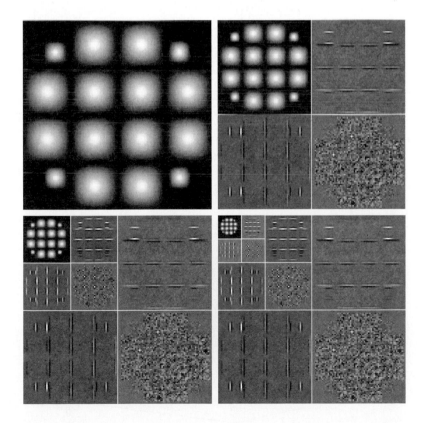

FIGURE 7.23 A three-scale FWT.

FIGURE 7.24
Fourth-order
symlets:
(a)–(b) decompo-
sition filters;
(c)–(d) recon-
struction filters;
(e) the one-
dimensional
wavelet; (f) the
one-dimensional
scaling function;
and (g) one of
three two-
dimensional
wavelets,
$\psi^H(x, y)$.

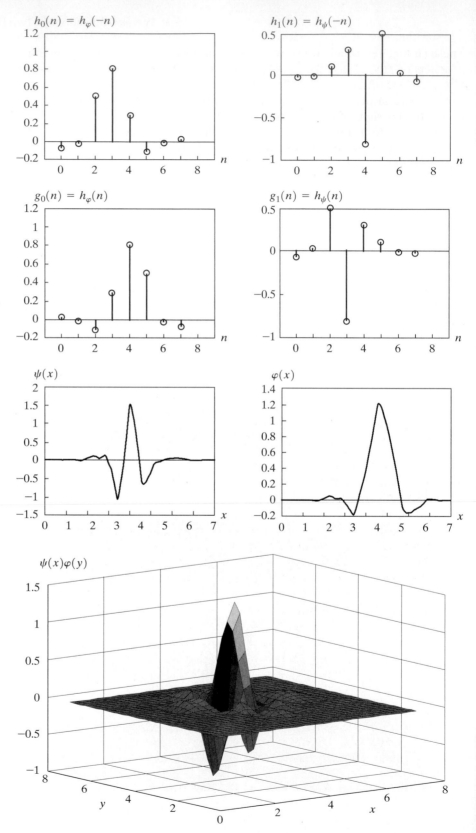

the previous decomposition—is the input. Each decomposition produces four quarter-size output images that are arranged as shown in Fig. 7.22(b) and substituted for the input from which they were derived. Note the directional nature of the wavelet-based subimages, W_ψ^H, W_ψ^V, and W_ψ^D, at each scale.   ▪

The decomposition filters used in the preceding example are part of a well-known family of wavelets called *symlets*, short for "symmetrical wavelets." Although they are not perfectly symmetrical, they are designed to have the least asymmetry and highest number of vanishing moments[†] for a given compact support (Daubechies [1992]). Figures 7.24(e) and (f) show the fourth-order one-dimensional symlets (i.e., wavelet and scaling functions). Figures 7.24(a)–(d) show the corresponding decomposition and reconstruction filters. The coefficients of lowpass reconstruction filter $g_0(n) = h_\varphi(n)$ are 0.0322, −0.0126, −0.0992, 0.2979, 0.8037, 0.4976, −0.0296, and −0.0758 for $0 \le n \le 7$. The coefficients of the remaining orthonormal filters can be obtained using Eq. (7.1-23). Figure 7.24(g), a low-resolution graphic depiction of wavelet $\psi^H(x, y)$, is provided as an illustration of how a one-dimensional scaling and wavelet function combine to form a separable, two-dimensional wavelet.

We conclude the section with two examples that demonstrate the usefulness of wavelets in image processing. As in the Fourier domain, the basic approach is to

1. Compute the two-dimensional wavelet transform of an image.
2. Alter the transform.
3. Compute the inverse transform.

Because the DWT's scaling and wavelet vectors are used as lowpass and highpass filters, most Fourier-based filtering techniques have an equivalent "wavelet domain" counterpart.

▪ Figure 7.25 provides a simple illustration. In Fig. 7.25(a), the lowest scale approximation component of the discrete wavelet transform shown in Fig. 7.23(c) has been eliminated by setting its values to zero. When the inverse transform is computed using these modified transform coefficients, as seen in Fig. 7.25(b), the effect is to emphasize or highlight the reconstructed image's edges. This provides considerable insight into the location of the original image's edges, despite the fact that they are relatively soft, sinusoidal transitions. By zeroing the horizontal details as well—see Figs. 7.25(c) and (d)—we can isolate the vertical edges.   ▪

EXAMPLE 7.13:
Wavelet-based
edge detection.

▪ As a second example, consider the magnetic resonance image (MRI) of a human head shown in Fig. 7.26(a). As can be seen in the background, the image has been uniformly corrupted with a form of additive or multiplicative white

EXAMPLE 7.14:
Wavelet-based
noise removal.

[†]The kth moment of wavelet $\psi(x)$ is $m(k) = \int x^k \psi(x)\, dx$. Zero moments impact the smoothness of the scaling and wavelet functions and our ability to represent them as polynomials. An order-N symlet has N vanishing moments.

FIGURE 7.25
Modifying a DWT
for edge
detection: (a) and
(c) two-scale
decompositions
with selected
coefficients
deleted; (b) and
(d) the
corresponding
reconstructions.

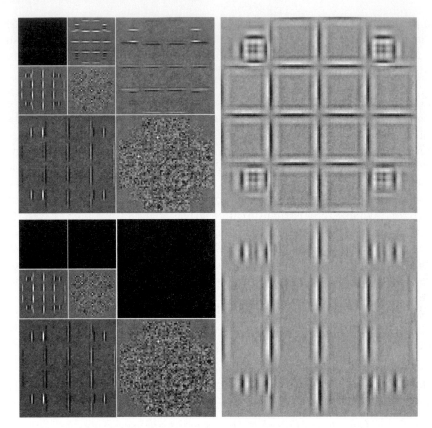

noise. The general wavelet-based procedure for *denoising* the image (i.e., suppressing the noise part) is as follows:

1. Choose a wavelet (e.g., Haar, symlet,…) and number of levels or scales, P, for the decomposition. Then compute the FWT of the noisy image.
2. Threshold the detail coefficients. That is, select and apply a threshold to the detail coefficients from scales $J - 1$ to $J - P$. This can be accomplished by *hard thresholding*, which means setting to zero the elements whose absolute values are lower than the threshold, or by *soft thresholding*, which involves first setting to zero the elements whose absolute values are lower than the threshold and then scaling the nonzero coefficients toward zero. Soft thresholding eliminates the discontinuity (at the threshold) that is inherent in hard thresholding.
3. Perform a wavelet reconstruction based on the original approximation coefficients at level $J - P$ and the modified detail coefficients for levels $J - 1$ to $J - P$.

Figure 7.26(b) shows the result of performing these operations with fourth-order symlets, two scales (i.e., $P = 2$), and a global threshold of 94.9093. Note the reduction in noise and corresponding loss of quality at the image edges. This loss of edge detail is greatly reduced in Fig. 7.26(c), which was generated by zeroing the highest-resolution detail coefficients (not the lower-resolution de-

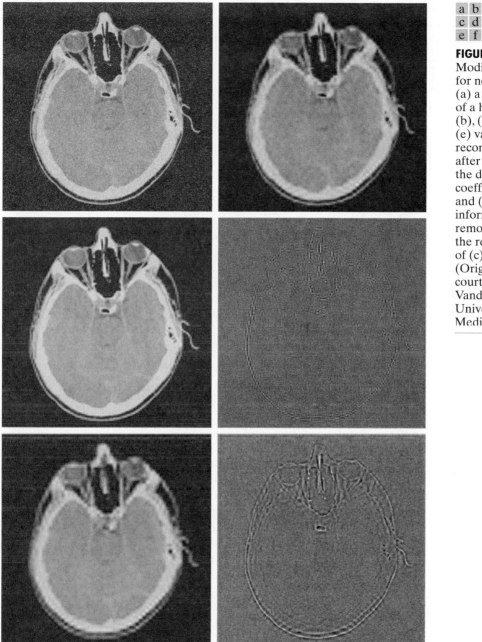

FIGURE 7.26
Modifying a DWT
for noise removal:
(a) a noisy MRI
of a human head;
(b), (c) and
(e) various
reconstructions
after thresholding
the detail
coefficients; (d)
and (f) the
information
removed during
the reconstruction
of (c) and (e).
(Original image
courtesy
Vanderbuilt
University
Medical Center.)

tails) and reconstructing the image. Here, almost all of the background noise has
been eliminated and the edges are only slightly disturbed. Figure 7.26(d) shows
the information that is lost in the process. This result was generated by computing the inverse FWT of the decomposed image with all but the highest-
resolution detail coefficients zeroed. As can be seen, it contains most of the

noise in the original image and some of the edge information. Figures 7.26(e) and (f) are a similar set of images involving all detail coefficients. That is, Fig. 7.26(e) is a reconstruction of the DWT in which the details at both levels of the decomposition have been zeroed; Fig. 7.26(f) is a reconstruction in which all but the detail coefficients (i.e., only the lowest level approximation coefficients) have been eliminated. Note the significant increase in edge information in Fig. 7.26(f) and the corresponding decrease in edge detail in Fig. 7.26(e). ■

7.6 Wavelet Packets

The fast wavelet transform decomposes a function into a series of logarithmically related frequency bands. That is, the low frequencies are grouped into narrow bands, while the high frequencies are grouped into wider bands. If you look along the frequency axis of the time-frequency plane in Fig. 7.21(c), this is immediately apparent. It is a defining characteristic of what are commonly called *constant-Q filters*. If we want greater control over the partitioning of the time-frequency plane (e.g., smaller bands at the higher frequencies), the FWT must be generalized to yield a more flexible decomposition—called a *wavelet packet* (Coifman and Wickerhauser [1992]). The cost of this generalization is an increase in computational complexity from $O(M)$ for the FWT to $O(M \log M)$.

Consider again the two-scale filter bank of Fig. 7.16(a)—but picture the decomposition as a *binary tree*. Figure 7.27(a) details the structure of the tree and links the appropriate FWT scaling and wavelet coefficients [from Fig. 7.16(a)] to its *nodes*. The *root node* is assigned the highest-scale approximation coefficients, which are samples of the function itself, while the *leaves* inherit the transform's approximation and detail coefficient outputs. The lone intermediate node, $W_\varphi(J - 1, n)$, is a filter bank approximation that is ultimately filtered to become two leaf nodes. Note that the coefficients of each node are the weights of a linear expansion that produces a band-limited "piece" of root node $f(n)$. Since any such piece is an element of a known scaling or wavelet subspace (see Sections 7.2.2 and 7.2.3), we can replace the generating coefficients in Fig. 7.27(a) by the corresponding subspace. The result is the *subspace analysis tree* of Fig. 7.27(b). Although the variable W is used to denote both coefficients and subspaces, the two quantities are distinguishable by the format of their subscripts.

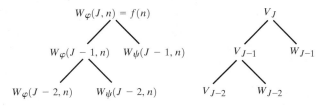

a b

FIGURE 7.27 A coefficient (a) and analysis (b) tree for the two-scale FWT analysis bank of Fig. 7.16.

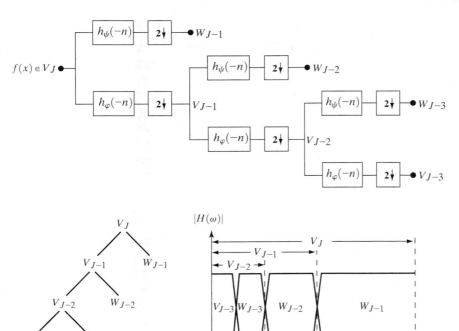

a
b c

FIGURE 7.28 A three-scale FWT filter bank: (a) block diagram; (b) decomposition space tree; and (c) spectrum splitting characteristics.

These concepts are further illustrated in Fig. 7.28, where a three-scale FWT analysis bank, analysis tree, and corresponding frequency spectrum are depicted. Unlike Fig. 7.16(a), the block diagram of Fig. 7.28(a) is labeled to resemble the analysis tree in Fig. 7.28(b)—as well as the spectrum in Fig. 7.28(c). Thus, while the output of the upper-left filter and subsampler is, to be accurate, $W_\psi(J - 1, n)$, it has been labeled W_{J-1}—the subspace of the function that is generated by the $W_\psi(J - 1, n)$ transform coefficients. This subspace corresponds to the upper-right leaf of the associated analysis tree, as well as the rightmost or widest bandwidth segment of the corresponding frequency spectrum.

Analysis trees provide a compact and informative way of representing multiscale wavelet transforms. They are simple to draw, take less space than their corresponding filter and subsampler-based block diagrams, and make it relatively easy to spot valid decompositions. The three-scale analysis tree of Fig. 7.28(b), for example, offers the following three expansion options:

$$V_J = V_{J-1} \oplus W_{J-1} \tag{7.6-1}$$

$$V_J = V_{J-2} \oplus W_{J-2} \oplus W_{J-1} \tag{7.6-2}$$

$$V_J = V_{J-3} \oplus W_{J-3} \oplus W_{J-2} \oplus W_{J-1}. \tag{7.6-3}$$

They correspond to the one-, two-, and three-scale FWT decompositions of Section 7.4 and may be obtained from Eq. (7.2-27) of Section 7.2.3 by letting $j_0 = J - P$ for $P = \{1, 2, 3\}$. In general, a P-scale FWT analysis tree supports P unique decompositions.

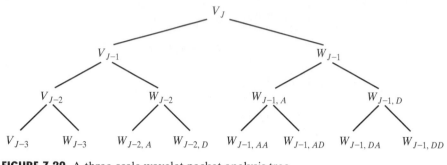

FIGURE 7.29 A three-scale wavelet packet analysis tree.

Analysis trees are also an efficient mechanism for representing *wavelet packets*, which are nothing more than *conventional wavelet transforms in which the details are iteratively filtered*. Thus, the three-scale FWT analysis tree of Fig. 7.28(b) becomes the three-scale *wavelet packet* tree of Fig. 7.29. Note the additional subscripting that is introduced. The first subscript of a double-subscripted node identifies the scale of the FWT *parent* node from which it descended. The second—a variable length string of A's and D's—encodes the path from the parent to the node. An A designates approximation filtering, while a D indicates detail filtering. Subspace $W_{J-1,DA}$, for example, is obtained by "filtering" the scale $J - 1$ FWT coefficients (i.e., parent W_{J-1} in Fig. 7.29) through an additional detail filter (yielding $W_{J-1,D}$), followed by an approximation filter (giving $W_{J-1,DA}$). Figures 7.30(a) and (b) are the filter bank and spectrum splitting characteristics of the analysis tree in Fig. 7.29. Note that the "naturally ordered" outputs of the filter bank in Fig. 7.30(a) have been reordered based on frequency content in Fig. 7.30(b) (see Problem 7.25 for more on "frequency ordered" wavelets).

The three-scale packet tree in Fig. 7.29 almost triples the number of decompositions (and associated time-frequency tilings) that are available from the three-scale FWT tree. Recall that in a normal FWT, we split, filter, and downsample the lowpass bands alone. This creates a fixed logarithmic relationship between frequency bands. Thus, while the three-scale FWT analysis tree of Fig. 7.28(a) offers three possible decompositions—see Eqs. (7.6-1) to (7.6-3)—the wavelet packet tree of Fig. 7.29 supports 26 different decompositions. For instance, V_J [and therefore function $f(n)$] can be expanded as

$$V_J = V_{J-3} \oplus W_{J-3} \oplus W_{J-2,A} \oplus W_{J-2,D} \oplus W_{J-1,AA} \oplus$$
$$W_{J-1,AD} \oplus W_{J-1,DA} \oplus W_{J-1,DD} \quad (7.6\text{-}4)$$

whose spectrum is shown in Fig. 7.30(b), or

$$V_J = V_{J-1} \oplus W_{J-1,A} \oplus W_{J-1,DA} \oplus W_{J-1,DD} \quad (7.6\text{-}5)$$

whose spectrum is depicted in Fig. 7.31. Note the difference between this last spectrum and the full packet spectrum of Fig. 7.30(b), or the three-scale FWT spectrum of Fig. 7.28(c). In general, P-scale, one-dimensional wavelet packet transforms (and associated $P + 1$-level analysis trees) support

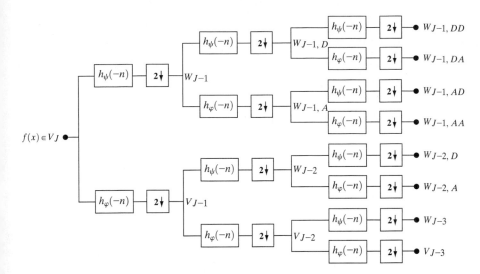

FIGURE 7.30 The (a) filter bank and (b) spectrum splitting characteristics of a three-scale full wavelet packet analysis tree.

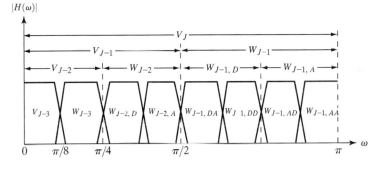

$$D(P + 1) = \left[D(P)\right]^2 + 1 \qquad (7.6\text{-}6)$$

unique decompositions, where $D(1) = 1$. With such a large number of valid expansions, packet-based transforms provide improved control of the partitioning of the decomposed function's spectrum. Of course, the cost of this control is an increase in computational complexity [compare the filter bank in Fig. 7.28(a) to that of Fig. 7.30(a)].

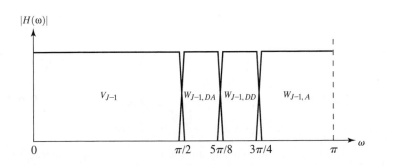

FIGURE 7.31 The spectrum of the decomposition in Eq. (7.6-5).

a b

FIGURE 7.32 The first decomposition of a two-dimensional FWT: (a) the spectrum and (b) the subspace analysis tree.

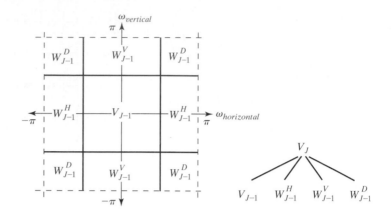

Now consider the two-dimensional, four-band filter bank of Fig. 7.22(a). As was noted in Section 7.5, it splits approximation $W_\varphi(j + 1, m, n)$ into outputs, $W_\varphi(j, m, n)$, $W_\psi^H(j, m, n)$, $W_\psi^V(j, m, n)$, and $W_\psi^D(j, m, n)$. As in the one-dimensional case, it can be "iterated" to generate P scale transforms for scales $j = J - 1, J - 2, \ldots, J - P$, with $W_\varphi(J, m, n) = f(m, n)$. The spectrum resulting from the first iteration [i.e., using $j + 1 = J$ in Fig. 7.22(a)] is shown in Fig. 7.32(a). Note that it divides the frequency plane into four equal areas. The low-frequency quarter band in the center of the plane coincides with transform coefficients $W_\varphi(J - 1, m, n)$ and scaling space V_{J-1}. (This nomenclature is consistent with the one-dimensional case.) To accommodate the two-dimensional nature of the input, however, we now have three (rather than one) wavelet subspaces. They are denoted W_{J-1}^H, W_{J-1}^V, and W_{J-1}^D and correspond to coefficients $W_\psi^H(J - 1, m, n)$, $W_\psi^V(J - 1, m, n)$, and $W_\psi^D(J - 1, m, n)$, respectively. Figure 7.32(b) shows the resulting four-band, single-scale *quaternary FWT analysis tree*. Note the superscripts that link the wavelet subspace designations to their transform coefficient counterparts.

Figure 7.33 shows a portion of a three-scale, two-dimensional wavelet packet analysis tree. Like its one-dimensional counterpart in Fig. 7.29, the first subscript of every node that is a descendent of a conventional FWT detail node is

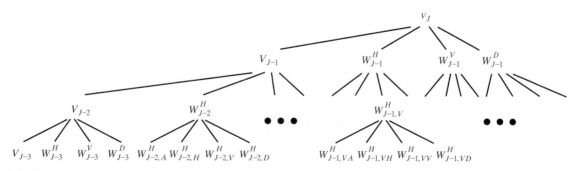

FIGURE 7.33 A three-scale, full wavelet packet decomposition tree. Only a portion of the tree is provided.

the scale of that *parent* detail node. The second subscript—a variable length string of A's, H's, V's and D's—encodes the path from the parent to the node under consideration. The node labeled $W^H_{J-1,VD}$, for example, is obtained by "row/column filtering" the scale $J - 1$ FWT horizontal detail coefficients (i.e., parent W^H_{J-1} in Fig. 7.33) through an additional detail/approximation filter (yielding $W^H_{J-1,V}$), followed by a detail/detail filter (giving $W^H_{J-1,VD}$). A P-scale, two-dimensional wavelet packet tree supports

$$D(P + 1) = \left[D(P) \right]^4 + 1 \qquad (7.6\text{-}7)$$

unique expansions, where $D(1) = 1$. Thus, the three-scale tree of Fig. 7.33 offers 83,522 possible decompositions!

EXAMPLE 7.15:
Two-dimensional wavelet packet decompositions.

■ A single wavelet packet tree presents numerous decomposition options. In fact, the number of possible decompositions is often so large that it is imprac-tical, if not impossible, to enumerate or examine them individually. An efficient algorithm for finding optimal decompositions with respect to application spe-cific criteria is highly desirable. As will be seen, classical entropy-based criteria are applicable in many situations and well suited to binary and quartenary tree searching algorithms.

Consider the problem of compressing the fingerprint image in Fig. 7.34(a). Using three-scale wavelet packet trees, there are 83,522 [see Eq. (7.6-7)]

a b

FIGURE 7.34 (a) A scanned fingerprint and (b) its three-scale, full wavelet packet decomposition. (Original image courtesy of the National Institute of Standards and Technology.)

potential decompositions that could serve as the starting point for the compression process. Figure 7.34(b) shows one of them—a full wavelet packet, 64-leaf decomposition like the analysis tree of Fig. 7.33. Note that the leaves of the tree correspond to the subbands of the 8 × 8 array of decomposed subimages in Fig. 7.34(b). The probability that this particular 64-leaf decomposition is in some way optimal for the purpose of compression, however, is relatively low. In the absence of a suitable optimality criterion, we can neither confirm nor deny it.

One reasonable criterion for selecting a decomposition for the compression of the image of Fig. 7.34(a) is the additive cost function

$$E(f) = \sum_{m,n} |f(m, n)|. \tag{7.6-8}$$

This function measures the entropy or information content of two-dimensional function f. Entropies near 0 indicate functions with little to no information. For example, the entropy of function $f(m, n) = 0$ for all m and n is 0. It provides no information. High entropy values, on the other hand, are indicative of functions with many nonzero values. Since most transform-based compression schemes work by truncating or thresholding the small coefficients to zero, a cost function that maximizes the number of near-zero values is a reasonable criterion for selecting a "best," from a compression viewpoint, decomposition.

The cost function just described is both computationally simple and easily adapted to tree optimization routines. The optimization algorithm must use the function to minimize the "cost" of the decomposition tree's leaf nodes. Minimal entropy leaf nodes should be favored because they have more near-zero values, which leads to greater compression. Because the cost function of Eq. (7.6-8) is a local measure that uses only the information available at the node under consideration, an efficient algorithm for finding minimal entropy solutions is easily constructed:

For each node of the analysis tree, beginning with the root and proceeding level by level to the leaves:

1. Compute both the entropy of the node, denoted E_P (for parent entropy), and the entropy of its four offspring—denoted E_A, E_H, E_V, and E_D. For two-dimensional wavelet packet decompositions, the parent is a two-dimensional array of approximation or detail coefficients; the offspring are the filtered approximation, horizontal, vertical, and diagonal details.

2. If the combined entropy of the offspring is less than the entropy of the parent—that is, $E_A + E_H + E_V + E_D < E_P$—include the offspring in the analysis tree. If the combined entropy of the offspring is greater than or equal to that of the parent, prune the offspring, keeping only the parent. It is a leaf of the optimized analysis tree.

The preceding algorithm can be used to (1) prune wavelet packet trees or (2) design procedures for computing optimal trees from scratch. In the latter case, nonessential siblings—descendants of nodes that would be eliminated in Step 2 of the algorithm—would not be computed. Figure 7.35 shows the optimized decomposition that results from applying the algorithm to the image of Fig. 7.34(a) with the cost function of Eq. (7.6-8). The corresponding analysis tree is given in Fig. 7.36. Note that many of the original full packet decomposi-

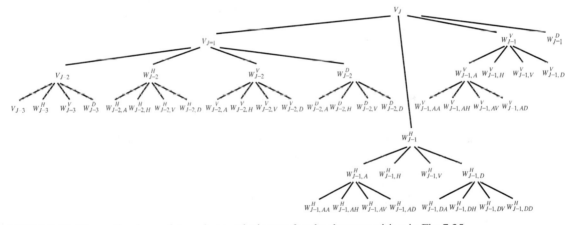

FIGURE 7.36 The optimal wavelet packet analysis tree for the decomposition in Fig. 7.35.

tion's 64 subbands in Fig. 7.34(b) (and corresponding analysis tree's 64 leaves in Fig. 7.33) have been eliminated. In addition, the subimages that are not split (or further decomposed) in Fig. 7.35 are relatively smooth and composed of pixels that are middle gray in value. Since all but the approximation subimage of this figure have been scaled so that gray level 128 indicates a zero-valued coefficient, these subimages contain little information. There would be no overall decrease in entropy realized by splitting them. ■

The preceding example is based on a real-world problem that was solved through the use of wavelets. The Federal Bureau of Investigation (FBI) currently maintains a large database of fingerprints and has established a wavelet-based national standard for the digitization and compression of fingerprint images (FBI [1993]). Using biorthogonal wavelets, the standard achieves a typical compression ratio of 15:1. The advantages of wavelet-based compression over the more traditional JPEG approach are examined in the next chapter.

The decomposition filters used in Example 7.15, as well as by the FBI, are part of a well-known family of wavelets called Cohen-Daubechies-Feauveau biorthogonal wavelets (Cohen, Daubechies, and Feauveau [1992]). Because the family's scaling and wavelet functions are symmetrical and have similar lengths, they are among the most widely used biorthogonal wavelets. Figures 7.37(e)–(h) show the dual scaling and wavelet functions that are characteristic of biorthogonal bases. Figures 7.37(a)–(d) are the corresponding decomposition and reconstruction filters. The coefficients of lowpass decomposition filter $h_0(n)$ are 0, 0.0019, −0.0019, −0.017, 0.0119, 0.0497, −0.0773, −0.0941, 0.4208, 0.8259, 0.4208, −0.0941, −0.0773, 0.0497, 0.0119, −0.017, −0.0019, and 0.0010 for $0 \le n \le 17$. Highpass decomposition filter $h_1(n)$ has coefficients 0, 0, 0, 0.0144, −0.0145, −0.0787, 0.0404, 0.4178, −0.7589, 0.4178, 0.0404, −0.0787, −0.0145, 0.0144, 0, 0, 0, and 0 for $0 \le n \le 17$. The corresponding coefficients of the biorthogonal synthesis filters can be computed using $g_0(n) = (-1)^{n+1}h_1(n)$ and $g_1(n) = (-1)^n h_0(n)$ of Eq. (7.1-15). That is, they are cross-modulated versions of the decomposition filters. Note that zero padding is employed to make the filters the same length.

Summary

The material of this chapter establishes a solid mathematical foundation for understanding and accessing the role of wavelets and multiresolution analysis in image processing. Wavelets and wavelet transforms are relatively new imaging tools that are being rapidly applied to a wide variety of image processing problems. Because of their similarity to the Fourier transform, many of the techniques in Chapter 4 have wavelet domain counterparts. A partial listing of the imaging applications that have been approached from a wavelet point of view includes image matching, registration, segmentation, denoising, restoration, enhancement, compression, morphological filtering, and computed tomography. Since it is impractical to cover all of these applications in a single chapter, the topics included were chosen for their value in introducing or clarifying fundamental concepts and preparing the reader for further study in the field. In Chapter 8, we will apply wavelets to the compression of images.

References and Further Reading

There are many good texts on wavelets and their application. Several complement our treatment and were relied on during the development of the core sections of the chapter. The material in Section 7.1.2 on subband coding and digital filtering follows the book by Vetterli and Kovacevic [1995], while Sections 7.2 and 7.4 on multiresolution expansions and the fast wavelet transform follow the treatment of these subjects in Burrus, Gopinath, and Guo [1998]. The remainder of the material in the chapter is based on the

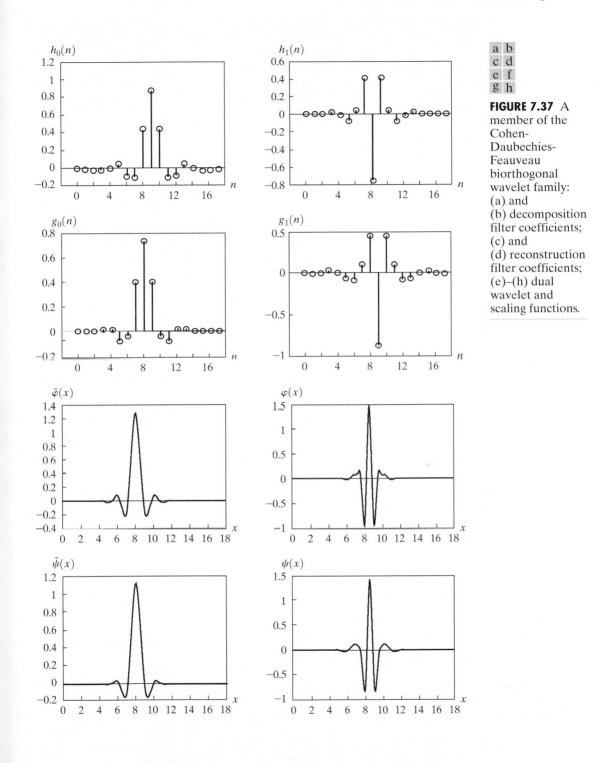

a b
c d
e f
g h

FIGURE 7.37 A member of the Cohen-Daubechies-Feauveau biorthogonal wavelet family: (a) and (b) decomposition filter coefficients; (c) and (d) reconstruction filter coefficients; (e)–(h) dual wavelet and scaling functions.

references cited in the text. Experimental work and many of the examples were done using the MATLAB wavelet toolbox (Misiti, Misiti, Oppenheim, and Poggi [1996]).

The history of wavelet analysis is recorded in a book by Hubbard [1998]. The early predecessors of wavelets were developed simultaneously in different fields and unified in a paper by Mallat [1987]. It brought a mathematical framework to the field. Much of the history of wavelets can be traced through the works of Meyer [1987] [1990] [1992a, b] [1993], Mallat [1987] [1989a–c] [1998], and Daubechies [1988] [1990] [1992] [1993] [1996]. The current interest in wavelets was stimulated by many of their publications. The book by Daubechies [1992] is a classic source for the mathematical details of wavelet theory.

The application of wavelets to image processing is addressed in general image processing texts, like Castleman [1996], and many application-specific books, some of which are conference proceedings. In this latter category, for example, are Rosenfeld [1984], Prasad and Iyengar [1997], and Topiwala [1998]. Recent articles that can serve as starting points for further research into specific imaging applications include Thévenaz and Unser [2000] for image registration; Chang and Kuo [1993] and Unser [1995] on texture-based classification; Heijmans and Goutsias [2000] on morphological wavelets; Banham et al. [1994], Wang, Zhang, and Pan [1995], and Banham and Kastaggelos [1996] on image restoration; Xu et al. [1994] and Chang, Yu, and Vetterli [2000] on image enhancement; Delaney and Bresler [1995] and Westenberg and Roerdink [2000] on computed tomography; and Lee, Sun, and Chen [1995], Liang and Kuo [1999], Wang, Lee, and Toraichi [1999], and You and Bhattacharya [2000] on image description and matching. One of the most important applications of wavelets is image compression—see, for example, Antonini et al. [1992], Wei et al. [1998], and the book by Topiwala [1998]. Finally, there have been a number of special issues devoted to wavelets, including a special issue on wavelet transforms and multiresolution signal anaysis in the *IEEE Transactions on Information Theory* [1992], a special issue on wavelets and signal processing in the *IEEE Transactions on Signal Processing* [1993], and a special section on multiresolution representation in the *IEEE Transactions on Pattern Analysis and Machine Intelligence* [1989].

Although the chapter focuses on the fundamentals of wavelets and their application to image processing, there is considerable interest in the construction of wavelets themselves. The interested reader is referred to the work of Battle [1987] [1988], Daubechies [1988] [1992], Cohen and Daubechies [1992], Meyer [1990], Mallat [1989b], Unser, Aldroubi, and Eden [1993], and Gröchenig and Madych [1992]. This is not an exhaustive list but should serve as a starting point for further reading. See also the general references on subband coding and filter banks, including Strang and Nguyen [1996] and Vetterli and Kovacevic [1995], and the references included in the chapter with respect to the wavelets we used as examples.

Problems

7.1 Design a system for decoding the prediction residual pyramid generated by the encoder of Fig. 7.2(b) and draw its block diagram. Assume there is no quantization error introduced by the encoder.

See inside front cover

Detailed solutions to the problems marked with a star can be found in the book web site. The site also contains suggested projects based on the material in this chapter.

★ **7.2** Construct a fully populated approximation pyramid and corresponding prediction residual pyramid for the image

$$f(x, y) = \begin{bmatrix} 1 & 2 & 3 & 4 \\ 5 & 6 & 7 & 8 \\ 9 & 10 & 11 & 12 \\ 13 & 14 & 15 & 16 \end{bmatrix}.$$

Use 2×2 block neighborhood averaging for the approximation filter in Fig. 7.2(b) and assume the interpolation filter is omitted.

★ **7.3**   Given a $2^J \times 2^J$ image, does a $J + 1$-level pyramid reduce or expand the amount of data required to represent the image? What is the compression or expansion ratio?

7.4   Prove that the following filters from Table 7.1 form perfect reconstruction filter banks:

★ **(a)** Quadrature mirror filter (QMF)

(b) Orthonormal filter

7.5   Are quadrature mirror filters biorthogonal or othonormal or both?

7.6   Compute the coefficients of the Daubechies synthesis filters $g_0(n)$ and $g_1(n)$ for Example 7.2. Using Eq. (7.1-22) with $m = 0$ only, show that the filters are orthonormal.

★ **7.7**   Draw a two-dimensional four-band filter bank decoder to reconstruct input $x(m, n)$ in Fig. 7.5.

7.8   Obtain the Haar transformation matrix for $N = 8$.

7.9   **(a)** Compute the Haar transform of the 2×2 image

$$\mathbf{F} = \begin{bmatrix} 3 & -1 \\ 6 & 2 \end{bmatrix}.$$

(b) The inverse Haar transform is $\mathbf{F} = \mathbf{H}^T \mathbf{T} \mathbf{H}$, where $\mathbf{T}$ is the Haar transform of $\mathbf{F}$ and $\mathbf{H}^T$ is the matrix inverse of $\mathbf{H}$. Show that $\mathbf{H}_2^{-1} = \mathbf{H}_2^T$ and use it to compute the inverse Haar transform of the result in (a).

7.10   Compute the expansion coefficients of 2-tuple $[3, 2]^T$ for the following bases and write the corresponding expansions:

★ **(a)** Basis $\varphi_0 = [1/\sqrt{2}, 1/\sqrt{2}]^T$ and $\varphi_1 = [1/\sqrt{2}, -1/\sqrt{2}]^T$ on $\mathbf{R}^2$, the set of real 2-tuples.

(b) Basis $\varphi_0 = [1, 0]^T$ and $\varphi_1 = [1, 1]^T$, and its dual, $\tilde{\varphi}_0 = [1, -1]^T$ and $\tilde{\varphi}_1 = [0, 1]^T$, on $\mathbf{R}^2$.

(c) Basis $\varphi_0 = [1, 0]^T, \varphi_1 = [-1/2, \sqrt{3}/2]^T$, and $\varphi_1 = [-1/2, -\sqrt{3}/2]^T$, and their duals, $\tilde{\varphi}_i = 2\varphi_i/3$ for $i = \{0, 1, 2\}$, on $\mathbf{R}^2$.

Hint: Vector inner products must be used in place of the integral inner products of Section 7.2.1.

7.11   Show that scaling function

$$\varphi(x) = \begin{cases} 1 & 0.25 \le x < 0.75 \\ 0 & \text{elsewhere} \end{cases}$$

does not satisfy the second requirement of a multiresolution analysis.

7.12   Write an expression for scaling space V_3 as a function of scaling function $\varphi(x)$. Use the Haar scaling function definition of Eq. (7.2-14) to draw the Haar V_3 scaling functions at translations $k = \{0, 1, 2\}$.

★ **7.13**   Draw wavelet $\psi_{3,3}(x)$ for the Haar wavelet function. Write an expression for $\psi_{3,3}(x)$ in terms of the Haar scaling function.

7.14   Suppose function $f(x)$ is a member of Haar scaling space V_3—that is, $f(x) \in V_3$. Use Eq. (7.2-22) to express V_3 as a function of scaling space V_0 and any required wavelet spaces. If $f(x)$ is 0 outside the interval $[0, 1)$, sketch the scaling and wavelet functions required for a linear expansion of $f(x)$ based on your expression.

7.15   Compute the first four terms of the wavelet series expansion of the function used in Example 7.7 with starting scale $j_0 = 1$. Write the resulting expansion in terms of the scaling and wavelet functions involved. How does your result compare to the example, where the starting scale was $j_0 = 0$?

7.16   The DWT in Eqs. (7.3-5) and (7.3-6) is a function of starting scale j_0.

(a) Recompute the one-dimensional DWT of function $f(n) = \{1, 4, -3, 0\}$ for $0 \le n \le 3$ in Example 7.8 with $j_0 = 1$ (rather than 0).

(b) Use the result from (a) to compute $f(1)$ from the transform values.

★ **7.17**   What does the following continuous wavelet transform reveal about the one-dimensional function upon which it was based?

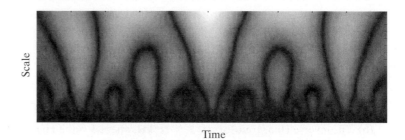

7.18   (a) The continuous wavelet transform of Problem 7.17 is computer generated. The function upon which it is based was first sampled at discrete intervals. What is continuous about the transform—or what distinguishes it from the discrete wavelet transform of the function?

★ (b) Under what circumstances is the DWT a better choice than the CWT? Are there times when the CWT is better than the DWT?

★ **7.19**   Draw the FWT filter bank required to compute the transform in Problem 7.16. Label all inputs and outputs with the appropriate sequences.

7.20   The computational complexity of an M-point fast wavelet transform is $O(M)$. That is, the number of operations is proportional M. What determines the constant of proportionality?

7.21 ★ (a) If the input to the three-scale FWT filter bank of Fig. 7.28(a) is the Haar scaling function $\varphi(n) = 1$ for $n = 0, 1, \ldots, 7$ and 0 elsewhere, what is the resulting transform with respect to Haar wavelets?

(b) What is the transform if the input is the corresponding Haar wavelet function $\psi(n) = \{1, 1, 1, 1, -1, -1, -1, -1\}$ for $n = 0, 1, \ldots, 7$?

(c) What input sequence produces transform $\{0, 0, 0, 0, 0, 0, B, 0\}$ with nonzero coefficient $W_\psi(2, 2) = B$?

★ **7.22**   The two-dimensional fast wavelet transform is similar to the pyramidal coding scheme of Section 7.2.1. How are they similar? Given the three-scale wavelet transform in Fig. 7.8(a), how would you construct the corresponding approximation pyramid? How many levels would it have?

7.23   Compute the two-dimensional wavelet transform with respect to Haar wavelets of the 2×2 image in Problem 7.9. Draw the required filter bank and label all inputs and outputs with the proper arrays.

★**7.24**   In the Fourier domain

$$f(x - x_0, y - y_0) \Leftrightarrow F(u, v)e^{-2\pi(ux_0/M + vy_0/N)}$$

and translation does not effect the display of $|F(u, v)|$. Using the following sequence of images, explain the translation property of wavelet transforms. The leftmost image contains two 32×32 white squares centered on a 128×128 gray background. The second image (from the left) is its single-scale wavelet transform with respect to Haar wavelets. The third is the wavelet transform of the original image after shifting it 32 pixels to the right and downward, and the final (rightmost) image is the wavelet transform of the original image after it has been shifted 1 pixel to the right and downward.

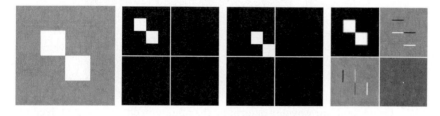

7.25   The following table shows the Haar wavelet and scaling functions for a four-scale fast wavelet transform. Sketch the additional basis functions needed for a full three-scale packet decomposition. Give the mathematical expression or expressions for determining them. Then order the basis functions according to frequency content and explain the results.

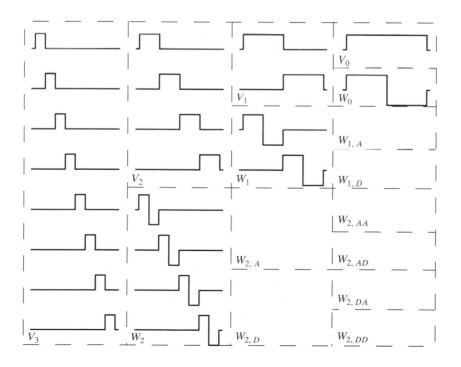

7.26 A wavelet packet decomposition of the vase from Fig. 7.1 is shown below.

    (a) Draw the corresponding decomposition analysis tree, labeling all nodes with the names of the proper scaling and wavelet spaces.

    (b) Draw and label the decomposition's frequency spectrum.

7.27 Using the Haar wavelet, determine the minimum entropy packet decomposition for the function $f(n) = 0.25$ for $n = 0, 1, 2, \ldots, 15$. Employ the nonnormalized Shannon entropy,

$$E[f(n)] = \sum_n f^2(n) \ln[f^2(n)]$$

as the minimization criterion. Draw the optimal tree, labeling the nodes with the computed entropy values.

8 *Image Compression*

But life is short and information endless ...
Abbreviation is a necessary evil and the abbreviator's
business is to make the best of a job which, although
intrinsically bad, is still better than nothing.

Aldous Huxley

Preview

Every day, an enormous amount of information is stored, processed, and transmitted digitally. Companies provide business associates, investors, and potential customers with financial data, annual reports, inventory, and product information over the Internet. Order entry and tracking, two of the most basic on-line transactions, are routinely performed from the comfort of one's own home. The U.S., as part of its *digital-* or *e-government* initiative, has made the entire catalog (and some of the holdings) of the Library of Congress, the world's largest library, electronically accessible; and cable television programming on demand is on the verge of becoming a reality. Because much of this on-line information is graphical or pictorial in nature, the storage (see Section 2.4.2) and communications requirements are immense. Methods of compressing the data prior to storage and/or transmission are of significant practical and commercial interest.

Image compression addresses the problem of reducing the amount of data required to represent a digital image. The underlying basis of the reduction process is the removal of redundant data. From a mathematical viewpoint, this amounts to transforming a 2-D pixel array into a statistically uncorrelated data set. The transformation is applied prior to storage or transmission of the image. At some later time, the compressed image is decompressed to reconstruct the original image or an approximation of it.

Interest in image compression dates back more than 35 years. The initial focus of research efforts in this field was on the development of analog methods for reducing video transmission bandwidth, a process called *bandwidth compression*. The advent of the digital computer and subsequent development of advanced integrated circuits, however, caused interest to shift from analog to digital compression approaches. With the relatively recent adoption of several key international image compression standards, the field has undergone significant growth through the practical application of the theoretic work that began in the 1940s, when C. E. Shannon and others first formulated the probabilistic view of information and its representation, transmission, and compression.

Currently, image compression is recognized as an "enabling technology." In addition to the areas just mentioned, image compression is the natural technology for handling the increased spatial resolutions of today's imaging sensors and evolving broadcast television standards. Furthermore, image compression plays a major role in many important and diverse applications, including televideo-conferencing, remote sensing (the use of satellite imagery for weather and other earth-resource applications), document and medical imaging, facsimile transmission (FAX), and the control of remotely piloted vehicles in military, space, and hazardous waste management applications. In short, an ever-expanding number of applications depend on the efficient manipulation, storage, and transmission of binary, gray-scale, and color images.

In this chapter, we examine both the theoretic and practical aspects of the image compression process. Sections 8.1 through 8.3 constitute an introduction to the fundamentals that collectively form the theory of this discipline. Section 8.1 describes the data redundancies that may be exploited by image compression algorithms. A model-based paradigm for the general compression-decompression process is presented in Section 8.2. Section 8.3 examines in some detail a number of basic concepts from information theory and their role in establishing fundamental limits on the representation of information.

Sections 8.4 through 8.6 cover the practical aspects of image compression, including both the principal techniques in use and the standards that have been instrumental in increasing the scope and acceptance of this discipline. Compression techniques fall into two broad categories: *information preserving* and *lossy*. Section 8.4 addresses methods in the first category, which are particularly useful in image archiving (as in the storage of legal or medical records). These methods allow an image to be compressed and decompressed without losing information. Section 8.5 describes methods in the second category, which provide higher levels of data reduction but result in a less than perfect reproduction of the original image. Lossy image compression is useful in applications such as broadcast television, videoconferencing, and facsimile transmission, in which a certain amount of error is an acceptable trade-off for increased compression performance. Finally, Section 8.6 deals with existing and proposed image compression standards.

8.1 Fundamentals

The term *data compression* refers to the process of reducing the amount of data required to represent a given quantity of information. A clear distinction must be made between *data* and *information*. They are not synonymous. In fact, data are the means by which information is conveyed. Various amounts of data may be used to represent the same amount of information. Such might be the case, for example, if a long-winded individual and someone who is short and to the point were to relate the same story. Here, the information of interest is the story; words are the data used to relate the information. If the two individuals use a different number of words to tell the same basic story, two different versions of the story are created, and at least one includes nonessential data. That is, it contains data (or words) that either provide no relevant information or simply restate that which is already known. It is thus said to contain *data redundancy*.

Data redundancy is a central issue in digital image compression. It is not an abstract concept but a mathematically quantifiable entity. If n_1 and n_2 denote the number of information-carrying units in two data sets that represent the same information, the *relative data redundancy* R_D of the first data set (the one characterized by n_1) can be defined as

$$R_D = 1 - \frac{1}{C_R} \qquad (8.1\text{-}1)$$

where C_R, commonly called the *compression ratio*, is

$$C_R = \frac{n_1}{n_2}. \qquad (8.1\text{-}2)$$

For the case $n_2 = n_1$, $C_R - 1$ and $R_D = 0$, indicating that (relative to the second data set) the first representation of the information contains no redundant data. When $n_2 \ll n_1$, $C_R \to \infty$ and $R_D \to 1$, implying significant compression and highly redundant data. Finally, when $n_2 \gg n_1$, $C_R \to 0$ and $R_D \to -\infty$, indicating that the second data set contains much more data than the original representation. This, of course, is the normally undesirable case of data expansion. In general, C_R and R_D lie in the open intervals $(0, \infty)$ and $(-\infty, 1)$, respectively. A practical compression ratio, such as 10 (or 10:1), means that the first data set has 10 information carrying units (say, bits) for every 1 unit in the second or compressed data set. The corresponding redundancy of 0.9 implies that 90% of the data in the first data set is redundant.

In digital image compression, three basic data redundancies can be identified and exploited: *coding* redundancy, *interpixel* redundancy, and *psychovisual* redundancy. Data compression is achieved when one or more of these redundancies are reduced or eliminated.

8.1.1 Coding Redundancy

In Chapter 3 we developed the technique for image enhancement by histogram processing on the assumption that the gray levels of an image are random quantities. We showed that a great deal of information about the appearance of an image could be obtained from a histogram of its gray levels. In this section, we utilize a similar formulation to show how the gray-level histogram of an image also can provide a great deal of insight into the construction of codes[†] to reduce the amount of data used to represent it.

Let us assume, once again, that a discrete random variable r_k in the interval $[0, 1]$ represents the gray levels of an image and that each r_k occurs with probability $p_r(r_k)$. As in Chapter 3,

$$p_r(r_k) = \frac{n_k}{n} \qquad k = 0, 1, 2, \ldots, L - 1 \tag{8.1-3}$$

where L is the number of gray levels, n_k is the number of times that the kth gray level appears in the image, and n is the total number of pixels in the image. If the number of bits used to represent each value of r_k is $l(r_k)$, then the average number of bits required to represent each pixel is

$$L_{\text{avg}} = \sum_{k=0}^{L-1} l(r_k) p_r(r_k). \tag{8.1-4}$$

That is, the average length of the code words assigned to the various gray-level values is found by summing the product of the number of bits used to represent each gray level and the probability that the gray level occurs. Thus the total number of bits required to code an $M \times N$ image is MNL_{avg}.

Representing the gray levels of an image with a natural m-bit binary code[‡] reduces the right-hand side of Eq. (8.1-4) to m bits. That is, $L_{\text{avg}} = m$ when m is substituted for $l(r_k)$. Then the constant m may be taken outside the summation, leaving only the sum of the $p_r(r_k)$ for $0 \le k \le L - 1$, which, of course, equals 1.

EXAMPLE 8.1:
A simple illustration of variable-length coding.

■ An 8-level image has the gray-level distribution shown in Table 8.1. If a natural 3-bit binary code [see code 1 and $l_1(r_k)$ in Table 8.1] is used to represent the 8 possible gray levels, L_{avg} is 3 bits, because $l_1(r_k) = 3$ bits for all r_k. If code 2 in Table 8.1 is used, however, the average number of bits required to code the image is reduced to

---

[†] A *code* is a system of symbols (letters, numbers, bits, and the like) used to represent a body of information or set of events. Each piece of information or event is assigned a sequence of *code symbols*, called a *code word*. The number of symbols in each code word is its *length*. One of the most famous codes was used by Paul Revere on April 18, 1775. The phrase "one if by land, two if by sea" is often used to describe that code, in which one or two lights were used to indicate whether the British were traveling by land or sea.

[‡] A *natural* (or *straight*) binary code is one in which each event or piece of information to be encoded (such as gray-level value) is assigned one of 2^m m-bit binary codes from an m-bit binary counting sequence.

r_k	$p_r(r_k)$	Code 1	$l_1(r_k)$	Code 2	$l_2(r_k)$
$r_0 = 0$	0.19	000	3	11	2
$r_1 = 1/7$	0.25	001	3	01	2
$r_2 = 2/7$	0.21	010	3	10	2
$r_3 = 3/7$	0.16	011	3	001	3
$r_4 = 4/7$	0.08	100	3	0001	4
$r_5 = 5/7$	0.06	101	3	00001	5
$r_6 = 6/7$	0.03	110	3	000001	6
$r_7 = 1$	0.02	111	3	000000	6

TABLE 8.1
Example of variable-length coding.

$$L_{avg} = \sum_{k=0}^{7} l_2(r_k) p_r(r_k)$$

$$= 2(0.19) + 2(0.25) + 2(0.21) + 3(0.16) + 4(0.08)$$
$$+ 5(0.06) + 6(0.03) + 6(0.02)$$

$$= 2.7 \text{ bits.}$$

From Eq. (8.1-2), the resulting compression ratio C_R is 3/2.7 or 1.11. Thus approximately 10% of the data resulting from the use of code 1 is redundant. The exact level of redundancy can be determined from Eq. (8.1-1):

$$R_D = 1 - \frac{1}{1.11} = 0.099.$$

Figure 8.1 illustrates the underlying basis for the compression achieved by code 2. It shows both the histogram of the image [a plot of $p_r(r_k)$ versus r_k] and $l_2(r_k)$. Because these two functions are inversely proportional—that is, $l_2(r_k)$ increases as $p_r(r_k)$ decreases—the shortest code words in code 2 are assigned to the gray levels that occur most frequently in an image.   ■

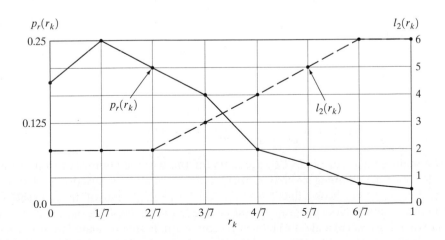

FIGURE 8.1
Graphic representation of the fundamental basis of data compression through variable-length coding.

In the preceding example, assigning fewer bits to the more probable gray levels than to the less probable ones achieves data compression. This process commonly is referred to as *variable-length coding*. If the gray levels of an image are coded in a way that uses more code symbols than absolutely necessary to represent each gray level [that is, the code fails to minimize Eq. (8.1-4)], the resulting image is said to contain *coding redundancy*. In general, coding redundancy is present when the codes assigned to a set of events (such as gray-level values) have not been selected to take full advantage of the probabilities of the events. It is almost always present when an image's gray levels are represented with a straight or natural binary code. In this case, the underlying basis for the coding redundancy is that images are typically composed of objects that have a regular and somewhat predictable morphology (shape) and reflectance, and are generally sampled so that the objects being depicted are much larger than the picture elements. The natural consequence is that, in most images, certain gray levels are more probable than others (that is, the histograms of most images are not uniform). A natural binary coding of their gray levels assigns the same number of bits to both the most and least probable values, thus failing to minimize Eq. (8.1-4) and resulting in coding redundancy.

8.1.2 Interpixel Redundancy

Consider the images shown in Figs. 8.2(a) and (b). As Figs. 8.2(c) and (d) show, these images have virtually identical histograms. Note also that both histograms are trimodal, indicating the presence of three dominant ranges of gray-level values. Because the gray levels in these images are not equally probable, variable-length coding can be used to reduce the coding redundancy that would result from a straight or natural binary encoding of their pixels. The coding process, however, would not alter the level of correlation between the pixels within the images. In other words, the codes used to represent the gray levels of each image have nothing to do with the correlation between pixels. These correlations result from the structural or geometric relationships between the objects in the image.

Figures 8.2(e) and (f) show the respective *autocorrelation coefficients* computed along one line of each image. These coefficients were computed using a normalized version of Eq. (4.6-30) in which

$$\gamma(\Delta n) = \frac{A(\Delta n)}{A(0)} \tag{8.1-5}$$

where

$$A(\Delta n) = \frac{1}{N - \Delta n} \sum_{y=0}^{N-1-\Delta n} f(x, y) f(x, y + \Delta n). \tag{8.1-6}$$

The scaling factor in Eq. (8.1-6) accounts for the varying number of sum terms that arise for each integer value of Δn. Of course, Δn must be strictly less than N, the number of pixels on a line. The variable x is the coordinate of the line used in the computation. Note the dramatic difference between the shape of the functions shown in Figs. 8.2(e) and (f). Their shapes can be qualitatively related to the structure in the images in Figs. 8.2(a) and (b). This relationship is particularly noticeable

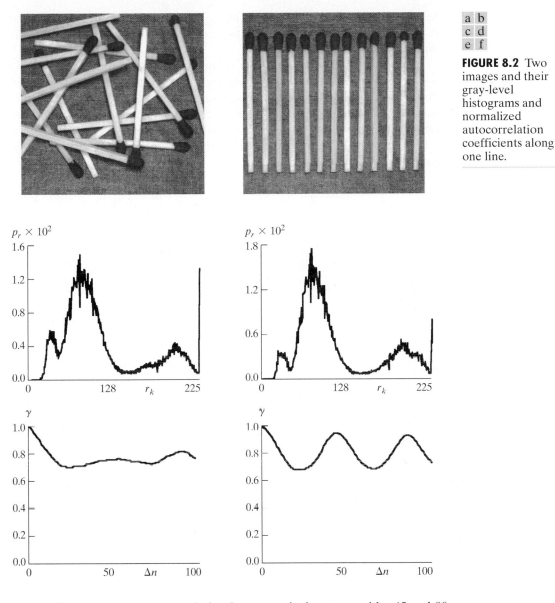

a b
c d
e f

FIGURE 8.2 Two images and their gray-level histograms and normalized autocorrelation coefficients along one line.

in Fig. 8.2(f), where the high correlation between pixels separated by 45 and 90 samples can be directly related to the spacing between the vertically oriented matches of Fig. 8.2(b). In addition, the adjacent pixels of both images are highly correlated. When Δn is 1, γ is 0.9922 and 0.9928 for the images of Figs. 8.2(a) and (b), respectively. These values are typical of most properly sampled television images.

These illustrations reflect another important form of data redundancy—one directly related to the interpixel correlations within an image. Because the value of any given pixel can be reasonably predicted from the value of its neighbors, the information carried by individual pixels is relatively small. Much of the visual contribution of a single pixel to an image is redundant; it could have been guessed on the basis of the values of its neighbors. A variety of names, including *spatial redundancy, geometric redundancy*, and *interframe redundancy*, have been coined to refer to these interpixel dependencies. We use the term *interpixel redundancy* to encompass them all.

In order to reduce the interpixel redundancies in an image, the 2-D pixel array normally used for human viewing and interpretation must be transformed into a more efficient (but usually "nonvisual") format. For example, the differences between adjacent pixels can be used to represent an image. Transformations of this type (that is, those that remove interpixel redundancy) are referred to as *mappings*. They are called *reversible mappings* if the original image elements can be reconstructed from the transformed data set.

EXAMPLE 8.2:
A simple illustration of run-length coding.

■ Figure 8.3 illustrates a simple mapping procedure. Figure 8.3(a) depicts a 1-in. by 3-in. section of an electrical assembly drawing that has been sampled at

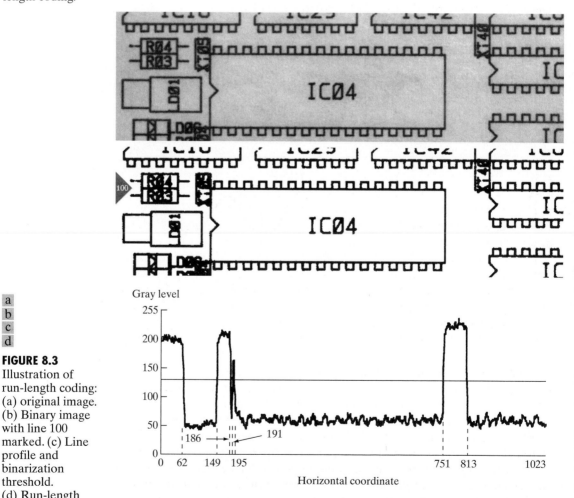

a
b
c
d

FIGURE 8.3
Illustration of run-length coding:
(a) original image.
(b) Binary image with line 100 marked. (c) Line profile and binarization threshold.
(d) Run-length code.

Line 100: (1, 63) (0, 87) (1, 37) (0, 5) (1, 4) (0, 556) (1, 62) (0, 210)

approximately 330 dpi (dots per inch). Figure 8.3(b) shows a binary version of this drawing, and Fig. 8.3(c) depicts the gray-level profile of one line of the image and the threshold used to obtain the binary version (see Section 3.1). Because the binary image contains many regions of constant intensity, a more efficient representation can be constructed by mapping the pixels along each scan line $f(x, 0), f(x, 1), \ldots, f(x, N-1)$ into a sequence of pairs (g_1, w_1), $(g_2, w_2), \ldots$, in which g_i denotes the ith gray level encountered along the line and w_i the run length of the ith run. In other words, the thresholded image can be more efficiently represented by the value and length of its constant gray-level runs (a nonvisual representation) than by a 2-D array of binary pixels.

Figure 8.3(d) shows the run-length encoded data corresponding to the thresholded line profile of Fig. 8.3(c). Only 88 bits are needed to represent the 1024 bits of binary data. In fact, the entire 1024×343 section shown in Fig. 8.3(b) can be reduced to 12,166 runs. As 11 bits are required to represent each run-length pair, the resulting compression ratio and corresponding relative redundancy are

$$C_R = \frac{(1024)(343)(1)}{(12166)(11)} = 2.63$$

and

$$R_D = 1 - \frac{1}{2.63} = 0.62.$$ ■

8.1.3 Psychovisual Redundancy

We noted in Section 2.1 that the brightness of a region, as perceived by the eye, depends on factors other than simply the light reflected by the region. For example, intensity variations (Mach bands) can be perceived in an area of constant intensity. Such phenomena result from the fact that the eye does not respond with equal sensitivity to all visual information. Certain information simply has less relative importance than other information in normal visual processing. This information is said to be *psychovisually redundant*. It can be eliminated without significantly impairing the quality of image perception.

That psychovisual redundancies exist should not come as a surprise, because human perception of the information in an image normally does not involve quantitative analysis of every pixel value in the image. In general, an observer searches for distinguishing features such as edges or textural regions and mentally combines them into recognizable groupings. The brain then correlates these groupings with prior knowledge in order to complete the image interpretation process.

Psychovisual redundancy is fundamentally different from the redundancies discussed earlier. Unlike coding and interpixel redundancy, psychovisual redundancy is associated with real or quantifiable visual information. Its elimination is possible only because the information itself is not essential for normal visual processing. Since the elimination of psychovisually redundant data results in a loss of quantitative information, it is commonly referred to as *quantization*. This terminology is consistent with normal usage of the word, which generally means the mapping of a broad range of input values to a limited number of output values, as discussed in Section 2.4. As it is an irreversible operation (visual information is lost), quantization results in lossy data compression.

EXAMPLE 8.3:
Compression by quantization.

■ Consider the images in Fig. 8.4. Figure 8.4(a) shows a monochrome image with 256 possible gray levels. Figure 8.4(b) shows the same image after uniform quantization to four bits or 16 possible levels. The resulting compression ratio is 2 : 1. Note, as discussed in Section 2.4, that false contouring is present in the previously smooth regions of the original image. This is the natural visual effect of more coarsely representing the gray levels of the image.

Figure 8.4(c) illustrates the significant improvements possible with quantization that takes advantage of the peculiarities of the human visual system. Although the compression ratio resulting from this second quantization procedure also is 2 : 1, false contouring is greatly reduced at the expense of some additional but less objectionable graininess. The method used to produce this result is known as *improved gray-scale* (IGS) *quantization*. It recognizes the eye's inherent sensitivity to edges and breaks them up by adding to each pixel a pseudorandom number, which is generated from the low-order bits of neighboring pixels, before quantizing the result. Because the low-order bits are fairly random (see the bit planes in Section 3.2.4), this amounts to adding a level of randomness, which depends on the local characteristics of the image, to the artificial edges normally associated with false contouring.

Table 8.2 illustrates this method. A sum—initially set to zero—is first formed from the current 8-bit gray-level value and the four least significant bits of a previously generated sum. If the four most significant bits of the current value are 1111_2, however, 0000_2 is added instead. The four most significant bits of the resulting sum are used as the coded pixel value. ■

a b c

FIGURE 8.4
(a) Original image.
(b) Uniform quantization to 16 levels. (c) IGS quantization to 16 levels.

Pixel	Gray Level	Sum	IGS Code
$i - 1$	N/A	0000 0000	N/A
i	0110 1100	0110 1100	0110
$i + 1$	1000 1011	1001 0111	1001
$i + 2$	1000 0111	1000 1110	1000
$i + 3$	1111 0100	1111 0100	1111

TABLE 8.2
IGS quantization
procedure.

Improved gray-scale quantization is typical of a large group of quantization procedures that operate directly on the gray levels of the image to be compressed. They usually entail a decrease in the image's spatial and/or gray-scale resolution. The resulting false contouring or other related effects necessitates the use of heuristic techniques to compensate for the visual impact of quantization. The normal 2:1 line interlacing approach used in commercial broadcast television, for example, is a form of quantization in which interleaving portions of adjacent frames allows reduced video scanning rates with little decrease in perceived image quality.

8.1.4 Fidelity Criteria

As noted previously, removal of psychovisually redundant data results in a loss of real or quantitative visual information. Because information of interest may be lost, a repeatable or reproducible means of quantifying the nature and extent of information loss is highly desirable. Two general classes of criteria are used as the basis for such an assessment: (1) objective fidelity criteria and (2) subjective fidelity criteria.

When the level of information loss can be expressed as a function of the original or input image and the compressed and subsequently decompressed output image, it is said to be based on an *objective fidelity criterion*. A good example is the root-mean-square (rms) error between an input and output image. Let $f(x, y)$ represent an input image and let $\hat{f}(x, y)$ denote an estimate or approximation of $f(x, y)$ that results from compressing and subsequently decompressing the input. For any value of x and y, the error $e(x, y)$ between $f(x, y)$ and $\hat{f}(x, y)$ can be defined as

$$e(x, y) = \hat{f}(x, y) - f(x, y) \tag{8.1-7}$$

so that the total error between the two images is

$$\sum_{x=0}^{M-1}\sum_{y=0}^{N-1}[\hat{f}(x, y) - f(x, y)]$$

where the images are of size $M \times N$. The *root-mean-square error*, e_{rms}, between $f(x, y)$ and $\hat{f}(x, y)$ then is the square root of the squared error averaged over the $M \times N$ array, or

$$e_{rms} = \left[\frac{1}{MN}\sum_{x=0}^{M-1}\sum_{y=0}^{N-1}[\hat{f}(x, y) - f(x, y)]^2\right]^{1/2} \tag{8.1-8}$$

A closely related objective fidelity criterion is the mean-square signal-to-noise ratio of the compressed-decompressed image. If $\hat{f}(x, y)$ is considered [by a simple rearrangement of the terms in Eq. (8.1-7)] to be the sum of the original image $f(x, y)$ and a noise signal $e(x, y)$, the *mean-square signal-to-noise ratio* of the output image, denoted SNR_{ms}, is

$$SNR_{ms} = \frac{\displaystyle\sum_{x=0}^{M-1}\sum_{y=0}^{N-1} \hat{f}(x, y)^2}{\displaystyle\sum_{x=0}^{M-1}\sum_{y=0}^{N-1} [\hat{f}(x, y) - f(x, y)]^2}. \tag{8.1-9}$$

The rms value of the signal-to-noise ratio, denoted SNR_{rms}, is obtained by taking the square root of Eq. (8.1-9).

Although objective fidelity criteria offer a simple and convenient mechanism for evaluating information loss, most decompressed images ultimately are viewed by humans. Consequently, measuring image quality by the subjective evaluations of a human observer often is more appropriate. This can be accomplished by showing a "typical" decompressed image to an appropriate cross section of viewers and averaging their evaluations. The evaluations may be made using an absolute rating scale or by means of side-by-side comparisons of $f(x, y)$ and $\hat{f}(x, y)$. Table 8.3 shows one possible absolute rating scale. Side-by-side comparisons can be done with a scale such as $\{-3, -2, -1, 0, 1, 2, 3\}$ to represent the subjective evaluations {*much worse, worse, slightly worse, the same, slightly better, better, much better*}, respectively. In either case, the evaluations are said to be based on *subjective fidelity criteria*.

EXAMPLE 8.4:
Comparisons of image quality.

■ The rms errors in the quantized images of Figs. 8.4(b) and (c) are 6.93 and 6.78 gray levels, respectively. The corresponding rms signal-to-noise ratios are 10.25 and 10.39. Although these values are quite similar, a subjective evaluation of the visual quality of the two coded images might result in a *marginal* rating for the image in Fig. 8.4(b) and a *passable* rating for that in Fig. 8.4(c).   ■

TABLE 8.3
Rating scale of the Television Allocations Study Organization. (Frendendall and Behrend.)

Value	Rating	Description
1	Excellent	An image of extremely high quality, as good as you could desire.
2	Fine	An image of high quality, providing enjoyable viewing. Interference is not objectionable.
3	Passable	An image of acceptable quality. Interference is not objectionable.
4	Marginal	An image of poor quality; you wish you could improve it. Interference is somewhat objectionable.
5	Inferior	A very poor image, but you could watch it. Objectionable interference is definitely present.
6	Unusable	An image so bad that you could not watch it.

8.2   Image Compression Models

In Section 8.1 we discussed individually three general techniques for reducing or compressing the amount of data required to represent an image. However, these techniques typically are combined to form practical image compression systems. In this section, we examine the overall characteristics of such a system and develop a general model to represent it.

As Fig. 8.5 shows, a compression system consists of two distinct structural blocks: an *encoder* and a *decoder*.[†] An input image $f(x, y)$ is fed into the encoder, which creates a set of symbols from the input data. After transmission over the *channel*, the encoded representation is fed to the decoder, where a reconstructed output image $\hat{f}(x, y)$ is generated. In general, $\hat{f}(x, y)$ may or may not be an exact replica of $f(x, y)$. If it is, the system is error free or information preserving; if not, some level of distortion is present in the reconstructed image.

Both the encoder and decoder shown in Fig. 8.5 consist of two relatively independent functions or subblocks. The encoder is made up of a *source encoder*, which removes input redundancies, and a *channel encoder*, which increases the noise immunity of the source encoder's output. As would be expected, the decoder includes a *channel decoder* followed by a *source decoder*. If the channel between the encoder and decoder is noise free (not prone to error), the channel encoder and decoder are omitted, and the general encoder and decoder become the source encoder and decoder, respectively.

8.2.1 The Source Encoder and Decoder

The source encoder is responsible for reducing or eliminating any coding, interpixel, or psychovisual redundancies in the input image. The specific application and associated fidelity requirements dictate the best encoding approach to use in any given situation. Normally, the approach can be modeled by a series of three independent operations. As Fig. 8.6(a) shows, each operation is designed to reduce one of the three redundancies described in Section 8.1. Figure 8.6(b) depicts the corresponding source decoder.

In the first stage of the source encoding process, the *mapper* transforms the input data into a (usually nonvisual) format designed to reduce interpixel redundancies in the input image. This operation generally is reversible and may or may not reduce directly the amount of data required to represent the image. Run-length coding (Sections 8.1.2 and 8.4.3) is an example of a mapping that

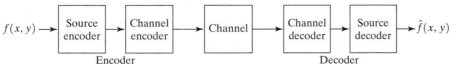

FIGURE 8.5 A general compression system model.

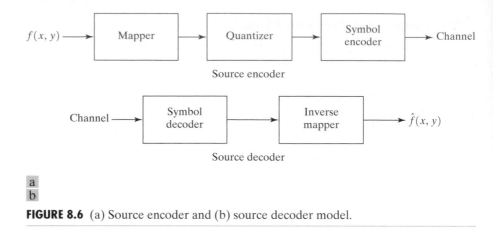

FIGURE 8.6 (a) Source encoder and (b) source decoder model.

directly results in data compression in this initial stage of the overall source encoding process. The representation of an image by a set of transform coefficients (Section 8.5.2) is an example of the opposite case. Here, the mapper transforms the image into an array of coefficients, making its interpixel redundancies more accessible for compression in later stages of the encoding process.

The second stage, or *quantizer* block in Fig. 8.6(a), reduces the accuracy of the mapper's output in accordance with some preestablished fidelity criterion. This stage reduces the psychovisual redundancies of the input image. As noted in Section 8.1.3, this operation is irreversible. Thus it must be omitted when error-free compression is desired.

In the third and final stage of the source encoding process, the *symbol coder* creates a fixed- or variable-length code to represent the quantizer output and maps the output in accordance with the code. The term *symbol coder* distinguishes this coding operation from the overall source encoding process. In most cases, a variable-length code is used to represent the mapped and quantized data set. It assigns the shortest code words to the most frequently occurring output values and thus reduces coding redundancy. The operation, of course, is reversible. Upon completion of the symbol coding step, the input image has been processed to remove each of the three redundancies described in Section 8.1.

Figure 8.6(a) shows the source encoding process as three successive operations, but all three operations are not necessarily included in every compression system. Recall, for example, that the quantizer must be omitted when error-free compression is desired. In addition, some compression techniques normally are modeled by merging blocks that are physically separate in Fig. 8.6(a). In the predictive compression systems of Section 8.5.1, for instance, the mapper and quantizer are often represented by a single block, which simultaneously performs both operations.

The source decoder shown in Fig. 8.6(b) contains only two components: a *symbol decoder* and an *inverse mapper*. These blocks perform, in reverse order, the inverse operations of the source encoder's symbol encoder and mapper blocks. Because quantization results in irreversible information loss, an inverse quantizer block is not included in the general source decoder model shown in Fig. 8.6(b).

8.2.2 The Channel Encoder and Decoder

The channel encoder and decoder play an important role in the overall encoding-decoding process when the channel of Fig. 8.5 is noisy or prone to error. They are designed to reduce the impact of channel noise by inserting a controlled form of redundancy into the source encoded data. As the output of the source encoder contains little redundancy, it would be highly sensitive to transmission noise without the addition of this "controlled redundancy."

One of the most useful channel encoding techniques was devised by R. W. Hamming (Hamming [1950]). It is based on appending enough bits to the data being encoded to ensure that some minimum number of bits must change between valid code words. Hamming showed, for example, that if 3 bits of redundancy are added to a 4-bit word, so that the *distance*[†] between any two valid code words is 3, all single-bit errors can be detected *and* corrected. (By appending additional bits of redundancy, multiple-bit errors can be detected and corrected.) The 7-bit *Hamming (7,4) code* word $h_1 h_2 \ldots h_5 h_6 h_7$ associated with a 4-bit binary number $b_3 b_2 b_1 b_0$ is

$$
\begin{aligned}
h_1 &= b_3 \oplus b_2 \oplus b_0 & h_3 &= b_3 \\
h_2 &= b_3 \oplus b_1 \oplus b_0 & h_5 &= b_2 \\
h_4 &= b_2 \oplus b_1 \oplus b_0 & h_6 &= b_1 \\
& & h_7 &= b_0
\end{aligned}
\tag{8.2-1}
$$

where $\oplus$ denotes the exclusive OR operation. Note that bits h_1, h_2, and h_4 are even-parity bits for the bit fields $b_3 b_2 b_0$, $b_3 b_1 b_0$, and $b_2 b_1 b_0$, respectively. (Recall that a string of binary bits has even parity if the number of bits with a value of 1 is even.)

To decode a Hamming encoded result, the channel decoder must check the encoded value for odd parity over the bit fields in which even parity was previously established. A single-bit error is indicated by a nonzero parity word $c_4 c_2 c_1$, where

$$
\begin{aligned}
c_1 &= h_1 \oplus h_3 \oplus h_5 \oplus h_7 \\
c_2 &= h_2 \oplus h_3 \oplus h_6 \oplus h_7 \\
c_4 &= h_4 \oplus h_5 \oplus h_6 \oplus h_7.
\end{aligned}
\tag{8.2-2}
$$

If a nonzero value is found, the decoder simply complements the code word bit position indicated by the parity word. The decoded binary value is then extracted from the corrected code word as $h_3 h_5 h_6 h_7$.

■ Consider the transmission of the 4-bit IGS data of Table 8.2 over a noisy communication channel. A single-bit error could cause a decompressed pixel to deviate from its correct value by as many as 128 gray levels.[‡] A Hamming

EXAMPLE 8.5:
Hamming
encoding.

[†]The *distance* between two code words is defined as the minimum number of digits that must change in one word so that the other word results. For example, the distance between 101101 and 011101 is 2. The *minimum distance* of a code is the smallest number of digits by which any two code words differ.

[‡]A simple procedure for decompressing 4-bit IGS data is to multiply the decimal equivalent of the IGS value by 16. For example, if the IGS value is 1110, the decompressed gray level is (14)(16) or 224. If the most significant bit of this IGS value was incorrectly transmitted as a 0, the decompressed gray level becomes 96. The resulting error is 128 gray levels.

channel encoder can be utilized to increase the noise immunity of this source encoded IGS data by inserting enough redundancy to allow the detection and correction of single-bit errors. From Eq. (8.2-1), the Hamming encoded value for the first IGS value in Table 8.2 is 1100110_2. Because the Hamming channel encoder increases the number of bits required to represent the IGS value from 4 to 7, the 2 : 1 compression ratio noted in the IGS example is reduced to 8/7 or 1.14 : 1. This reduction in compression is the price paid for increased noise immunity.   ■

8.3 Elements of Information Theory

In Section 8.1 we introduced several ways to reduce the amount of data used to represent an image. The question that naturally arises is: How few data actually are needed to represent an image? That is, is there a minimum amount of data that is sufficient to describe completely an image without loss of information? Information theory provides the mathematical framework to answer this and related questions.

8.3.1 Measuring Information

See inside front cover
Consult the book web site for a brief review of probability theory.

The fundamental premise of information theory is that the generation of information can be modeled as a probabilistic process that can be measured in a manner that agrees with intuition. In accordance with this supposition, a random event E that occurs with probability $P(E)$ is said to contain

$$I(E) = \log \frac{1}{P(E)} = -\log P(E) \tag{8.3-1}$$

units of information. The quantity $I(E)$ often is called the *self-information* of E. Generally speaking, the amount of self-information attributed to event E is inversely related to the probability of E. If $P(E) = 1$ (that is, the event always occurs), $I(E) = 0$ and no information is attributed to it. That is, because no uncertainty is associated with the event, no information would be transferred by communicating that the event has occurred. However, if $P(E) = 0.99$, communicating that E has occurred conveys some small amount of information. Communicating that E has *not* occurred conveys more information, because this outcome is less likely.

The base of the logarithm in Eq. (8.3-1) determines the unit used to measure information.[†] If the base m logarithm is used, the measurement is said to be in m-ary units. If the base 2 is selected, the resulting unit of information is called a *bit*. Note that if $P(E) = \frac{1}{2}$, $I(E) = -\log_2 \frac{1}{2}$, or 1 bit. That is, 1 bit is the amount of information conveyed when one of two possible equally likely events occurs. A simple example of such a situation is flipping a coin and communicating the result.

[†] When we do not explicitly specify the base of the log used in an expression, the result may be interpreted in any base and corresponding information unit.

8.3.2 The Information Channel

When self-information is transferred between an information source and a user of the information, the source of information is said to be connected to the user of information by an *information channel*. The information channel is the physical medium that links the source to the user. It may be a telephone line, an electromagnetic energy propagation path, or a wire in a digital computer. Figure 8.7 shows a simple mathematical model for a discrete information system. Here, the parameter of particular interest is the system's *capacity*, defined as its ability to transfer information.

Let us assume that the information source in Fig. 8.7 generates a random sequence of symbols from a finite or countably infinite set of possible symbols. That is, the output of the source is a discrete random variable. The set of source symbols $\{a_1, a_2, \ldots, a_J\}$ is referred to as the *source alphabet A*, and the elements of the set, denoted a_j, are called *symbols* or *letters*. The probability of the event that the source will produce symbol a_j is $P(a_j)$, and

$$\sum_{j=1}^{J} P(a_j) = 1. \tag{8.3-2}$$

A $J \times 1$ vector $\mathbf{z} = \left[P(a_1), P(a_2), \ldots, P(a_J) \right]^T$ customarily is used to represent the set of all source symbol probabilities $\{P(a_1), P(a_2), \ldots, P(a_J)\}$. The finite *ensemble* $(A, \mathbf{z})$ describes the information source completely.

The probability that the discrete source will emit symbol a_j is $P(a_j)$, so the self-information generated by the production of a single source symbol is, in accordance with Eq. (8.3-1), $I(a_j) = -\log P(a_j)$. If k source symbols are generated, the law of large numbers stipulates that, for a sufficiently large value of k, symbol a_j will (on average) be output $kP(a_j)$ times. Thus the average self-information obtained from k outputs is

$$-kP(a_1) \log P(a_1) - kP(a_2) \log P(a_2) - \ldots - kP(a_J) kP(a_J)$$

or

$$-k \sum_{j=1}^{J} P(a_j) \log P(a_j).$$

The average information per source output, denoted $H(\mathbf{z})$, is

$$H(\mathbf{z}) = - \sum_{j=1}^{J} P(a_j) \log P(a_j). \tag{8.3-3}$$

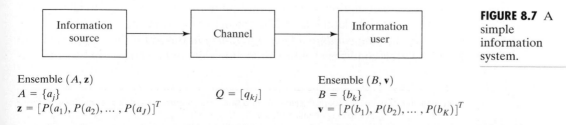

Ensemble $(A, \mathbf{z})$
$A = \{a_j\}$
$\mathbf{z} = \left[P(a_1), P(a_2), \ldots, P(a_J) \right]^T$

$Q = \left[q_{kj} \right]$

Ensemble $(B, \mathbf{v})$
$B = \{b_k\}$
$\mathbf{v} = \left[P(b_1), P(b_2), \ldots, P(b_K) \right]^T$

FIGURE 8.7 A simple information system.

This quantity is called the *uncertainty* or *entropy* of the source. It defines the average amount of information (in *m*-ary units per symbol) obtained by observing a single source output. As its magnitude increases, more uncertainty and thus more information is associated with the source. If the source symbols are equally probable, the entropy or uncertainty of Eq. (8.3-3) is maximized and the source provides the greatest possible average information per source symbol.

Having modeled the information source, we can develop the input-output characteristics of the information channel rather easily. Because we modeled the input to the channel in Fig. 8.7 as a discrete random variable, the information transferred to the output of the channel is also a discrete random variable. Like the source random variable, it takes on values from a finite or countably infinite set of symbols $\{b_1, b_2, \ldots, b_K\}$ called the *channel alphabet, B*. The probability of the event that symbol b_k is presented to the *information user* is $P(b_k)$. The finite ensemble $(B, \mathbf{v})$, where $\mathbf{v} = \left[P(b_1), P(b_2), \ldots, P(b_K)\right]^T$, describes the channel output completely and thus the information received by the user.

The probability $P(b_k)$ of a given channel output and the probability distribution of the source $\mathbf{z}$ are related by the expression[†]

$$P(b_k) = \sum_{j=1}^{J} P(b_k \mid a_j)P(a_j) \qquad (8.3\text{-}4)$$

where $P(b_k \mid a_j)$ is the conditional probability that output symbol b_k is received, given that source symbol a_j was generated. If the conditional probabilities referenced in Eq. (8.3-4) are arranged in a matrix $K \times J$ matrix $\mathbf{Q}$, such that

$$\mathbf{Q} = \begin{bmatrix} P(b_1 \mid a_1) & P(b_1 \mid a_2) & \cdots & P(b_1 \mid a_J) \\ P(b_2 \mid a_1) & \cdot & \cdots & \cdot \\ \cdot & \cdot & & \cdot \\ \cdot & \cdot & \cdots & \cdot \\ \cdot & & & \\ P(b_K \mid a_1) & P(b_K \mid a_2) & \cdots & P(b_K \mid a_J) \end{bmatrix} \qquad (8.3\text{-}5)$$

then the probability distribution of the complete output alphabet can be computed from

$$\mathbf{v} = \mathbf{Q}\mathbf{z}. \qquad (8.3\text{-}6)$$

Matrix $\mathbf{Q}$, with elements $q_{kj} = P(b_k \mid a_j)$, is referred to as the *forward channel transition matrix* or by the abbreviated term *channel matrix*.

To determine the capacity of an information channel with forward channel transition matrix $\mathbf{Q}$, the entropy of the information source must first be computed under the assumption that the information user observes a particular output b_k. Equation (8.3-4) defines a distribution of source symbols for any observed b_k, so each b_k has one *conditional entropy function*. Based on the steps leading to Eq. (8.3-3), this conditional entropy function, denoted $H(\mathbf{z} \mid b_k)$, can be written as

---

[†]One of the fundamental laws of probability theory is that, for an arbitrary event D and t mutually exclusive events $C_1, C_2, \ldots, C_t$, the total probability of D is $P(D) = P(D \mid C_1)P(C_1) + \ldots + P(D \mid C_t)P(C_t)$.

$$H(\mathbf{z}|b_k) = -\sum_{j=1}^{J} P(a_j|b_k)\log P(a_j|b_k) \qquad (8.3\text{-}7)$$

where $P(a_j|b_k)$ is the probability that symbol a_j was transmitted by the source, given that the user received b_k. The expected (average) value of this expression over all b_k is

$$H(\mathbf{z}|\mathbf{v}) = \sum_{k=1}^{K} H(\mathbf{z}|b_k)P(b_k), \qquad (8.3\text{-}8)$$

which, after substitution of Eq. (8.3-7) for $H(\mathbf{z}|b_k)$ and some minor rearrangement,[†] can be written as

$$H(\mathbf{z}|\mathbf{v}) = -\sum_{j=1}^{J}\sum_{k=1}^{K} P(a_j, b_k)\log P(a_j|b_k). \qquad (8.3\text{-}9)$$

Here, $P(a_j, b_k)$ is the joint probability of a_j and b_k. That is, $P(a_j, b_k)$ is the probability that a_j is transmitted *and* b_k is received.

The term $H(\mathbf{z}|\mathbf{v})$ is called the *equivocation* of $\mathbf{z}$ with respect to $\mathbf{v}$. It represents the average information of one source symbol, assuming observation of the output symbol that resulted from its generation. Because $H(\mathbf{z})$ is the average information of one source symbol, assuming no knowledge of the resulting output symbol, the difference between $H(\mathbf{z})$ and $H(\mathbf{z}|\mathbf{v})$ is the average information received upon observing a single output symbol. This difference, denoted $I(\mathbf{z}, \mathbf{v})$ and called the *mutual information* of $\mathbf{z}$ and $\mathbf{v}$, is

$$I(\mathbf{z}, \mathbf{v}) = H(\mathbf{z}) - H(\mathbf{z}|\mathbf{v}). \qquad (8.3\text{-}10)$$

Substituting Eqs. (8.3-3) and (8.3-9) for $H(\mathbf{z})$ and $H(\mathbf{z}|\mathbf{v})$, and recalling that $P(a_j) = P(a_j, b_1) + P(a_j, b_2) + \cdots + P(a_j, b_K)$, yields

$$I(\mathbf{z}, \mathbf{v}) = \sum_{j=1}^{J}\sum_{k=1}^{K} P(a_j, b_k)\log \frac{P(a_j, b_k)}{P(a_j)P(b_k)}, \qquad (8.3\text{-}11)$$

which, after further manipulation, can be written as

$$I(\mathbf{z}, \mathbf{v}) = \sum_{j=1}^{J}\sum_{k=1}^{K} P(a_j)q_{kj}\log \frac{q_{kj}}{\sum_{i=1}^{J} P(a_i)q_{ki}}. \qquad (8.3\text{-}12)$$

Thus the average information received upon observing a single output of the information channel is a function of the input or source symbol probability vector $\mathbf{z}$ and channel matrix $\mathbf{Q}$. The minimum possible value of $I(\mathbf{z}, \mathbf{v})$ is zero and occurs when the input and output symbols are statistically independent, in which case $P(a_j, b_k) = P(a_j)P(b_k)$ and the log term in Eq. (8.3-11) is 0 for all j and k. The maximum value of $I(\mathbf{z}, \mathbf{v})$ over all possible choices of source probabilities in vector $\mathbf{z}$ is the *capacity*, C, of the channel described by channel matrix $\mathbf{Q}$. That is,

$$C = \max_{\mathbf{z}}\left[I(\mathbf{z}, \mathbf{v})\right] \qquad (8.3\text{-}13)$$

[†] Use is made of the fact that the joint probability of two events, C and D, is $P(C, D) = P(C)P(D|C) = P(D)P(C|D)$.

where the maximum is taken over all possible input symbol probabilities. The capacity of the channel defines the maximum rate (in m-ary information units per source symbol) at which information can be transmitted reliably through the channel. Moreover, the capacity of a channel does not depend on the input probabilities of the source (that is, on how the channel is used) but is a function of the conditional probabilities defining the channel alone.

EXAMPLE 8.6:
The binary case.

■ Consider a binary information source with source alphabet $A = \{a_1, a_2\} = \{0, 1\}$. The probabilities that the source will produce symbols a_1 and a_2 are $P(a_1) = p_{bs}$ and $P(a_2) = 1 - p_{bs} = \bar{p}_{bs}$, respectively. From Eq. (8.3-3), the entropy of the source is

$$H(\mathbf{z}) = -p_{bs} \log_2 p_{bs} - \bar{p}_{bs} \log_2 \bar{p}_{bs}.$$

Because $\mathbf{z} = [P(a_1), P(a_2)]^T = [p_{bs}, 1 - p_{bs}]^T$, $H(\mathbf{z})$ depends on the single parameter p_{bs}, and the right-hand side of the equation is called the *binary entropy function*, denoted $H_{bs}(\cdot)$. Thus, for example, $H_{bs}(t)$ is the function $-t \log_2 t - \bar{t} \log_2 \bar{t}$. Figure 8.8(a) shows a plot of $H_{bs}(p_{bs})$ for $0 \leq p_{bs} \leq 1$. Note that H_{bs} obtains its maximum value (of 1 bit) when p_{bs} is $\frac{1}{2}$. For all other values of p_{bs}, the source provides less than 1 bit of information.

Now assume that the information is to be transmitted over a noisy binary information channel and let the probability of an error during the transmission of any symbol be p_e. Such a channel is called a *binary symmetric channel* (BSC) and is defined by the channel matrix

$$\mathbf{Q} = \begin{bmatrix} 1 - p_e & p_e \\ p_e & 1 - p_e \end{bmatrix} = \begin{bmatrix} \bar{p}_e & p_e \\ p_e & \bar{p}_e \end{bmatrix}.$$

For each input or source symbol, the BSC produces one output b_j from the output alphabet $B = \{b_1, b_2\} = \{0, 1\}$. The probabilities of receiving output symbols b_1 and b_2 can be determined from Eq. (8.3-6):

$$\mathbf{v} = \mathbf{Qz} = \begin{bmatrix} \bar{p}_e & p_e \\ p_e & \bar{p}_e \end{bmatrix} \begin{bmatrix} p_{bs} \\ \bar{p}_{bs} \end{bmatrix} = \begin{bmatrix} \bar{p}_e p_{bs} + p_e \bar{p}_{bs} \\ p_e p_{bs} + \bar{p}_e \bar{p}_{bs} \end{bmatrix}.$$

Consequently, because $\mathbf{v} = [P(b_1), P(b_2)]^T = [P(0), P(1)]^T$, the probability that the output is a 0 is $\bar{p}_e p_{bs} + p_e \bar{p}_{bs}$, and the probability that it is a 1 is $p_e p_{bs} + \bar{p}_e \bar{p}_{bs}$.

The mutual information of the BSC can now be computed from Eq. (8.3-12). Expanding the summations of this equation and collecting the appropriate terms gives

$$I(\mathbf{z}, \mathbf{v}) = H_{bs}(p_{bs} p_e + \bar{p}_{bs} \bar{p}_e) - H_{bs}(p_e)$$

where $H_{bs}(\cdot)$ is the binary entropy function of Fig. 8.8(a). For a fixed value of p_e, $I(\mathbf{z}, \mathbf{v})$ is 0 when p_{bs} is 0 or 1. Moreover, $I(\mathbf{z}, \mathbf{v})$ achieves its maximum value when the binary source symbols are equally probable. Figure 8.8(b) shows $I(\mathbf{z}, \mathbf{v})$ for all values of p_{bs} and a given channel error p_e.

In accordance with Eq. (8.3-13), the capacity of the BSC is obtained from the maximum of the mutual information over all possible source distributions. From

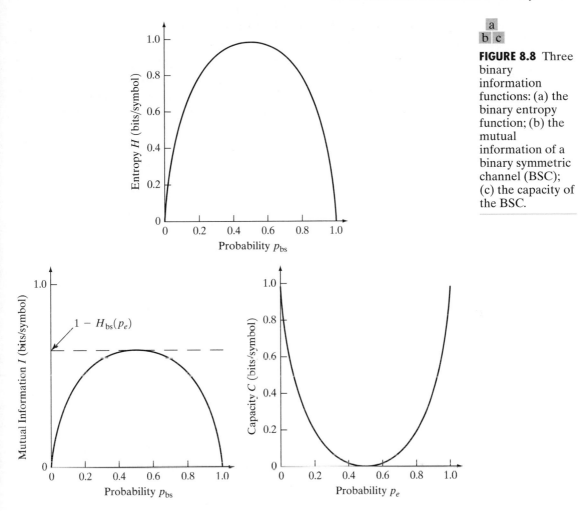

FIGURE 8.8 Three binary information functions: (a) the binary entropy function; (b) the mutual information of a binary symmetric channel (BSC); (c) the capacity of the BSC.

Fig. 8.8(b), which plots $I(\mathbf{z}, \mathbf{v})$ for all possible binary source distributions (that is, for $0 \leq p_{bs} \leq 1$ or for $\mathbf{z} = [0, 1]^T$ to $\mathbf{z} = [1, 0]^T$), we see that $I(\mathbf{z}, \mathbf{v})$ is maximum (for any p_e) when $p_{bs} = \frac{1}{2}$. This value of p_{bs} corresponds to source probabilities vector $\mathbf{z} = [\frac{1}{2}, \frac{1}{2}]^T$. The corresponding value of $I(\mathbf{z}, \mathbf{v})$ is $1 - H_{bs}(p_e)$. Thus the capacity of the BSC, plotted in Fig. 8.8(c), is

$$C = 1 - H_{bs}(p_e).$$

Note that when there is no possibility of a channel error $(p_e = 0)$—as well as when a channel error is a certainty $(p_e = 1)$—the capacity of the channel obtains its maximum value of 1 bit/symbol. In either case, maximum information transfer is possible because the channel's output is completely predictable. However, when $p_e = \frac{1}{2}$, the channel's output is completely unpredictable and no information can be transferred through it. ■

8.3.3 Fundamental Coding Theorems

The overall mathematical framework introduced in Section 8.3.2 is based on the model shown in Fig. 8.7, which contains an information source, channel, and user. In this section, we add a communication system to the model and examine three basic theorems regarding the coding or representation of information. As Fig. 8.9 shows, the communication system is inserted between the source and the user and consists of an encoder and decoder.

The noiseless coding theorem

When both the information channel and communication system are error free, the principal function of the communication system is to represent the source as compactly as possible. Under these circumstances, the *noiseless coding theorem*, also called *Shannon's first theorem* (Shannon [1948]), defines the minimum average code word length per source symbol that can be achieved.

A source of information with finite ensemble $(A, \mathbf{z})$ and statistically independent source symbols is called a *zero-memory* source. If we consider its output to be an n-tuple of symbols from the source alphabet (rather than a single symbol), the source output then takes on one of J^n possible values, denoted α_i, from the set of all possible n element sequences $A' = \{\alpha_1, \alpha_2, \dots, \alpha_{J^n}\}$. In other words, each α_i (called a *block random variable*) is composed of n symbols from A. (The notation A' distinguishes the set of block symbols from A, the set of single symbols.) The probability of a given α_i is $P(\alpha_i)$, ...which is related to the single-symbol probabilities $P(a_j)$ by

$$P(\alpha_i) = P(a_{j1})P(a_{j2})\cdots P(a_{jn}) \tag{8.3-14}$$

where subscripts $j1, j2, \dots, jn$ are used to index the n symbols from A that make up an α_i. As before, the vector $\mathbf{z}'$ (the prime is added to indicate the use of the block random variable) denotes the set of all source probabilities $\{P(\alpha_1), P(\alpha_2), \dots, P(\alpha_{J^n})\}$, and the entropy of the source is

$$H(\mathbf{z}') = -\sum_{i=1}^{J^n} P(\alpha_i) \log P(\alpha_i).$$

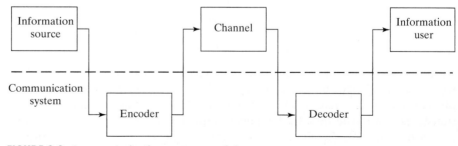

FIGURE 8.9 A communication system model.

Substituting Eq. (8.3-14) for $P(\alpha_i)$ and simplifying yields

$$H(\mathbf{z}') = nH(\mathbf{z}). \tag{8.3-15}$$

Thus the entropy of the zero-memory information source (which produces the block random variable) is n times the entropy of the corresponding single symbol source. Such a source is referred to as the *nth extension* of the single symbol or nonextended source. Note that the first extension of any source is the nonextended source itself.

Because the self-information of source output α_i is $\log\left[1/P(\alpha_i)\right]$, it seems reasonable to code α_i with a code word of integer length $l(\alpha_i)$ such that

$$\log \frac{1}{P(\alpha_i)} \leq l(\alpha_i) < \log \frac{1}{P(\alpha_i)} + 1. \tag{8.3-16}$$

Intuition suggests that the source output α_i be represented by a code word whose length is the smallest integer exceeding the self-information of α_i.[†] Multiplying this result by $P(\alpha_i)$ and summing over all i gives

$$\sum_{i=1}^{J^n} P(\alpha_i) \log \frac{1}{P(\alpha_i)} \leq \sum_{i=1}^{J^n} P(\alpha_i) l(\alpha_i) < \sum_{i=1}^{J^n} P(\alpha_i) \log \frac{1}{P(\alpha_i)} + 1$$

or

$$H(\mathbf{z}') \leq L'_{\text{avg}} < H(\mathbf{z}') + 1 \tag{8.3-17}$$

where L'_{avg} represents the average word length of the code corresponding to the nth extension of the nonextended source. That is,

$$L'_{\text{avg}} = \sum_{i=1}^{J^n} P(\alpha_i) l(\alpha_i). \tag{8.3-18}$$

Dividing Eq. (8.3-17) by n and noting from Eq. (8.3-15) that $H(\mathbf{z}')/n$ is equal to $H(\mathbf{z})$ yields

$$H(\mathbf{z}) \leq \frac{L'_{\text{avg}}}{n} < H(\mathbf{z}) + \frac{1}{n} \tag{8.3-19}$$

which, in the limiting case, becomes

$$\lim_{n \to \infty} \left[\frac{L'_{\text{avg}}}{n} \right] = H(\mathbf{z}). \tag{8.3-20}$$

Equation (8.3-19) states Shannon's first theorem for a zero-memory source. It shows that it is possible to make L'_{avg}/n arbitrarily close to $H(\mathbf{z})$ by coding infinitely long extensions of the source. Although derived under the assumption of statistically independent source symbols, the result is easily extended to more general sources, where the occurrence of source symbol a_j may depend on a finite number of preceding symbols. These types of sources (called *Markov sources*) commonly are used to model interpixel correlations in an image. Because $H(\mathbf{z})$

[†] A uniquely decodable code can be constructed subject to this constraint.

is a lower bound on L'_{avg}/n [that is, the limit of L'_{avg}/n as n becomes large in Eq. (8.3-20) is $H(\mathbf{z})$], the *efficiency* η of any encoding strategy can be defined as

$$\eta = n\,\frac{H(\mathbf{z})}{L'_{avg}}. \tag{8.3-21}$$

EXAMPLE 8.7:
Extension coding.

■ A zero-memory information source with source alphabet $A = \{a_1, a_2\}$ has symbol probabilities $P(a_1) = \frac{2}{3}$ and $P(a_2) = \frac{1}{3}$. From Eq. (8.3-3), the entropy of this source is 0.918 bits/symbol. If symbols a_1 and a_2 are represented by the binary code words 0 and 1, $L'_{avg} = 1$ bit/symbol and the resulting code efficiency is $\eta = (1)(0.918)/1$, or 0.918.

Table 8.4 summarizes the code just described and an alternative encoding based on the second extension of the source. The lower portion of Table 8.4 lists the four block symbols ($\alpha_1, \alpha_2, \alpha_3,$ and α_4) in the second extension of the source. From Eq. (8.3-14) their probabilities are $\frac{4}{9}$, $\frac{2}{9}$, $\frac{2}{9}$, and $\frac{1}{9}$, respectively. In accordance with Eq. (8.3-18), the average word length of the second encoding is $\frac{17}{9}$ or 1.89 bits/symbol. The entropy of the second extension is twice the entropy of the nonextended source, or 1.83 bits/symbol, so the efficiency of the second encoding is $\eta = 1.83/1.89 = 0.97$. It is slightly better than the nonextended coding efficiency of 0.92. Encoding the second extension of the source reduces the average number of code bits per source symbol from 1 bit/symbol to 1.89/2 or 0.94 bits/symbol. ■

The noisy coding theorem

If the channel of Fig. 8.9 is noisy or prone to error, interest shifts from representing the information as compactly as possible to encoding it so that reliable communication is possible. The question that naturally arises is: How small can the error in communication be made?

EXAMPLE 8.8:
Noisy binary channel.

■ Suppose that a BSC has a probability of error $p_e = 0.01$ (that is, 99% of all source symbols are transmitted through the channel correctly). A simple method for increasing the reliability of the communication is to repeat each message or binary symbol several times. Suppose, for example, that rather than transmitting a 0 or a 1, the coded messages 000 and 111 are used. The

TABLE 8.4
Extension coding example.

α_i	Source Symbols	$P(\alpha_i)$ Eq. (8.3-14)	$I(\alpha_i)$ Eq. (8.3-1)	$l(\alpha_i)$ Eq. (8.3-16)	Code Word	Code Length
First Extension						
α_1	a_1	2/3	0.59	1	0	1
α_2	a_2	1/3	1.58	2	1	1
Second Extension						
α_1	$a_1 a_1$	4/9	1.17	2	0	1
α_2	$a_1 a_2$	2/9	2.17	3	10	2
α_3	$a_2 a_1$	2/9	2.17	3	110	3
α_4	$a_2 a_2$	1/9	3.17	4	111	3

probability that no errors will occur during the transmission of a three-symbol message is or $(1 - p_e)^3$ or $\bar{p}_e^3$. The probability of a single error is $3p_e\bar{p}_e^2$, the probability of two errors is $3p_e^2\bar{p}_e$, and the probability of three errors is p_e^3. Because the probability of a single symbol transmission error is less than 50%, received messages can be decoded by using a majority vote of the three received symbols. Thus the probability of incorrectly decoding a three-symbol code word is the sum of the probabilities of two symbol errors and three symbol errors, or $p_e^3 + 3p_e^2\bar{p}_e$. When no errors or a single error occurs, the majority vote decodes the message correctly. For $p_e = 0.01$, the probability of a communication error is reduced to 0.0003.   ■

By extending the repetitive coding scheme just described, we can make the overall error in communication as small as desired. In the general case, we do so by encoding the nth extension of the source using K-ary code sequences of length r, where $K^r \geq J^n$. The key to this approach is to select only φ of the K^r possible code sequences as valid code words and devise a decision rule that optimizes the probability of correct decoding. In the preceding example, repeating each source symbol three times is equivalent to encoding the nonextended binary source using two out of 2^3, or 8, possible 3-bit code words. The two valid code words are 000 and 111. If a nonvalid code word is presented to the decoder, a majority vote of the three code bits determines the output.

A zero-memory information source generates information at a *rate* (in information units per source symbol) equal to its entropy $H(\mathbf{z})$. The nth extension of the source provides information at a rate of $H(\mathbf{z}')/n$ information units per symbol. If the information is coded, as in the preceding example, the maximum rate of coded information is $r^{-1} \log \varphi$ (information units per code symbol) and occurs when the φ valid code words used to code the source are equally probable. Hence, a code of size φ and block length r is said to have a rate of

$$R = \frac{1}{r} \log \varphi \tag{8.3-22}$$

information units per symbol. *Shannon's second theorem* (Shannon [1948]), also called the *noisy coding theorem*, tells us that for any $R < C$, where C is the capacity of the *zero-memory channel* with matrix $\mathbf{Q}$,[†] there exists an integer r, and code of block length r and rate R such that the probability of a block decoding error is less than or equal to ε for any $\varepsilon > 0$. Thus the probability of error can be made arbitrarily small so long as the coded message rate is less than the capacity of the channel.

The source coding theorem

The theorems described thus far establish fundamental limits on error-free communication over both reliable and unreliable channels. In this section, we turn to the case in which the channel is error free but the communication process itself is lossy. Under these circumstances, the principal function of the communication

---

[†] A zero-memory channel is one in which the channel's response to the current input symbol is independent of its response to previous input symbols.

system is "information compression." In most cases, the average error introduced by the compression is constrained to some maximum allowable level D. We want to determine the smallest rate, subject to a given fidelity criterion, at which information about the source can be conveyed to the user. This problem is specifically addressed by a branch of information theory known as *rate distortion theory*.

Let the information source and decoder outputs in Fig. 8.9 be defined by the finite ensembles $(A, \mathbf{z})$ and $(B, \mathbf{z})$, respectively. The assumption now is that the channel of Fig. 8.9 is error free, so a channel matrix $\mathbf{Q}$, which relates $\mathbf{z}$ to $\mathbf{v}$ in accordance with Eq. (8.3-6), can be thought of as modeling the encoding-decoding process alone. Because the encoding-decoding process is deterministic, $\mathbf{Q}$ describes an artificial zero-memory channel that models the effect of the information compression and decompression. Each time the source produces source symbol a_j, it is represented by a code symbol that is then decoded to yield output symbol b_k with probability q_{kj} (see Section 8.3.2).

Addressing the problem of encoding the source so that the average distortion is less than D requires that a rule be formulated to assign quantitatively a distortion value to every possible approximation at the source output. For the simple case of a nonextended source, a nonnegative cost function $\rho(a_j, b_k)$, called a *distortion measure*, can be used to define the penalty associated with reproducing source output a_j with decoder output b_k. The output of the source is random, so the distortion also is a random variable whose average value, denoted $d(\mathbf{Q})$, is

$$
d(\mathbf{Q}) = \sum_{j=1}^{J} \sum_{k=1}^{K} \rho(a_j, b_k) P(a_j, b_k)
$$

$$
= \sum_{j=1}^{J} \sum_{k=1}^{K} \rho(a_j, b_k) P(a_j) q_{kj}. \tag{8.3-23}
$$

The notation $d(\mathbf{Q})$ emphasizes that the average distortion is a function of the encoding-decoding procedure, which (as noted previously) is modeled by $\mathbf{Q}$. A particular encoding-decoding procedure is said to be *D-admissible* if and only if the average distortion associated with $\mathbf{Q}$ is less than or equal to D. The set of all D-admissible encoding-decoding procedures therefore is

$$
\mathbf{Q}_D = \{ q_{kj} \mid d(\mathbf{Q}) \le D \}. \tag{8.3-24}
$$

Because every encoding-decoding procedure is defined by an artificial channel matrix $\mathbf{Q}$, the average information obtained from observing a single decoder output can be computed in accordance with Eq. (8.3-12). Hence, we can define a *rate distortion function*

$$
R(D) = \min_{\mathbf{Q} \in \mathbf{Q}_D} \left[I(\mathbf{z}, \mathbf{v}) \right], \tag{8.3-25}
$$

which assumes the minimum value of Eq. (8.3-12) over all D-admissible codes. Note that the minimum can be taken over $\mathbf{Q}$, because $I(\mathbf{z}, \mathbf{v})$ is a function of the probabilities in vector $\mathbf{z}$ and elements in matrix $\mathbf{Q}$. If $D = 0$, $R(D)$ is less than or equal to the entropy of the source, or $R(0) \le H(\mathbf{z})$.

Equation (8.3-25) defines the minimum rate at which information about the source can be conveyed to the user subject to the constraint that the average distortion be less than or equal to D. To compute this rate [that is, $R(D)$], we simply minimize $I(\mathbf{z}, \mathbf{v})$ [Eq. (8.3-12)] by appropriate choice of $\mathbf{Q}$ (or q_{kj}) subject to the constraints

$$q_{kj} \geq 0, \tag{8.3-26}$$

$$\sum_{k=1}^{K} q_{kj} = 1, \tag{8.3-27}$$

and

$$d(\mathbf{Q}) = D. \tag{8.3-28}$$

Equations (8.3-26) and (8.3-27) are fundamental properties of channel matrix $\mathbf{Q}$. The elements of $\mathbf{Q}$ must be positive and, because some output must be received for any input symbol generated, the terms in any one column of $\mathbf{Q}$ must sum to 1. Equation (8.3-28) indicates that the minimum information rate occurs when the maximum possible distortion is allowed.

▪ Consider a zero-memory binary source with equally probable source symbols $\{0, 1\}$ and the simple distortion measure

$$\rho(a_j, b_k) = 1 - \delta_{jk}$$

where δ_{jk} is the unit delta function. Because $\rho(a_j, b_k)$ is 1 if $a_j \neq b_k$ but is 0 otherwise, each encoding-decoding error is counted as one unit of distortion. The calculus of variations can be used to compute $R(D)$. Letting $\mu_1, \mu_2, \ldots, \mu_{J+1}$ be Lagrangian multipliers, we form the augmented criterion function

$$J(\mathbf{Q}) = I(\mathbf{z}, \mathbf{v}) - \sum_{j=1}^{J} \mu_j \sum_{k=1}^{K} q_{kj} - \mu_{J+1} d(\mathbf{Q}),$$

equate its JK derivatives with respect to q_{kj} to 0 (that is, $dJ/dq_{kj} = 0$), and solve the resulting equations, together with the $J + 1$ equations associated with Eqs. (8.3-27) and (8.3-28), for unknowns q_{kj} and $\mu_1, \mu_2, \ldots, \mu_{J+1}$. If the resulting q_{kj} are nonnegative [or satisfy Eq. (8.3-26)], a valid solution is found. For the source and distortion pair defined above, we get the following 7 equations (with 7 unknowns):

$$2q_{11} = (q_{11} + q_{12}) \exp[2\mu_1] \qquad 2q_{22} = (q_{21} + q_{22}) \exp[2\mu_2]$$
$$2q_{12} = (q_{11} + q_{12}) \exp[2\mu_1 + \mu_3] \qquad 2q_{21} = (q_{21} + q_{22}) \exp[2\mu_2 + \mu_3]$$
$$q_{11} + q_{21} = 1 \qquad\qquad\qquad\qquad q_{12} + q_{22} = 1$$
$$q_{21} + q_{12} = 2D.$$

A series of tedious but straightforward algebraic steps then yields

$$q_{12} = q_{21} = D$$
$$q_{11} = q_{22} = 1 - D$$
$$\mu_1 = \mu_2 = \log \sqrt{2(1 - D)}$$
$$\mu_3 = \log \frac{D}{1 - D}$$

EXAMPLE 8.9:
Computing the rate distortion function for a zero-memory binary source.

so that

$$Q = \begin{bmatrix} 1 - D & D \\ D & 1 - D \end{bmatrix}.$$

It is given that the source symbols are equally probable, so the maximum possible distortion is $\frac{1}{2}$. Thus $0 \le D \le \frac{1}{2}$ and the elements of Q satisfy Eq. (8.3-12) for all D. The mutual information associated with Q and the previously defined binary source is computed by using Eq. (8.3-12). Noting the similarity between Q and the binary symmetric channel matrix, however, we can immediately write

$$I(z, v) = 1 - H_{bs}(D).$$

This result follows from Example 8.6 by substituting $p_{bs} = \frac{1}{2}$ and $p_e = D$ into $I(z, v) = H_{bs}(p_{bs} p_e + \bar{p}_{bs} \bar{p}_e) - H_{bs}(p_e)$. The rate distortion function follows immediately from Eq. (8.3-25):

$$R(D) = \min_{Q \in Q_D} \left[1 - H_{bs}(D) \right] = 1 - H_{bs}(D).$$

The final simplification is based on the fact that, for a given D, $1 - H_{bs}(D)$ assumes a single value, which, by default, is the minimum. The resulting function is plotted in Fig. 8.10. Its shape is typical of most rate distortion functions. Note the maximum value of D, denoted D_{max}, such that $R(D) = 0$ for all $D \ge D_{max}$. In addition, $R(D)$ is always positive, monotonically decreasing, and convex in the interval $(0, D_{max})$.   ■

Rate distortion functions can be computed analytically for simple sources and distortion measures, as in the preceding example. Moreover, convergent iterative algorithms suitable for implementation on digital computers can be used when

FIGURE 8.10 The rate distortion function for a binary symmetric source.

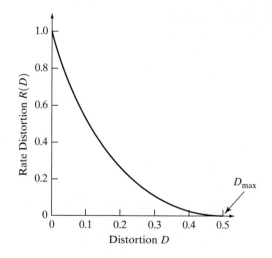

analytical methods fail or are impractical. After $R(D)$ is computed (for any zero-memory source and *single-letter* distortion measure[†]), the *source coding theorem* tells us that, for any $\varepsilon > 0$, there exists an r, and code of block length r and rate $R < R(D) + \varepsilon$, such that the average per-letter distortion satisfies the condition $d(\mathbf{Q}) \leq D + \varepsilon$. An important practical consequence of this theorem and the noisy coding theorem is that the source output can be recovered at the decoder with an arbitrarily small probability of error provided that the channel has capacity $C > R(D) + \varepsilon$. This latter result is known as the *information transmission theorem*.

8.3.4 Using Information Theory

Information theory provides the basic tools needed to deal with information representation and manipulation directly and quantitatively. In this section we explore the application of these tools to the specific problem of image compression. Because the fundamental premise of information theory is that the generation of information can be modeled as a probabilistic process, we first develop a statistical model of the image generation process.

■ Consider the problem of estimating the information content (or entropy) of the simple 8-bit image:

21	21	21	95	169	243	243	243
21	21	21	95	169	243	243	243
21	21	21	95	169	243	243	243
21	21	21	95	169	243	243	243

One relatively simple approach is to assume a particular source model and compute the entropy of the image based on that model. For example, we can assume that the image was produced by an imaginary "8-bit gray-level source" that sequentially emitted statistically independent pixels in accordance with a predefined probability law. In this case, the source symbols are gray levels, and the source alphabet is composed of 256 possible symbols. If the symbol probabilities are known, the average information content or entropy of each pixel in the image can be computed by using Eq. (8.3-3). In the case of a uniform probability density, for instance, the source symbols are equally probable, and the source is characterized by an entropy of 8 bits/pixel. That is, the average information per source output (pixel) is 8 bits. Then the total entropy of the preceding 4×8 image is 256 bits. This particular image is but one of $2^{8 \times 4 \times 8}$, or 2^{256} ($\sim 10^{77}$), equally probable 4×8 images that can be produced by the source.

An alternative method of estimating information content is to construct a source model based on the relative frequency of occurrence of the gray levels in the image under consideration. That is, an observed image can be interpreted as a sample of the behavior of the gray-level source that generated it. Because

---

[†] A single-letter distortion measure is one in which the distortion associated with a block of letters (or symbols) is the sum of the distortions for each letter (or symbol) in the block.

the observed image is the only available indicator of source behavior, modeling the probabilities of the source symbols using the gray-level histogram of the sample image is reasonable:

Gray Level	Count	Probability
21	12	3/8
95	4	1/8
169	4	1/8
243	12	3/8

An estimate, called the *first-order estimate*, of the entropy of the source can be computed with Eq. (8.3-3). The first-order estimate in this example is 1.81 bits/pixel. The entropy of the source and/or image thus is approximately 1.81 bits/pixel, or 58 total bits.

Better estimates of the entropy of the gray-level source that generated the sample image can be computed by examining the relative frequency of pixel blocks in the sample image, where a block is a grouping of adjacent pixels. As block size approaches infinity, the estimate approaches the source's true entropy. (This result can be shown with the procedure utilized to prove the validity of the noiseless coding theorem in Section 8.3.3.) Thus by assuming that the sample image is connected from line to line and end to beginning, we can compute the relative frequency of pairs of pixels (that is, the second extension of the source):

Gray-level Pair	Count	Probability
(21, 21)	8	1/4
(21, 95)	4	1/8
(95, 169)	4	1/8
(169, 243)	4	1/8
(243, 243)	8	1/4
(243, 21)	4	1/8

The resulting entropy estimate [again using Eq. (8.3-3)] is 2.5/2, or 1.25 bits/pixel, where division by 2 is a consequence of considering two pixels at a time. This estimate is called the *second-order estimate* of the source entropy, because it was obtained by computing the relative frequencies of 2-pixel blocks. Although the third-, fourth-, and higher-order estimates would provide even better approximations of source entropy, convergence of these estimates to the true source entropy is slow and computationally involved. For instance, a general 8-bit image has $(2^8)^2$, or 65,536, possible symbol pairs whose relative frequency must be computed. If 5-pixel blocks are considered, the number of possible 5-tuples is $(2^8)^5$, or $\sim 10^{12}$. ■

Although computing the actual entropy of an image is difficult, estimates such as those in the preceding example provide insight into image compressibility.

The first-order estimate of entropy, for example, is a lower bound on the compression that can be achieved through variable-length coding alone. (Recall from Section 8.1.1 that variable-length coding is used to reduce coding redundancies.) In addition, the differences between the higher-order estimates of entropy and the first-order estimate indicate the presence or absence of interpixel redundancies. That is, they reveal whether the pixels in an image are statistically independent. If the pixels are statistically independent (that is, there is no interpixel redundancy), the higher-order estimates are equivalent to the first-order estimate, and variable-length coding provides optimal compression. For the image considered in the preceding example, the numerical difference between the first- and second-order estimates indicates that a mapping can be created that allows an additional $1.81 - 1.25 = 0.56$ bits/pixel to be eliminated from the image's representation.

■ Consider mapping the pixels of the image in the preceding example to create the representation:

EXAMPLE 8.11:
Using mappings to reduce entropy.

21	0	0	74	74	74	0	0
21	0	0	74	74	74	0	0
21	0	0	74	74	74	0	0
21	0	0	74	74	74	0	0

Here, we construct a difference array by replicating the first column of the original image and using the arithmetic difference between adjacent columns for the remaining elements. For example, the element in the first row, second column of the new representation is $(21 - 21)$, or 0. The resulting difference distribution is

Gray Level or Difference	Count	Probability
0	12	1/2
21	4	1/8
74	12	3/8

If we now consider the mapped array to be generated by a "difference source," we can again use Eq. (8.3-3) to compute a first-order estimate of the entropy of the array, which is 1.41 bits/pixel. Thus by variable-length coding the mapped difference image, the original image can be represented with only 1.41 bits/pixel or a total of about 46 bits. This value is greater than the 1.25 bits/pixel second-order estimate of entropy computed in the preceding example, so we know that we can find an even better mapping. ■

The preceding examples illustrate that the first-order estimate of the entropy of an image is not necessarily the minimum code rate for the image. The reason is that pixels in an image generally are not statistically independent. As noted in Section 8.2, the process of minimizing the actual entropy of an image is called source coding. In the error-free case it encompasses the two operations of mapping and symbol coding. If information loss can be tolerated, it also includes the third step of quantization.

The slightly more complicated problem of lossy image compression can also be approached using the tools of information theory. In this case, however, the principal result is the source coding theorem. As indicated in Section 8.3.3, this theorem reveals that any zero-memory source can be encoded by using a code of rate $R < R(D)$ such that the average per symbol distortion is less than D. To apply this result correctly to lossy image compression requires identifying an appropriate source model, devising a meaningful distortion measure, and computing the resulting rate distortion function $R(D)$. The first step of this process has already been considered. The second step can be conveniently approached through the use of an objective fidelity criterion from Section 8.1.4. The final step involves finding a matrix $\mathbf{Q}$ whose elements minimize Eq. (8.3-12), subject to the constraints imposed by Eqs. (8.3-24) through (8.3-28). Unfortunately, this task is particularly difficult—and only a few cases of any practical interest have been solved. One is when the images are Gaussian random fields and the distortion measure is a weighted square error function. In this case, the optimal encoder must expand the image into its Karhunen-Loève components (see Section 11.4) and represent each component with equal mean-square error (Davisson [1972]).

8.4 Error-Free Compression

In numerous applications error-free compression is the only acceptable means of data reduction. One such application is the archival of medical or business documents, where lossy compression usually is prohibited for legal reasons. Another is the processing of satellite imagery, where both the use and cost of collecting the data makes any loss undesirable. Yet another is digital radiography, where the loss of information can compromise diagnostic accuracy. In these and other cases, the need for error-free compression is motivated by the intended use or nature of the images under consideration.

In this section, we focus on the principal error-free compression strategies currently in use. They normally provide compression ratios of 2 to 10. Moreover, they are equally applicable to both binary and gray-scale images. As indicated in Section 8.2, error-free compression techniques generally are composed of two relatively independent operations: (1) devising an alternative representation of the image in which its interpixel redundancies are reduced; and (2) coding the representation to eliminate coding redundancies. These steps correspond to the mapping and symbol coding operations of the source coding model discussed in connection with Fig. 8.6.

8.4.1 Variable-Length Coding

The simplest approach to error-free image compression is to reduce *only* coding redundancy. Coding redundancy normally is present in any natural binary encoding of the gray levels in an image. As we noted in Section 8.1.1, it can be eliminated by coding the gray levels so that Eq. (8.1-4) is minimized. To do so requires construction of a variable-length code that assigns the shortest possible code words to the most probable gray levels. Here, we examine several optimal and near optimal techniques for constructing such a code. These techniques are formulated in the language of information theory. In practice, the source symbols may be either the gray levels of an image or the output of a gray-level mapping operation (pixel differences, run lengths, and so on).

Huffman coding

The most popular technique for removing coding redundancy is due to Huffman (Huffman [1952]). When coding the symbols of an information source individually, *Huffman coding* yields the smallest possible number of code symbols per source symbol. In terms of the noiseless coding theorem (see Section 8.3.3), the resulting code is optimal for a fixed value of n, subject to the constraint that the source symbols be coded *one at a time*.

The first step in Huffman's approach is to create a series of source reductions by ordering the probabilities of the symbols under consideration and combining the lowest probability symbols into a single symbol that replaces them in the next source reduction. Figure 8.11 illustrates this process for binary coding (*K*-ary Huffman codes can also be constructed). At the far left, a hypothetical set of source symbols and their probabilities are ordered from top to bottom in terms of decreasing probability values. To form the first source reduction, the bottom two probabilities, 0.06 and 0.04, are combined to form a "compound symbol" with probability 0.1. This compound symbol and its associated probability are placed in the first source reduction column so that the probabilities of the reduced source are also ordered from the most to the least probable. This process is then repeated until a reduced source with two symbols (at the far right) is reached.

The second step in Huffman's procedure is to code each reduced source, starting with the smallest source and working back to the original source. The minimal length binary code for a two-symbol source, of course, is the symbols 0 and 1. As Fig. 8.12 shows, these symbols are assigned to the two symbols on the right (the assignment is arbitrary; reversing the order of the 0 and 1 would work just as well). As the reduced source symbol with probability 0.6 was generated by combining two symbols in the reduced source to its left, the 0 used to code it is now assigned to *both* of these symbols, and a 0 and 1 are arbitrarily

Original source		Source reduction			
Symbol	Probability	1	2	3	4
a_2	0.4	0.4	0.4	0.4	0.6
a_6	0.3	0.3	0.3	0.3	0.4
a_1	0.1	0.1	0.2	0.3	
a_4	0.1	0.1	0.1		
a_3	0.06	0.1			
a_5	0.04				

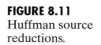

FIGURE 8.11
Huffman source reductions.

FIGURE 8.12
Huffman code
assignment
procedure.

	Original source		Source reduction			
Sym.	Prob.	Code	1	2	3	4
a_2	0.4	1	0.4　1	0.4　1	0.4　1	┌0.6　0
a_6	0.3	00	0.3　00	0.3　00	0.3　00 ◄┐	0.4　1
a_1	0.1	011	0.1　011	┌0.2　010 ◄	└0.3　01 ◄┘	
a_4	0.1	0100	0.1　0100 ◄┐	└0.1　011 ◄┘		
a_3	0.06	01010 ◄┐	└0.1　0101 ◄┘			
a_5	0.04	01011 ◄┘				

appended to each to distinguish them from each other. This operation is then repeated for each reduced source until the original source is reached. The final code appears at the far left in Fig. 8.12. The average length of this code is

$$L_{avg} = (0.4)(1) + (0.3)(2) + (0.1)(3) + (0.1)(4) + (0.06)(5) + (0.04)(5)$$
$$= 2.2 \text{ bits/symbol}$$

and the entropy of the source is 2.14 bits/symbol. In accordance with Eq. (8.3-21), the resulting Huffman code efficiency is 0.973.

Huffman's procedure creates the optimal code for a set of symbols and probabilities *subject to* the constraint that the symbols be coded one at a time. After the code has been created, coding and/or decoding is accomplished in a simple lookup table manner. The code itself is an instantaneous uniquely decodable *block code*. It is called a *block code* because each source symbol is mapped into a fixed sequence of code symbols. It is *instantaneous* because each code word in a string of code symbols can be decoded without referencing succeeding symbols. It is *uniquely decodable* because any string of code symbols can be decoded in only one way. Thus, any string of Huffman encoded symbols can be decoded by examining the individual symbols of the string in a left to right manner. For the binary code of Fig. 8.12, a left-to-right scan of the encoded string 010100111100 reveals that the first valid code word is 01010, which is the code for symbol a_3. The next valid code is 011, which corresponds to symbol a_1. Continuing in this manner reveals the completely decoded message to be $a_3 a_1 a_2 a_2 a_6$.

Other near optimal variable length codes

When a large number of symbols is to be coded, the construction of the optimal binary Huffman code is a nontrivial task. For the general case of J source symbols, $J - 2$ source reductions must be performed (see Fig. 8.11) and $J - 2$ code assignments made (see Fig. 8.12). Thus construction of the optimal Huffman code for an image with 256 gray levels requires 254 source reductions and 254 code assignments. In view of the computational complexity of this task, sacrificing coding efficiency for simplicity in code construction sometimes is necessary.

Table 8.5 illustrates four variable-length codes that provide such a trade-off. Note that the average length of the Huffman code—the last row of the table—is lower than the other codes listed. The natural binary code has the greatest average length. In addition, the 4.05 bits/symbol code rate achieved by Huffman's technique approaches the 4.0 bits/symbol entropy bound of the source, com-

TABLE 8.5
Variable-length
codes.

Source symbol	Probability	Binary Code	Huffman	Truncated Huffman	B₂-Code	Binary Shift	Huffman Shift
Block 1							
a_1	0.2	00000	10	11	C00	000	10
a_2	0.1	00001	110	011	C01	001	11
a_3	0.1	00010	111	0000	C10	010	110
a_4	0.06	00011	0101	0101	C11	011	100
a_5	0.05	00100	00000	00010	C00C00	100	101
a_6	0.05	00101	00001	00011	C00C01	101	1110
a_7	0.05	00110	00010	00100	C00C10	110	1111
Block 2							
a_8	0.04	00111	00011	00101	C00C11	111 000	00 10
a_9	0.04	01000	00110	00110	C01C00	111 001	00 11
a_{10}	0.04	01001	00111	00111	C01C01	111 010	00 110
a_{11}	0.04	01010	00100	01000	C01C10	111 011	00 100
a_{12}	0.03	01011	01001	01001	C01C11	111 100	00 101
a_{13}	0.03	01100	01110	10 0000	C10C00	111 101	00 1110
a_{14}	0.03	01101	01111	10 0001	C10C01	111 110	00 1111
Block 3							
a_{15}	0.03	01110	01100	10 0010	C10C10	111 111 000	00 00 10
a_{16}	0.02	01111	010000	10 0011	C10C11	111 111 001	00 00 11
a_{17}	0.02	10000	010001	10 0100	C11C00	111 111 010	00 00 110
a_{18}	0.02	10001	001010	10 0101	C11C01	111 111 011	00 00 100
a_{19}	0.02	10010	001011	10 0110	C11C10	111 111 100	00 00 101
a_{20}	0.02	10011	011010	10 0111	C11C11	111 111 101	00 00 1110
a_{21}	0.01	10100	011011	10 1000	C00C00C00	111 111 110	00 00 1111
Entropy	4.0						
Average length		5.0	4.05	4.24	4.65	4.59	4.13

puted by using Eq. (8.3-3) and given at the bottom of the table. Although none of the remaining codes in Table 8.5 achieve the Huffman coding efficiency, all are easier to construct. Like Huffman's technique, they assign the shortest code words to the most likely source symbols.

Column 5 of Table 8.5 illustrates a simple modification of the basic Huffman coding strategy known as *truncated Huffman coding*. A truncated Huffman code is generated by Huffman coding only the most probable ψ symbols of the source, for some positive integer ψ less than J. A prefix code followed by a suitable fixed-length code is used to represent all other source symbols. In Table 8.5, ψ arbitrarily was selected as 12 and the prefix code was generated as the 13th Huffman code word. That is, a "prefix symbol" whose probability was the sum of the probabilities of symbols a_{13} through a_{21} was included as a 13th symbol during the Huffman coding of the 12 most probable source symbols. The remaining 9 symbols were then coded using the prefix code, which turned out to be 10, and a 4-bit binary value equal to the symbol subscript minus 13.

Column 6 of Table 8.5 illustrates a second, near optimal, and variable-length code known as a B-code. It is close to optimal when the source symbol probabilities obey a power law of the form

$$P(a_j) = cj^{-\beta} \tag{8.4-1}$$

for some positive constant β and normalizing constant $c = 1/\sum_{j=0}^{J} j^{-\beta}$. For example, the distribution of run lengths in a binary representation of a typical typewritten text document is nearly exponential. As Table 8.5 shows, each code word is made up of *continuation* bits, denoted C, and *information* bits, which are natural binary numbers. The only purpose of the continuation bits is to separate individual code words, so they simply alternate between 0 and 1 for each code word in a string. The B-code shown in Table 8.5 is called a B_2-code, because two information bits are used per continuation bit. The sequence of B_2-codes corresponding to the source symbol string $a_{11}a_2a_7$ is 001 010 101 000 010 or 101 110 001 100 110, depending on whether the first continuation bit is assumed to be 0 or 1.

The two remaining variable-length codes in Table 8.5 are referred to as *shift codes*. A shift code is generated by (1) arranging the source symbols so that their probabilities are monotonically decreasing, (2) dividing the total number of symbols into symbol blocks of equal size, (3) coding the individual elements within all blocks identically, and (4) adding special *shift-up* and/or *shift-down* symbols to identify each block. Each time a shift-up or shift-down symbol is recognized at the decoder, it moves one block up or down with respect to a predefined reference block.

To generate the 3-bit binary shift code in column 7 of Table 8.5, the 21 source symbols are first ordered in accordance with their probabilities of occurrence and divided into three blocks of seven symbols. The individual symbols (a_1 through a_7) of the upper block—considered the reference block—are then coded with the binary codes 000 through 110. The eighth binary code (111) is not included in the reference block; instead, it is used as a single shift-up control that identifies the remaining blocks (in this case, a shift-down symbol is not used). The symbols in the remaining two blocks are then coded by one or two shift-up symbols in combination with the binary codes used to code the reference block. For example, source symbol a_{19} is coded as 111 111 100.

The Huffman shift code in column 8 of Table 8.5 is generated in a similar manner. The principal difference is in the assignment of a probability to the shift symbol prior to Huffman coding the reference block. Normally, this assignment is accomplished by summing the probabilities of all the source symbols outside the reference block; that is, by using the same concept utilized to define the prefix symbol in the truncated Huffman code. Here, the sum is taken over symbols a_8 through a_{21} and is 0.39. The shift symbol is thus the most probable symbol and is assigned one of the shortest Huffman code words (00).

Arithmetic coding

Unlike the variable-length codes described previously, *arithmetic coding* generates nonblock codes. In arithmetic coding, which can be traced to the work of Elias (see Abramson [1963]), a one-to-one correspondence between source symbols and code words does not exist. Instead, an entire sequence of source symbols (or message) is assigned a single arithmetic code word. The code word itself defines an interval of real numbers between 0 and 1. As the number of symbols in the message increases, the interval used to represent it becomes smaller

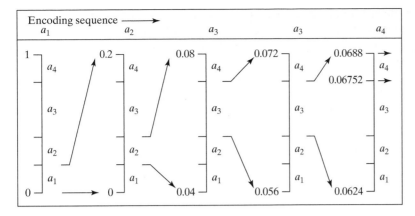

FIGURE 8.13
Arithmetic coding
procedure.

and the number of information units (say, bits) required to represent the interval becomes larger. Each symbol of the message reduces the size of the interval in accordance with its probability of occurrence. Because the technique does not require, as does Huffman's approach, that each source symbol translate into an integral number of code symbols (that is, that the symbols be coded one at a time), it achieves (but only in theory) the bound established by the noiseless coding theorem of Section 8.3.3.

Figure 8.13 illustrates the basic arithmetic coding process. Here, a five-symbol sequence or message, $a_1a_2a_3a_3a_4$, from a four-symbol source is coded. At the start of the coding process, the message is assumed to occupy the entire half-open interval $[0, 1)$. As Table 8.6 shows, this interval is initially subdivided into four regions based on the probabilities of each source symbol. Symbol a_1, for example, is associated with subinterval $[0, 0.2)$. Because it is the first symbol of the message being coded, the message interval is initially narrowed to $[0, 0.2)$. Thus in Fig. 8.13 $[0, 0.2)$ is expanded to the full height of the figure and its end points labeled by the values of the narrowed range. The narrowed range is then subdivided in accordance with the original source symbol probabilities and the process continues with the next message symbol. In this manner, symbol a_2 narrows the subinterval to $[0.04, 0.08)$, a_3 further narrows it to $[0.056, 0.072)$, and so on. The final message symbol, which must be reserved as a special end-of-message indicator, narrows the range to $[0.06752, 0.0688)$. Of course, any number within this subinterval—for example, 0.068—can be used to represent the message.

Source Symbol	Probability	Initial Subinterval
a_1	0.2	$[0.0, 0.2)$
a_2	0.2	$[0.2, 0.4)$
a_3	0.4	$[0.4, 0.8)$
a_4	0.2	$[0.8, 1.0)$

TABLE 8.6
Arithmetic coding
example.

In the arithmetically coded message of Fig. 8.13, three decimal digits are used to represent the five-symbol message. This translates into $^3/_5$ or 0.6 decimal digits per source symbol and compares favorably with the entropy of the source, which, from Eq. (8.3-3), is 0.58 decimal digits or 10-ary units/symbol. As the length of the sequence being coded increases, the resulting arithmetic code approaches the bound established by the noiseless coding theorem. In practice, two factors cause coding performance to fall short of the bound: (1) the addition of the end-of-message indicator that is needed to separate one message from another; and (2) the use of finite precision arithmetic. Practical implementations of arithmetic coding address the latter problem by introducing a scaling strategy and a rounding strategy (Langdon and Rissanen [1981]). The scaling strategy renormalizes each subinterval to the $[0, 1)$ range before subdividing it in accordance with the symbol probabilities. The rounding strategy guarantees that the truncations associated with finite precision arithmetic do not prevent the coding subintervals from being represented accurately.

8.4.2 LZW Coding

Having examined the principal methods for removing coding redundancy, we now consider one of several error-free compression techniques that also attack an image's interpixel redundancies. The technique, called Lempel-Ziv-Welch (LZW) coding, assigns fixed-length code words to variable length sequences of source symbols but requires no a priori knowledge of the probability of occurrence of the symbols to be encoded. Recall from Section 8.3.3 that Shannon's first theorem states that the nth extension of a zero-memory source can be coded with fewer average bits per source symbol than the nonextended source itself. Despite the fact that it must be licensed under United States Patent No. 4,558,302, LZW compression has been integrated into a variety of mainstream imaging file formats, including the *graphic interchange format* (GIF), *tagged image file format* (TIFF), and the *portable document format* (PDF).

LZW coding is conceptually very simple (Welch [1984]). At the onset of the coding process, a codebook or "dictionary" containing the source symbols to be coded is constructed. For 8-bit monochrome images, the first 256 words of the dictionary are assigned to the gray values $0, 1, 2, \ldots, 255$. As the encoder sequentially examines the image's pixels, gray-level sequences that are not in the dictionary are placed in algorithmically determined (e.g., the next unused) locations. If the first two pixels of the image are white, for instance, sequence "255-255" might be assigned to location 256, the address following the locations reserved for gray levels 0 through 255. The next time that two consecutive white pixels are encountered, code word 256, the address of the location containing sequence 255-255, is used to represent them. If a 9-bit, 512-word dictionary is employed in the coding process, the original $(8 + 8)$ bits that were used to represent the two pixels are replaced by a single 9-bit code word. Cleary, the size of the dictionary is an important system parameter. If it is too small, the detection of matching gray-level sequences will be less likely; if it is too large, the size of the code words will adversely affect compression performance.

■ Consider the following 4 × 4, 8-bit image of a vertical edge:

$$
\begin{array}{cccc}
39 & 39 & 126 & 126 \\
39 & 39 & 126 & 126 \\
39 & 39 & 126 & 126 \\
39 & 39 & 126 & 126
\end{array}
$$

Table 8.7 details the steps involved in coding its 16 pixels. A 512-word dictionary with the following starting content is assumed:

Dictionary Location	Entry
0	0
1	1
⋮	⋮
255	255
256	—
⋮	⋮
511	—

Locations 256 through 511 are initially unused.

The image is encoded by processing its pixels in a left-to-right, top-to-bottom manner. Each successive gray-level value is concatenated with a variable—column 1 of Table 8.7—called the "currently recognized sequence." As can be seen, this variable is initially null or empty. The dictionary is searched for each

Currently Recognized Sequence	Pixel Being Processed	Encoded Output	Dictionary Location (Code Word)	Dictionary Entry
	39			
39	39	39	256	39-39
39	126	39	257	39-126
126	126	126	258	126-126
126	39	126	259	126-39
39	39			
39-39	126	256	260	39-39-126
126	126			
126-126	39	258	261	126-126-39
39	39			
39-39	126			
39-39-126	126	260	262	39-39-126-126
126	39			
126-39	39	259	263	126-39-39
39	126			
39-126	126	257	264	39-126-126
126		126		

TABLE 8.7
LZW coding
example.

concatenated sequence and if found, as was the case in the first row of the table, is replaced by the newly concatenated and recognized (i.e., located in the dictionary) sequence. This was done in column 1 of row 2. No output codes are generated, nor is the dictionary altered. If the concatenated sequence is not found, however, the address of the currently recognized sequence is output as the next encoded value, the concatenated but unrecognized sequence is added to the dictionary, and the currently recognized sequence is initialized to the current pixel value. This occurred in row 2 of the table. The last two columns detail the gray-level sequences that are added to the dictionary when scanning the entire 4 × 4 image. Nine additional code words are defined. At the conclusion of coding, the dictionary contains 265 code words and the LZW algorithm has successfully identified several repeating gray-level sequences—leveraging them to reduce the original 128-bit image to 90 bits (i.e., 10 9-bit codes). The encoded output is obtained by reading the third column from top to bottom. The resulting compression ratio is 1.42:1.   ■

A unique feature of the LZW coding just demonstrated is that the coding dictionary or code book is created while the data are being encoded. Remarkably, an LZW decoder builds an identical decompression dictionary as it decodes simultaneously the encoded data stream. It is left as an exercise to the reader (see Problem 8.16) to decode the output of the preceding example and reconstruct the code book. Although not needed in this example, most practical applications require a strategy for handling dictionary overflow. A simple solution is to flush or reinitialize the dictionary when it becomes full and continue coding with a new initialized dictionary. A more complex option is to monitor compression performance and flush the dictionary when it becomes poor or unacceptable. Alternately, the least used dictionary entries can be tracked and replaced when necessary.

8.4.3 Bit-Plane Coding

Another effective technique for reducing an image's interpixel redundancies is to process the image's bit planes individually. The technique, called *bit-plane coding*, is based on the concept of decomposing a multilevel (monochrome or color) image into a series of binary images and compressing each binary image via one of several well-known binary compression methods. In this section, we describe the most popular decomposition approaches and review several of the more commonly used compression methods.

Bit-plane decomposition

The gray levels of an *m*-bit gray-scale image can be represented in the form of the base 2 polynomial

$$a_{m-1}2^{m-1} + a_{m-2}2^{m-2} + \cdots + a_1 2^1 + a_0 2^0. \tag{8.4-2}$$

Based on this property, a simple method of decomposing the image into a collection of binary images is to separate the *m* coefficients of the polynomial into *m* 1-bit *bit planes*. As noted in Chapter 3, the zeroth-order bit plane is generated by collecting the a_0 bits of each pixel, while the $(m - 1)$st-order bit plane contains the a_{m-1} bits or coefficients. In general, each bit plane is numbered from 0 to $m - 1$ and is constructed by setting its pixels equal to the values of the appropriate bits or polynomial coefficients from each pixel in the original image. The inherent disadvantage of this approach is that small changes in gray level can have a significant impact on the complexity of the bit planes. If a pixel of intensity 127 (01111111) is adjacent to a pixel of intensity 128 (10000000), for instance, every bit plane will contain a corresponding 0 to 1 (or 1 to 0) transition. For example, as the most significant bits of the two binary codes for 127 and 128 are different, bit plane 7 will contain a zero-valued pixel next to a pixel of value 1, creating a 0 to 1 (or 1 to 0) transition at that point.

An alternative decomposition approach (which reduces the effect of small gray-level variations) is to first represent the image by an m-bit *Gray code*. The m-bit Gray code $g_{m-1} \cdots g_2 g_1 g_0$ that corresponds to the polynomial in Eq. (8.4-2) can be computed from

$$g_i = a_i \oplus a_{i+1} \qquad 0 \le i \le m - 2 \qquad (8.4\text{-}3)$$

$$g_{m-1} = a_{m-1}.$$

Here, $\oplus$ denotes the exclusive OR operation. This code has the unique property that successive code words differ in only one bit position. Thus, small changes in gray level are less likely to affect all m bit planes. For instance, when gray levels 127 and 128 are adjacent, only the 7th bit plane will contain a 0 to 1 transition, because the Gray codes that correspond to 127 and 128 are 11000000 and 01000000, respectively.

■ The 1024 × 1024 images shown in Figs. 8.14(a) and (b) are used to illustrate the compression techniques described in the remainder of this section. The 8-bit monochrome image of a child was generated with a high-resolution CCD camera. The binary image of a warranty deed prepared by President Andrew Jackson in 1796 was produced on a flatbed document scanner. Figures 8.15 and 8.16 show the eight binary and Gray-coded bit planes of the image of the child.

EXAMPLE 8.13:
Bit-plane coding.

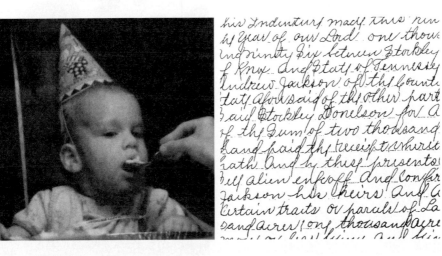

a b

FIGURE 8.14 A 1024 × 1024 (a) 8-bit monochrome image and (b) binary image.

FIGURE 8.15 The four most significant binary (left column) and Gray-coded (right column) bit planes of the image in Fig. 8.14(a).

Bit 7

Bit 6

Bit 5

Bit 4

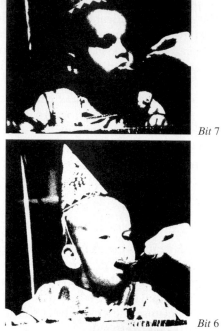

Bit 7

Bit 6

Bit 5

Bit 4

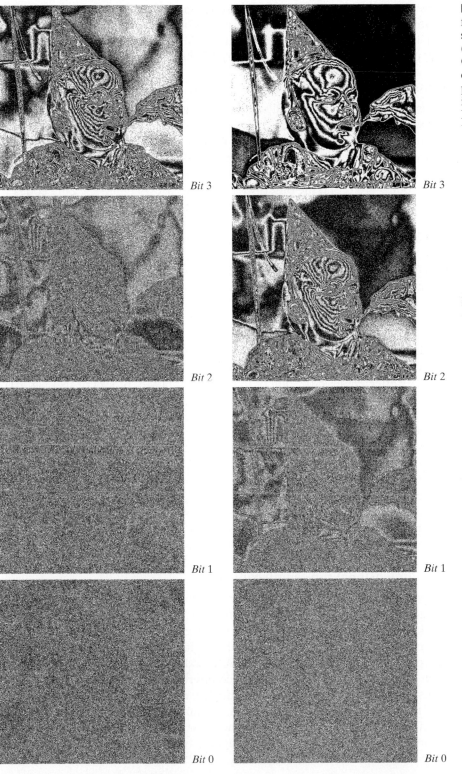

Bit 3   *Bit* 3

Bit 2   *Bit* 2

Bit 1   *Bit* 1

Bit 0   *Bit* 0

FIGURE 8.16 The four least significant binary (left column) and Gray-coded (right column) bit planes of the image in Fig. 8.14(a).

Note that the high-order bit planes are far less complex than their low-order counterparts. That is, they contain large uniform areas of significantly less detail, busyness, or randomness. In addition, the Gray-coded bit planes are less complex than the corresponding binary bit planes. ■

Constant area coding

A simple but effective method of compressing a binary image or bit plane is to use special code words to identify large areas of contiguous 1's or 0's. In one such approach, called *constant area coding* (CAC), the image is divided into blocks of size $p \times q$ pixels, which are classified as all white, all black, or mixed intensity. The most probable or frequently occurring category is then assigned the 1-bit code word 0, and the other two categories are assigned the 2-bit codes 10 and 11. Compression is achieved because the pq bits that normally would be used to represent each constant area are replaced by a 1-bit or 2-bit code word. Of course, the code assigned to the mixed intensity category is used as a prefix, which is followed by the pq-bit pattern of the block.

When predominantly white text documents are being compressed, a slightly simpler approach is to code the solid white areas as 0 and all other blocks (including the solid black blocks) by a 1 followed by the bit pattern of the block. This approach, called *white block skipping* (WBS), takes advantage of the anticipated structural tendencies of the image to be compressed. As few solid black areas are expected, they are grouped with the mixed intensity regions, allowing a 1-bit code word to be used for the highly probable white blocks. A particularly effective modification of this procedure (with blocks of size $1 \times q$) is to code the solid white lines as 0's and all other lines with a 1 followed by the normal WBS code sequence. Another is to employ an iterative approach in which the binary image or bit plane is decomposed into successively smaller and smaller subblocks. For 2-D blocks, a solid white image is coded as a 0, and all other images are divided into subblocks that are assigned a prefix of 1 and similarly coded. That is, if a subblock is all white, it is represented by the prefix 1, indicating that it is a first iteration subblock, followed by a 0, indicating that it is solid white. If the subblock is not solid white, the decomposition process is repeated until a predefined subblock size is reached and coded as either a 0 (if it is all white) or a 1 followed by the block bit pattern.

One-dimensional run-length coding

An effective alternative to constant area coding is to represent each row of an image or bit plane by a sequence of lengths that describe successive runs of black and white pixels. This technique, referred to as *run-length coding*, was developed in the 1950s and has become, along with its 2-D extensions, the standard compression approach in facsimile (FAX) coding. The basic concept is to code each contiguous group of 0's or 1's encountered in a left to right scan of a row by its length and to establish a convention for determining the value of the run. The most common approaches for determining the value of a run are (1) to specify the value of the first run of each row, or (2) to assume that each row begins with a white run, whose run length may in fact be zero.

Although run-length coding is in itself an effective method of compressing an image (see the example in Section 8.1.2), additional compression usually can be realized by variable-length coding the run lengths themselves. In fact, the black and white run lengths may be coded separately using variable-length codes that are specifically tailored to their own statistics. For example, letting symbol a_j represent a black run of length j, we can estimate the probability that symbol a_j was emitted by an imaginary black run-length source by dividing the number of black run lengths of length j in the entire image by the total number of black runs. An estimate of the entropy of this black run-length source, denoted H_0, follows by substituting these probabilities into Eq. (8.3-3). A similar argument holds for the entropy of the white runs, denoted H_1. The approximate run-length entropy of the image is

$$H_{\text{RL}} = \frac{H_0 + H_1}{L_0 + L_1} \qquad (8.4\text{-}4)$$

where the variables L_0 and L_1 denote the average values of black and white run lengths, respectively. Equation (8.4-4) provides an estimate of the average number of bits per pixel required to code the run lengths in a binary image using a variable-length code.

Two-dimensional run-length coding

One-dimensional run-length coding concepts are easily extended to create a variety of 2-D coding procedures. One of the better known results is *relative address coding* (RAC), which is based on the principle of tracking the binary transitions that begin and end each black and white run. Figure 8.17(a) illustrates one implementation of this approach. Note that ec is the distance from the current transition c to the last transition of the current line e, whereas cc' is the distance from c to the first similar (in the same direction) transition past e, denoted c', on the previous line. If $ec \leq cc'$, the RAC coded distance d is set equal to ec and used to represent the current transition at c. But if $cc' < ec$, d is set equal to cc'.

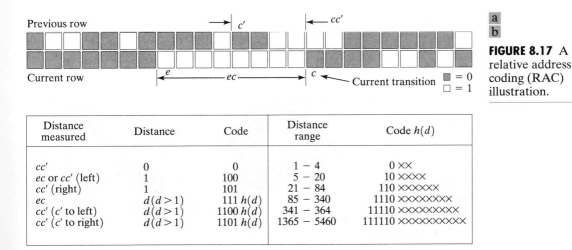

a
b

FIGURE 8.17 A relative address coding (RAC) illustration.

Distance measured	Distance	Code	Distance range	Code $h(d)$
cc'	0	0	1 – 4	0 ××
ec or cc' (left)	1	100	5 – 20	10 ××××
cc' (right)	1	101	21 – 84	110 ××××××
ec	$d\,(d>1)$	111 $h(d)$	85 – 340	1110 ××××××××
cc' (c' to left)	$d\,(d>1)$	1100 $h(d)$	341 – 364	11110 ××××××××××
cc' (c' to right)	$d\,(d>1)$	1101 $h(d)$	1365 – 5460	111110 ××××××××××××

Like run-length coding, relative address coding requires the adoption of a convention for determining run values. In addition, imaginary transitions at the start and end of each line, as well as an imaginary starting line (say, an all-white line), must be assumed so that image boundaries can be handled properly. Finally, since the probability distributions of the RAC distances of most images are not uniform in practice (see Section 8.1.1), the final step of the RAC process is to code the RAC distance selected (that is, the shortest) and its distance, d, by using a suitable variable-length code. As Fig. 8.17(b) shows, a code similar to a B_1-code can be utilized. The smallest distances are assigned the shortest code words, and all other distances are coded by using a prefix to indicate the shortest RAC distance, a second prefix that assigns d to a specific range of distances, and the binary representation [denoted $\times\times\times \ldots \times$ in Fig. 8.17(b)] of d minus the base distance of the range itself. If ec and cc' are +8 and +4, as in Fig. 8.17(a), the proper RAC code word is 1100011. Finally, if $d = 0$, c is directly below c', whereas if $d = 1$, the decoder may have to determine the closest transition point, because the 100 code does not specify whether the measurement is relative to the current row or to the previous row.

Contour tracing and coding

Relative address coding is one approach for representing intensity transitions that make up the contours in a binary image. Another approach is to represent each contour by a set of boundary points or by a single boundary point and a set of directionals. The latter sometimes is referred to as *direct contour tracing*. In this section, we describe yet another method, called *predictive differential quantizing* (PDQ), which demonstrates the essential characteristics of both approaches. It is a scan-line-oriented contour tracing procedure.

In predictive differential quantizing, the front and back contours (Fig. 8.18) of each object of an image are traced simultaneously to generate a sequence of pairs (Δ', Δ''). The term Δ' is the difference between the starting coordinates of the front contours on adjacent lines, and Δ'' is the difference between the front-to-back contour lengths. These differences, together with special messages that indicate the start of new contours (the *new start* message) and the end of old contours (the *merge* message), represent each object. If Δ'' is replaced by the

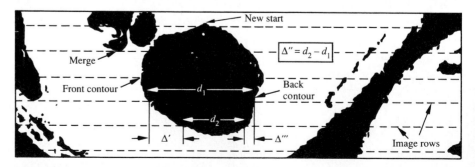

FIGURE 8.18 Parameters of the PDQ algorithm.

difference between the back contour coordinates of adjacent lines, denoted Δ''', the technique is referred to as *double delta coding* (DDC).

The new start and merge messages allow the (Δ', Δ'') or (Δ', Δ''') pairs generated on a scan line basis to be linked properly to the corresponding pairs in the previous and next rows. Without these messages, the decoder would be unable to reference one difference pair to another or to position correctly the contours within the image. To avoid encoding both the row and column coordinates of each new start and merge message, a unique code often is used to identify scan lines that do not contain object pixels. The final step in both PDQ and DDC coding is to represent Δ', Δ'' or Δ''', and the coordinates of the new starts and merges with a suitable variable-length code.

■ We conclude this section by comparing the binary compression techniques described previously. Each approach was used to compress the images of Fig. 8.14. The resulting code rates and compression ratios are provided in Tables 8.8 and 8.9. When interpreting these results, note that first-order estimates (see Section 8.3.4) of the entropies of the RLC run lengths and PDQ and DDC distances were computed and used as an approximation of the compression performance that could be achieved under the variable-length coding approaches of Section 8.4.1.

The results in Tables 8.8 and 8.9 show that all the techniques were able to eliminate some amount of interpixel redundancy. That is, the resulting code rates

EXAMPLE 8.14:
Comparison of binary compression techniques.

Method	Bit-plane code rate (bits/pixel)								Code Rate	Compression Ratio
	7	6	5	4	3	2	1	0		
Binary Bit-Plane Coding										
CBC (4 × 4)	0.14	0.24	0.60	0.79	0.99	—	—	—	5.75	1.4:1
RLC	0.09	0.19	0.51	0.68	0.87	1.00	1.00	1.00	5.33	1.5:1
PDQ	0.07	0.18	0.79					—	6.04	1.3:1
DDC	0.07	0.18	0.79	—	—	—	—	—	6.03	1.3:1
RAC	0.06	0.15	0.62	0.91	—	—	—	—	5.17	1.4:1
Gray Bit-Plane Coding										
CBC (4 × 4)	0.14	0.18	0.48	0.40	0.61	0.98	—	—	4.80	1.7:1
RLC	0.09	0.13	0.40	0.33	0.51	0.85	1.00	1.00	4.29	1.9:1
PDQ	0.07	0.12	0.61	0.40	0.82	—	—	—	5.02	1.6:1
DDC	0.07	0.11	0.61	0.40	0.81	—	—	—	5.00	1.6:1
RAC	0.06	0.10	0.49	0.31	0.62	—	—	—	4.05	1.8:1

TABLE 8.8
Error-free bit-plane coding results for Fig. 8.14(a): $H \approx 6.82$ bits/pixel

	WBS (1 × 8)	WBS (4 × 4)	RLC	PDQ	DDC	RAC
Code rate (bits/pixel)	0.48	0.39	0.32	0.23	0.22	0.23
Compression ratio	2.1:1	2.6:1	3.1:1	4.4:1	4.7:1	4.4:1

TABLE 8.9
Error-free binary image compression results for Fig. 8.14(b): $H \approx 0.55$ bits/pixel.

were less than the first-order entropy estimate of each image. Run-length coding proved to be the best coding method for the bit-plane coded images, whereas the 2-D techniques (such as PDQ, DDC, and RAC) performed better when compressing the binary image. Furthermore, the relatively straightforward procedure of Gray coding the image of Fig. 8.14(a) improved the achievable coding performance by about 1 bit/pixel. Finally, note that the five compression methods were able to compress the monochrome image only by a factor of 1 to 2, while compressing the binary image of Fig. 8.14(b) by a factor of 2 to 5. As Table 8.8 shows, the reason for this performance difference is that the algorithms were unable to compress the lower-order bit planes of the bit-plane coded images. In fact, the dashed entries of the table indicate instances in which the algorithms caused data expansion. In these cases, the raw data were used to represent the bit plane, and only 1 bit/pixel was added to the total code rate.   ■

8.4.4 Lossless Predictive Coding

Let us now turn to an error-free compression approach that does not require decomposition of an image into a collection of bit planes. The approach, commonly referred to as *lossless predictive coding*, is based on eliminating the interpixel redundancies of closely spaced pixels by extracting and coding *only* the new information in each pixel. The *new information* of a pixel is defined as the difference between the actual and predicted value of that pixel.

Figure 8.19 shows the basic components of a lossless predictive coding system. The system consists of an encoder and a decoder, each containing an identical *predictor*. As each successive pixel of the input image, denoted f_n, is introduced to the encoder, the predictor generates the anticipated value of that pixel based on some number of past inputs. The output of the predictor is then rounded to the nearest integer, denoted $\hat{f}_n$, and used to form the difference or *prediction error*

$$e_n = f_n - \hat{f}_n, \tag{8.4-5}$$

a
b

FIGURE 8.19 A lossless predictive coding model: (a) encoder; (b) decoder.

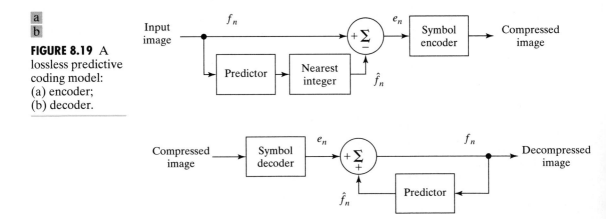

which is coded using a variable-length code (by the symbol encoder) to generate the next element of the compressed data stream. The decoder of Fig. 8.19(b) reconstructs e_n from the received variable-length code words and performs the inverse operation

$$f_n = e_n + \hat{f}_n. \tag{8.4-6}$$

Various local, global, and adaptive (see Section 8.5.1) methods can be used to generate $\hat{f}_n$. In most cases, however, the prediction is formed by a linear combination of m previous pixels. That is,

$$\hat{f}_n = \text{round}\left[\sum_{i=1}^{m} \alpha_i f_{n-i} \right] \tag{8.4-7}$$

where m is the order of the linear predictor, round is a function used to denote the rounding or nearest integer operation, and the α_i for $i = 1, 2, \ldots, m$ are prediction coefficients. In raster scan applications, the subscript n indexes the predictor outputs in accordance with their time of occurrence. That is, $f_n, \hat{f}_n$, and e_n in Eqs. (8.4-5) through (8.4-7) could be replaced with the more explicit notation $f(t), \hat{f}(t)$, and $e(t)$, where t represents time. In other cases, n is used as an index on the spatial coordinates and/or frame number (in a time sequence of images) of an image. In 1-D linear predictive coding, for example, Eq. (8.4-7) can be written

$$\hat{f}(x, y) = \text{round}\left[\sum_{i=1}^{m} \alpha_i f(x, y - i) \right] \tag{8.4-8}$$

where each subscripted variable is now expressed explicitly as a function of spatial coordinates x and y. Note that Eq. (8.4-8) indicates that the 1-D linear prediction $\hat{f}(x, y)$ is a function of the previous pixels on the current line alone. In 2-D predictive coding, the prediction is a function of the previous pixels in a left-to-right, top-to-bottom scan of an image. In the 3-D case, it is based on these pixels and the previous pixels of preceding frames. Equation (8.4-8) cannot be evaluated for the first m pixels of each line, so these pixels must be coded by using other means (such as a Huffman code) and considered as an overhead of the predictive coding process. A similar comment applies to the higher-dimensional cases.

■ Consider encoding the monochrome image of Fig. 8.14(a) using the simple first-order linear predictor

$$\hat{f}(x, y) = \text{round}\left[\alpha f(x, y - 1) \right]. \tag{8.4-9}$$

EXAMPLE 8.15:
Predictive coding.

A predictor of this general form commonly is called a *previous pixel* predictor, and the corresponding predictive coding procedure is referred to as *differential coding* or *previous pixel coding*. Figure 8.20(a) shows the prediction error image that results from Eq. (8.4-9) with $\alpha = 1$. In this image, gray-level 128 represents a prediction error of zero, whereas all nonzero positive and negative prediction errors (under and over estimates) are multiplied by 8 and displayed as lighter and darker shades of gray, respectively. The mean value of the prediction image is 128.02, which corresponds to an average prediction error of only 0.02 bits.

FIGURE 8.20
(a) The prediction error image resulting from Eq. (8.4-9).
(b) Gray-level histogram of the original image.
(c) Histogram of the prediction error.

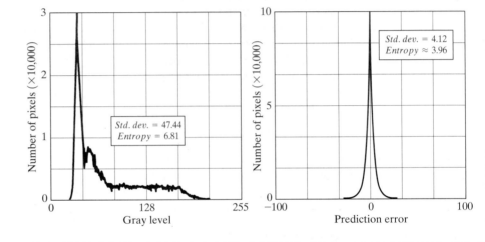

Figures 8.20(b) and (c) show the gray-level histogram of the image in Fig. 8.14(a) and the histogram of the prediction error resulting from Eq. (8.4-9). Note that the variance of the prediction error in Fig. 8.20(c) is much smaller than the variance of the gray levels in the original image. Moreover, the first-order estimate of the entropy of the prediction error image is significantly less than the corresponding first-order estimate for the original image (3.96 bits/pixel as opposed to 6.81 bits/pixel). This decrease in entropy reflects removal of a great deal of redundancy by the predictive coding process, despite the fact that for m-bit images, $(m + 1)$-bit numbers are needed to represent accurately the error sequence that results from Eq. (8.4-5). Although any of the variable-length coding procedures of Section 8.4.1 can be used to code this error sequence, the resulting compression will be limited to approximately 8/3.96, or about 2:1. In general, an estimate of the maximum compression of any lossless predictive coding approach may be obtained by dividing the average number of bits used to represent each pixel in the original image by a first-order estimate of the entropy of the prediction error data. ■

The preceding example emphasizes that the amount of compression achieved in lossless predictive coding is related directly to the entropy reduction that results from mapping the input image into the prediction error sequence. Because a great deal of interpixel redundancy is removed by the prediction and differencing process, the probability density function of the prediction error is, in general, highly peaked at zero and characterized by a relatively small (in comparison to the input gray-level distribution) variance. In fact, the density function of the prediction error often is modeled by the zero mean uncorrelated Laplacian pdf

$$p_e(e) = \frac{1}{\sqrt{2}\sigma_e} e^{\frac{-\sqrt{2}|e|}{\sigma_e}} \tag{8.4-10}$$

where σ_e is the standard deviation of e.

8.5 Lossy Compression

Unlike the error-free approaches outlined in the previous section, lossy encoding is based on the concept of compromising the accuracy of the reconstructed image in exchange for increased compression. If the resulting distortion (which may or may not be visually apparent) can be tolerated, the increase in compression can be significant. In fact, many lossy encoding techniques are capable of reproducing recognizable monochrome images from data that have been compressed by more than 100 : 1 and images that are virtually indistinguishable from the originals at 10 : 1 to 50 : 1. Error-free encoding of monochrome images, however, seldom results in more than a 3 : 1 reduction in data. As indicated in Section 8.2, the principal difference between these two approaches is the presence or absence of the quantizer block of Fig. 8.6.

8.5.1 Lossy Predictive Coding

In this section, we add a quantizer to the model introduced in Section 8.4.4 and examine the resulting trade-off between reconstruction accuracy and compression performance. As Fig. 8.21 shows, the quantizer, which absorbs the nearest integer function of the error free encoder, is inserted between the symbol

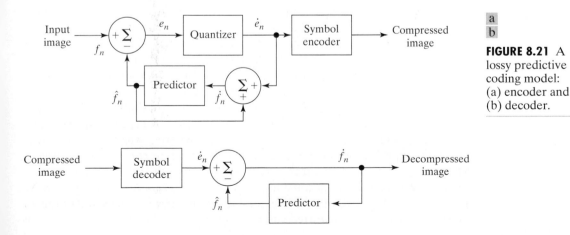

a
b

FIGURE 8.21 A lossy predictive coding model: (a) encoder and (b) decoder.

encoder and the point at which the prediction error is formed. It maps the prediction error into a limited range of outputs, denoted $\dot{e}_n$, which establish the amount of compression and distortion associated with lossy predictive coding.

In order to accommodate the insertion of the quantization step, the error-free encoder of Fig. 8.19(a) must be altered so that the predictions generated by the encoder and decoder are equivalent. As Fig. 8.21(a) shows, this is accomplished by placing the lossy encoder's predictor within a feedback loop, where its input, denoted $\dot{f}_n$, is generated as a function of past predictions and the corresponding quantized errors. That is,

$$\dot{f}_n = \dot{e}_n + \hat{f}_n \tag{8.5-1}$$

where $\hat{f}_n$ is as defined in Section 8.4.4. This closed loop configuration prevents error buildup at the decoder's output. Note from Fig. 8.21(b) that the output of the decoder also is given by Eq. (8.5-1).

EXAMPLE 8.16:
Delta modulation.

■ *Delta modulation* (DM) is a simple but well-known form of lossy predictive coding in which the predictor and quantizer are defined as

$$\hat{f}_n = \alpha \dot{f}_{n-1} \tag{8.5-2}$$

and

$$\dot{e}_n = \begin{cases} +\zeta & \text{for } e_n > 0 \\ -\zeta & \text{otherwise} \end{cases} \tag{8.5-3}$$

where α is a prediction coefficient (normally less than 1) and ζ is a positive constant. The output of the quantizer, $\dot{e}_n$, can be represented by a single bit (Fig. 8.22a), so the symbol encoder of Fig. 8.21(a) can utilize a 1-bit fixed-length code. The resulting DM code rate is 1 bit/pixel.

Figure 8.22(c) illustrates the mechanics of the delta modulation process, where the calculations needed to compress and reconstruct the input sequence $\{14, 15, 14, 15, 13, 15, 15, 14, 20, 26, 27, 28, 27, 27, 29, 37, 47, 62, 75, 77, 78, 79, 80,$ $81, 81, 82, 82\}$ with $\alpha = 1$ and $\zeta = 6.5$ are tabulated. The process begins with the error-free transfer of the first input pixel to the decoder. With the initial condition $\dot{f}_0 = f_0 = 14$ established at both the encoder and decoder, the remaining outputs can be computed by repeatedly evaluating Eqs. (8.5-2), (8.4-5), (8.5-3), and (8.5-1). Thus, when $n = 1$, for example, $\hat{f}_1 = (1)(14) = 14$, $e_1 = 15 -$ $14 = 1$, $\dot{e}_1 = +6.5$ (because $e_1 > 0$), $\dot{f}_1 = 6.5 + 14 = 20.5$, and the resulting reconstruction error is $(15 - 20.5)$, or -5.5 gray levels.

Figure 8.22(b) shows graphically the tabulated data shown in Fig. 8.22(c). Both the input and completely decoded output (f_n and $\dot{f}_n$) are shown. Note that in the rapidly changing area from $n = 14$ to 19, where ζ was too small to represent the input's largest changes, a distortion known as *slope overload* occurs. Moreover, when ζ was too large to represent the input's smallest changes, as in the relatively smooth region from $n = 0$ to $n = 7$, *granular noise* appears. In most images, these two phenomena lead to blurred object edges and grainy or noisy surfaces (that is, distorted smooth areas). ■

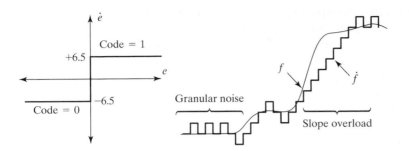

FIGURE 8.22 An example of delta modulation.

Input		Encoder				Decoder		Error
n	f	$\hat{f}$	e	$\dot{e}$	$\dot{f}$	$\hat{f}$	$\dot{f}$	$[f - \dot{f}]$
0	14	—	—	—	14.0	—	14.0	0.0
1	15	14.0	1.0	6.5	20.5	14.0	20.5	−5.5
2	14	20.5	−6.5	−6.5	14.0	20.5	14.0	0.0
3	15	14.0	1.0	6.5	20.5	14.0	20.5	−5.5
.	.	.	.	.	.	.	.	.
14	29	20.5	8.5	6.5	27.0	20.5	27.0	2.0
15	37	27.0	10.0	6.5	33.5	27.0	33.5	3.5
16	47	33.5	13.5	6.5	40.0	33.5	40.0	7.0
17	62	40.0	22.0	6.5	46.5	40.0	46.5	15.5
18	75	46.5	28.5	6.5	53.0	46.5	53.0	22.0
19	77	53.0	24.0	6.5	59.6	53.0	59.6	17.5
.	.	.	.	.	.	.	.	.

The distortions noted in the preceding example are common to all forms of lossy predictive coding. The severity of these distortions depends on a complex set of interactions between the quantization and prediction methods employed. Despite these interactions, the predictor normally is designed under the assumption of no quantization error, and the quantizer is designed to minimize its own error. That is, the predictor and quantizer are designed independently of each other.

Optimal predictors

The optimal predictor used in most predictive coding applications minimizes the encoder's mean-square prediction error[†]

$$E\{e_n^2\} = E\{[f_n - \hat{f}_n]^2\} \tag{8.5-4}$$

subject to the constraint that

$$\dot{f}_n = \dot{e}_n + \hat{f}_n \approx e_n + \hat{f}_n = f_n \tag{8.5-5}$$

and

$$\hat{f}_n = \sum_{i=1}^{m} \alpha_i f_{n-i}. \tag{8.5-6}$$

[†] The notation $E\{\cdot\}$ denotes the statistical expectation operator.

That is, the optimization criterion is chosen to minimize the mean-square prediction error, the quantization error is assumed to be negligible $(\dot{e}_n \approx e_n)$, and the prediction is constrained to a linear combination of m previous pixels.[†] These restrictions are not essential, but they simplify the analysis considerably and, at the same time, decrease the computational complexity of the predictor. The resulting predictive coding approach is referred to as *differential pulse code modulation* (DPCM).

Under these conditions, the optimal predictor design problem is reduced to the relatively straightforward exercise of selecting the m prediction coefficients that minimize the expression

$$E\{e_n^2\} = E\left\{\left[f_n - \sum_{i=1}^{m} \alpha_i f_{n-1}\right]^2\right\}. \tag{8.5-7}$$

Differentiating Eq. (8.5-7) with respect to each coefficient, equating the derivatives to zero, and solving the resulting set of simultaneous equations under the assumption that f_n has mean zero and variance σ^2 yields

$$\boldsymbol{\alpha} = \mathbf{R}^{-1}\mathbf{r} \tag{8.5-8}$$

where $\mathbf{R}^{-1}$ is the inverse of the $m \times m$ autocorrelation matrix

$$\mathbf{R} = \begin{bmatrix} E\{f_{n-1}f_{n-1}\} & E\{f_{n-1}f_{n-2}\} & \cdots & E\{f_{n-1}f_{n-m}\} \\ E\{f_{n-2}f_{n-1}\} & \cdot & \cdots & \cdot \\ \cdot & \cdot & \cdots & \cdot \\ \cdot & & & \\ E\{f_{n-m}f_{n-1}\} & E\{f_{n-m}f_{n-2}\} & \cdots & E\{f_{n-m}f_{n-m}\} \end{bmatrix} \tag{8.5-9}$$

and $\mathbf{r}$ and $\boldsymbol{\alpha}$ are the m-element vectors

$$\mathbf{r} = \begin{bmatrix} E\{f_n f_{n-1}\} \\ E\{f_n f_{n-2}\} \\ \vdots \\ E\{f_n f_{n-m}\} \end{bmatrix} \quad \text{and} \quad \boldsymbol{\alpha} = \begin{bmatrix} \alpha_1 \\ \alpha_2 \\ \vdots \\ \alpha_m \end{bmatrix}. \tag{8.5-10}$$

Thus for any input image, the coefficients that minimize Eq. (8.5-7) can be determined via a series of elementary matrix operations. Moreover, the coefficients depend only on the autocorrelations of the pixels in the original image. The variance of the prediction error that results from the use of these optimal coefficients is

$$\sigma_e^2 = \sigma^2 - \boldsymbol{\alpha}^T\mathbf{r} = \sigma^2 - \sum_{i=1}^{m} E\{f_n f_{n-i}\}\alpha_i. \tag{8.5-11}$$

Although the mechanics of evaluating Eq. (8.5-8) are quite simple, computation of the autocorrelations needed to form $\mathbf{R}$ and $\mathbf{r}$ is so difficult in practice

---

[†] In general, the optimal predictor for a non-Gaussian image is a nonlinear function of the pixels used to form the estimate.

that *local* predictions (those in which the prediction coefficients are computed image by image) are almost never used. In most cases, a set of *global* coefficients is computed by assuming a simple image model and substituting the corresponding autocorrelations into Eqs. (8.5-9) and (8.5-10). For instance, when a 2-D Markov source (see Section 8.3.3) with separable autocorrelation function

$$E\{f(x, y)f(x - i, y - i)\} = \sigma^2 \rho_v^i \rho_h^i \qquad (8.5\text{-}12)$$

and generalized fourth-order linear predictor

$$\hat{f}(x, y) = \alpha_1 f(x, y - 1) + \alpha_2 f(x - 1, y - 1) \qquad (8.5\text{-}13)$$
$$+ \alpha_3 f(x - 1, y) + \alpha_4 f(x - 1, y + 1)$$

are assumed, the resulting optimal coefficients (Jain [1989]) are

$$\alpha_1 = \rho_h \quad \alpha_2 = -\rho_v \rho_h \quad \alpha_3 = \rho_v \quad \alpha_4 = 0 \qquad (8.5\text{-}14)$$

where ρ_h and ρ_v are the horizontal and vertical correlation coefficients, respectively, of the image under consideration.

Finally, the sum of the prediction coefficients in Eq. (8.5-6) normally is required to be less than or equal to one. That is,

$$\sum_{i=1}^{m} \alpha_i \leq 1. \qquad (8.5\text{-}15)$$

This restriction is made to ensure that the predictor's output falls within the allowed range of gray levels and to reduce the impact of transmission noise, which is generally seen as horizontal streaks in the reconstructed image. Reducing the DPCM decoder's susceptibility to input noise is important, because a single error (under the right circumstances) can propagate to all future outputs. That is, the decoder's output may become unstable. By further restricting Eq. (8.5-15) to be strictly less than 1 confines the impact of an input error to a small number of outputs.

■ Consider the prediction error that results from DPCM coding the monochrome image of Fig. 8.23 under the assumption of zero quantization error and with each of four predictors:

EXAMPLE 8.17:
Comparison of prediction techniques.

FIGURE 8.23 A 512 × 512 8-bit monochrome image.

$$\hat{f}(x, y) = 0.97f(x, y - 1) \tag{8.5-16}$$

$$\hat{f}(x, y) = 0.5f(x, y - 1) + 0.5f(x - 1, y) \tag{8.5-17}$$

$$\hat{f}(x, y) = 0.75f(x, y - 1) + 0.75f(x - 1, y) - 0.5f(x - 1, y - 1) \tag{8.5-18}$$

$$\hat{f}(x, y) = \begin{cases} 0.97f(x, y - 1) & \text{if } \Delta h \le \Delta v \\ 0.97f(x - 1, y) & \text{otherwise} \end{cases} \tag{8.5-19}$$

where $\Delta h = |f(x - 1, y) - f(x - 1, y - 1)|$ and $\Delta v = |f(x, y - 1) - f(x - 1, y - 1)|$ denote the horizontal and vertical gradients at point (x, y). Equations (8.5-16) through (8.5-18) define a relatively robust set of α_i, which provide satisfactory performance over a wide range of images. The adaptive predictor of Eq. (8.5-19) is designed to improve edge rendition by computing a local measure of the directional properties of an image (Δh and Δv) and selecting a predictor specifically tailored to the measured behavior.

Figures 8.24(a) through (d) show the prediction error images that result from using the predictors of Eqs. (8.5-16) through (8.5-19). Note that the visually

a b
c d

FIGURE 8.24 A comparison of four linear prediction techniques.

perceptible error decreases as the order of the predictor increases.[†] The standard deviations of the prediction error distributions follow a similar pattern. They are 4.9, 3.7, 3.3, and 4.1 gray levels, respectively.   ■

Optimal quantization

The staircase quantization function $t = q(s)$ shown in Fig. 8.25 is an odd function of s [that is, $q(-s) = -q(s)$] that can be completely described by the $L/2$ values of s_i and t_i shown in the first quadrant of the graph. These break points define function discontinuities and are called the *decision* and *reconstruction* *levels* of the quantizer. As a matter of convention, s is considered to be mapped to t_i if it lies in the half-open interval $(s_i, s_{i+1}]$.

The quantizer design problem is to select the best s_i and t_i for a particular optimization criterion and input probability density function $p(s)$. If the optimization criterion, which could be either a statistical or psychovisual measure,[‡] is the minimization of the mean-square quantization error (that is, $E\{(s - t_i)^2\}$) and $p(s)$ is an even function, the conditions for minimal error (Max [1960]) are

$$\int_{s_{i-1}}^{s_i} (s - t_i)p(s) \, ds = 0 \qquad i = 1, 2, \ldots, \frac{L}{2} \qquad (8.5\text{-}20)$$

$$s_i = \begin{cases} 0 & i = 0 \\ \dfrac{t_i + t_{i+1}}{2} & i = 1, 2, \ldots, \dfrac{L}{2} - 1 \\ \infty & i = \dfrac{L}{2} \end{cases} \qquad (8.5\text{-}21)$$

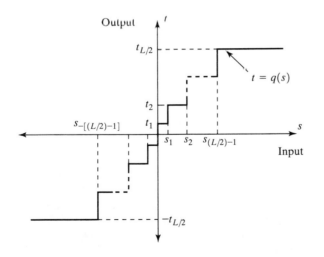

Output

$t = q(s)$

FIGURE 8.25 A typical quantization function.

[†]Predictors that use more than three or four previous pixels provide little compression gain for the added predictor complexity (Habibi [1971]).
[‡]See Netravali [1977] and Limb and Rubinstein [1978] for more on psychovisual measures.

TABLE 8.10
Lloyd-Max
quantizers for a
Laplacian
probability
density function
of unit variance.

Levels	2		4		8	
i	s_i	t_i	s_i	t_i	s_i	t_i
1	∞	0.707	1.102	0.395	0.504	0.222
2			∞	1.810	1.181	0.785
3					2.285	1.576
4					∞	2.994
θ		1.414		1.087		0.731

and

$$s_{-i} = -s_i \quad t_{-i} = -t_i. \tag{8.5-22}$$

Equation (8.5-20) indicates that the reconstruction levels are the centroids of the areas under $p(s)$ over the specified decision intervals, whereas Eq. (8.5-21) indicates that the decision levels are halfway between the reconstruction levels. Equation (8.5-22) is a consequence of the fact that q is an odd function. For any L, the s_i and t_i that satisfy Eqs. (8.5-20) through (8.5-22) are optimal in the mean-square error sense; the corresponding quantizer is called an L-level *Lloyd-Max* quantizer.

Table 8.10 lists the 2-, 4-, and 8-level Lloyd-Max decision and reconstruction levels for a unit variance Laplacian probability density function [see Eq. (8.4-10)]. Because obtaining an explicit or closed-form solution to Eqs. (8.5-20) through (8.5-22) for most nontrivial $p(s)$ is difficult, these values were generated numerically (Paez and Glisson [1972]). The three quantizers shown provide fixed output rates of 1, 2, and 3 bits/pixel, respectively. As Table 8.10 was constructed for a unit variance distribution, the reconstruction and decision levels for the case of $\sigma \neq 1$ are obtained by multiplying the tabulated values by the standard deviation of the probability density function under consideration. The final row of the table lists the step size, θ, that simultaneously satisfies Eqs. (8.5-20) through (8.5-22) *and* the additional constraint that

$$t_i - t_{i-1} = s_i - s_{i-1} = \theta. \tag{8.5-23}$$

If a symbol encoder that utilizes a variable-length code is used in the general lossy predictive encoder of Fig. 8.21(a), an *optimum uniform quantizer* of step size θ will provide a lower code rate (for a Laplacian pdf) than a fixed-length coded Lloyd-Max quantizer with the same output fidelity (O'Neil [1971]).

Although the Lloyd-Max and optimum uniform quantizers are not adaptive, much can be gained from adjusting the quantization levels based on the local behavior of an image. In theory, slowly changing regions can be finely quantized, while the rapidly changing areas are quantized more coarsely. This approach simultaneously reduces both granular noise and slope overload, while requiring only a minimal increase in code rate. The trade-off is increased quantizer complexity.

EXAMPLE 8.18:
Illustration of
quantization and
reconstruction.

■ Figures 8.26(a), (c), and (e) show the DPCM reconstructed images that resulted from combining the 2-, 4-, and 8-level Lloyd-Max quantizers in Table 8.10 with the planar predictor of Eq. (8.5-18). The quantizers were generated by multiplying the tabulated Lloyd-Max decision and reconstruction levels by the standard deviation of the nonquantized planar prediction error from the preceding example (that is, 3.3 gray levels). Note that the edges of the decoded images are blurred from slope overload. This result is particularly noticeable in Fig. 8.26(a), which was generated using a two-level quantizer, but is less apparent in Figs. 8.26(c) and (e), where four and eight quantization levels were applied. Figures 8.27(a), (c), and (e) show the scaled differences between these decoded images and the original image of Fig. 8.23.

In order to generate the decoded images in Figs. 8.26(b), (d), and (f), and the resulting error images in Figs. 8.27(b), (d), and (f), we used an adaptive quantization method in which the best (in a mean-square error sense) of four possible quantizers was selected for each block of 16 pixels. The four quantizers were scaled versions of the optimal Lloyd-Max quantizers previously described. The scaling factors were 0.5, 1.0, 1.75, and 2.5. Because a 2-bit code was appended to each block in order to specify the selected quantizer, the overhead associated with the quantizer switching was $^2\!/_{16}$ or 0.125 bits/pixel. Note the substantial decrease in perceived error that resulted from this relatively small increase in code rate.

Table 8.11 lists the rms errors of the difference images in Figs. 8.27(a) through (f), as well as for a number of other combinations of predictors and quantizers. Note that in a mean-square error sense, the two-level adaptive quantizers performed about as well as the four-level nonadaptive versions. Moreover, the four-level adaptive quantizers outperformed the eight-level nonadaptive approaches. In general, the numerical results indicate that the predictors of Eqs. (8.5-15), (8.5-17), and (8.5-19) exhibit the same overall characteristics as the predictor of Eq. (8.5-18). The compression that resulted under each of the quantization methods is listed in the last row of Table 8.11. Note that the substantial decrease in rms error [Eq. (8.1-8)] achieved by the adaptive approaches did not significantly affect compression performance.   ■

8.5.2 Transform Coding

The predictive coding techniques discussed in Section 8.5.1 operate directly on the pixels of an image and thus are *spatial domain methods*. In this section, we consider compression techniques that are based on modifying the transform of an image. In *transform coding*, a reversible, linear transform (such as the Fourier

Predictor	Lloyd-Max Quantizer			Adaptive Quantizer		
	2-level	**4-level**	**8-level**	**2-level**	**4-level**	**8-level**
Eq. (8.5-16)	30.88	6.86	4.08	7.49	3.22	1.55
Eq. (8.5-17)	14.59	6.94	4.09	7.53	2.49	1.12
Eq. (8.5-18)	9.90	4.30	2.31	4.61	1.70	0.76
Eq. (8.5-19)	38.18	9.25	3.36	11.46	2.56	1.14
Compression	8.00:1	4.00:1	2.70:1	7.11:1	3.77:1	2.56:1

TABLE 8.11
Lossy DPCM root-mean-square error summary.

FIGURE 8.26 DPCM result images: (a) 1.0; (b) 1.125; (c) 2.0; (d) 2.125; (e) 3.0; (f) 3.125 bits/pixel.

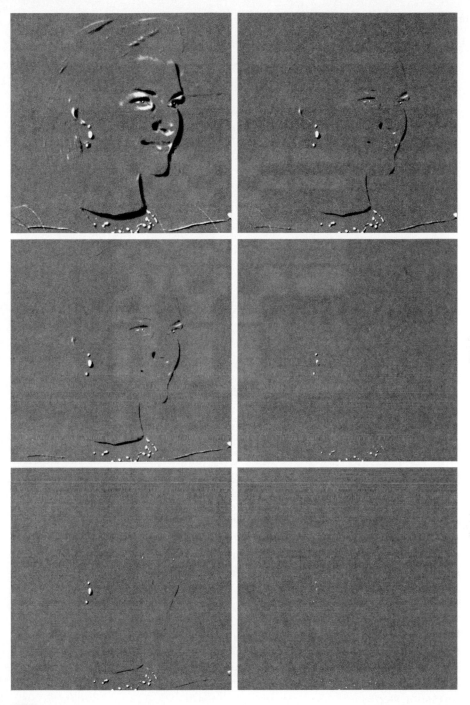

a b
c d
e f

FIGURE 8.27 The scaled (×8) DPCM error images that correspond to Figs. 8.26(a) through (f).

469

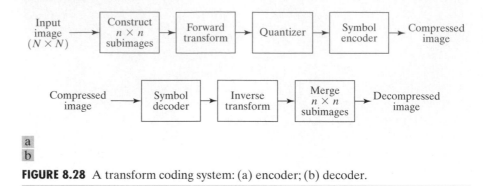

a
b

FIGURE 8.28 A transform coding system: (a) encoder; (b) decoder.

transform) is used to map the image into a set of transform coefficients, which are then quantized and coded. For most natural images, a significant number of the coefficients have small magnitudes and can be coarsely quantized (or discarded entirely) with little image distortion. A variety of transformations, including the discrete Fourier transform (DFT) of Chapter 4, can be used to transform the image data.

Figure 8.28 shows a typical transform coding system. The decoder implements the inverse sequence of steps (with the exception of the quantization function) of the encoder, which performs four relatively straightforward operations: subimage decomposition, transformation, quantization, and coding. An $N \times N$ input image first is subdivided into subimages of size $n \times n$, which are then transformed to generate $(N/n)^2$ subimage transform arrays, each of size $n \times n$. The goal of the transformation process is to decorrelate the pixels of each subimage, or to pack as much information as possible into the smallest number of transform coefficients. The quantization stage then selectively eliminates or more coarsely quantizes the coefficients that carry the least information. These coefficients have the smallest impact on reconstructed subimage quality. The encoding process terminates by coding (normally using a variable-length code) the quantized coefficients. Any or all of the transform encoding steps can be adapted to local image content, called *adaptive transform coding*, or fixed for all subimages, called *nonadaptive transform coding*.

Transform selection

Transform coding systems based on a variety of discrete 2-D transforms have been constructed and/or studied extensively. The choice of a particular transform in a given application depends on the amount of reconstruction error that can be tolerated and the computational resources available. Compression is achieved during the quantization of the transformed coefficients (not during the transformation step).

Consider an image $f(x, y)$ of size $N \times N$ whose forward, discrete transform, $T(u, v)$, can be expressed in terms of the general relation

$$T(u, v) = \sum_{x=0}^{N-1}\sum_{y=0}^{N-1} f(x, y)g(x, y, u, v) \qquad (8.5\text{-}24)$$

for $u, v = 0, 1, 2, \ldots, N - 1$. Given $T(u, v), f(x, y)$ similarly can be obtained using the generalized inverse discrete transform

$$f(x, y) = \sum_{u=0}^{N-1}\sum_{v=0}^{N-1} T(u, v)h(x, y, u, v) \qquad (8.5\text{-}25)$$

for $x, y = 0, 1, 2, \ldots, N - 1$. In these equations, $g(x, y, u, v)$ and $h(x, y, u, v)$ are called the *forward* and *inverse transformation kernels*, respectively. For reasons that will become clear later in the section, they also are referred to as *basis functions* or *basis images*. The $T(u, v)$ for $u, v = 0, 1, 2, \ldots, N - 1$ in Eq. (8.5-25) are called *transform coefficients*; they can be viewed as the expansion coefficients—see Section 7.2.1—of a series expansion of $f(x, y)$ with respect to basis functions $h(x, y, u, v)$.

The forward kernel in Eq. (8.5-24) is said to be *separable* if

$$g(x, y, u, v) = g_1(x, u)g_2(y, v). \qquad (8.5\text{-}26)$$

In addition, the kernel is *symmetric* if g_1 is functionally equal to g_2. In this case, Eq. (8.5-26) can be expressed in the form

$$g(x, y, u, v) = g_1(x, u)g_1(y, v). \qquad (8.5\text{-}27)$$

Identical comments apply to the inverse kernel if $g(x, y, u, v)$ is replaced by $h(x, y, u, v)$ in Eqs. (8.5-26) and (8.5-27). It is not difficult to show that a 2-D transform with a separable kernel can be computed using row-column or column-row passes of the corresponding 1-D transform, in the manner explained in Section 4.6.1.

The forward and inverse transformation kernels in Eqs. (8.5-24) and (8.5-25) determine the type of transform that is computed and the overall computational complexity and reconstruction error of the transform coding system in which they are employed. The most well-known transform kernel pair is

$$g(x, y, u, v) = \frac{1}{N^2} e^{-j2\pi(ux+vy)/N} \qquad (8.5\text{-}28)$$

and

$$h(x, y, u, v) = e^{j2\pi(ux+vy)/N} \qquad (8.5\text{-}29)$$

where $j = \sqrt{-1}$. Substituting these kernels into Eqs. (8.5-24) and (8.5-25) yields a simplified version (in which $M = N$) of the discrete Fourier transform pair introduced in Section 4.2.2.

A computationally simpler transformation that is also useful in transform coding, called the *Walsh-Hadamard transform* (WHT), is derived from the functionally identical kernels

$$g(x, y, u, v) = h(x, y, u, v) = \frac{1}{N} (-1)^{\sum_{i=0}^{m-1}[b_i(x)p_i(u)+b_i(y)p_i(v)]} \qquad (8.5\text{-}30)$$

where $N = 2^m$. The summation in the exponent of this expression is performed in modulo 2 arithmetic and $b_k(z)$ is the kth bit (from right to left) in the binary representation of z. If $m = 3$ and $z = 6$ (110 in binary), for example, $b_0(z) = 0$, $b_1(z) = 1$, and $b_2(z) = 1$. The $p_i(u)$ in Eq. (8.5-30) are computed using:

$$
\begin{aligned}
p_0(u) &= b_{m-1}(u) \\
p_1(u) &= b_{m-1}(u) + b_{m-2}(u) \\
p_2(u) &= b_{m-2}(u) + b_{m-3}(u) \\
&\;\;\vdots \\
p_{m-1}(u) &= b_1(u) + b_0(u)
\end{aligned}
\tag{8.5-31}
$$

where the sums, as noted previously, are performed in modulo 2 arithmetic. Similar expressions apply to $p_i(v)$.

Unlike the kernels of the DFT, which are sums of sines and cosines [see Eqs. (8.5-28) and (8.5-29)], the Walsh-Hadamard kernels consist of alternating plus and minus 1's arranged in a checkerboard pattern. Figure 8.29 shows the kernel for $N = 4$. Each block consists of $4 \times 4 = 16$ elements (subsquares). White denotes +1 and black denotes −1. To obtain the top, left block, we let $u = v = 0$ and plot values of $g(x, y, 0, 0)$ for $x, y = 0, 1, 2, 3$. All values in this case are +1. The second block on the top row is a plot of values of $g(x, y, 0, 1)$ for $x, y = 0$, 1, 2, 3, and so on. As already noted, the importance of the Walsh-Hadamard transform is its simplicity of implementation—all kernel values are +1 or −1.

One of the most frequently used transformations for image compression is the *discrete cosine transform* (DCT). It is obtained by substituting the following (equal) kernels into Eqs. (8.5-24) and (8.5-25)

$$
\begin{aligned}
g(x, y, u, v) &= h(x, y, u, v) \\
&= \alpha(u)\alpha(v) \cos\left[\frac{(2x + 1)u\pi}{2N}\right] \cos\left[\frac{(2y + 1)v\pi}{2N}\right]
\end{aligned}
\tag{8.5-32}
$$

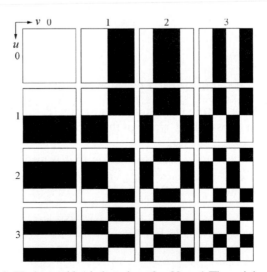

FIGURE 8.29 Walsh-Hadamard basis functions for $N = 4$. The origin of each block is at its top left.

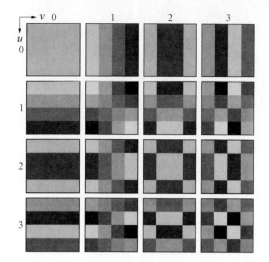

FIGURE 8.30 Discrete-cosine basis functions for $N = 4$. The origin of each block is at its top left.

where

$$\alpha(u) - \begin{cases} \sqrt{\dfrac{1}{N}} & \text{for } u = 0 \\ \sqrt{\dfrac{2}{N}} & \text{for } u = 1, 2, \ldots, N - 1 \end{cases} \qquad (8.5\text{-}33)$$

and similarly for $\alpha(v)$. Figure 8.30 shows $g(x, y, u, v)$ for the case $N = 4$. The computation follows the same format as explained for Fig. 8.29, with the difference that the values of g are not integers. In Fig. 8.30, the lighter gray levels correspond to larger values of g.

■ Figures 8.31(a), (c), and (e) show three approximations of the 512×512 monochrome image in Fig. 8.23. These pictures were obtained by dividing the original image into subimages of size 8×8, representing each subimage using one of the transforms just described (i.e., the DFT, WHT, or DCT transform), truncating 50% of the resulting coefficients, and taking the inverse transform of the truncated coefficient arrays.

In each case, the 32 retained coefficients were selected on the basis of maximum magnitude. When we disregard any quantization or coding issues, this process amounts to compressing the original image by a factor of 2. Note that in all cases, the 32 discarded coefficients had little visual impact on reconstructed image quality. Their elimination, however, was accompanied by some mean-square error, which can be seen in the scaled error images of Figs. 8.31(b), (d), and (f). The actual rms errors were 1.28, 0.86, and 0.68 gray levels, respectively. ■

EXAMPLE 8.19:
Transform coding with the DFT, WHT, and DCT.

a b
c d
e f

FIGURE 8.31 Approximations of Fig. 8.23 using the (a) Fourier, (c) Hadamard, and (e) cosine transforms, together with the corresponding scaled error images.

The small differences in mean-square reconstruction error noted in the preceding example are related directly to the energy or information packing properties of the transforms employed. In accordance with Eq. (8.5-25), an $n \times n$ image $f(x, y)$ can be expressed as a function of its 2-D transform $T(u, v)$:

$$f(x, y) = \sum_{u=0}^{n-1} \sum_{v=0}^{n-1} T(u, v) h(x, y, u, v) \tag{8.5-34}$$

for $x, y = 0, 1, \dots, n - 1$. Note that we merely replaced N in Eq. (8.5-25) with n and now consider $f(x, y)$ to represent a subimage of the image being compressed. Since the inverse kernel $h(x, y, u, v)$ in Eq. (8.5-34) depends only on the indices x, y, u, v, and not on the values of $f(x, y)$ or $T(u, v)$, it can be viewed as defining a set of *basis functions* or *basis images* for the series defined by Eq. (8.5-34). This interpretation becomes clearer if the notation used in Eq. (8.5-34) is modified to obtain

$$\mathbf{F} = \sum_{u=0}^{n-1} \sum_{v=0}^{n-1} T(u, v) \mathbf{H}_{uv} \tag{8.5-35}$$

where $\mathbf{F}$ is an $n \times n$ matrix containing the pixels of $f(x, y)$ and

$$\mathbf{H}_{uv} = \begin{bmatrix} h(0, 0, u, v) & h(0, 1, u, v) & \cdots & h(0, n - 1, u, v) \\ h(1, 0, u, v) & \cdot & \cdots & \cdot \\ \cdot & \cdot & \cdots & \cdot \\ \cdot & \cdot & \cdots & \cdot \\ \cdot & & & \\ h(n - 1, 0, u, v) & h(n - 1, 1, u, v) & \cdots & h(n - 1, n - 1, u, v) \end{bmatrix}. \tag{8.5-36}$$

Then $\mathbf{F}$, the matrix containing the pixels of the input subimage, is explicitly defined as a linear combination of n^2 matrices of size $n \times n$; that is, the $\mathbf{H}_{uv}$ for u, $v = 0, 1, \dots, n - 1$ in Eq. (8.5-36). These matrices in fact are the basis images (or functions) of the series expansion in Eq. (8.5-35); the associated $T(u, v)$ are the expansion coefficients. Figures 8.29 and 8.30 illustrate graphically the WHT and DCT basis images for the case of $n = 4$.

If we now define a transform coefficient *masking function*

$$\gamma(u, v) = \begin{cases} 0 & \text{if } T(u, v) \text{ satisfies a specified truncation criterion} \\ 1 & \text{otherwise} \end{cases} \tag{8.5-37}$$

for $u, v = 0, 1, \dots, n - 1$, an approximation of $\mathbf{F}$ can be obtained from the truncated expansion

$$\hat{\mathbf{F}} = \sum_{u=0}^{n-1} \sum_{v=0}^{n-1} \gamma(u, v) T(u, v) \mathbf{H}_{uv} \tag{8.5-38}$$

where $\gamma(u, v)$ is constructed to eliminate the basis images that make the smallest contribution to the total sum in Eq. (8.5-35). The mean-square error between subimage $\mathbf{F}$ and approximation $\hat{\mathbf{F}}$ then is

$$
\begin{aligned}
e_{\mathrm{ms}} &= E\{\|\mathbf{F} - \hat{\mathbf{F}}\|^2\} \\
&= E\left\{\left\|\sum_{u=0}^{n-1}\sum_{v=0}^{n-1} T(u, v)\mathbf{H}_{uv} - \sum_{u=0}^{n-1}\sum_{v=0}^{n-1}\gamma(u, v)T(u, v)\mathbf{H}_{uv}\right\|^2\right\} \\
&= E\left\{\left\|\sum_{u=0}^{n-1}\sum_{v=0}^{n-1} T(u, v)\mathbf{H}_{uv}[1 - \gamma(u, v)]\right\|^2\right\} \\
&= \sum_{u=0}^{n-1}\sum_{v=0}^{n-1}\sigma_{T(u,v)}^2[1 - \gamma(u, v)]
\end{aligned}
\tag{8.5-39}
$$

where $\|\mathbf{F} - \hat{\mathbf{F}}\|$ is the norm of matrix $(\mathbf{F} - \hat{\mathbf{F}})$, and $\sigma_{T(u,v)}^2$ is the variance of the coefficient at transform location (u, v). The final simplification is based on the orthonormal nature of the basis images and the assumption that the pixels of $\mathbf{F}$ are generated by a random process with zero mean and known covariance. The total mean-square approximation error thus is the sum of the variances of the discarded transform coefficients; that is, the coefficients for which $\gamma(u, v) = 0$, so that $[1 - \gamma(u, v)]$ in Eq. (8.5-39) is 1. Transformations that redistribute or pack the most information into the fewest coefficients provide the best subimage approximations and, consequently, the smallest reconstruction errors. Finally, under the assumptions that led to Eq. (8.5-39), the mean-square error of the $(N/n)^2$ subimages of an $N \times N$ image are identical. Thus the mean-square error (being a measure of *average* error) of the $N \times N$ image equals the mean-square error of a single subimage.

The earlier example showed that the information packing ability of the DCT is superior to that of the DFT and WHT. Although this condition usually holds for most natural images, the Karhunen-Loève transform (the KLT is discussed in Chapter 11), not the DCT, is the optimal transform in an information packing sense. That is, the KLT minimizes the mean-square error in Eq. (8.5-39) for any input image and any number of retained coefficients (Kramer and Mathews [1956]).[†] However, because the KLT is data dependent, obtaining the KLT basis images for each subimage, in general, is a nontrivial computational task. For this reason, the KLT is seldom used in practice for image compression. Instead, a transform, such as the DFT, WHT, or DCT, whose basis images are fixed (input independent), normally is used. Of the possible input independent transforms, the nonsinusoidal transforms (such as the WHT transform) are the simplest to implement. The sinusoidal transforms (such as the DFT or DCT) more closely approximate the information packing ability of the optimal KLT.

Hence many transform coding systems are based on the DCT, which provides a good compromise between information packing ability and computational complexity. In fact, the properties of the DCT have proved to be of such practical value that the DCT has become an international standard for transform coding

---

[†] An additional condition for optimality is that the masking function of Eq. (8.5-37) selects the KLT coefficients of maximum variance.

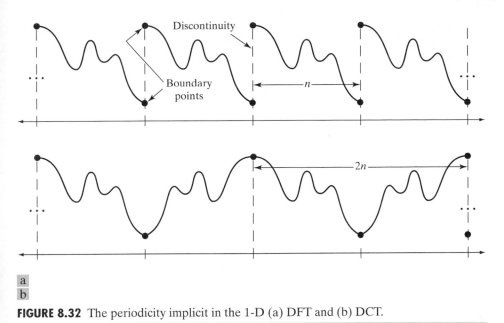

a
b

FIGURE 8.32 The periodicity implicit in the 1-D (a) DFT and (b) DCT.

systems (see Section 8.6). Compared to the other input independent transforms, it has the advantages of having been implemented in a single integrated circuit, packing the most information into the fewest coefficients[†] (for most natural images), and minimizing the blocklike appearance, called *blocking artifact*, that results when the boundaries between subimages become visible. This last property is particularly important in comparisons with the other sinusoidal transforms. As Fig. 8.32(a) shows, the implicit n-point periodicity (see Section 4.6) of the DFT gives rise to boundary discontinuities that result in substantial high-frequency transform content. When the DFT transform coefficients are truncated or quantized, the Gibbs phenomenon[‡] causes the boundary points to take on erroneous values, which appear in an image as blocking artifact. That is, the boundaries between adjacent subimages become visible because the boundary pixels of the subimages assume the mean values of discontinuities formed at the boundary points [see Fig. 8.32(a)]. The DCT of Fig. 8.32(b) reduces this effect, because its implicit $2n$-point periodicity does not inherently produce boundary discontinuities.

Subimage size selection

Another significant factor affecting transform coding error and computational complexity is subimage size. In most applications, images are subdivided so that the correlation (redundancy) between adjacent subimages is reduced to some

[†] Ahmed et al. [1974] first noticed that the KLT basis images of a first-order Markov image source closely resemble the DCT's basis images. As the correlation between adjacent pixels approaches one, the input dependent KLT basis images become identical to the input independent DCT basis images (Clarke [1985]).

[‡] This phenomenon, described in most electrical engineering texts on circuit analysis, occurs because the Fourier transform fails to converge uniformly at discontinuities. At discontinuities, Fourier expansions take the mean values.

acceptable level and so that n is an integer power of 2 where, as before, n is the subimage dimension. The latter condition simplifies the computation of the subimage transforms (see the base-2 successive doubling methods in Section 4.6.6). In general, both the level of compression and computational complexity increase as the subimage size increases. The most popular subimage sizes are 8×8 and 16×16.

EXAMPLE 8.20:
Effects of subimage size on transform coding.

■ Figure 8.33 illustrates graphically the impact of subimage size on transform coding reconstruction error. The data plotted were obtained by dividing the monochrome image of Fig. 8.23 into subimages of size $n \times n$, for $n = 2, 4, 8, 16,$ and 32, computing the transform of each subimage, truncating 75% of the resulting coefficients, and taking the inverse transform of the truncated arrays. Note that the Hadamard and cosine curves flatten as the size of the subimage becomes greater than 8×8, whereas the Fourier reconstruction error decreases even more rapidly in this region. Extrapolation of these curves to larger values of n suggests that the Fourier reconstruction error will cross the Walsh-Hadamard curve and converge to the cosine result. In fact, this result is consistent with the theoretical and experimental findings reported by Netravali and Limb [1980] and by Pratt [1991] for a 2-D Markov image source.

All three curves intersect when 2×2 subimages are used. In this case, only one of the four coefficients (25%) of each transformed array was retained. The coefficient in all cases was the dc component, so the inverse transform simply replaced the four subimage pixels by their average value [see Eq. (4.2-22)]. This condition is evident in Fig. 8.34(d), which shows a zoomed portion of the 2×2 DCT result. Note that the blocking artifact that is prevalent in this result decreases as the subimage size increases to 4×4 and 8×8 in Figs. 8.34(e) and (f). Figure 8.34(c) shows a zoomed portion of the original image for reference. In addition, Figs. 8.34(a) and (b) facilitate comparison of these results to those of the preceding example. ■

FIGURE 8.33
Reconstruction error versus subimage size.

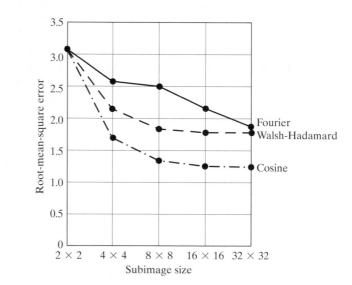

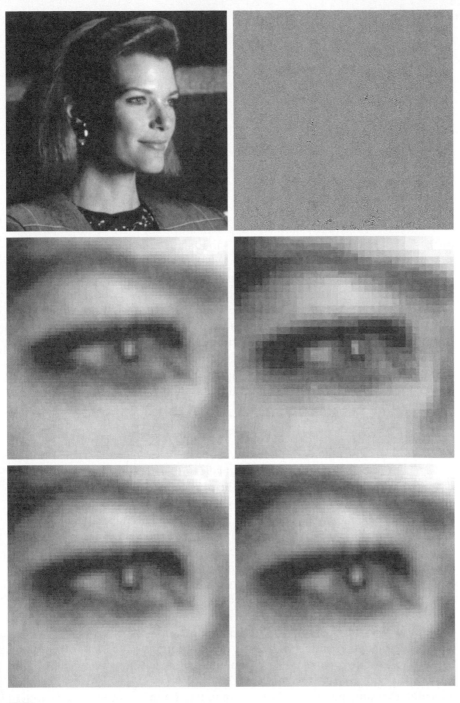

FIGURE 8.34 Approximations of Fig. 8.23 using 25% of the DCT coefficients: (a) and (b) 8 × 8 subimage results; (c) zoomed original; (d) 2 × 2 result; (e) 4 × 4 result; and (f) 8 × 8 result.

Bit allocation

The reconstruction error associated with the truncated series expansion of Eq. (8.5-38) is a function of the number and relative importance of the transform coefficients that are discarded, as well as the precision that is used to represent the retained coefficients. In most transform coding systems, the retained coefficients are selected [that is, the masking function of Eq. (8.5-37) is constructed] on the basis of maximum variance, called *zonal coding*, or on the basis of maximum magnitude, called *threshold coding*. The overall process of truncating, quantizing, and coding the coefficients of a transformed subimage is commonly called *bit allocation*.

EXAMPLE 8.21:
Bit allocation.

■ Figures 8.35(a) and (b) show two approximations of Fig. 8.23 in which 87.5% of the DCT coefficients of each 8 × 8 subimage were discarded. The first result was obtained via threshold coding by keeping the eight largest transform coefficients, and the second image was generated by using a zonal coding approach. In the latter case, each DCT coefficient was considered a random variable whose distribution could be computed over the ensemble of all transformed subimages. The 8 distributions of largest variance (12.5% of the 64 coefficients in the transformed 8 × 8 subimage) were located and used to determine the coordinates, u and v, of the coefficients, $T(u, v)$, that were retained for all subimages. Note that the threshold coding difference image of Fig. 8.35(c) contains far less error than the zonal coding result in Fig. 8.35(d). Figures 8.35(e) and (f) provide a closer view of a small portion of the reconstructed images in (a) and (c). ■

Zonal coding implementation   Zonal coding is based on the information theory concept of viewing information as uncertainty. Therefore the transform coefficients of maximum variance carry the most image information and should be retained in the coding process. The variances themselves can be calculated directly from the ensemble of $(N/n)^2$ transformed subimage arrays, as in the preceding example, or based on an assumed image model (say, a Markov autocorrelation function). In either case, the zonal sampling process can be viewed, in accordance with Eq. (8.5-38), as multiplying each $T(u, v)$ by the corresponding element in a *zonal mask*, which is constructed by placing a 1 in the locations of maximum variance and a 0 in all other locations. Coefficients of maximum variance usually are located around the origin of an image transform, resulting in the typical zonal mask shown in Fig. 8.36(a).

The coefficients retained during the zonal sampling process must be quantized and coded, so zonal masks are sometimes depicted showing the number of bits used to code each coefficient [Fig. 8.36(b)]. In most cases, the coefficients are allocated the same number of bits, or some fixed number of bits is distributed among them unequally. In the first case, the coefficients generally are normalized by their standard deviations and uniformly quantized. In the second case, a quantizer, such as an optimal Lloyd-Max quantizer, is designed for each coefficient. To construct the required quantizers, the zeroth or dc coefficient normally is

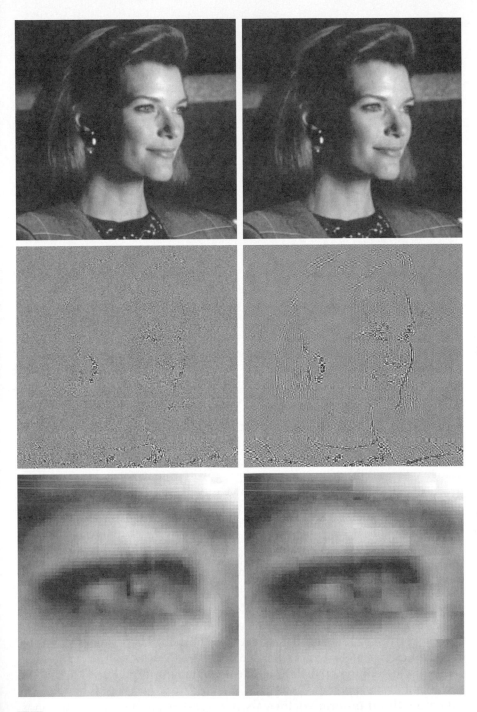

FIGURE 8.35 Approximations of Fig. 8.23 using 12.5% of the 8 × 8 DCT coefficients: (a), (c), and (e) threshold coding results; (b), (d), and (f) zonal coding results.

a b
c d

FIGURE 8.36 A typical (a) zonal mask, (b) zonal bit allocation, (c) threshold mask, and (d) thresholded coefficient ordering sequence. Shading highlights the coefficients that are retained.

1	1	1	1	1	0	0	0	8	7	6	4	3	2	1	0
1	1	1	1	0	0	0	0	7	6	5	4	3	2	1	0
1	1	1	0	0	0	0	0	6	5	4	3	3	1	1	0
1	1	0	0	0	0	0	0	4	4	3	3	2	1	0	0
1	0	0	0	0	0	0	0	3	3	3	2	1	1	0	0
0	0	0	0	0	0	0	0	2	2	1	1	1	0	0	0
0	0	0	0	0	0	0	0	1	1	1	0	0	0	0	0
0	0	0	0	0	0	0	0	0	0	0	0	0	0	0	0

1	1	0	1	1	0	0	0	0	1	5	6	14	15	27	28
1	1	1	1	0	0	0	0	2	4	7	13	16	26	29	42
1	1	0	0	0	0	0	0	3	8	12	17	25	30	41	43
1	0	0	0	0	0	0	0	9	11	18	24	31	40	44	53
0	0	0	0	0	0	0	0	10	19	23	32	39	45	52	54
0	1	0	0	0	0	0	0	20	22	33	38	46	51	55	60
0	0	0	0	0	0	0	0	21	34	37	47	50	56	59	61
0	0	0	0	0	0	0	0	35	36	48	49	57	58	62	63

modeled by a Rayleigh density function, whereas the remaining coefficients are modeled by a Laplacian or Gaussian density.[†] The number of quantization levels (and thus the number of bits) allotted to each quantizer is made proportional to $\log_2 \sigma^2_{T(u,v)}$. This allocation is consistent with rate distortion theory, which indicates that a Gaussian random variable of variance σ^2 cannot be represented by less than $\frac{1}{2}\log_2(\sigma^2/D)$ bits and be reproduced with mean-square error less than D (see Problem 8.11). The intuitive conclusion is that the information content of a Gaussian random variable is proportional to $\log_2(\sigma^2/D)$. Thus the retained coefficients in Eq. (8.5-38)—which (in the context of the current discussion) are selected on the basis of maximum variance—should be assigned bits in proportion to the logarithm of the coefficient variances.

Threshold coding implementation   Zonal coding usually is implemented by using a single fixed mask for all subimages. Threshold coding, however, is inherently adaptive in the sense that the location of the transform coefficients retained for each subimage vary from one subimage to another. In fact, threshold coding is the adaptive transform coding approach most often used in practice because of its computational simplicity. The underlying concept is that, for any subimage, the transform coefficients of largest magnitude make the most significant contribution to reconstructed subimage quality, as demonstrated in the

---

[†] As each coefficient is a linear combination of the pixels in its subimage [see Eq. (8.5-24)], the central limit theorem suggests that, as subimage size increases, the coefficients tend to become Gaussian. This result does not apply to the dc coefficient, however, because nonnegative images always have positive dc coefficients.

last example. Because the locations of the maximum coefficients vary from one subimage to another, the elements of $\gamma(u, v)T(u, v)$ normally are reordered (in a predefined manner) to form a 1-D, run-length coded sequence. Figure 8.36(c) shows a typical *threshold mask* for one subimage of a hypothetical image. This mask provides a convenient way to visualize the threshold coding process for the corresponding subimage, as well as to mathematically describe the process using Eq. (8.5-38). When the mask is applied [via Eq. (8.5-38)] to the subimage for which it was derived, and the resulting $n \times n$ array is reordered to form an n^2-element coefficient sequence in accordance with the zigzag ordering pattern of Fig. 8.34(d), the reordered 1-D sequence contains several long runs of 0's [the zigzag pattern becomes evident by starting at 0 in Fig. 8.36(d) and following the numbers in sequence]. These runs normally are run-length coded. The nonzero or retained coefficients, corresponding to the mask locations that contain a 1, are represented using one of the variable-length codes of Section 8.4.

There are three basic ways to threshold a transformed subimage or, stated differently, to create a subimage threshold masking function of the form given in Eq. (8.5-37): (1) A single global threshold can be applied to all subimages; (2) a different threshold can be used for each subimage; or (3) the threshold can be varied as a function of the location of each coefficient within the subimage. In the first approach, the level of compression differs from image to image, depending on the number of coefficients that exceed the global threshold. In the second, called *N-largest coding*, the same number of coefficients is discarded for each subimage. As a result, the code rate is constant and known in advance. The third technique, like the first, results in a variable code rate, but offers the advantage that thresholding *and* quantization can be combined by replacing $\gamma(u, v)T(u, v)$ in Eq. (8.5-38) with

$$\hat{T}(u, v) - \text{round}\left[\frac{T(u, v)}{Z(u, v)}\right] \qquad (8.5\text{-}40)$$

where $\hat{T}(u, v)$ is a thresholded and quantized approximation of $T(u, v)$, and $Z(u, v)$ is an element of the transform normalization array

$$\mathbf{Z} = \begin{bmatrix} Z(0, 0) & Z(0, 1) & \cdots & Z(0, n-1) \\ Z(1, 0) & & & \\ \cdot & \cdot & \cdots & \cdot \\ \cdot & \cdot & \cdots & \cdot \\ \cdot & \cdot & & \cdot \\ Z(n-1, 0) & Z(n-1, 1) & \cdots & Z(n-1, n-1) \end{bmatrix} . \qquad (8.5\text{-}41)$$

Before a normalized (thresholded and quantized) subimage transform, $\hat{T}(u, v)$, can be inverse transformed to obtain an approximation of subimage $f(x, y)$, it must be multiplied by $Z(u, v)$. The resulting denormalized array, denoted $\dot{T}(u, v)$, is an approximation of $\hat{T}(u, v)$:

$$\dot{T}(u, v) = \hat{T}(u, v)Z(u, v). \qquad (8.5\text{-}42)$$

The inverse transform of $\dot{T}(u, v)$ yields the decompressed subimage approximation.

a b
FIGURE 8.37
(a) A threshold
coding
quantization
curve [see
Eq. (8.5-40)].
(b) A typical
normalization
matrix.

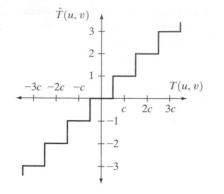

Figure 8.37(a) depicts Eq. (8.5-40) graphically for the case in which $Z(u, v)$ is assigned a particular value c. Note that $\hat{T}(u, v)$ assumes integer value k if and only if

$$kc - \frac{c}{2} \leq T(u, v) < kc + \frac{c}{2}.$$

If $Z(u, v) > 2T(u, v)$, then $\hat{T}(u, v) = 0$ and the transform coefficient is completely truncated or discarded. When $\hat{T}(u, v)$ is represented with a variable-length code that increases in length as the magnitude of k increases, the number of bits used to represent $T(u, v)$ is controlled by the value of c. Thus the elements of **Z** can be scaled to achieve a variety of compression levels. Figure 8.37(b) shows a typical normalization array. This array, which has been used extensively in the JPEG[†] standardization efforts (see Section 8.6.2), weighs each coefficient of a transformed subimage according to heuristically determined perceptual or psychovisual importance.

EXAMPLE 8.22:
Illustration of
threshold coding.

■ Figures 8.38(a) and (b) show two threshold-coded approximations of the monochrome image in Fig. 8.23. Both images were generated using an 8×8 DCT and the normalization array of Fig. 8.37(b). The first result, which provides a compression ratio of about 34 to 1, was obtained by direct application of that normalization array. The second, which compresses the original image by a ratio of 67 to 1, was generated after multiplying (scaling) the normalization array by 4. By comparison, the *average* compression ratio obtained by using all the error-free methods discussed in Section 8.4 was only 2.62 to 1.

The differences between the original image of Fig. 8.23 and the reconstructed images of Figs. 8.38(a) and (b) are shown in Figs. 8.38(c) and (d), respectively. The corresponding rms errors [see Eq. (8.1-8)] are 3.42 and 6.33 gray levels. The precise nature of the errors is more visible in the zoomed images of Figures 8.38(e) and (f). These images show a magnified section of Figs. 8.38(a) and (b), respectively. They allow a better assessment of the subtle differences between the reconstructed images.   ■

[†] JPEG is an abbreviation for the Joint Photographic Experts Group.

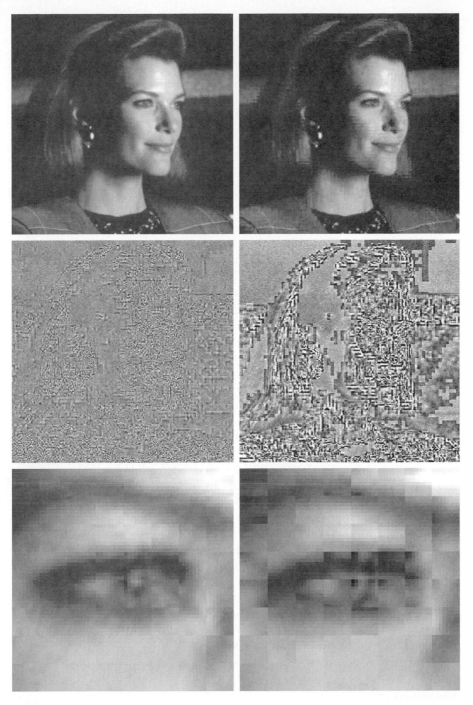

FIGURE 8.38 Left column: Approximations of Fig. 8.23 using the DCT and normalization array of Fig. 8.37(b). Right column: Similar results for 4**Z**.

a
b

FIGURE 8.39 A wavelet coding system: (a) encoder; (b) decoder.

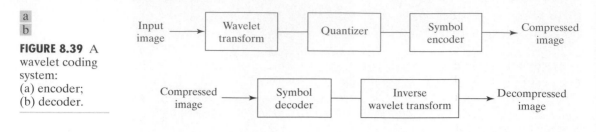

8.5.3 Wavelet Coding

Like the transform coding techniques of the previous section, wavelet coding is based on the idea that the coefficients of a transform that decorrelates the pixels of an image can be coded more efficiently than the original pixels themselves. If the transform's basis functions—in this case wavelets—pack most of the important visual information into a small number of coefficients, the remaining coefficients can be quantized coarsely or truncated to zero with little image distortion.

Figure 8.39 shows a typical wavelet coding system. To encode a $2^J \times 2^J$ image, an analyzing wavelet, ψ, and minimum decomposition level, $J - P$, are selected and used to compute the image's discrete wavelet transform. If the wavelet has a complimentary scaling function φ, the fast wavelet transform (see Sections 7.4 and 7.5) can be used. In either case, the computed transform converts a large portion of the original image to horizontal, vertical, and diagonal decomposition coefficients with zero mean and Laplacian-like distributions. Recall the image of Fig. 7.1 and the dramatically simpler statistics of its wavelet transform in Fig. 7.8(a). Since many of the computed coefficients carry little visual information, they can be quantized and coded to minimize intercoefficient and coding redundancy. Moreover, the quantization can be adapted to exploit any positional correlation across the P decomposition levels. One or more of the lossless coding methods of Section 8.4, including run-length, Huffman, arithmetic, and bit-plane coding, can be incorporated into the final symbol coding step. Decoding is accomplished by inverting the encoding operations—with the exception of quantization, which cannot be reversed exactly.

The principal difference between the wavelet-based system of Fig. 8.39 and the transform coding system of Fig. 8.28 is the omission of the transform coder's subimage processing stages. Because wavelet transforms are both computationally efficient and inherently local (i.e., their basis functions are limited in duration), subdivision of the original image is unnecessary. As will be seen in the following example, the removal of the subdivision step eliminates the blocking artifact that characterizes DCT-based approximations at high compression ratios.

EXAMPLE 8.23:
Comparing wavelet and DCT-based coding.

■ Figure 8.40 shows two wavelet-based approximations of the monochrome image in Figure 8.23. Figure 8.40(a) was reconstructed from an encoding that compressed the original image by 34:1; Fig. 8.40(b) was generated from a 67:1 encoding. Since these ratios are identical to the compression levels of

Example 8.22, Figs. 8.40(a) through (f) can be compared—both qualitatively and quantitatively—to the transform-based results in Figs. 8.38(a) through (f).

A visual comparison of Figs. 8.40(c) and (d) with 8.38(c) and (d), respectively, reveals a noticeable decrease of error in the wavelet coding results. In fact, the rms error of the wavelet-based image in Fig. 8.40(a) is 2.29 gray levels, as opposed to 3.42 gray levels for the corresponding transform-based result in Fig. 8.38(a). In a similar manner, the rms errors of the approximations in Figs. 8.38(b) and 8.40(b) are 6.33 and 2.96, respectively. The computed errors favor the wavelet-based results at both compression levels.

Besides decreasing the reconstruction error for a given level of compression, wavelet coding—see Figs. 8.40(e) and (f)—dramatically increases (in a subjective sense) image quality. This is particularly evident in Fig. 8.40(f). Note that the blocking artifact that dominated the corresponding transform-based result in Fig. 8.38(f) is no longer present.   ■

When the level of compression exceeds 67:1, the largest ratio examined in the previous two examples, there is an increased loss of texture in the woman's clothing and blurring of her eyes. Both effects are visible in Figs. 8.41(a) and (b), which were reconstructed from 108:1 and 167:1 wavelet-based encodings of the original image in Fig. 8.23. The increased blurring is particularly evident in Figs. 8.41(e) and (f). The rms errors for Figs. 8.41(a) and (b) are 3.72 and 4.73 gray levels, respectively. A subjective evaluation of either image reveals its obvious superiority to the 67:1 transform-based result in Fig. 8.38(b). Its rms error was 6.33 gray levels. Thus, at more than twice the level of compression, the most highly compressed wavelet-based reconstruction has only 75% of the error of the less compressed transform-based result—and superior perceived quality as well.

To conclude our discussion of wavelet-based compression, we end with a brief overview of the major factors influencing coding complexity, performance, and reconstruction error.

Wavelet selection

The wavelets chosen as the basis of the forward and inverse transforms in Fig. 8.39 affect all aspects of wavelet coding system design and performance. They impact directly the computational complexity of the transforms and, less directly, the system's ability to compress and reconstruct images of acceptable error. When the transforming wavelet has a companion scaling function, the transformation can be implemented as a sequence of digital filtering operations, with the number of filter taps equal to the number of nonzero wavelet and scaling vector coefficients. The ability of the wavelet to pack information into a small number of transform coefficients determines its compression and reconstruction performance.

The most widely used expansion functions for wavelet-based compression are the Daubechies wavelets and biorthogonal wavelets. The latter allow useful analysis properties, like number of zero moments (see Section 7.5), to be incorporated into the decomposition filters, while important synthesis properties, like smoothness of reconstruction, are built into the reconstruction filters.

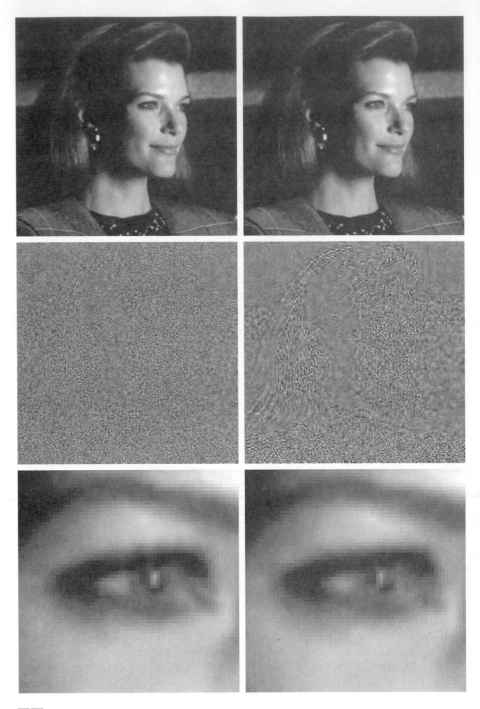

FIGURE 8.40 (a), (c), and (e) Wavelet coding results comparable to the transform-based results in Figs. 8.38(a), (c), and (e); (b), (d), and (f) similar results for Figs. 8.38(b), (d), and (f).

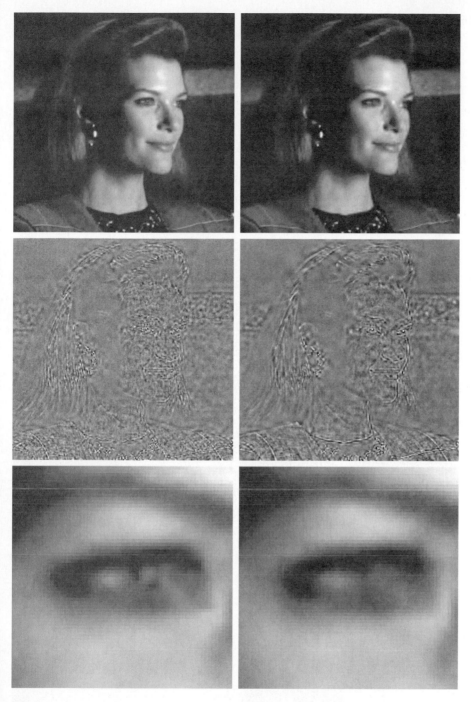

FIGURE 8.41 (a), (c), and (e) Wavelet coding results with a compression ratio of 108 to 1; (b), (d), and (f) similar results for a compression of 167 to 1.

EXAMPLE 8.24:
Wavelet bases in
wavelet coding.

■ Figure 8.42 contains four discrete wavelet transforms of the woman in Fig. 8.23. Haar wavelets, the simplest and only discontinuous wavelets considered here, were used as the expansion or basis functions in Fig. 8.42(a). Daubechies wavelets, among the most popular imaging wavelets, were used in Fig. 8.42(b), and symlets, which are an extension of the Daubechies wavelets with increased symmetry, were used in Fig. 8.42(c). The Cohen-Daubechies-Feauveau wavelets that were employed in Fig. 8.42(d) are included to illustrate the capabilities of biorthogonal wavelets. As in previous results of this type, all detail coefficients have been scaled to make the underlying structure more visible—with gray-level 128 corresponding to coefficient value 0.

As can be seen in Table 8.12, the number of operations involved in the computation of the transforms in Fig. 8.42 increases from 4 to 28 multiplications

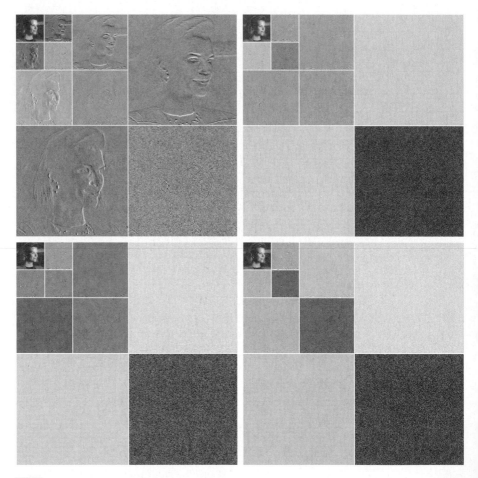

a b
c d

FIGURE 8.42 Wavelet transforms of Fig. 8.23 with respect to (a) Haar wavelets, (b) Daubechies wavelets, (c) symlets, and (d) Cohen-Daubechies-Feauveau biorthogonal wavelets.

Wavelet	Filter Taps (Scaling + Wavelet)	Zeroed Coefficients
Haar (see Ex. 7.10)	2 + 2	46%
Daubechies (see Fig. 7.6)	8 + 8	51%
Symlet (see Fig. 7.24)	8 + 8	51%
Biorthogonal (see Fig. 7.37)	17 + 11	55%

TABLE 8.12
Wavelet transform filter taps and zeroed coefficients when truncating the transforms in Fig. 8.42 below 1.5.

and additions per coefficient (for each decomposition level) as you move from Fig. 8.42(a) to (d). All four transforms were computed using a fast wavelet transform (i.e., filter bank) formulation. Note that as the computational complexity (i.e., the number of filter taps) increases, the information packing ability does as well. When Haar wavelets are employed and the detail coefficients below 1.5 are truncated to zero, 46% of the total transform is zeroed. With the more complex biorthogonal wavelets, the number of zeroed coefficients rises to 55%, increasing the potential compression by almost 10%. ■

Decomposition level selection

Another factor affecting wavelet coding computational complexity and reconstruction error is the number of transform decomposition levels. Since a P-scale fast wavelet transform involves P filter bank iterations, the number of operations in the computation of the forward and inverse transforms increases with the number of decomposition levels. Moreover, quantizing the increasingly lower-scale coefficients that result with more decomposition levels impacts increasingly larger areas of the reconstructed image. In many applications, like searching image databases or transmitting images for progressive reconstruction, the resolution of the stored or transmitted images and the scale of the lowest useful approximations normally determine the number of transform levels.

■ Table 8.13 illustrates the effect of decomposition level selection on the coding of Fig. 8.23 with a fixed global threshold of 25. As in previous wavelet coding examples, only detail coefficients are truncated. The table lists both the percentage of zeroed coefficients and the resulting rms reconstruction errors. Note that the initial decompositions are responsible for the majority of the data compression. There is little change in the number of truncated coefficients above three decomposition levels. ■

EXAMPLE 8.25:
Decomposition levels in wavelet coding.

Scales and Filter Bank Iterations	Approximation Coefficient Image	Truncated Coefficients (%)	Reconstruction Error (rms)
1	256 × 256	75%	1.93
2	128 × 128	93%	2.69
3	64 × 64	97%	3.12
4	32 × 32	98%	3.25
5	16 × 16	98%	3.27

TABLE 8.13
Decomposition level impact on wavelet coding the 512 × 512 image of Fig. 8.23.

FIGURE 8.43 The impact of dead zone interval selection on wavelet coding.

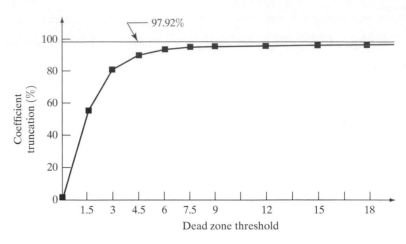

Quantizer design

The largest factor effecting wavelet coding compression and reconstruction error is coefficient quantization. Although the most widely used quantizers are uniform, the effectiveness of the quantization can be improved significantly by (1) introducing an enlarged quantization interval around zero, called a *dead zone*, or (2) adapting the size of the quantization interval from scale to scale. In either case, the selected quantization intervals must be transmitted to the decoder with the encoded image bit stream. The intervals themselves may be determined heuristically or computed automatically based on the image being compressed. For example, a global coefficient threshold could be computed as the median of the absolute values of the first-level detail coefficients or as a function of the number of zeroes that are truncated and the amount of energy that is retained in the reconstructed image.

EXAMPLE 8.26: Dead zone interval selection in wavelet coding.

■ Figure 8.43 illustrates the impact of dead zone interval size on the percentage of truncated detail coefficients for a three-scale biorthogonal wavelet-based encoding of the woman in Fig. 8.23. As the size of the dead zone increases, the number of truncated coefficients does as well. Above the knee of the curve (i.e., beyond 4.5), there is little gain. This is due to the fact that the histogram of the detail coefficients is highly peaked around zero (see, for example, Fig. 7.8).

The rms reconstruction errors corresponding to the dead zone thresholds in Fig. 8.43 grow from 0 to 1.77 gray levels at a threshold of 4.5 and to 2.79 gray levels for a threshold of 18, where the number of zeroes reaches 96.43%. If every detail coefficient were eliminated, that percentage would increase by about 1.5%, but the reconstruction error would grow to 7.6 gray levels.   ■

8.6 Image Compression Standards

Many of the lossy and error-free compression methods described so far play important roles in popular image compression standards. In this section we examine a few of these standards and use them to demonstrate the methods presented earlier. Most of the standards discussed are sanctioned by the International Standardization Organization (ISO) and the Consultative Committee of the International Telephone and Telegraph (CCITT). They address both binary and continuous-tone (monochrome and color) image compression, as well as both still-frame and video (i.e., sequential-frame) applications.

8.6.1 Binary Image Compression Standards

Two of the most widely used image compression standards are the CCITT Group 3 and 4 standards for binary image compression. Although they are currently utilized in a wide variety of computer applications, they were originally designed as facsimile (FAX) coding methods for transmitting documents over telephone networks. The Group 3 standard applies a nonadaptive, 1-D run-length coding technique in which the last $K - 1$ lines of each group of K lines (for $K = 2$ or 4) are optionally coded in a 2-D manner. The Group 4 standard is a simplified or streamlined version of the Group 3 standard in which only 2-D coding is allowed. Both standards use the same nonadaptive 2-D coding approach. This approach is quite similar to the relative address coding (RAC) technique described in Section 8.4.2.

During the development of the CCITT standards, eight representative "test" documents were selected and used as a baseline for evaluating various binary compression alternatives. The existing Group 3 and 4 standards compress these documents, which include both typed and handwritten text (in several languages) as well as a few line drawings, by about 15:1. Because the Group 3 and 4 standards are based on nonadaptive techniques, however, they sometimes result in data expansion (e.g., with half-tone images). To overcome this and related problems, the Joint Bilevel Imaging Group (JBIG)—a joint committee of the CCITT and ISO— has adopted and/or proposed several other binary compression standards. These include JBIG1, an adaptive arithmetic compression technique that provides both the best average and worst-case binary compression available currently, and JBIG2 (now a final committee draft), which achieves compressions that are typically 2 to 4 times greater than JBIG1. These standards can be used to compress both binary and gray-scale images of up to 6 gray-coded bits/pixel (on a bit plane basis.)

One-dimensional compression

In the 1-D CCITT Group 3 compression method, each line of an image[†] is encoded as a series of variable-length code words that represent the run lengths of the alternating white and black runs in a left-to-right scan of the line. The code words themselves are of two types. If the run length is less than 63, a terminating code from the modified Huffman code in Table 8.14 is used. If the run length is greater than 63, the largest possible makeup code (not exceeding the run length) from Table 8.15 is used in conjunction with a terminating code that represents the difference between the makeup code and the actual run length. The standard requires that each line begins with a white run-length code word, which may in fact be 00110101, the code for a white run of length zero. Finally,

---

[†] In the standard, images are referred to as *pages* and sequences of images are referred to as *documents*.

TABLE 8.14
CCITT
terminating codes.

Run Length	White Code Word	Black Code Word	Run Length	White Code Word	Black Code Word
0	00110101	0000110111	32	00011011	000001101010
1	000111	010	33	00010010	000001101011
2	0111	11	34	00010011	000011010010
3	1000	10	35	00010100	000011010011
4	1011	011	36	00010101	000011010100
5	1100	0011	37	00010110	000011010101
6	1110	0010	38	00010111	000011010110
7	1111	00011	39	00101000	000011010111
8	10011	000101	40	00101001	000001101100
9	10100	000100	41	00101010	000001101101
10	00111	0000100	42	00101011	000011011010
11	01000	0000101	43	00101100	000011011011
12	001000	0000111	44	00101101	000001010100
13	000011	00000100	45	00000100	000001010101
14	110100	00000111	46	00000101	000001010110
15	110101	000011000	47	00001010	000001010111
16	101010	0000010111	48	00001011	000001100100
17	101011	0000011000	49	01010010	000001100101
18	0100111	0000001000	50	01010011	000001010010
19	0001100	00001100111	51	01010100	000001010011
20	0001000	00001101000	52	01010101	000000100100
21	0010111	00001101100	53	00100100	000000110111
22	0000011	00000110111	54	00100101	000000111000
23	0000100	00000101000	55	01011000	000000100111
24	0101000	00000010111	56	01011001	000000101000
25	0101011	00000011000	57	01011010	000001011000
26	0010011	000011001010	58	01011011	000001011001
27	0100100	000011001011	59	01001010	000000101011
28	0011000	000011001100	60	01001011	000000101100
29	00000010	000011001101	61	00110010	000001011010
30	00000011	000001101000	62	00110011	000001100110
31	00011010	000001101001	63	00110100	000001100111

a unique end-of-line (EOL) code word 000000000001 is used to terminate each line, as well as to signal the first line of each new image. The end of a sequence of images is indicated by six consecutive EOLs.

Two-dimensional compression

The 2-D compression approach adopted for both the CCITT Group 3 and 4 standards is a line-by-line method in which the position of each black-to-white or white-to-black run transition is coded with respect to the position of a *reference element* a_0 that is situated on the current *coding line*. The previously coded line is called the *reference line*; the reference line for the first line of each new image is an imaginary white line.

TABLE 8.15
CCITT makeup
codes.

Run Length	White Code Word	Black Code Word	Run Length	White Code Word	Black Code Word
64	11011	0000001111	960	011010100	0000001110011
128	10010	000011001000	1024	011010101	0000001110100
192	010111	000011001001	1088	011010110	0000001110101
256	0110111	000001011011	1152	011010111	0000001110110
320	00110110	000000110011	1216	011011000	0000001110111
384	00110111	000000110100	1280	011011001	0000001010010
448	01100100	000000110101	1344	011011010	0000001010011
512	01100101	0000001101100	1408	011011011	0000001010100
576	01101000	0000001101101	1472	010011000	0000001010101
640	01100111	0000001001010	1536	010011001	0000001011010
704	011001100	0000001001011	1600	010011010	0000001011011
768	011001101	0000001001100	1664	011000	0000001100100
832	011010010	0000001001101	1728	010011011	0000001100101
896	011010011	0000001110010			

	Code Word			Code Word
1792	00000001000		2240	000000010110
1856	00000001100		2304	000000010111
1920	00000001101		2368	000000011100
1984	000000010010		2432	000000011101
2048	000000010011		2496	000000011110
2112	000000010100		2560	000000011111
2176	000000010101			

Figure 8.44 shows the basic coding process for a single scan line. Note that the initial steps of the procedure are directed at locating several key transition or *changing elements*: a_0, a_1, a_2, b_1, and b_2. A changing element is defined as a pixel whose value is different from that of the previous pixel on the same line. The most important changing element is a_0 (the reference element), which is either set to the location of an imaginary white changing element to the left of the first pixel of each new coding line or determined from the previous coding mode. After a_0 is located, a_1 is identified as the location of the next changing element to the right of a_0 on the current coding line, a_2 as the next changing element to the right of a_1 on the coding line, b_1 as the changing element of the opposite value (of a_0) and to the right of a_0 on the reference (or previous) line, and b_2 as the next changing element to the right of b_1 on the reference line. If any of these changing elements are not detected, they are set to the location of an imaginary pixel to the right of the last pixel on the appropriate line. Figure 8.45 provides two illustrations of the general relationships between the various changing elements.

After identification of the current reference element and associated changing elements, two simple tests are performed to select one of three possible coding modes: *pass mode, vertical mode*, or *horizontal mode*. The inital test, which

FIGURE 8.44
CCITT 2-D
coding procedure.
The notation
$|a_1 b_1|$ denotes the
absolute value of
the distance
between changing
elements a_1
and b_1.

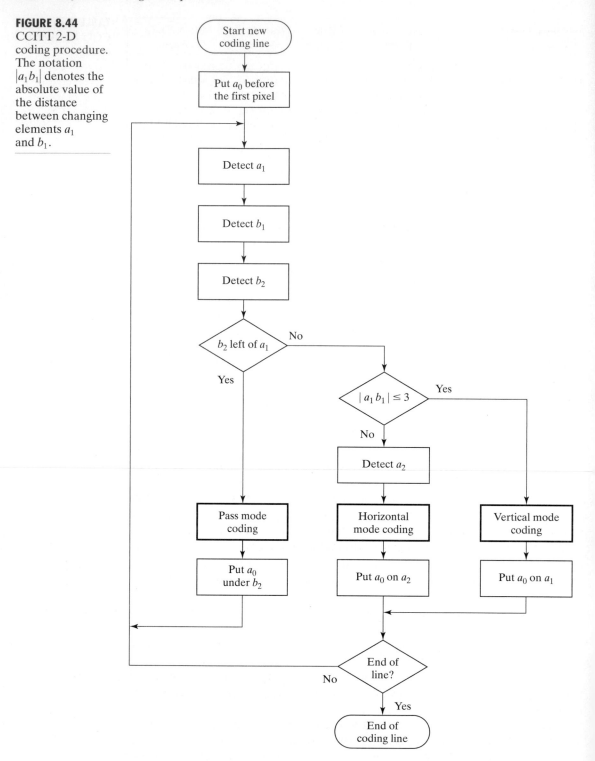

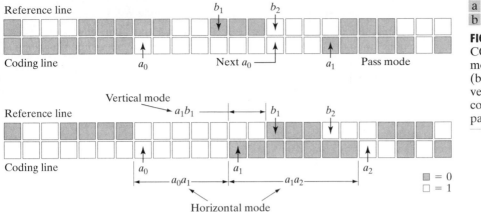

FIGURE 8.45
CCITT (a) pass mode and (b) horizontal and vertical mode coding parameters.

corresponds to the first branch point in the flow chart shown in Fig. 8.44, compares the location of b_2 to that of a_1. The second test, which corresponds to the second branch point in Fig. 8.44, computes the distance (in pixels) between the locations of a_1 and b_1 and compares it against 3. Depending on the outcome of these tests, one of the three outlined coding blocks of Fig. 8.44 is entered and the appropriate coding procedure is executed. A new reference element is then established, as per the flow chart, in preparation for the next coding iteration.

Table 8.16 defines the specific codes utilized for each of the three possible coding modes. In pass mode, which specifically excludes the case in which b_2 is directly above a_1, only the pass mode code word 0001 is needed. As Fig. 8.45(a) shows, this mode identifies white or black reference line runs that do not overlap the current white or black coding line runs. In horizontal coding mode, the distances from a_0 to a_1 and a_1 to a_2 must be coded in accordance with the termination and makeup codes of Tables 8.14 and 8.15 and then appended to the horizontal mode code word 001. This is indicated in Table 8.16 by the notation $001 + M(a_0a_1) + M(a_1a_2)$, where a_0a_1 and a_1a_2 denote the distances from a_0 to a_1 and a_1 to a_2, respectively. Finally, in vertical coding mode, one of six special variable-length codes is assigned to the distance between a_1 and b_1. Figure 8.45(b) illustrates the parameters involved in both horizontal and vertical mode coding. The *extension mode* code word at the bottom of Table 8.16 is used to enter an optional facsimile coding mode. For example, the 0000001111 code is used to initiate an uncompressed mode of transmission.

■ Although Fig. 8.45(b) is annotated with the parameters for both horizontal and vertical mode coding, the depicted condition in reality is a case for vertical mode coding. That is, as b_2 is to the right of a_1, the first (or pass mode) test in Fig. 8.44 fails. The second test, which determines whether the vertical or horizontal coding mode is entered, indicates that vertical mode coding should be used, because the distance from a_1 to b_1 is less than 3. In accordance with Table 8.16, the appropriate code word is 000010, implying that a_1 is two pixels left of b_1. In preparation for the next coding iteration, a_0 is moved to the location of a_1. ■

EXAMPLE 8.27:
CCITT vertical mode coding example.

TABLE 8.16
CCITT two-
dimensional code
table.

Mode	Code Word
Pass	0001
Horizontal	$001 + M(a_0a_1) + M(a_1a_2)$
Vertical	
a_1 below b_1	1
a_1 one to the right of b_1	011
a_1 two to the right of b_1	000011
a_1 three to the right of b_1	0000011
a_1 one to the left of b_1	010
a_1 two to the left of b_1	000010
a_1 three to the left of b_1	0000010
Extension	0000001×××

8.6.2 Continuous Tone Still Image Compression Standards

The CCITT and ISO have defined several continuous tone (as opposed to bi-nary) image compression standards. These standards, which are in various phases of the adoption process, address both monochrome and color image compression. In contrast to the binary compression standards described in Section 8.6.1, continuous tone standards are based principally on the lossy transform coding techniques of Sections 8.5.2 and 8.5.3. To develop the standards, CCITT and ISO committees solicited algorithm recommendations from a large number of companies, universities, and research laboratories. The best of those submitted were selected on the basis of image quality and compression performance. The resulting standards, which include the original DCT-based JPEG standard, the recently proposed wavelet-based JPEG 2000 standard, and the JPEG-LS standard, a lossless to near lossless adaptive prediction scheme that includes a mechanism for flat region detection and run-length coding (ISO/IEC [1999]), represent the state of the art in continuous tone image compression.

JPEG

One of the most popular and comprehensive continuous tone, still frame compression standards is the JPEG standard. It defines three different coding systems: (1) a lossy *baseline coding system*, which is based on the DCT and is adequate for most compression applications; (2) an *extended coding system* for greater compression, higher precision, or progressive reconstruction applications; and (3) a lossless *independent coding system* for reversible compression. To be JPEG compatible, a product or system must include support for the baseline system. No particular file format, spatial resolution, or color space model is specified.

In the baseline system, often called the *sequential baseline system*, the input and output data precision is limited to 8 bits, whereas the quantized DCT values are restricted to 11 bits. The compression itself is performed in three sequential steps: DCT computation, quantization, and variable-length code assignment. The image is first subdivided into pixel blocks of size 8×8, which are processed left to right, top to bottom. As each 8×8 block or subimage is encountered, its 64 pixels are level shifted by subtracting the quantity 2^{n-1}, where 2^n is the maximum number of gray levels. The 2-D discrete cosine transform of the block is then computed, quantized in accordance with Eq. (8.5-40), and reordered, using the zigzag pattern of Fig. 8.36(d), to form a 1-D sequence of quantized coefficients.

Since the one-dimensionally reordered array generated under the zigzag pattern of Fig. 8.36(d) is qualitatively arranged according to increasing spatial frequency, the JPEG coding procedure is designed to take advantage of the long runs of zeros that normally result from the reordering. In particular, the nonzero AC[†] coefficients are coded using a variable-length code that defines the coefficient's value and number of preceding zeros. The DC coefficient is difference coded relative to the DC coefficient of the previous subimage. Tables 8.17, 8.18, and 8.19 provide the default JPEG Huffman codes for luminance imagery. The JPEG recommended luminance quantization array is given in Fig. 8.37(b) and can be scaled to provide a variety of compression levels. Although default coding tables and proven quantization arrays are provided for both luminance and chrominance processing, the user is free to construct custom tables and/or arrays, which may in fact be adapted to the characteristics of the image being compressed.

■ Consider compression and reconstruction of the following 8×8 subimage with the JPEG baseline standard:

52	55	61	66	70	61	64	73
63	59	66	90	109	85	69	72
62	59	68	113	144	104	66	73
63	58	71	122	154	106	70	69
67	61	68	104	126	88	68	70
79	65	60	70	77	68	58	75
85	71	64	59	55	61	65	83
87	79	69	68	65	76	78	94

EXAMPLE 8.28:
JPEG baseline coding and decoding.

The original image consists of 256 or 2^8 possible gray levels, so the coding process begins by level shifting the pixels of the original subimage by -2^7 or -128 gray levels. The resulting shifted array is

−76	−73	−67	−62	−58	−67	−64	−55
−65	−69	−62	−38	−19	−43	−59	−56
−66	−69	−60	−15	16	−24	−62	−55
−65	−70	−57	−6	26	−22	−58	−59
−61	−67	−60	−24	−2	−40	−60	−58
−49	−63	−68	−58	−51	−65	−70	−53
−43	−57	−64	−69	−73	−67	−63	−45
−41	−49	−59	−60	−63	−52	−50	−34

[†] In the standard, the term AC denotes all transform coefficients with the exception of the zeroth or DC coefficient.

TABLE 8.17
JPEG coefficient
coding categories.

Range	DC Difference Category	AC Category
0	0	N/A
−1, 1	1	1
−3, −2, 2, 3	2	2
−7, …, −4, 4, …, 7	3	3
−15, …, −8, 8, …, 15	4	4
−31, …, −16, 16, …, 31	5	5
−63, …, −32, 32, …, 63	6	6
−127, …, −64, 64, …, 127	7	7
−255, …, −128, 128, …, 255	8	8
−511, …, −256, 256, …, 511	9	9
−1023, …, −512, 512, …, 1023	A	A
−2047, …, −1024, 1024, …, 2047	B	B
−4095, …, −2048, 2048, …, 4095	C	C
−8191, …, −4096, 4096, …, 8191	D	D
−16383, …, −8192, 8192, …, 16383	E	E
−32767, …, −16384, 16384, …, 32767	F	N/A

TABLE 8.18
JPEG default DC
code (luminance).

Category	Base Code	Length	Category	Base Code	Length
0	010	3	6	1110	10
1	011	4	7	11110	12
2	100	5	8	111110	14
3	00	5	9	1111110	16
4	101	7	A	11111110	18
5	110	8	B	111111110	20

which, when transformed in accordance with the forward DCT of Eqs. (8.5-24) and (8.5-32) for $N = 8$, becomes

$$
\begin{array}{rrrrrrrr}
-415 & -29 & -62 & 25 & 55 & -20 & -1 & 3 \\
7 & -21 & -62 & 9 & 11 & -7 & -6 & 6 \\
-46 & 8 & 77 & -25 & -30 & 10 & 7 & -5 \\
-50 & 13 & 35 & -15 & -9 & 6 & 0 & 3 \\
11 & -8 & -13 & -2 & -1 & 1 & -4 & 1 \\
-10 & 1 & 3 & -3 & -1 & 0 & 2 & -1 \\
-4 & -1 & 2 & -1 & 2 & -3 & 1 & -2 \\
-1 & -1 & -1 & -2 & -1 & -1 & 0 & -1
\end{array}
$$

TABLE 8.19
JPEG default AC
code (luminance)
(continues on next page).

Run/Category	Base Code	Length	Run/Category	Base Code	Length
0/0	**1010 (= EOB)**	**4**			
0/1	00	3	8/1	11111010	9
0/2	01	4	8/2	111111111000000	17
0/3	100	6	8/3	1111111110110111	19
0/4	1011	8	8/4	1111111110111000	20
0/5	11010	10	8/5	1111111110111001	21
0/6	111000	12	8/6	1111111110111010	22
0/7	1111000	14	8/7	1111111110111011	23
0/8	1111110110	18	8/8	1111111110111100	24
0/9	1111111110000010	25	8/9	1111111110111101	25
0/A	1111111110000011	26	8/A	1111111110111110	26
1/1	1100	5	9/1	111111000	10
1/2	111001	8	9/2	1111111110111111	18
1/3	1111001	10	9/3	1111111111000000	19
1/4	111110110	13	9/4	1111111111000001	20
1/5	11111110110	16	9/5	1111111111000010	21
1/6	1111111110000100	22	9/6	1111111111000011	22
1/7	1111111110000101	23	9/7	1111111111000100	23
1/8	1111111110000110	24	9/8	1111111111000101	24
1/9	1111111110000111	25	9/9	1111111111000110	25
1/A	1111111110001000	26	9/A	1111111111000111	26
2/1	11011	6	Λ/1	111111001	10
2/2	11111000	10	A/2	1111111111001000	18
2/3	1111110111	13	A/3	1111111111001001	19
2/4	1111111110001001	20	A/4	1111111111001010	20
2/5	1111111110001010	21	A/5	1111111111001011	21
2/6	1111111110001011	22	A/6	1111111111001100	22
2/7	1111111110001100	23	A/7	1111111111001101	23
2/8	1111111110001101	24	A/8	1111111111001110	24
2/9	1111111110001110	25	A/9	1111111111001111	25
2/A	1111111110001111	26	A/A	1111111111010000	26
3/1	111010	7	B/1	111111010	10
3/2	111110111	11	B/2	1111111111010001	18
3/3	1111110111	14	B/3	1111111111010010	19
3/4	1111111110010000	20	B/4	1111111111010011	20
3/5	1111111110010001	21	B/5	1111111111010100	21
3/6	1111111110010010	22	B/6	1111111111010101	22
3/7	1111111110010011	23	B/7	1111111111010110	23
3/8	1111111110010100	24	B/8	1111111111010111	24
3/9	1111111110010101	25	B/9	1111111111011000	25
3/A	1111111110010110	26	B/A	1111111111011001	26
4/1	111011	7	C/1	1111111010	11
4/2	1111111000	12	C/2	1111111111011010	18
4/3	1111111110010111	19	C/3	1111111111011011	19
4/4	1111111110011000	20	C/4	1111111111011100	20
4/5	1111111110011001	21	C/5	1111111111011101	21
4/6	1111111110011010	22	C/6	1111111111011110	22
4/7	1111111110011011	23	C/7	1111111111011111	23
4/8	1111111110011100	24	C/8	1111111111100000	24
4/9	1111111110011101	25	C/9	1111111111100001	25
4/A	1111111110011110	26	C/A	1111111111100010	26

TABLE 8.19
JPEG default AC
code (luminance)
(continued).

Run/Category	Base Code	Length	Run/Category	Base Code	Length
5/1	1111010	8	D/1	11111111010	12
5/2	1111111001	12	D/2	1111111111100011	18
5/3	1111111110011111	19	D/3	1111111111100100	19
5/4	1111111110100000	20	D/4	1111111111100101	20
5/5	1111111110100001	21	D/5	1111111111100110	21
5/6	1111111110100010	22	D/6	1111111111100111	22
5/7	1111111110100011	23	D/7	1111111111101000	23
5/8	1111111110100100	24	D/8	1111111111101001	24
5/9	1111111110100101	25	D/9	1111111111101010	25
5/A	1111111110100110	26	D/A	1111111111101011	26
6/1	1111011	8	E/1	111111110110	13
6/2	11111111000	13	E/2	1111111111101100	18
6/3	1111111110100111	19	E/3	1111111111101101	19
6/4	1111111110101000	20	E/4	1111111111101110	20
6/5	1111111110101001	21	E/5	1111111111101111	21
6/6	1111111110101010	22	E/6	1111111111110000	22
6/7	1111111110101011	23	E/7	1111111111110001	23
6/8	1111111110101100	24	E/8	1111111111110010	24
6/9	1111111110101101	25	E/9	1111111111110011	25
6/A	1111111110101110	26	E/A	1111111111110100	26
7/1	11111001	9	**F/0**	**111111110111**	**12**
7/2	1111111001	13	F/1	1111111111110101	17
7/3	1111111110101111	19	F/2	1111111111110110	18
7/4	1111111110110000	20	F/3	1111111111110111	19
7/5	1111111110110001	21	F/4	1111111111111000	20
7/6	1111111110110010	22	F/5	1111111111111001	21
7/7	1111111110110011	23	F/6	1111111111111010	22
7/8	1111111110110100	24	F/7	1111111111111011	23
7/9	1111111110110101	25	F/8	1111111111111100	24
7/A	1111111110110110	26	F/9	1111111111111101	25
			F/A	1111111111111110	26

If the JPEG recommended normalization array of Fig. 8.37(b) is used to quantize the transformed array, the scaled and truncated [that is, normalized in accordance with Eq. (8.5-40)] coefficients are

$$
\begin{array}{rrrrrrrr}
-26 & -3 & -6 & 2 & 2 & 0 & 0 & 0 \\
1 & -2 & -4 & 0 & 0 & 0 & 0 & 0 \\
-3 & 1 & 5 & -1 & -1 & 0 & 0 & 0 \\
-4 & 1 & 2 & -1 & 0 & 0 & 0 & 0 \\
1 & 0 & 0 & 0 & 0 & 0 & 0 & 0 \\
0 & 0 & 0 & 0 & 0 & 0 & 0 & 0 \\
0 & 0 & 0 & 0 & 0 & 0 & 0 & 0 \\
0 & 0 & 0 & 0 & 0 & 0 & 0 & 0
\end{array}
$$

where, for instance, the DC coefficient is computed as

$$\hat{T}(0,0) = \text{round}\left[\frac{T(0,0)}{Z(0,0)}\right]$$

$$= \text{round}\left[\frac{-415}{16}\right] = -26.$$

Note that the transformation and normalization process produces a large number of zero-valued coefficients. When the coefficients are reordered in accordance with the zigzag ordering pattern of Fig. 8.36(d), the resulting 1-D coefficient sequence is

$$[-26 \ -3 \ 1 \ -3 \ -2 \ -6 \ 2 \ -4 \ 1 \ -4 \ 1 \ 1 \ 5 \ 0 \ 2 \ 0 \ 0 \ -1 \ 2 \ 0 \ 0 \ 0 \ 0 \ 0 \ -1 \ -1 \ \text{EOB}]$$

where the EOB symbol denotes the end-of-block condition. A special EOB Huffman code word (see category 0 and run-length 0 in Table 8.19) is provided to indicate that the remainder of the coefficients in a reordered sequence are zeros.

The construction of the default JPEG code for the reordered coefficient sequence begins with the computation of the difference between the current DC coefficient and that of the previously encoded subimage. As the subimage here was taken from Fig. 8.23 and the DC coefficient of the transformed and quantized subimage to its immediate left was −17, the resulting DPCM difference is $[-26 - (-17)]$ or −9, which lies in DC difference category 4 of Table 8.17. In accordance with the default Huffman difference code of Table 8.18, the proper base code for a category 4 difference is 101 (a 3-bit code), while the total length of a completely encoded category 4 coefficient is 7 bits. The remaining 4 bits must be generated from the least significant bits (LSBs) of the difference value. For a general DC difference category (say, category K), an additional K bits are needed and computed as either the K LSBs of the positive difference or the K LSBs of the negative difference minus 1. For a difference of −9, the appropriate LSBs are $(0111) - 1$ or 0110, and the complete DPCM coded DC code word is 1010110.

The nonzero AC coefficients of the reordered array are coded similarly from Tables 8.17 and 8.19. The principal difference is that each default AC Huffman code word depends on the number of zero-valued coefficients preceding the nonzero coefficient to be coded, as well as the magnitude category of the nonzero coefficient. (See the column labeled Run/Category in Table 8.19.) Thus the first nonzero AC coefficient of the reordered array (−3) is coded as 0100. The first 2 bits of this code indicate that the coefficient was in magnitude category 2 and preceded by no zero-valued coefficients (see Table 8.17); the last 2 bits are generated by the same process used to arrive at the LSBs of the DC difference code. Continuing in this manner, the completely coded (reordered) array is

1010110 0100 001 0100 0101 100001 0110 100011 001 100011 001
    001 100101 11100110 110110 0110 11110100 000 1010

where the spaces have been inserted solely for readability. Although it was not needed in this example, the default JPEG code contains a special code word for a run of 15 zeros followed by a zero (see category 0 and run-length F in Table 8.19).

The total number of bits in the completely coded reordered array (and thus the number of bits required to represent the entire 8×8, 8-bit subimage of this example) is 92. The resulting compression ratio is 512/92, or about 5.6:1.

To decompress a JPEG compressed subimage, the decoder must first re-create the normalized transform coefficients that led to the compressed bit stream. Because a Huffman-coded binary sequence is instantaneous and uniquely decodable, this step is easily accomplished in a simple lookup table manner. Here the regenerated array of quantized coefficients is

−26	−3	−6	2	2	0	0	0
1	−2	−4	0	0	0	0	0
−3	1	5	−1	−1	0	0	0
−4	1	2	−1	0	0	0	0
1	0	0	0	0	0	0	0
0	0	0	0	0	0	0	0
0	0	0	0	0	0	0	0
0	0	0	0	0	0	0	0

After denormalization in accordance with Eq. (8.5-42), the array becomes

−416	−33	−60	32	48	0	0	0
12	−24	−56	0	0	0	0	0
−42	13	80	−24	−40	0	0	0
−56	17	44	−29	0	0	0	0
18	0	0	0	0	0	0	0
0	0	0	0	0	0	0	0
0	0	0	0	0	0	0	0
0	0	0	0	0	0	0	0

where, for example, the DC coefficient is computed as

$$\dot{T}(0,0) = \hat{T}(0,0)Z(0,0) = (-26)(16) = -416.$$

The completely reconstructed subimage is obtained by taking the inverse DCT of the denormalized array in accordance with Eqs. (8.5-25) and (8.5-32) to obtain

−70	−64	−61	−64	−69	−66	−58	−50
−72	−73	−61	−39	−30	−40	−54	−59
−68	−78	−58	−9	13	−12	−48	−64
−59	−77	−57	0	22	−13	−51	−60
−54	−75	−64	−23	−13	−44	−63	−56
−52	−71	−72	−54	−54	−71	−71	−54
−45	−59	−70	−68	−67	−67	−61	−50
−35	−47	−61	−66	−60	−48	−44	−44

and level shifting each inverse transformed pixel by $+2^7$ (or $+128$) to yield

58	64	67	64	59	62	70	78
56	55	67	89	98	88	74	69
60	50	70	119	141	116	80	64
69	51	71	128	149	115	77	68
74	53	64	105	115	84	65	72
76	57	56	74	75	57	57	74
83	69	59	60	61	61	67	78
93	81	67	62	69	80	84	84

Any differences between the original and reconstructed subimage are a result of the lossy nature of the JPEG compression and decompression process. In this example, the errors range from -14 to $+11$ and are distributed as follows:

-6	-9	-6	2	11	-1	-6	-5
7	4	-1	1	11	-3	-5	3
2	9	-2	-6	-3	-12	-14	9
-6	7	0	-4	-5	-9	7	1
-7	8	4	-1	11	4	3	-2
3	8	4	-4	2	11	1	1
2	2	5	-1	-6	0	-2	5
-6	-2	2	6	-4	-4	-6	10

The root-mean-square error of the overall compression and reconstruction process is approximately 5.9 gray levels.   ■

   The reconstructed subimage in the preceding example is located physically at about the center of the woman's right eye in Fig. 8.38(a). Note that both the original subimage and reconstructed result contain a local gray-level peak in the fourth row and fifth column, where a light is reflected in the woman's pupil. This local peak causes the root-mean-square error of the reconstructed subimage to exceed substantially the overall error of the completely decompressed image. In fact, it is approximately twice as great as the error associated with Fig. 8.38(a), which also was compressed with the baseline JPEG algorithm. The reason is that many of the subimages of the original image are nearly constant and can be represented with little distortion. Figure 8.38(b) provides an additional JPEG baseline compression result.

JPEG 2000

Although not yet formally adopted, JPEG 2000 extends the initial JPEG standard to provide increased flexibility in both the compression of continuous tone still images and access to the compressed data. For example, portions of a JPEG 2000 compressed image can be extracted for retransmission, storage, display, and/or editing. The standard is based on the wavelet coding techniques of Section 8.5.3. Coefficient quantization is adapted to individual scales and subbands and the quantized coefficients are arithmetically coded on a bit-plane basis (see Section 8.4). Using the notation of the standard, an image is encoded as follows (ISO/IEC [2000]).

The first step of the encoding process is to DC level shift the samples of the *Ssiz*-bit unsigned image to be coded by subtracting 2^{Ssiz-1}. If the image has more than one *component*—like the red, green, and blue planes of a color image—each component is individually shifted. If there are exactly three components, they may be optionally decorrelated using a reversible or nonreversible linear combination of the components. The *irreversible component transform* of the standard, for example, is

$$Y_0(x, y) = 0.299I_0(x, y) + 0.587I_1(x, y) + 0.114I_2(x, y)$$

$$Y_1(x, y) = -0.16875I_0(x, y) - 0.33126I_1(x, y) + 0.5I_2(x, y) \quad (8.6\text{-}1)$$

$$Y_2(x, y) = 0.5I_0(x, y) - 0.41869I_1(x, y) - 0.08131I_2(x, y)$$

where I_0, I_1, and I_2 are the level-shifted input components and Y_0, Y_1, and Y_2 are the corresponding decorrelated components. If the input components are the red, green, and blue planes of a color image, Eq. (8.6-1) approximates the $R'G'B'$ to $Y'C_bC_r$ color video transform (Poynton [1996]).[†] The goal of the transformation is to improve compression efficiency; transformed components Y_1 and Y_2 are difference images whose histograms are highly peaked around zero.

After the image has been level shifted and optionally decorrelated, its components are optionally divided into *tiles*. Tiles are rectangular arrays of pixels that contain the same relative portion of all components. Thus, the tiling process creates *tile components* that can be extracted and reconstructed independently, providing a simple mechanism for accessing and/or manipulating a limited region of a coded image.

The one-dimensional discrete wavelet transform of the rows and columns of each tile component is then computed. For error-free compression, the transform is based on a biorthogonal, 5-3 coefficient scaling and wavelet vector (Le Gall and Tabatabai [1988]). A rounding procedure is defined for non-integer-valued transform coefficients. In lossy applications, a 9-7 coefficient scaling-wavelet vector (Antonini, Barlaud, Mathieu, and Daubechies [1992]) is employed. In either case, the transform is computed using the fast wavelet transform of Section 7.4 or via a *lifting-based* approach (Mallat [1999]). The coefficients needed to construct a 9-7 FWT analysis filter bank are given in Table 8.20. The complementary lifting-based implementation involves six sequential "lifting" and "scaling" operations:

[†] $R'G'B'$ is a gamma corrected, nonlinear version of a linear CIE (International Commission on Illumination) RGB colorimetry value. Y' is luminance and C_b and C_r are color differences (i.e., scaled $B' - Y'$ and $R' - Y'$ values).

Filter Tap	Highpass Wavelet Coefficient	Lowpass Scaling Coefficient
0	−1.115087052456994	0.6029490182363579
±1	0.5912717631142470	0.2668641184428723
±2	0.05754352622849957	−0.07822326652898785
±3	−0.09127176311424948	−0.01686411844287495
±4	0	0.02674875741080976

TABLE 8.20
Impulse responses of the low and highpass analysis filters for an irreversible 9-7 wavelet transform.

$$Y(2n+1) = X(2n+1) + \alpha[X(2n) + X(2n+2)],$$
$$i_0 - 3 \leq 2n + 1 < i_1 + 3$$
$$Y(2n) = X(2n) + \beta[Y(2n-1) + Y(2n+1)], \quad i_0 - 2 \leq 2n < i_1 + 2$$
$$Y(2n+1) = Y(2n+1) + \gamma[Y(2n) + Y(2n+2)],$$
$$i_0 - 1 \leq 2n + 1 < i_1 + 1 \quad (8.6\text{-}2)$$
$$Y(2n) = Y(2n) + \delta[Y(2n-1) + Y(2n+1)], \quad i_0 \leq 2n < i_1$$
$$Y(2n+1) = -K \cdot Y(2n+1), \quad i_0 \leq 2n + 1 < i_1$$
$$Y(2n) = Y(2n)/K, \quad i_0 \leq 2n < i_1.$$

Here, X is the tile component being transformed, Y is the resulting transform, and i_0 and i_1 define the position of the tile component within a component. That is, they are the indices of the first sample of the tile-component row or column being transformed and the one immediately following the last sample. Variable n assumes values based on i_0, i_1, and which of the six operations is being performed. If $n < i_0$ or $n \geq i_1$, $X(n)$ is obtained by symmetrically extending X. For example, $X(i_0 - 1) = X(i_0 + 1), X(i_0 - 2) = X(i_0 + 2), X(i_1) = X(i_1 - 2)$, and $X(i_1 + 1) = X(i_1 - 3)$. At the conclusion of the lifting and scaling operations, the even-indexed values of Y are equivalent to the FWT lowpass filtered output; the odd-indexed values of Y correspond to the highpass FWT filtered result. Lifting parameters α, β, γ, and δ are −1.586134342, −0.052980118, 0.882911075, and 0.433506852, respectively. Scaling factor K is 1.230174105.

The transformation just described produces four subbands—a low-resolution approximation of the tile component and the component's horizontal, vertical, and diagonal frequency characteristics. Repeating the transformation N_L times, with subsequent iterations restricted to the previous decomposition's approximation coefficients, produces an N_L-scale wavelet transform. Adjacent scales are related spatially by powers of 2 and the lowest scale contains the only explicitly defined approximation of the original tile component. As can be surmised from Fig. 8.46, where the notation of the standard is summarized for the case of $N_L = 2$, a general N_L-scale transform contains $3N_L + 1$ subbands whose coefficients are denoted a_b, for $b = N_L LL, N_L HL, \ldots, 1HL, 1LH, 1HH$. The standard does not specify the number of scales to be computed.

When each of the tile components has been processed, the total number of transform coefficients is equal to the number of samples in the original image—

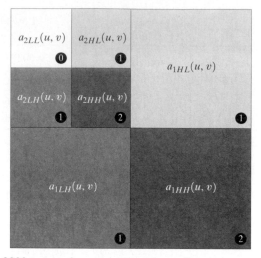

FIGURE 8.46 JPEG 2000 two-scale wavelet transform tile-component coefficient notation and analysis gain.

but the important visual information is concentrated in a few coefficients. To reduce the number of bits needed to represent the transform, coefficient $a_b(u, v)$ of subband b is quantized to value $q_b(u, v)$ using

$$q_b(u, v) = \text{sign}[a_b(u, v)] \cdot \text{floor}\left[\frac{|a_b(u, v)|}{\Delta_b}\right] \tag{8.6-3}$$

where *quantiztion step size* Δ_b is

$$\Delta_b = 2^{R_b - \varepsilon_b}\left(1 + \frac{\mu_b}{2^{11}}\right), \tag{8.6-4}$$

R_b is the *nominal dynamic range* of subband b, and ε_b and μ_b are the number of bits allotted to the *exponent* and *mantissa* of the subband's coefficients. The nominal dynamic range of subband b is the sum of the number of bits used to represent the original image and the *analysis gain* bits for subband b. Subband analysis gain bits follow the simple pattern shown in Fig. 8.46. For example, there are two analysis gain bits for subband $b = 1HH$.

For error-free compression, $\mu_b = 0$, $R_b = \varepsilon_b$, and $\Delta_b = 1$. For irreversible compression, no particular quantization step size is specified in the standard. Instead, the number of exponent and mantissa bits must be provided to the decoder on a subband basis, called *explicit quantization*, or for the $N_L LL$ subband only, called *implicit quantization*. In the latter case, the remaining subbands are quantized using extrapolated $N_L LL$ subband parameters. Letting ε_0 and μ_0 be the number of bits allocated to the $N_L LL$ subband, the extrapolated parameters for subband b are

$$\mu_b = \mu_0 \tag{8.6-5}$$
$$\varepsilon_b = \varepsilon_0 + nsd_b - nsd_0$$

where nsd_b denotes the number of subband decomposition levels from the original image tile component to subband b.

The final steps of the encoding process are *coefficient bit modeling*, arithmetic coding, bit-stream layering, and packetizing. The coefficients of each transformed tile-component's subbands are arranged into rectangular blocks called *code blocks*, which are individually coded a bit plane at a time. Starting from the most significant bit plane with a nonzero element, each bit plane is processed in three passes. Each bit of a bit plane is coded in only one of the three passes, which are called *significance propagation*, *magnitude refinement*, and *cleanup*. The outputs are then arithmetically coded and grouped with similar passes from other code blocks to form *layers*. A layer is an arbitrary number of groupings of coding passes from each code block. The resulting layers are finally partitioned into *packets*, providing an additional method of extracting a spatial region of interest from the total code stream. Packets are the fundamental unit of the encoded code stream.

JPEG 2000 decoders simply invert the operations described previously. After decoding the bit modeled, arithmetically coded, layered, and packetized codestream, a user-selected number of the original image's tile-component subbands are reconstructed. Although the encoder may have encoded M_b bit planes for a particular subband, the user—due to the embedded nature of the code stream—may choose to decode only N_b bit planes. This amounts to quantizing the code block's coefficients using a step size of $2^{M_b - N_b} \cdot \Delta_b$. Any nondecoded bits are set to zero and the resulting coefficients, denoted $\bar{q}_b(u, v)$, are dequantized using

$$R_{q_b}(u, v) = \begin{cases} \left(\bar{q}_b(u, v) + 2^{M_b - N_b(u, v)}\right) \cdot \Delta_b & \bar{q}_b(u, v) > 0 \\ \left(\bar{q}_b(u, v) - 2^{M_b - N_b(u, v)}\right) \cdot \Delta_b & \bar{q}_b(u, v) < 0 \\ 0 & \bar{q}_b(u, v) = 0 \end{cases} \quad (8.6\text{-}6)$$

where $R_{q_b}(u, v)$ denotes a dequantized transform coefficient and $N_b(u, v)$ is the number of decoded bit planes for $\bar{q}_b(u, v)$. The dequantized coefficients are then inverse transformed by column and by row using an FWT^{-1} filter bank whose coefficients are obtained from Table 8.20 and Eq. (7.1-15) or via the following lifting-based operations:

$$X(2n) = K \cdot Y(2n), \quad i_0 - 3 \le 2n < i_1 + 3$$
$$X(2n + 1) = (-1/K) \cdot Y(2n + 1), \quad i_0 - 2 \le 2n + 1 < i_1 + 2$$
$$X(2n) = X(2n) - \delta[X(2n - 1) + X(2n + 1)],$$
$$i_0 - 3 \le 2n < i_1 + 3 \quad (8.6\text{-}7)$$
$$X(2n + 1) = X(2n + 1) - \gamma[X(2n) + X(2n + 2)],$$
$$i_0 - 2 \le 2n + 1 < i_1 + 2$$
$$X(2n) = X(2n) - \beta[X(2n - 1) + X(2n + 1)], \quad i_0 - 1 \le 2n < i_1 + 1$$
$$X(2n + 1) = X(2n + 1) - \alpha[X(2n) + X(2n + 2)], \quad i_0 \le 2n + 1 < i_1$$

where parameters $\alpha, \beta, \gamma, \delta$, and K are as defined for Eq. (8.6-2). Dequantized coefficient row or column element $Y(n)$ is symmetrically extended when necessary. The final decoding steps are the assembly of the component tiles, inverse component transformation (if required), and DC level shifting. For irreversible coding, the inverse component transformation is

$$
\begin{aligned}
I_0(x, y) &= Y_0(x, y) + 1.402 Y_2(x, y) \\
I_1(x, y) &= Y_0(x, y) - 0.34413 Y_1(x, y) - 0.71414 Y_2(x, y) \qquad (8.6\text{-}8) \\
I_2(x, y) &= Y_0(x, y) + 1.772 Y_1(x, y)
\end{aligned}
$$

and the transformed pixels are shifted by $+2^{Ssiz-1}$. Figures 8.40 and 8.41 of Section 8.5.3, which illustrate compression ratios from $34:1$ to $167:1$, were generated using the irreversible JPEG 2000 algorithm.

8.6.3 Video Compression Standards

Video compression standards extend the transform-based, still image compression techniques of the previous section to include methods for reducing temporal or frame-to-frame redundancies. Although there are a variety of video coding standards in use today, most rely on similar video compression techniques. Depending on the intended application, the standards can be grouped into two broad categories: (1) video teleconferencing standards and (2) multimedia standards.

A number of video teleconferencing standards, including H.261 (also referred to as PX64), H.262, H.263, and H.320, have been defined by the International Telecommunications Union (ITU), the successor to the CCITT. H.261 is intended for operation at affordable telecom bit rates and to support full motion video transmission over T1[†] lines with delays of less than 150 ms. Delays exceeding 150 ms do not provide viewers the "feeling" of direct visual feedback. H.263, on the other hand, is designed for very low bit rate video, in the range of 10 to 30 kbit/s, and H.320, a superset of H.261, is constructed for Integrated Services Digital Network[‡] (ISDN) bandwidths. Each standard uses a motion-compensated, DCT-based coding scheme. Since motion estimation is difficult to perform in the transform domain, blocks of pixels, called *macroblocks*, are compared to neighboring blocks of the previous frame and used to compute a motion compensated prediction error. The prediction error is then discrete cosine transformed in 8×8 pixel blocks, quantized, and coded for transmission or storage.

Multimedia video compression standards for video on demand, digital HDTV broadcasting, and image/video database services use similar motion estimation and coding techniques. The principal standards—MPEG-1, MPEG-2, and MPEG-4—were developed under the auspices of the Motion Picture Experts Group of the CCITT and ISO. MPEG-1 is an "entertainment quality" coding standard for the storage and retrieval of video on digital media like compact disk read-only memories (CD-ROMs). It supports bit rates on the order of

---

[†] The T1 line was introduced by the Bell system for digital voice communications over short distances of 10 to 50 miles. Twenty-four telephone channels are time-multiplexed, sampled, and coded into a 1.544 Mbit/s PCM (pulse code modulation) signal for transmission over a single T1 line.

[‡] Two ISDN "B" channels provide sufficient bandwidth (i.e., 128 kbit/s) to transmit compressed 320×240 images at 15 frames per second.

1.5 Mbit/s. MPEG-2 addresses applications involving video quality between NTSC/PAL[†] and CCIR 601[‡]; bit rates from 2 to 10 Mbit/s, a range that is suitable for cable TV distribution and narrow-channel satellite broadcasting, are supported. The goal of both MPEG-1 and MPEG-2 is to make the storage and transmission of digital audio and video (AV) material efficient. MPEG-4, on the other hand, provides (1) improved video compression efficiency; (2) content-based interactivity, such as AV object-based access and efficient integration of natural and synthetic data; and (3) universal access, including increased robustness in error-prone environments, the ability to add or drop AV objects, and object resolution scalability. Although these functionalities create a need to segment arbitrarily shaped video objects, segmentation is not a part of the standard. A great deal of video content—like computer games—is produced and readily available in the form of video objects. MPEG-4 targets bit rates between 5 and 64 kbit/s for mobile and public switched telephone network (PSTN) applications and up to 4 Mbit/s for TV and film applications. In addition, it supports both constant bit rate and variable bit rate coding.

Like the ITU teleconferencing standards, MPEG standards are built around a hybrid block-based DPCM/DCT coding scheme. Figure 8.47 shows a typical MPEG encoder. It exploits redundancies within and between adjacent video frames, motion uniformity between frames, and the psychovisual properties of the human visual system. The input of the encoder is an 8×8 array of pixels, called an *image block*. The standards define a *macroblock* as a 2×2 array of image blocks (i.e., a 16×16 array of image elements) and a *slice* as a row of nonoverlapping macroblocks. For color video, a macroblock, is composed of four luminance blocks, denoted Y_1 through Y_4, and two chrominance blocks, C_b and C_r. Recall that color difference signal C_b is blue minus luminance and C_r is red minus luminance. Because the eye has far less spatial acuity for color than for luminance, these two components are often sampled at half the horizontal and vertical resolution of the luminance signal, resulting in a $4:1:1$ sampling ratio between $Y' : C_b : C_r$.

The grayed elements of the primary input-to-output path in Fig. 8.47 parallel the transformation, quantization, and variable-length coding operations of a JPEG encoder. The principal difference is the input, which may be a conventional block of image data or the difference between a conventional block and a prediction of it based on similar blocks in previous and/or subsequent video frames. This leads to three basic types of encoded output frames:

1. *Intraframe* or *independent frame* (*I-frame*). An I-frame is compressed independently of all previous and future video frames. Of the three possible encoded output frames, it most highly resembles a JPEG encoded image. Moreover, it is the reference point for the motion estimation needed to generate subsequent *P-* and *B-frames*. I-frames provide the highest degree of

---

[†] NTSC and PAL are acronyms for the National Television System Committee and Phase Alternate Line, respectively. Both are composite color video standards.

[‡] CCIR is an acronym for the International Radio Consultive Committee.

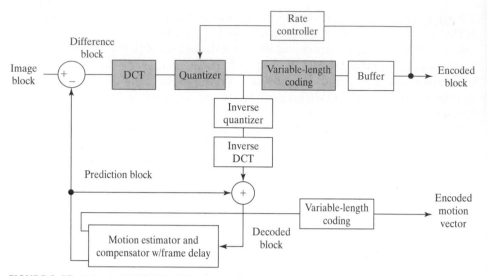

FIGURE 8.47 A basic DPCM/DCT encoder for motion compensated video compression.

random access, ease of editing, and greatest resistance to the propagation of transmission error. As a result, all standards require their periodic insertion into the compressed codestream.

2. *Predictive frame (P-frame).* A P-frame is the compressed difference between the current frame and a prediction of it based on the previous I- or P-frame. The difference is formed in the leftmost summer of Fig. 8.47. The prediction is motion compensated and typically involves sliding the decoded block in the lower part of Fig. 8.47 around its immediate neighborhood in the current frame and computing a measure of correlation (such as the sum of the square of the pixel-by-pixel differences). In fact, the process is often carried out in subpixel increments (such as sliding the subimage $\frac{1}{4}$ pixels at a time), which necessitates interpolating pixel values prior to computing the correlation measure. The computed motion vector is variable-length coded and transmitted as an integral part of the encoded data stream. Motion estimation is carried out on the macroblock level.

3. *Bidirectional frame (B-frame).* A B-frame is the compressed difference between the current frame and a prediction of it based on the previous I- or P-frame and next P-frame. Accordingly, the decoder must have access to both past and future reference frames. The encoded frames are therefore reordered before transmission; the decoder reconstructs and displays them in the proper sequence.

The encoder of Fig. 8.47 is designed to generate a bit stream that matches the capacity of the intended video channel. To accomplish this, the quantization parameters are adjusted by the *rate controller* as a function of the occupancy of the output *buffer*. As the buffer becomes fuller, the quantization is made coarser, so that fewer bits stream into the buffer.

Summary

The principal objectives of this chapter were to present the theoretic foundation of digital image compression and to describe the most commonly used compression methods that form the core of the technology as it exists currently. Although the level of the presentation is introductory in nature, the depth and breadth of the material covered are sufficient to serve as the basis for independent reading in this subject area. The references provide an entry into the extensive body of literature dealing with image compression and related topics. In addition to extensive uses involving gray-scale imagery, compression methods are playing an increasingly important role in document image storage and transmission, as evidenced by the emergence of the international standards discussed in Section 8.6. In addition to medical imaging, compression is one of the few areas of image processing that has received a sufficiently broad commercial appeal to warrant the adoption of widely accepted standards.

References and Further Reading

The introductory material of the chapter, which is generally confined to Sections 8.1 and 8.2, is basic to image compression and may be found in one form or another in most of the general image processing books cited at the end of Chapter 1. The material in Section 8.1.3 on improved gray-scale quantization is based on Bisignani, Richards, and Whelan [1966]. For additional information on the human visual system, see Netravali and Limb [1980], as well as in Huang [1966], Schreiber and Knapp [1958], and the references cited at the end of Chapter 2. Subjective fidelity criteria are discussed in Frendendall and Behrend [1960]. Error detecting and correcting codes are covered in most introductory texts on switching or finite automata theory, as well as in general information theory texts.

The material in Section 8.3 is based on several excellent books on information theory. Noteworthy are Abramson [1963], Blahut [1987], and Berger [1971]. Shannon's classic paper, "A Mathematical Theory of Communication" [1948], lays the foundation for most of the material in the section and is another excellent reference.

The descriptions of the error-free encoding techniques of Section 8.4 are, for the most part, based on the original papers cited in the text or as follows. The algorithms covered are representative of the work in this area, but are by no means exhaustive. The material on LZW coding has its origins in the work of Ziv and Lempel [1977, 1978]. The material on arthmetic coding follows the development in Witten, Neal, and Cleary [1987]. One of the more important implementations of arithmetic coding is summarized in Pennebaker et al. [1988]. For additional information on bit-plane coding, see Schwartz and Barker [1966] and the tutorial by Rabbani and Jones [1991], which also contains a good discussion of lossless predictive coding. Huang and Hussian [1975] first published the details of white-block skipping. Relative address coding and predictive differential quantizing were first reported by Yamazaki, Wakahara, and Teramura [1976] and Huang and Tretiak [1972], respectively. The adaptive predictor of Eq. (8.5-19) is from Graham [1958].

The material in Section 8.5 covers the principal lossy encoding approaches. Various other methods are directly based on these techniques. Noteworthy among them are *hybrid encoding* (Habibi [1974]), a scheme that combines 1-D transform coding and DPCM to obtain about the same performance as 2-D transform coding using fewer computations; *subband coding* (Woods and O'Neil [1986]), in which an image is filtered into a set of images (with different spatial frequencies) that may be individually DPCM coded; and *interframe coding* (Roese et al. [1977]), where the redundancy between successive frames in a time sequence of images is reduced by using a predictive or transform coding approach. In addition, a variety of lossy techniques are closely related to the techniques described. These include, among others, *block truncation coding* (Delp and Mitchell [1979]), in which a 1-bit quantizer is designed for each $n \times n$ block of a subdivided image; *vector quantization* (Linde et al. [1980]), in which an image is decomposed into vectors (containing pixels, transform coefficients, and so on) that are matched against a code book of possible vectors and coded to indicate the best fit; and *hierarchical coding* (Knowlton [1980]), which usually involves the generation of a pyramid-structured data set that can be progressively accessed to obtain better and better representations of the original image. These references do not necessarily cite the inventor of the techniques; they provide a starting point for additional reading on the methods. Other articles or books of interest include Tasto and Wintz [1971], Gharavi and Tabatabai [1988], Baylon and Lim [1990], Candy et al. [1971], Jain and Jain [1981], Healy and Mitchell [1981], Lema and Mitchell [1984], Udpikar and Raina [1987], Gray [1984], Equitz [1989], Sezan et al. [1989], Tanimoto [1979], Blume and Fand [1989], Rabbani and Jones [1991], and Storer and Reif [1991]. Almost every issue of the *IEEE Transactions on Image Processing* includes several articles on video and still image compression, with many of the articles related to wavelet- and fractal-based compression, vector quantization, and video motion compensation. See, for example, Boulgouris et al. [2001], Martin and Bell [2001], Chen and Wilson [2000], Hartenstein et al. [2000], Yang and Ramchandran [2000], and Meyer et al. [2000] as a starting point for further reading and references.

Section 8.6 is based primarily on the published drafts and formal standards of the International Standards Organization and the Consultative Committee of International Telephone and Telegraph. These documents are available from the standards organizations or the American National Standards Institute (ANSI). Additional references on compression standards include Hunter and Robinson [1980], Ang et al. [1991], Fox [1991], Pennebaker and Mitchell [1992], Bhatt et al. [1997], Sikora [1997], Bhaskaran and Konstantinos [1997], Ngan et al. [1999], Weinberger et al. [2000], and Symes [2001].

Several survey articles have been devoted to the field of image compression. Noteworthy are Netravali and Limb [1980], A. K. Jain [1981], a special issue on picture communication systems in the *IEEE Transactions on Communications* [1981], a special issue on the encoding of graphics in the *Proceedings of IEEE* [1980], a special issue on visual communication systems in the *Proceedings of the IEEE* [1985], a special issue on image sequence compression in the *IEEE Transactions on Image Processing* [1994], and a special issue on vector quantization in the *IEEE Transactions on Image Processing* [1996].

Problems

8.1   (a) Can variable-length coding procedures be used to compress a histogram equalized image with 2^n gray levels? Explain.

(b) Can such an image contain interpixel redundancies that could be exploited for data compression?

8.2   One variation of the run-length coding procedure described in Section 8.1.2 involves (1) coding only the runs of 0's or 1's (not both) and (2) assigning a special code to the start of each line to reduce the effect of transmission errors. One possible code pair is (x_k, r_k), where x_k and r_k represent the kth run's starting coordinate and run length, respectively. The code $(0, 0)$ is used to signal each new line.

(a) Derive a general expression for the maximum average runs per scan line required to guarantee data compression when run-length coding a $2^n \times 2^n$ binary image.

(b) Compute the maximum allowable value for $n = 10$.

★**8.3**   Consider an 8-pixel line of gray-scale data, {12, 12, 13, 13, 10, 13, 57, 54}, which has been uniformly quantized with 6-bit accuracy. Construct its 3-bit IGS code.

8.4   Compute the rms error and rms signal-to-noise ratios for the decoded IGS data of Problem 8.3.

8.5   **(a)** Use the Hamming (7, 4) code to code the IGS quantized data of Table 8.2.

   ★ **(b)** Determine which bit, if any, is in error in the Hamming encoded messages 1100111, 1100110, and 1100010. What are the decoded values?

★**8.6**   The base e unit of information is commonly called a *nat*, and the base 10 information unit is called a *Hartley*. Compute the conversion factors needed to relate these units to the base 2 unit of information (the bit).

★**8.7**   Prove that, for a zero-memory source with q symbols, the maximum value of the entropy is $\log q$, which is achieved if and only if all source symbols are equiprobable. *Hint*: Consider the quantity $\log q - H(\mathbf{z})$ and note the inequality $\ln x \le x - 1$.

8.8   Calculate the various probabilities associated with the information channel in which $A = \{0, 1\}$, $B = \{0, 1\}$, $\mathbf{z} = [0.75, 0.25]^T$ and

$$\mathbf{Q} = \begin{bmatrix} \dfrac{2}{3} & \dfrac{1}{3} \\ \dfrac{1}{10} & \dfrac{9}{10} \end{bmatrix}.$$

Include $P(a = 0)$, $P(a = 1)$, $P(b = 0)$, $P(b = 1)$, $P(b = 0|a = 0)$, $P(b = 0|a = 1)$, $P(b = 1|a = 0)$, $P(b = 1|a = 1)$, $P(a = 0|b = 0)$, $P(a = 0|b = 1)$, $P(a = 1|b = 0)$, $P(a = 1|b = 1)$, $P(a = 0,b = 0)$, $P(a = 0, b = 1)$, $P(a = 1, b = 0)$, and $P(a = 1, b = 1)$.

★**8.9**   Consider the binary information source and BSC of the example in Section 8.3.2 and let $p_{bs} = 3/4$ and $p_e = 1/3$.

   (a) What is the entropy of the source?

   (b) How much less uncertainty about the input is there when the output has been observed?

   (c) What is this difference in uncertainty called and how does it compare numerically to the channel's capacity?

8.10   A *binary erasure channel* is one in which there is a finite probability β that a transmitted symbol will not be received. The channel has three possible outputs: a 0, an erasure (no received symbol), and a 1. These three outcomes form the three rows of the binary erasure channel matrix

$$\mathbf{Q} = \begin{bmatrix} 1 - \beta & 0 \\ \beta & \beta \\ 0 & 1 - \beta \end{bmatrix}.$$

   (a) Find the capacity of the channel.

   ★ **(b)** Would you prefer a binary symmetric channel with a 0.125 probability of error or an erasure channel with probability of erasure $\beta = 0.5$?

8.11   The rate distortion function of a zero-memory Gaussian source of arbitrary mean and variance σ^2 with respect to the mean-square error criterion (Berger [1971]) is

$$R(D) = \begin{cases} \frac{1}{2} \log \frac{\sigma^2}{D} & \text{for } 0 \leq D \leq \sigma^2 \\ 0 & \text{for } D \geq \sigma^2. \end{cases}$$

★ **(a)** Plot this function.

(b) What is $D_{\max}$?

(c) If a distortion of no more than 75% of the source's variance is allowed, what is the maximum compression that can be achieved?

8.12   **(a)** How many unique Huffman codes are there for a three-symbol source?

(b) Construct them.

8.13   **(a)** Compute the entropy of the source whose symbol probabilities are defined in Table 8.1.

(b) Construct a Huffman code for the source symbols and explain any differences between the constructed code and Code 2 of the table.

(c) Construct the best B_1-code for this distribution.

(d) Construct the best 2-bit binary shift code.

(e) Divide the symbols into two blocks of four and construct the best Huffman shift code.

(f) Compute the average word lengths for each code and compare them to the entropy from part (a).

★ **8.14**   The arithmetic decoding process is the reverse of the encoding procedure. Decode the message 0.23355 given the coding model

Symbol	Probability
a	0.2
e	0.3
i	0.1
o	0.2
u	0.1
!	0.1

8.15   Use the LZW coding algorithm of Section 8.4.2 to encode the 7-bit ASCII string "aaaaaaaaaa".

★ **8.16**   Devise an algorithm for decoding the LZW encoded output of Example 8.12. Since the dictionary that was used during the encoding is not available, the code book must be reproduced as the output is decoded.

8.17   **(a)** Construct the entire 4-bit Gray code.

(b) Create a general procedure for converting a Gray-coded number to its binary equivalent and use it to decode 0111010100111.

8.18   A 64 × 64 pixel binary image has been coded using 1-D WBS with blocks of four pixels. The WBS code for one line of the image was 0110010000001000010010000000, where a 0 is used to represent a black pixel.

(a) Decode the line.

(b) Create a 1-D iterative WBS procedure that begins by looking for all white lines (a 64-pixel block) and successively halves nonwhite intervals until four pixel blocks are reached.

(c) Use your algorithm to code the previously decoded line. It should require fewer bits.

8.19 ★ (a) Explain why the first similar transition past e on the previous line is used as c' in relative address coding.

(b) Can you devise an alternate approach?

8.20 An image whose autocorrelation function is of the form of Eq. (8.5-12) with $\rho_h = 0$ is to be DPCM coded using a second-order predictor.

(a) Form the autocorrelation matrix $\mathbf{R}$ and vector $\mathbf{r}$.

(b) Find the optimal prediction coefficients.

(c) Compute the variance of the prediction error that would result from using the optimal coefficients.

★8.21 Derive the Lloyd-Max decision and reconstruction levels for $L = 4$ and the uniform probability density function

$$p(s) = \begin{cases} \dfrac{1}{2A} & -A \le s \le A \\ 0 & \text{otherwise.} \end{cases}$$

8.22 Use the CCITT Group 4 compression algorithm to code the second line of the following two-line segment:

```
0 1 1 0 0 1 1 1 0 0 1 1 1 1 1 1 1 1 1 0 0 0 0 1
1 1 1 1 1 1 1 0 0 0 1 1 1 0 0 0 0 1 1 1 1 1 1 1
```

Assume that the initial reference element a_0 is located on the first pixel of the second line segment.

★8.23 (a) List all the members of the JPEG DC coefficient difference category 3.

(b) Compute their default Huffman codes using Table 8.18.

8.24 A radiologist from a well-known research hospital recently attended a medical conference at which a system that could transmit 4096×4096 12-bit digitized X-ray images over standard T1 phone lines was exhibited. The system transmitted the images in a compressed form using a *progressive* technique in which a reasonably good approximation of the X-ray was first reconstructed at the viewing station and then refined gradually to produce an error-free display. The transmission of the data needed to generate the first approximation took approximately 5 or 6 s. Refinements were made every 5 or 6 s (on the average) for the next 1 min, with the first and last refinements having the most and least significant impact on the reconstructed X-ray, respectively. The physician was favorably impressed with the system, because she could begin her diagnosis by using the first approximation of the X-ray and complete it as the error-free reconstruction of the X-ray was being generated. Upon returning to her office, she submitted a purchase request to the hospital administrator. Unfortunately, the hospital was on a relatively tight budget, which recently had been stretched thinner by the hiring of an aspiring young electrical engineering graduate. To appease the radiologist, the administrator gave the young engineer the task of designing such a system. (He thought it might be cheaper to design and build a similar system in-house. The hospital currently owned some of the elements of such a system, but the transmission of the raw X-ray data took more than 2 min.) The administrator asked the engineer to have an initial block diagram by the afternoon staff meeting. With little time and only a copy of *Digital Image Processing* (this text, of course) from his recent school days in hand, the engineer was able to devise conceptually a system to satisfy the transmission and associated compression requirements. Construct a conceptual block diagram of such a system, specifying the compression techniques you would recommend.

8.25   Show that the lifting-based wavelet transform defined by Eq. (8.6-2) is equivalent to the traditional FWT filter bank implementation using the coefficients in Table 8.20. Define the filter coefficients in terms of $\alpha, \beta, \gamma, \delta$, and K.

8.26   Compute the quantization step sizes of the subbands for a JPEG 2000 encoded image in which implicit quantization is used and 8 bits are allotted to the mantissa and exponent of the $2LL$ subband.

★ **8.27**   Draw the block diagram of the companion MPEG decoder for the encoder in Fig. 8.47.

9 Morphological Image Processing

> In form and feature, face and limb,
> I grew so like my brother
> That folks got taking me for him
> And each for one another.
>
> *Henry Sambrooke Leigh, Carols of Cockayne, The Twins*

Preview

The word *morphology* commonly denotes a branch of biology that deals with the form and structure of animals and plants. We use the same word here in the context of *mathematical morphology* as a tool for extracting image components that are useful in the representation and description of region shape, such as boundaries, skeletons, and the convex hull. We are interested also in morphological techniques for pre- or postprocessing, such as morphological filtering, thinning, and pruning.

The language of mathematical morphology is set theory. As such, morphology offers a unified and powerful approach to numerous image processing problems. Sets in mathematical morphology represent objects in an image. For example, the set of all black pixels in a binary image is a complete morphological description of the image. In binary images, the sets in question are members of the 2-D integer space Z^2 (see Section 2.4.2), where each element of a set is a tuple (2-D vector) whose coordinates are the (x, y) coordinates of a black (or white, depending on convention) pixel in the image. Gray-scale digital images can be represented as sets whose components are in Z^3. In this case, two components of each element of the set refer to the coordinates of a pixel, and the third corresponds to its discrete gray-level value. Sets in higher dimensional spaces can contain other image attributes, such as color and time varying components.

In the following sections we develop and illustrate several important concepts in mathematical morphology. Many of the ideas introduced here can be formulated in terms of n-dimensional Euclidean space, E^n. However, our interest initially is on binary images whose components are elements of Z^2. We discuss extensions to gray-scale images in Section 9.6.

The material in this chapter begins a transition from a focus on purely image processing methods, whose input and output are images, to processes in which the inputs are images, but the outputs are attributes extracted from those images, in the sense defined in Section 1.1. Tools like morphology and related concepts are a cornerstone of the mathematical foundation that is utilized for extracting "meaning" from an image. Other approaches are developed and applied in the remaining chapters of this book.

9.1 Preliminaries

In this section we introduce some basic concepts from set theory that are needed as foundation for the remaining sections of this chapter.

9.1.1 Some Basic Concepts from Set Theory

Let A be a set in Z^2. If $a = (a_1, a_2)$ is an element of A, then we write

$$a \in A. \qquad (9.1\text{-}1)$$

Similarly, if a is not an element of A, we write

$$a \notin A. \qquad (9.1\text{-}2)$$

The set with no elements is called the *null* or *empty set* and is denoted by the symbol $\varnothing$.

A set is specified by the contents of two braces: $\{\cdot\}$. The elements of the sets with which we are concerned in this chapter are the *coordinates* of pixels representing objects or other features of interest in an image. For example, when we write an expression of the form $C = \{w \mid w = -d, \text{ for } d \in D\}$ we mean that set C is the set of elements, w, such that w is formed by multiplying each of the two coordinates of all the elements of set D by -1.

If every element of a set A is also an element of another set B, then A is said to be a *subset* of B, denoted as

$$A \subseteq B. \qquad (9.1\text{-}3)$$

The *union* of two sets A and B, denoted by

$$C = A \cup B, \qquad (9.1\text{-}4)$$

is the set of all elements belonging to either A, B, or both. Similarly, the *intersection* of two sets A and B, denoted by

$$D = A \cap B, \qquad (9.1\text{-}5)$$

is the set of all elements belonging to both A and B.

Two sets A and B are said to be *disjoint* or *mutually exclusive* if they have no common elements. In this case,

$$A \cap B = \varnothing. \qquad (9.1\text{-}6)$$

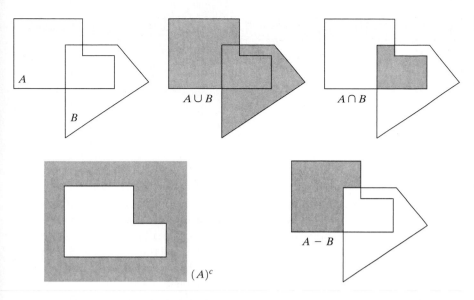

FIGURE 9.1
(a) Two sets A and B. (b) The union of A and B. (c) The intersection of A and B. (d) The complement of A. (e) The difference between A and B.

The *complement* of a set A is the set of elements not contained in A:

$$A^c = \{w \mid w \notin A\}. \qquad (9.1\text{-}7)$$

The *difference* of two sets A and B, denoted $A - B$, is defined as

$$A - B = \{w \mid w \in A, w \notin B\} = A \cap B^c. \qquad (9.1\text{-}8)$$

We see that this is the set of elements that belong to A, but not to B. Figure 9.1 illustrates the preceding concepts. The result of the set operation indicated in each figure is shown in gray.

We need two additional definitions that are used extensively in morphology but generally are not found in basic texts on set theory. The *reflection* of set B, denoted $\hat{B}$, is defined as

$$\hat{B} = \{w \mid w = -b, \quad \text{for} \quad b \in B\}. \qquad (9.1\text{-}9)$$

The *translation* of set A by point $z = (z_1, z_2)$, denoted $(A)_z$, is defined as

$$(A)_z = \{c \mid c = a + z, \quad \text{for} \quad a \in A\}. \qquad (9.1\text{-}10)$$

Figure 9.2 illustrates these two definitions using the sets from Fig. 9.1. The black dot identifies the origin of the sets shown in the figure.

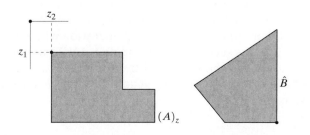

FIGURE 9.2
(a) Translation of A by z.
(b) Reflection of B. The sets A and B are from Fig. 9.1.

TABLE 9.1
The three basic logical operations.

p	q	p AND q (also $p \cdot q$)	p OR q (also $p + q$)	NOT (p) (also $\bar{p}$)
0	0	0	0	1
0	1	0	1	1
1	0	0	1	0
1	1	1	1	0

9.1.2 Logic Operations Involving Binary Images

The majority of applications based on the morphological concepts discussed in this chapter involve binary images. Logic operations, although simple in nature, provide a powerful complement to implementation of image processing algorithms based on morphology. Logic operations were introduced in Section 3.4 in connection with masking. In the following discussion, we are interested in logic operations involving binary pixels and images.

The principal logic operations used in image processing are AND, OR, and NOT (COMPLEMENT). Their properties are summarized in Table 9.1. These operations are *functionally complete* in the sense that they can be combined to form any other logic operation.

Logic operations are performed on a pixel by pixel basis between corresponding pixels of two or more images (except NOT, which operates on the pixels of a single image). Because the AND operation of two binary variables is 1 only when both variables are 1, the result at any location in a resulting AND image is 1 only if the corresponding pixels in the two input images are 1. Figure 9.3 shows various examples of logic operations involving images, where black indicates a binary 1 and white indicates a 0. (We use both conventions in this chapter, sometimes reversing the binary meaning of dark [black or gray] and light [white], depending on which is clearer in a given situation.) Other logic operations are easily constructed using the definitions in Table 9.1. For instance, the XOR (exclusive OR) operation yields a 1 when one or the other pixel (but *not* both) is 1, and it yields a 0 otherwise. This operation is unlike the OR operation, which is 1 when one or the other pixel is 1, or when both pixels are 1. Similarly, the NOT-AND operation selects the black pixels that simultaneously are in B, and not in A.

It is important to note that the logic operations just described have a one-to-one correspondence with the set operations discussed in Section 9.1.1, with the limitation that logic operations are restricted to binary variables, which is not the case in general for set operations. Thus, for example, the intersection operation in set theory reduces to the AND operation when the variables involved are binary. Terms such as *intersection* and *AND* (and even their notation) often are used interchangeably in the literature to denote general or binary set operations, with the meaning generally being clear from the context of the discussion.

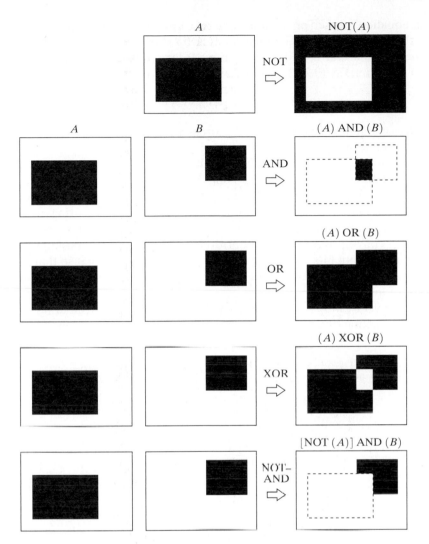

FIGURE 9.3 Some logic operations between binary images. Black represents binary 1s and white binary 0s in this example.

9.2 Dilation and Erosion

We begin the discussion of morphological operations by treating in some detail two operations: *dilation* and *erosion*. These operations are fundamental to morphological processing. In fact, many of the morphological algorithms discussed in this chapter are based on these two primitive operations.

9.2.1 Dilation

With A and B as sets in Z^2, the *dilation* of A by B, denoted $A \oplus B$, is defined as

$$A \oplus B = \{z \mid (\hat{B})_z \cap A \neq \varnothing\}. \tag{9.2-1}$$

This equation is based on obtaining the reflection of B about its origin and shifting this reflection by z. The dilation of A by B then is the set of all *displacements*, z, such that $\hat{B}$ and A overlap by at least one element. Based on this interpretation, Eq. (9.2-1) may be rewritten as

$$A \oplus B = \{z | [(\hat{B})_z \cap A] \subseteq A\}. \tag{9.2-2}$$

Set B is commonly referred to as the *structuring element* in dilation, as well as in other morphological operations.

Equation (9.2-1) is not the only definition of dilation in the current literature on morphology (see Problems 9.10 and 9.11 for two different, yet equivalent, definitions). However, the preceding definition has a distinct advantage over other formulations in that it is more intuitive when the structuring element B is viewed as a convolution mask. Although dilation is based on set operations, whereas convolution is based on arithmetic operations, the basic process of "flipping" B about its origin and then successively displacing it so that it slides over set (image) A is analogous to the convolution process discussed in Sections 3.5 and 4.2.4.

Figure 9.4(a) shows a simple set, and Fig. 9.4(b) shows a structuring element and its reflection (the dark dot denotes the origin of the element). In this case the

a b c
d   e

FIGURE 9.4
(a) Set A.
(b) Square structuring element (dot is the center).
(c) Dilation of A by B, shown shaded.
(d) Elongated structuring element.
(e) Dilation of A using this element.

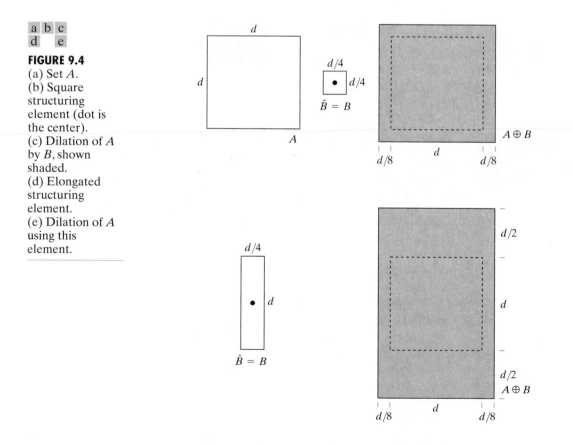

structuring element and its reflection are equal because B is symmetric with respect to its origin. The dashed line in Fig. 9.4(c) shows the original set for reference, and the solid line shows the limit beyond which any further displacements of the origin of $\hat{B}$ by z would cause the intersection of $\hat{B}$ and A to be empty. Therefore, all points inside this boundary constitute the dilation of A by B. Figure 9.4(d) shows a structuring element designed to achieve more dilation vertically than horizontally. Figure 9.4(e) shows the dilation achieved with this element.

■ One of the simplest applications of dilation is for bridging gaps. Figure 9.5(a) shows the same image with broken characters that we studied in Fig. 4.19 in connection with lowpass filtering. The maximum length of the breaks is known to be two pixels. A simple structuring element that can be used for repairing the gaps is shown in Fig. 9.5(b). The result of dilating the original image with this structuring element is shown in Fig. 9.5(c). The gaps have been bridged. One immediate advantage of the morphological approach over the lowpass filtering method we used to bridge the gaps in Fig. 4.19 is that the morphological method resulted directly in a binary image. Lowpass filtering, on the other hand, started with a binary image and produced a gray-scale image, which would require a pass with a thresholding function to convert it back to binary form. ■

EXAMPLE 9.1: Use of morphological dilation for bridging gaps.

9.2.2 Erosion

For sets A and B in Z^2 the erosion of A by B, denoted $A \ominus B$, is defined as

$$A \ominus B = \{z \mid (B)_z \subseteq A\}. \qquad (9.2\text{-}3)$$

In words, this equation indicates that the erosion of A by B is the set of all points z such that B, translated by z, is contained in A. As in the case of dilation,

Historically, certain computer
programs were written using
only two digits rather than
four to define the applicable
year. Accordingly, the
company's software may
recognize a date using "00"
as 1900 rather than the year
2000.

Historically, certain computer
programs were written using
only two digits rather than
four to define the applicable
year. Accordingly, the
company's software may
recognize a date using "00"
as 1900 rather than the year
2000.

0	1	0
1	1	1
0	1	0

a c
b

FIGURE 9.5
(a) Sample text of poor resolution with broken characters (magnified view).
(b) Structuring element.
(c) Dilation of (a) by (b). Broken segments were joined.

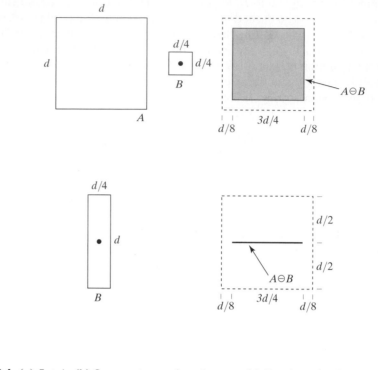

a b c
d   e

FIGURE 9.6 (a) Set A. (b) Square structuring element. (c) Erosion of A by B, shown shaded. (d) Elongated structuring element. (e) Erosion of A using this element.

Eq. (9.2-3) is not the only definition of erosion. (See Problems 9.12 and 9.13 for two different, yet equivalent, definitions.) However, Eq. (9.2-3) usually is favored in practical implementations of morphology for the same reasons stated earlier in connection with Eq. (9.2-1).

Figure 9.6 shows a process similar to that shown in Fig. 9.4. As before, set A is shown as a dashed line for reference in Fig. 9.6(c). The boundary of the shaded region shows the limit beyond which further displacement of the origin of B would cause this set to cease being completely contained in A. Thus, the locus of points within this boundary (i.e., the shaded region) constitutes the erosion of A by B. Figure 9.6(d) shows an elongated structuring element, and Fig. 9.6(e) shows the erosion of A by this element. Note that the original set was eroded down to a line.

Dilation and erosion are duals of each other with respect to set complementation and reflection. That is,

$$(A \ominus B)^c = A^c \oplus \hat{B}. \tag{9.2-4}$$

We proceed to prove this result formally in order to illustrate a typical approach for establishing the validity of morphological expressions. Starting with the definition of erosion, we have

$$(A \ominus B)^c = \left\{ z \,|\, (B)_z \subseteq A \right\}^c.$$

If set $(B)_z$ is contained in set A, then $(B)_z \cap A^c = \varnothing$, in which case the preceding equation becomes

$$(A \ominus B)^c = \{z \mid (B)_z \cap A^c = \varnothing\}^c.$$

But the complement of the set of z's that satisfy $(B)_z \cap A^c = \varnothing$ is the set of z's such that $(B)_z \cap A^c \neq \varnothing$. Thus

$$(A \ominus B)^c = \{z \mid (B)_z \cap A^c \neq \varnothing\}$$
$$= A^c \oplus \hat{B}$$

where the last step follows from Eq. (9.2-1). This concludes the proof.

■ One of the simplest uses of erosion is for eliminating irrelevant detail (in terms of size) from a binary image. Figure 9.7(a) shows a binary image composed of squares of sizes 1, 3, 5, 7, 9, and 15 pixels on the side. Suppose that we want to eliminate all the squares except the largest ones. We can do this by eroding the image with a structuring element of a size somewhat smaller than the objects we wish to keep. In this example we chose a structuring element of size 13×13 pixels. The result of eroding the original image with this structuring element is shown in Fig. 9.7(b). Only portions of the largest squares remain. As shown in Fig. 9.7(c), we can restore these three squares to their original 15×15 size by dilating them with the same structuring element we used for erosion (dilation does not fully restore eroded objects in general; see Problem 9.18). Note in all three images in this example that objects are represented by white pixels, rather than by black pixels as in the previous example. As noted earlier, both representations are used in practice. Unless it is stated otherwise, it is generally understood that the "active" elements of structuring elements assume the same binary values as the objects of interest. The concepts presented in this example are the basis for morphological filtering, as discussed in the following section. ■

EXAMPLE 9.2:
Use of morphological erosion for removing image components.

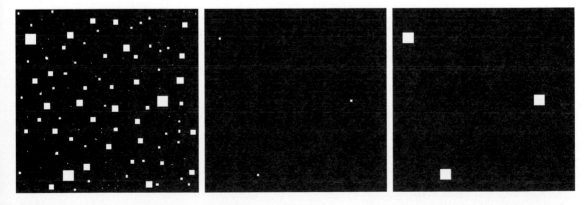

a b c

FIGURE 9.7 (a) Image of squares of size 1, 3, 5, 7, 9, and 15 pixels on the side. (b) Erosion of (a) with a square structuring element of 1's, 13 pixels on the side. (c) Dilation of (b) with the same structuring element.

9.3 Opening and Closing

As we have seen, dilation expands an image and erosion shrinks it. In this section we discuss two other important morphological operations: opening and closing. *Opening* generally smoothes the contour of an object, breaks narrow isthmuses, and eliminates thin protrusions. *Closing* also tends to smooth sections of contours but, as opposed to opening, it generally fuses narrow breaks and long thin gulfs, eliminates small holes, and fills gaps in the contour.

The opening of set A by structuring element B, denoted $A \circ B$, is defined as

$$A \circ B = (A \ominus B) \oplus B. \tag{9.3-1}$$

Thus, the opening A by B is the erosion of A by B, followed by a dilation of the result by B.

Similarly, the closing of set A by structuring element B, denoted $A \bullet B$, is defined as

$$A \bullet B = (A \oplus B) \ominus B, \tag{9.3-2}$$

which, in words, says that the closing of A by B is simply the dilation of A by B, followed by the erosion of the result by B.

The opening operation has a simple geometric interpretation (Fig. 9.8). Suppose that we view the structuring element B as a (flat) "rolling ball." The *boundary* of $A \circ B$ is then established by the points in B that reach the *farthest* into the boundary of A as B is rolled around the *inside* of this boundary. This geometric *fitting* property of the opening operation leads to a set-theoretic formulation, which states that the opening of A by B is obtained by taking the union of all translates of B that fit into A. That is, opening can be expressed as a fitting process such that

$$A \circ B = \cup \{(B)_z | (B)_z \subseteq A\} \tag{9.3-3}$$

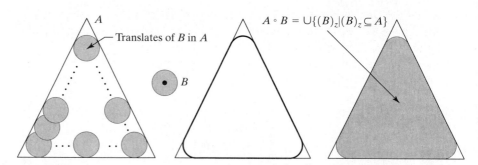

a b c d

FIGURE 9.8 (a) Structuring element B "rolling" along the inner boundary of A (the dot indicates the origin of B). (c) The heavy line is the outer boundary of the opening. (d) Complete opening (shaded).

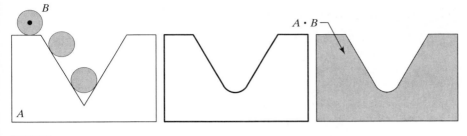

a b c

FIGURE 9.9 (a) Structuring element B "rolling" on the outer boundary of set A. (b) Heavy line is the outer boundary of the closing. (c) Complete closing (shaded).

where $\cup\{\cdot\}$ denotes the union of all the sets inside the braces.

Closing has a similar geometric interpretation, except that now we roll B on the outside of the boundary (Fig. 9.9). It will be shown shortly that opening and closing are duals of each other, so having to roll the ball on the outside is not unexpected. Geometrically, a point w is an element of $A \bullet B$ if and only if $(B)_z \cap A \neq \varnothing$ for any translate of $(B)_z$ that contains w. Figure 9.9 illustrates the basic geometrical properties of closing.

■ Figure 9.10 further illustrates the opening and closing operations. Figure 9.10(a) shows a set A, and Fig. 9.10(b) shows various positions of a disk structuring element during the erosion process. When completed, this process resulted in the disjoint figure shown in Fig. 9.10(c). Note the elimination of the bridge between the two main sections. Its width was thin in relation to the diameter of the structuring element; that is, the structuring element could not be completely contained in this part of the set, thus violating the conditions of Eq. (9.2-3). The same also was true of the two rightmost members of the object. Protruding elements where the disk did not fit were eliminated. Figure 9.10(d) shows the process of dilating the eroded set, and Fig. 9.10(e) shows the final result of opening. Note that outward pointing corners were rounded, whereas inward pointing corners were not affected.

Similarly, Figs. 9.10(f) through (i) show the results of closing A with the same structuring element. We note that the inward pointing corners were rounded, whereas the outward pointing corners remained unchanged. The leftmost intrusion on the boundary of A was reduced in size significantly, because the disk did not fit there. Note also the smoothing that resulted in parts of the object from both opening and closing the set A with a circular structuring element.   ■

EXAMPLE 9.3: A simple illustration of morphological opening and closing.

As in the case of dilation and erosion, opening and closing are duals of each other with respect to set complementation and reflection. That is,

$$(A \bullet B)^c = (A^c \circ \hat{B}).\qquad(9.3\text{-}4)$$

FIGURE 9.10
Morphological
opening and
closing. The
structuring
element is the
small circle shown
in various
positions in (b).
The dark dot is
the center of the
structuring
element.

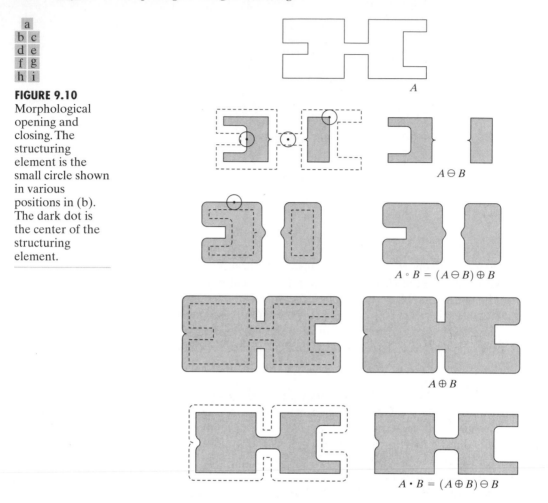

$$A$$

$$A \ominus B$$

$$A \circ B = (A \ominus B) \oplus B$$

$$A \oplus B$$

$$A \cdot B = (A \oplus B) \ominus B$$

We leave the proof of this result as an exercise (Problem 9.14).
The opening operation satisfies the following properties:

(i) $A \circ B$ is a subset (subimage) of A.
(ii) If C is a subset of D, then $C \circ B$ is a subset of $D \circ B$.
(iii) $(A \circ B) \circ B = A \circ B$.

Similarly, the closing operation satisfies the following properties:

(i) A is a subset (subimage) of $A \cdot B$.
(ii) If C is a subset of D, then $C \cdot B$ is a subset of $D \cdot B$.
(iii) $(A \cdot B) \cdot B = A \cdot B$.

Note from condition (*iii*) in both cases that multiple openings or closings of a
set have no effect after the operator has been applied once.

■ Morphological operations can be used to construct filters similar in concept to the spatial filters discussed in Chapter 3. The binary image shown in Fig. 9.11(a) shows a section of a fingerprint corrupted by noise. Here the noise manifests itself as light elements on a dark background and as dark elements on the light components of the fingerprint. The objective is to eliminate the noise and its effects on the print while distorting it as little as possible. A morphological filter consisting of opening followed by closing can be used to accomplish this objective.

The structuring element used is shown in Fig. 9.11(b). The rest of Fig. 9.11 shows a step-by-step sequence of the filtering operation. Figure 9.11(c) shows the result of eroding A with the structuring element. The background noise was completely eliminated in the erosion stage of opening because in this case all noise components are physically smaller than the structuring element. The size of the noise elements (dark spots) contained within the fingerprint actually increased in size. The reason is that these elements actually are inner boundaries that should increase in size as the object is eroded. This enlargement is countered by performing dilation on Fig. 9.11(c). Figure 9.11(d) shows the result. The noise components contained in the fingerprint were reduced in size or deleted completely.

EXAMPLE 9.4:
Use of opening and closing for morphological filtering.

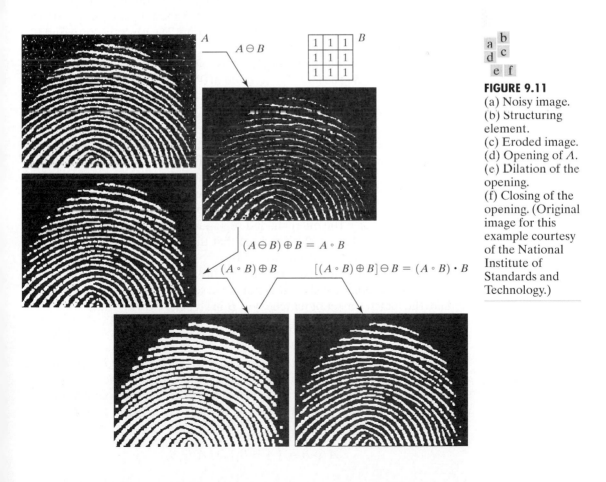

$$A \ominus B$$

$$(A \ominus B) \oplus B = A \circ B$$

$$(A \circ B) \oplus B$$

$$[(A \circ B) \oplus B] \ominus B = (A \circ B) \cdot B$$

a b
a c
d c
  e f

FIGURE 9.11
(a) Noisy image.
(b) Structuring element.
(c) Eroded image.
(d) Opening of A.
(e) Dilation of the opening.
(f) Closing of the opening. (Original image for this example courtesy of the National Institute of Standards and Technology.)

The two operations just described constitute the opening of A by B. We note in Fig. 9.11(d) that the net effect of opening was to eliminate virtually all noise components in both the background and the fingerprint itself. However, new gaps between the fingerprint ridges were created. To counter this undesirable effect, we perform a dilation on the opening, as shown in Fig. 9.11(e). Most of the breaks were restored, but the ridges were thickened, a condition that can be remedied by erosion. The result, shown in Fig. 9.11(f), constitutes the closing of the opening in Fig. 9.11(d). This final result is remarkably clean of noise specks, but it has the disadvantage that some of the print ridges were not fully repaired, and thus contain breaks. This is not totally unexpected, because no conditions were built into the procedure for maintaining connectivity (we discuss this issue again in Example 9.8 and give ways to address it in Section 11.1.5).   ■

9.4 The Hit-or-Miss Transformation

The morphological hit-or-miss transform is a basic tool for shape detection. We introduce this concept with the aid of Fig. 9.12, which shows a set A consisting of three shapes (subsets), denoted X, Y, and Z. The shading in Figs. 9.12(a) through (c) indicates the original sets, whereas the shading in Figs. 9.12(d) and (e) indicates the result of morphological operations. The objective is to find the location of one of the shapes, say, X.

Let the origin of each shape be located at its center of gravity. Let X be enclosed by a small window, W. The *local background* of X with respect to W is defined as the set difference $(W - X)$, as shown in Fig. 9.12(b). Figure 9.12(c) shows the complement of A, which is needed later. Figure 9.12(d) shows the erosion of A by X (the dashed lines are included for reference). Recall that the erosion of A by X is the set of locations of the *origin* of X, such that X is completely contained in A. Interpreted another way, $A \ominus X$ may be viewed geometrically as the set of all locations of the origin of X at which X found a match (hit) in A. Keep in mind that in Fig. 9.12 A consists only of the three *disjoint* sets X, Y, and Z.

Figure 9.12(e) shows the erosion of the complement of A by the local background set $(W - X)$. The outer shaded region in Fig. 9.12(e) is part of the erosion. We note from Figs. 9.12(d) and (e) that the set of locations for which X *exactly* fits inside A is the *intersection* of the erosion of A by X and the erosion of A^c by $(W - X)$ as shown in Fig. 9.12(f). This intersection is precisely the location sought. In other words, if B denotes the set composed of X and its background, the match (or set of matches) of B in A, denoted $A \circledast B$, is

$$A \circledast B = (A \ominus X) \cap [A^c \ominus (W - X)]. \qquad (9.4\text{-}1)$$

We can generalize the notation somewhat by letting $B = (B_1, B_2)$, where B_1 is the set formed from elements of B associated with an object and B_2 is the set of elements of B associated with the corresponding background. From the preceding discussion, $B_1 = X$ and $B_2 = (W - X)$. With this notation, Eq. (9.4-1) becomes

$$A \circledast B = (A \ominus B_1) \cap (A^c \ominus B_2). \qquad (9.4\text{-}2)$$

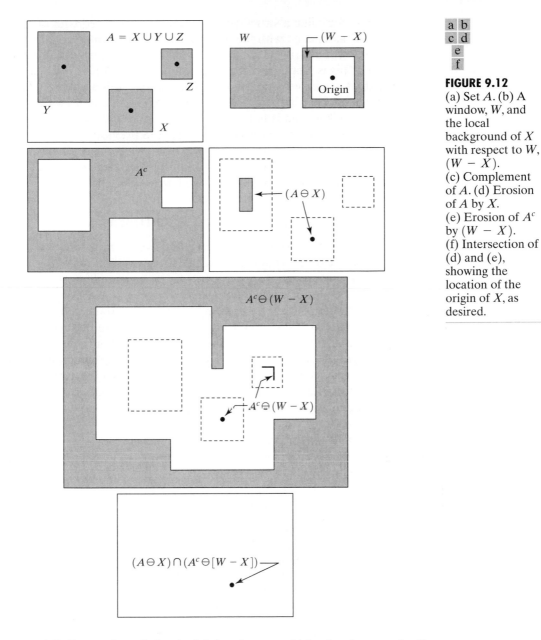

FIGURE 9.12
(a) Set A. (b) A window, W, and the local background of X with respect to W, $(W - X)$. (c) Complement of A. (d) Erosion of A by X. (e) Erosion of A^c by $(W - X)$. (f) Intersection of (d) and (e), showing the location of the origin of X, as desired.

Thus, set $A \circledast B$ contains all the (origin) points at which, simultaneously, B_1 found a match ("hit") in A and B_2 found a match in A^c. By using the definition of set differences given in Eq. (9.1-8) and the dual relationship between erosion and dilation given in Eq. (9.2-4), we can write Eq. (9.4-2) as

$$A \circledast B = \left(A \ominus B_1\right) - \left(A \oplus \hat{B}_2\right). \qquad (9.4\text{-}3)$$

However, Eq. (9.4-2) is considerably more intuitive. We refer to any of the preceding three equations as the *morphological hit-or-miss transform*.

The reason for using a structuring element B_1 associated with objects and an element B_2 associated with the background is based on an assumed definition that two or more objects are distinct only if they form disjoint (disconnected) sets. This is guaranteed by requiring that each object have at least a one-pixel-thick background around it. In some applications, we may be interested in detecting certain patterns (combinations) of 1's and 0's within a set, in which case a background is not required. In such an instance, the hit-or-miss transform reduces to simple erosion. As indicated previously, erosion is still a set of matches, but without the additional requirement of a background match for detecting individual objects. This simplified pattern detection scheme is used in some of the algorithms developed in the following section.

9.5 Some Basic Morphological Algorithms

With the preceding discussion as background, we are now ready to consider some practical uses of morphology. When dealing with binary images, the principal application of morphology is extracting image components that are useful in the representation and description of shape. In particular, we consider morphological algorithms for extracting boundaries, connected components, the convex hull, and the skeleton of a region. We also develop several methods (for region filling, thinning, thickening, and pruning) that are used frequently in conjunction with these algorithms as pre- or postprocessing steps. We make extensive use in this section of "mini-images," designed to clarify the mechanics of each morphological process as we introduce it. The images are binary, with 1's shown shaded and 0's shown in white.

9.5.1 Boundary Extraction

The boundary of a set A, denoted by $\beta(A)$, can be obtained by first eroding A by B and then performing the set difference between A and its erosion. That is,

$$\beta(A) = A - (A \ominus B) \qquad (9.5\text{-}1)$$

where B is a suitable structuring element.

Figure 9.13 illustrates the mechanics of boundary extraction. It shows a simple binary object, a structuring element B, and the result of using Eq. (9.5-1).

a	b
c	d

FIGURE 9.13 (a) Set A. (b) Structuring element B. (c) A eroded by B. (d) Boundary, given by the set difference between A and its erosion.

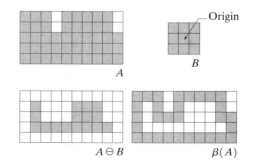

A

B

$A \ominus B$

$\beta(A)$

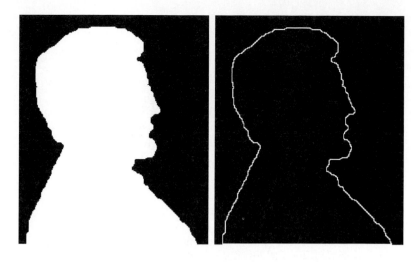

FIGURE 9.14
(a) A simple
binary image, with
1's represented in
white. (b) Result
of using
Eq. (9.5-1) with
the structuring
element in
Fig. 9.13(b).

Although the structuring element shown in Fig. 9.13(b) is among the most frequently used, it is by no means unique. For example, using a 5×5 structuring element of 1's would result in a boundary between 2 and 3 pixels thick. Note that when the origin of B is on the edges of the set, part of the structuring element may be outside the image. The normal treatment of this condition is to assume that values outside the borders of the image are 0.

■ Figure 9.14 further illustrates the use of Eq. (9.5-1) with the structuring element of Fig. 9.13(b). In this example, binary 1's are shown in white and 0's in black, so the elements of the structuring element, which are 1's, also are treated as white. Because of the structuring element used, the boundary shown in Fig. 9.14(b) is one pixel thick. ■

EXAMPLE 9.5:
Boundary
extraction by
morphological
processing.

9.5.2 Region Filling

Next we develop a simple algorithm for region filling based on set dilations, complementation, and intersections. In Fig. 9.15 A denotes a set containing a subset whose elements are 8-connected boundary points of a region. Beginning with a point p inside the boundary, the objective is to fill the entire region with 1's.

If we adopt the convention that all nonboundary (background) points are labeled 0, then we assign a value of 1 to p to begin. The following procedure then fills the region with 1's:

$$X_k = (X_{k-1} \oplus B) \cap A^c \quad k = 1, 2, 3, \ldots \tag{9.5-2}$$

where $X_0 = p$, and B is the symmetric structuring element shown in Fig. 9.15(c). The algorithm terminates at iteration step k if $X_k = X_{k-1}$. The set union of X_k and A contains the filled set and its boundary.

The dilation process of Eq. (9.5-2) would fill the entire area if left unchecked. However, the intersection at each step with A^c limits the result to inside the region of interest. This is our first example of how a morphological process can be conditioned to meet a desired property. In the current application, it is

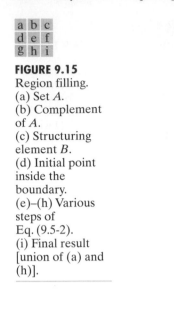

FIGURE 9.15
Region filling.
(a) Set A.
(b) Complement
of A.
(c) Structuring
element B.
(d) Initial point
inside the
boundary.
(e)–(h) Various
steps of
Eq. (9.5-2).
(i) Final result
[union of (a) and
(h)].

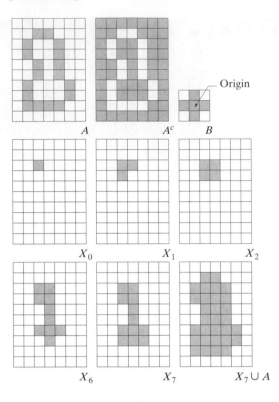

appropriately called *conditional dilation*. The rest of Fig. 9.15 illustrates further
the mechanics of Eq. (9.5-2). Although this example has only one subset, the concept clearly applies to any finite number of such subsets, assuming that a point
inside each boundary is given.

EXAMPLE 9.6:
Morphological
region filling.

■ Figure 9.16(a) shows an image composed of white circles with black inner
spots. An image such as this might result from thresholding into two levels a
scene containing polished spheres (e.g., ball bearings). The dark spots inside
the spheres are the result of reflections. The objective is to eliminate the reflections by region filling. Figure 9.16(a) shows one point selected inside one of
the spheres, and Fig. 9.16(b) shows the result of filling that component. Finally,
Fig. 9.16(c) shows the result of filling all the spheres. Because it must be known
whether black points are background points or sphere inner points, fully automating this procedure requires that additional "intelligence" be built into the
algorithm (see Problem 9.23).   ■

9.5.3 Extraction of Connected Components

The concepts of connectivity and connected components were introduced in
Section 2.5.2. In practice, extraction of connected components in a binary image
is central to many automated image analysis applications. Let Y represent a
connected component contained in a set A and assume that a point p of Y is

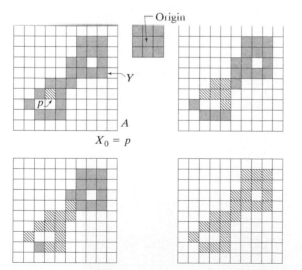

a b c

FIGURE 9.16 (a) Binary image (the white dot inside one of the regions is the starting point for the region-filling algorithm). (b) Result of filling that region (c) Result of filling all regions.

known. Then the following iterative expression yields all the elements of Y:

$$X_k = (X_{k-1} \oplus B) \cap A \quad k = 1, 2, 3, \dots \quad (9.5\text{-}3)$$

where $X_0 = p$, and B is a suitable structuring element, as shown in Fig. 9.17. If $X_k = X_{k-1}$, the algorithm has converged and we let $Y = X_k$.

Equation (9.5-3) is similar in form to Eq. (9.5-2). The only difference is the use of A instead of its complement. This difference arises because all the elements sought (that is, the elements of the connected component) are labeled 1.

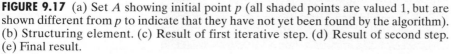

a b c
d   e

FIGURE 9.17 (a) Set A showing initial point p (all shaded points are valued 1, but are shown different from p to indicate that they have not yet been found by the algorithm). (b) Structuring element. (c) Result of first iterative step. (d) Result of second step. (e) Final result.

The intersection with A at each iterative step eliminates dilations centered on elements labeled 0. Figure 9.17 illustrates the mechanics of Eq. (9.5-3). Note that the shape of the structuring element assumes 8-connectivity between pixels. As in the region-filling algorithm, the results just discussed are applicable to any finite number of sets of connected components contained in A, assuming that a point is known in each connected component.

EXAMPLE 9.7:
Using connected components to detect foreign objects in packaged food.

■ Connected components are used frequently for automated inspection. Figure 9.18(a) shows an X-ray image of a chicken breast that contains bone fragments. It is of considerable interest to be able to detect such foreign objects in processed food before packaging and/or shipping. In this particular case, the density of the bones is such that their nominal gray-level values are different from the background. This make extraction of the bones from the background a simple matter by using a single threshold (thresholding was introduced in Section 3.1 and is discussed in considerable more detail in Section 10.3). The result is the binary image shown in Fig. 9.18(b).

The most significant feature in this figure is the fact that the points that remain are clustered into objects (bones), rather than being isolated, irrelevant points. We can make sure that only objects of "significant" size remain by erod-

a
b
c d

FIGURE 9.18
(a) X-ray image of chicken filet with bone fragments.
(b) Thresholded image. (c) Image eroded with a 5 × 5 structuring element of 1's.
(d) Number of pixels in the connected components of (c). (Image courtesy of NTB Elektronische Geraete GmbH, Diepholz, Germany, www.ntbxray.com.)

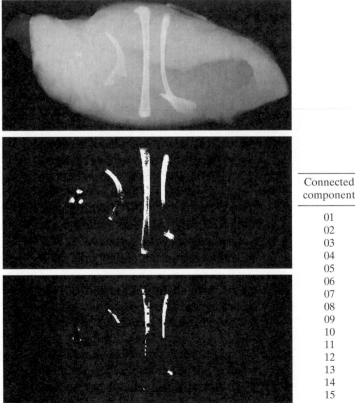

Connected component	No. of pixels in connected comp
01	11
02	9
03	9
04	39
05	133
06	1
07	1
08	743
09	7
10	11
11	11
12	9
13	9
14	674
15	85

ing the thresholded image. In this example, we define as significant any object that remains after erosion with a 5×5 structuring element of 1's. The result of erosion is shown in Fig. 9.18(c). The next step is to analyze the size of the objects that remain. We label (identify) these objects by extracting the connected components in the image. The table in Fig. 9.18(d) lists the results of the extraction. There are a total of 15 connected components, with four of them being dominant in size. This is enough to determine that significant foreign objects are contained in the original image. If desired, further characterization (such as shape) is possible using the techniques discussed in Chapter 11.   ■

9.5.4 Convex Hull

A set A is said to be *convex* if the straight line segment joining any two points in A lies entirely within A. The *convex hull* H of an arbitrary set S is the smallest convex set containing S. The set difference $H - S$ is called the *convex deficiency* of S. As discussed in more detail in Sections 11.1.4 and 11.3.2, the convex hull and convex deficiency are useful for object description. Here, we present a simple morphological algorithm for obtaining the convex hull, $C(A)$, of a set A.

Let $B^i, i = 1, 2, 3, 4$, represent the four structuring elements in Fig. 9.19(a). The procedure consists of implementing the equation:

$$X^i_k = \left(X_{k-1} \circledast B^i\right) \cup A \quad i = 1, 2, 3, 4 \quad \text{and} \quad k = 1, 2, 3, \dots \quad (9.5\text{-}4)$$

with $X^i_0 = A$. Now let $D^i = X^i_{\text{conv}}$, where the subscript "conv" indicates convergence in the sense that $X^i_k = X^i_{k-1}$. Then the convex hull of A is

$$C(A) = \bigcup_{i=1}^{4} D^i. \qquad (9.5\text{-}5)$$

In other words, the procedure consists of iteratively applying the hit-or-miss transform to A with B^1; when no further changes occur, we perform the union with A and call the result D^1. The procedure is repeated with B^2 (applied to A) until no further changes occur, and so on. The union of the four resulting D's constitutes the convex hull of A. Note that we are using the simplified implementation of the hit-or-miss transform in which no background match is required, as discussed at the end of Section 9.4.

Figure 9.19 illustrates the procedure given in Eqs. (9.5-4) and (9.5-5). Figure 9.19(a) shows the structuring elements used to extract the convex hull. The origin of each element is at its center. The $\times$ entries indicate "don't care" conditions. This means that a structuring element is said to have found a match in A if the 3-by-3 region of A under the structuring element mask at that location matches the pattern of the mask. For a particular mask, a pattern match occurs when the center of the 3-by-3 region in A is 0, and the three pixels under the shaded mask elements are 1. The values of the other pixels in the 3-by-3 region do not matter. Also, with respect to the notation in Fig. 9.19(a), B^i is a clockwise rotation of B^{i-1} by 90°.

Figure 9.19(b) shows a set A for which the convex hull is sought. Starting with $X^1_0 = A$ resulted in the set shown in Fig. 9.19(c) after four iterations of Eq. (9.5-4). Then, letting $X^2_0 = A$ and again using Eq. (9.5-4) resulted in the set

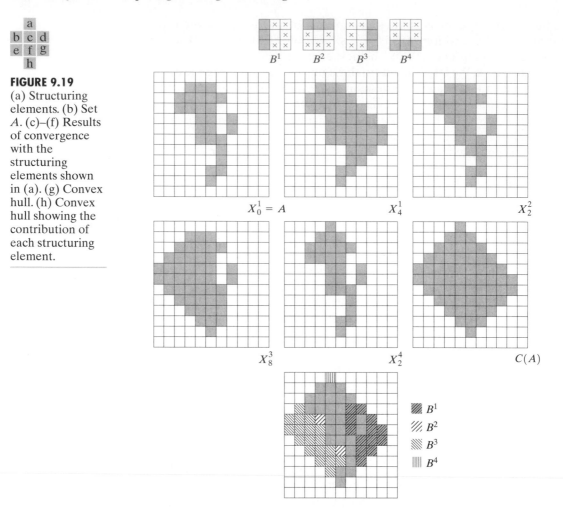

FIGURE 9.19
(a) Structuring elements. (b) Set A. (c)–(f) Results of convergence with the structuring elements shown in (a). (g) Convex hull. (h) Convex hull showing the contribution of each structuring element.

shown in Fig. 9.19(d) (convergence was achieved in only two steps in this case). The next two results were obtained in the same way. Finally, forming the union of the sets in Figs. 9.19(c), (d), (e), and (f) resulted in the convex hull shown in Fig. 9.19(g). The contribution of each structuring element is highlighted in the composite set shown in Fig. 9.19(h).

One obvious shortcoming of the procedure just outlined is that the convex hull can grow beyond the minimum dimensions required to guarantee convexity. One simple approach to reduce this effect is to limit growth so that it does not extend past the vertical and horizontal dimensions of the original set of points. Imposing this limitation on the example in Fig. 9.19 resulted in the image shown in Fig. 9.20. Boundaries of greater complexity can be used to limit growth even further in images with more detail. For example, we could use the maximum dimensions of the original set of points along the vertical, horizontal, and diagonal directions. The price paid for refinements such as this is additional complexity (and increased computational requirements) of the algorithm.

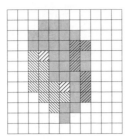

FIGURE 9.20 Result of limiting growth of convex hull algorithm to the maximum dimensions of the original set of points along the vertical and horizontal directions.

9.5.5 Thinning

The thinning of a set A by a structuring element B, denoted $A \otimes B$, can be defined in terms of the hit-or-miss transform:

$$A \otimes B = A - (A \circledast B)$$
$$= A \cap (A \circledast B)^c. \qquad (9.5\text{-}6)$$

As in the previous section, we are interested only in pattern matching with the structuring elements, so no background operation is required in the hit-or-miss transform. A more useful expression for thinning A symmetrically is based on a *sequence* of structuring elements:

$$\{B\} = \{B^1, B^2, B^3, \ldots, B^n\} \qquad (9.5\text{-}7)$$

where B^i is a rotated version of B^{i-1}. Using this concept, we now define thinning by a sequence of structuring elements as

$$A \otimes \{B\} = ((\ldots((A \otimes B^1) \otimes B^2)\ldots) \otimes B^n). \qquad (9.5\text{-}8)$$

The process is to thin A by *one pass* with B^1, then thin the result with one pass of B^2, and so on, until A is thinned with one pass of B^n. The entire process is repeated until no further changes occur. Each individual thinning pass is performed using Eq. (9.5-6).

Figure 9.21(a) shows a set of structuring elements commonly used for thinning, and Fig. 9.21(b) shows a set A to be thinned by using the procedure just discussed. Figure 9.21(c) shows the result of thinning with one raster pass of A with B^1, and Figs. 9.21(d) through (k) show the results of passes with the other structuring elements. Convergence was achieved after the second pass of B^6. Figure 9.21(l) shows the thinned result. Finally, Fig. 9.21(m) shows the thinned set converted to m-connectivity (see Section 2.5.2) to eliminate multiple paths.

9.5.6 Thickening

Thickening is the morphological dual of thinning and is defined by the expression

$$A \odot B = A \cup (A \circledast B) \qquad (9.5\text{-}9)$$

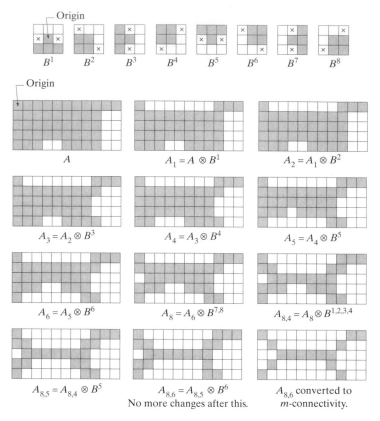

FIGURE 9.21 (a) Sequence of rotated structuring elements used for thinning. (b) Set A. (c) Result of thinning with the first element. (d)–(i) Results of thinning with the next seven elements (there was no change between the seventh and eighth elements). (j) Result of using the first four elements again. (l) Result after convergence. (m) Conversion to m-connectivity.

where B is a structuring element suitable for thickening. As in thinning, thickening can be defined as a sequential operation:

$$A \odot \{B\} = ((\ldots ((A \odot B^1) \odot B^2)\ldots) \odot B^n). \qquad (9.5\text{-}10)$$

The structuring elements used for thickening have the same form as those shown in Fig. 9.21(a) in connection with thinning, but with all 1's and 0's interchanged. However, a separate algorithm for thickening is seldom used in practice. Instead, the usual procedure is to thin the background of the set in question and then complement the result. In other words, to thicken a set A, we form $C = A^c$, thin C, and then form C^c. Figure 9.22 illustrates this procedure.

Depending on the nature of A, this procedure may result in some disconnected points, as Fig. 9.22(d) shows. Hence thickening by this method usually is followed by a simple postprocessing step to remove disconnected points. Note from Fig. 9.22(c) that the thinned background forms a boundary for the thick-

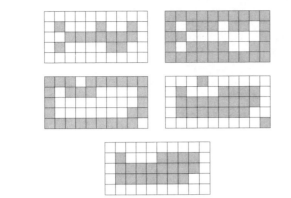

a b
c d
e

FIGURE 9.22 (a) Set A. (b) Complement of A. (c) Result of thinning the complement of A. (d) Thickened set obtained by complementing (c). (e) Final result, with no disconnected points.

ening process. This useful feature is not present in the direct implementation of thickening using Eq. (9.5-10), and it is one of the principal reasons for using background thinning to accomplish thickening.

9.5.7 Skeletons

As shown in Fig. 9.23, the notion of a skeleton, $S(A)$, of a set A is intuitively simple. We deduce from this figure that

(a) If z is a point of $S(A)$ and $(D)_z$ is the largest disk centered at z and contained in A, one cannot find a larger disk (not necessarily centered at z) containing $(D)_z$ and included in A. The disk $(D)_z$ is called a *maximum disk*.
(b) The disk $(D)_z$ touches the boundary of A at two or more different places.

The skeleton of A can be expressed in terms of erosions and openings. That is, it can be shown (Serra [1982]) that

$$S(A) = \bigcup_{k=0}^{K} S_k(A) \tag{9.5-11}$$

with

$$S_k(A) = (A \ominus kB) - (A \ominus kB) \circ B \tag{9.5-12}$$

where B is a structuring element, and $(A \ominus kB)$ indicates k successive erosions of A:

$$(A \ominus kB) = (\ldots (A \ominus B) \ominus B) \ominus \ldots) \ominus B \tag{9.5-13}$$

k times, and K is the last iterative step before A erodes to an empty set. In other words,

$$K = \max\{k \mid (A \ominus kB) \neq \varnothing\}. \tag{9.5-14}$$

FIGURE 9.23
(a) Set A.
(b) Various positions of maximum disks with centers on the skeleton of A.
(c) Another maximum disk on a different segment of the skeleton of A.
(d) Complete skeleton.

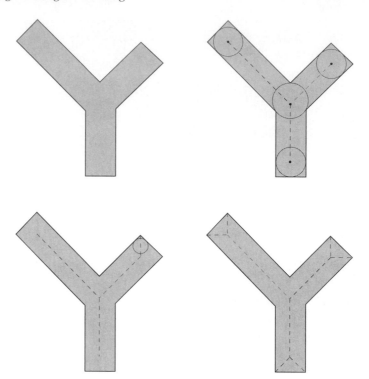

The formulation given in Eqs. (9.5-11) and (9.5-12) states that $S(A)$ can be obtained as the union of the *skeleton subsets* $S_k(A)$. Also, it can be shown that A can be *reconstructed* from these subsets by using the equation

$$A = \bigcup_{k=0}^{K} \left(S_k(A) \oplus kB \right) \tag{9.5-15}$$

where $\left(S_k(A) \oplus kB \right)$ denotes k successive dilations of $S_k(A)$; that is,

$$\left(S_k(A) \oplus kB \right) = \left(\left(\ldots \left(S_k(A) \oplus B \right) \oplus B \right) \oplus \ldots \right) \oplus B. \tag{9.5-16}$$

EXAMPLE 9.8:
Computing the skeleton of a simple figure.

■ Figure 9.24 illustrates the concepts just discussed. The first column shows the original set (at the top) and two erosions by the structuring element B. Note that one more erosion of A would yield the empty set, so $K = 2$ in this case. The second column shows the opening of the sets in the first column by B. These results are easily explained by the fitting characterization of the opening operation discussed in connection with Fig. 9.8. The third column simply contains the set differences between the first and second columns.

The fourth column contains two partial skeletons and the final result (at the bottom of the column). The final skeleton not only is thicker than it needs to be but, more important, it is not connected. This result is not unexpected, as nothing in the preceding formulation of the morphological skeleton guarantees connectivity. Morphology produces an elegant formulation in terms of erosions and openings of the given set. However, heuristic formulations such as the algo-

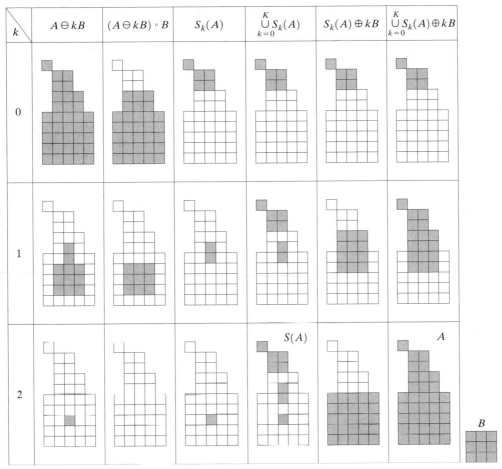

FIGURE 9.24 Implementation of Eqs. (9.5-11) through (9.5-15). The original set is at the top left, and its morphological skeleton is at the bottom of the fourth column. The reconstructed set is at the bottom of the sixth column.

rithm developed in Section 11.1.5 are needed if, as is usually the case, the skeleton must be maximally thin, connected, and minimally eroded.

The fifth column shows $S_0(A)$, $S_1(A) \oplus B$, and $(S_2(A) \oplus 2B) = (S_2(A) \oplus B) \oplus B$. Finally, the last column shows reconstruction of set A, which, according to Eq. (9.5-15), is the union of the dilated skeleton subsets shown in the fifth column. ∎

9.5.8 Pruning

Pruning methods are an essential complement to thinning and skeletonizing algorithms because these procedures tend to leave parasitic components that need to be "cleaned up" by postprocessing. We begin the discussion with a pruning problem and then develop a morphological solution based on the material introduced in the preceding sections. Thus we take this opportunity to illustrate how to go about solving a problem by combining several of the techniques discussed up to this point.

A common approach in the automated recognition of hand-printed characters is to analyze the shape of the skeleton of each character. These skeletons often are characterized by "spurs" (parasitic components). Spurs are caused during erosion by non uniformities in the strokes composing the characters. We develop a morphological technique for handling this problem, starting with the assumption that the length of a parasitic component does not exceed a specified number of pixels.

Figure 9.25(a) shows the skeleton of a hand-printed "a." The parasitic component on the leftmost part of the character is illustrative of what we are interested in removing. The solution is based on suppressing a parasitic branch by successively eliminating its end point. Of course, this also shortens (or eliminates) other branches in the character but, in the absence of other structural information, the assumption in this example is that any branch with three or less pixels is to be eliminated. Thinning of an input set A with a sequence of structuring elements designed to detect only end points achieves the desired result. That is, let

$$X_1 = A \otimes \{B\} \qquad (9.5\text{-}17)$$

where $\{B\}$ denotes the structuring element sequence shown in Figs. 9.25(b) and (c) [see Eq. (9.5-7) regarding structuring-element sequences]. The sequence of

a	b
	c
d	e
f	g

FIGURE 9.25
(a) Original image. (b) and (c) Structuring elements used for deleting end points. (d) Result of three cycles of thinning. (e) End points of (d). (f) Dilation of end points conditioned on (a). (g) Pruned image.

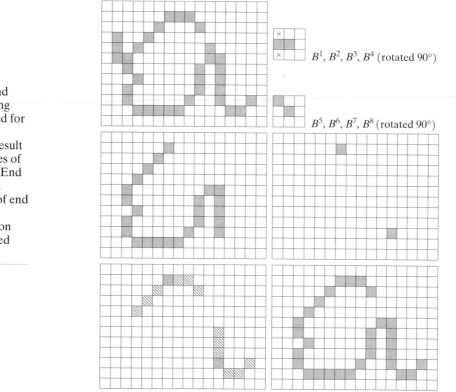

B^1, B^2, B^3, B^4 (rotated 90°)

B^5, B^6, B^7, B^8 (rotated 90°)

structuring elements consists of two different structures, each of which is ro-tated 90° for a total of eight elements. The × in Fig. 9.25(b) signifies a "don't care" condition, in the sense that it does not matter whether the pixel in that lo-cation has a value of 0 or 1. Numerous results reported in the literature on mor-phology are based on the use of a *single* structuring element, similar to the one in Fig. 9.25(b), but having "don't care" conditions along the entire first column. This is incorrect. For example, this element would identify the point located in the eighth row, fourth column of Fig. 9.25(a) as an end point, thus eliminating it and breaking connectivity in the stroke.

Applying Eq. (9.5-17) to A three times yields the set X_1 shown in Fig. 9.25(d). The next step is to "restore" the character to its original form, but with the par-asitic branches removed. To do so first requires forming a set X_2 containing all end points in X_1 [Fig. 9.25(e)]:

$$X_2 = \bigcup_{k=1}^{8} (X_1 \circledast B^k) \qquad (9.5\text{-}18)$$

where the B^k are the same end-point detectors shown in Figs. 9.25(b) and (c). The next step is dilation of the end points three times, using set A as a delimiter:

$$X_3 = (X_2 \oplus H) \cap A \qquad (9.5\text{-}19)$$

where H is a 3×3 structuring element of 1's. As in the case of region filling and ex-traction of connected components, this type of conditional dilation prevents the creation of 1-valued elements outside the region of interest, as evidenced by the result shown in Fig. 9.25(f). Finally, the union of X_3 and X_1 yields the desired result,

$$X_4 = X_1 \cup X_3 \qquad (9.5\text{-}20)$$

as shown in Fig. 9.25(g).

In more complex scenarios, use of Eq. (9.5-19) sometimes picks up the "tips" of some parasitic branches. This condition can occur when the end points of these branches are near the skeleton. Although Eq. (9.5-17) may eliminate them, they can be picked up again during dilation because they are valid points in A. Unless entire parasitic elements are picked up again (a rare case if these ele-ments are short with respect to valid strokes), detecting and eliminating them is easy because they are disconnected regions.

A natural thought at this juncture is that there must be easier ways to solve this problem. For example, we could just keep track of all deleted points and sim-ply reconnect the appropriate points to all end points left after application of Eq. (9.5-17). This option is valid, but the advantage of the formulation just pre-sented is that the use of simple morphological constructs solved the entire prob-lem. In practical situations when a set of such tools is available, the advantage is that no new algorithms have to be written. We simply combine the necessary morphological functions into a sequence of operations.

9.5.9 Summary of Morphological Operations on Binary Images

Table 9.2 summarizes the morphological results developed in the preceding sec-tions, and Fig. 9.26 summarizes the basic types of structuring elements used in the various morphological processes discussed thus far. Note that the Roman nu-merals in the third column of Table 9.2 refer to the structuring elements in Fig. 9.26.

TABLE 9.2
Summary of morphological operations and their properties.

Operation	Equation	Comments (The Roman numerals refer to the structuring elements shown in Fig. 9.26).
Translation	$(A)_z = \{w \mid w = a + z, \quad \text{for } a \in A\}$	Translates the origin of A to point z.
Reflection	$\hat{B} = \{w \mid w = -b, \quad \text{for } b \in B\}$	Reflects all elements of B about the origin of this set.
Complement	$A^c = \{w \mid w \notin A\}$	Set of points not in A.
Difference	$A - B = \{w \mid w \in A, w \notin B\}$ $= A \cap B^c$	Set of points that belong to A but not to B.
Dilation	$A \oplus B = \{z \mid (\hat{B})_z \cap A \neq \varnothing\}$	"Expands" the boundary of A. (I)
Erosion	$A \ominus B = \{z \mid (B)_z \subseteq A\}$	"Contracts" the boundary of A. (I)
Opening	$A \circ B = (A \ominus B) \oplus B$	Smoothes contours, breaks narrow isthmuses, and eliminates small islands and sharp peaks. (I)
Closing	$A \bullet B = (A \oplus B) \ominus B$	Smoothes contours, fuses narrow breaks and long thin gulfs, and eliminates small holes. (I)
Hit-or-miss transform	$A \circledast B = (A \ominus B_1) \cap (A^c \ominus B_2)$ $= (A \ominus B_1) - (A \oplus \hat{B}_2)$	The set of points (coordinates) at which, simultaneously, B_1 found a match ("hit") in A and B_2 found a match in A^c.
Boundary extraction	$\beta(A) = A - (A \ominus B)$	Set of points on the boundary of set A. (I)
Region filling	$X_k = (X_{k-1} \oplus B) \cap A^c; X_0 = p$ and $k = 1, 2, 3, \ldots$	Fills a region in A, given a point p in the region. (II)
Connected components	$X_k = (X_{k-1} \oplus B) \cap A; X_0 = p$ and $k = 1, 2, 3, \ldots$	Finds a connected component Y in A, given a point p in Y. (I)
Convex hull	$X_k^i = (X_{k-1}^i \circledast B^i) \cup A; i = 1, 2, 3, 4;$ $k = 1, 2, 3, \ldots; X_0^i = A;$ and $D^i = X_{\text{conv}}^i.$	Finds the convex hull $C(A)$ of set A, where "conv" indicates convergence in the sense that $X_k^i = X_{k-1}^i.$ (III)

Operation	Equation	**Comments** (The Roman numerals refer to the structuring elements shown in Fig. 9.26).
Thinning	$A \otimes B = A - (A \circledast B)$ $= A \cap (A \circledast B)^c$ $A \otimes \{B\} =$ $((\dots((A \otimes B^1) \otimes B^2)\dots) \otimes B^n)$ $\{B\} = \{B^1, B^2, B^3, \dots, B^n\}$	Thins set A. The first two equations give the basic definition of thinning. The last two equations denote thinning by a sequence of structuring elements. This method is normally used in practice. (IV)
Thickening	$A \odot B = A \cup (A \circledast B)$ $A \odot \{B\} =$ $((\dots(A \odot B^1) \odot B^2 \dots) \odot B^n)$	Thickens set A. (See preceding comments on sequences of structuring elements.) Uses IV with 0's and 1's reversed.
Skeletons	$S(A) = \bigcup_{k=0}^{K} S_k(A)$ $S_k(A) = \bigcup_{k=0}^{K} \{(A \ominus kB)$ $- [(A \ominus kB) \circ B]\}$ Reconstruction of A: $A = \bigcup_{k=0}^{K} (S_k(A) \oplus kB)$	Finds the skeleton $S(A)$ of set A. The last equation indicates that A can be reconstructed from its skeleton subsets $S_k(A)$. In all three equations, K is the value of the iterative step after which the set A erodes to the empty set. The notation $(A \ominus kB)$ denotes the kth iteration of successive erosion of A by B. (I)
Pruning	$X_1 = A \otimes \{B\}$ $X_2 = \bigcup_{k=1}^{8}(X_1 \circledast B^k)$ $X_3 = (X_2 \oplus H) \cap A$ $X_4 = X_1 \cup X_3$	X_4 is the result of pruning set A. The number of times that the first equation is applied to obtain X_1 must be specified. Structuring elements V are used for the first two equations. In the third equation H denotes structuring element I.

TABLE 9.2
Summary of morphological results and their properties.
(continued)

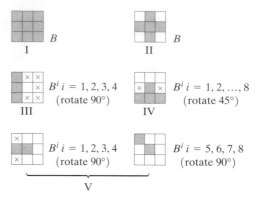

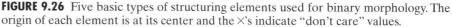

FIGURE 9.26 Five basic types of structuring elements used for binary morphology. The origin of each element is at its center and the ×'s indicate "don't care" values.

9.6 Extensions to Gray-Scale Images

In this section we extend to gray-scale images the basic operations of dilation, erosion, opening, and closing. We then use these operations to develop several basic gray-scale morphological algorithms. In particular, we develop algorithms for boundary extraction via a morphological gradient operation, and for region partitioning based on texture content. We also discuss algorithms for smoothing and sharpening, which often are useful as pre- or postprocessing steps.

Throughout the discussions that follow, we deal with digital image functions of the form $f(x, y)$ and $b(x, y)$, where $f(x, y)$ is the input image and $b(x, y)$ is a structuring element, itself a subimage function. The assumption is that these functions are discrete in the sense introduced in Section 2.4.2. That is, if Z denotes the set of real integers, the assumption is that (x, y) are integers from $Z \times Z$ and that f and b are functions that assign a gray-level value (a real number from the set of real numbers, R) to each distinct pair of coordinates (x, y). If the gray levels also are integers, Z replaces R.

9.6.1 Dilation

Gray-scale dilation of f by b, denoted $f \oplus b$, is defined as

$$(f \oplus b)(s, t) =$$
$$\max\{f(s - x, t - y) + b(x, y) \mid (s - x), (t - y) \in D_f; (x, y) \in D_b\} \quad (9.6\text{-}1)$$

where D_f and D_b are the domains of f and b, respectively. Keep in mind that f and b are functions, rather than sets, as is the case in binary morphology.

The condition that $(s - x)$ and $(t - y)$ have to be in the domain of f, and x and y have to be in the domain of b, is analogous to the condition in the binary definition of dilation, where the two sets have to overlap by at least one element. Note also that the form of Eq. (9.6-1) is similar to 2-D convolution [Eq. (4.2-30)], with the max operation replacing the sums of convolution and the addition replacing the products of convolution.

We illustrate the notation and mechanics of Eq. (9.6-1) by means of simple 1-D functions. For functions of one variable, Eq. (9.6-1) reduces to the expression

$$(f \oplus b)(s) = \max\{f(s - x) + b(x) \,|\, (s - x) \in D_f \quad \text{and} \quad x \in D_b\}.$$

Recall from the discussion of convolution that $f(-x)$ is simply $f(x)$ mirrored with respect to the origin of the x axis. As in convolution, the function $f(s - x)$ moves to the right for positive s, and to the left for negative s. The requirements that the value of $(s - x)$ has to be in the domain of f and that the value of x has to be in the domain of b imply that f and b overlap. As noted in the previous paragraph, these conditions are analogous to the requirement in the binary definition of dilation, where the two sets have to overlap by at least one element. Finally, unlike the binary case, f, rather than the structuring element b, is shifted. Equation (9.6-1) could be written so that b undergoes translation instead of f. However, if D_b is smaller than D_f (a condition almost always found in practice), the form given in Eq. (9.6-1) is simpler in terms of indexing and achieves the same result. Conceptually, f sliding by b is really no different than b sliding by f. In fact, although this equation is easier to implement, the actual mechanics of gray-scale dilation are easier to visualize if b is the function that slides past f.

An example is shown in Fig. 9.27. Note that at each position of the structuring element the value of dilation at that point is the maximum of the sum of f

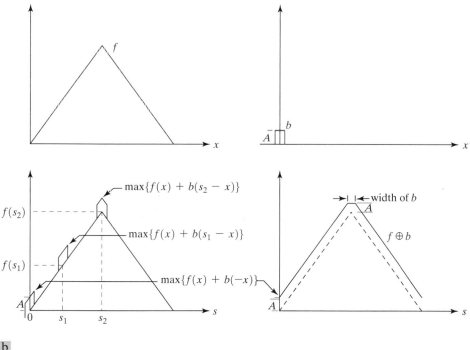

a b
c d

FIGURE 9.27 (a) A simple function. (b) Structuring element of height A. (c) Result of dilation for various positions of sliding b past f. (d) Complete result of dilation (shown solid).

and b in the interval spanned by b. The general effect of performing dilation on a gray-scale image is twofold: (1) If all the values of the structuring element are positive, the output image tends to be brighter than the input. (2) Dark details either are reduced or eliminated, depending on how their values and shapes relate to the structuring element used for dilation.

9.6.2 Erosion

Gray-scale erosion, denoted $f \ominus b$, is defined as

$$
(f \ominus b)(s, t) = \\
\min \{ f(s + x, t + y) - b(x, y) \,|\, (s + x), (t + y) \in D_f; (x, y) \in D_b \} \quad (9.6\text{-}2)
$$

where D_f and D_b are the domains of f and b, respectively. The condition that $(s + x)$ and $(t + y)$ have to be in the domain of f, and x and y have to be in the domain of b, is analogous to the condition in the binary definition of erosion, where the structuring element has to be completely contained by the set being eroded. Note that the form of Eq. (9.6-2) is similar in form to 2-D correlation [Eq. (4.6-30)], with the min operation replacing the sums of correlation and subtraction replacing the products of correlation.

We illustrate the mechanics of Eq. (9.6-2) by eroding a simple 1-D function. For functions of one variable, the expression for erosion reduces to

$$
(f \ominus b)(s) = \min \{ f(s + x) - b(x) \,|\, (s + x) \in D_f \text{ and } x \in D_b \}.
$$

As in correlation, the function $f(s + x)$ moves to the left for positive s and to the right for negative s. The requirements that $(s + x) \in D_f$ and $x \in D_b$ imply that the range of b is completely contained within the range of the displaced f. As noted in the previous paragraph, these requirements are analogous to those in the binary definition of erosion, where the structuring element has to be contained completely in the set being eroded.

Finally, unlike the binary definition of erosion, f, rather than the structuring element b, is shifted. Equation (9.6-2) could be written so that b is the function translated, resulting in a more complicated expression in terms of indexing. Because f sliding past b conceptually is the same as b sliding past f, the form of Eq. (9.6-2) is used for the reasons stated at the end of the discussion on dilation. Figure 9.28 shows the result of eroding the function of Fig. 9.27(a) by the structuring element of Fig. 9.27(b).

FIGURE 9.28
Erosion of the function shown in Fig. 9.27(a) by the structuring element shown in Fig. 9.27(b).

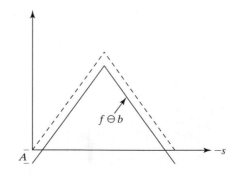

Equation (9.6-2) indicates that erosion is based on choosing the minimum value of $(f - b)$ in the interval defined by the shape of the structuring element. The general effect of performing erosion on a gray-scale image is twofold: (1) If all the elements of the structuring element are positive, the output image tends to be darker than the input image. (2) The effect of bright details in the input image that are smaller in area than the structuring element is reduced, with the degree of reduction being determined by the gray-level values surrounding the bright detail and by the shape and amplitude values of the structuring element itself.

Gray-scale dilation and erosion are duals with respect to function complementation and reflection. That is,

$$(f \ominus b)^c(s, t) = (f^c \oplus \hat{b})(s, t) \tag{9.6-3}$$

where $f^c = -f(x, y)$ and $\hat{b} = b(-x, -y)$. Except as needed for clarity, we simplify the notation in the following discussions by omitting the arguments of all functions.

■ Figure 9.29(a) shows a simple 512×512 gray-scale image, and Fig. 9.29(b) shows the result of dilating this image with a "flat-top" structuring element in the shape of a parallelepiped of unit height and size 5×5 pixels. Based on the preceding discussion, dilation is expected to produce an image that is brighter than the original and in which small, dark details have been reduced or eliminated.

EXAMPLE 9.9:
Illustration of dilation and erosion on a gray-scale image.

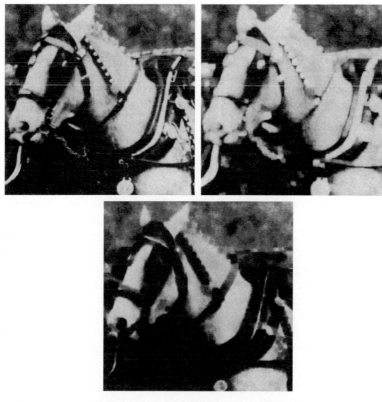

a b
c

FIGURE 9.29
(a) Original image. (b) Result of dilation. (c) Result of erosion. (Courtesy of Mr. A. Morris, Leica Cambridge, Ltd.)

These effects clearly are visible in Fig. 9.29(b). Not only does the image appear brighter than the original, but the sizes of dark features, such as the nostrils and the dark components of the studded rein extending from the ears down to the neck, have been reduced. Figure 9.29(c) shows the result of eroding the original image. Note the opposite effect to dilation. The eroded image is darker, and the sizes of small, bright features (such as the studs on the rein) were reduced.   ■

9.6.3 Opening and Closing

The expressions for opening and closing of gray-scale images have the same form as their binary counterparts. The opening of image f by subimage (structuring element) b, denoted $f \circ b$, is

$$f \circ b = (f \ominus b) \oplus b. \tag{9.6-4}$$

As in the binary case, opening is simply the erosion of f by b, followed by a dilation of the result by b. Similarly, the closing of f by b, denoted $f \bullet b$, is

$$f \bullet b = (f \oplus b) \ominus b. \tag{9.6-5}$$

The opening and closing for gray-scale images are duals with respect to complementation and reflection. That is,

$$(f \bullet b)^c = f^c \circ \hat{b}. \tag{9.6-6}$$

Because $f^c = -f(x, y)$, Eq. (9.6-6) can be written also as $-(f \bullet b) = (-f \circ \hat{b})$.

Opening and closing of images have a simple geometric interpretation. Suppose that we view an image function $f(x, y)$ in 3-D perspective (like a relief map), with the x- and y-axes being the usual spatial coordinates and the third axis being gray-level values. In this representation, the image appears as a discrete surface whose value at any point (x, y) is the value of f at those coordinates. Suppose that we open f by a spherical structuring element, b, viewing this element as a "rolling ball." Then the mechanics of opening f by b may be interpreted geometrically as the process of pushing the ball against the *underside* of the surface, while at the same time rolling it so that the entire underside of the surface is traversed. The opening, $f \circ b$, then is the surface of the highest points reached by any part of the sphere as it slides over the *entire* undersurface of f.

Figure 9.30 illustrates this concept. Figure 9.30(a) shows a scan line of a gray-scale image as a continuous function to simplify the illustration. Figure 9.30(b) shows the rolling ball in various positions, and Fig. 9.30(c) shows the complete result of opening f by b along the scan line. All the peaks that were narrow with respect to the diameter of the ball were reduced in amplitude and sharpness. In practical applications, opening operations usually are applied to remove small (with respect to the size of the structuring element) light details, while leaving the overall gray levels and larger bright features relatively undisturbed. The initial erosion removes the small details, but it also darkens the image. The subsequent dilation again increases the overall intensity of the image without reintroducing the details totally removed by erosion.

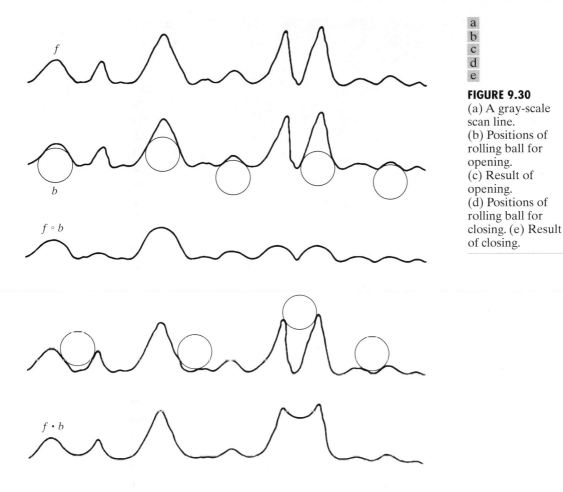

FIGURE 9.30
(a) A gray-scale scan line.
(b) Positions of rolling ball for opening.
(c) Result of opening.
(d) Positions of rolling ball for closing. (e) Result of closing.

Figures 9.30(d) and (e) show the result of closing f by b. Here, the ball slides on top of the surface, and peaks essentially are left in their original form (assuming that their separation at the narrowest point exceeds the diameter of the ball). In practice, closing is generally used to remove dark details from an image, while leaving bright features relatively undisturbed. The initial dilation removes the dark details and brightens the image, and the subsequent erosion darkens the image without reintroducing the details removed by dilation. It is of interest to compare Fig. 9.30 with Figs. 9.8 and 9.9.

The gray-scale opening operation satisfies the following properties:

(i) $(f \circ b) \lrcorner f$.
(ii) If $f_1 \lrcorner f_2$, then $(f_1 \circ b) \lrcorner (f_2 \circ b)$.
(iii) $(f \circ b) \circ b = f \circ b$.

The notation $e \lrcorner r$ is used to indicate that the domain of e is a subset of the domain of r, and also that $e(x, y) \le r(x, y)$ for any (x, y) in the domain of e.

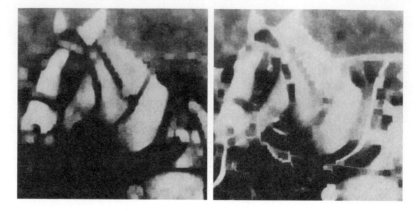

a b

FIGURE 9.31 (a) Opening and (b) closing of Fig. 9.29(a). (Courtesy of Mr. A. Morris, Leica Cambridge, Ltd.)

Similarly, the closing operation satisfies the following properties:

(i) $f \lrcorner (f \bullet b)$.
(ii) If $f_1 \lrcorner f_2$, then $(f_1 \bullet b) \lrcorner (f_2 \bullet b)$.
(iii) $(f \bullet b) \bullet b = f \bullet b$.

The usefulness of these expressions is similar to that of their binary counterparts.

EXAMPLE 9.10:
Illustration of gray-scale opening and closing.

■ Figure 9.31(a) shows the result of opening the image in Fig. 9.29(a) with the same structuring element used there. Note the decreased sizes of the small, bright details, with no appreciable effect on the darker gray levels. Figure 9.31(b) shows the closing of Fig. 9.29(a). Note the decreased sizes of the small, dark details, with relatively little effect on the bright features.   ■

9.6.4 Some Applications of Gray-Scale Morphology

We conclude the discussion of morphological techniques by presenting in some detail various applications of gray-scale morphology. Unless stated otherwise, all the images shown are of size 512×512 and were processed by using the structuring element discussed in connection with Fig. 9.29.

Morphological smoothing

One way to achieve smoothing is to perform a morphological opening followed by a closing. The net result of these two operations is to remove or attenuate both bright and dark artifacts or noise. Figure 9.32 shows a smoothed version of the image shown in Fig. 9.29(a).

Morphological gradient

In addition to the operations discussed earlier in connection with the removal of small dark and bright artifacts, dilation and erosion often are used to com-

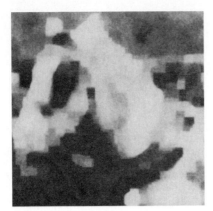

FIGURE 9.32 Morphological smoothing of the image in Fig. 9.29(a). (Courtesy of Mr. A. Morris, Leica Cambridge, Ltd.)

pute the *morphological gradient* of an image, denoted g:

$$g = (f \oplus b) - (f \ominus b). \tag{9.6-7}$$

Figure 9.33 shows the result of computing the morphological gradient of the image shown in Fig. 9.29(a). As expected, the morphological gradient highlights sharp gray-level transitions in the input image. As opposed to gradients obtained using the methods discussed in Section 3.7.3, morphological gradients obtained using symmetrical structuring elements tend to depend less on edge directionality.

Top-hat transformation

The so-called morphological *top-hat* transformation of an image, denoted h, is defined as

$$h = f - (f \circ b) \tag{9.6-8}$$

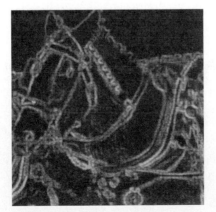

FIGURE 9.33 Morphological gradient of the image in Fig. 9.29(a). (Courtesy of Mr. A. Morris, Leica Cambridge, Ltd.)

FIGURE 9.34 Result of performing a top-hat transformation on the image of Fig. 9.29(a). (Courtesy of Mr. A. Morris, Leica Cambridge, Ltd.)

where, as before, f is the input image and b is the structuring element function. This transformation—which owes its original name to the use of a cylindrical or parallelepiped structuring element function with a flat top—is useful for enhancing detail in the presence of shading. Figure 9.34 shows the result of performing a top-hat transformation on the image of Fig. 9.29(a). Note the enhancement of detail in the background region below the lower part of the horse's head.

Textural segmentation

Figure 9.35(a) shows a simple gray-scale image composed of two texture regions. The region on the right consists of circular blobs of larger diameter than those on the left. The objective is to find the boundary between the two regions based on their textural content.

Because closing tends to remove dark details from an image, the procedure in this particular case is to close the input image by using successively larger structuring elements. When the size of the structuring element corresponds to that of

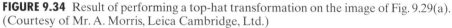

a b

FIGURE 9.35
(a) Original image. (b) Image showing boundary between regions of different texture. (Courtesy of Mr. A. Morris, Leica Cambridge, Ltd.)

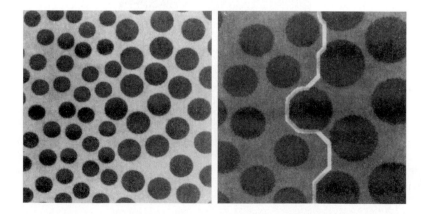

the small blobs, they are removed from the image, leaving only a light background in the area previously occupied by them. At this point in the process, only the larger blobs and the light background on the left and between the large blobs themselves, remain. Next, a single opening is performed with a structuring element that is large in relation to the separation between the large blobs. This operation removes the light patches between the blobs, leaving a dark region on the right consisting of the large dark blobs and the now equally dark patches between these blobs. At this point the process has produced a light region on the left and a dark region on the right. A simple threshold then yields the boundary between the two textural regions. Figure 9.35(b) shows the resulting boundary superimposed on the original image. It is instructive to work through this example in more detail using the rolling ball analogy described in Fig. 9.30.

Granulometry

Granulometry is a field that deals principally with determining the size distribution of particles in an image. Figure 9.36(a) shows an image consisting of light objects of three different sizes. The objects not only are overlapping, but they also are too cluttered to enable detection of individual particles. Because the particles are lighter than the background, the following morphological approach can be used to determine size distribution. Opening operations with structuring elements of increasing size are performed on the original image. The difference between the original image and its opening is computed after each pass when a different structuring element is completed. At the end of the process, these differences are normalized and then used to construct a histogram of particle-size distribution. This approach is based on the idea that opening operations of a particular size have the most effect on regions of the input image that contain particles of similar size. Thus, a measure of the relative number of such particles is obtained by computing the difference between the input and output images. Figure 9.36(b) shows the resulting size distribution in this case. The histogram indicates the presence of three predominant particle sizes in the input image. This type of processing is useful for describing regions with a predominant particle-like character.

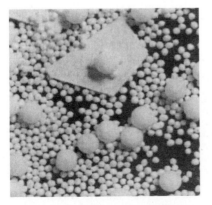

Size Dist'n

FIGURE 9.36
(a) Original image consisting of overlapping particles; (b) size distribution. (Courtesy of Mr. A. Morris, Leica Cambridge, Ltd.)

Summary

The morphological concepts and techniques introduced in this chapter constitute a powerful set of tools for extracting features of interest in an image. One of the most appealing aspects of morphological image processing is the extensive set-theoretical foundation from which morphological techniques have evolved. A significant advantage in terms of implementation is the fact that dilation and erosion are primitive operations that are the basis for a broad class of morphological algorithms. As shown in the following chapter, morphology can be used as the basis for developing image segmentation procedures with a wide range of applications. As discussed in Chapter 11, morphological techniques also play a major role in procedures for image description.

References and Further Reading

The book by Serra [1982] is a fundamental reference on morphological image processing. See also Serra [1988], Giardina and Dougherty [1988], and Haralick and Shapiro [1992]. Additional early references relevant to our discussion include Blum [1967], Lantuéjoul [1980], Maragos [1987], and Haralick et al. [1987]. For an overview of both binary and gray-scale morphology, see Basart and Gonzalez [1992] and Basart et al. [1992]. This set of references provides ample background for the material covered in Sections 9.1 through 9.4.

Important issues of implementing morphological algorithms such as the ones given in Section 9.5 and 9.6 are exemplified in the papers by Jones and Svalbe [1994], Park and Chin [1995], Sussner and Ritter [1997], Anelli et al. [1998], and Shaked and Bruckstein [1998]. Current work in the theory and applications of morphological image processing is summarized in the book by Goutsias and Bloomberg [2000], and in a special issue of *Pattern Recognition* [2000]. See also a recent compilation of references by Rosenfeld [2000]. The books by Marchand-Maillet and Sharaiha [2000] on binary image processing, and by Ritter and Wilson [2001] on image algebra, also are on interest.

Problems

See inside front cover

Detailed solutions to the problems marked with a star can be found in the book web site. The site also contains suggested projects based on the material in this chapter.

9.1   Digital images in this book are embedded in square grid arrangements and pixels are allowed to be 4-, 8-, or *m*-connected. However, other grid arrangements are possible. Specifically, a hexagonal grid arrangement that leads to 6-connectivity, is sometimes used (see the following figure).

   (a) How would you convert an image from a square grid to a hexagonal grid?

   (b) Discuss the shape invariance to rotation of objects represented in a square grid as opposed to a hexagonal grid.

   (c) Is it possible to have ambiguous diagonal configurations in a hexagonal grid, as is the case with 8-connectivity? (See Section 2.5.2.)

9.2 ★ **(a)** With reference to Fig. 9.1(a), sketch the set $(A \cap B) \cup (A \cup B)^c$.

(b) Give expressions for the sets shown shaded in the following figures. The shaded areas in each image constitute one set, so give an expression for each of the three images:

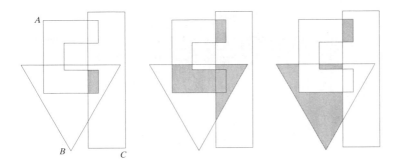

9.3 ★ **(a)** Give a morphological algorithm for converting an 8-connected binary boundary to an m-connected boundary (see Section 2.5.2). You may assume that the boundary is fully connected and that it is one pixel thick.

(b) Does the operation of your algorithm require more than one iteration with each structuring element? Explain your reasoning.

(c) Is the performance of your algorithm independent of the order in which the structuring elements are applied? If your answer is yes, prove it; otherwise give an example that illustrates the dependence of your procedure on the order of application of the structuring elements.

9.4 Erosion of a set A by structuring element B is a subset of A as long as the origin of B is contained by B. Give an example in which the erosion $A \ominus B$ lies outside, or partially outside, A.

9.5 The following four statements are true. Advance an argument that establishes the reason(s) for their validity. Part (a) is true in general. Parts (b) through (d) are true only for *digital* sets. To show the validity of (b) through (d), draw a discrete, square grid (as shown in Problem 9.1) and give an example for each case using sets composed of points on this grid. *Hint:* Keep the number of points in each case as small as possible while still establishing the validity of the statements.

★ **(a)** The erosion of a convex set by a convex structuring element is a convex set.

★ **(b)** The dilation of a convex set by a convex structuring element is not necessarily always convex.

(c) The points in a convex digital set are not always connected.

(d) It is possible to have a set of points in which a line joining every pair of points in the set lies within the set but the set is not convex.

★**9.6** With reference to the image shown, give the structuring element and morphological operation(s) that produced each of the results shown in images (a) through (d). Show the origin of each structuring element clearly. The dashed lines show the boundary of the original set and are included only for reference. Note that in (d) all corners are rounded.

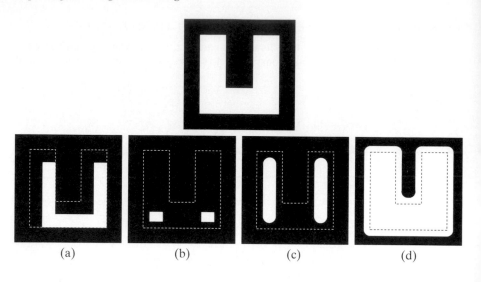

(a)   (b)   (c)   (d)

9.7   Let A denote the set shown shaded in the following figure. Refer to the structuring elements shown (the black dots denote the origin). Sketch the result of the following morphological operations:

(a) $(A \ominus B^4) \oplus B^2$

(b) $(A \ominus B^1) \oplus B^3$

(c) $(A \oplus B^1) \oplus B^3$

(d) $(A \oplus B^3) \ominus B^2$

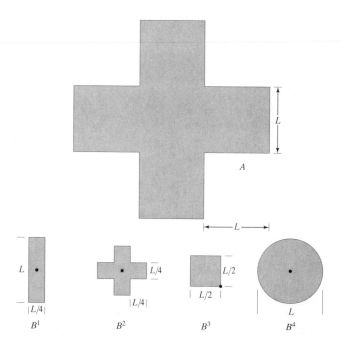

★9.8   (a) What is the limiting effect of repeatedly dilating an image? Assume that a trivial (one point) structuring element is not used.

(b) What is the smallest image from which you can start in order for your answer in part (a) to hold?

9.9   (a) What is the limiting effect of repeatedly eroding an image? Assume that a trivial (one point) structuring element is not used.

(b) What is the smallest image from which you can start in order for your answer in part (a) to hold?

★9.10   An alternative definition of dilation is

$$A \oplus B = \{w \in Z^2 | w = a + b, \text{ for some } a \in A \text{ and } b \in B\}.$$

Show that this definition and the definition in Eq. (9.2-1) are equivalent.

9.11   (a) Show that the definition of dilation given in Problem 9.10 is equivalent to yet another definition of dilation:

$$A \oplus B = \bigcup_{b \in B} (A)_b$$

(This expression also is called the *Minkowsky addition* of two sets.)

(b) Show that the expression in (a) also is equivalent to the definition in Eq. (9.2-1).

★9.12   An alternative definition of erosion is

$$A \ominus B = \{w \in Z^2 | w + b \in A, \text{ for every } b \in B\}.$$

Show that this definition is equivalent to the definition in Eq. (9.2-3).

9.13   (a) Show that the definition of erosion given in Problem 9.12 is equivalent to yet another definition of erosion:

$$A \ominus B = \bigcap_{b \in B} (A)_{-b}$$

(If $-b$ is replaced with b, this expression is called the *Minkowsky subtraction* of two sets.)

(b) Show that the expression in (a) also is equivalent to the definition in Eq. (9.2-3).

★9.14   Prove the validity of the duality expression $(A \bullet B)^c = (A^c \circ \hat{B})$.

9.15   Prove the validity of the following expressions:

★ (a) $A \circ B$ is a subset (subimage) of A.

(b) If C is a subset of D, then $C \circ B$ is a subset of $D \circ B$.

(c) $(A \circ B) \circ B = A \circ B$.

9.16   Prove the validity of the following expressions (assume that the origin of B is contained in B and that Problem 9.16(a) is true):

(a) A is a subset (subimage) of $A \bullet B$.

(b) If C is a subset of D, then $C \bullet B$ is a subset of $D \bullet B$.

(c) $(A \bullet B) \bullet B = A \bullet B$.

9.17   Refer to the image and structuring element shown. Sketch what the sets C, D, E, and F would look like in the following sequence of operations: $C = A \ominus B$; $D = C \oplus B$; $E = D \oplus B$; and $F = E \ominus B$. The initial set A consists of all the image components shown in white, with the exception of the structuring element *B*. Note that this sequence of operations is simply the opening of *A* by *B*, followed by the closing of that opening by *B*. You may assume that *B* is just large enough to enclose each of the noise components.

★**9.18**   In the example given in connection with Fig. 9.7, it is demonstrated that the squares surviving the erosion step could be fully restored by dilating them with the same structuring element used for erosion. Erosion followed by dilation is the opening of the image, and we know that, in general, opening does not result in identical reconstruction of the objects to which it is applied. Explain why full reconstruction of the remaining squares was possible in Fig. 9.7.

 9.19   **(a)** Sketch the result of applying the hit-or-miss transform to the image and structuring element shown. Indicate clearly the origin and border you selected for the structuring element.

     (b) Compare your result with the result of matching by correlation (Fig. 4.46), stating differences and similarities between the two.

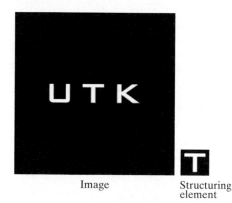

Image   Structuring
                                     element

★**9.20**   Three features (lake, bay, and line segment) useful for differentiating thinned objects in an image are shown in the following figures. Develop a morphological/logical algorithm for differentiating among these shapes. The input to your algorithm would be one of these three shapes. The output must be the identity of the input. You may assume that the features are 1 pixel thick and that each is fully connected. However, they can appear in any orientation.

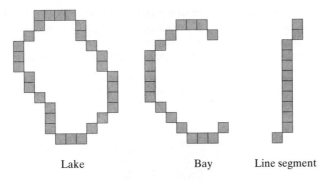

Lake   Bay   Line segment

9.21 Discuss what you would expect the result to be in each of the following cases:

(a) The starting point of the region filling algorithm of section 9.5.2 is a point on the boundary of the object.

(b) The starting point in the region filling algorithm is outside of the boundary.

(c) Sketch what the convex hull of the figure in Problem 9.7 would look like as computed with the algorithm given in Section 9.5.4. Assume that $L = 3$ pixels.

★**9.22** (a) Discuss the effect of using the structuring element in Fig. 9.15(c) for boundary extraction, instead of the one shown in Fig. 9.13(b).

(b) What would be the effect of using a 3×3 structuring element composed of all 1's in the region filling algorithm of Eq. (9.5-2), instead of the structuring element shown in Fig. 9.15(c)?

9.23 Propose a method for fully automating the example shown in Fig. 9.16. This implies determining which black points are background points and which are points contained within the spheres (i.e., black regions completely contained within white regions). Assume that binary 1's are represented in white.

★**9.24** The algorithm given in Section 9.5.3 for extracting connected components requires that a point be known in each connected component in order to extract them all. Suppose that you are given a binary image containing an arbitrary (unknown) number of connected components. Propose a completely automated procedure for extracting all connected components. Assume that points belonging to connected components are labeled 1 and background points are labeled 0.

9.25 Suppose that the image $f(x, y)$ and structuring element $b(x, y)$ in Eq. (9.6-1) both are rectangular, with domains D_f and D_b denoted $([F_{x1}, F_{x2}], [F_{y1}, F_{y2}])$ and $([B_{x1}, B_{x2}], [B_{y1}, B_{y2}])$ respectively. For example, the closed intervals $[F_{x1}, F_{x2}]$ and $[F_{y1}, F_{y2}]$ are the ranges of x and y in the x- and y-axes of the xy-plane where the function $f(x, y)$ is defined.

★ (a) Assume that $(x, y) \in D_b$, and derive expressions for the intervals over which the displacement variables s and t can range to satisfy Eq. (9.6-1). These intervals in the s- and t-axes define the rectangular domain of $(f \oplus b)(s, t)$ in the st-plane.

(b) Repeat for erosion, as defined in Eq. (9.6-2).

9.26 A gray-scale image, $f(x, y)$, is corrupted by nonoverlapping noise spikes that can be modeled as small, cylindrical artifacts of radii $R_{min} \le r \le R_{max}$ and amplitude $A_{min} \le a \le A_{max}$.

★ (a) Develop a morphological filtering approach for cleaning up the image.

(b) Repeat (a), but now assume that there is overlapping of, at most, four noise spikes.

9.27   A preprocessing step in an application of microscopy is concerned with the issue of isolating individual round particles from similar particles that overlap in groups of two or more particles (see following image). Assuming that all particles are of the same size, propose a morphological algorithm that produces three images consisting respectively of

★ **(a)** Only of particles that have merged with the boundary of the image.

(b) Only overlapping particles.

(c) Only nonoverlapping particles.

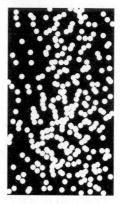

9.28   A high-technology manufacturing plant wins a government contract to manufacture high-precision washers of the form shown in the following figure. The terms of the contract require that the shape of all washers be inspected by an imaging system. In this context, shape inspection refers to deviations from round on the inner and outer edges of the washers. You may assume the following: (1) A "golden" (perfect with respect to the problem) image of an acceptable washer is available; and (2) the imaging and positioning systems ultimately used in the system will have an accuracy high enough to allow you to ignore errors due to digitalization and positioning. You are hired as a consultant to help specify the visual inspection part of the system. Propose a solution based on morphological/logic operations. Your answer should be in the form of a block diagram.

10 *Image Segmentation*

The whole is equal to the sum of its parts.

Euclid

The whole is greater than the sum of its parts.

Max Wertheimer

Preview

The material in the previous chapter began a transition from image processing methods whose input and output are images, to methods in which the inputs are images, but the outputs are attributes extracted from those images (in the sense defined in Section 1.1). Segmentation is another major step in that direction.

Segmentation subdivides an image into its constituent regions or objects. The level to which the subdivision is carried depends on the problem being solved. That is, segmentation should stop when the objects of interest in an application have been isolated. For example, in the automated inspection of electronic assemblies, interest lies in analyzing images of the products with the objective of determining the presence or absence of specific anomalies, such as missing components or broken connection paths. There is no point in carrying segmentation past the level of detail required to identify those elements.

Segmentation of nontrivial images is one of the most difficult tasks in image processing. Segmentation accuracy determines the eventual success or failure of computerized analysis procedures. For this reason, considerable care should be taken to improve the probability of rugged segmentation. In some situations, such as industrial inspection applications, at least some measure of control over the environment is possible at times. The experienced image processing system designer invariably pays considerable attention to such opportunities. In other applications, such as autonomous target acquisition, the system designer has no control of the environment. Then the usual approach is to focus on selecting the types of sensors most likely to enhance the objects of interest while diminishing the contribution of irrelevant image detail. A good example is the use of infrared imaging by the military to detect objects with strong heat signatures, such as equipment and troops in motion.

Image segmentation algorithms generally are based on one of two basic properties of intensity values: discontinuity and similarity. In the first category, the approach is to partition an image based on abrupt changes in intensity, such as edges in an image. The principal approaches in the second category are based on partitioning an image into regions that are similar according to a set of predefined criteria. Thresholding, region growing, and region splitting and merging are examples of methods in this category.

In this chapter we discuss a number of approaches in the two categories just mentioned. We begin the development with methods suitable for detecting gray-level discontinuities such as points, lines, and edges. Edge detection in particular has been a staple of segmentation algorithms for many years. In addition to edge detection per se, we also discuss methods for connecting edge segments and for "assembling" edges into region boundaries. The discussion on edge detection is followed by the introduction of various thresholding techniques. Thresholding also is a fundamental approach to segmentation that enjoys a significant degree of popularity, especially in applications where speed is an important factor. The discussion on thresholding is followed by the development of several region-oriented segmentation approaches. We then discuss a morphological approach to segmentation called *watershed segmentation*. This approach is particularly attractive because it combines several of the positive attributes of segmentation based on the techniques presented in the first part of the chapter. We conclude the chapter with a discussion on the use of motion cues for image segmentation.

10.1 Detection of Discontinuities

In this section we present several techniques for detecting the three basic types of gray-level discontinuities in a digital image: points, lines, and edges. The most common way to look for discontinuities is to run a mask through the image in the manner described in Section 3.5. For the 3×3 mask shown in Fig. 10.1, this procedure involves computing the sum of products of the coefficients with the gray

FIGURE 10.1 A general 3×3 mask.

w_1	w_2	w_3
w_4	w_5	w_6
w_7	w_8	w_9

levels contained in the region encompassed by the mask. That is, with reference to Eq. (3.5-3), the response of the mask at any point in the image is given by

$$R = w_1 z_1 + w_2 z_2 + \cdots + w_9 z_9$$
$$= \sum_{i=1}^{9} w_i z_i \qquad (10.1\text{-}1)$$

where z_i is the gray level of the pixel associated with mask coefficient w_i. As usual, the response of the mask is defined with respect to its center location. The details for implementing mask operations are discussed in Section 3.5.

10.1.1 Point Detection

The detection of isolated points in an image is straightforward in principle. Using the mask shown in Fig. 10.2(a), we say that a point has been detected at the location on which the mask is centered if

$$|R| \geq T \qquad (10.1\text{-}2)$$

where T is a nonnegative threshold and R is given by Eq. (10.1-1). Basically, this formulation measures the weighted differences between the center point and its neighbors. The idea is that an *isolated* point (a point whose gray level is significantly different from its background and which is located in a homogeneous or nearly homogeneous area) will be quite different from its surroundings, and thus be easily detectable by this type of mask. Note that the mask in Fig. 10.2(a) is the same as the mask shown in Fig. 3.39(d) in connection with Laplacian operations. However, the emphasis here is strictly on the detection of points. That is, the only differences that are considered of interest are those

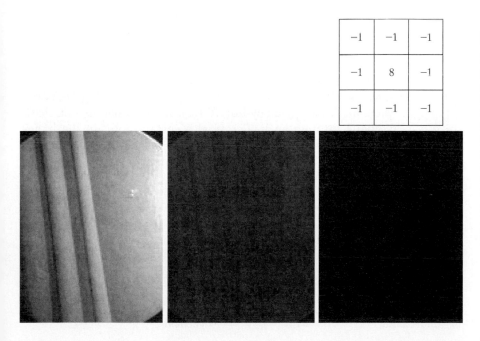

a
b c d

FIGURE 10.2
(a) Point detection mask.
(b) X-ray image of a turbine blade with a porosity.
(c) Result of point detection.
(d) Result of using Eq. (10.1-2). (Original image courtesy of X-TEK Systems Ltd.)

large enough (as determined by T) to be considered isolated points. Note that the mask coefficients sum to zero, indicating that the mask response will be zero in areas of constant gray level.

EXAMPLE 10.1:
Detection of isolated points in an image.

■ We illustrate segmentation of isolated points from an image with the aid of Fig. 10.2(b), which shows an X-ray image of a jet-engine turbine blade with a porosity in the upper, right quadrant of the image. There is a single black pixel embedded within the porosity. Figure 10.2(c) is the result of applying the point detector mask to the X-ray image, and Fig. 10.2(d) shows the result of using Eq. (10.1-2) with T equal to 90% of the highest absolute pixel value of the image in Fig. 10.2(c). (Threshold selection is discussed in detail in Section 10.3.) The single pixel is clearly visible in this image (the pixel was enlarged manually so that it would be visible after printing). This type of detection process is rather specialized because it is based on single-pixel discontinuities that have a homogeneous background in the area of the detector mask. When this condition is not satisfied, other methods discussed in this chapter are more suitable for detecting gray-level discontinuities.   ■

10.1.2 Line Detection

The next level of complexity is line detection. Consider the masks shown in Fig. 10.3. If the first mask were moved around an image, it would respond more strongly to lines (one pixel thick) oriented horizontally. With a constant background, the maximum response would result when the line passed through the middle row of the mask. This is easily verified by sketching a simple array of 1's with a line of a different gray level (say, 5's) running horizontally through the array. A similar experiment would reveal that the second mask in Fig. 10.3 responds best to lines oriented at $+45°$; the third mask to vertical lines; and the fourth mask to lines in the $-45°$ direction. These directions can be established also by noting that the preferred direction of each mask is weighted with a larger coefficient (i.e., 2) than other possible directions. Note that the coefficients in each mask sum to zero, indicating a zero response from the masks in areas of constant gray level.

Let R_1, R_2, R_3, and R_4 denote the responses of the masks in Fig. 10.3, from left to right, where the R's are given by Eq. (10.1-1). Suppose that the four masks are run individually through an image. If, at a certain point in the image, $|R_i| > |R_j|$, for all $j \neq i$, that point is said to be more likely associated with a line in the direction of mask i. For example, if at a point in the image, $|R_1| > |R_j|$ for

FIGURE 10.3 Line masks.

-1	-1	-1		-1	-1	2		-1	2	-1		2	-1	-1
2	2	2		-1	2	-1		-1	2	-1		-1	2	-1
-1	-1	-1		2	-1	-1		-1	2	-1		-1	-1	2

Horizontal	$+45°$	Vertical	$-45°$

$j = 2, 3, 4$, that particular point is said to be more likely associated with a horizontal line. Alternatively, we may be interested in detecting lines in a specified direction. In this case, we would use the mask associated with that direction and threshold its output, as in Eq. (10.1-2). In other words, if we are interested in detecting all the lines in an image in the direction defined by a given mask, we simply run the mask through the image and threshold the absolute value of the result. The points that are left are the strongest responses, which, for lines one pixel thick, correspond closest to the direction defined by the mask. The following example illustrates this procedure.

■ Figure 10.4(a) shows a digitized (binary) portion of a wire-bond mask for an electronic circuit. Suppose that we are interested in finding all the lines that are one pixel thick and are oriented at −45°. For this purpose, we use the last mask shown in Fig. 10.3. The absolute value of the result is shown in Fig. 10.4(b). Note that all vertical and horizontal components of the image were eliminated, and that the components of the original image that tend toward a −45° direction

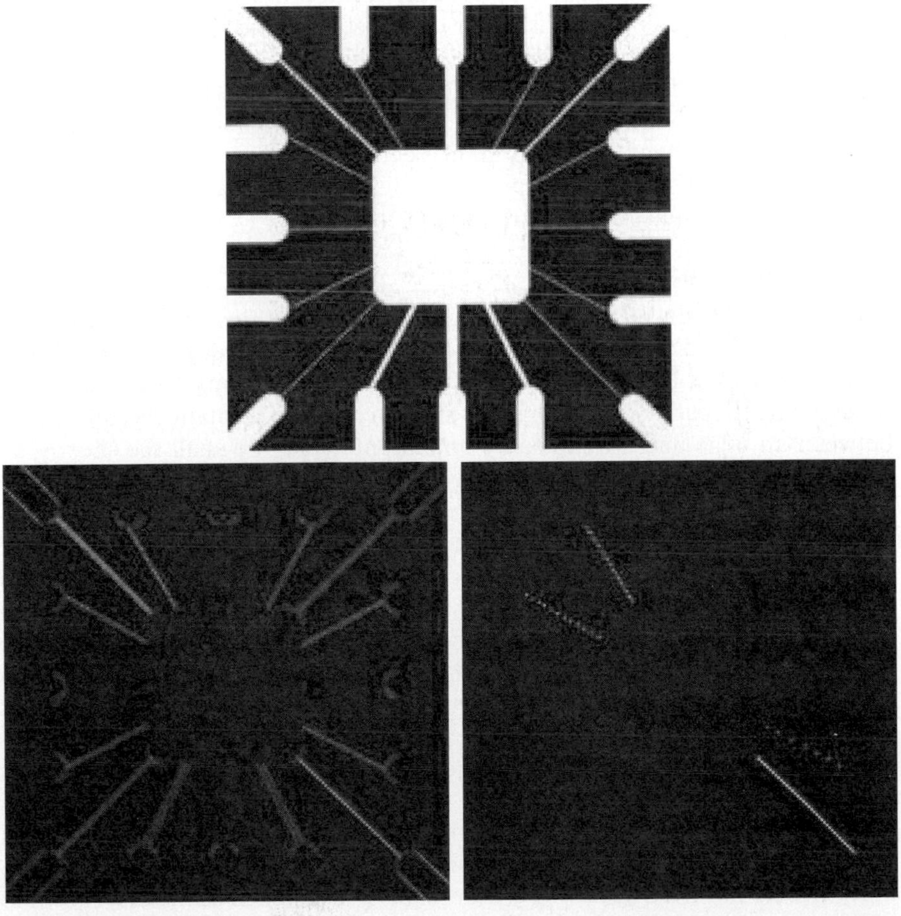

EXAMPLE 10.2:
Detection of lines in a specified direction.

a
b c

FIGURE 10.4
Illustration of line detection.
(a) Binary wire-bond mask.
(b) Absolute value of result after processing with −45° line detector.
(c) Result of thresholding image (b).

produced the strongest responses in Fig. 10.4(b). In order to determine which lines best fit the mask, we simply threshold this image. The result of using a threshold equal to the maximum value in the image is shown in Fig. 10.4(c). The maximum value is a good choice for a threshold in applications such as this because the input image is binary and we are looking for the strongest responses. Figure 10.4(c) shows in white all points that passed the threshold test. In this case, the procedure extracted the only line segment that was one pixel thick and oriented at −45° (the other component of the image oriented in this direction in the top, left quadrant is not one pixel thick). The isolated points shown in Fig. 10.4(c) are points that also had similarly strong responses to the mask. In the original image, these points and their immediate neighbors are oriented in such as way that the mask produced a maximum response at those isolated locations. These isolated points can be detected using the mask in Fig. 10.2(a) and then deleted, or they could be deleted using morphological erosion, as discussed in the last chapter.   ■

10.1.3 Edge Detection

Although point and line detection certainly are important in any discussion on segmentation, edge detection is by far the most common approach for detecting meaningful discontinuities in gray level. In this section we discuss approaches for implementing first- and second-order digital derivatives for the detection of edges in an image. We introduced these derivatives in Section 3.7 in the context of image enhancement. The focus in this section is on their properties for edge detection. Some of the concepts previously introduced are restated briefly here for the sake continuity in the discussion.

Basic formulation

Edges were introduced informally in Section 3.7.1. In this section we look at the concept of a digital edge a little closer. Intuitively, an edge is a set of connected pixels that lie on the boundary between two regions. However, we already went through some length in Section 2.5.2 to explain the difference between an edge and a boundary. Fundamentally, as we shall see shortly, an edge is a "local" concept whereas a region boundary, owing to the way it is defined, is a more global idea. A reasonable definition of "edge" requires the ability to measure gray-level transitions in a meaningful way.

We start by modeling an edge intuitively. This will lead us to a formalism in which "meaningful" transitions in gray levels can be measured. Intuitively, an ideal edge has the properties of the model shown in Fig. 10.5(a). An ideal edge according to this model is a set of connected pixels (in the vertical direction here), each of which is located at an orthogonal step transition in gray level (as shown by the horizontal profile in the figure).

In practice, optics, sampling, and other image acquisition imperfections yield edges that are blurred, with the degree of blurring being determined by factors such as the quality of the image acquisition system, the sampling rate, and illumination conditions under which the image is acquired. As a result, edges are more closely modeled as having a "ramplike" profile, such as the one shown in

Model of an ideal digital edge

Model of a ramp digital edge

FIGURE 10.5
(a) Model of an
ideal digital edge.
(b) Model of a
ramp edge. The
slope of the ramp
is proportional to
the degree of
blurring in the
edge.

Gray-level profile
of a horizontal line
through the image

Gray-level profile
of a horizontal line
through the image

Fig. 10.5(b). The slope of the ramp is inversely proportional to the degree of blurring in the edge. In this model, we no longer have a thin (one pixel thick) path. Instead, an edge point now is any point contained in the ramp, and an edge would then be a set of such points that are connected. The "thickness" of the edge is determined by the length of the ramp, as it transitions from an initial to a final gray level. This length is determined by the slope, which, in turn, is determined by the degree of blurring. This makes sense: Blurred edges tend to be thick and sharp edges tend to be thin.

Figure 10.6(a) shows the image from which the close-up in Fig. 10.5(b) was extracted. Figure 10.6(b) shows a horizontal gray-level profile of the edge between the two regions. This figure also shows the first and second derivatives of the gray-level profile. The first derivative is positive at the points of transition into and out of the ramp as we move from left to right along the profile; it is constant for points in the ramp; and is zero in areas of constant gray level. The second derivative is positive at the transition associated with the dark side of the edge, negative at the transition associated with the light side of the edge, and zero along the ramp and in areas of constant gray level. The signs of the derivatives in Fig. 10.6(b) would be reversed for an edge that transitions from light to dark.

We conclude from these observations that the magnitude of the first derivative can be used to detect the presence of an edge at a point in an image (i.e., to determine if a point is on a ramp). Similarly, the sign of the second derivative can be used to determine whether an edge pixel lies on the dark or light side of an edge. We note two additional properties of the second derivative around an edge: (1) It produces two values for every edge in an image (an undesirable feature); and (2) an imaginary straight line joining the extreme positive and negative values of the second derivative would cross zero near the midpoint of the edge. This *zero-crossing* property of the second derivative is quite useful

FIGURE 10.6
(a) Two regions
separated by a
vertical edge.
(b) Detail near
the edge, showing
a gray-level
profile, and the
first and second
derivatives of the
profile.

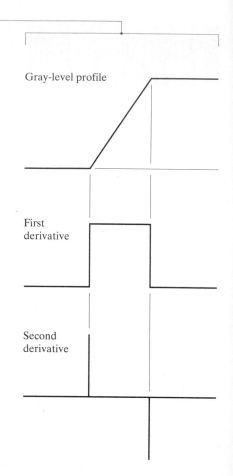

for locating the centers of thick edges, as we show later in this section. Finally,
we note that some edge models make use of a smooth transition into and out
of the ramp (Problem 10.5). However, the conclusions at which we arrive in the
following discussion are the same. Also, it is evident from this discussion that we
are dealing here with local measures (thus the comment made in Section 2.5.2
about the local nature of edges).

Although attention thus far has been limited to a 1-D horizontal profile, a
similar argument applies to an edge of any orientation in an image. We simply
define a profile perpendicular to the edge direction at any desired point and
interpret the results as in the preceding discussion.

EXAMPLE 10.3:
Behavior of the
first and second
derivatives
around a noisy
edge.

■ The edges shown in Fig. 10.5 and 10.6 are free of noise. The image segments
in the first column in Fig. 10.7 show close-ups of four ramp edges separating a
black region on the left and a white region on the right. It is important to keep
in mind that the entire transition from black to white is a *single* edge. The image
segment at the top, left is free of noise. The other three images in the first column of Fig. 10.7 are corrupted by additive Gaussian noise with zero mean and

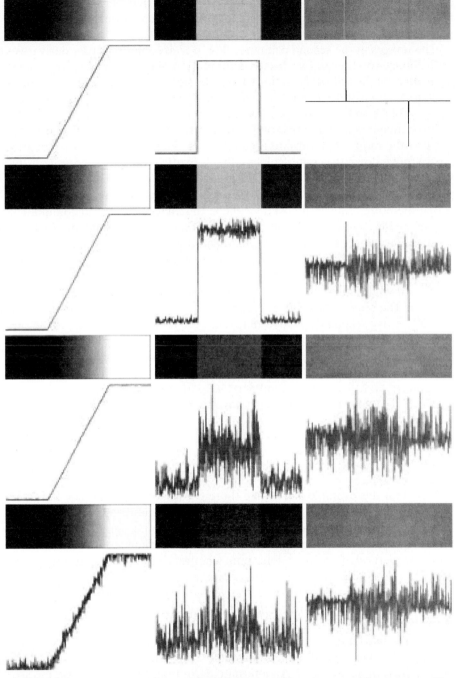

FIGURE 10.7 First column: images and gray-level profiles of a ramp edge corrupted by random Gaussian noise of mean 0 and σ = 0.0, 0.1, 1.0, and 10.0, respectively. Second column: first-derivative images and gray-level profiles. Third column: second-derivative images and gray-level profiles.

a
b
c
d

standard deviation of 0.1, 1.0, and 10.0 gray levels, respectively. The graph shown below each of these images is a gray-level profile of a horizontal scan line passing through the image.

The images in the second column of Fig. 10.7 are the first-order derivatives of the images on the left (we discuss computation of the first and second image derivatives in the following section). Consider, for example, the center image at the top. As discussed in connection with Fig. 10.6(b), the derivative is zero in the constant black and white regions. These are the two black areas shown in the derivative image. The derivative of a constant ramp is a constant, equal to the slope of the ramp. This constant area in the derivative image is shown in gray. As we move down the center column, the derivatives become increasingly different from the noiseless case. In fact, it would be difficult to associate the last profile in that column with a ramp edge. What makes these results interesting is that the noise really is almost invisible in the images on the left column. The last image is a slightly grainy, but this corruption is almost imperceptible. These examples are good illustrations of the sensitivity of derivatives to noise.

As expected, the second derivative is even more sensitive to noise. The second derivative of the noiseless image is shown in the top, right image. The thin black and white lines are the positive and negative components explained in Fig. 10.6. The gray in these images represents zero due to scaling. We note that the only noisy second derivative that resembles the noiseless case is the one corresponding to noise with a standard deviation of 0.1 gray levels. The other two second-derivative images and profiles clearly illustrate that it would be difficult indeed to detect their positive and negative components, which are the truly useful features of the second derivative in terms of edge detection.

The fact that fairly little noise can have such a significant impact on the two key derivatives used for edge detection in images is an important issue to keep in mind. In particular, image smoothing should be a serious consideration prior to the use of derivatives in applications where noise with levels similar to those we have just discussed is likely to be present.   ■

Based on this example and on the three paragraphs that precede it, we are led to the conclusion that, to be classified as a meaningful edge point, the transition in gray level associated with that point has to be significantly stronger than the background at that point. Since we are dealing with local computations, the method of choice to determine whether a value is "significant" or not is to use a threshold. Thus, we *define* a point in an image as being an *edge point* if its two-dimensional first-order derivative is greater than a specified threshold. A set of such points that are connected according to a predefined criterion of connectedness (see Section 2.5.2) is by definition an *edge*. The term *edge segment* generally is used if the edge is short in relation to the dimensions of the image. A key problem in segmentation is to assemble edge segments into longer edges, as explained in Section 10.2. An alternate definition if we elect to use the second-derivative is simply to define the edge points in an image as the zero crossings of its second derivative. The definition of an edge in this case is the same as above. It is important to note that these definitions do not guarantee success in finding edges in an image. They simply give us a formalism to look for them.

As in Chapter 3, first-order derivatives in an image are computed using the gradient. Second-order derivatives are obtained using the Laplacian.

Gradient operators

First-order derivatives of a digital image are based on various approximations of the 2-D gradient. The gradient of an image $f(x, y)$ at location (x, y) is defined as the *vector*

See inside front cover

Consult the book web site for a brief review of vector analysis.

$$\nabla \mathbf{f} = \begin{bmatrix} G_x \\ G_y \end{bmatrix} = \begin{bmatrix} \dfrac{\partial f}{\partial x} \\ \dfrac{\partial f}{\partial y} \end{bmatrix}. \tag{10.1-3}$$

It is well known from vector analysis that the gradient vector points in the direction of maximum rate of change of f at coordinates (x, y).

An important quantity in edge detection is the magnitude of this vector, denoted ∇f, where

$$\nabla f = \text{mag}(\nabla \mathbf{f}) = \left[G_x^2 + G_y^2 \right]^{1/2}. \tag{10.1-4}$$

This quantity gives the maximum rate of increase of $f(x, y)$ per unit distance in the direction of $\nabla \mathbf{f}$. It is a common (although not strictly correct) practice to refer to ∇f also as the *gradient*. We will adhere to convention and also use this term interchangeably, differentiating between the vector and its magnitude only in cases in which confusion is likely.

The *direction* of the gradient vector also is an important quantity. Let $\alpha(x, y)$ represent the direction angle of the vector $\nabla \mathbf{f}$ at (x, y). Then, from vector analysis,

$$\alpha(x, y) = \tan^{-1}\left(\frac{G_y}{G_x} \right) \tag{10.1-5}$$

where the angle is measured with respect to the x-axis. The direction of an edge at (x, y) is *perpendicular* to the direction of the gradient vector at that point. Computation of the gradient of an image is based on obtaining the partial derivatives $\partial f / \partial x$ and $\partial f / \partial y$ at every pixel location. Let the 3×3 area shown in Fig. 10.8(a) represent the gray levels in a neighborhood of an image. As discussed in Section 3.7.3, one of the simplest ways to implement a first-order partial derivative at point z_5 is to use the following *Roberts cross-gradient operators*:

$$G_x = (z_9 - z_5) \tag{10.1-6}$$

and

$$G_y = (z_8 - z_6). \tag{10.1-7}$$

These derivatives can be implemented for an entire image by using the masks shown in Fig. 10.8(b) with the procedure discussed in Section 3.5.

Masks of size 2×2 are awkward to implement because they do not have a clear center. An approach using masks of size 3×3 is given by

$$G_x = (z_7 + z_8 + z_9) - (z_1 + z_2 + z_3) \tag{10.1-8}$$

z_1	z_2	z_3
z_4	z_5	z_6
z_7	z_8	z_9

−1	0
0	1

0	−1
1	0

Roberts

−1	−1	−1
0	0	0
1	1	1

−1	0	1
−1	0	1
−1	0	1

Prewitt

−1	−2	−1
0	0	0
1	2	1

−1	0	1
−2	0	2
−1	0	1

Sobel

a
b c
d e
f g

FIGURE 10.8 A 3 × 3 region of an image (the z's are gray-level values) and various masks used to compute the gradient at point labeled z_5.

and

$$G_y = (z_3 + z_6 + z_9) - (z_1 + z_4 + z_7). \tag{10.1-9}$$

In this formulation, the difference between the first and third rows of the 3 × 3 image region approximates the derivative in the x-direction, and the difference between the third and first columns approximates the derivative in the y-direction. The masks shown in Figs. 10.8(d) and (e), called the *Prewitt operators*, can be used to implement these two equations.

A slight variation of these two equations uses a weight of 2 in the center coefficient:

$$G_x = (z_7 + 2z_8 + z_9) - (z_1 + 2z_2 + z_3) \tag{10.1-10}$$

and

$$G_y = (z_3 + 2z_6 + z_9) - (z_1 + 2z_4 + z_7). \tag{10.1-11}$$

A weight value of 2 is used to achieve some smoothing by giving more importance to the center point (Problem 10.8). Figures 10.8(f) and (g), called the *Sobel* operators, are used to implement these two equations. The Prewitt and Sobel operators are among the most used in practice for computing digital gradients. The Prewitt masks are simpler to implement than the Sobel masks, but the latter have slightly superior noise-suppression characteristics, an important issue when dealing with derivatives. Note that the coefficients in all the masks shown in Fig. 10.8 sum to 0, indicating that they give a response of 0 in areas of constant gray level, as expected of a derivative operator.

The masks just discussed are used to obtain the gradient components G_x and G_y. Computation of the gradient requires that these two components be combined in the manner shown in Eq. (10.1-4). However, this implementation is not always desirable because of the computational burden required by squares and square roots. An approach used frequently is to approximate the gradient by absolute values:

$$\nabla f \approx |G_x| + |G_y|. \tag{10.1-12}$$

This equation is much more attractive computationally, and it still preserves relative changes in gray levels. As discussed in Section 3.7.3, the price paid for this advantage is that the resulting filters will not be isotropic (invariant to rotation) in general. However, this is not an issue when masks such as the Prewitt and Sobel masks are used to compute G_x and G_y. These masks give isotropic results only for vertical and horizontal edges, so even if we used Eq. (10.1-4) to compute the gradient, the results would be isotropic only for edges in those directions. In this case, Eqs. (10.1-4) and (10.1-12) give the same result (Problem 10.6).

It is possible to modify the 3×3 masks in Fig. 10.8 so that they have their strongest responses along the diagonal directions. The two additional Prewitt and Sobel masks for detecting discontinuities in the diagonal directions are shown in Fig. 10.9.

■ Figure 10.10 illustrates the response of the two components of the gradient, $|G_x|$ and $|G_y|$, as well as the gradient image formed from the sum of these two

EXAMPLE 10.4:
Illustration of the gradient and its components.

0	1	1	−1	−1	0
−1	0	1	−1	0	1
−1	−1	0	0	1	1

Prewitt

0	1	2	−2	−1	0
−1	0	1	−1	0	1
−2	−1	0	0	1	2

Sobel

a b
c d

FIGURE 10.9 Prewitt and Sobel masks for detecting diagonal edges.

a b
c d

FIGURE 10.10
(a) Original
image. (b) $|G_x|$,
component of the
gradient in the
x-direction.
(c) $|G_y|$,
component in the
y-direction.
(d) Gradient
image, $|G_x| + |G_y|$.

components. The directionality of the two components is evident in Figs. 10.10(b) and (c). Note in particular how strong the roof tile, horizontal brick joints, and horizontal segments of the windows are in Fig. 10.10(b). By contrast, Fig. 10.10(c) favors the vertical components, such as the corner of the near wall, the vertical components of the window, the vertical joints of the brick, and the lamppost on the right side of the picture.

The original image is of reasonably high resolution (1200 × 1600 pixels) and, at the distance the image was taken, the contribution made to image detail by the wall bricks is still significant. This level of detail often is undesirable, and one way to reduce it is to smooth the image. Figure 10.11 shows the same sequence of images as in Fig. 10.10, but with the original image being smoothed first using a 5 × 5 averaging filter. The response of each mask now shows almost no contribution due to the bricks, with the result being dominated mostly by the principal edges. Note that averaging caused the response of all edges to be weaker.

In Figs. 10.10 and 10.11, it is evident that the horizontal and vertical Sobel masks respond about equally well to edges oriented in the minus and plus 45° directions. If it is important to emphasize edges along the diagonal directions, then one of the mask pairs in Fig. 10.9 should be used. The absolute responses of the diagonal Sobel masks are shown in Fig. 10.12. The stronger diagonal response of these masks is evident in this figure. Both diagonal masks have similar response to horizontal and vertical edges but, as expected, their response in these directions is weaker than the response of the horizontal and vertical Sobel masks shown in Figs. 10.10(b) and 10.10(c).   ■

a b
c d

FIGURE 10.11
Same sequence as
in Fig. 10.10, but
with the original
image smoothed
with a 5 × 5
averaging filter.

a b

FIGURE 10.12
Diagonal edge
detection.
(a) Result of using
the mask in
Fig. 10.9(c).
(b) Result of using
the mask in
Fig. 10.9(d). The
input in both cases
was Fig. 10.11(a).

The Laplacian

The Laplacian of a 2-D function $f(x, y)$ is a second-order derivative defined as

$$\nabla^2 f = \frac{\partial^2 f}{\partial x^2} + \frac{\partial^2 f}{\partial y^2}. \tag{10.1-13}$$

Digital approximations to the Laplacian were introduced in Section 3.7.2. For
a 3 × 3 region, one of the two forms encountered most frequently in practice is

$$\nabla^2 f = 4z_5 - (z_2 + z_4 + z_6 + z_8) \tag{10.1-14}$$

FIGURE 10.13
Laplacian masks
used to
implement
Eqs. (10.1-14) and
(10.1-15),
respectively.

0	−1	0
−1	4	−1
0	−1	0

−1	−1	−1
−1	8	−1
−1	−1	−1

where the z's are defined in Fig. 10.8(a). A digital approximation including the diagonal neighbors is given by

$$\nabla^2 f = 8z_5 - (z_1 + z_2 + z_3 + z_4 + z_6 + z_7 + z_8 + z_9). \quad (10.1\text{-}15)$$

Masks for implementing these two equations are shown in Fig. 10.13. We note from these masks that the implementations of Eqs. (10.1-14) and (10.1-15) are isotropic for rotation increments of 90° and 45°, respectively.

The Laplacian generally is not used in its original form for edge detection for several reasons: As a second-order derivative, the Laplacian typically is unacceptably sensitive to noise (Fig. 10.7). The magnitude of the Laplacian produces double edges (see Figs. 10.6 and 10.7), an undesirable effect because it complicates segmentation. Finally, the Laplacian is unable to detect edge direction. For these reasons, the role of the Laplacian in segmentation consists of (1) using its zero-crossing property for edge location, as mentioned earlier in this section, or (2) using it for the complementary purpose of establishing whether a pixel is on the dark or light side of an edge, as we show in Section 10.3.6.

In the first category, the Laplacian is combined with smoothing as a precursor to finding edges via zero-crossings. Consider the function

$$h(r) = -e^{-\frac{r^2}{2\sigma^2}} \quad (10.1\text{-}16)$$

where $r^2 = x^2 + y^2$ and σ is the standard deviation. Convolving this function with an image blurs the image, with the degree of blurring being determined by the value of σ. The Laplacian of h (the second derivative of h with respect to r) is

$$\nabla^2 h(r) = -\left[\frac{r^2 - \sigma^2}{\sigma^4}\right] e^{-\frac{r^2}{2\sigma^2}}. \quad (10.1\text{-}17)$$

This function is commonly referred to as the *Laplacian of a Gaussian* (LoG) because Eq. (10.1-16) is in the form of a Gaussian function. Figure 10.14 shows a 3-D plot, image, and cross section of the LoG function. Also shown is a 5 × 5 mask that approximates $\nabla^2 h$. This approximation is not unique. Its purpose is to capture the essential *shape* of $\nabla^2 h$; that is, a positive central term, surrounded by an adjacent negative region that increases in value as a function of distance from the origin, and a zero outer region. The coefficients also must sum to zero, so that the response of the mask is zero in areas of constant gray level. A mask this small is useful only for images that are essentially noise free. Due to its shape, the Laplacian of a Gaussian sometimes is called the *Mexican hat* function.

Because the second derivative is a linear operation, convolving an image with $\nabla^2 h$ is the same as convolving the image with the Gaussian smoothing function of Eq. (10.1-16) first and then computing the Laplacian of the result.

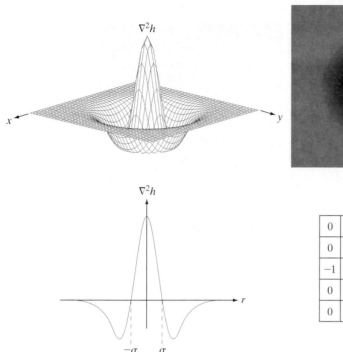

FIGURE 10.14
Laplacian of a Gaussian (LoG).
(a) 3-D plot.
(b) Image (black is negative, gray is the zero plane, and white is positive).
(c) Cross section showing zero crossings.
(d) 5 × 5 mask approximation to the shape of (a).

0	0	−1	0	0
0	−1	−2	−1	0
−1	−2	16	−2	−1
0	−1	−2	−1	0
0	0	−1	0	0

Thus, we see that the purpose of the Gaussian function in the LoG formulation is to smooth the image, and the purpose of the Laplacian operator is to provide an image with zero crossings used to establish the location of edges. Smoothing the image reduces the effect of noise and, in principle, it counters the increased effect of noise caused by the second derivatives of the Laplacian. It is of interest to note that neurophysiological experiments carried out in the early 1980s (Ullman [1981], Marr [1982]) provide evidence that certain aspects of human vision can be modeled mathematically in the basic form of Eq. (10.1-17).

■ Figure 10.15(a) shows the angiogram image discussed in Section 1.3.2. Figure 10.15(b) shows the Sobel gradient of this image, included here for comparison. Figure 10.15(c) is a spatial Gaussian function (with a standard deviation of five pixels) used to obtain a 27 × 27 spatial smoothing mask. The mask was obtained by sampling this Gaussian function at equal intervals. Figure 10.15(d) is the spatial mask used to implement Eq. (10.1-15). Figure 10.15(e) is the LoG image obtained by smoothing the original image with the Gaussian smoothing mask, followed by application of the Laplacian mask (this image was cropped to eliminate the border effects produced by the smoothing mask). As noted in the preceding paragraph, $\nabla^2 h$ can be computed by application of (c) followed by (d). Employing this procedure provides more control over the smoothing function, and often results in two masks that are much smaller when compared with a single composite mask that implements Eq. (10.1-17) directly. A composite mask usually is larger because it must incorporate the more complex shape shown in Fig. 10.14(a).

EXAMPLE 10.5:
Edge finding by zero crossings.

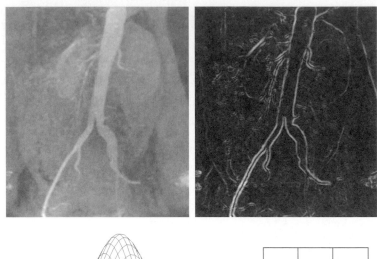

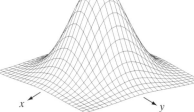

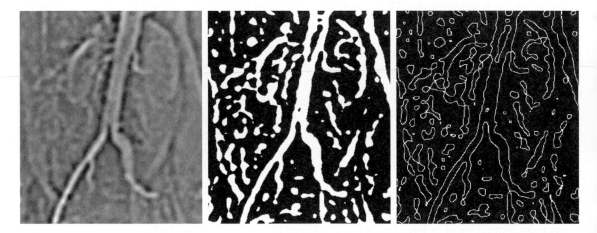

FIGURE 10.15 (a) Original image. (b) Sobel gradient (shown for comparison). (c) Spatial Gaussian smoothing function. (d) Laplacian mask. (e) LoG. (f) Thresholded LoG. (g) Zero crossings. (Original image courtesy of Dr. David R. Pickens, Department of Radiology and Radiological Sciences, Vanderbilt University Medical Center.)

The LoG result shown in Fig. 10.15(e) is the image from which zero crossings are computed to find edges. One straightforward approach for approximating zero crossings is to threshold the LoG image by setting all its positive values to, say, white, and all negative values to black. The result is shown in Fig. 10.15(f). The logic behind this approach is that zero crossings occur between positive and negative values of the Laplacian. Finally, Fig. 10.15(g) shows the estimated zero crossings, obtained by scanning the thresholded image and noting the transitions between black and white.

Comparing Figs. 10.15(b) and (g) reveals several interesting and important differences. First, we note that the edges in the zero-crossing image are thinner than the gradient edges. This is a characteristic of zero crossings that makes this approach attractive. On the other hand, we see in Fig. 10.15(g) that the edges determined by zero crossings form numerous closed loops. This so-called spaghetti effect is one of the most serious drawbacks of this method. Another major drawback is the computation of zero crossings, which is the foundation of the method. Although it was reasonably straightforward in this example, the computation of zero crossings presents a challenge in general, and considerably more sophisticated techniques often are required to obtain acceptable results (Huertas and Medione [1986]).

Zero-crossing methods are of interest because of their noise reduction capabilities and potential for rugged performance. However, the limitations just noted present a significant barrier in practical applications. For this reason, edge-finding techniques based on various implementations of the gradient still are used more frequently than zero crossings in the implementation of segmentation algorithms. ■

10.2 Edge Linking and Boundary Detection

Ideally, the methods discussed in the previous section should yield pixels lying only on edges. In practice, this set of pixels seldom characterizes an edge completely because of noise, breaks in the edge from nonuniform illumination, and other effects that introduce spurious intensity discontinuities. Thus edge detection algorithms typically are followed by linking procedures to assemble edge pixels into meaningful edges. Several basic approaches are suited to this purpose.

10.2.1 Local Processing

One of the simplest approaches for linking edge points is to analyze the characteristics of pixels in a small neighborhood (say, 3×3 or 5×5) about every point (x, y) in an image that has been labeled an edge point by one of the techniques discussed in the previous section. All points that are similar according to a set of predefined criteria are linked, forming an edge of pixels that share those criteria.

The two principal properties used for establishing similarity of edge pixels in this kind of analysis are (1) the strength of the response of the gradient operator used to produce the edge pixel; and (2) the direction of the gradient vector. The first property is given by the value of ∇f, as defined in Eq. (10.1-4) or (10.1-12). Thus an edge pixel with coordinates (x_0, y_0) in a predefined neighborhood of

(x, y), is similar in magnitude to the pixel at (x, y) if

$$|\nabla f(x, y) - \nabla f(x_0, y_0)| \leq E \qquad (10.2\text{-}1)$$

where E is a nonnegative threshold.

The direction (angle) of the gradient vector is given by Eq. (10.1-5). An edge pixel at (x_0, y_0) in the predefined neighborhood of (x, y) has an angle similar to the pixel at (x, y) if

$$|\alpha(x, y) - \alpha(x_0, y_0)| < A \qquad (10.2\text{-}2)$$

where A is a nonnegative angle threshold. As noted in Eq. (10.1-5), the direction of the edge at (x, y) is *perpendicular* to the direction of the gradient vector at that point.

A point in the predefined neighborhood of (x, y) is linked to the pixel at (x, y) if both magnitude and direction criteria are satisfied. This process is repeated at every location in the image. A record must be kept of linked points as the center of the neighborhood is moved from pixel to pixel. A simple bookkeeping procedure is to assign a different gray level to each set of linked edge pixels.

EXAMPLE 10.6:
Edge-point
linking based on
local processing.

■ To illustrate the foregoing procedure, consider Fig. 10.16(a), which shows an image of the rear of a vehicle. The objective is to find rectangles whose sizes makes them suitable candidates for license plates. The formation of these rectangles can be accomplished by detecting strong horizontal and vertical edges. Figures 10.16(b) and (c) show vertical and horizontal edges obtained by using

a b
c d

FIGURE 10.16
(a) Input image.
(b) G_y component
of the gradient.
(c) G_x component
of the gradient.
(d) Result of edge
linking. (Courtesy
of Perceptics
Corporation.)

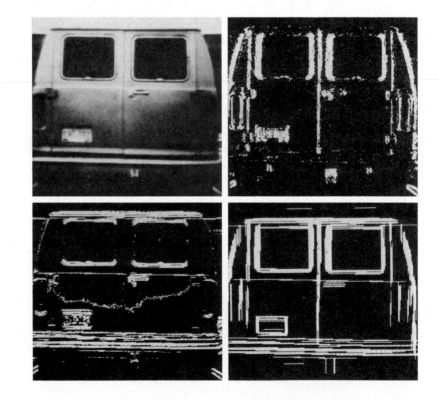

the horizontal and vertical Sobel operators. Figure 10.16(d) shows the result of linking all points that simultaneously had a gradient value greater than 25 and whose gradient directions did not differ by more than 15°. The horizontal lines were formed by sequentially applying these criteria to every row of Fig. 10.16(c). A sequential column scan of Fig. 10.16(b) yielded the vertical lines. Further processing consisted of linking edge segments separated by small breaks and deleting isolated short segments. As Fig. 10.16(d) shows, the rectangle corresponding to the license plate was one of the few rectangles detected in the image. It would be a simple matter to locate the license plate based on these rectangles (the width-to-height ratio of the license plate rectangle has a distinctive 2:1 proportion for U.S. plates). ■

10.2.2 Global Processing via the Hough Transform

In this section, points are linked by determining first if they lie on a curve of specified shape. Unlike the local analysis method discussed in Section 10.2.1, we now consider global relationships between pixels.

Given n points in an image, suppose that we want to find subsets of these points that lie on straight lines. One possible solution is to first find all lines determined by every pair of points and then find all subsets of points that are close to particular lines. The problem with this procedure is that it involves finding $n(n - 1)/2 \sim n^2$ lines and then performing $(n)(n(n - 1))/2 \sim n^3$ comparisons of every point to all lines. This approach is computationally prohibitive in all but the most trivial applications.

Hough [1962] proposed an alternative approach, commonly referred to as the *Hough transform.* Consider a point (x_i, y_i) and the general equation of a straight line in slope-intercept form, $y_i = ax_i + b$. Infinitely many lines pass through (x_i, y_i), but they all satisfy the equation $y_i = ax_i + b$ for varying values of a and b. However, writing this equation as $b = -x_i a + y_i$ and considering the *ab*-plane (also called *parameter space*) yields the equation of a *single* line for a fixed pair (x_i, y_i). Furthermore, a second point (x_j, y_j) also has a line in parameter space associated with it, and this line intersects the line associated with (x_i, y_i) at (a', b'), where a' is the slope and b' the intercept of the line containing both (x_i, y_i) and (x_j, y_j) in the *xy*-plane. In fact, all points contained on this line have lines in parameter space that intersect at (a', b'). Figure 10.17 illustrates these concepts.

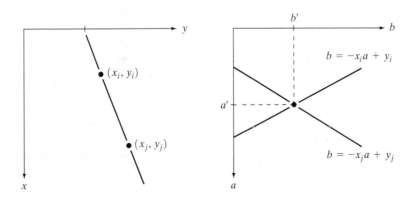

FIGURE 10.17
(a) *xy*-plane.
(b) Parameter space.

FIGURE 10.18
Subdivision of the
parameter plane
for use in the
Hough transform.

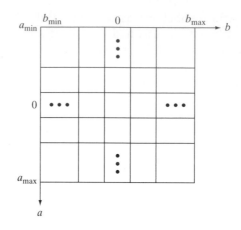

The computational attractiveness of the Hough transform arises from subdividing the parameter space into so-called *accumulator cells*, as illustrated in Fig. 10.18, where (a_{max}, a_{min}) and (b_{max}, b_{min}) are the expected ranges of slope and intercept values. The cell at coordinates (i, j), with accumulator value $A(i, j)$, corresponds to the square associated with parameter space coordinates (a_i, b_j). Initially, these cells are set to zero. Then, for every point (x_k, y_k) in the image plane, we let the parameter a equal each of the allowed subdivision values on the a-axis and solve for the corresponding b using the equation $b = -x_k a + y_k$. The resulting b's are then rounded off to the nearest allowed value in the b-axis. If a choice of a_p results in solution b_q, we let $A(p, q) = A(p, q) + 1$. At the end of this procedure, a value of Q in $A(i, j)$ corresponds to Q points in the xy-plane lying on the line $y = a_i x + b_j$. The number of subdivisions in the ab-plane determines the accuracy of the colinearity of these points.

Note that subdividing the a axis into K increments gives, for every point (x_k, y_k), K values of b corresponding to the K possible values of a. With n image points, this method involves nK computations. Thus the procedure just discussed is *linear* in n, and the product nK does not approach the number of computations discussed at the beginning of this section unless K approaches or exceeds n.

A problem with using the equation $y = ax + b$ to represent a line is that the slope approaches infinity as the line approaches the vertical. One way around this difficulty is to use the normal representation of a line:

$$x \cos \theta + y \sin \theta = \rho. \tag{10.2-3}$$

Figure 10.19(a) illustrates the geometrical interpretation of the parameters used in Eq. (10.2-3). The use of this representation in constructing a table of accumulators is identical to the method discussed for the slope-intercept representation. Instead of straight lines, however, the loci are sinusoidal curves in the $\rho\theta$-plane. As before, Q collinear points lying on a line $x \cos \theta_j + y \sin \theta_j = \rho_i$ yield Q sinusoidal curves that intersect at (ρ_i, θ_j) in the parameter space. Incrementing θ and solving for the corresponding ρ gives Q entries in accumulator $A(i, j)$ associated with the cell determined by (ρ_i, θ_j). Figure 10.19(b) illustrates the subdivision of the parameter space.

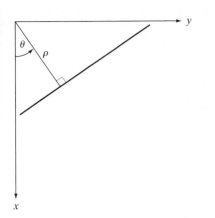

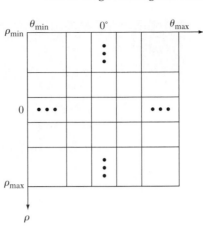

FIGURE 10.19
(a) Normal
representation of
a line.
(b) Subdivision of
the $\rho\theta$-plane into
cells.

The range of angle θ is $\pm 90°$, measured with respect to the x-axis. Thus with reference to Fig. 10.19(a), a horizontal line has $\theta = 0°$, with ρ being equal to the positive x-intercept. Similarly, a vertical line has $\theta = 90°$, with ρ being equal to the positive y-intercept, or $\theta = -90°$, with ρ being equal to the negative y-intercept.

■ Figure 10.20 illustrates the Hough transform based on Eq. (10.2-3). Figure 10.20(a) shows an image with five labeled points. Each of these points is mapped onto the $\rho\theta$-plane, as shown in Fig. 10.20(b). The range of θ values is $\pm 90°$, and the range of the ρ-axis is $\pm\sqrt{2}D$, where D is the distance between corners in the image. Unlike the transform based on using the slope intercept, each of these curves has a different sinusoidal shape. The horizontal line resulting from the mapping of point 1 is a special case of a sinusoid with zero amplitude.

The colinearity detection property of the Hough transform is illustrated in Fig. 10.20(c). Point A (not to be confused with accumulator values) denotes the intersection of the curves corresponding to points 1, 3, and 5 in the xy-image plane. The location of point A indicates that these three points lie on a straight line passing through the origin ($\rho = 0$) and oriented at $-45°$. Similarly, the curves intersecting at point B in the parameter space indicate that points 2, 3, and 4 lie on a straight line oriented at $45°$ and whose distance from the origin is one-half the diagonal distance from the origin of the image to the opposite corner.

Finally, Fig. 10.20(d) indicates the fact that the Hough transform exhibits a reflective adjacency relationship at the right and left edges of the parameter space. This property, shown by the points marked A, B, and C in Fig. 10.20(d), is the result of the manner in which θ and ρ change sign at the $\pm 90°$ boundaries. ■

EXAMPLE 10.7:
Illustration of the
Hough transform.

Although the focus so far has been on straight lines, the Hough transform is applicable to any function of the form $g(\mathbf{v}, \mathbf{c}) = 0$, where $\mathbf{v}$ is a vector of coordinates and $\mathbf{c}$ is a vector of coefficients. For example, the points lying on the circle

$$(x - c_1)^2 + (y - c_2)^2 = c_3^2 \qquad (10.2\text{-}4)$$

can be detected by using the approach just discussed. The basic difference is the presence of three parameters (c_1, c_2, and c_3), which results in a 3-D parameter

a b
c d

FIGURE 10.20
Illustration of the
Hough transform.
(Courtesy of Mr.
D. R. Cate, Texas
Instruments, Inc.)

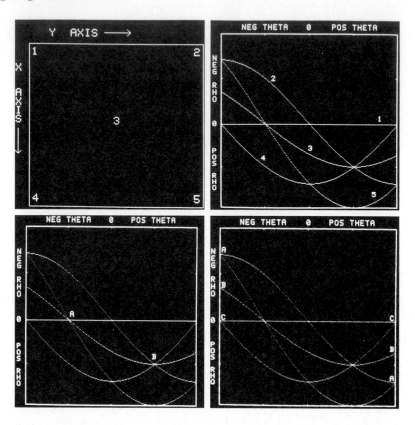

space with cubelike cells and accumulators of the form $A(i, j, k)$. The procedure is to increment c_1 and c_2, solve for the c_3 that satisfies Eq. (10.2-4), and update the accumulator corresponding to the cell associated with the triplet (c_1, c_2, c_3). Clearly, the complexity of the Hough transform is proportional to the number of coordinates and coefficients in a given functional representation. Further generalizations of the Hough transform to detect curves with no simple analytic representations are possible, as is the application of the transform to gray-scale images. Several references dealing with these extensions are included at the end of this chapter.

We now return to the edge-linking problem. An approach based on the Hough transform is as follows:

1. Compute the gradient of an image and threshold it to obtain a binary image.
2. Specify subdivisions in the $\rho\theta$-plane.
3. Examine the counts of the accumulator cells for high pixel concentrations.
4. Examine the relationship (principally for continuity) between pixels in a chosen cell.

The concept of continuity in this case usually is based on computing the distance between disconnected pixels identified during traversal of the set of pixels corresponding to a given accumulator cell. A gap at any point is significant if the distance

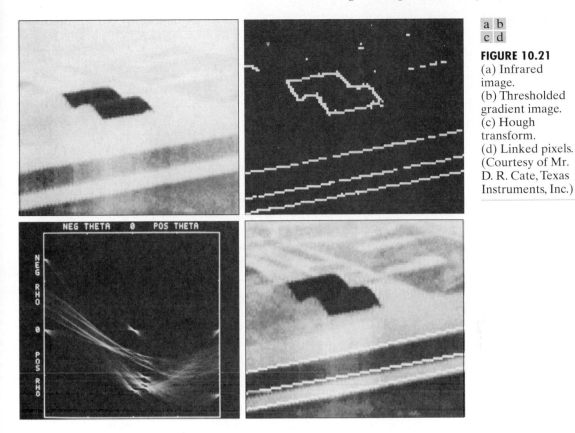

a b
c d

FIGURE 10.21
(a) Infrared
image.
(b) Thresholded
gradient image.
(c) Hough
transform.
(d) Linked pixels.
(Courtesy of Mr.
D. R. Cate, Texas
Instruments, Inc.)

between that point and its closest neighbor exceeds a certain threshold. (See Section 2.5 for a discussion of connectivity, neighborhoods, and distance measures.)

■ Figure 10.21(a) shows an aerial infrared image containing two hangars and a runway. Figure 10.21(b) is a thresholded gradient image obtained using the Sobel operators discussed in Section 10.1.3 (note the small gaps in the borders of the runway). Figure 10.21(c) shows the Hough transform of the gradient image. Figure 10.21(d) shows (in white) the set of pixels linked according to the criteria that (1) they belonged to one of the three accumulator cells with the highest count, and (2) no gaps were longer than five pixels. Note the disappearance of the gaps as a result of linking.   ■

EAMPLE 10.8:
Using the Hough
transform for
edge linking.

10.2.3 Global Processing via Graph-Theoretic Techniques

In this section we discuss a global approach for edge detection and linking based on representing edge segments in the form of a graph and searching the graph for low-cost paths that correspond to significant edges. This representation provides a rugged approach that performs well in the presence of noise. As might be expected, the procedure is considerably more complicated and requires more processing time than the methods discussed so far.

FIGURE 10.22
Edge element
between pixels p
and q.

We begin the development with some basic definitions. A *graph* $G = (N, U)$ is a finite, nonempty set of nodes N, together with a set U of unordered pairs of distinct elements of N. Each pair (n_i, n_j) of U is called an *arc*. A graph in which the arcs are directed is called a *directed graph*. If an arc is directed from node n_i to node n_j, then n_j is said to be a *successor* of the *parent* node n_i. The process of identifying the successors of a node is called *expansion* of the node. In each graph we define levels, such that level 0 consists of a single node, called the *start* or *root* node, and the nodes in the last level are called *goal* nodes. A *cost* $c(n_i, n_j)$ can be associated with every arc (n_i, n_j). A sequence of nodes $n_1, n_2, \ldots, n_k$, with each node n_i being a successor of node n_{i-1}, is called a *path* from n_1 to n_k. The cost of the entire path is

$$c = \sum_{i=2}^{k} c(n_{i-1}, n_i). \tag{10.2-5}$$

The following discussion is simplified if we define an *edge element* as the boundary between two pixels p and q, such that p and q are 4-neighbors, as Fig. 10.22 illustrates. Edge elements are identified by the xy-coordinates of points p and q. In other words, the edge element in Fig. 10.22 is defined by the pairs $(x_p, y_p)(x_q, y_q)$. Consistent with the definition given in Section 10.1.3, an *edge* is a sequence of connected edge elements.

We can illustrate how the concepts just discussed apply to edge detection using the 3×3 image shown in Fig. 10.23(a). The outer numbers are pixel

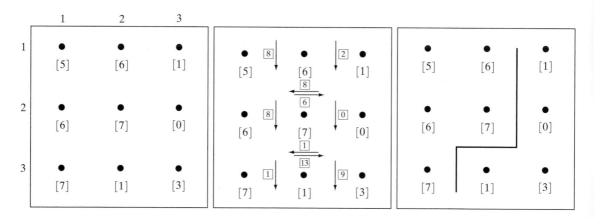

a b c

FIGURE 10.23 (a) A 3×3 image region. (b) Edge segments and their costs. (c) Edge corresponding to the lowest-cost path in the graph shown in Fig. 10.24.

coordinates and the numbers in brackets represent gray-level values. Each edge element, defined by pixels p and q, has an associated cost, defined as

$$c(p, q) = H - \left[f(p) - f(q)\right] \qquad (10.2\text{-}6)$$

where H is the highest gray-level value in the image (7 in this case), and $f(p)$ and $f(q)$ are the gray-level values of p and q, respectively. By convention, the point p is on the right-hand side of the direction of travel along edge elements. For example, the edge segment $(1, 2)(2, 2)$ is between points $(1, 2)$ and $(2, 2)$ in Fig. 10.23(b). If the direction of travel is to the right, then p is the point with coordinates $(2, 2)$ and q is point with coordinates $(1, 2)$; therefore, $c(p, q) = 7 - [7 - 6] = 6$. This cost is shown in the box below the edge segment. If, on the other hand, we are traveling to the *left* between the same two points, then p is point $(1, 2)$ and q is $(2, 2)$. In this case the cost is 8, as shown above the edge segment in Fig. 10.23(b). To simplify the discussion, we assume that edges start in the top row and terminate in the last row, so that the first element of an edge can be only between points $(1, 1)$, $(1, 2)$ or $(1, 2)$, $(1, 3)$. Similarly, the last edge element has to be between points $(3, 1)$, $(3, 2)$ or $(3, 2)$, $(3, 3)$. Keep in mind that p and q are 4-neighbors, as noted earlier.

Figure 10.24 shows the graph for this problem. Each node (rectangle) in the graph corresponds to an edge element from Fig. 10.23. An arc exists between two nodes if the two corresponding edge elements taken in succession can be part

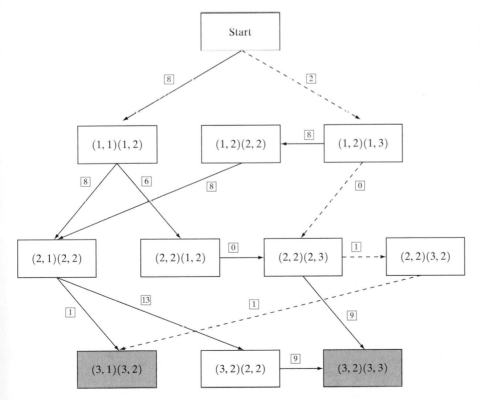

FIGURE 10.24
Graph for the image in Fig. 10.23(a). The lowest-cost path is shown dashed.

of an edge. As in Fig. 10.23(b), the cost of each edge segment, computed using Eq. (10.2-6), is shown in a box on the side of the arc leading into the corresponding node. Goal nodes are shown shaded. The minimum cost path is shown dashed, and the edge corresponding to this path is shown in Fig. 10.23(c).

In general, the problem of finding a minimum-cost path is not trivial in terms of computation. Typically, the approach is to sacrifice optimality for the sake of speed, and the following algorithm represents a class of procedures that use heuristics in order to reduce the search effort. Let $r(n)$ be an estimate of the cost of a minimum-cost path from the start node s to a goal node, where the path is constrained to go through n. This cost can be expressed as the estimate of the cost of a minimum-cost path from s to n plus an estimate of the cost of that path from n to a goal node; that is,

$$r(n) = g(n) + h(n). \tag{10.2-7}$$

Here, $g(n)$ can be chosen as the lowest-cost path from s to n found so far, and $h(n)$ is obtained by using any available heuristic information (such as expanding only certain nodes based on previous costs in getting to that node). An algorithm that uses $r(n)$ as the basis for performing a graph search is as follows:

Step 1: Mark the start node OPEN and set $g(s) = 0$.
Step 2: If no node is OPEN exit with failure; otherwise, continue.
Step 3: Mark CLOSED the OPEN node n whose estimate $r(n)$ computed from Eq. (10.2-7) is smallest. (Ties for minimum r values are resolved arbitrarily, but always in favor of a goal node.)
Step 4: If n is a goal node, exit with the solution path obtained by tracing back through the pointers; otherwise, continue.
Step 5: Expand node n, generating all of its successors. (If there are no successors go to step 2.)
Step 6: If a successor n_i is not marked, set

$$r(n_i) = g(n) + c(n, n_i),$$

mark it OPEN, and direct pointers from it back to n.
Step 7: If a successor n_i is marked CLOSED or OPEN, update its value by letting

$$g'(n_i) = \min[g(n_i), g(n) + c(n, n_i)].$$

Mark OPEN those CLOSED successors whose g' values were thus lowered and redirect to n the pointers from all nodes whose g' values were lowered. Go to step 2.

This algorithm does not guarantee a minimum-cost path; its advantage is speed via the use of heuristics. However, if $h(n)$ is a lower bound on the cost of the minimal-cost path from node n to a goal node, the procedure indeed yields an optimal path to a goal (Hart et al. [1968]). If no heuristic information is available (that is, $h \equiv 0$), the procedure reduces to the *uniform-cost algorithm* of Dijkstra [1959].

EXAMPLE 10.9:
Edge finding by graph search.

■ Figure 10.25 shows an image of a noisy chromosome silhouette and an edge found using a heuristic graph search based on the algorithm developed in this

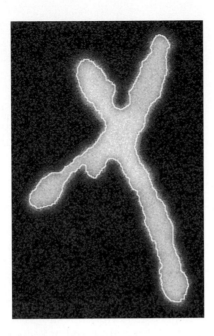

FIGURE 10.25
Image of noisy
chromosome
silhouette and
edge boundary
(in white)
determined by
graph search.

section. The edge is shown in white, superimposed on the original image. Note that in this case the edge and the boundary of the object are approximately the same. The cost was based on Eq. (10.2-6), and the heuristic used at any point on the graph was to determine and use the optimum path for five levels down from that point. Considering the amount of noise present in this image, the graph-search approach yielded a reasonably accurate result.   ■

10.3 Thresholding

Because of its intuitive properties and simplicity of implementation, image thresholding enjoys a central position in applications of image segmentation. Simple thresholding was first introduced in Section 3.1, and we have used it in various discussions in the preceding chapters. In this section, we introduce thresholding in a more formal way and extend it to techniques that are considerably more general than what has been presented thus far.

10.3.1 Foundation

Suppose that the gray-level histogram shown in Fig. 10.26(a) corresponds to an image, $f(x, y)$, composed of light objects on a dark background, in such a way that object and background pixels have gray levels grouped into two dominant modes. One obvious way to extract the objects from the background is to select a threshold T that separates these modes. Then any point (x, y) for which $f(x, y) > T$ is called an *object point*; otherwise, the point is called a *background point*. This is the type of thresholding introduced in Section 3.1.

Figure 10.26(b) shows a slightly more general case of this approach, where three dominant modes characterize the image histogram (for example, two types

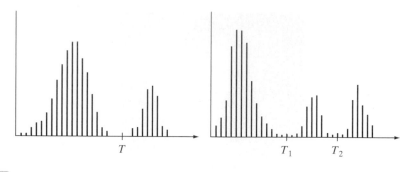

a b

FIGURE 10.26 (a) Gray-level histograms that can be partitioned by (a) a single threshold, and (b) multiple thresholds.

of light objects on a dark background). Here, *multilevel thresholding* classifies a point (x, y) as belonging to one object class if $T_1 < (x, y) \le T_2$, to the other object class if $f(x, y) > T_2$, and to the background if $f(x, y) \le T_1$. In general, segmentation problems requiring multiple thresholds are best solved using region growing methods, such as those discussed in Section 10.4.

Based on the preceding discussion, thresholding may be viewed as an operation that involves tests against a function T of the form

$$T = T[x, y, p(x, y), f(x, y)] \tag{10.3-1}$$

where $f(x, y)$ is the gray level of point (x, y) and $p(x, y)$ denotes some local property of this point—for example, the average gray level of a neighborhood centered on (x, y). A thresholded image $g(x, y)$ is defined as

$$g(x, y) = \begin{cases} 1 & \text{if } f(x, y) > T \\ 0 & \text{if } f(x, y) \le T. \end{cases} \tag{10.3-2}$$

Thus, pixels labeled 1 (or any other convenient gray level) correspond to objects, whereas pixels labeled 0 (or any other gray level not assigned to objects) correspond to the background.

When T depends only on $f(x, y)$ (that is, only on gray-level values) the threshold is called *global*. If T depends on both $f(x, y)$ and $p(x, y)$, the threshold is called *local*. If, in addition, T depends on the spatial coordinates x and y, the threshold is called *dynamic* or *adaptive*.

10.3.2 The Role of Illumination

In Section 2.3.4 we introduced a simple model in which an image $f(x, y)$ is formed as the product of a reflectance component $r(x, y)$ and an illumination component $i(x, y)$. The purpose of this section is to use this model to discuss briefly the effect of illumination on thresholding, especially on global thresholding.

Consider the computer generated reflectance function shown in Fig. 10.27(a). The histogram of this function, shown in Fig. 10.27(b), is clearly bimodal and could be partitioned easily by placing a single global threshold, T, in the histogram

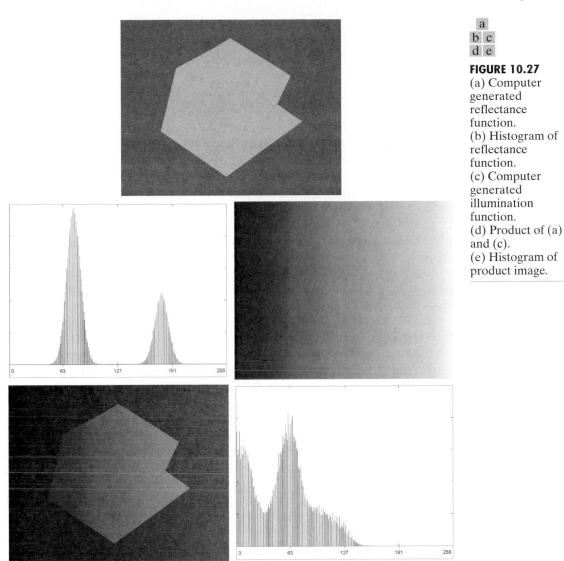

FIGURE 10.27
(a) Computer
generated
reflectance
function.
(b) Histogram of
reflectance
function.
(c) Computer
generated
illumination
function.
(d) Product of (a)
and (c).
(e) Histogram of
product image.

valley. Multiplying the reflectance function in Fig. 10.27(a) by the illumination function shown in Fig. 10.27(c) yields the image shown in Fig. 10.27(d). Figure 10.27(e) shows the histogram of this image. Note that the original valley was virtually eliminated, making segmentation by a single threshold an impossible task. Although we seldom have the reflectance function by itself to work with, this simple illustration shows that the reflective nature of objects and background could be such that they are easily separable. However, the image resulting from poor (in this case nonuniform) illumination could be quite difficult to segment.

The reason why the histogram in Fig. 10.27(e) is so distorted can be explained with aid of the discussion in Section 4.5. From Eq. (4.5-1),

$$f(x, y) = i(x, y)r(x, y). \tag{10.3-3}$$

Taking the natural logarithm of this equation yields a sum:

$$
\begin{aligned}
z(x, y) &= \ln f(x, y) \\
&= \ln i(x, y) + \ln r(x, y) \\
&= i'(x, y) + r'(x, y).
\end{aligned}
\tag{10.3-4}
$$

From probability theory (Papoulis [1991]), if $i'(x, y)$ and $r'(x, y)$ are independent random variables, the histogram of $z(x, y)$ is given by the convolution of the histograms of $i'(x, y)$ and $r'(x, y)$. If $i(x, y)$ were constant, $i'(x, y)$ would be constant also, and its histogram would be a simple spike (like an impulse). The convolution of this impulselike function with the histogram of $r'(x, y)$ would leave the basic shape of this histogram unchanged (recall from the discussion in Section 4.2.4 that convolution of a function with an impulse copies the function at the location of the impulse). But if $i'(x, y)$ had a broader histogram (resulting from nonuniform illumination), the convolution process would smear the histogram of $r'(x, y)$, yielding a histogram for $z(x, y)$ whose shape could be quite different from that of the histogram of $r'(x, y)$. The degree of distortion depends on the broadness of the histogram of $i'(x, y)$, which in turn depends on the nonuniformity of the illumination function.

We have dealt with the logarithm of $f(x, y)$, instead of dealing with the image function directly, but the essence of the problem is clearly explained by using the logarithm to separate the illumination and reflectance components. This approach allows histogram formation to be viewed as a convolution process, thus explaining why a distinct valley in the histogram of the reflectance function could be smeared by improper illumination.

When access to the illumination source is available, a solution frequently used in practice to compensate for nonuniformity is to project the illumination pattern onto a constant, white reflective surface. This yields an image $g(x, y) = ki(x, y)$, where k is a constant that depends on the surface and $i(x, y)$ is the illumination pattern. Then, for any image $f(x, y) = i(x, y)r(x, y)$ obtained with the same illumination function, simply dividing $f(x, y)$ by $g(x, y)$ yields a normalized function $h(x, y) = f(x, y)/g(x, y) = r(x, y)/k$. Thus, if $r(x, y)$ can be segmented by using a single threshold T, then $h(x, y)$ can be segmented by using a single threshold of value T/k.

10.3.3 Basic Global Thresholding

With reference to the discussion in Section 10.3.1, the simplest of all thresholding techniques is to partition the image histogram by using a single global threshold, T, as illustrated in Fig. 10.26(a). Segmentation is then accomplished by scanning the image pixel by pixel and labeling each pixel as object or background, depending on whether the gray level of that pixel is greater or less than the value of T. As indicated earlier, the success of this method depends entirely on how well the histogram can be partitioned.

EXAMPLE 10.10:
Global
thresholding.

■ Figure 10.28(a) shows a simple image, and Fig. 10.28(b) shows its histogram. Figure 10.28(c) shows the result of segmenting Fig. 10.28(a) by using a threshold T midway between the maximum and minimum gray levels. This threshold

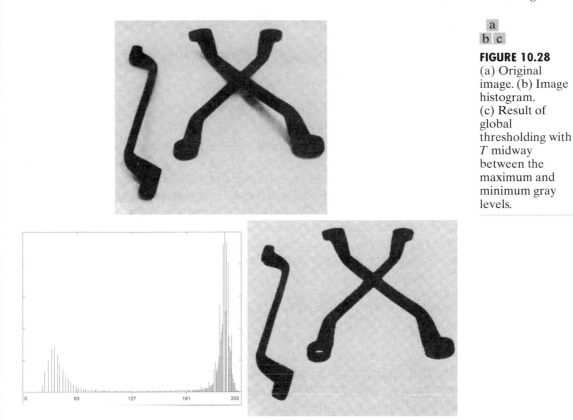

FIGURE 10.28
(a) Original
image. (b) Image
histogram.
(c) Result of
global
thresholding with
T midway
between the
maximum and
minimum gray
levels.

achieved a "clean" segmentation by eliminating the shadows and leaving only the objects themselves. The objects of interest in this case are darker than the background, so any pixel with a gray level $\leq T$ was labeled black (0), and any pixel with a gray level $>T$ was labeled white (255). The key objective is merely to generate a binary image, so the black–white relationship could be reversed.

The type of global thresholding just described can be expected to be successful in highly controlled environments. One of the areas in which this often is possible is in industrial inspection applications, where control of the illumination usually is feasible. ■

The threshold in the preceding example was specified by using a heuristic approach, based on visual inspection of the histogram. The following algorithm can be used to obtain T automatically:

1. Select an initial estimate for T.
2. Segment the image using T. This will produce two groups of pixels: G_1 consisting of all pixels with gray level values $>T$ and G_2 consisting of pixels with values $\leq T$.
3. Compute the average gray level values μ_1 and μ_2 for the pixels in regions G_1 and G_2.

4. Compute a new threshold value:

$$T = \frac{1}{2}(\mu_1 + \mu_2).$$

5. Repeat steps 2 through 4 until the difference in T in successive iterations is smaller than a predefined parameter T_o.

When there is reason to believe that the background and object occupy comparable areas in the image, a good initial value for T is the average gray level of the image. When objects are small compared to the area occupied by the background (or vice versa), then one group of pixels will dominate the histogram and the average gray level is not as good an initial choice. A more appropriate initial value for T in cases such as this is a value midway between the maximum and minimum gray levels. The parameter T_o is used to stop the algorithm after changes become small in terms of this parameter. This is used when speed of iteration is an important issue.

EXAMPLE 10.11:
Image segmentation using an estimated global threshold.

■ Figure 10.29 shows an example of segmentation based on a threshold estimated using the preceding algorithm. Figure 10.29(a) is the original image, and Fig. 10.29(b) is the image histogram. Note the clear valley of the histogram. Application of the iterative algorithm resulted in a value of 125.4 after three iterations starting with the average gray level and $T_o = 0$. The result obtained using $T = 125$ to segment the original image is shown in Fig. 10.29(c). As expected from the clear separation of modes in the histogram, the segmentation between object and background was very effective.   ■

10.3.4 Basic Adaptive Thresholding

As illustrated in Fig. 10.27, imaging factors such as uneven illumination can transform a perfectly segmentable histogram into a histogram that cannot be partitioned effectively by a single global threshold. An approach for handling such a situation is to divide the original image into subimages and then utilize a different threshold to segment each subimage. The key issues in this approach are how to subdivide the image and how to estimate the threshold for each resulting subimage. Since the threshold used for each pixel depends on the location of the pixel in terms of the subimages, this type of thresholding is adaptive. We illustrate adaptive thresholding with a simple example. A more comprehensive example is given in the next section.

EXAMPLE 10.12:
Basic adaptive thresholding.

■ Figure 10.30(a) shows the image from Fig. 10.27(d), which we concluded could not be thresholded effectively with a single global threshold. In fact, Fig. 10.30(b) shows the result of thresholding the image with a global threshold manually placed in the valley of its histogram [see Fig. 10.27(e)]. One approach to reduce the effect of nonuniform illumination is to subdivide the image into smaller subimages, such that the illumination of each subimage is approximately uniform. Figure 10.30(c) shows such a partition, obtained by subdividing the image into four equal parts, and then subdividing each part by four again.

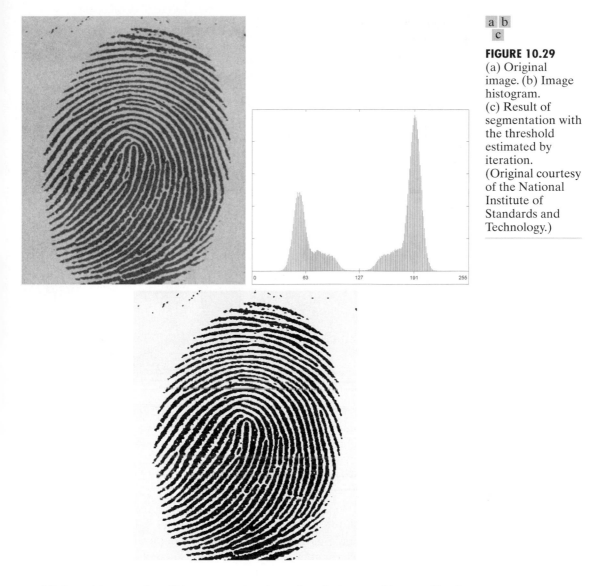

FIGURE 10.29
(a) Original
image. (b) Image
histogram.
(c) Result of
segmentation with
the threshold
estimated by
iteration.
(Original courtesy
of the National
Institute of
Standards and
Technology.)

All the subimages that did not contain a boundary between object and background had variances of less than 75. All subimages containing boundaries had variances in excess of 100. Each subimage with variance greater than 100 was segmented with a threshold computed for that subimage using the algorithm discussed in the previous section. The initial value for T in each case was selected as the point midway between the minimum and maximum gray levels in the subimage. All subimages with variance less than 100 were treated as one composite image, which was segmented using a single threshold estimated using the same algorithm.

The result of segmentation using this procedure is shown in Fig. 10.30(d). With the exception of two subimages, the improvement over Fig. 10.30(b) is

a b
c d

FIGURE 10.30
(a) Original
image. (b) Result
of global
thresholding.
(c) Image
subdivided into
individual
subimages.
(d) Result of
adaptive
thresholding.

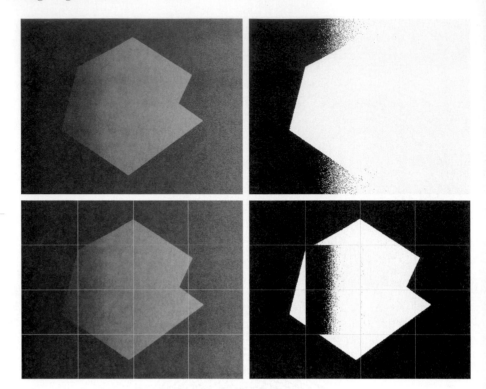

evident. The boundary between object and background in each of the improperly segmented subimages was small and dark, and the resulting histogram was almost unimodal. Figure 10.31(a) shows the top improperly segmented subimage from Fig. 10.30(c) and the subimage directly above it, which was segmented properly. The histogram of the subimage that was properly segmented is clearly bimodal, with well-defined peaks and valley. The other histogram is almost unimodal, with no clear distinction between object and background.

Figure 10.31(d) shows the failed subimage further subdivided into much smaller subimages, and Fig. 10.31(e) shows the histogram of the top, left small subimage. This subimage contains the transition between object and background. This smaller subimage has a clearly bimodal histogram and should be easily segmentable. This, in fact, is the case, as shown in Fig. 10.31(f). This figure also shows the segmentation of all the other small subimages. All these subimages had a nearly unimodal histogram, and their average gray level was closer to the object than to the background, so they were all classified as object. It is left as a project for the reader to show that considerably more accurate segmentation can be achieved by subdividing the entire image in Fig. 10.30(a) into subimages of the size shown in Fig. 10.31(d).   ■

10.3.5 Optimal Global and Adaptive Thresholding

In this section we discuss a method for estimating thresholds that produce the minimum average segmentation error. As an illustration, the method is applied

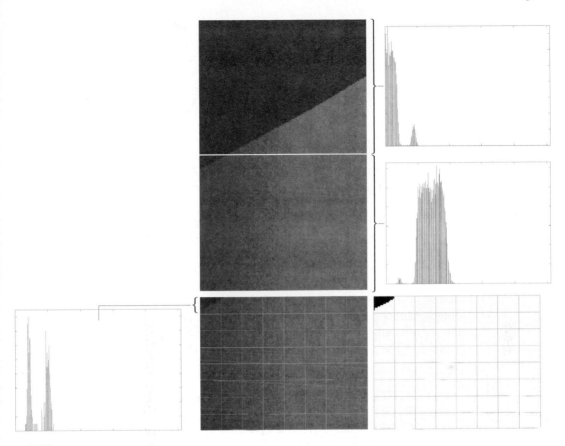

a	b
	c

| e | d | f |

FIGURE 10.31 (a) Properly and improperly segmented subimages from Fig 10.30. (b)–(c) Corresponding histograms. (d) Further subdivision of the improperly segmented subimage. (e) Histogram of small subimage at top, left. (f) Result of adaptively segmenting (d).

to a problem that requires solution of several important issues found frequently in the practical application of thresholding.

   Suppose that an image contains only two principal gray-level regions. Let z denote gray-level values. We can view these values as random quantities, and their histogram may be considered an estimate of their probability density function (PDF), $p(z)$. This overall density function is the sum or mixture of two densities, one for the light and the other for the dark regions in the image. Furthermore, the mixture parameters are proportional to the relative areas of the dark and light regions. If the form of the densities is known or assumed, it is possible to determine an optimal threshold (in terms of minimum error) for segmenting the image into the two distinct regions.

   Figure 10.32 shows two probability density functions. Assume that the larger of the two PDFs corresponds to the background levels while the smaller one

See inside front cover

Consult the book web site for a brief review of probability theory.

FIGURE 10.32
Gray-level
probability
density functions
of two regions in
an image.

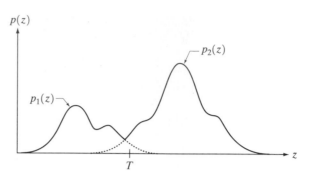

describes the gray levels of objects in the image. The mixture probability density function describing the overall gray-level variation in the image is

$$p(z) = P_1 p_1(z) + P_2 p_2(z). \tag{10.3-5}$$

Here, P_1 and P_2 are the probabilities of occurrence of the two classes of pixels; that is, P_1 is the probability (a number) that a random pixel with value z is an object pixel. Similarly, P_2 is the probability that the pixel is a background pixel. We are assuming that any given pixel belongs either to an object or to the background, so that

$$P_1 + P_2 = 1. \tag{10.3-6}$$

An image is segmented by classifying as background all pixels with gray levels greater than a threshold T (see Fig. 10.32). All other pixels are called object pixels. Our main objective is to select the value of T that minimizes the average error in making the decisions that a given pixel belongs to an object or to the background.

Recall that the probability of a random variable having a value in the interval $[a, b]$ is the integral of its probability density function from a to b, which is the area of the PDF curve between these two limits. Thus, the probability of *erroneously* classifying a background point as an object point is

$$E_1(T) = \int_{-\infty}^{T} p_2(z)\, dz. \tag{10.3-7}$$

This is the area under the curve of $p_2(z)$ to the left of the threshold. Similarly, the probability of erroneously classifying an object point as background is

$$E_2(T) = \int_{T}^{\infty} p_1(z)\, dz, \tag{10.3-8}$$

which is the area under the curve of $p_1(z)$ to the right of T. Then the overall probability of error is

$$E(T) = P_2 E_1(T) + P_1 E_2(T). \tag{10.3-9}$$

Note how the quantities E_1 and E_2 are weighted (given importance) by the probability of occurrence of object or background pixels. Note also that the subscripts are opposites. This is simple to explain. Consider, for example, the extreme case in which background points are known never to occur. In this case $P_2 = 0$. The contribution to the overall error (E) of classifying a background point as an object point (E_1) should be zeroed out because background points are known never to occur. This is accomplished by multiplying E_1 by $P_2 = 0$. If background and object points are equally likely to occur, then the weights are $P_1 = P_2 = 0.5$.

To find the threshold value for which this error is minimal requires differentiating $E(T)$ with respect to T (using Leibniz's rule) and equating the result to 0. The result is

$$P_1 p_1(T) = P_2 p_2(T). \tag{10.3-10}$$

This equation is solved for T to find the optimum threshold. Note that if $P_1 = P_2$, then the optimum threshold is where the curves for $p_1(z)$ and $p_2(z)$ intersect (see Fig. 10.32).

Obtaining an analytical expression for T requires that we know the equations for the two PDFs. Estimating these densities in practice is not always feasible, and an approach used often is to employ densities whose parameters are reasonably simple to obtain. One of the principal densities used in this manner is the Gaussian density, which is completely characterized by two parameters: the mean and the variance. In this case,

$$p(z) = \frac{P_1}{\sqrt{2\pi}\sigma_1} e^{-\frac{(z-\mu_1)^2}{2\sigma_1^2}} + \frac{P_2}{\sqrt{2\pi}\sigma_2} e^{-\frac{(z-\mu_2)^2}{2\sigma_2^2}} \tag{10.3-11}$$

where μ_1 and σ_1^2 are the mean and variance of the Gaussian density of one class of pixels (say, objects) and μ_2 and σ_2^2 are the mean and variance of the other class. Using this equation in the general solution of Eq. (10.3-10) results in the following solution for the threshold T:

$$AT^2 + BT + C = 0 \tag{10.3-12}$$

where

$$\begin{aligned}
A &= \sigma_1^2 - \sigma_2^2 \\
B &= 2(\mu_1\sigma_2^2 - \mu_2\sigma_1^2) \\
C &= \sigma_1^2\mu_2^2 - \sigma_2^2\mu_1^2 + 2\sigma_1^2\sigma_2^2 \ln(\sigma_2 P_1/\sigma_1 P_2).
\end{aligned} \tag{10.3-13}$$

Since a quadratic equation has two possible solutions, two threshold values may be required to obtain the optimal solution.

If the variances are equal, $\sigma^2 = \sigma_1^2 = \sigma_2^2$, a single threshold is sufficient:

$$T = \frac{\mu_1 + \mu_2}{2} + \frac{\sigma^2}{\mu_1 - \mu_2} \ln\left(\frac{P_2}{P_1}\right). \tag{10.3-14}$$

If $P_1 = P_2$, the optimal threshold is the average of the means. The same is true if $\sigma = 0$. Determining the optimal threshold may be similarly accomplished for other densities of known form, such as the Raleigh and log-normal densities.

Instead of assuming a functional form for $p(z)$, a minimum mean-square-error approach may be used to estimate a composite gray-level PDF of an image from the image histogram. For example, the mean square error between the (continuous) mixture density $p(z)$ and the (discrete) image histogram $h(z_i)$ is

$$e_{ms} = \frac{1}{n} \sum_{i=1}^{n} \left[p(z_i) - h(z_i) \right]^2 \qquad (10.3\text{-}15)$$

where an n-point histogram is assumed. The principal reason for estimating the complete density is to determine the presence or absence dominant modes in the PDF. For example, two dominant modes typically indicate the presence of edges in the image (or region) over which the PDF is computed.

In general, determining analytically the parameters that minimize this mean square error is not a simple matter. Even for the Gaussian case, the straightforward computation of equating the partial derivatives to 0 leads to a set of simultaneous transcendental equations that usually can be solved only by numerical procedures, such as a conjugate gradients or Newton's method for simultaneous nonlinear equations.

EXAMPLE 10.13:
Use of optimum thresholding for image segmentation.

■ The following is one of the earliest (and still one of the most instructive) examples of segmentation by optimum thresholding in image processing. This example is particularly interesting at this junction because it shows how segmentation results can be improved by employing preprocessing techniques based on methods developed in our discussion of image enhancement. In addition, the example also illustrates the use of local histogram estimation and adaptive thresholding. The general problem is to outline automatically the boundaries of heart ventricles in cardioangiograms (X-ray images of a heart that has been injected with a contrast medium). The approach discussed here was developed by Chow and Kaneko [1972] for outlining boundaries of the left ventricle of the heart.

Prior to segmentation, all images were preprocessed as follows: (1) Each pixel was mapped with a log function (see Section 3.2.2) to counter exponential effects caused by radioactive absorption. (2) An image obtained before application of the contrast medium was subtracted from each image captured after the medium was injected in order to remove the spinal column present in both images (see Section 3.4.1). (3) Several angiograms were summed in order to reduce random noise (see Section 3.4.2). Figure 10.33 shows a cardioangiogram

a b

FIGURE 10.33 A cardioangiogram before and after preprocessing. (Chow and Kaneko.)

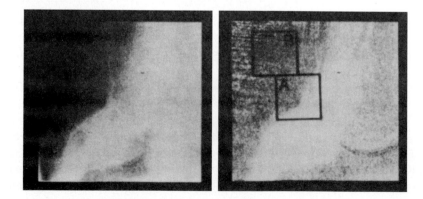

before and after preprocessing (an explanation of the regions marked A and B is given in the following paragraph).

In order to compute the optimal thresholds, each preprocessed image was subdivided into 49 regions by placing a 7×7 grid with 50% overlap over each image (all original images shown in this example are of size 256×256 pixels). Each of the 49 resulting overlapped regions contained 64×64 pixels. Figures 10.34(a) and 10.34(b) are the histograms of the regions marked A and B in Fig. 10.33(b). Note that the histogram for region A clearly is bimodal, indicating the presence of a boundary. The histogram for region B, however, is unimodal, indicating the absence of two markedly distinct regions.

After all 49 histograms were computed, a test of bimodality was performed to reject the unimodal histograms. The remaining histograms were then fitted by bimodal Gaussian density curves [see Eq. (10.3-11)] using a conjugate gradient hill-climbing method to minimize the error function given in Eq. (10.3-15). The $\times$'s and $\bigcirc$'s in Fig. 10.34(a) are two fits to the histogram shown in black dots. The optimum thresholds were then obtained by using Eqs. (10.3-12) and (10.3-13).

At this stage of the process only the regions with bimodal histograms were assigned thresholds. The thresholds for the remaining regions were obtained by interpolating these thresholds. Then a second interpolation was carried out point by point by using neighboring threshold values so that, at the end of the procedure, every point in the image had been assigned a threshold. Finally, a binary decision was carried out for each pixel using the rule

$$f(x, y) = \begin{cases} 1 & \text{if } f(x, y) \geq T_{xy} \\ 0 & \text{otherwise} \end{cases}$$

where T_{xy} was the threshold assigned to location (x, y) in the image [note that these are adaptive thresholds, because they depend on the spatial coordinates (x, y)]. Boundaries were obtained by taking the gradient of the binary picture. Figure 10.35 shows the boundaries superimposed on the original image. Considering the variability and complexity of the images involved, this procedure yielded excellent segmentation results.   ■

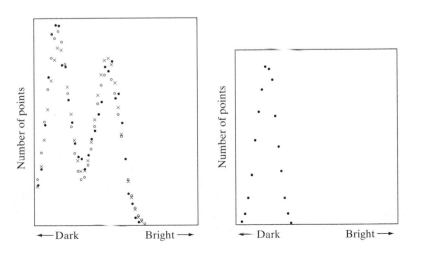

←—Dark   Bright —→   ← Dark   Bright —→

FIGURE 10.34
Histograms (black dots) of (a) region A, and (b) region B in Fig. 10.33(b). (Chow and Kaneko.)

FIGURE 10.35
Cardioangiogram showing superimposed boundaries. (Chow and Kaneko.)

10.3.6 Use of Boundary Characteristics for Histogram Improvement and Local Thresholding

Based on the discussions in the previous five sections, it is intuitively evident that the chances of selecting a "good" threshold are enhanced considerably if the histogram peaks are tall, narrow, symmetric, and separated by deep valleys. One approach for improving the shape of histograms is to consider only those pixels that lie on or near the edges between objects and the background. An immediate and obvious improvement is that histograms would be less dependent on the relative sizes of objects and the background. For instance, the histogram of an image composed of a small object on a large background area (or vice versa) would be dominated by a large peak because of the high concentration of one type of pixels. Figures 10.30 and 10.31 are a good illustration of how segmentation performance is affected by this condition.

If only the pixels on or near the edge between object and the background were used, the resulting histogram would have peaks of approximately the same height. In addition, the probability that any of those given pixels lies on an object would be approximately equal to the probability that it lies on the background, thus improving the symmetry of the histogram peaks. Finally, as indicated in the following paragraph, using pixels that satisfy some simple measures based on gradient and Laplacian operators has a tendency to deepen the valley between histogram peaks.

The principal problem with the approach just discussed is the implicit assumption that the edges between objects and background are known. This information clearly is not available during segmentation, as finding a division between objects and background is precisely what segmentation is all about. However, from the discussion in Section 10.1.3, an indication of whether a pixel is on an edge may be obtained by computing its gradient. In addition, use of the Laplacian can yield information regarding whether a given pixel lies on the dark or light side of an edge. The average value of the Laplacian is 0 at the transition of an edge (see Fig. 10.6), so in practice the valleys of histograms formed from the pixels selected by a gradient/Laplacian criterion can be expected to be sparsely populated. This property produces the highly desirable deep valleys discussed previously.

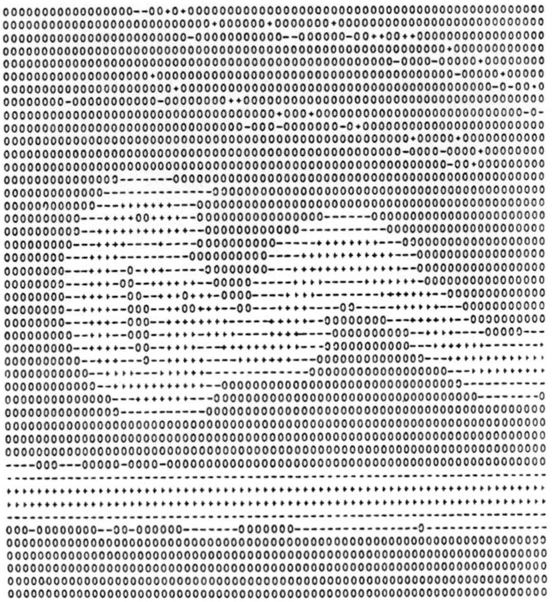

```
00000000000000000---00+0+0000000000000000000000000000000000000000
000000000000000000000000000+000000+0000000+000000000000000000000000
0000000000000000000000000000-0000000000---00000-00++00++00000000000000
0000000000000000000000000000000000000000000000000000000-0000+00000000000
00000000000000000000000000000000000000000000000000---0000-00000+0000000000
0000000000000000000+000000000000000000000000000000000000-00000+00000
00000000-00000000000-000000000++00000000000000000000000000000-0-00+0
00000000-0000000000-0000000++00000000000000000000000000000000000-0-
0000000000000000000000000000000000000+000+00000000000000000000000000-0+
00000000000000000000000000000000-000-0000000-0+000000000000000000000
00000000000000000000000000000000000000000000000-0000-0000+00000000000
00000000000000000000000000000000000000000000000000-00-000000000000000
000000000000000003-------000000000000000000000000000000000000000000000
00000000000---------------0000000000000000000000000000000000000000000000
00000000000---+++++++++---0000000000000000000000000000000000000000000000
000000000---+++++00+++++---0000000000000000------00000000000000000000000
000000000---+++++++++++++---000000000000---------00000000000000000000
00000000---++++---------+---00000000000-----++++++++---00000000000000000
0000000---++++---------------000000000---++++++++++++---00000000000000
0000000---+++++-0-+++++----0000000---+++++++++++++++---0000000000000
0000000---++++-00---+++++---00000------++++++++++++++---0000000000000
0000000---+++++--00--++++++--0000-------+++++++++++++---00000000000
0000000---+++++-00-+0+++--0000----+++++++++++---000000000000
0000000---++++---000--++00++---00------00--------++++++---0000000000
0000000---++++-+-000-++++++--------++++++--00-------++++++--0000000---
0000000---++++++--00--+++++---------++++++--00000-----+++++----00000---
0000000---++++++---00-++++---------+++++++---00000000000---+++++----0000000
0000000---+++++---0-+++++-----++++++++--------00000000000---++++++++++
0000000---++++++--0-+++++++----------000000000000---++++++++++
000000000---++++++---00000000000000000000000000---------
00000000000---++++++++++---0000000000000000000000000-------0
000000000000---+++++++++---0000000000000000000000000000000000000
000000000000003-----------0000000000000000000000000000000000
000000000000000000000000000000000000000000000000000000000000000000
000000000000000000000000000000000000000000000000000000000000000
000000000000000000000000000000000000000000000000000000000000000
----000---00000-0000-00000000000000000000-------------------------
-----------------------------------------------------------------
+++++++++++++++++++++++++++++++++++++++++++++++++++++++++++++++++
+++++++++++++++++++++++++++++++++++++++++++++++++++++++++++++++++
-----------------------------------------------------------------
000-00000000--00-000000-------0000000----------------0-----------
0000000000000000000000000000000000000000000000000000000000000005
00000000000000000000000000000000000000000000000000000000000000000
0000000000000000000000000000000000000000000000000000000000000000
00000000000000000000000000000000000000000000000000000000000000000
00000000000000000000000000000000000000000000000000000000000000000
```

FIGURE 10.36
Image of a handwritten stroke coded by using Eq. (10.3-16). (Courtesy of IBM Corporation.)

The gradient ∇f at any point (x, y) in an image is given by Eq. (10.1-4) or (10.1-12). Similarly, the Laplacian $\nabla^2 f$ is given by Eq. (10.1-14) or (10.1-15). These two quantities may be used to form a three-level image, as follows:

$$s(x, y) = \begin{cases} 0 & \text{if } \nabla f < T \\ + & \text{if } \nabla f \geq T \quad \text{and} \quad \nabla^2 f \geq 0 \\ - & \text{if } \nabla f \geq T \quad \text{and} \quad \nabla^2 f < 0 \end{cases} \qquad (10.3\text{-}16)$$

where the symbols $0, +,$ and $-$ represent any three distinct gray levels, T is a threshold, and the gradient and Laplacian are computed at every point (x, y). For a dark object on a light background, and with reference to Fig. 10.6, the use of Eq. (10.3-16) produces an image $s(x, y)$ in which (1) all pixels that are not on an edge (as determined by ∇f being less than T) are labeled 0; (2) all pixels on the dark side of an edge are labeled $+$; and (3) all pixels on the light side of an edge are labeled $-$. The symbols $+$ and $-$ in Eq. (10.3-16) are reversed for a light object on a dark background. Figure 10.36 shows the labeling produced by Eq. (10.3-16) for an image of a dark, underlined stroke written on a light background.

The information obtained with this procedure can be used to generate a segmented, binary image in which 1's correspond to objects of interest and 0's correspond to the background. The transition (along a horizontal or vertical scan line) from a light background to a dark object must be characterized by the occurrence of a $-$ followed by a $+$ in $s(x, y)$. The interior of the object is composed of pixels that are labeled either 0 or $+$. Finally, the transition from the object back to the background is characterized by the occurrence of a $+$ followed by a $-$. Thus a horizontal or vertical scan line containing a section of an object has the following structure:

$$(\cdots)(-, +)(0 \text{ or } +)(+, -)(\cdots)$$

a
b

FIGURE 10.37
(a) Original image. (b) Image segmented by local thresholding. (Courtesy of IBM Corporation.)

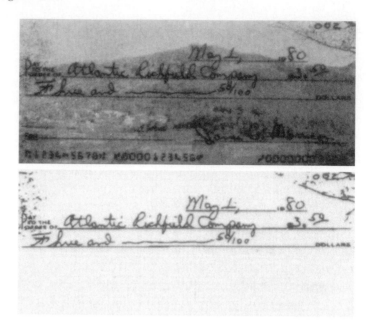

where $(\cdots)$ represents any combination of $+$, $-$, and 0. The innermost parentheses contain object points and are labeled 1. All other pixels along the same scan line are labeled 0, with the exception of any other sequence of (0 or $+$) bounded by $(-, +)$ and $(+, -)$.

EXAMPLE 10.14:
Image segmentation by local thresholding.

■ Figure 10.37(a) shows an image of an ordinary scenic bank check. Figure 10.38 shows the histogram as a function of gradient values for pixels with gradients greater than 5. Note that this histogram has two dominant modes that are symmetric, nearly of the same height, and are separated by a distinct valley. Finally, Fig. 10.37(b) shows the segmented image obtained by using Eq. (10.3-16) with T at or near the midpoint of the valley. The result was made binary by using the sequence analysis just discussed. Note that this example is an illustration of local thresholding, as defined in Eq. (10.3-1), because the value of T was determined from a histogram of the gradient and Laplacian, which are local properties. ■

FIGURE 10.38
Histogram of pixels with gradients greater than 5. (Courtesy of IBM Corporation.)

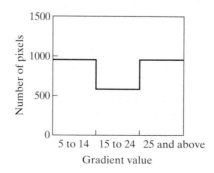

10.3.7 Thresholds Based on Several Variables

So far we have been concerned with thresholding gray levels. In some cases, a sensor can make available more than one variable to characterize each pixel in an image, and thus allow *multispectral thresholding*. As discussed in some detail in Section 6.7, color imaging is a good example, in which each pixel is characterized by three RGB values. In this case, constructing a 3-D "histogram" becomes possible. The basic procedure is analogous to the method used for one variable. For example, for an image with three variables (RGB components), each having 16 possible levels, a $16 \times 16 \times 16$ grid (cube) is formed. Inserted in each cell of the cube is the number of pixels whose RGB components have values corresponding to the coordinates defining the location of that particular cell. Each entry is then divided by the total number of pixels in the image to form a normalized histogram.

The concept of thresholding now becomes one of finding clusters of points in 3-D space. Suppose, for example, that K significant clusters of points are found in the histogram. The image can be segmented by assigning one arbitrary value (say, white) to pixels whose RGB components are closer to one cluster and another value (say, black) to the other pixels in the image. This concept is easily extendable to more components and certainly to more clusters. The principal difficulty is that cluster seeking becomes an increasingly complex task as the number of variables increases. Cluster-seeking methods can be found, for example, in the books by Duda, Hart, and Stork [2001], and Tou and Gonzalez [1974].

■ The image shown in Fig. 10.39(a) is a monochrome picture of a color photograph. The original color image is composed of three 16-level RGB images. The scarf is a vivid red, and the hair and facial colors are light and different in spectral characteristics from the window and other background features.

Figure 10.39(b) was obtained by thresholding about one of the histogram clusters corresponding to facial tones. Note that the window, which in the monochrome

EXAMPLE 10.15: Multispectral thresholding.

a b c

FIGURE 10.39 (a) Original color image shown as a monochrome picture. (b) Segmentation of pixels with colors close to facial tones. (c) Segmentation of red components.

picture is close in gray-level value to the hair, does not appear in the segmented image because of the use of multispectral characteristics to separate these two regions. Figure 10.39(c) was obtained by thresholding about a cluster close to the red axis. In this case only the scarf and part of a flower (which is red also) appeared in the segmented result. The threshold used to obtain both results was a distance of one cell. Thus any pixel whose components were outside the cell enclosing the center of the cluster in question was classified as background (black). Pixels whose components placed them inside the cell were coded white.   ■

As discussed in Section 6.7, color segmentation can be based on any of the color models introduced in Chapter 6. For instance, hue and saturation are important properties in numerous applications dealing with the use of imaging for automated inspection. These properties are particularly important in attempts to emulate the equivalent function performed by humans, such as in the inspection of fruits for ripeness or in the inspection of manufactured goods. As mentioned in Chapter 6, the Hue, Saturation, Intensity (HSI) model is ideal for these types of applications because it is closely related to the way in which humans describe the perception of color. A segmentation approach using the hue and saturation components of a color signal also is particularly attractive, because it involves 2-D data clusters that are easier to analyze than, say, the 3-D clusters needed for RGB segmentation.

10.4 Region-Based Segmentation

The objective of segmentation is to partition an image into regions. In Sections 10.1 and 10.2 we approached this problem by finding boundaries between regions based on discontinuities in gray levels, whereas in Section 10.3 segmentation was accomplished via thresholds based on the distribution of pixel properties, such as gray-level values or color. In this section we discuss segmentation techniques that are based on finding the regions directly.

10.4.1 Basic Formulation

Let R represent the entire image region. We may view segmentation as a process that partitions R into n subregions, $R_1, R_2, \ldots, R_n$, such that

(a) $\bigcup_{i=1}^{n} R_i = R.$

(b) R_i is a connected region, $i = 1, 2, \ldots, n.$

(c) $R_i \cap R_j = \varnothing$ for all i and $j, i \neq j.$

(d) $P(R_i) = \text{TRUE}$ for $i = 1, 2, \ldots, n.$

(e) $P(R_i \cup R_j) = \text{FALSE}$ for any adjacent regions R_i and $R_j.$

Here, $P(R_k)$ is a logical predicate defined over the points in set R_k and $\varnothing$ is the null set.

Condition (a) indicates that the segmentation must be complete; that is, every pixel must be in a region. Condition (b) requires that points in a region must be connected in some predefined sense (see Section 2.5.2 regarding connectivity).

Condition (c) indicates that the regions must be disjoint. Condition (d) deals with the properties that must be satisfied by the pixels in a segmented region—for example $P(R_i)$ = TRUE if all pixels in R_i have the same gray level. Finally, condition (e) indicates that regions R_i and R_j are different in the sense of predicate P.

10.4.2 Region Growing

As its name implies, *region growing* is a procedure that groups pixels or subregions into larger regions based on predefined criteria. The basic approach is to start with a set of "seed" points and from these grow regions by appending to each seed those neighboring pixels that have properties similar to the seed (such as specific ranges of gray level or color).

Selecting a set of one or more starting points often can be based on the nature of the problem, as will be shown in Example 10.16. When a priori information is not available, the procedure is to compute at every pixel the same set of properties that ultimately will be used to assign pixels to regions during the growing process. If the result of these computations shows clusters of values, the pixels whose properties place them near the centroid of these clusters can be used as seeds.

The selection of similarity criteria depends not only on the problem under consideration, but also on the type of image data available. For example, the analysis of land-use satellite imagery depends heavily on the use of color. This problem would be significantly more difficult, or even impossible, to handle without the inherent information available in color images. When the images are monochrome, region analysis must be carried out with a set of descriptors based on gray levels and spatial properties (such as moments or texture). We discuss descriptors useful for region characterization in Chapter 11.

Descriptors alone can yield misleading results if connectivity or adjacency information is not used in the region-growing process. For example, visualize a random arrangement of pixels with only three distinct gray-level values. Grouping pixels with the same gray level to form a "region" without paying attention to connectivity would yield a segmentation result that is meaningless in the context of this discussion.

Another problem in region growing is the formulation of a stopping rule. Basically, growing a region should stop when no more pixels satisfy the criteria for inclusion in that region. Criteria such as gray level, texture, and color, are local in nature and do not take into account the "history" of region growth. Additional criteria that increase the power of a region-growing algorithm utilize the concept of size, likeness between a candidate pixel and the pixels grown so far (such as a comparison of the gray level of a candidate and the average gray level of the grown region), and the shape of the region being grown. The use of these types of descriptors is based on the assumption that a model of expected results is at least partially available.

■ Figure 10.40(a) shows an X-ray image of a weld (the horizontal dark region) containing several cracks and porosities (the bright, white streaks running horizontally through the middle of the image). We wish to use region growing to segment the regions of the weld failures. These segmented features could be used

EXAMPLE 10.16:
Application of region growing in weld inspection.

a b
c d

FIGURE 10.40
(a) Image
showing defective
welds. (b) Seed
points. (c) Result
of region growing.
(d) Boundaries of
segmented
defective welds
(in black).
(Original image
courtesy of
X-TEK Systems,
Ltd.).

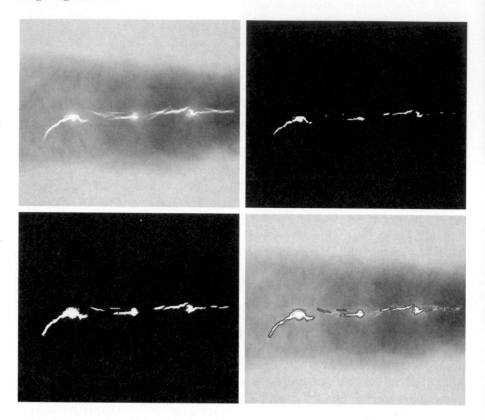

for inspection, for inclusion in a database of historical studies, for controlling an automated welding system, and for other numerous applications.

The first order of business is to determine the initial seed points. In this application, it is known that pixels of defective welds tend to have the maximum allowable digital value (255 in this case). Based on this information, we selected as starting points all pixels having values of 255. The points thus extracted from the original image are shown in Fig. 10.40(b). Note that many of the points are clustered into *seed regions*.

The next step is to choose criteria for region growing. In this particular example we chose two criteria for a pixel to be annexed to a region: (1) The absolute gray-level difference between any pixel and the seed had to be less than 65. This number is based on the histogram shown in Fig. 10.41 and represents the difference between 255 and the location of the first major valley to the left, which is representative of the highest gray level value in the dark weld region. (2) To be included in one of the regions, the pixel had to be 8-connected to at least one pixel in that region. If a pixel was found to be connected to more than one region, the regions were merged.

Figure 10.40(c) shows the regions that resulted by starting with the seeds in Fig. 10.40(b) and utilizing the criteria defined in the previous paragraph. Superimposing the boundaries of these regions on the original image [Fig. 10.40(d)] reveals that the region-growing procedure did indeed segment the defective

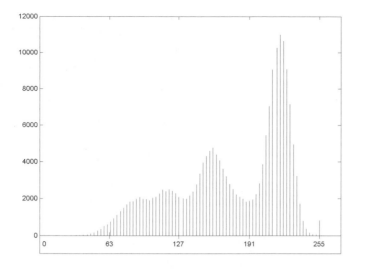

FIGURE 10.41
Histogram of
Fig. 10.40(a).

welds with an acceptable degree of accuracy. It is of interest to note that it was not necessary to specify any stopping rules in this case because the criteria for region growing were sufficient to isolate the features of interest.   ■

It was mentioned in Section 10.3.1 in connection with Fig. 10.26(b) that problems having multimodal histograms generally are best solved using region-based approaches. The histogram shown in Fig. 10.41 is an excellent example of a "clean" multimodal histogram. This histogram and the results in Example 10.16 confirm the assertion that, even with well-behaved histograms, multilevel thresholding is a difficult proposition. Based on the results of Example 10.16, it should be intuitively obvious that this problem cannot be solved effectively by any reasonably general method of automatic threshold selection based on gray levels alone. The use of *connectivity* was fundamental in solving the problem.

10.4.3 Region Splitting and Merging

The procedure just discussed grows regions from a set of seed points. An alternative is to subdivide an image initially into a set of arbitrary, disjoint regions and then merge and/or split the regions in an attempt to satisfy the conditions stated in Section 10.4.1. A split and merge algorithm that iteratively works toward satisfying these constraints is developed next.

Let R represent the entire image region and select a predicate P. One approach for segmenting R is to subdivide it successively into smaller and smaller quadrant regions so that, for any region R_i, $P(R_i) = $ TRUE. We start with the entire region. If $P(R) = $ FALSE, we divide the image into quadrants. If P is FALSE for any quadrant, we subdivide that quadrant into subquadrants, and so on. This particular splitting technique has a convenient representation in the form of a so-called *quadtree* (that is, a tree in which nodes have exactly four descendants), as illustrated in Fig. 10.42. Note that the root of the tree corresponds to the entire image and that each node corresponds to a subdivision. In this case, only R_4 was subdivided further.

FIGURE 10.42
(a) Partitioned
image.
(b) Corresponding
quadtree.

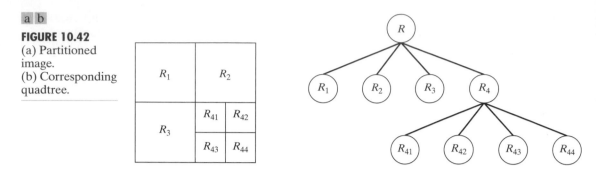

If only splitting were used, the final partition likely would contain adjacent regions with identical properties. This drawback may be remedied by allowing merging, as well as splitting. Satisfying the constraints of Section 10.4.1 requires merging only adjacent regions whose combined pixels satisfy the predicate P. That is, two adjacent regions R_j and R_k are merged only if $P(R_j \cup R_k) = $ TRUE.

The preceding discussion may be summarized by the following procedure, in which, at any step we

1. Split into four disjoint quadrants any region R_i for which $P(R_i) = $ FALSE.
2. Merge any adjacent regions R_j and R_k for which $P(R_j \cup R_k) = $ TRUE.
3. Stop when no further merging or splitting is possible.

Several variations of the preceding basic theme are possible. For example, one possibility is to split the image initially into a set of blocks. Further splitting is carried out as described previously, but merging is initially limited to groups of four blocks that are descendants in the quadtree representation and that satisfy the predicate P. When no further mergings of this type are possible, the procedure is terminated by one final merging of regions satisfying step 2. At this point, the merged regions may be of different sizes. The principal advantage of this approach is that it uses the same quadtree for splitting and merging, until the final merging step.

EXAMPLE 10.17:
Split and merge.

■ Figure 10.43(a) shows a simple image. We define $P(R_i) = $ TRUE if at least 80% of the pixels in R_i have the property $|z_j - m_i| \le 2\sigma_i$, where z_j is the gray level of the jth pixel in R_i, m_i is the mean gray level of that region, and σ_i is the

FIGURE 10.43
(a) Original
image. (b) Result
of split and merge
procedure.
(c) Result of
thresholding (a).

standard deviation of the gray levels in R_i. If $P(R_i)$ = TRUE under this condition, the values of all the pixels in R_i were set equal to m_i. Splitting and merging was done using the algorithm outlined previously. The result of applying this technique to the image in Fig. 10.43(a) is shown in Fig. 10.43(b). Note that the image was segmented perfectly. The image shown in Fig. 10.43(c) was obtained by thresholding Fig. 10.43(a), with a threshold placed midway between the two principal peaks of the histogram. The shading (and the stem of the leaf) were erroneously eliminated by the thresholding procedure.   ■

As used in the preceding example, properties based on the mean and standard deviation of pixels in a region attempt to quantify the *texture* of a region (see Section 11.3.3 for a discussion on texture). The concept of *texture segmentation* is based on using measures of texture for the predicates $P(R_i)$. That is, we can perform texture segmentation by any of the methods discussed in this section by specifying predicates based on texture content.

10.5 Segmentation by Morphological Watersheds

Thus far, we have discussed segmentation based on three principal concepts: (a) detection of discontinuities, (b) thresholding, and (c) region processing. Each of these approaches was found to have advantages (for example, speed in the case of global thresholding) and disadvantages (for example, the need for postprocessing, such as edge linking, in methods based on detecting discontinuities in gray levels). In this section we discuss an approach based on the concept of so-called *morphological watersheds*. As will become evident in the following discussion, segmentation by watersheds embodies many of the concepts of the other three approaches and, as such, often produces more stable segmentation results, including continuous segmentation boundaries. This approach also provides a simple framework for incorporating knowledge-based constraints (see Fig. 1.23) in the segmentation process.

10.5.1 Basic Concepts

The concept of watersheds is based on visualizing an image in three dimensions: two spatial coordinates versus gray levels. In such a "topographic" interpretation, we consider three types of points: (a) points belonging to a regional minimum; (b) points at which a drop of water, if placed at the location of any of those points, would fall with certainty to a single minimum; and (c) points at which water would be equally likely to fall to more than one such minimum. For a particular regional minimum, the set of points satisfying condition (b) is called the *catchment basin* or *watershed* of that minimum. The points satisfying condition (c) form crest lines on the topographic surface and are termed *divide lines* or *watershed lines*.

The principal objective of segmentation algorithms based on these concepts is to find the watershed lines. The basic idea is simple: Suppose that a hole is punched in each regional minimum and that the entire topography is flooded from below by letting water rise through the holes at a uniform rate. When the rising water in distinct catchment basins is about to merge, a dam is built to prevent the merging. The flooding will eventually reach a stage when only the tops of the dams are visible above the water line. These dam boundaries correspond to the divide lines of the watersheds. Therefore, they are the (continuous) boundaries extracted by a watershed segmentation algorithm.

These ideas can be explained further with the aid of Fig. 10.44. Figure 10.44(a) shows a simple gray-scale image and Fig. 10.44(b) is a topographic view, in which the height of the "mountains" is proportional to gray-level values in the input image. For ease of interpretation, the backsides of structures are shaded. This is not to be confused with gray-level values; only the general topography of the three-dimensional representation is of interest. In order to prevent the rising water from spilling out through the edges of the structure, we imagine the perimeter of the entire topography (image) being enclosed by dams of height greater than the highest possible mountain, whose value is determined by the highest possible gray-level value in the input image.

Suppose that a hole is punched in each regional minimum [shown as dark areas in Fig. 10.44(b)] and that the entire topography is flooded from below

a b
c d

FIGURE 10.44
(a) Original image.
(b) Topographic view. (c)–(d) Two stages of flooding.

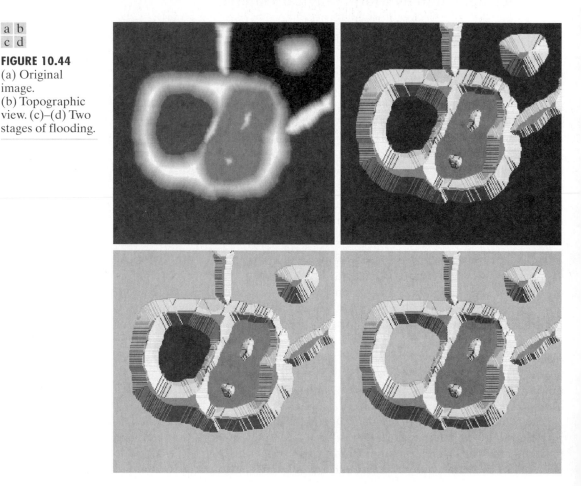

by letting water rise through the holes at a uniform rate. Figure 10.44(c) shows the first stage of flooding, where the "water," shown in light gray, has covered only areas that correspond to the very dark background in the image. In Figs. 10.44(d) and (e) we see that the water now has risen into the first and second catchment basins, respectively. As the water continues to rise, it will eventually overflow from one catchment basin into another. The first indication of this is shown in 10.44(f). Here, water from the left basin actually overflowed into the basin on the right and a short "dam" (consisting of single pixels) was built to prevent water from merging at that level of flooding (the details of dam building are discussed in the following section). The effect is more pronounced as water continues to rise, as shown in Fig. 10.44(g). This figure shows a longer dam between the two catchment basins and another dam in the top part of the right basin. The latter dam was built to prevent merging of water from that basin with water from areas corresponding to the background. This process is continued until the maximum level of flooding (corresponding to the highest gray-level value in the image) is reached. The final dams correspond to the watershed lines, which are the desired segmentation result. The

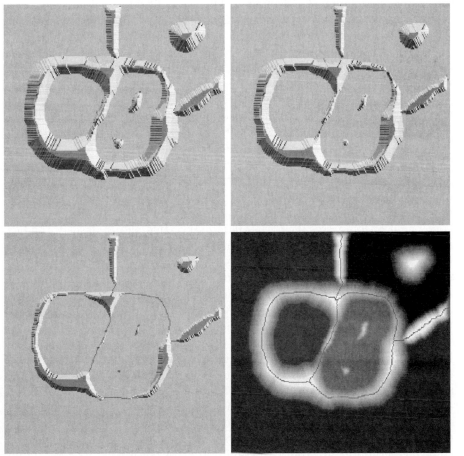

e f
g h

FIGURE 10.44
(Continued)
(e) Result of further flooding. (f) Beginning of merging of water from two catchment basins (a short dam was built between them). (g) Longer dams. (h) Final watershed (segmentation) lines. (Courtesy of Dr. S. Beucher, CMM/Ecole des Mines de Paris.)

result for this example is shown in Fig. 10.44(h) as a dark, one-pixel-thick path superimposed on the original image. Note the important property that the watershed lines form a connected path, thus giving continuous boundaries between regions.

One of the principal applications of watershed segmentation is in the extraction of nearly uniform (bloblike) objects from the background. Regions characterized by small variations in gray levels have small gradient values. Thus, in practice, we often see watershed segmentation applied to the gradient of an image, rather than to the image itself. In this formulation, the regional minima of catchment basins correlate nicely with the small value of the gradient corresponding to the objects of interest.

10.5.2 Dam Construction

Before proceeding, let us consider how to construct the dams or watershed lines required by watershed segmentation algorithms. Dam construction is based on binary images, which are members of 2-D integer space Z^2 (see Section 2.4.2). The simplest way to construct dams separating sets of binary points is to use morphological dilation (see Section 9.2.1).

The basics of how to construct dams using dilation are illustrated in Fig. 10.45. Figure 10.45(a) shows portions of two catchment basins at flooding step $n - 1$ and Fig. 10.45(b) shows the result at the next flooding step, n. The water has spilled from one basin to the other and, therefore, a dam must be built to keep this from happening. In order to be consistent with notation to be introduced shortly, let M_1 and M_2 denote the sets of coordinates of points in two regional minima. Then let the set of coordinates of points in the *catchment basin* associated with these two minima at stage $n - 1$ of flooding be denoted by $C_{n-1}(M_1)$ and $C_{n-1}(M_2)$, respectively. These are the two black regions shown in Fig. 10.45(a).

Let the union of these two sets be denoted by $C[n - 1]$. There are two connected components in Fig. 10.45(a) (see Section 2.5.2 regarding connected components) and only one connected component in Fig. 10.45(b). This connected component encompasses the earlier two components, shown dashed. The fact that two connected components have become a *single* component indicates that water between the two catchment basins has merged at flooding step n. Let this connected component be denoted q. Note that the two components from step $n - 1$ can be extracted from q by performing the simple AND operation $q \cap C[n - 1]$. We note also that all points belonging to an individual catchment basin form a single connected component.

Suppose that each of the connected components in Fig. 10.45(a) is dilated by the structuring element shown in Fig. 10.45(c), subject to two conditions: (1) The dilation has to be constrained to q (this means that the center of the structuring element can be located only at points in q during dilation), and (2) the dilation cannot be performed on points that would cause the sets being dilated to merge (become a single connected component). Figure 10.45(d) shows that a first dilation pass (in light gray) expanded the boundary of each original connected component. Note that condition (1) was satisfied by every point

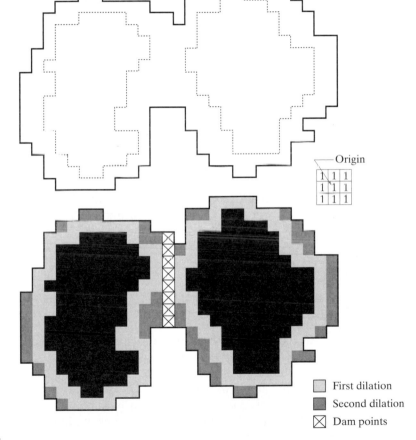

a
b c
d

FIGURE 10.45 (a) Two partially flooded catchment basins at stage $n - 1$ of flooding. (b) Flooding at stage n, showing that water has spilled between basins (for clarity, water is shown in white rather than black). (c) Structuring element used for dilation. (d) Result of dilation and dam construction.

during dilation, and condition (2) did not apply to any point during the dilation process; thus the boundary of each region was expanded uniformly.

In the second dilation (shown in medium gray), several points failed condition (1) while meeting condition (2), resulting in the broken perimeter shown in the figure. It also is evident that the only points in q that satisfy the two conditions under consideration describe the one-pixel-thick connected path shown crossed-hatched in Fig. 10.45(d). This path constitutes the desired separating dam at stage n of flooding. Construction of the dam at this level of flooding is completed by setting all the points in the path just determined to a value greater than the maximum gray-level value of the image. The height of all dams is generally set at 1 plus the maximum allowed value in the image. This will prevent water from crossing over the part of the completed dam as the level of flooding is increased. It is important to note that dams built by this procedure, which are the desired segmentation boundaries, are connected components. In other words, this method eliminates the problems of broken segmentation lines.

Although the procedure just described is based on a simple example, the method used for more complex situations is exactly the same, including the use of the 3×3 symmetric structuring element shown in Fig. 10.45(c).

10.5.3 Watershed Segmentation Algorithm

Let $M_1, M_2, \ldots, M_R$ be sets denoting the *coordinates* of the points in the regional minima of an image $g(x, y)$. As indicated at the end of Section 10.5.1, this typically will be a gradient image. Let $C(M_i)$ be a set denoting the coordinates of the points in the catchment basin associated with regional minimum M_i (recall that the points in any catchment basin form a connected component). The notation min and max will be used to denote the minimum and maximum values of $g(x, y)$. Finally, let $T[n]$ represent the set of coordinates (s, t) for which $g(s, t) < n$. That is,

$$T[n] = \{(s, t) \mid g(s, t) < n\}. \tag{10.5-1}$$

Geometrically, $T[n]$ is the set of coordinates of points in $g(x, y)$ lying below the plane $g(x, y) = n$.

The topography will be flooded in *integer* flood increments, from $n = \text{min} + 1$ to $n = \text{max} + 1$. At any step n of the flooding process, the algorithm needs to know the number of points below the flood depth. Conceptually, suppose that the coordinates in $T[n]$ that are below the plane $g(x, y) = n$ are "marked" black, and all other coordinates are marked white. Then when we look "down" on the xy-plane at any increment n of flooding, we will see a binary image in which black points correspond to points in the function that are below the plane $g(x, y) = n$. This interpretation is quite useful in helping understand the following discussion.

Let $C_n(M_i)$ denote the set of coordinates of points in the catchment basin associated with minimum M_i that are flooded at stage n. With reference to the discussion in the previous paragraph, $C_n(M_i)$ may be viewed as a binary image given by

$$C_n(M_i) = C(M_i) \cap T[n]. \tag{10.5-2}$$

In other words, $C_n(M_i) = 1$ at location (x, y) if $(x, y) \in C(M_i)$ AND $(x, y) \in T[n]$; otherwise $C_n(M_i) = 0$. The geometrical interpretation of this result is straightforward. We are simply using the AND operator to isolate at stage n of flooding the portion of the binary image in $T[n]$ that is associated with regional minimum M_i.

Next, we let $C[n]$ denote the union of the flooded catchment basins portions at stage n:

$$C[n] = \bigcup_{i=1}^{R} C_n(M_i). \qquad (10.5\text{-}3)$$

Then $C[\max + 1]$ is the union of all catchment basins:

$$C[\max + 1] = \bigcup_{i=1}^{R} C(M_i). \qquad (10.5\text{-}4)$$

It can be shown (Problem 10.29) that the elements in both $C_n(M_i)$ and $T[n]$ are never replaced during execution of the algorithm, and that the number of elements in these two sets either increases or remains the same as n increases. Thus, it follows that $C[n - 1]$ is a subset of $C[n]$. According to Eqs. (10.5-2) and (10.5-3), $C[n]$ is a subset of $T[n]$, so it follows that $C[n - 1]$ is a subset of $T[n]$. From this we have the important result that each connected component of $C[n - 1]$ is contained in exactly one connected component of $T[n]$.

The algorithm for finding the watershed lines is initialized with $C[\min + 1] = T[\min + 1]$. The algorithm then proceeds recursively, assuming at step n that $C[n - 1]$ has been constructed. A procedure for obtaining $C[n]$ from $C[n - 1]$ is as follows. Let Q denote the set of connected components in $T[n]$. Then, for each connected component $q \in Q[n]$, there are three possibilities:

(a) $q \cap C[n - 1]$ is empty.
(b) $q \cap C[n - 1]$ contains one connected component of $C[n - 1]$.
(c) $q \cap C[n - 1]$ contains more than one connected component of $C[n - 1]$.

Construction of $C[n]$ from $C[n - 1]$ depends on which of these three conditions holds. Condition (a) occurs when a new minimum is encountered, in which case connected component q is incorporated into $C[n - 1]$ to form $C[n]$. Condition (b) occurs when q lies within the catchment basin of some regional minimum, in which case q is incorporated into $C[n - 1]$ to form $C[n]$. Condition (c) occurs when all, or part, of a ridge separating two or more catchment basins is encountered. Further flooding would cause the water level in these catchment basins to merge. Thus a dam (or dams if more than two catchment basins are involved) must be built within q to prevent overflow between the catchment basins. As explained in the previous section, a one-pixel-thick dam can be constructed when needed by dilating $q \cap C[n - 1]$ with a 3×3 structuring element of 1's, and constraining the dilation to q.

Algorithm efficiency is improved by using only values of n that correspond to existing gray-level values in $g(x, y)$; we can determine these values, as well as the values of min and max, from the histogram of $g(x, y)$.

FIGURE 10.46
(a) Image of
blobs. (b) Image
gradient.
(c) Watershed
lines.
(d) Watershed
lines
superimposed on
original image.
(Courtesy of Dr.
S. Beucher,
CMM/Ecole des
Mines de Paris.)

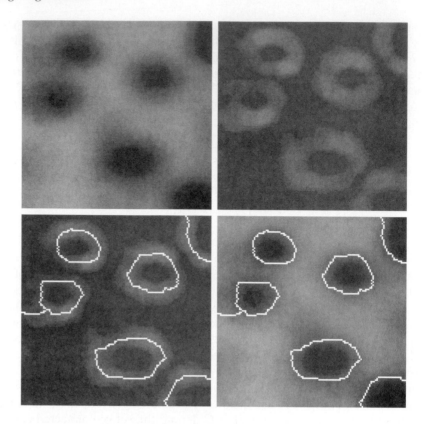

EXAMPLE 10.18:
Illustration of the
watershed
segmentation
algorithm.

■ Consider the image and its gradient, shown in Figs. 10.46(a) and (b), respectively. Application of the watershed algorithm just described yielded the watershed lines (white paths) of the gradient image shown in Fig. 10.46(c). These segmentation boundaries are shown superimposed on the original image in Fig. 10.46(d). As noted at the beginning of this section, the segmentation boundaries have the important property of being connected paths. ■

10.5.4 The Use of Markers

Direct application of the watershed segmentation algorithm in the form discussed in the previous section generally leads to *oversegmentation* due to noise and other local irregularities of the gradient. As shown in Fig. 10.47, oversegmentation can be serious enough to render the result of the algorithm virtually useless. In this case, this means a large number of segmented regions. A practical solution to this problem is to limit the number of allowable regions by incorporating a preprocessing stage designed to bring additional knowledge into the segmentation procedure.

An approach used to control oversegmentation is based on the concept of markers. A *marker* is a connected component belonging to an image. We have *internal* markers, associated with objects of interest, and *external* markers, associated with the background. A procedure for marker selection typically will

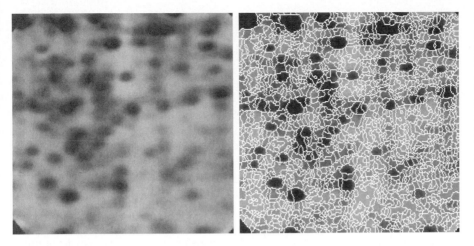

FIGURE 10.47
(a) Electrophoresis image. (b) Result of applying the watershed segmentation algorithm to the gradient image. Oversegmentation is evident. (Courtesy of Dr. S. Beucher, CMM/Ecole des Mines de Paris.)

consist of two principal steps: (1) preprocessing; and (2) definition of a set of criteria that markers must satisfy. To illustrate, consider Fig. 10.47(a) again. Part of the problem that led to the oversegmented result in Fig. 10.47(b) is the large number of potential minima. Because of their size, many of these minima really are irrelevant detail. As has been pointed out several times in earlier discussions, an effective method for minimizing the effect of small spatial detail is to filter the image with a smoothing filter. This is an appropriate preprocessing scheme in this particular case.

Suppose that we define an internal marker in this case as (1) a region that is surrounded by points of higher "altitude;" (2) such that the points in the region form a connected component; and (3) in which all the points in the connected component have the same gray-level value. After the image was smoothed, the internal markers resulting from this definition are shown as light gray, bloblike regions in Fig. 10.48(a). Next, the watershed algorithm was applied to the smoothed

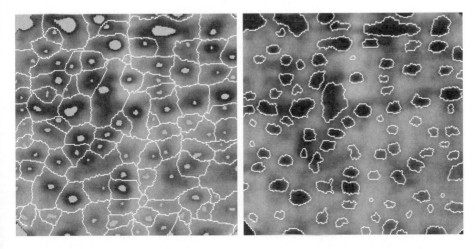

FIGURE 10.48
(a) Image showing internal markers (light gray regions) and external markers (watershed lines). (b) Result of segmentation. Note the improvement over Fig. 10.47(b). (Courtesy of Dr. S. Beucher, CMM/Ecole des Mines de Paris.)

image, under the restriction that these internal markers be the *only* allowed regional minima. Figure 10.48(a) shows the resulting watershed lines. These watershed lines are defined as the external markers. Note that the points along the watershed lines are good candidates for the background because they pass along the highest points between neighboring markers.

The external markers shown in Fig. 10.48(a) effectively partition the image into regions, with each region containing a single internal marker and part of the background. The problem is thus reduced to partitioning each of these regions into two: a single object and its background. We can bring to bear on this simplified problem many of the segmentation techniques discussed earlier in this chapter. Another approach is simply to apply the watershed segmentation algorithm to each individual region. In other words, we simply take the gradient of the smoothed image [as in Fig. 10.46(b)] and then restrict the algorithm to operate on a single watershed that contains the marker in that particular region. The result obtained using this approach is shown in 10.48(b). The improvement over the image in 10.47(b) is evident.

Marker selection can range from simple procedures based on gray-level values and connectivity, as was just illustrated, to more complex descriptions involving size, shape, location, relative distances, texture content, and so on (see Chapter 11 regarding descriptors). The point is that using markers brings a priori knowledge to bear on the segmentation problem. The reader is reminded that humans often aid segmentation and higher-level tasks in every-day vision by using a priori knowledge, one of the most familiar being the use of context. Thus, the fact that segmentation by watersheds offers a framework that can make effective use of this type of knowledge is a significant advantage of this method.

10.6 The Use of Motion in Segmentation

Motion is a powerful cue used by humans and many animals to extract objects of interest from a background of irrelevant detail. In imaging applications, motion arises from a relative displacement between the sensing system and the scene being viewed, such as in robotic applications, autonomous navigation, and dynamic scene analysis. In the following sections we consider the use of motion in segmentation both spatially and in the frequency domain.

10.6.1 Spatial Techniques

Basic approach

One of the simplest approaches for detecting changes between two image frames $f(x, y, t_i)$ and $f(x, y, t_j)$ taken at times t_i and t_j, respectively, is to compare the two images pixel by pixel. One procedure for doing this is to form a difference image. Suppose that we have a reference image containing only stationary components. Comparing this image against a subsequent image of the same scene, but including a moving object, results in the difference of the two images canceling the stationary elements, leaving only nonzero entries that correspond to the nonstationary image components.

A difference image between two images taken at times t_i and t_j may be defined as

$$d_{ij}(x, y) = \begin{cases} 1 & \text{if } |f(x, y, t_i) - f(x, y, t_j)| > T \\ 0 & \text{otherwise} \end{cases} \qquad (10.6\text{-}1)$$

where T is a specified threshold. Note that $d_{ij}(x, y)$ has a value of 1 at spatial coordinates (x, y) only if the gray-level difference between the two images is appreciably different at those coordinates, as determined by the specified threshold T. It is assumed that all images are of the same size. Finally, we note that the values of the coordinates (x, y) in Eq. (10.6-1) span the dimensions of these images, so that the difference image $d_{ij}(x, y)$ also is of same size as the images in the sequence.

In dynamic image processing, all pixels in $d_{ij}(x, y)$ with value 1 are considered the result of object motion. This approach is applicable only if the two images are registered spatially and if the illumination is relatively constant within the bounds established by T. In practice, 1-valued entries in $d_{ij}(x, y)$ often arise as a result of noise. Typically, these entries are isolated points in the difference image, and a simple approach to their removal is to form 4- or 8-connected regions of 1's in $d_{ij}(x, y)$ and then ignore any region that has less than a predetermined number of entries. Although it may result in ignoring small and/or slow-moving objects, this approach improves the chances that the remaining entries in the difference image actually are the result of motion.

Accumulative differences

Isolated entries resulting from noise is not an insignificant problem when trying to extract motion components from a sequence of images. Although the number of these entries can be reduced by a thresholded connectivity analysis, this filtering process can also remove small or slow-moving objects as noted in the previous section. One way to address this problem is by considering changes at a pixel location over several frames, thus introducing a "memory" into the process. The idea is to ignore changes that occur only sporadically over a frame sequence and can therefore be attributed to random noise.

Consider a sequence of image frames $f(x, y, t_1), f(x, y, t_2), \ldots, f(x, y, t_n)$ and let $f(x, y, t_1)$ be the reference image. An *accumulative difference image* (ADI) is formed by comparing this reference image with every subsequent image in the sequence. A counter for each pixel location in the accumulative image is incremented every time a difference occurs at that pixel location between the reference and an image in the sequence. Thus when the kth frame is being compared with the reference, the entry in a given pixel of the accumulative image gives the number of times the gray level at that position was different from the corresponding pixel value in the reference image. Differences are established, for example, by using Eq. (10.6-1).

Often useful is consideration of three types of accumulative difference images: *absolute*, *positive*, and *negative* ADIs. Assuming that the gray-level values of the moving objects are larger than the background, these three types of ADIs are defined as follows. Let $R(x, y)$ denote the reference image and, to simplify the notation, let k denote t_k, so that $f(x, y, k) = f(x, y, t_k)$. We assume that $R(x, y) = f(x, y, 1)$. Then, for any $k > 1$, and keeping in mind that the values of the ADIs are *counts*, we define the following for all relevant values of (x, y):

$$A_k(x, y) = \begin{cases} A_{k-1}(x, y) + 1 & \text{if } |R(x, y) - f(x, y, k)| > T \\ A_{k-1}(x, y) & \text{otherwise} \end{cases} \quad (10.6\text{-}2)$$

$$P_k(x, y) = \begin{cases} P_{k-1}(x, y) + 1 & \text{if } [R(x, y) - f(x, y, k)] > T \\ P_{k-1}(x, y) & \text{otherwise} \end{cases} \quad (10.6\text{-}3)$$

and

$$N_k(x, y) = \begin{cases} N_{k-1}(x, y) + 1 & \text{if } [R(x, y) - f(x, y, k)] < -T \\ N_{k-1}(x, y) & \text{otherwise} \end{cases} \quad (10.6\text{-}4)$$

where $A_k(x, t)$, $P_k(x, y)$, and $N_k(x, y)$ are the absolute, positive, and negative ADIs, respectively, after the kth image in the sequence is encountered.

It is understood that these ADIs start out with all zero values (counts). Note also that the ADIs are the same size as the images in the sequence. As noted previously, the images in the sequence are all assumed to be of the same size. Finally, we note that the order of the inequalities and signs of the thresholds in Eqs. (10.6-3) and (10.6-4) are reversed if the gray-level values of the background pixels are greater than the levels of the moving objects.

EXAMPLE 10.19:
Computation of the absolute, positive, and negative accumulative difference images.

■ Figure 10.49 shows the three ADIs displayed as intensity images for a rectangular object of dimension 75×50 pixels that is moving in a southeasterly direction at a speed of $5\sqrt{2}$ pixels per frame. The images are of size 256×256 pixels. We note the following: (1) The nonzero area of the positive ADI is equal to the size of the moving object. (2) The location of the positive ADI corresponds to the location of the moving object in the reference frame. (3) The number of counts in the positive ADI stops increasing when the moving object is displaced completely with respect to the same object in the reference frame.

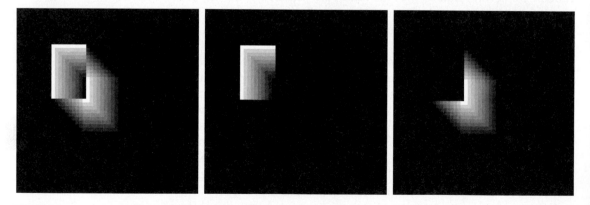

a b c

FIGURE 10.49 ADIs of a rectangular object moving in a southeasterly direction. (a) Absolute ADI. (b) Positive ADI. (c) Negative ADI.

(4) The absolute ADI contains the regions of the positive and negative ADI. (5) The direction and speed of the moving object can be determined from the entries in the absolute and negative ADIs.   ■

Establishing a reference image

A key to the success of the techniques discussed in the preceding two sections is having a reference image against which subsequent comparisons can be made. As indicated, the difference between two images in a dynamic imaging problem has the tendency to cancel all stationary components, leaving only image elements that correspond to noise and to the moving objects. The noise problem can be handled by the filtering approach mentioned earlier or by forming an accumulative difference image, as discussed in the preceding section.

In practice, obtaining a reference image with only stationary elements is not always possible, and building a reference from a set of images containing one or more moving objects becomes necessary. This necessity applies particularly to situations describing busy scenes or in cases where frequent updating is required. One procedure for generating a reference image is as follows. Consider the first image in a sequence to be the reference image. When a nonstationary component has moved completely out of its position in the reference frame, the corresponding background in the present frame can be duplicated in the location originally occupied by the object in the reference frame. When all moving objects have moved completely out of their original positions, a reference image containing only stationary components will have been created. Object displacement can be established by monitoring the changes in the positive ADI, as indicated in the preceding section.

■ Figures 10.50(a) and (b) show two image frames of a traffic intersection. The first image is considered the reference, and the second depicts the same scene some time later. The objective is to remove the principal moving objects in the reference image in order to create a static image. Although there are other smaller moving objects, the principal moving feature is the automobile at the intersection moving from left to right. For illustrative purposes we focus on this object. By monitoring the changes in the positive ADI, it is possible to determine the initial position of a moving object, as explained previously. Once the area

EXAMPLE 10.20:
Building a reference image.

a b c

FIGURE 10.50 Building a static reference image. (a) and (b) Two frames in a sequence. (c) Eastbound automobile subtracted from (a) and the background restored from the corresponding area in (b). (Jain and Jain.)

occupied by this object is identified, the object can be removed from the image by subtraction. By looking at the frame in the sequence at which the positive ADI stopped changing, we can copy from this image the area previously occupied by the moving object in the initial frame. This area then is pasted onto the image from which the object was cut out, thus restoring the background of that area. If this is done for all moving objects, the result is a reference image with only static components against which we can compare subsequent frames for motion detection, as explained in the previous two sections. The result of removing the east-bound moving vehicle in this case is shown in Fig. 10.50(c). ■

10.6.2 Frequency Domain Techniques

In this section we consider the problem of determining motion estimates via a Fourier transform formulation. Consider a sequence $f(x, y, t), t = 0, 1, \ldots,$ $K - 1$, of K digital image frames of size $M \times N$ generated by a stationary camera. We begin the development by assuming that all frames have a homogeneous background of zero intensity. The exception is a single, 1-pixel object of unit intensity that is moving with constant velocity. Suppose that for frame one ($t = 0$) the image plane is *projected* onto the x-axis; that is, the pixel intensities are summed across the columns. This operation yields a 1-D array with M entries that are 0, except at the location where the object is projected. Multiplying the components of the array by $\exp\left[j2\pi a_1 x \Delta t\right], x = 0, 1, 2, \ldots,$ $M - 1$, with the object at coordinates (x', y') at that instant of time, produces a sum equal to $\exp\left[j2\pi a_1 x' \Delta t\right]$. In this notation a_1 is a positive integer, and Δt is the time interval between frames.

Suppose that in frame two ($t = 1$) the object has moved to coordinates $(x' + 1, y')$; that is, it has moved 1 pixel parallel to the x-axis. Then repeating the projection procedure discussed in the previous paragraph yields the sum $\exp\left[j2\pi a_1 (x' + 1)\Delta t\right]$. If the object continues to move 1 pixel location per frame, then, at any integer instant of time, the result is $\exp\left[j2\pi a_1 (x' + t)\Delta t\right]$, which, using Euler's formula, may be expressed as

$$e^{j2\pi a_1(x'+t)\Delta t} = \cos\left[2\pi a_1(x' + t)\Delta t\right] + j\sin\left[2\pi a_1(x' + t)\Delta t\right] \quad (10.6\text{-}5)$$

for $t = 0, 1, \ldots, K - 1$. In other words, this procedure yields a complex sinusoid with frequency a_1. If the object were moving v_1 pixels (in the x-direction) between frames, the sinusoid would have frequency $v_1 a_1$. Because t varies between 0 and $K - 1$ in integer increments, restricting a_1 to integer values causes the discrete Fourier transform of the complex sinusoid to have two peaks—one located at frequency $v_1 a_1$ and the other at $K - v_1 a_1$. This latter peak is the result of symmetry in the discrete Fourier transform, as discussed in Section 4.6, and may be ignored. Thus a peak search in the Fourier spectrum yields $v_1 a_1$. Division of this quantity by a_1 yields v_1, which is the velocity component in the x-direction, as the frame rate is assumed to be known. A similar argument would yield v_2, the component of velocity in the y-direction.

A sequence of frames in which no motion takes place produces identical exponential terms, whose Fourier transform would consist of a single peak at a frequency of 0 (a single dc term). Therefore, because the operations discussed so far are linear, the general case involving one or more moving objects in an arbitrary static background would have a Fourier transform with a peak at dc corresponding to static image components and peaks at locations proportional to the velocities of the objects.

These concepts may be summarized as follows. For a sequence of K digital images of size $M \times N$, the sum of the weighted projections onto the x axis at any integer instant of time is

$$g_x(t, a_1) = \sum_{x=0}^{M-1} \sum_{y=0}^{N-1} f(x, y, t) e^{j2\pi a_1 x \Delta t} \qquad t = 0, 1, \ldots, K - 1. \qquad (10.6\text{-}6)$$

Similarly, the sum of the projections onto the y-axis is

$$g_y(t, a_2) = \sum_{y=0}^{N-1} \sum_{x=0}^{M-1} f(x, y, t) e^{j2\pi a_2 y \Delta t} \qquad t = 0, 1, \ldots, K - 1 \qquad (10.6\text{-}7)$$

where, as noted already, a_1 and a_2 are positive integers.

The 1-D Fourier transforms of Eqs. (10.6-6) and (10.6-7), respectively, are

$$G_x(u_1, a_1) = \frac{1}{K} \sum_{t=0}^{K-1} g_x(t, a_1) e^{-j2\pi u_1 t / K} \qquad u_1 = 0, 1, \ldots, K - 1 \qquad (10.6\text{-}8)$$

and

$$G_y(u_2, a_2) = \frac{1}{K} \sum_{t=0}^{K-1} g_y(t, a_2) e^{j2\pi u_2 t / K} \qquad u_2 = 0, 1, \ldots, K - 1. \qquad (10.6\text{-}9)$$

In practice, computation of these transforms is carried out using an FFT algorithm, as discussed in Section 4.6.

The frequency-velocity relationship is

$$u_1 = a_1 v_1 \qquad (10.6\text{-}10)$$

and

$$u_2 = a_2 v_2. \qquad (10.6\text{-}11)$$

In this formulation the unit of velocity is in pixels per total frame time. For example, $v_1 = 10$ is interpreted as a motion of 10 pixels in K frames. For frames that are taken uniformly, the actual physical speed depends on the frame rate and the distance between pixels. Thus if $v_1 = 10$, $K = 30$, the frame rate is two images per second, and the distance between pixels is 0.5 m, then the actual physical speed in the x-direction is

$$v_1 = (10 \text{ pixels})(0.5 \text{ m/pixel})(2 \text{ frames/s})/(30 \text{ frames})$$
$$= 1/3 \text{ m/s}.$$

The sign of the x-component of the velocity is obtained by computing

$$S_{1x} = \frac{d^2 \text{Re}[g_x(t, a_1)]}{dt^2} \Bigg|_{t=n} \qquad (10.6\text{-}12)$$

and

$$S_{2x} = \frac{d^2 \mathrm{Im}\left[g_x(t, a_1)\right]}{dt^2}\Bigg|_{t=n} \qquad (10.6\text{-}13)$$

Because g_x is sinusoidal, it can be shown that S_{1x} and S_{2x} will have the same sign at an arbitrary point in time, n, if the velocity component v_1 is positive. Conversely, opposite signs in S_{1x} and S_{2x} indicate a negative component. If either S_{1x} or S_{2x} is zero, we consider the next closest point in time, $t = n \pm \Delta t$. Similar comments apply to computing the sign of v_2.

EXAMPLE 10.21:
Detection of a small moving object via the frequency domain.

■ Figures 10.51 through 10.54 illustrate the effectiveness of the approach just derived. Figure 10.51 shows one of a 32-frame sequence of LANDSAT images generated by adding white noise to a reference image. The sequence contains a superimposed target moving at 0.5 pixel per frame in the x-direction and 1 pixel per frame in the y-direction. The target, shown circled in Fig. 10.52, has a Gaussian intensity distribution spread over a small (9-pixel) area and is not easily discernible by eye. The results of computing Eqs. (10.6-8) and (10.6-9) with $a_1 = 6$ and $a_2 = 4$ are shown in Figs. 10.53 and 10.54, respectively. The peak at $u_1 = 3$ in Fig. 10.53 yields $v_1 = 0.5$ from Eq. (10.6-10). Similarly, the peak at $u_2 = 4$ in Fig. 10.54 yields $v_2 = 1.0$ from Eq. (10.6-11). ■

Guidelines for the selection of a_1 and a_2 can be explained with the aid of Figs. 10.53 and 10.54. For instance, suppose that we had used $a_2 = 15$ instead of $a_2 = 4$. In that case the peaks in Fig. 10.54 would now be at $u_2 = 15$ and 17 because $v_2 = 1.0$, which would be a seriously aliased result. As discussed in Section 2.4.4, aliasing is caused by undersampling (too few frames in the present discussion, as the range of u is determined by K). Because $u = av$, one possibility is to select a as the integer closest to $a = u_{\max}/v_{\max}$, where $u_{\max}$ is the aliasing frequency limitation established by K by and $v_{\max}$ is the maximum expected object velocity.

FIGURE 10.51
LANDSAT frame. (Cowart, Snyder, and Ruedger.)

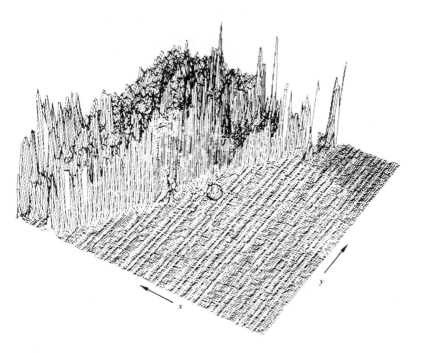

FIGURE 10.52
Intensity plot of
the image in
Fig. 10.51, with
the target circled.
(Rajala, Riddle,
and Snyder.)

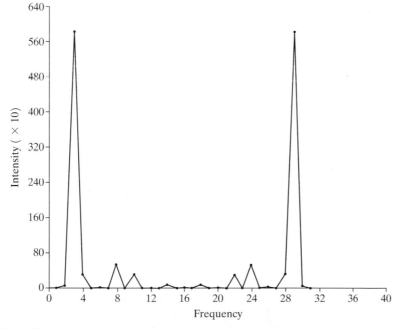

FIGURE 10.53 Spectrum of Eq. (10.6-8) showing a peak at $u_1 = 3$. (Rajala, Riddle, and Snyder.)

FIGURE 10.54
Spectrum of
Eq. (10.6-9)
showing a peak at
$u_2 = 4$. (Rajala,
Riddle, and
Snyder.)

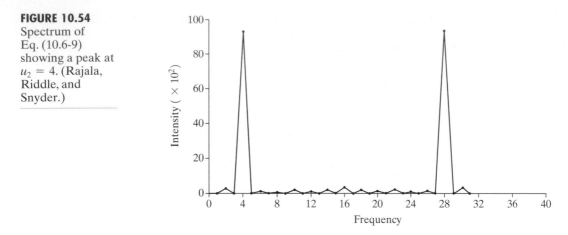

Summary

Image segmentation is an essential preliminary step in most automatic pictorial pattern recognition and scene analysis problems. As indicated by the range of examples presented in the previous sections, the choice of one segmentation technique over another is dictated mostly by the peculiar characteristics of the problem being considered. The methods discussed in this chapter, although far from exhaustive, are representative of techniques commonly used in practice. The following references can be used as the basis for further study of this topic.

References and Further Reading

Because of its central role in autonomous image processing, segmentation is a topic covered in most books dealing with image processing, image analysis, and computer vision. The following books provide complementary and/or supplementary reading for our coverage of this topic: Shapiro and Stockman [2001], Sonka et al. [1999], Petrou and Bosdogianni [1999], and Umbaugh [1998].

Work dealing with the use of masks to detect gray-level discontinuities (Section 10.1) has a long history. Numerous masks have been proposed over the years: Roberts [1965], Prewitt [1970], Kirsh [1971], Robinson [1976], Frei and Chen [1977], and Canny [1986]. A review article by Fram and Deutsch [1975] contains numerous masks and an evaluation of their performance. The issue of mask performance, especially for edge detection, still is an area of considerable interest, as exemplified by Qian and Huang [1996], Wang et al. [1996], Heath et al. [1997, 1998], and Ando [2000]. Edge detection on color images has been increasing in popularity for a number of multisensing applications. See, for example, Salinas, Abidi and Gonzalez [1996]; Zugaj and Lattuati [1998]; Mirmehdi and Petrou [2000]; and Plataniotis and Venetsanopoulos [2000]. The interplay between image characteristic and mask performance also is a topic of current interest, as exemplified by Ziou [2001]. Our presentation of the zero-crossing properties of the Laplacian is based on a paper by Marr and Hildredth [1980] and on the book by Marr [1982]. See also a paper by Clark [1989] on authenticating edges produced by zero-crossing algorithms. (Corrections of parts of the Clark paper are given by Piech [1990].) As mentioned in Section 10.1, zero crossing via the Laplacian of a Gaussian is an important approach whose relative performance is still an active topic of research (Gunn [1998, 1999]).

The Hough transform (Hough [1962]) has emerged over the past decade as a method of choice for global pixel linking and curve detection. Numerous generalizations to the basic transform discussed in this chapter have been proposed over the years. For example, Lo and Tsai [1995] discuss an approach for detecting thick lines, Guil et al. [1995, 1997] deal with fast implementations of the Hough transform and detection of primitive curves, Daul at al. [1998] discuss further generalizations for detecting elliptical arcs, and Shapiro [1996] deals with implementation of the Hough transform for gray-scale images. The algorithm presented in Section 10.2.3 is based on Martelli [1972, 1976]. For additional reading on heuristic graph searching, see Nilsson [1980], Umeyama [1988], and Sonka et al. [1999].

As mentioned at the beginning of Section 10.3, thresholding techniques enjoy a significant degree of popularity because they are simple to implement. It is not surprising that there is a considerable body of work reported in the literature on this topic. A good appreciation of the extent of this literature can be gained from the review papers by Sahoo et al. [1988] and by Lee et al. [1990]. We spent a significant level of effort in Section 10.3.2 dealing with the effects of illumination on thresholding. The types of approaches used to deal with this problem are illustrated by the work of Perez and Gonzalez [1987], Parker [1991], Murase and Nayar [1994], Bischsel [1998], and Drew et al. [1999]. For additional reading on the material in Sections 10.3.3 and 10.3.4, see Jain et al. [1995]. The early work of Chow and Kaneko [1972] discussed in Section 10.3.5 is still a standard in terms of illustrating the key aspects of a threshold-based image segmentation solution. Essentially the same can be said for the material presented in Section 10.3.6 (due to White and Rohrer [1983]), which combines thresholding, the gradient, and the Laplacian in the solution of a difficult segmentation problem. It is interesting to compare the fundamental similarities in terms of image segmentation capability between these two articles and work on thresholding done almost twenty years later (Cheriet et al. [1998], Sauvola and Pietikainen [2000]). See also Liang et al. [2000] and Chan et al. [2000] for alternate approaches to the problem of detecting boundaries in images similar in concept to those studied by Chow and Kaneko.

See Fu and Mui [1981] for an early survey on the topic of region-oriented segmentation. The works of Haddon and Boyce [1990] and of Pavlidis and Liow [1990] are among the earliest efforts to integrate region and boundary information for the purpose of segmentation. A newer region-growing approach proposed by Hojjatoleslami and Kittler [1998] also is of interest. For current basic coverage of region-oriented segmentation concepts, see Shapiro and Stockman [2001] and Sonka et al. [1999].

Segmentation by watersheds was shown in Section 10.5 to be a powerful concept. Early references dealing with segmentation by watersheds are Serra [1988], Beucher [1990], and Beucher and Meyer [1992]. The paper by Baccar et al. [1996] discusses segmentation based on data fusion and morphological watersheds. The progress in this field in a little more than one decade is evident in a special issue of *Pattern Recognition* [2000], devoted entirely to this topic. As indicated in our discussion in Section 10.5, one of the key issues with watersheds is the problem of over segmentation. The papers by Najmanand and Schmitt [1996], Haris et al. [1998], and Bleau and Leon [2000] are illustrative of approaches for dealing with this problem. Bieniek and Moga [2000] discuss a watershed segmentation algorithm based on connected components.

The material in Section 10.6.1 is from Jain, R. [1981]. See also Jain, Kasturi, and Schunck [1995]. The material in Section 10.6.2 is from Rajala, Riddle, and Snyder [1983]. See also the papers by Shariat and Price [1990] and by Cumani et al. [1991]. The books by Shapiro and Stockman [2001] and by Sonka et al. [1999] provide additional reading regarding motion estimation.

Problems

★**10.1**   A binary image contains straight lines oriented horizontally, vertically, at 45°, and at −45°. Give a set of 3 × 3 masks that can be used to detect 1-pixel-long breaks in these lines. Assume that the gray level of the lines is 1 and that the gray level of the background is 0.

10.2   Propose a technique for detecting gaps of length ranging between 1 and L pixels in line segments of a binary image. Assume that the lines are 1 pixel thick. Base your technique on 8-neighbor connectivity analysis, rather than attempting to construct masks for detecting the gaps.

10.3   Refer to Fig. 10.4 in answering the following questions.

★ **(a)** Some of the lines joining the pads and center element in Fig. 10.4(b) are single lines, while some others are double lines. Explain why.

(b) How would you go about eliminating the components in Fig. 10.4(c) that are not part of the line oriented at −45°?

10.4   Consider a horizontal intensity profile through the middle of a binary image that contains a step edge running vertically through the center of the image. Draw what the profile would look like after the image has been blurred by an averaging mask of size $n \times n$, with coefficients equal to $1/n^2$. For simplicity, assume that the image was scaled so that its gray levels are 0 on the left of the edge and 1 to its right. Also, assume that the size of the mask is much smaller than the image, so that image border effects are not a concern near the center of the horizontal intensity profile.

★**10.5**   Suppose that we had used the edge model shown, instead of the ramp model of an edge shown in Fig. 10.6. Sketch the gradient and Laplacian of each profile.

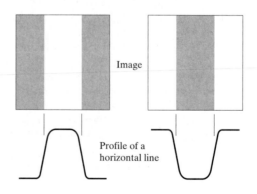

Image

Profile of a
horizontal line

10.6   Refer to Fig. 10.8 in answering the following questions.

(a) Assume that the Sobel masks are used to obtain G_x and G_y. Show that in this case the gradient computed by using Eqs. (10.1-4) and (10.1-12) give identical results for edges oriented in the horizontal and vertical directions.

(b) Show that this is true also for the Prewitt masks.

★**10.7**   Show that the Sobel and Prewitt gradient masks of Figs. 10.8 and 10.9 give isotropic results only for horizontal and vertical edges, and for edges oriented at ±45°.

10.8   The results obtained by a single pass through an image of some 2-D masks can be achieved also by two passes using 1-D masks. For example, the same result of using a 3 × 3 smoothing mask with coefficients 1/9 can be obtained by a pass of the mask [1 1 1] through an image. The result of this pass is then followed by a pass of the mask

$$\begin{bmatrix} 1 \\ 1 \\ 1 \end{bmatrix}.$$

The final result is then scaled by 1/9. Show that the Sobel masks (Fig. 10.8) can be implemented by one pass of a *differencing* mask of the form [−1 0 1] (or its vertical counterpart) followed by a *smoothing* mask of the form [1 2 1] (or its vertical counterpart).

★**10.9** The so-called *compass gradient operators* of size 3 × 3 are designed to measure gradients of edges oriented in eight directions: E, NE, N, NW, W, SW, S, and SE. Give the form of these eight operators using coefficients valued 0, 1, or −1. Specify the gradient direction of each mask, keeping in mind that the gradient direction is orthogonal to the edge direction.

10.10 The center rectangle in the binary image shown is of size $m \times n$ pixels.

(a) Sketch the gradient of this image using the approximation given in Eq. (10.1-12). Assume that G_x and G_y are obtained by using the Sobel operators. Show all relevant different pixel values in the gradient image.

(b) Sketch the histogram of edge directions computed from Eq. (10.1-5). Be precise in labeling the height of each peak of the histogram.

(c) Sketch the Laplacian of the image for the approximation given in Eq. (10.1-14). Show all relevant different pixel values in the Laplacian image.

10.11 With reference to Eq. (10.1-17),

★ (a) Show that the average value of the Laplacian operator $\nabla^2 h$ is zero.

(b) Prove that the average value of any image convolved with this operator also is zero.

(c) Would (b) be true in general for the approximations to the Laplacian given in Eqs. (10.1-14) and (10.1-15)? Explain.

10.12 Refer to Fig. 10.15.

(a) Explain why the edges in Fig. 10.15(g) form closed contours.

(b) Does the zero-crossings method for edge finding always result in edges that are closed contours? Give a reason for your answer.

★**10.13** Refer to the discussion in Section 10.2.2.

   (a) Explain why the Hough mapping of point 1 in Fig. 10.20(b) is a straight line.

   (b) Is this the only point that would produce this result?

   (c) Explain the reflective adjacency relationship illustrated in Fig. 10.20(d).

10.14 Refer to the discussion in Section 10.2.2.

   (a) Develop a general procedure for obtaining the normal representation of a line from its slope-intercept equation $y = ax + b$.

   (b) Find the normal representation of the line $y = -2x + 1$.

10.15 An important area of application for image segmentation techniques is in processing images resulting from so-called bubble chamber events. These images arise from experiments in high-energy physics in which a beam of particles of known properties is directed onto a target of known nuclei. A typical event consists of incoming tracks, any one of which, in the event of a collision, branches out into secondary tracks of particles emanating from the point of collision. Propose a segmentation approach for detecting all tracks that contain at least 100 pixels and are angled at any of the following six directions off the horizontal: $\pm 25°$, $\pm 50°$, and $\pm 75°$. The allowable estimation error in any of these six directions is $\pm 5°$. For a track to be valid it must be at least 100 pixels long and not have more than three gaps, any of which cannot exceed 10 pixels. You may assume that the images have been preprocessed so that they are binary and that all tracks are 1 pixel wide, except at the point of collision from which they emanate. Your procedure should be able to differentiate between tracks that have the same direction but different origins. (*Hint:* Base your solution on the Hough transform.)

★**10.16** Refer to Figs. 10.22 and 10.24.

   (a) Superimpose on Fig. 10.22 all the possible edges given by the graph in Fig. 10.24.

   (b) Compute the cost of the minimum-cost path.

10.17 Find the edge corresponding to the minimum-cost path in the subimage shown. The numbers in brackets are gray levels and the outer numbers are spatial coordinates. Assume that the edge starts in the first column and ends in the last column.

	1	2	3
1	• [2]	• [1]	• [0]
2	• [1]	• [1]	• [7]
3	• [6]	• [8]	• [2]

★**10.18** The images shown are quite different, but their histograms are identical. Suppose that each image is blurred with a 3 × 3 smoothing mask.

    (a) Would the histograms still be identical after blurring?

    (b) If your answer is no, sketch the two histograms.

10.19 Consider a noiseless image of size $N \times N$ whose first $N/2$ columns have gray level L_A and its remaining columns have gray level L_B, where $L_B > L_A$. The histogram of this image has only two peaks, of identical height, one located at L_A and the other at L_B. Segmenting this image into two halves based on its gray-level content is a trivial task that can be accomplished by a single global threshold located between L_A and L_B. Suppose, however, that you multiply the image by a gray-scale wedge that varies from 0 on the left to K on the right, with $K > L_B$. What would the histogram of this new image look like? Label the various parts of the histogram clearly.

10.20 Refer to the threshold finding algorithm introduced in Section 10.3.3. Assume that in a given problem the histogram is bimodal and, furthermore, that the shape of the modes can be approximated by Gaussian curves of the form $A_1 e^{-(z-m_1)^2/2\sigma_1^2}$ and $A_2 e^{-(z-m_2)^2/2\sigma_2^2}$. Assume that $m_1 < m_2$ and give conditions (in terms of the parameters of these curves) for the following to be true when the algorithm converges:

    ★ **(a)** The threshold is equal to $(m_1 + m_2)/2$.

    ★ **(b)** The threshold is to the left of m_1.

    (c) The threshold is in the interval: $(m_1 + m_2)/2 < T < m_2$.

    If it is impossible for any one of these conditions to exist, so state, and give a reason.

10.21 The illumination, $i(x, y)$, of a scene is known to vary as a function of spatial coordinates according to the equation $i(x, y) = A(1 + e^{-B[(x-N/2)^2+(y-N/2)^2]})$, measured in some appropriately normalized units. The digitized scene, denoted $f(x, y)$, is of size $N \times N$, with $N = 1000$. The constants have values $A = 0.5$ and $B = 10^{-4}$. It is known that any subimage of $f(x, y)$ having a size greater than 10×10 pixels can be properly segmented, as long as the *illumination* spanning the area of the subimage did not vary by more than 0.1 units between any two points in the subimage when $f(x, y)$ was acquired. It is known also that the thresholds that work for the segmentation of the subimages under these conditions can be obtained using the algorithm introduced in Section 10.3.3. Propose an adaptive technique capable of thresholding this image. Not all subimages need to be of the same size, but you are required to use the largest possible subimage at any location in the image.

★**10.22** Suppose that an image has the gray-level probability density functions shown. Here, $p_1(z)$ corresponds to objects and $p_2(z)$ corresponds to the background. Assume that $P_1 = P_2$ and find the optimal threshold between object and background pixels.

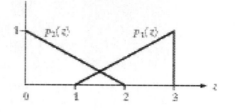

10.23 The functional form of the Rayleigh probability density function and a typical plot are shown in the following figure. This density is well suited for approximating skewed histograms, such as the ones shown in Fig. 10.29. Use the Rayleigh density to set up a two-category bimodal problem, as in Problem 10.22. Find the optimal threshold in terms of the a priori probabilities P_1 and P_2, and the parameters of the Rayleigh density. The long "tails" of the two densities should be in opposite directions, as in Fig. 10.29.

$$p(z) = \begin{cases} \dfrac{2}{b}(z - a)e^{-(z-a)^2/b} & z \geq a \\ 0 & z < a \end{cases}$$

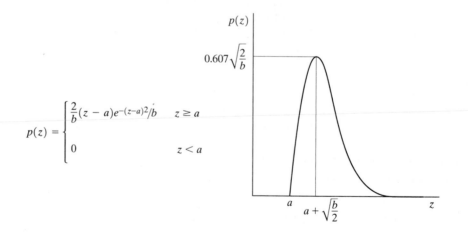

★**10.24** Start with Eq. (10.3-10) and derive Eqs. (10.3-12) and (10.3-13).

10.25 Derive Eq. (10.3-14) starting with Eqs. (10.3-12) and (10.3-13).

★**10.26** The mean and standard deviation of the background pixels in the image shown are 110 and 15, respectively. The object pixels have mean and standard deviation values of 200 and 40, respectively. Propose a thresholding solution for segmenting the objects out of the image. State clearly any assumptions that you make in solving this problem.

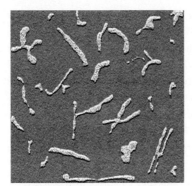

10.27 Propose an approach based on region growing for segmenting the image shown in Problem 10.26. State clearly any assumptions that you make in solving this problem.

10.28 Segment the image shown by using the split and merge procedure discussed in Section 10.4.3. Let $P(R_i) = $ TRUE if all pixels in R_i have the same gray level. Show the quadtree corresponding to your segmentation.

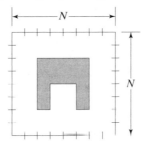

10.29 Refer to the discussion in Section 10.5.3.

★ **(a)** Show that the elements in sets $C_n(M_i)$ and $T[n]$ are never replaced during execution of the watershed algorithm.

(b) Show that the number of elements in sets $C_n(M_i)$ and $T[n]$ either increases or remains the same as n increases.

10.30 The boundaries illustrated in Section 10.5, obtained using the watershed segmentation algorithm, form closed loops. Advance an argument that establishes whether or not closed boundaries always result from application of this algorithm.

★ **10.31** Give a step-by-step implementation of the dam building procedure for the one-dimensional gray-level cross section shown. Show a drawing of the cross section at each step, showing "water" levels and dams constructed.

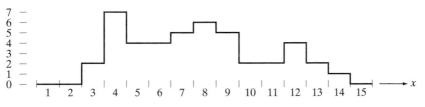

10.32 What would the negative ADI image in Fig. 10.49(c) look like if we tested against T, instead of testing against $-T$, in Eq. (10.6-4)?

10.33 Determine if the following statements are true of false. Explain the reason for your answer.

(a) The nonzero entries of the absolute ADI continue to grow in dimension as long as the object is moving.

(b) The nonzero entries in the positive ADI always occupy the same area, regardless of the motion undergone by the object.

(c) The nonzero entries in the negative ADI continue to grow in dimension as long as the object is moving.

10.34 Suppose that in Example 10.21 motion along the x axis is set to zero. The object now moves only along the y axis at 1 pixel per frame for 32 frames and then (instantaneously) reverses direction and moves in exactly the opposite direction for another 32 frames. What would Figs. 10.53 and 10.54 look like under these conditions?

10.35 The speed of a bullet in flight is to be estimated by using high-speed imaging techniques. The method of choice involves the use of a TV camera and flash that exposes the scene for K s. The bullet is 2.5 cm long, 1 cm wide, and its range of speed is 750 ± 250 m/s. The camera optics produce an image in which the bullet occupies 10% of the horizontal resolution of a 256×256 digital image.

(a) Determine the maximum value of K that will guarantee that the blur from motion does not exceed 1 pixel.

(b) Determine the minimum number of frames per second that would have to be taken in order to guarantee that at least two complete images of the bullet are obtained during its path through the field of view of the camera.

(c) Propose a segmentation procedure for automatically extracting the bullet from a sequence of frames.

(d) Propose a method for automatically determining the speed of the bullet.

11 Representation and Description

Well, but reflect; have we not several times
acknowledged that names rightly given are the
likenesses and images of the things which they name?

Socrates

Preview

After an image has been segmented into regions by methods such as those discussed in Chapter 10, the resulting aggregate of segmented pixels usually is represented and described in a form suitable for further computer processing. Basically, representing a region involves two choices: (1) We can represent the region in terms of its external characteristics (its boundary), or (2) we can represent it in terms of its internal characteristics (the pixels comprising the region). Choosing a representation scheme, however, is only part of the task of making the data useful to a computer. The next task is to *describe* the region based on the chosen representation. For example, a region may be *represented* by its boundary, and the boundary *described* by features such as its length, the orientation of the straight line joining its extreme points, and the number of concavities in the boundary.

An external representation is chosen when the primary focus is on shape characteristics. An internal representation is selected when the primary focus is on regional properties, such as color and texture. Sometimes it may be necessary to use both types of representation. In either case, the features selected as descriptors should be as insensitive as possible to variations in size, translation, and rotation. For the most part, the descriptors discussed in this chapter satisfy one or more of these properties.

11.1 Representation

The segmentation techniques discussed in Chapter 10 yield raw data in the form of pixels along a boundary or pixels contained in a region. Although these data sometimes are used directly to obtain descriptors (as in determining the texture of a region), standard practice is to use schemes that compact the data into representations that are considerably more useful in the computation of descriptors. In this section we discuss various representation approaches.

11.1.1 Chain Codes

Chain codes are used to represent a boundary by a connected sequence of straight-line segments of specified length and direction. Typically, this representation is based on 4- or 8-connectivity of the segments. The direction of each segment is coded by using a numbering scheme such as the ones shown in Fig. 11.1.

Digital images usually are acquired and processed in a grid format with equal spacing in the x- and y-directions, so a chain code could be generated by following a boundary in, say, a clockwise direction and assigning a direction to the segments connecting every pair of pixels. This method generally is unacceptable for two principal reasons: (1) The resulting chain of codes tends to be quite long, and (2) any small disturbances along the boundary due to noise or imperfect segmentation cause changes in the code that may not be related to the shape of the boundary.

An approach frequently used to circumvent the problems just discussed is to resample the boundary by selecting a larger grid spacing, as illustrated in Fig. 11.2(a). Then, as the boundary is traversed, a boundary point is assigned to each node of the large grid, depending on the proximity of the original boundary to that node, as shown in Fig. 11.2(b). The resampled boundary obtained in this way then can be represented by a 4- or 8-code, as shown in Figs. 11.2(c) and (d), respectively. The starting point in Fig. 11.2(c) is (arbitrarily) at the top, left dot, and the boundary is the shortest allowable 4- or 8-path in the grid of Fig. 11.2(b). The boundary representation in Fig. 11.2(c) is the chain code 0033 … 01, and in Fig. 11.2(d) it is the code 0766…12. As might be expected, the accuracy of the resulting code representation depends on the spacing of the sampling grid.

FIGURE 11.1
Direction numbers for (a) 4-directional chain code, and (b) 8-directional chain code.

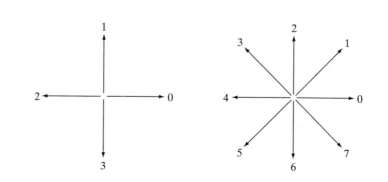

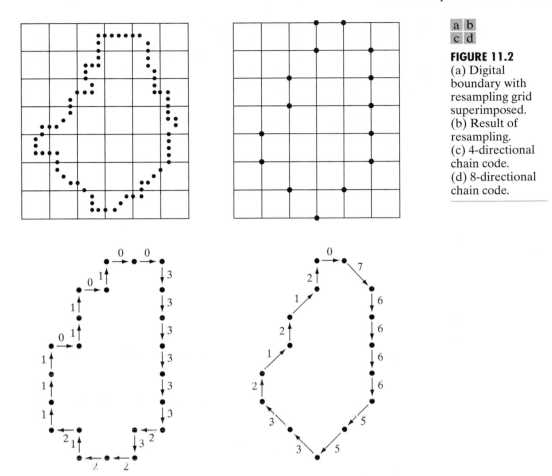

FIGURE 11.2
(a) Digital
boundary with
resampling grid
superimposed.
(b) Result of
resampling.
(c) 4-directional
chain code.
(d) 8-directional
chain code.

The chain code of a boundary depends on the starting point. However, the code can be normalized with respect to the starting point by a straightforward procedure: We simply treat the chain code as a circular sequence of direction numbers and redefine the starting point so that the resulting sequence of numbers forms an integer of minimum magnitude. We can normalize also for rotation by using the *first difference* of the chain code instead of the code itself. This difference is obtained by counting the number of direction changes (in a counterclockwise direction) that separate two adjacent elements of the code. For instance, the first difference of the 4-direction chain code 10103322 is 3133030. If we elect to treat the code as a circular sequence, then the first element of the difference is computed by using the transition between the last and first components of the chain. Here, the result is 33133030. Size normalization can be achieved by altering the size of the resampling grid.

These normalizations are exact only if the boundaries themselves are invariant to rotation and scale change, which, in practice, is seldom the case. For instance, the same object digitized in two different orientations will in general have different boundary shapes, with the degree of dissimilarity being proportional to image resolution. This effect can be reduced by selecting chain elements that are large in proportion to the distance between pixels in the digitized image and/or by orienting the resampling grid along the principal axes of the object to be coded, as discussed in Section 11.2.2, or along its eigen axes, as discussed in Section 11.4.

11.1.2 Polygonal Approximations

A digital boundary can be approximated with arbitrary accuracy by a polygon. For a closed curve, the approximation is exact when the number of segments in the polygon is equal to the number of points in the boundary so that each pair of adjacent points defines a segment in the polygon. In practice, the goal of polygonal approximation is to capture the "essence" of the boundary shape with the fewest possible polygonal segments. This problem in general is not trivial and can quickly turn into a time-consuming iterative search. However, several polygonal approximation techniques of modest complexity and processing requirements are well suited for image processing applications.

Minimum perimeter polygons

We begin the discussion of polygonal approximations with a method for finding *minimum perimeter polygons*. The procedure is best explained by an example. Suppose that we enclose a boundary by a set of concatenated cells, as shown in Fig. 11.3(a). It helps to visualize this enclosure as two walls corresponding to the outside and inside boundaries of the strip of cells, and think of the object boundary as a rubber band contained within the walls. If the rubber band is allowed to shrink, it takes the shape shown in Fig. 11.3(b), producing a polygon of minimum perimeter that fits the geometry established by the cell strip. If each cell encompasses only one point on the boundary, the error in each cell between the original boundary and the rubber-band approximation at most would be $\sqrt{2}d$, where d is the minimum possible distance between different pixels (i.e., the distance between lines in the sampling grid used to produce the digital image). This error can be reduced by half by forcing each cell to be centered on its corresponding pixel.

a b

FIGURE 11.3
(a) Object boundary enclosed by cells.
(b) Minimum perimeter polygon.

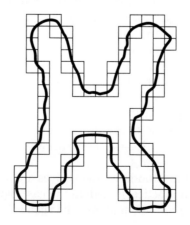
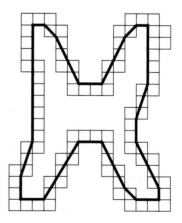

Merging techniques

Merging techniques based on average error or other criteria have been applied to the problem of polygonal approximation. One approach is to merge points along a boundary until the least square error line fit of the points merged so far exceeds a preset threshold. When this condition occurs, the parameters of the line are stored, the error is set to 0, and the procedure is repeated, merging new points along the boundary until the error again exceeds the threshold. At the end of the procedure the intersections of adjacent line segments form the vertices of the polygon. One of the principal difficulties with this method is that vertices in the resulting approximation do not always correspond to inflections (such as corners) in the original boundary, because a new line is not started until the error threshold is exceeded. If, for instance, a long straight line were being tracked and it turned a corner, a number (depending on the threshold) of points past the corner would be absorbed before the threshold was exceeded. However, splitting (discussed next) along with merging may be used to alleviate this difficulty.

Splitting techniques

One approach to boundary segment *splitting* is to subdivide a segment successively into two parts until a specified criterion is satisfied. For instance, a requirement might be that the maximum perpendicular distance from a boundary segment to the line joining its two end points not exceed a preset threshold. If it does, the farthest point from the line becomes a vertex, thus subdividing the initial segment into two subsegments. This approach has the advantage of seeking prominent inflection points. For a closed boundary, the best starting points usually are the two farthest points in the boundary. For example, Fig. 11.4(a) shows an object boundary, and Fig. 11.4(b) shows a subdivision of this boundary (solid line) about its farthest points. The point marked c is the farthest point (in terms of perpendicular distance) from the top boundary segment to line ab. Similarly, point d is the farthest point in the bottom segment. Figure 11.4(c) shows the result of using the splitting procedure with a threshold equal to 0.25 times the length of line ab. As no point in the new

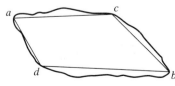

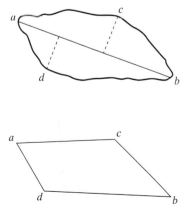

a	b
c	d

FIGURE 11.4
(a) Original boundary.
(b) Boundary divided into segments based on extreme points. (c) Joining of vertices.
(d) Resulting polygon.

boundary segments has a perpendicular distance (to its corresponding straight-line segment) that exceeds this threshold, the procedure terminates with the polygon shown in Fig. 11.4(d).

11.1.3 Signatures

A signature is a 1-D functional representation of a boundary and may be generated in various ways. One of the simplest is to plot the distance from the centroid to the boundary as a function of angle, as illustrated in Fig. 11.5. Regardless of how a signature is generated, however, the basic idea is to reduce the boundary representation to a 1-D function, which presumably is easier to describe than the original 2-D boundary.

Signatures generated by the approach just described are invariant to translation, but they do depend on rotation and scaling. Normalization with respect to rotation can be achieved by finding a way to select the same starting point to generate the signature, regardless of the shape's orientation. One way to do so is to select the starting point as the point farthest from the centroid, if this point happens to be unique and independent of rotational aberrations for each shape of interest. Another way is to select the point on the eigen axis (see Section 11.4) that is farthest from the centroid. This method requires more computation but is more rugged because the direction of the eigen axis is determined by using all contour points. Yet another way is to obtain the chain code of the boundary and then use the approach discussed in Section 11.1.1, assuming that the coding is coarse enough so that rotation does not affect its circularity.

Based on the assumptions of uniformity in scaling with respect to both axes and that sampling is taken at equal intervals of θ, changes in size of a shape result in changes in the amplitude values of the corresponding signature. One way to normalize for this result is to scale all functions so that they always span the

FIGURE 11.5
Distance-versus-angle signatures.
In (a) $r(\theta)$ is constant. In (b), the signature consists of repetitions of the pattern
$r(\theta) = A \sec\theta$ for $0 \leq \theta \leq \pi/4$ and $r(\theta) = A \csc\theta$ for $\pi/4 < \theta \leq \pi/2$.

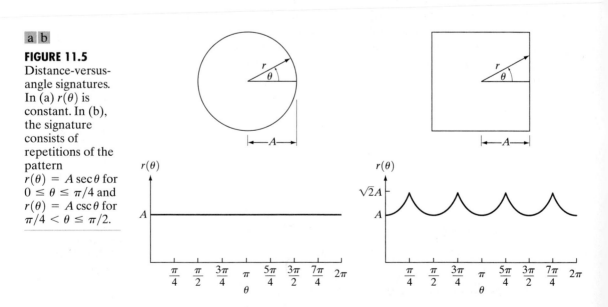

same range of values, say, $[0, 1]$. The main advantage of this method is simplicity, but it has the potentially serious disadvantage that scaling of the entire function depends on only two values: the minimum and maximum. If the shapes are noisy, this dependence can be a source of error from object to object. A more rugged (but also more computationally intensive) approach is to divide each sample by the variance of the signature, assuming that the variance is not zero—as in the case of Fig. 11.5(a)—or so small that it creates computational difficulties. Use of the variance yields a variable scaling factor that is inversely proportional to changes in size and works much as automatic gain control does. Whatever the method used, keep in mind that the basic idea is to remove dependency on size while preserving the fundamental shape of the waveforms.

Of course, distance versus angle is not the only way to generate a signature. For example, another way is to traverse the boundary and, corresponding to each point on the boundary, plot the angle between a line tangent to the boundary at that point and a reference line. The resulting signature, although quite different from the $r(\theta)$ curve, would carry information about basic shape characteristics. For instance, horizontal segments in the curve would correspond to straight lines along the boundary, because the tangent angle would be constant there. A variation of this approach is to use the so-called *slope density function* as a signature. This function is simply a histogram of tangent-angle values. As a histogram is a measure of concentration of values, the slope density function responds strongly to sections of the boundary with constant tangent angles (straight or nearly straight segments) and has deep valleys in sections producing rapidly varying angles (corners or other sharp inflections).

11.1.4 Boundary Segments

Decomposing a boundary into segments often is useful. Decomposition reduces the boundary's complexity and thus simplifies the description process. This approach is particularly attractive when the boundary contains one or more significant concavities that carry shape information. In this case use of the convex hull of the region enclosed by the boundary is a powerful tool for robust decomposition of the boundary.

As defined in Section 9.5.4, the *convex hull H* of an arbitrary set S is the smallest convex set containing S. The set difference $H - S$ is called the *convex deficiency D* of the set S. To see how these concepts might be used to partition a boundary into meaningful segments, consider Fig. 11.6(a), which shows an

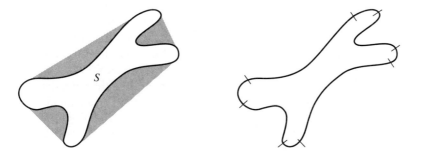

a b

FIGURE 11.6
(a) A region, S, and its convex deficiency (shaded).
(b) Partitioned boundary.

object (set S) and its convex deficiency (shaded regions). The region boundary can be partitioned by following the contour of S and marking the points at which a transition is made into or out of a component of the convex deficiency. Figure 11.6(b) shows the result in this case. Note that in principle, this scheme is independent of region size and orientation.

In practice, digital boundaries tend to be irregular because of digitization, noise, and variations in segmentation. These effects usually result in convex deficiencies that have small, meaningless components scattered randomly throughout the boundary. Rather than attempt to sort out these irregularities by postprocessing, a common approach is to smooth a boundary prior to partitioning. There are a number of ways to do so. One way is to traverse the boundary and replace the coordinates of each pixel by the average coordinates of k of its neighbors along the boundary. This approach works for small irregularities, but it is time-consuming and difficult to control. Large values of k can result in excessive smoothing, whereas small values of k might not be sufficient in some segments of the boundary. A more rugged technique is to use a polygonal approximation, as discussed in Section 11.1.2, prior to finding the convex deficiency of a region. Most digital boundaries of interest are simple polygons (polygons without self-intersection). Graham and Yao [1983] give an algorithm for finding the convex hull of such polygons.

The concepts of a convex hull and its deficiency are equally useful for describing an entire region, as well as just its boundary. For example, description of a region might be based on its area and the area of its convex deficiency, the number of components in the convex deficiency, the relative location of these components, and so on. Recall that a morphological algorithm for finding the convex hull was developed in Section 9.5.4. References cited at the end of this chapter contain other formulations.

11.1.5 Skeletons

An important approach to representing the structural shape of a plane region is to reduce it to a graph. This reduction may be accomplished by obtaining the *skeleton* of the region via a thinning (also called *skeletonizing*) algorithm. Thinning procedures play a central role in a broad range of problems in image processing, ranging from automated inspection of printed circuit boards to counting of asbestos fibers in air filters. We already discussed in Section 9.5.7 the basics of skeletonizing using morphology. However, as noted in that section, the procedure discussed there made no provisions for keeping the skeleton connected. The algorithm developed here corrects that problem.

The skeleton of a region may be defined via the medial axis transformation (MAT) proposed by Blum [1967]. The MAT of a region R with border B is as follows. For each point p in R, we find its closest neighbor in B. If p has more than one such neighbor, it is said to belong to the *medial axis* (skeleton) of R. The concept of "closest" (and the resulting MAT) depend on the definition of a distance (see Section 2.5.3). Figure 11.7 shows some examples using the Euclidean distance. The same results would be obtained with the maximum disk of Section 9.5.7.

The MAT of a region has an intuitive definition based on the so-called "prairie fire concept." Consider an image region as a prairie of uniform, dry

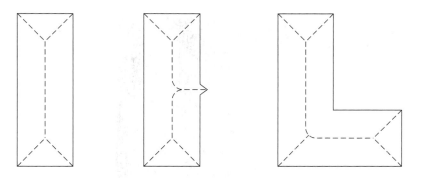

a b c
FIGURE 11.7
Medial axes
(dashed) of three
simple regions.

grass, and suppose that a fire is lit along its border. All fire fronts will advance
into the region at the same speed. The MAT of the region is the set of points
reached by more than one fire front at the same time.

Although the MAT of a region yields an intuitively pleasing skeleton, direct
implementation of this definition typically is expensive computationally. Implementation potentially involves calculating the distance from every interior point
to every point on the boundary of a region. Numerous algorithms have been proposed for improving computational efficiency while at the same time attempting
to produce a medial axis representation of a region. Typically, these are thinning
algorithms that iteratively delete edge points of a region subject to the constraints
that deletion of these points (1) does not remove end points, (2) does not break
connectivity, and (3) does not cause excessive erosion of the region.

In this section we present an algorithm for thinning binary regions. Region
points are assumed to have value 1 and background points to have value 0. The
method consists of successive passes of two basic steps applied to the contour
points of the given region, where, based on the definition given in Section 2.5.2,
a *contour point* is any pixel with value 1 and having at least one 8 neighbor valued 0. With reference to the 8-neighborhood notation shown in Fig. 11.8, step
1 flags a contour point p_1 for deletion if the following conditions are satisfied:

(a) $2 \leq N(p_1) \leq 6$
(b) $T(p_1) = 1$
(c) $p_2 \cdot p_4 \cdot p_6 = 0$
(d) $p_4 \cdot p_6 \cdot p_8 = 0$ (11.1-1)

where $N(p_1)$ is the number of nonzero neighbors of p_1; that is,

$$N(p_1) = p_2 + p_3 + \cdots + p_8 + p_9 \qquad (11.1\text{-}2)$$

p_9	p_2	p_3
p_8	p_1	p_4
p_7	p_6	p_5

FIGURE 11.8
Neighborhood
arrangement used
by the thinning
algorithm.

FIGURE 11.9
Illustration of
conditions (a)
and (b) in
Eq. (11.1-1). In
this case
$N(p_1) = 4$ and
$T(p_1) = 3$.

0	0	1
1	p_1	0
1	0	1

and $T(p_1)$ is the number of 0-1 transitions in the ordered sequence $p_2, p_3, \ldots,$ p_8, p_9, p_2. For example, $N(p_1) = 4$ and $T(p_1) = 3$ in Fig. 11.9.

In step 2, conditions (a) and (b) remain the same, but conditions (c) and (d) are changed to

$$(c')\ p_2 \cdot p_4 \cdot p_8 = 0$$
$$(d')\ p_2 \cdot p_6 \cdot p_8 = 0 \qquad\qquad (11.1\text{-}3)$$

Step 1 is applied to every border pixel in the binary region under consideration. If one or more of conditions (a)–(d) are violated, the value of the point in question is not changed. If all conditions are satisfied the point is flagged for deletion. However, the point is not deleted until all border points have been processed. This delay prevents changing the structure of the data during execution of the algorithm. After step 1 has been applied to all border points, those that were flagged are deleted (changed to 0). Then step 2 is applied to the resulting data in exactly the same manner as step 1.

Thus one iteration of the thinning algorithm consists of (1) applying step 1 to flag border points for deletion; (2) deleting the flagged points; (3) applying step 2 to flag the remaining border points for deletion; and (4) deleting the flagged points. This basic procedure is applied iteratively until no further points are deleted, at which time the algorithm terminates, yielding the skeleton of the region.

Condition (a) is violated when contour point p_1 only has one or seven 8-neighbors valued 1. Having only one such neighbor implies that p_1 is the end point of a skeleton stroke and obviously should not be deleted. Deleting p_1 if it had seven such neighbors would cause erosion into the region. Condition (b) is violated when it is applied to points on a stroke 1 pixel thick. Hence this condition prevents disconnection of segments of a skeleton during the thinning operation. Conditions (c) and (d) are satisfied simultaneously by the minimum set of values: $(p_4 = 0\ or\ p_6 = 0)$ or $(p_2 = 0\ and\ p_8 = 0)$. Thus with reference to the neighborhood arrangement in Fig. 11.8, a point that satisfies these conditions, as well as conditions (a) and (b), is an east or south boundary point or a northwest corner point in the boundary. In either case, p_1 is not part of the skeleton and should be removed. Similarly, conditions (c') and (d') are satisfied simultaneously by the following minimum set of values: $(p_2 = 0\ or\ p_8 = 0)$ or $(p_4 = 0$ $and\ p_6 = 0)$. These correspond to north or west boundary points, or a southeast corner point. Note that northeast corner points have $p_2 = 0$ and $p_4 = 0$, and thus satisfy conditions (c) and (d), as well as (c') and (d'). The same is true for southwest corner points, which have $p_6 = 0$ and $p_8 = 0$.

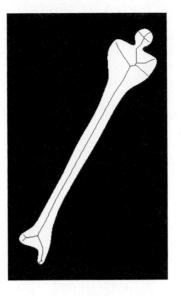

FIGURE 11.10
Human leg bone
and skeleton of
the region shown
superimposed.

■ Figure 11.10 shows a segmented image of a human leg bone and, superimposed, the skeleton of the region computed using the algorithm just discussed. For the most part, the skeleton looks intuitively correct. There is a double branch on the right side of the "shoulder" of the bone that at first glance one would expect to be a single branch, as on the corresponding left side. Note, however, that the right shoulder is somewhat broader (in the long direction) than the left shoulder. That is what caused the branch to be created by the algorithm. This type of unpredictable behavior is not unusual in skeletonizing algorithms. ■

EXAMPLE 11.1:
The skeleton of a
region.

11.2 Boundary Descriptors

In this section we consider several approaches to describing the boundary of a region, and in Section 11.3 we focus on regional descriptors. Parts of Sections 11.4 and 11.5 are applicable to both boundaries and regions.

11.2.1 Some Simple Descriptors

The *length* of a boundary is one of its simplest descriptors. The number of pixels along a boundary gives a rough approximation of its length. For a chain-coded curve with unit spacing in both directions, the number of vertical and horizontal components plus $\sqrt{2}$ times the number of diagonal components gives its exact length.

The *diameter* of a boundary B is defined as

$$\text{Diam}(B) = \max_{i,j}\left[D(p_i, p_j)\right] \qquad (11.2\text{-}1)$$

where D is a distance measure (see Section 2.5.3) and p_i and p_j are points on the boundary. The value of the diameter and the orientation of a line segment connecting the two extreme points that comprise the diameter (this line is called the *major axis* of the boundary) are useful descriptors of a boundary. The *minor axis* of a boundary is defined as the line perpendicular to the major axis, and of such length that a box passing through the outer four points of intersection of the boundary with the two axes completely encloses the boundary.[†] The box just described is called the *basic rectangle*, and the ratio of the major to the minor axis is called the *eccentricity* of the boundary. This also is a useful descriptor.

Curvature is defined as the rate of change of slope. In general, obtaining reliable measures of curvature at a point in a digital boundary is difficult because these boundaries tend to be locally "ragged." However, using the difference between the slopes of adjacent boundary segments (which have been represented as straight lines) as a descriptor of curvature at the point of intersection of the segments sometimes proves useful. For example, the vertices of boundaries such as those shown in Figs. 11.3(b) and 11.4(d) lend themselves well to curvature descriptions. As the boundary is traversed in the clockwise direction, a vertex point *p* is said to be part of a *convex* segment if the change in slope at *p* is nonnegative; otherwise, *p* is said to belong to a segment that is *concave*. The description of curvature at a point can be refined further by using ranges in the change of slope. For instance, *p* could be part of a nearly straight segment if the change is less than $10°$ or a *corner* point if the change exceeds $90°$. Note, however, that these descriptors must be used with care because their interpretation depends on the length of the individual segments relative to the overall length of the boundary.

11.2.2 Shape Numbers

As explained in Section 11.1.1, the first difference of a chain-coded boundary depends on the starting point. The *shape number* of such a boundary, based on the 4-directional code of Fig. 11.1(a), is defined as the first difference of smallest magnitude. The *order n* of a shape number is defined as the number of digits in its representation. Moreover, *n* is even for a closed boundary, and its value limits the number of possible different shapes. Figure 11.11 shows all the shapes of order 4, 6, and 8, along with their chain-code representations, first differences, and corresponding shape numbers. Note that the first difference is computed by treating the chain code as a circular sequence, as discussed in Section 11.1.1. Although the first difference of a chain code is independent of rotation, in general the coded boundary depends on the orientation of the grid. One way to normalize the grid orientation is by aligning the chain-code grid with the sides of the basic rectangle defined in the previous section.

In practice, for a desired shape order, we find the rectangle of order *n* whose eccentricity (defined in the previous section) best approximates that of the basic rectangle and use this new rectangle to establish the grid size. For example, if $n = 12$, all the rectangles of order 12 (that is, those whose perimeter length is 12) are 2×4, 3×3, and 1×5. If the eccentricity of the 2×4 rectangle best matches the eccentricity of the basic rectangle for a given boundary, we establish a 2×4 grid centered on the basic rectangle and use the procedure outlined

---

[†]Do not confuse this definition of major and minor axes with the eigen axes, which are defined in - Section 11.4.

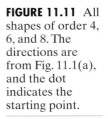

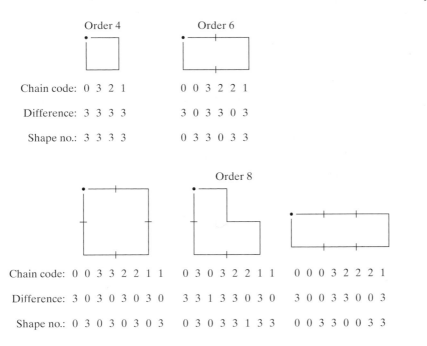

Order 4

Chain code: 0 3 2 1

Difference: 3 3 3 3

Shape no.: 3 3 3 3

Order 6

0 0 3 2 2 1

3 0 3 3 0 3

0 3 3 0 3 3

Order 8

Chain code: 0 0 3 3 2 2 1 1   0 3 0 3 2 2 1 1   0 0 0 3 2 2 2 1

Difference: 3 0 3 0 3 0 3 0   3 3 1 3 3 0 3 0   3 0 0 3 3 0 0 3

Shape no.: 0 3 0 3 0 3 0 3   0 3 0 3 3 1 3 3   0 0 3 3 0 0 3 3

in Section 11.1.1 to obtain the chain code. The shape number follows from the first difference of this code. Although the order of the resulting shape number usually equals n because of the way the grid spacing was selected, boundaries with depressions comparable to this spacing sometimes yield shape numbers of order greater than n. In this case, we specify a rectangle of order lower than n and repeat the procedure until the resulting shape number is of order n.

■ Suppose that $n = 18$ is specified for the boundary shown in Fig. 11.12(a). To obtain a shape number of this order requires following the steps just discussed. The first step is to find the basic rectangle, as shown in Fig. 11.12(b). The closest rectangle of order 18 is a 3×6 rectangle, requiring subdivision of the basic rectangle as shown in Fig. 11.12(c), where the chain-code directions are aligned with the resulting grid. The final step is to obtain the chain code and use its first difference to compute the shape number, as shown in Fig. 11.12(d).   ■

EXAMPLE 11.2: Computing shape numbers.

11.2.3 Fourier Descriptors

Figure 11.13 shows a K-point digital boundary in the xy-plane. Starting at an arbitrary point (x_0, y_0), coordinate pairs $(x_0, y_0), (x_1, y_1), (x_2, y_2), \ldots, (x_{K-1}, y_{K-1})$ are encountered in traversing the boundary, say, in the counterclockwise direction. These coordinates can be expressed in the form $x(k) = x_k$ and $y(k) = y_k$. With this notation, the boundary itself can be represented as the sequence of coordinates $s(k) = [x(k), y(k)]$, for $k = 0, 1, 2, \ldots, K - 1$. Moreover, each coordinate pair can be treated as a complex number so that

$$s(k) = x(k) + jy(k) \tag{11.2-2}$$

a b
c d

FIGURE 11.12
Steps in the
generation of a
shape number.

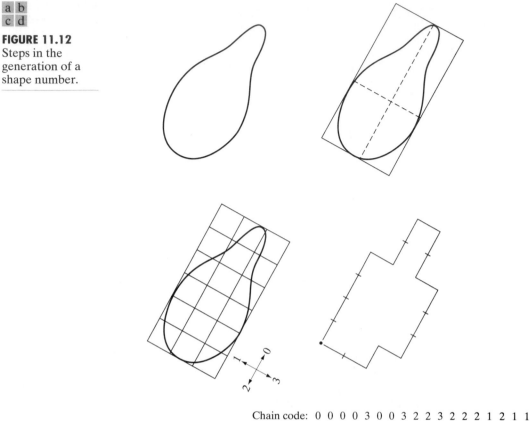

Chain code: 0 0 0 0 3 0 0 3 2 2 3 2 2 2 1 2 1 1

Difference: 3 0 0 0 3 1 0 3 3 0 1 3 0 0 3 1 3 0

Shape no.: 0 0 0 3 1 0 3 3 0 1 3 0 0 3 1 3 0 3

for $k = 0, 1, 2, \ldots, K - 1$. That is, the x-axis is treated as the real axis and the y-axis as the imaginary axis of a sequence of complex numbers. Although the interpretation of the sequence was recast, the nature of the boundary itself was not changed. Of course, this representation has one great advantage: It reduces a 2-D to a 1-D problem.

From Section 4.2.1, the discrete Fourier transform (DFT) of $s(k)$ is

$$a(u) = \frac{1}{K} \sum_{k=0}^{K-1} s(k)e^{-j2\pi uk/K} \qquad (11.2\text{-}3)$$

for $u = 0, 1, 2, \ldots, K - 1$. The complex coefficients $a(u)$ are called the *Fourier descriptors* of the boundary. The inverse Fourier transform of these coefficients restores $s(k)$. That is,

$$s(k) = \sum_{u=0}^{K-1} a(u)e^{j2\pi uk/K} \qquad (11.2\text{-}4)$$

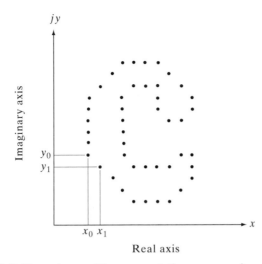

FIGURE 11.13 A digital boundary and its representation as a complex sequence. The points (x_0, y_0) and (x_1, y_1) shown are (arbitrarily) the first two points in the sequence.

for $k = 0, 1, 2, \ldots, K - 1$. Suppose, however, that instead of all the Fourier coefficients, only the first P coefficients are used. This is equivalent to setting $a(u) = 0$ for $u > P - 1$ in Eq. (11.2-4). The result is the following *approximation* to $s(k)$:

$$\hat{s}(k) = \sum_{u=0}^{P-1} a(u)e^{j2\pi uk/K} \qquad (11.2\text{-}5)$$

tor $k = 0, 1, 2, \ldots, K - 1$. Although only P terms are used to obtain each component of $\hat{s}(k)$, k still ranges from 0 to $K - 1$. That is, the *same* number of points exists in the approximate boundary, but not as many terms are used in the reconstruction of each point. Recall from discussions of the Fourier transform in Chapter 4 that high-frequency components account for fine detail, and low-frequency components determine global shape. Thus the smaller P becomes, the more detail that is lost on the boundary. The following example demonstrates this clearly.

EXAMPLE 11.3: Illustration of Fourier descriptors.

■ Figure 11.14 shows a square boundary consisting of $K = 64$ points and the results of using Eq. (11.2-5) to reconstruct this boundary for various values of P. Note that the value of P has to be about 8 before the reconstructed boundary looks more like a square than a circle. Next, note that little in the way of corner definition occurs until P is about 56, at which time the corner points begin to "break out" of the sequence. Finally, note that, when $P = 61$, the curves begin to straighten, which leads to an almost exact replica of the original one additional coefficient later. Thus, a few low-order coefficients are able to capture gross shape, but many more high-order terms are required to define accurately sharp features such as corners and straight lines. This result is not unexpected in view of the role played by low- and high-frequency components in defining the shape of a region. ■

FIGURE 11.14
Examples of reconstruction from Fourier descriptors. P is the number of Fourier coefficients used in the reconstruction of the boundary.

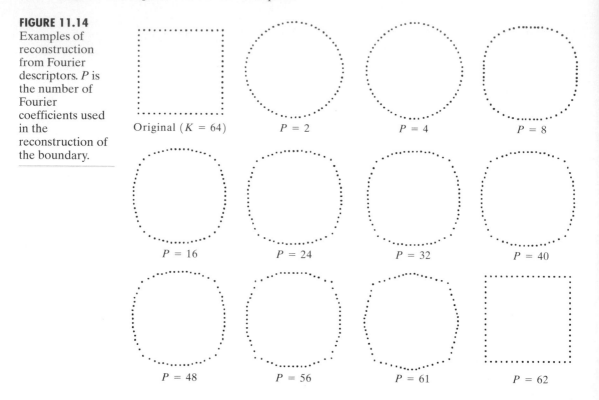

Original ($K = 64$) $P = 2$ $P = 4$ $P = 8$

$P = 16$ $P = 24$ $P = 32$ $P = 40$

$P = 48$ $P = 56$ $P = 61$ $P = 62$

As demonstrated in the preceding example, a few Fourier descriptors can be used to capture the gross essence of a boundary. This property is valuable, because these coefficients carry shape information. Thus they can be used as the basis for differentiating between distinct boundary shapes, as we discuss in some detail in Chapter 12.

We have stated several times that descriptors should be as insensitive as possible to translation, rotation, and scale changes. In cases where results depend on the order in which points are processed, an additional constraint is that descriptors should be insensitive to starting point. Fourier descriptors are not directly insensitive to these geometrical changes, but the changes in these parameters can be related to simple transformations on the descriptors. For example, consider rotation, and recall from elementary mathematical analysis that rotation of a point by an angle θ about the origin of the complex plane is accomplished by multiplying the point by $e^{j\theta}$. Doing so to every point of $s(k)$ rotates the entire sequence about the origin. The rotated sequence is $s(k)e^{j\theta}$, whose Fourier descriptors are

$$a_r(u) = \frac{1}{K} \sum_{k=0}^{K-1} s(k)e^{j\theta}e^{-j2\pi uk/K}$$
$$= a(u)e^{j\theta} \qquad (11.2\text{-}6)$$

for $u = 0, 1, 2, \ldots, K - 1$. Thus rotation simply affects all coefficients equally by a multiplicative *constant* term $e^{j\theta}$.

TABLE 11.1
Some basic
properties of
Fourier
descriptors.

Transformation	Boundary	Fourier Descriptor
Identity	$s(k)$	$a(u)$
Rotation	$s_r(k) = s(k)e^{j\theta}$	$a_r(u) = a(u)e^{j\theta}$
Translation	$s_t(k) = s(k) + \Delta_{xy}$	$a_t(u) = a(u) + \Delta_{xy}\delta(u)$
Scaling	$s_s(k) = \alpha s(k)$	$a_s(u) = \alpha a(u)$
Starting point	$s_p(k) = s(k - k_0)$	$a_p(u) = a(u)e^{-j2\pi k_0 u/K}$

Table 11.1 summarizes the Fourier descriptors for a boundary sequence $s(k)$ that undergoes rotation, translation, scaling, and changes in starting point. The symbol Δ_{xy} is defined as $\Delta_{xy} = \Delta x + j\Delta y$, so the notation $s_t(k) = s(k) + \Delta_{xy}$ indicates redefining (translating) the sequence as

$$s_t(k) = [x(k) + \Delta x] + j[y(k) + \Delta y]. \tag{11.2-7}$$

In other words, translation consists of adding a constant displacement to all coordinates in the boundary. Note that translation has no effect on the descriptors, except for $u = 0$, which has the impulse function $\delta(u)$.[†] Finally, the expression $s_p(k) = s(k - k_0)$ means redefining the sequence as

$$s_p = x(k - k_0) + jy(k - k_0), \tag{11.2-8}$$

which merely changes the starting point of the sequence to $k = k_0$ from $k = 0$. The last entry in Table 11.1 shows that a change in starting point affects all descriptors in a different (but known) way, in the sense that the term multiplying $a(u)$ depends on u.

11.2.4 Statistical Moments

The shape of boundary segments (and of signature waveforms) can be described quantitatively by using simple statistical moments, such as the mean, variance, and higher-order moments. To see how this can be accomplished, consider Fig. 11.15(a), which shows the segment of a boundary, and Fig. 11.15(b), which shows the segment represented as a 1-D function $g(r)$ of an arbitrary variable r. This function is obtained by connecting the two end points of the segment and rotating the line segment until it is horizontal. The coordinates of the points are rotated by the same angle.

Let us treat the amplitude of g as a discrete random variable v and form an amplitude histogram $p(v_i), i = 0, 1, 2, \ldots, A - 1$, where A is the number of discrete amplitude increments in which we divide the amplitude scale. Then, keeping in mind that $p(v_i)$ is an estimate of the probability of value v_i occurring, it follows from Eq. (3.3-18) that the nth moment of v about its mean is

$$\mu_n(v) = \sum_{i=0}^{A-1} (v_i - m)^n p(v_i) \tag{11.2-9}$$

See inside front cover
Consult the book web site for a brief review of probability theory.

[†] Recall from Chapter 4 that the Fourier transform of a constant is an impulse located at the origin. Recall also that the impulse function is zero everywhere else.

a b

FIGURE 11.15
(a) Boundary
segment.
(b) Representation
as a 1-D function.

where

$$m = \sum_{i=0}^{A-1} v_i p(v_i). \tag{11.2-10}$$

The quantity m is recognized as the mean or average value of v and μ_2 as its variance. Generally, only the first few moments are required to differentiate between signatures of clearly distinct shapes.

An alternative approach is to normalize $g(r)$ to unit area and treat it as a histogram. In other words, $g(r_i)$ is now treated as the probability of value r_i occurring. In this case, r is treated as the random variable and the moments are

$$\mu_n(r) = \sum_{i=0}^{K-1} (r_i - m)^n g(r_i) \tag{11.2-11}$$

where

$$m = \sum_{i=0}^{K-1} r_i g(r_i). \tag{11.2-12}$$

In this notation, K is the number of points on the boundary, and $\mu_n(r)$ is directly related to the shape of $g(r)$. For example, the second moment $\mu_2(r)$ measures the spread of the curve about the mean value of r and the third moment $\mu_3(r)$ measures its symmetry with reference to the mean.

Basically, what we have accomplished is to reduce the description task to that of describing 1-D functions. Although moments are by far the most popular method, they are not the only descriptors that could be used for this purpose. For instance, another method involves computing the 1-D discrete Fourier transform, obtaining its spectrum, and using the first q components of the spectrum to describe $g(r)$. The advantage of moments over other techniques is that implementation of moments is straightforward and they also carry a "physical" interpretation of boundary shape. The insensitivity of this approach to rotation is clear from Fig. 11.15. Size normalization, if desired, can be achieved by scaling the range of values of g and r.

11.3 Regional Descriptors

In this section we consider various approaches for describing image regions. Keep in mind that it is common practice to use of both boundary and regional descriptors combined.

11.3.1 Some Simple Descriptors

The *area* of a region is defined as the number of pixels in the region. The *perimeter* of a region is the length of its boundary. Although area and perimeter are sometimes used as descriptors, they apply primarily to situations in which the size of the regions of interest is invariant. A more frequent use of these two descriptors is in measuring *compactness* of a region, defined as (perimeter)^2/area. Compactness is a dimensionless quantity (and thus is insensitive to uniform scale changes) and is minimal for a disk-shaped region. With the exception of errors introduced by rotation of a digital region, compactness also is insensitive to orientation.

Other simple measures used as region descriptors include the mean and median of the gray levels, the minimum and maximum gray-level values, and the number of pixels with values above and below the mean.

■ Even a simple region descriptor such as normalized area can be quite useful in extracting information from images. For instance, Fig. 11.16 shows a satellite infrared image of the Americas. As discussed in more detail in Section 1.3.4, images such as these provide a global inventory of human settlements. The sensor used to collect these images has the capability to detect visible and near-infrared emissions, such as lights, fires, and flares. The table alongside the images shows (by region from top to bottom) the ratio of the area occupied by white (the lights) to the total light area in all four regions. A simple measurement like this can give, for example, a relative estimate by region of electrical energy consumed. The data can be refined by normalizing it with respect to land mass per region, with respect to population numbers, and so on. ■

EXAMPLE 11.4:
Using area computations to extract information from images.

11.3.2 Topological Descriptors

Topological properties are useful for global descriptions of regions in the image plane. Simply defined, *topology* is the study of properties of a figure that are unaffected by any deformation, as long as there is no tearing or joining of the figure (sometimes these are called *rubber-sheet* distortions). For example, Fig. 11.17 shows a region with two holes. Thus if a topological descriptor is defined by the number of holes in the region, this property obviously will not be affected by a stretching or rotation transformation. In general, however, the number of holes will change if the region is torn or folded. Note that, as stretching affects distance, topological properties do not depend on the notion of distance or any properties implicitly based on the concept of a distance measure.

Another topological property useful for region description is the number of connected components. A *connected component* of a region was defined in Section 2.5.2. Figure 11.18 shows a region with three connected components. (See Section 9.5.3 regarding an algorithm for computing connected components.)

The number of holes H and connected components C in a figure can be used to define the *Euler number* E:

$$E = C - H \qquad (11.3-1)$$

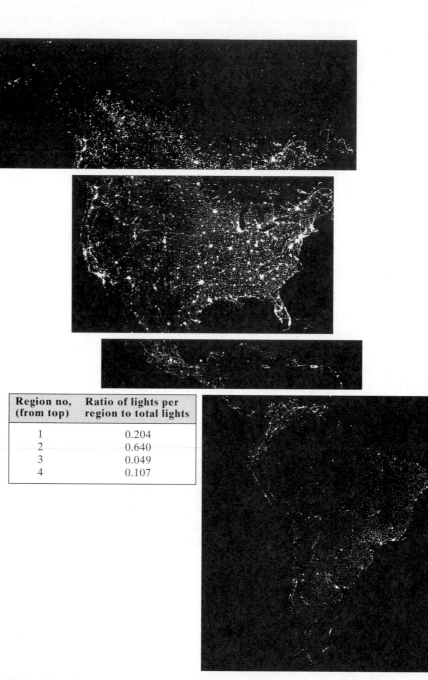

Region no, (from top)	Ratio of lights per region to total lights
1	0.204
2	0.640
3	0.049
4	0.107

FIGURE 11.16 Infrared images of the Americas at night. (Courtesy of NOAA.)

FIGURE 11.17 A region with two holes.

The Euler number is also a topological property. The regions shown in Fig. 11.19, for example, have Euler numbers equal to 0 and −1, respectively, because the "A" has one connected component and one hole and the "B" one connected component but two holes.

Regions represented by straight-line segments (referred to as *polygonal networks*) have a particularly simple interpretation in terms of the Euler number. Figure 11.20 shows a polygonal network. Classifying interior regions of such a network into faces and holes often is important. Denoting the number of vertices by V, the number of edges by Q, and the number of faces by F gives the following relationship, called the *Euler formula*:

$$V - Q + F = C - H \tag{11.3-2}$$

which, in view of Eq. (11.3-1), is equal to the Euler number:

$$V - Q + F = C - H$$
$$= E. \tag{11.3-3}$$

The network shown in Fig. 11.20 has 7 vertices, 11 edges, 2 faces, 1 connected region, and 3 holes; thus the Euler number is −2:

$$7 - 11 + 2 = 1 - 3 = -2.$$

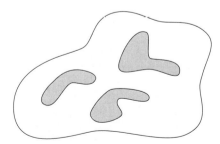

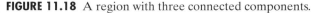

FIGURE 11.18 A region with three connected components.

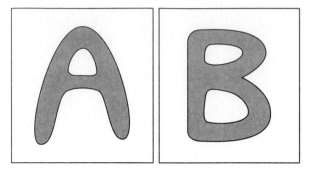

FIGURE 11.19 Regions with Euler number equal to 0 and −1, respectively.

Topological descriptors provide an additional feature that is often useful in characterizing regions in a scene.

EXAMPLE 11.5:
Use of connected components for extracting the largest features in a segmented image.

■ Figure 11.21(a) shows a 512 × 512, 8-bit image of Washington, D.C. taken by a NASA LANDSAT satellite. This particular image is in the near infrared band (see Fig. 1.10 for details). Suppose that we want to segment the river using only this image (as opposed to using several multispectral images, which would simplify the task). Since the river is a rather dark, uniform region of the image, thresholding is an obvious thing to try. The result of thresholding the image with the highest possible threshold value before the river became a disconnected region is shown in Fig. 11.21(b). The threshold was selected manually to illustrate the point that it would be impossible in this case to segment the river by itself without other regions of the image also appearing in the thresholded result. The objective of this example is to illustrate how connected components can be used to "finish" the segmentation.

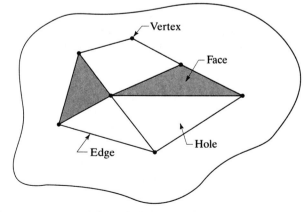

FIGURE 11.20 A region containing a polygonal network.

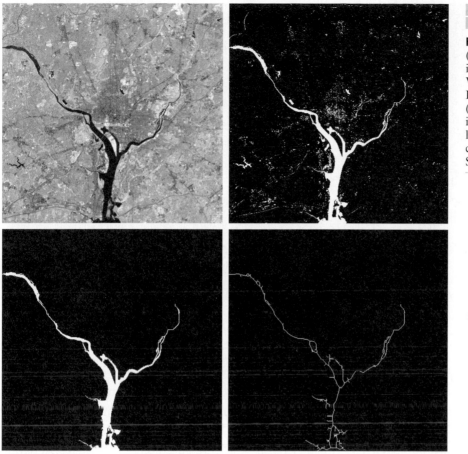

FIGURE 11.21
(a) Infrared
image of the
Washington,
D.C. area.
(b) Thresholded
image. (c) The
largest connected
component of (b).
Skeleton of (c).

The image in Fig. 11.21(b) has 1591 connected components (obtained using 8-connectivity) and its Euler number is 1552, from which we deduce that the number of holes is 39. Figure 11.21(c) shows the connected component with the largest number of elements (8479). This is the desired result, which we already know cannot be segmented by itself from the image. Note how clean this result is. If we wanted to perform measurements, like the length of each branch of the river, we could use the skeleton of the connected component [Fig. 11.21(d)] to do so. In other words, the length of each branch in the skeleton would be a reasonably close approximation to the length of the river branch it represents. ■

11.3.3 Texture

An important approach to region description is to quantify its *texture* content. Although no formal definition of texture exists, intuitively this descriptor provides measures of properties such as smoothness, coarseness, and regularity (Fig. 11.22 shows some examples). The three principal approaches used in image processing to describe the texture of a region are statistical, structural, and spectral. Statistical approaches yield characterizations of textures as smooth, coarse,

a b c

FIGURE 11.22 The white squares mark, from left to right, smooth, coarse, and regular textures. These are optical microscope images of a superconductor, human cholesterol, and a microprocessor. (Courtesy of Dr. Michael W. Davidson, Florida State University.)

grainy, and so on. Structural techniques deal with the arrangement of image primitives, such as the description of texture based on regularly spaced parallel lines. Spectral techniques are based on properties of the Fourier spectrum and are used primarily to detect global periodicity in an image by identifying high-energy, narrow peaks in the spectrum.

Statistical approaches

One of the simplest approaches for describing texture is to use statistical moments of the gray-level histogram of an image or region. Let z be a random variable denoting gray levels and let $p(z_i), i = 0, 1, 2, \ldots, L - 1$, be the corresponding histogram, where L is the number of distinct gray levels. From Eq. (3.3-18), the nth moment of z about the mean is

$$\mu_n(z) = \sum_{i=0}^{L-1} (z_i - m)^n p(z_i) \qquad (11.3\text{-}4)$$

where m is the mean value of z (the average gray level):

$$m = \sum_{i=0}^{L-1} z_i p(z_i). \qquad (11.3\text{-}5)$$

Note from Eq. (11.3-4) that $\mu_0 = 1$ and $\mu_1 = 0$. The second moment [the *variance* $\sigma^2(z) = \mu_2(z)$] is of particular importance in texture description. It is a measure of gray-level contrast that can be used to establish descriptors of relative smoothness. For example, the measure

$$R = 1 - \frac{1}{1 + \sigma^2(z)} \qquad (11.3\text{-}6)$$

is 0 for areas of constant intensity (the variance is zero there) and approaches 1 for large values of $\sigma^2(z)$. Because variance values tend to be large for gray-scale images with values, for example, in the range 0 to 255, it is a good idea to normalize the variance to the interval $[0, 1]$ for use in Eq. (11.3-6). This is done simply by dividing $\sigma^2(z)$ by $(L - 1)^2$ in Eq. (11.3-6). The standard deviation, $\sigma(z)$, also is used frequently as a measure of texture because values of the standard deviation tend to be more intuitive to many people.

The third moment,

$$\mu_3(z) = \sum_{i=0}^{L-1} (z_i - m)^3 p(z_i), \qquad (11.3\text{-}7)$$

is a measure of the skewness of the histogram while the fourth moment is a measure of its relative flatness. The fifth and higher moments are not so easily related to histogram shape, but they do provide further quantitative discrimination of texture content. Some useful additional texture measures based on histograms include a measure of "uniformity," given by

$$U = \sum_{i=0}^{L-1} p^2(z_i), \qquad (11.3\text{-}8)$$

and an *average entropy* measure, which the reader might recall from basic information theory, or from our discussion in Chapter 8, is defined as

$$e = -\sum_{i=0}^{L-1} p(z_i) \log_2 p(z_i). \qquad (11.3\text{-}9)$$

Because the p's have values in the range $[0, 1]$ and their sum equals 1, measure U is maximum for an image in which all gray levels are equal (maximally uniform), and decreases from there. Entropy is a measure of variability and is 0 for a constant image.

■ Table 11.2 summarizes the values of the preceding measures for the three types of textures highlighted in Fig. 11.22. The mean just tells us the average gray level of each region and is useful only as a rough idea of intensity, not really texture. The standard deviation is much more informative; the numbers clearly show that the first texture has significantly less variability in gray level (it is smoother) than the other two textures. The coarse texture shows up clearly in this measure. As expected, the same comments hold for R, because it measures essentially the same thing as the standard deviation. The third moment generally is useful for determining the degree of symmetry of histograms and whether they are skewed to the left (negative value) or the right (positive value).

EXAMPLE 11.6:
Texture measures based on histograms.

TABLE 11.2
Texture measures
for the subimages
shown in
Fig. 11.22.

Texture	Mean	Standard deviation	R (normalized)	Third moment	Uniformity	Entropy
Smooth	82.64	11.79	0.002	−0.105	0.026	5.434
Coarse	143.56	74.63	0.079	−0.151	0.005	7.783
Regular	99.72	33.73	0.017	0.750	0.013	6.674

This gives a rough idea of whether the gray levels are biased toward the dark or light side of the mean. In terms of texture, the information derived from the third moment is useful only when variations between measurements are large. Looking at the measure of uniformity, we again conclude that the first subimage is smoother (more uniform than the rest) and that the most random (lowest uniformity) corresponds to the coarse texture. This is not surprising. Finally, the entropy values are in the opposite order and thus lead us to the same conclusions as the uniformity measure did. The first subimage has the lowest variation in gray level and the coarse image the most. The regular texture is in between the two extremes with respect to both these measures.   ■

Measures of texture computed using only histograms suffer from the limitation that they carry no information regarding the relative position of pixels with respect to each other. One way to bring this type of information into the texture-analysis process is to consider not only the distribution of intensities, but also the positions of pixels with equal or nearly equal intensity values.

Let P be a position operator and let $\mathbf{A}$ be a $k \times k$ matrix whose element a_{ij} is the number of times that points with gray level z_i occur (in the position specified by P) relative to points with gray level z_j, with $1 \le i, j \le k$. For instance, consider an image with three gray levels, $z_1 = 0, z_2 = 1$, and $z_3 = 2$, as follows:

$$
\begin{array}{ccccc}
0 & 0 & 0 & 1 & 2 \\
1 & 1 & 0 & 1 & 1 \\
2 & 2 & 1 & 0 & 0 \\
1 & 1 & 0 & 2 & 0 \\
0 & 0 & 1 & 0 & 1
\end{array}
$$

Defining the position operator P as "one pixel to the right and one pixel below" yields the following 3×3 matrix $\mathbf{A}$:

$$
\mathbf{A} = \begin{bmatrix} 4 & 2 & 1 \\ 2 & 3 & 2 \\ 0 & 2 & 0 \end{bmatrix}
$$

where, for example, a_{11} (top left) is the number of times that a point with level $z_1 = 0$ appears one pixel location below and to the right of a pixel with the same gray level, and a_{13} (top right) is the number of times that a point with level $z_1 = 0$ appears one pixel location below and to the right of a point with gray level $z_3 = 2$. The size of $\mathbf{A}$ is determined by the number of distinct gray levels in the input image. Thus application of the concepts discussed in this section usually requires that intensities be requantized into a few gray-level bands in order to keep the size of $\mathbf{A}$ manageable.

Let n be the total number of point pairs in the image that satisfy P (in the preceding example $n = 16$, the sum of all values in matrix $\mathbf{A}$). If a matrix $\mathbf{C}$ is formed by dividing every element of $\mathbf{A}$ by n, then c_{ij} is an estimate of the joint probability that a pair of points satisfying P will have values (z_i, z_j). The matrix $\mathbf{C}$ is called the *gray-level co-occurrence matrix*. Because $\mathbf{C}$ depends on P, the presence of given texture patterns may be detected by choosing an appropriate position operator. For instance, the operator used in the preceding example is sensitive to bands of constant intensity running at $-45°$. (Note that the highest value in $\mathbf{A}$ was $a_{11} = 4$, partially due to a streak of points with intensity 0 and running at $-45°$.) More generally, the problem is to analyze a given $\mathbf{C}$ matrix in order to categorize the texture of the region over which $\mathbf{C}$ was computed. A set of descriptors useful for this purpose includes the following:

1. Maximum probability

$$\max_{i,j}(c_{ij})$$

2. Element difference moment of order k

$$\sum_i \sum_j (i - j)^k c_{ij}$$

3. Inverse element difference moment of order k

$$\sum_i \sum_j c_{ij}/(i - j)^k \qquad i \neq j$$

4. Uniformity

$$\sum_i \sum_j c_{ij}^2$$

5. Entropy

$$-\sum_i \sum_j c_{ij} \log_2 c_{ij}$$

The basic idea is to characterize the "content" of $\mathbf{C}$ via these descriptors. For example, the first property gives an indication of the strongest response to P. The second descriptor has a relatively low value when the high values of $\mathbf{C}$ are near the main diagonal, because the differences $(i - j)$ are smaller there. The third descriptor has the opposite effect. The fourth descriptor is highest when the c_{ij}s are all equal. As noted previously, the fifth descriptor is a measure of randomness, achieving its highest value when all elements of $\mathbf{C}$ are maximally random.

One approach for using these descriptors is to "teach" a system representative descriptor values for a set of different textures. The texture of an unknown region is then subsequently determined by how closely its descriptors match those stored in the system memory. We discuss matching in more detail in Chapter 12.

Structural approaches

As mentioned at the beginning of this section, a second major category of texture description is based on structural concepts. Suppose that we have a rule of the form $S \rightarrow aS$, which indicates that the symbol S may be rewritten as aS (for example, three applications of this rule would yield the string $aaaS$). If a represents a circle [Fig. 11.23(a)] and the meaning of "circles to the right" is assigned to a string of the form $aaa\ldots$, the rule $S \rightarrow aS$ allows generation of the texture pattern shown in Fig. 11.23(b).

Suppose next that we add some new rules to this scheme: $S \rightarrow bA$, $A \rightarrow cA$, $A \rightarrow c$, $A \rightarrow bS$, $S \rightarrow a$, where the presence of a b means "circle down" and the presence of a c means "circle to the left." We can now generate a string of the form $aaabccbaa$ that corresponds to a 3×3 matrix of circles. Larger texture patterns, such as the one shown in Fig. 11.23(c), can be generated easily in the same way. (Note, however, that these rules can also generate structures that are not rectangular.)

The basic idea in the foregoing discussion is that a simple "texture primitive" can be used to form more complex texture patterns by means of some rules that limit the number of possible arrangements of the primitive(s). These concepts lie at the heart of relational descriptions, a topic that we treat in more detail in Section 11.5.

Spectral approaches

As indicated in Section 5.4, the Fourier spectrum is ideally suited for describing the directionality of periodic or almost periodic 2-D patterns in an image. These global texture patterns, although easily distinguishable as concentrations of high-energy bursts in the spectrum, generally are quite difficult to detect with spatial methods because of the local nature of these techniques.

Here, we consider three features of the Fourier spectrum that are useful for texture description: (1) Prominent peaks in the spectrum give the principal

a
b
c

FIGURE 11.23
(a) Texture primitive.
(b) Pattern generated by the rule $S \rightarrow aS$.
(c) 2-D texture pattern generated by this and other rules.

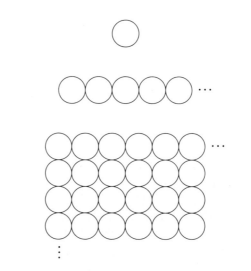

direction of the texture patterns. (2) The location of the peaks in the frequency plane gives the fundamental spatial period of the patterns. (3) Eliminating any periodic components via filtering leaves nonperiodic image elements, which can then be described by statistical techniques. Recall that the spectrum is symmetric about the origin, so only half of the frequency plane needs to be considered. Thus for the purpose of analysis, every periodic pattern is associated with only one peak in the spectrum, rather than two.

Detection and interpretation of the spectrum features just mentioned often are simplified by expressing the spectrum in polar coordinates to yield a function $S(r, \theta)$, where S is the spectrum function and r and θ are the variables in this coordinate system. For each direction θ, $S(r, \theta)$ may be considered a 1-D function $S_\theta(r)$. Similarly, for each frequency r, $S_r(\theta)$ is a 1-D function. Analyzing $S_\theta(r)$ for a fixed value of θ yields the behavior of the spectrum (such as the presence of peaks) along a radial direction from the origin, whereas analyzing $S_r(\theta)$ for a fixed value of r yields the behavior along a circle centered on the origin.

A more global description is obtained by integrating (summing for discrete variables) these functions:

$$S(r) = \sum_{\theta=0}^{\pi} S_\theta(r) \qquad (11.3\text{-}10)$$

and

$$S(\theta) = \sum_{r=1}^{R_0} S_r(\theta) \qquad (11.3\text{-}11)$$

where R_0 is the radius of a circle centered at the origin.

The results of Eqs. (11.3-10) and (11.3-11) constitute a pair of values $[S(r), S(\theta)]$ for *each* pair of coordinates (r, θ). By varying these coordinates, we can generate two 1-D functions, $S(r)$ and $S(\theta)$, that constitute a spectral-energy description of texture for an entire image or region under consideration. Furthermore, descriptors of these functions themselves can be computed in order to characterize their behavior quantitatively. Descriptors typically used for this purpose are the location of the highest value, the mean and variance of both the amplitude and axial variations, and the distance between the mean and the highest value of the function.

■ Figure 11.24 illustrates the use of Eqs. (11.3-10) and (11.3-11) for global texture description. Figure 11.24(a) shows an image with periodic texture, and Fig. 11.24(b) shows its spectrum. Figures 11.24(c) and (d) show plots of $S(r)$ and $S(\theta)$, respectively. The plot of $S(r)$ is a typical structure, having high energy content near the origin and progressively lower values for higher frequencies. The plot of $S(\theta)$ shows prominent peaks at intervals of 45°, which clearly correspond to the periodicity in the texture content of the image.

As an illustration of how a plot of $S(\theta)$ could be used to differentiate between two texture patterns, Fig. 11.24(e) shows another image whose texture pattern is predominantly in the horizontal and vertical directions. Figure 11.24(f) shows the plot of $S(\theta)$ for the spectrum of this image. As expected, this plot shows peaks at 90° intervals. Discriminating between the two texture patterns by analyzing their corresponding $S(\theta)$ waveforms would be straightforward.   ■

EXAMPLE 11.7:
Spectral texture.

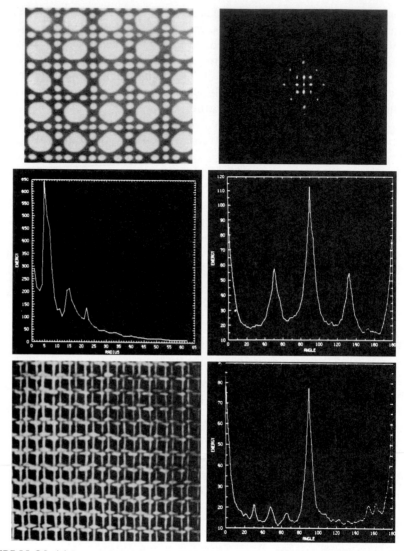

FIGURE 11.24 (a) Image showing periodic texture. (b) Spectrum. (c) Plot of $S(r)$. (d) Plot of $S(\theta)$. (e) Another image with a different type of periodic texture. (f) Plot of $S(\theta)$. (Courtesy of Dr. Dragana Brzakovic, University of Tennessee.)

11.3.4 Moments of Two-Dimensional Functions

For a 2-D continuous function $f(x, y)$, the moment of order $(p + q)$ is defined as

$$m_{pq} = \int_{-\infty}^{\infty} \int_{-\infty}^{\infty} x^p y^q f(x, y) \, dx \, dy \qquad (11.3\text{-}12)$$

for $p, q = 0, 1, 2, \ldots$ A uniqueness theorem (Papoulis [1991]) states that if $f(x, y)$ is piecewise continuous and has nonzero values only in a finite part of the xy-plane, moments of all orders exist, and the moment sequence (m_{pq}) is uniquely determined by $f(x, y)$. Conversely, (m_{pq}) uniquely determines $f(x, y)$.

The *central moments* are defined as

$$\mu_{pq} = \int_{-\infty}^{\infty} \int_{-\infty}^{\infty} (x - \bar{x})^p (y - \bar{y})^q f(x, y) \, dx \, dy \qquad (11.3\text{-}13)$$

where

$$\bar{x} = \frac{m_{10}}{m_{00}} \quad \text{and} \quad \bar{y} = \frac{m_{01}}{m_{00}}.$$

If $f(x, y)$ is a digital image, then Eq. (11.3-13) becomes

$$\mu_{pq} = \sum_x \sum_y (x - \bar{x})^p (y - \bar{y})^q f(x, y). \qquad (11.3\text{-}14)$$

The central moments of order up to 3 are

$$\mu_{00} = \sum_x \sum_y (x - \bar{x})^0 (y - \bar{y})^0 f(x, y)$$

$$= \sum_x \sum_y f(x, y)$$

$$= m_{00}$$

$$\mu_{10} = \sum_x \sum_y (x - \bar{x})^1 (y - \bar{y})^0 f(x, y)$$

$$= m_{10} - \frac{m_{10}}{m_{00}} (m_{00})$$

$$= 0$$

$$\mu_{01} = \sum_x \sum_y (x - \bar{x})^0 (y - y)^1 f(x, y)$$

$$= m_{01} - \frac{m_{01}}{m_{00}} (m_{00})$$

$$= 0$$

$$\mu_{11} = \sum_x \sum_y (x - \bar{x})^1 (y - \bar{y})^1 f(x, y)$$

$$= m_{11} - \frac{m_{10} m_{01}}{m_{00}}$$

$$= m_{11} - \bar{x} m_{01} = m_{11} - \bar{y} m_{10}$$

$$\mu_{20} = \sum_x \sum_y (x - \bar{x})^2 (y - \bar{y})^0 f(x, y)$$

$$= m_{20} - \frac{2m_{10}^2}{m_{00}} + \frac{m_{10}^2}{m_{00}}$$

$$= m_{20} - \frac{m_{10}^2}{m_{00}}$$

$$= m_{20} - \bar{x} m_{10}$$

$$\mu_{02} = \sum_x \sum_y (x - \bar{x})^0 (y - \bar{y})^2 f(x, y)$$

$$= m_{02} - \frac{m_{01}^2}{m_{00}}$$

$$= m_{02} - \bar{y} m_{01}$$

$$\mu_{21} = \sum_x \sum_y (x - \bar{x})^2 (y - \bar{y})^1 f(x, y)$$

$$= m_{21} - 2\bar{x} m_{11} - \bar{y} m_{20} + 2\bar{x}^2 m_{01}$$

$$\mu_{12} = \sum_x \sum_y (x - \bar{x})^1 (y - \bar{y})^2 f(x, y)$$

$$= m_{12} - 2\bar{y} m_{11} - \bar{x} m_{02} + 2\bar{y}^2 m_{10}$$

$$\mu_{30} = \sum_x \sum_y (x - \bar{x})^3 (y - \bar{y})^0 f(x, y)$$

$$= m_{30} - 3\bar{x} m_{20} + 2\bar{x}^2 m_{10}$$

$$\mu_{03} = \sum_x \sum_y (x - \bar{x})^0 (y - \bar{y})^3 f(x, y)$$

$$= m_{03} - 3\bar{y} m_{02} + 2\bar{y}^2 m_{01}.$$

In summary,

$$\mu_{00} = m_{00} \qquad\qquad \mu_{02} = m_{02} - \bar{y} m_{01}$$
$$\mu_{10} = 0 \qquad\qquad \mu_{30} = m_{30} - 3\bar{x} m_{20} + 2\bar{x}^2 m_{10}$$
$$\mu_{01} = 0 \qquad\qquad \mu_{03} = m_{03} - 3\bar{y} m_{02} + 2\bar{y}^2 m_{01}$$
$$\mu_{11} = m_{11} - \bar{y} m_{10} \qquad \mu_{21} = m_{21} - 2\bar{x} m_{11} - \bar{y} m_{20} + 2\bar{x}^2 m_{01}$$
$$\mu_{20} = m_{20} - \bar{x} m_{10} \qquad \mu_{12} = m_{12} - 2\bar{y} m_{11} - \bar{x} m_{02} + 2\bar{y}^2 m_{10}.$$

The *normalized central moments*, denoted η_{pq}, are defined as

$$\eta_{pq} = \frac{\mu_{pq}}{\mu_{00}^\gamma} \tag{11.3-15}$$

where

$$\gamma = \frac{p + q}{2} + 1 \tag{11.3-16}$$

for $p + q = 2, 3, \ldots$.

A set of seven *invariant moments* can be derived from the second and third moments.[†]

$$\phi_1 = \eta_{20} + \eta_{02} \tag{11.3-17}$$
$$\phi_2 = (\eta_{20} - \eta_{02})^2 + 4\eta_{11}^2 \tag{11.3-18}$$

[†] Derivation of these results involves concepts that are beyond the scope of this discussion. The book by Bell [1965] and the paper by Hu [1962] contain detailed discussions of these concepts. Moment invariants can be generalized to n dimensions (Mamistvalov [1998]).

$$\phi_3 = (\eta_{30} - 3\eta_{12})^2 + (3\eta_{21} - \eta_{03})^2 \qquad (11.3\text{-}19)$$

$$\phi_4 = (\eta_{30} + \eta_{12})^2 + (\eta_{21} + \eta_{03})^2 \qquad (11.3\text{-}20)$$

$$\phi_5 = (\eta_{30} - 3\eta_{12})(\eta_{30} + \eta_{12})[(\eta_{30} + \eta_{12})^2$$
$$- 3(\eta_{21} + \eta_{03})^2] + (3\eta_{21} - \eta_{03})(\eta_{21} + \eta_{03}) \qquad (11.3\text{-}21)$$
$$[3(\eta_{30} + \eta_{12})^2 - (\eta_{21} + \eta_{03})^2]$$

$$\phi_6 = (\eta_{20} - \eta_{02})[(\eta_{30} + \eta_{12})^2 - (\eta_{21} + \eta_{03})^2] \qquad (11.3\text{-}22)$$
$$+ 4\eta_{11}(\eta_{30} + \eta_{12})(\eta_{21} + \eta_{03})$$

$$\phi_7 = (3\eta_{21} - \eta_{03})(\eta_{30} + \eta_{12})[(\eta_{30} + \eta_{12})^2$$
$$- 3(\eta_{21} + \eta_{03})^2] + (3\eta_{12} - \eta_{30})(\eta_{21} + \eta_{03}) \qquad (11.3\text{-}23)$$
$$[3(\eta_{30} + \eta_{12})^2 - (\eta_{21} + \eta_{03})^2].$$

This set of moments is invariant to translation, rotation, and scale change.

■ The image shown in Fig. 11.25(a) was reduced to half size in Fig. 11.25(b), mirror-imaged in Fig. 11.25(c), and rotated by 2° and 45°, as shown in Figs. 11.25(d) and (e). The seven moment invariants given in Eqs. (11.3-17) through (11.3-23) were then computed for each of these images, and the logarithm of the results were taken to reduce the dynamic range. As Table 11.3 shows, the results for Figs. 11.25(b) through (e) are in reasonable agreement with the invariants computed for the original image. The major cause of error can be attributed to the digital nature of the data, especially for the rotated images.   ■

EXAMPLE 11.8:
Two-dimensional moment invariants.

11.4 Use of Principal Components for Description

The material discussed in this section is applicable to boundaries and regions. In addition, it can be used as the basis for describing sets of images that are registered spatially, but whose corresponding pixel values are different (e.g., the three component images of a color RGB image). Suppose that we are given the three component images of such a color image. The three images can be treated as a unit by expressing each group of three corresponding pixels as a vector. For example, let x_1, x_2, and x_3, respectively, be the values of the first pixel in each of the three images. These three elements can be expressed in the form of a 3-D *column* vector, $\mathbf{x}$, where

See inside front cover
Consult the book web site for a brief review of vectors and matrices.

$$\mathbf{x} = \begin{bmatrix} x_1 \\ x_2 \\ x_3 \end{bmatrix}.$$

This one vector represents *one* common pixel in all three images. If the images are of size $M \times N$, there will be a total of $K = MN$ three-dimensional vectors after all the pixels are represented in this manner. If we have n registered images, the vectors will be n-dimensional:

$$\mathbf{x} = \begin{bmatrix} x_1 \\ x_2 \\ \vdots \\ x_n \end{bmatrix}. \qquad (11.4\text{-}1)$$

FIGURE 11.25
Images used to
demonstrate
properties of
moment
invariants (see
Table 11.3).

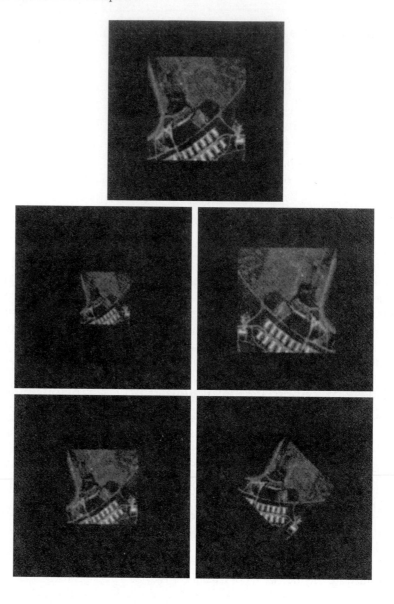

Throughout this section, the assumption is that all vectors are column vectors (i.e., matrices of order $n \times 1$). We can write them on a line of text simply by expressing them as $\mathbf{x} = (x_1, x_2, \ldots, x_n)^T$, where "$T$" indicates transpose.

We can treat the vectors as random quantities, just like we did when constructing a gray-level histogram. The only difference is that, instead of talking about quantities like the mean and variance of the random variables, we now talk about *mean vectors* and *covariance matrices* of the random vectors. The mean vector of the population is defined as

$$\mathbf{m_x} = E\{\mathbf{x}\} \tag{11.4-2}$$

Invariant (Log)	Original	Half Size	Mirrored	Rotated 2°	Rotated 45°
ϕ_1	6.249	6.226	6.919	6.253	6.318
ϕ_2	17.180	16.954	19.955	17.270	16.803
ϕ_3	22.655	23.531	26.689	22.836	19.724
ϕ_4	22.919	24.236	26.901	23.130	20.437
ϕ_5	45.749	48.349	53.724	46.136	40.525
ϕ_6	31.830	32.916	37.134	32.068	29.315
ϕ_7	45.589	48.343	53.590	46.017	40.470

TABLE 11.3
Moment invariants for the images in Figs. 11.25(a)-(e).

where $E\{\cdot\}$ is the expected value of the argument, and the subscript denotes that **m** is associated with the population of **x** vectors. Recall that the expected value of a vector or matrix is obtained by taking the expected value of each element.

The *covariance matrix* of the vector population is defined as

$$\mathbf{C_x} = E\{(\mathbf{x} - \mathbf{m_x})(\mathbf{x} - \mathbf{m_x})^T\}. \tag{11.4-3}$$

Because **x** is n dimensional, $\mathbf{C_x}$ and $(\mathbf{x} - \mathbf{m_x})(\mathbf{x} - \mathbf{m_x})^T$ are matrices of order $n \times n$. Element c_{ii} of $\mathbf{C_x}$ is the variance of x_i, the ith component of the **x** vectors in the population, and element c_{ij} of $\mathbf{C_x}$ is the covariance[†] between elements x_i and x_j of these vectors. The matrix $\mathbf{C_x}$ is real and symmetric. If elements x_i and x_j are uncorrelated, their covariance is zero and, therefore, $c_{ij} = c_{ji} = 0$. Note that all these definitions reduce to their familiar one-dimensional counterparts when $n = 1$.

For K vector samples from a random population, the mean vector can be approximated from the samples by using the familiar averaging expression

$$\mathbf{m_x} = \frac{1}{K} \sum_{k=1}^{K} \mathbf{x}_k. \tag{11.4-4}$$

Similarly, by expanding the product $(\mathbf{x} - \mathbf{m_x})(\mathbf{x} - \mathbf{m_x})^T$ and using Eqs. (11.4-2) and (11.4-4) we would find that the covariance matrix can be approximated from the samples as follows:

$$\mathbf{C_x} = \frac{1}{K} \sum_{k=1}^{K} \mathbf{x}_k \mathbf{x}_k^T - \mathbf{m_x} \mathbf{m_x}^T. \tag{11.4-5}$$

■ To illustrate the mechanics of Eqs. (11.4-4) and (11.4-5), consider the four vectors $\mathbf{x}_1 = (0, 0, 0)^T, \mathbf{x}_2 = (1, 0, 0)^T, \mathbf{x}_3 = (1, 1, 0)^T$, and $\mathbf{x}_4 = (1, 0, 1)^T$, where the transpose is used so that column vectors may be conveniently written horizontally on a line of text, as noted previously. Applying Eq. (11.4-4) yields the following mean vector:

$$\mathbf{m_x} = \frac{1}{4} \begin{bmatrix} 3 \\ 1 \\ 1 \end{bmatrix}.$$

EXAMPLE 11.9:
Computation of the mean vector and covariance matrix.

[†] Recall that the variance of a random variable x with mean m is defined as $E\{(x - m)^2\}$. The covariance of two random variables x_i and x_j is defined as $E\{(x_i - m_i)(x_j - m_j)\}$. If the variables are *uncorrelated*, their covariance is 0.

Similarly, use of Eq. (11.4-5) yields the following covariance matrix:

$$\mathbf{C_x} = \frac{1}{16} \begin{bmatrix} 3 & 1 & 1 \\ 1 & 3 & -1 \\ 1 & -1 & 3 \end{bmatrix}.$$

All the elements along the main diagonal are equal, which indicates that the three components of the vectors in the population have the same variance. Also, elements x_1 and x_2, as well as x_1 and x_3, are positively correlated; elements x_2 and x_3 are negatively correlated. ■

Because $\mathbf{C_x}$ is real and symmetric, finding a set of n orthonormal eigenvectors always is possible (Noble and Daniel [1988]). Let $\mathbf{e}_i$ and $\lambda_i, i = 1, 2, \ldots, n$, be the eigenvectors and corresponding eigenvalues of $\mathbf{C_x}$,[†] arranged (for convenience) in descending order so that $\lambda_j \geq \lambda_{j+1}$ for $j = 1, 2, \ldots, n - 1$. Let $\mathbf{A}$ be a matrix whose rows are formed from the eigenvectors of $\mathbf{C_x}$, ordered so that the first row of $\mathbf{A}$ is the eigenvector corresponding to the largest eigenvalue, and the last row is the eigenvector corresponding to the smallest eigenvalue.

Suppose that we use $\mathbf{A}$ as a transformation matrix to map the $\mathbf{x}$'s into vectors denoted by $\mathbf{y}$'s, as follows:

$$\mathbf{y} = \mathbf{A}(\mathbf{x} - \mathbf{m_x}). \qquad (11.4\text{-}6)$$

This expression is called the *Hotelling transform*, which, as will be shown shortly, has some interesting and useful properties.

It is not difficult to show that the mean of the $\mathbf{y}$ vectors resulting from this transformation is zero; that is,

$$\mathbf{m_y} = E\{\mathbf{y}\} = \mathbf{0}. \qquad (11.4\text{-}7)$$

It follows from basic matrix theory that the covariance matrix of the $\mathbf{y}$'s is given in terms of $\mathbf{A}$ and $\mathbf{C_x}$ by the expression

$$\mathbf{C_y} = \mathbf{A}\mathbf{C_x}\mathbf{A}^T. \qquad (11.4\text{-}8)$$

Furthermore, because of the way $\mathbf{A}$ was formed, $\mathbf{C_y}$ is a diagonal matrix whose elements along the main diagonal are the eigenvalues of $\mathbf{C_x}$; that is,

$$\mathbf{C_y} = \begin{bmatrix} \lambda_1 & & & 0 \\ & \lambda_2 & & \\ & & \ddots & \\ 0 & & & \lambda_n \end{bmatrix}. \qquad (11.4\text{-}9)$$

The off-diagonal elements of this covariance matrix are 0, so the elements of the $\mathbf{y}$ vectors are uncorrelated. Keep in mind that the λ_j's are the eigenvalues of $\mathbf{C_x}$ and that the elements along the main diagonal of a diagonal matrix are its eigenvalues (Noble and Daniel [1988]). Thus $\mathbf{C_x}$ and $\mathbf{C_y}$ have the same eigenvalues. In fact, the same is true for the eigenvectors.

[†] By definition, the eigenvectors and eigenvalues of an $n \times n$ matrix, $\mathbf{C}$, satisfy the relation $\mathbf{C}\mathbf{e}_i = \lambda_i \mathbf{e}_i$, for $i = 1, 2, \ldots, n$.

Another important property of the Hotelling transform deals with the reconstruction of **x** from **y**. Because the rows of **A** are orthonormal vectors, it follows that $\mathbf{A}^{-1} = \mathbf{A}^T$, and any vector **x** can be recovered from its corresponding **y** by using the expression

$$\mathbf{x} = \mathbf{A}^T\mathbf{y} + \mathbf{m_x}. \tag{11.4-10}$$

Suppose, however, that instead of using all the eigenvectors of $\mathbf{C_x}$ we form matrix $\mathbf{A}_k$ from the k eigenvectors corresponding to the k largest eigenvalues, yielding a transformation matrix of order $k \times n$. The **y** vectors would then be k dimensional, and the reconstruction given in Eq. (11.4-10) would no longer be exact (this is somewhat analogous to the procedure we used in Section 11.2.3 to describe a boundary with a few Fourier coefficients).

The vector reconstructed by using $\mathbf{A}_k$ is

$$\hat{\mathbf{x}} = \mathbf{A}_k^T\mathbf{y} + \mathbf{m_x}. \tag{11.4-11}$$

It can be shown that the mean square error between **x** and $\hat{\mathbf{x}}$ is given by the expression

$$\begin{aligned} e_{ms} &= \sum_{j=1}^{n} \lambda_j - \sum_{j=1}^{k} \lambda_j \\ &= \sum_{j=k+1}^{n} \lambda_j. \end{aligned} \tag{11.4-12}$$

The first line of Eq. (11.4-12) indicates that the error is zero if $k = n$ (that is, if all the eigenvectors are used in the transformation). Because the λ_j's decrease monotonically, Eq. (11.4-12) also shows that the error can be minimized by selecting the k eigenvectors associated with the largest eigenvalues. Thus the Hotelling transform is optimal in the sense that it minimizes the mean square error between the vectors **x** and their approximations $\hat{\mathbf{x}}$. Due to this idea of using the eigenvectors corresponding to the largest eigenvalues, the Hotelling transform also is known as the *principal components* transform.

■ Figure 11.26 shows six images generated by a 6-band multispectral scanner operating in the wavelengths shown in Table 11.4. Viewing the images as shown in Fig. 11.27 allows formation of a 6-dimensional vector $\mathbf{x} = (x_1, x_2, \ldots, x_6)^T$ from each set of corresponding pixels in the images, as discussed at the beginning of this section. The images in this particular application are of resolution 384×239 so the population consists of 91,776 vectors from which to compute the mean vector and covariance matrix. Table 11.5 shows the eigenvalues of $\mathbf{C_y}$. Note the dominance of the first two eigenvalues.

EXAMPLE 11.10:
Use of principal components to describe images.

Use of Equation (11.4-6) generated a set of transformed **y** vectors corresponding to the **x** vectors. From them, six principal component images were assembled (images are constructed from vectors simply by applying Fig. 11.27 in reverse). Figure 11.28 shows the results. Component 1 denotes the image formed from all the y_1 components of the transformed vectors, and so on for the other five images. Recall from basic matrix theory that y_1, for example, is obtained by performing the inner (dot) product of the first row of **A** with the column vector $(\mathbf{x} - \mathbf{m_x})^T$.

FIGURE 11.26 Six spectral images from an airborne scanner. (Courtesy of the Laboratory for Applications of Remote Sensing, Purdue University.)

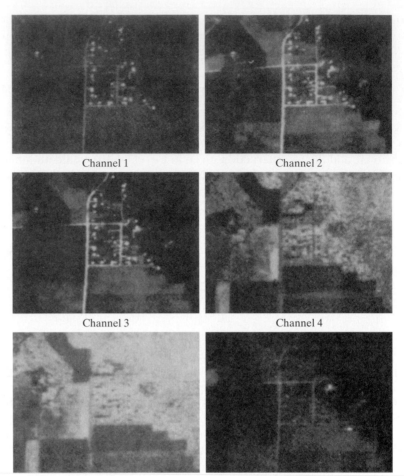

Channel 1   Channel 2

Channel 3   Channel 4

Channel 5   Channel 6

TABLE 11.4
Channel numbers and wavelengths.

Channel	Wavelength band (microns)
1	0.40–0.44
2	0.62–0.66
3	0.66–0.72
4	0.80–1.00
5	1.00–1.40
6	2.00–2.60

The first row of $\mathbf{A}$ is the eigenvector corresponding to the largest eigenvalue of the covariance matrix of the population, and this eigenvalue gives the variance of the gray levels of the first transformed image. Thus based on the numbers shown in Table 11.5, this image should have the highest contrast. That such is the case is quite evident in Fig. 11.28. Because the first two images account for about

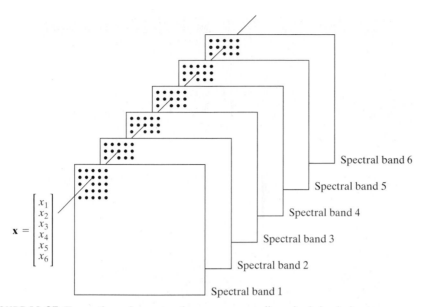

$$\mathbf{x} = \begin{bmatrix} x_1 \\ x_2 \\ x_3 \\ x_4 \\ x_5 \\ x_6 \end{bmatrix}$$

Spectral band 6

Spectral band 5

Spectral band 4

Spectral band 3

Spectral band 2

Spectral band 1

FIGURE 11.27 Formation of a vector from corresponding pixels in six images.

λ_1	λ_2	λ_3	λ_4	λ_5	λ_6
3210	931.4	118.5	83.88	64.00	13.40

TABLE 11.5
Eigenvalues of the covariance matrix obtained from the images in Fig. 11.26.

94% of the total variance, the fact that the other four principal-component images have low contrast is not unexpected. Thus if instead of storing all six images for posterity, only the first two transformed images, along with $\mathbf{m}_x$ and the first two rows of $\mathbf{A}$, were stored, a credible job of reconstructing an approximation to the six original images could be done at a later date. This capability for performing *data compression*, although not impressive by today's standards is a useful byproduct of the Hotelling transform. In terms of description, this means describing the content of six images with two, plus the mean vector and first two rows of the transformation matrix. The same argument would apply if instead of entire images we were discussing regions.   ■

■ In the preceding discussion we showed how to apply the principal components transformation to sets of images or regions. In this example we illustrate how to use principal components for describing boundaries and regions in a single image. The approach is to form two-dimensional vectors from the *coordinates* of the boundary or region. Consider the object shown in Fig. 11.29(a). Vectors are formed from the coordinates of the pixels in the object if we wish

EXAMPLE 11.11:
Use of principal components for describing boundaries and regions in a single image.

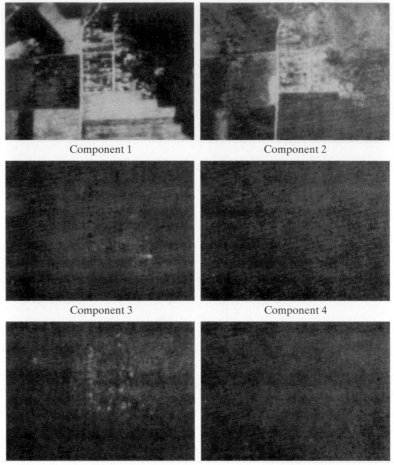

Component 1   Component 2

Component 3   Component 4

Component 5   Component 6

FIGURE 11.28 Six principal-component images computed from the data in Fig. 11.26. (Courtesy of the Laboratory for Applications of Remote Sensing, Purdue University.)

to describe the region. If we wish to describe the boundary, we only use the coordinates of the points on the boundary.

The resulting vectors then are treated as a 2-D population of random vectors. In other words, each pixel in the object is treated as a 2-D vector $\mathbf{x} = (a, b)^T$, where a and b are the coordinate values of that pixel with respect to the x_1- and x_2-axes. These vectors are used to compute the mean vector and covariance matrix of the population (object). The problem is much simpler than before because we are working in only two dimensions.

The net effect of using Eq. (11.4-6) is to establish a new coordinate system whose origin is at the centroid of the population (the coordinates of the mean vector) and whose axes are in the direction of the eigenvectors of $\mathbf{C_x}$, as shown in Fig. 11.29(b). This coordinate system clearly shows that the transformation in Eq. (11.4-6) is a rotation transformation that aligns the data with the eigenvectors,

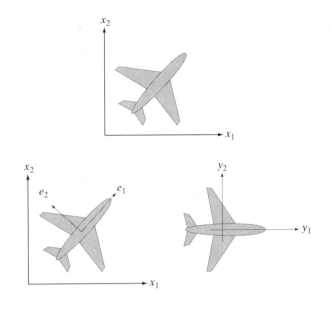

FIGURE 11.29 (a) An object. (b) Eigenvectors. (c) Object rotated by using Eq. (11.4-6). The net effect is to align the object along its eigen axes.

as shown in Fig. 11.29(c). In fact, this alignment is precisely the mechanism that decorrelates the data. Furthermore, as the eigenvalues appear along the main diagonal of $\mathbf{C_y}$, λ_i is the variance of component y_i along eigenvector $\mathbf{e}_i$. The two eigenvectors are perpendicular. The y-axes sometimes are called the *eigen axes*, for obvious reasons. ∎

The concept of aligning a 2-D object with its principal eigenvectors plays an important role in description. As noted earlier, description should be as independent as possible to variations in size, translation, and rotation. The ability to *align* the object with its principal axes provides a reliable means for removing the effects of rotation. The eigenvalues are the variances along the eigen axes, and can be used for size normalization. The effects of translation are accounted for by centering the object about its mean, as shown in Eq. (11.4-6). Keep in mind the fact that the method of description derived in this section is equally applicable to both regions and boundaries.

11.5 Relational Descriptors

We introduced in Section 11.3.3 the concept of rewriting rules for describing texture. In this section we expand that concept in the context of relational descriptors. These apply equally well to boundaries or regions, and their main purpose is to capture in the form of rewriting rules basic repetitive patterns in a boundary or region.

FIGURE 11.30
(a) A simple
staircase
structure.
(b) Coded
structure.

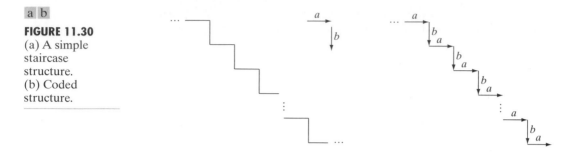

Consider the simple staircase structure shown in Fig. 11.30(a). Assume that this structure has been segmented out of an image and that we want to describe it in some formal way. By defining the two *primitive elements a* and *b* shown, we may code Fig. 11.30(a) in the form shown in Fig. 11.30(b). The most obvious property of the coded structure is the repetitiveness of the elements *a* and *b*. Therefore, a simple description approach is to formulate a recursive relationship involving these primitive elements. One possibility is to use the *rewriting rules*:

(a) $S \rightarrow aA$,
(b) $A \rightarrow bS$, and
(c) $A \rightarrow b$,

where S and A are variables and the elements a and b are constants corresponding to the primitives just defined. Rule 1 indicates that S, called the *starting symbol*, can be replaced by primitive a and variable A. This variable, in turn, can be replaced by b and S or by b alone. Replacing A with bS, leads back to the first rule and the procedure can be repeated. Replacing A with b terminates the procedure, because no variables remain in the expression. Figure 11.31 illustrates some sample derivations of these rules, where the numbers below the structures represent the order in which rules 1, 2, and 3 were applied. The relationship between a and b is preserved, because these rules force an a always to be followed by a b. Notably, these three simple rewriting rules can be used to generate (or describe) infinitely many "similar" structures. As we show in Chapter 12, this approach also has the advantage of a solid theoretical foundation.

FIGURE 11.31
Sample
derivations for
the rules $S \rightarrow aA$,
$A \rightarrow bS$, and
$A \rightarrow b$.

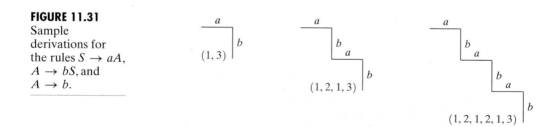

Because strings are 1-D structures, their application to image description requires establishing an appropriate method for reducing 2-D positional relations to 1-D form. Most applications of strings to image description are based on the idea of extracting connected line segments from the objects of interest. One approach is to follow the contour of an object and code the result with segments of specified direction and/or length. Figure 11.32 illustrates this procedure.

Another, somewhat more general, approach is to describe sections of an image (such as small homogeneous regions) by directed line segments, which can be joined in other ways besides head-to-tail connections. Figure 11.33(a) illustrates this approach, and Fig. 11.33(b) shows some typical operations that can be defined on abstracted primitives. Figure 11.33(c) shows a set of specific primitives consisting of line segments defined in four directions, and Fig. 11.33(d) shows a step-by-step generation of a specific shape, where $(\sim d)$ indicates the primitive d with its direction reversed. Note that each composite structure has a single head and a single tail. The result of interest is the last string, which describes the complete structure.

String descriptions are best suited for applications in which connectivity of primitives can be expressed in a head-to-tail or other continuous manner. Sometimes regions that are similar in terms of texture or other descriptor may not be contiguous, and techniques are required for describing such situations. One of the most useful approaches for doing so is to use tree descriptors.

A *tree T* is a finite set of one or more nodes for which

(a) there is a unique node $ designated the *root*, and
(b) the remaining nodes are partitioned into m disjointed sets $T_1, \ldots, T_m$, each of which in turn is a tree called a *subtree* of T.

The *tree frontier* is the set of nodes at the bottom of the tree (the *leaves*), taken in order from left to right. For example, the tree shown in Fig. 11.34 has root $ and frontier *xy*.

Generally, two types of information in a tree are important: (1) information about a node stored as a set of words describing the node, and (2) information relating a node to its neighbors, stored as a set of pointers to those neighbors.

FIGURE 11.32
Coding a region boundary with directed line segments.

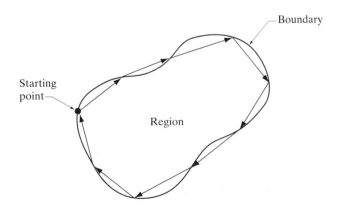

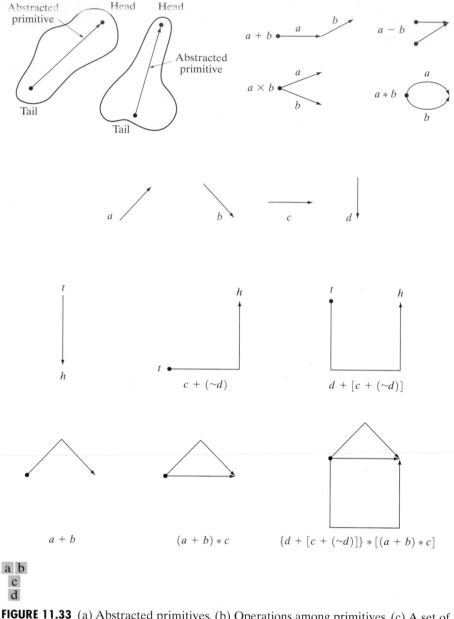

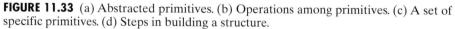

FIGURE 11.33 (a) Abstracted primitives. (b) Operations among primitives. (c) A set of specific primitives. (d) Steps in building a structure.

As used in image description, the first type of information identifies an image substructure (e.g., region or boundary segment), whereas the second type defines the physical relationship of that substructure to other substructures. For example, Fig. 11.35(a) can be represented by a tree by using the relationship "inside of." Thus, if the root of the tree is denoted $, Fig. 11.35(a) shows that the first

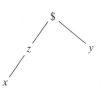

FIGURE 11.34 A simple tree with root $ and frontier *xy*.

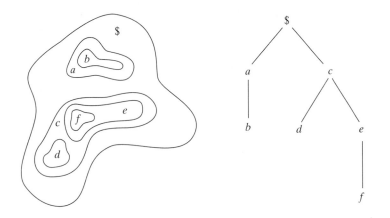

a b

FIGURE 11.35 (a) A simple composite region. (b) Tree representation obtained by using the relationship "inside of."

level of complexity involves *a* and *c* inside $, which produces two branches emanating from the root, as shown in Fig. 11.35(b). The next level involves *b* inside *a*, and *d* and *e* inside *c*. Finally, *f* inside *e* completes the tree.

Summary

The representation and description of objects or regions that have been segmented out of an image are early steps in the operation of most automated processes involving images. These descriptions, for example, constitute the input to the object recognition methods developed in the following chapter. As indicated by the range of description techniques covered in this chapter, the choice of one method over another is determined by the problem under consideration. The objective is to choose descriptors that "capture" essential differences between objects, or classes of objects, while maintaining as much independence as possible to changes in factors such as location, size, and orientation.

References and Further Reading

The chain-code representation discussed in Section 11.1.1 was first proposed by Freeman [1961, 1974]. For current work using chain codes see Bribiesca [1999], who also has extended chain codes to 3-D (Bribiesca [2000]). For a detailed discussion and algorithm to compute minimum-perimeter polygons (Section 11.1.2) see Sklansky et al. [1972]. Typical work on polygonal approximations a decade ago is illustrated in the papers by Bengt-soon and Eklundh [1991] and by Sato [1992]. The paper by Zhu and Chirlian [1995] presents an interesting approach to the detection of point inflections along a curve. See also Hu and Yan [1997]. More recent work in this area focuses on invariant polygonal fitting (Voss and Suesse [1997]), on methods for evaluating the performance of polygonal approximation algorithms (Rosin [1997]), on generic implementations (Huang and Sun [1999]), and on computational speed (Davis [1999]).

References for the discussion of signatures (Section 11.1.3) are Ballard and Brown [1982] and Gupta and Srinath [1988]. See Preparata and Shamos [1985] regarding fundamental formulations for finding the convex hull and convex deficiency (Section 11.1.4). See also the paper by Liu-Yu and Antipolis [1993]. Katzir et al. [1994] discuss the detection of partially occluded curves. Zimmer et al. [1997] discuss an improved algorithm for computing the convex hull, and Latecki and Lakämper [1999] discuss a convexity rule for shape decomposition.

The skeletonizing algorithm discussed in Section 11.1.5 is based on Zhang and Suen [1984]. Some useful additional comments on the properties and implementation of this algorithm are included in a paper by Lu and Wang [1986]. A paper by Jang and Chin [1990] provides an interesting tie between the discussion in Section 11.1.5 and the morphological concept of thinning introduced in Section 9.5.5. For thinning approaches in the presence of noise see Shi and Wong [1994] and Chen and Yu [1996]. Shaked and Bruckstein [1998] discuss a pruning algorithm useful for removing spurs from a skeleton. Fast computation of the medial axis transform is discussed by Sahni and Jenq [1992] and by Ferreira and Ubéda [1999]. The survey paper by Loncaric [1998] is of interest regarding many of the approaches discussed in Section 11.1.

Freeman and Shapira [1975] give an algorithm for finding the basic rectangle of a closed, chain-coded curve (Section 11.2.1). The discussion on shape numbers in Section 11.2.2 is based on the work of Bribiesca and Guzman [1980] and Bribiesca [1981]. For additional reading on Fourier descriptors (Section 11.2.3), see the early papers by Zahn and Roskies [1972] and by Persoon and Fu [1977]. See also Aguado et al. [1998] and Sonka et al. [1999]. Reddy and Chatterji [1996] discuss an interesting approach using the FFT to achieve invariance to translation, rotation, and scale change. The material in Section 11.2.4 is based on elementary probability theory (see, for example, Peebles [1993] and Popoulis [1991]).

For additional reading on Section 11.3.2, see Rosenfeld and Kak [1982] and Ballard and Brown [1982]. For an excellent introduction to texture (Section 11.3.3), see Haralick and Shapiro [1992]. For an early survey on texture, see Wechsler [1980]. The papers by Murino et al. [1998] and Garcia [1999], and the discussion by Shapiro and Stockman [2001], are representative of current work in this field.

The moment invariant approach discussed in Section 11.3.4 is from Hu [1962]. Also see Bell [1965]. To get an idea of the range of applications of moment invariants, see Hall [1979] regarding image matching and Cheung and Teoh [1999] regarding the use of moments for describing symmetry. Moment invariants were generalized to n dimensions by Mamistvalov [1998].

Hotelling [1933] was the first to derive and publish the approach that transforms discrete variables into uncorrelated coefficients. He referred to this technique as *the method of principal components*. His paper gives considerable insight into the method and is worth reading. Hotelling's transformation was rediscovered by Kramer and Mathews [1956] and by Huang and Schultheiss [1963]. Principal components are still a basic tool for image description used in numerous applications, as exemplified by Swets and Weng [1996] and by Duda, Heart, and Stork [2001]. References for the material in Section 11.5 are Gonzalez and Thomason [1978] and Fu [1982]. See also Sonka et al. [1999].

Problems

11.1 ★ (a) Show that redefining the starting point of a chain code so that the resulting sequence of numbers forms an integer of minimum magnitude makes the code independent of the initial starting point on the boundary.

(b) Find the normalized starting point of the code 11076765543322.

11.2 (a) Show that the first difference of a chain code normalizes it to rotation, as explained in Section 11.1.1.

(b) Compute the first difference of the code 0101030303323232212111.

11.3 ★ (a) Show that the rubber-band polygonal approximation approach discussed in Section 11.1.2 yields a polygon with minimum perimeter.

(b) Show that if each cell corresponds to a pixel on the boundary, the maximum possible error in that cell is $\sqrt{2}d$, where d is the minimum possible horizontal or vertical distance between adjacent pixels (i.e., the distance between lines in the sampling grid used to produce the digital image).

11.4 ★ (a) Discuss the effect on the resulting polygon if the error threshold is set to zero in the merging method discussed in Section 11.1.2.

(b) What would be the effect on the splitting method?

11.5 ★ (a) Plot the signature of a square boundary using the tangent angle method discussed in Section 11.1.3.

(b) Repeat for the slope density function.

Assume that the square is aligned with the x- and y-axes, and let the x-axis be the reference line. Start at the corner closest to the origin.

11.6 Find an expression for the signature of each of the following boundaries, and plot the signatures.

★ (a) An equilateral triangle

(h) A rectangle

(c) An ellipse

11.7 Draw the medial axis of

★ (a) A circle

★ (b) A square

(c) A rectangle

(d) An equilateral triangle

11.8 For each of the figures shown,

★ (a) Discuss the action taken at point p by step 1 of the skeletonizing algorithm presented in Section 11.1.5.

(b) Repeat for step 2 of the algorithm. Assume that $p = 1$ in all cases.

1	1	0	0	0	0	0	1	0	1	1	0
1	p	0	1	p	0	1	p	1	0	p	1
1	1	0	0	0	0	0	1	0	0	0	0

11.9   With reference to the skeletonizing algorithm in Section 11.1.5, what would the figure shown look like after

★ **(a)** One pass of step 1 of the algorithm?

(b) One pass of step 2 (on the result of step 1, not the original image)?

```
·  ·  ·  ·  ·  ·  ·  ·  ·  ·  ·  ·
·  ·  ·  ·  ·  ·  ·  ·  ·  ·  ·  ·
·  ·  ●  ●  ●  ·  ·  ●  ●  ●  ·  ·
·  ·  ●  ●  ●  ·  ·  ●  ●  ●  ·  ·
·  ·  ●  ●  ●  ●  ●  ●  ●  ●  ·  ·
·  ·  ●  ●  ●  ●  ●  ●  ●  ●  ·  ·
·  ·  ●  ●  ●  ●  ●  ●  ●  ●  ·  ·
·  ·  ●  ●  ●  ●  ●  ●  ●  ●  ·  ·
·  ·  ●  ●  ●  ●  ●  ●  ●  ●  ·  ·
·  ·  ●  ●  ●  ·  ·  ●  ●  ●  ·  ·
·  ·  ·  ·  ·  ·  ·  ·  ·  ·  ·  ·
·  ·  ·  ·  ·  ·  ·  ·  ·  ·  ·  ·
```

11.10 ★ **(a)** What is the order of the shape number for the figure shown?

(b) Obtain the shape number.

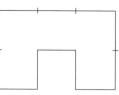

11.11   The procedure discussed in Section 11.2.3 for using Fourier descriptors consists of expressing the coordinates of a contour as complex numbers, taking the DFT of these numbers, and keeping only a few components of the DFT as descriptors of the boundary shape. The inverse DFT is then an approximation to the original contour. What class of contour shapes would have a DFT consisting of real numbers and how would the axis system in Fig. 11.13 have to be set up to obtain these real numbers?

★ **11.12**   Give the smallest number of statistical moment descriptors needed to differentiate between the signatures of the figures shown in Fig. 11.5.

11.13   Give two boundary shapes that have the same mean and third statistical moment descriptors, but different second moments.

★ **11.14**   Propose a set of descriptors capable of differentiating between the shapes of the characters $0, 1, 8, 9$, and X. (*Hint:* Use topological descriptors in conjunction with the convex hull.)

11.15   Consider a checkerboard image composed of alternating black and white squares, each of size $m \times m$. Give a position operator that would yield a diagonal co-occurrence matrix.

11.16   Obtain the gray-level co-occurrence matrix of a 5×5 image composed of a checkerboard of alternating 1's and 0's if

★ **(a)** the position operator P is defined as "one pixel to the right," and

(b) "two pixels to the right."

Assume that the top left pixel has value 0.

11.17   Prove the validity of Eqs. (11.4-7), (11.4-8), and (11.4-9).

★ **11.18**   It was mentioned in Example 11.10 that a credible job could be done of reconstructing approximations to the six original images by using only the two principal-component images associated with the largest eigenvalues. What would be the mean square error incurred in doing so? Express your answer as a percentage of the maximum possible error.

11.19 For a set of images of size 64×64, assume that the covariance matrix given in Eq. (11.4-9) turns out to be the identity matrix. What would be the mean square error between the original images and images reconstructed using Eq. (11.4-11) with only half of the original eigenvectors?

★**11.20** Under what conditions would you expect the major axes of a boundary, defined in Section 11.2.1, to be equal to the eigen axes of that boundary?

11.21 Give a spatial relationship and corresponding tree representation for a checkerboard pattern of black and white squares. Assume that the top left element is black and that the root of the tree corresponds to that element. Your tree can have no more than two branches emanating from each node.

★**11.22** You are contracted to design an image processing system for detecting imperfections on the inside of certain solid plastic wafers. The wafers are examined using an X-ray imaging system, which yields 8-bit images of 512×512 resolution. In the absence of imperfections, the images appear "bland," having a mean gray level of 100 and variance of 400. The imperfections appear as bloblike regions in which about 70% of the pixels have excursions in intensity of 50 gray levels or less about a mean of 100. A wafer is considered defective if such a region occupies an area exceeding 20×20 pixels in size. Propose a system based on texture analysis.

11.23 A company that bottles a variety of industrial chemicals has heard of your success solving imaging problems and hires you to design an approach for detecting when bottles are not full. The bottles appear as shown in the following figure as they move along a conveyor line past an automatic filling and capping station. A bottle is considered imperfectly filled when the level of the liquid is below the midway point between the bottom of the neck and the shoulder of the bottle. The shoulder is defined as the region of the bottle where the sides and slanted portion of the bottle intersect. The bottles are moving, but the company has an imaging system equipped with a illumination flash front end that effectively stops motion, so you will be given images that look very close to the sample shown here. Based on the material you have learned up to this point, propose a solution for detecting bottles that are not filled properly. State clearly all assumptions that you make and that are likely to impact the solution you propose.

11.24 Having heard about your success with the bottling problem, you are contacted by a fluids company that wishes to automate bubble-counting in certain processes for quality control. The company has solved the imaging problem and can obtain 8-bit images of resolution 700×700 pixels, such as the ones shown. Each image represents an area of 7 cm^2. The company wishes to do two things with each image: (1) Determine the ratio of the area occupied by bubbles to the total area of the image, and (2) count the number of distinct bubbles. Based on the material you have learned up to this point, propose a solution to this problem. In your solution, make sure to state the physical dimensions of the smallest bubble your solution can detect. State clearly all assumptions that you make and that are likely to impact the solution you propose.

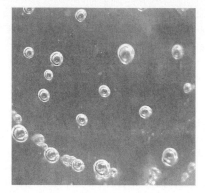

12 *Object Recognition*

> One of the most interesting aspects of the world
> is that it can be considered to be made up of
> patterns.
> A pattern is essentially an arrangement. It is
> characterized by the order of the elements of which
> it is made, rather than by the intrinsic nature of these
> elements.
>
> *Norbert Wiener*

Preview

We conclude our coverage of digital image processing with an introduction to techniques for object recognition. As noted in Section 1.1, we have defined the scope covered by our treatment of digital image processing to include recognition of *individual* image regions, which in this chapter we call *objects* or *patterns*.

The approaches to pattern recognition developed in this chapter are divided into two principal areas: decision-theoretic and structural. The first category deals with patterns described using quantitative descriptors, such as length, area, and texture. The second category deals with patterns best described by qualitative descriptors, such as the relational descriptors discussed in Section 11.5.

Central to the theme of recognition is the concept of "learning" from sample patterns. Learning techniques for both decision-theoretic and structural approaches are developed and illustrated in the material that follows.

12.1 Patterns and Pattern Classes

A *pattern* is an *arrangement of descriptors*, such as those discussed in Chapter 11. The name *feature* is used often in the pattern recognition literature to denote a descriptor. A *pattern class* is a family of patterns that share some common properties. Pattern classes are denoted $\omega_1, \omega_2, \ldots, \omega_W$, where W is the number of classes. Pattern recognition by machine involves techniques for assigning patterns to their respective classes—automatically and with as little human intervention as possible.

Three common pattern arrangements used in practice are vectors (for quantitative descriptions) and strings and trees (for structural descriptions). Pattern vectors are represented by bold lowercase letters, such as **x**, **y**, and **z**, and take the form

$$
\mathbf{x} = \begin{bmatrix} x_1 \\ x_2 \\ \vdots \\ x_n \end{bmatrix}
\tag{12.1-1}
$$

See inside front cover

Consult the book web site for a brief review of vectors and matrices.

where each component, x_i, represents the ith descriptor and n is the total number of such descriptors associated with the pattern. Pattern vectors are represented as columns (that is, $n \times 1$ matrices). Hence a pattern vector can be expressed in the form shown in Eq. (12.1-1) or in the equivalent form $\mathbf{x} = (x_1, x_2, \ldots, x_n)^T$, where T indicates transposition. The reader will recognize this notation from Section 11.4.

The nature of the components of a pattern vector **x** depends on the approach used to describe the physical pattern itself. Let us illustrate with an example that is both simple and gives a sense of history in the area of classification of measurements. In a classic paper, Fisher [1936] reported the use of what then was a new technique called *discriminant analysis* (discussed in Section 12.2) to recognize three types of iris flowers (*Iris setosa*, *virginica*, and *versicolor*) by measuring the widths and lengths of their petals (Fig. 12.1). In our present

FIGURE 12.1
Three types of iris flowers described by two measurements.

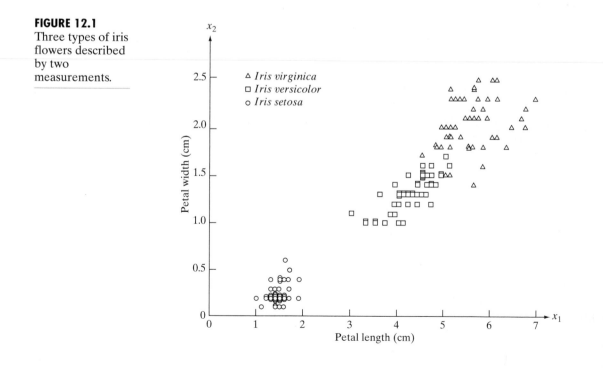

terminology, each flower is *described* by two measurements, which leads to a 2-D pattern vector of the form

$$\mathbf{x} = \begin{bmatrix} x_1 \\ x_2 \end{bmatrix} \qquad (12.1\text{-}2)$$

where x_1 and x_2 correspond to petal length and width, respectively. The three pattern classes in this case, denoted ω_1, ω_2, and ω_3, correspond to the varieties *setosa*, *virginica*, and *versicolor*, respectively.

Because the petals of flowers vary in width and length, the pattern vectors describing these flowers also will vary, not only between different classes, but also within a class. Figure 12.1 shows length and width measurements for several samples of each type of iris. After a set of measurements has been selected (two in this case), the components of a pattern vector become the entire description of each physical sample. Thus each flower in this case becomes a point in 2-D Euclidean space. We also note that measurements of petal width and length in this case adequately separated the class of *Iris setosa* from the other two but did not separate as successfully the *virginica* and *versicolor* types from each other. This result illustrates the classic *feature selection* problem, in which the degree of class separability depends strongly on the choice of descriptors selected for an application. We say considerably more about this issue in Sections 12.2 and 12.3.

Figure 12.2 shows another example of pattern vector generation. In this case, we are interested in different types of noisy shapes, a sample of which is shown in Fig. 12.2(a). If we elect to represent each object by its signature (see Section 11.1.3), we would obtain 1-D signals of the form shown in Fig. 12.2(b). Suppose that we elect to describe each signature simply by its sampled amplitude values; that is, we sample the signatures at some specified interval values of θ, denoted $\theta_1, \theta_2, \ldots, \theta_n$. Then we can form pattern vectors by letting $x_1 = r(\theta_1)$, $x_2 = r(\theta_2), \ldots, x_n = r(\theta_n)$. These vectors become points in n dimensional Euclidean space, and pattern classes can be imagined to be "clouds" in n dimensions.

Instead of using signature amplitudes directly, we could compute, say, the first n statistical moments of a given signature (Section 11.2.4) and use these descriptors as components of each pattern vector. In fact, as may be evident by now, pattern vectors can be generated in numerous other ways. We present some

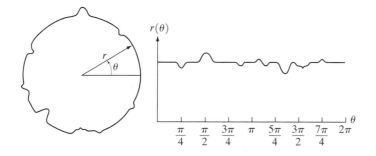

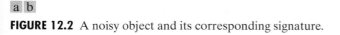

FIGURE 12.2 A noisy object and its corresponding signature.

of them throughout this chapter. For the moment, the key concept to keep in mind is that selecting the descriptors on which to base each component of a pattern vector has a profound influence on the eventual performance of object recognition based on the pattern vector approach.

The techniques just described for generating pattern vectors yield pattern classes characterized by quantitative information. In some applications, pattern characteristics are best described by structural relationships. For example, fingerprint recognition is based on the interrelationships of print features called *minutiae*. Together with their relative sizes and locations, these features are primitive components that describe fingerprint ridge properties, such as abrupt endings, branching, merging, and disconnected segments. Recognition problems of this type, in which not only quantitative measures about each feature but also the spatial relationships between the features determine class membership, generally are best solved by structural approaches. This subject was introduced in Section 11.5. We revisit it briefly here in the context of pattern descriptors.

Figure 12.3(a) shows a simple staircase pattern. This pattern could be sampled and expressed in terms of a pattern vector, similar to the approach used in Fig. 12.2. However, the basic structure, consisting of repetitions of two simple primitive elements, would be lost in this method of description. A more meaningful description would be to define the elements *a* and *b* and let the pattern be the string of symbols $w = \ldots ababab ab \ldots$, as shown in Fig. 12.3(b). The structure of this particular class of patterns is captured in this description by requiring that connectivity be defined in a head-to-tail manner, and by allowing only alternating symbols. This structural construct is applicable to staircases of any length but excludes other types of structures that could be generated by other combinations of the primitives *a* and *b*.

String descriptions adequately generate patterns of objects and other entities whose structure is based on relatively simple connectivity of primitives, usually associated with boundary shape. A more powerful approach for many applications is the use of tree descriptions, as defined in Section 11.5. Basically, most hierarchical ordering schemes lead to tree structures. For example, Fig. 12.4 is a satellite image of a heavily built downtown area and surrounding

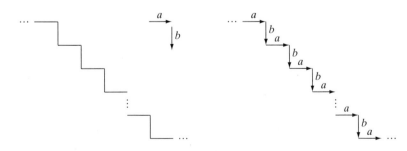

a b

FIGURE 12.3 (a) Staircase structure. (b) Structure coded in terms of the primitives *a* and *b* to yield the string description . . . *ababab* . . . .

FIGURE 12.4
Satellite image of
a heavily built
downtown area
(Washington,
D.C.) and
surrounding
residential areas.
(Courtesy of
NASA.)

residential areas. Let us define the entire image area by the symbol $. The (up-
side down) tree representation shown in Fig. 12.5 was obtained by using the
structural relationship "composed of." Thus the root of the tree represents the
entire image. The next level indicates that the image is composed of a downtown
and residential area. The residential area, in turn is composed of housing, highways, and shopping malls. The next level down further describes the housing
and highways. We can continue this type of subdivision until we reach the limit
of our ability to resolve different regions in the image.

We develop in the following sections recognition approaches for objects
described by all the techniques discussed in the preceding paragraphs.

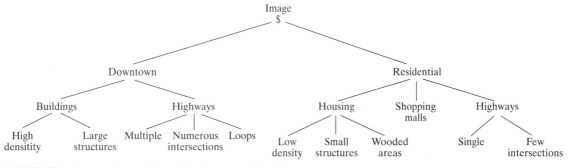

FIGURE 12.5 A tree description of the image in Fig. 12.4.

12.2 Recognition Based on Decision-Theoretic Methods

Decision-theoretic approaches to recognition are based on the use of *decision* (or *discriminant*) *functions*. Let $\mathbf{x} = (x_1, x_2, \ldots, x_n)^T$ represent an n-dimensional pattern vector, as discussed in Section 12.1. For W pattern classes $\omega_1, \omega_2, \ldots, \omega_W$, the basic problem in decision-theoretic pattern recognition is to find W decision functions $d_1(\mathbf{x}), d_2(\mathbf{x}), \ldots, d_W(\mathbf{x})$ with the property that, if a pattern $\mathbf{x}$ belongs to class ω_i, then

$$d_i(\mathbf{x}) > d_j(\mathbf{x}) \qquad j = 1, 2, \ldots, W; j \neq i. \qquad (12.2\text{-}1)$$

In other words, an unknown pattern $\mathbf{x}$ is said to belong to the ith pattern class if, upon substitution of $\mathbf{x}$ into all decision functions, $d_i(\mathbf{x})$ yields the largest numerical value. Ties are resolved arbitrarily.

The *decision boundary* separating class ω_i from ω_j is given by values of $\mathbf{x}$ for which $d_i(\mathbf{x}) = d_j(\mathbf{x})$ or, equivalently, by values of $\mathbf{x}$ for which

$$d_i(\mathbf{x}) - d_j(\mathbf{x}) = 0. \qquad (12.2\text{-}2)$$

Common practice is to identify the decision boundary between two classes by the single function $d_{ij}(\mathbf{x}) = d_i(\mathbf{x}) - d_j(\mathbf{x}) = 0$. Thus $d_{ij}(\mathbf{x}) > 0$ for patterns of class ω_i and $d_{ij}(\mathbf{x}) < 0$ for patterns of class ω_j. The principal objective of the discussion in this section is to develop various approaches for finding decision functions that satisfy Eq. (12.2-1).

12.2.1 Matching

Recognition techniques based on matching represent each class by a prototype pattern vector. An unknown pattern is assigned to the class to which it is closest in terms of a predefined metric. The simplest approach is the minimum-distance classifier, which, as its name implies, computes the (Euclidean) distance between the unknown and each of the prototype vectors. It chooses the smallest distance to make a decision. We also discuss an approach based on correlation, which can be formulated directly in terms of images and is quite intuitive.

Minimum distance classifier

Suppose that we define the prototype of each pattern class to be the mean vector of the patterns of that class:

$$\mathbf{m}_j = \frac{1}{N_j} \sum_{\mathbf{x} \in \omega_j} \mathbf{x}_j \qquad j = 1, 2, \ldots, W \qquad (12.2\text{-}3)$$

where N_j is the number of pattern vectors from class ω_j and the summation is taken over these vectors. As before, W is the number of pattern classes. One way to determine the class membership of an unknown pattern vector $\mathbf{x}$ is to assign it to the class of its closest prototype, as noted previously. Using the Euclidean distance to determine closeness reduces the problem to computing the distance measures:

$$D_j(\mathbf{x}) = \|\mathbf{x} - \mathbf{m}_j\| \qquad j = 1, 2, \ldots, W \qquad (12.2\text{-}4)$$

where $\|\mathbf{a}\| = (\mathbf{a}^T\mathbf{a})^{1/2}$ is the Euclidean norm. We then assign $\mathbf{x}$ to class ω_i if $D_i(\mathbf{x})$ is the smallest distance. That is, the smallest distance implies the best match in this formulation. It is not difficult to show (Problem 12.2) that selecting the smallest distance is equivalent to evaluating the functions

$$d_j(\mathbf{x}) = \mathbf{x}^T\mathbf{m}_j - \frac{1}{2}\mathbf{m}_j^T\mathbf{m}_j \qquad j = 1, 2, \dots, W \qquad (12.2\text{-}5)$$

and assigning $\mathbf{x}$ to class ω_i if $d_i(\mathbf{x})$ yields the largest numerical value. This formulation agrees with the concept of a decision function, as defined in Eq. (12.2-1).

From Eqs. (12.2-2) and (12.2-5), the decision boundary between classes ω_i and ω_j for a minimum distance classifier is

$$d_{ij}(\mathbf{x}) = d_i(\mathbf{x}) - d_j(\mathbf{x})$$

$$= \mathbf{x}^T(\mathbf{m}_i - \mathbf{m}_j) - \frac{1}{2}(\mathbf{m}_i - \mathbf{m}_j)^T(\mathbf{m}_i + \mathbf{m}_j) = 0. \qquad (12.2\text{-}6)$$

The surface given by Eq. (12.2-6) is the perpendicular bisector of the line segment joining $\mathbf{m}_i$ and $\mathbf{m}_j$ (see Problem 12.3). For $n = 2$, the perpendicular bisector is a line, for $n = 3$ it is a plane, and for $n > 3$ it is called a *hyperplane*.

■ Figure 12.6 shows two pattern classes extracted from the iris samples in Fig. 12.1. The two classes, *Iris versicolor* and *Iris setosa*, denoted ω_1 and ω_2, respectively, have sample mean vectors $\mathbf{m}_1 = (4.3, 1.3)^T$ and $\mathbf{m}_2 = (1.5, 0.3)^T$. From Eq. (12.2-5), the decision functions are

EXAMPLE 12.1:
Illustration of the minimum-distance classifier.

$$d_1(\mathbf{x}) = \mathbf{x}^T\mathbf{m}_1 - \frac{1}{2}\mathbf{m}_1^T\mathbf{m}_1$$

$$= 4.3x_1 + 1.3x_2 - 10.1$$

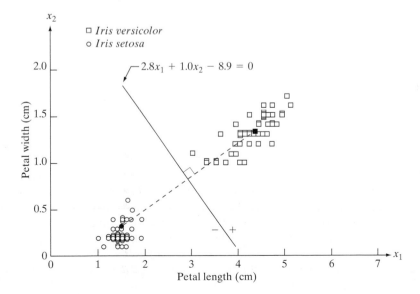

FIGURE 12.6
Decision boundary of minimum distance classifier for the classes of *Iris versicolor* and *Iris setosa*. The dark dot and square are the means.

and

$$d_2(\mathbf{x}) = \mathbf{x}^T\mathbf{m}_2 - \frac{1}{2}\mathbf{m}_2^T\mathbf{m}_2$$
$$= 1.5x_1 + 0.3x_2 - 1.17.$$

From Eq. (12.2-6), the equation of the boundary is

$$d_{12}(\mathbf{x}) = d_1(\mathbf{x}) - d_2(\mathbf{x})$$
$$= 2.8x_1 + 1.0x_2 - 8.9 = 0.$$

Figure 12.6 shows a plot of this boundary (note that the axes are not to the same scale). Substitution of any pattern vector from class ω_1 would yield $d_{12}(\mathbf{x}) > 0$. Conversely, any pattern from class ω_2 would yield $d_{12}(\mathbf{x}) < 0$. In other words, given an unknown pattern belonging to one of these two classes, the sign of $d_{12}(\mathbf{x})$ would be sufficient to determine the pattern's class membership.   ■

In practice, the minimum distance classifier works well when the distance between means is large compared to the spread or randomness of each class with respect to its mean. In Section 12.2.2 we show that the minimum distance classifier yields optimum performance (in terms of minimizing the average loss of misclassification) when the distribution of each class about its mean is in the form of a spherical "hypercloud" in n-dimensional pattern space.

The simultaneous occurrence of large mean separations and relatively small class spread occur seldomly in practice unless the system designer controls the nature of the input. An excellent example is provided by systems designed to read stylized character fonts, such as the familiar American Banker's Association E-13B font character set. As Fig. 12.7 shows, this particular font set consists of 14 characters that were purposely designed on a 9 × 7 grid in order to facilitate their reading. The characters usually are printed in ink that contains finely ground magnetic material. Prior to being read, the ink is subjected to a magnetic field, which accentuates each character to simplify detection. In other words, the segmentation problem is solved by artificially highlighting the key characteristics of each character.

The characters typically are scanned in a horizontal direction with a single-slit reading head that is narrower but taller than the characters. As the head moves across a character, it produces a 1-D electrical signal (a signature) that is conditioned to be proportional to the rate of increase or decrease of the character area under the head. For example, consider the waveform associated with the number 0 in Fig. 12.7. As the reading head moves from left to right, the area seen by the head begins to increase, producing a positive derivative (a positive rate of change). As the head begins to leave the left leg of the 0, the area under the head begins to decrease, producing a negative derivative. When the head is in the middle zone of the character, the area remains nearly constant, producing a zero derivative. This pattern repeats itself as the head enters the right leg of the character. The design of the font ensures that the waveform of each character is distinct from that of all others. It also ensures that the peaks and zeros of each waveform occur approximately on the vertical lines of the background

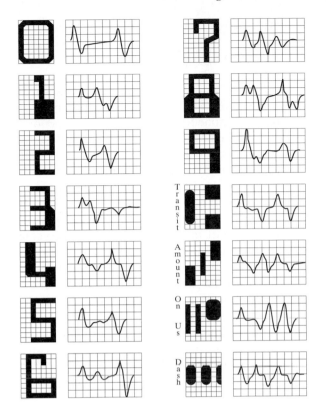

FIGURE 12.7
American Bankers Association E-13B font character set and corresponding waveforms.

grid on which these waveforms are displayed, as shown in Fig. 12.7. The E-13B font has the property that sampling the waveforms only at these points yields enough information for their proper classification. The use of magnetized ink aids in providing clean waveforms, thus minimizing scatter.

Designing a minimum distance classifier for this application is straightforward. We simply store the sample values of each waveform and let each set of samples be represented as a prototype vector $\mathbf{m}_j$, $j = 1, 2, \ldots, 14$. When an unknown character is to be classified, the approach is to scan it in the manner just described, express the grid samples of the waveform as a vector, $\mathbf{x}$, and identify its class by selecting the class of the prototype vector that yields the highest value in Eq. (12.2-5). High classification speeds can be achieved with analog circuits composed of resistor banks (see Problem 12.4).

Matching by correlation

We introduced the basic concept of image correlation in Section 4.6.4. Here, we consider it as the basis for finding matches of a subimage $w(x, y)$ of size $J \times K$ within an image $f(x, y)$ of size $M \times N$, where we assume that $J \leq M$ and $K \leq N$. Although the correlation approach can be expressed in vector form (see Problem 12.5), working directly with an image or subimage format is more intuitive (and traditional).

In its simplest form, the correlation between $f(x, y)$ and $w(x, y)$ is

$$c(x, y) = \sum_s \sum_t f(s, t)w(x + s, y + t) \tag{12.2-7}$$

for $x = 0, 1, 2, \ldots, M - 1, y = 0, 1, 2, \ldots, N - 1$, and the summation is taken over the image region where w and f overlap. Note by comparing this equation with Eq. (4.6-30) that it is implicitly assumed that the functions are real quantities and that we left out the MN constant. The reason is that we are going to use a normalized function in which these constants cancel out, and the definition given in Eq. (12.2-7) is used commonly in practice. We also used the symbols s and t in Eq. (12.2-7) to avoid confusion with m and n, which are used for other purposes in this chapter.

Figure 12.8 illustrates the procedure, where we assume that the origin of f is at its top left and the origin of w is at its center. For *one* value of (x, y), say, (x_0, y_0) inside f, application of Eq. (12.2-7) yields *one* value of c. As x and y are varied, w moves around the image area, giving the function $c(x, y)$. The maximum value(s) of c indicates the position(s) where w best matches f. Note that accuracy is lost for values of x and y near the edges of f, with the amount of error being in the correlation proportional to the size of w. This is the familiar border problem that we encountered numerous times in Chapter 3.

The correlation function given in Eq. (12.2-7) has the disadvantage of being sensitive to changes in the amplitude of f and w. For example, doubling all values of f doubles the value of $c(x, y)$. An approach frequently used to over-

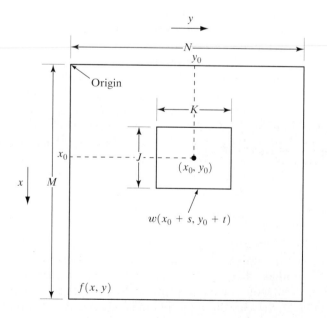

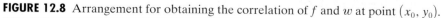

FIGURE 12.8 Arrangement for obtaining the correlation of f and w at point (x_0, y_0).

come this difficulty is to perform matching via the *correlation coefficient*, which is defined as

$$\gamma(x, y) = \frac{\sum_s \sum_t [f(s, t) - \bar{f}(s, t)][w(x + s, y + t) - \bar{w}]}{\left\{ \sum_s \sum_t [f(s, t) - \bar{f}(s, t)]^2 \sum_s \sum_t [w(x + s, y + t) - \bar{w}]^2 \right\}^{\frac{1}{2}}}$$

(12.2-8)

where $x = 0, 1, 2, \ldots, M - 1$, $y - 0, 1, 2, \ldots, N - 1$, $\bar{w}$ is the average value of the pixels in w (computed only once), $\bar{f}$ is the average value of f in the region coincident with the current location of w, and the summations are taken over the coordinates common to both f and w. The correlation coefficient $\gamma(x, y)$ is scaled in the range -1 to 1, independent of scale changes in the amplitude of f and w (see Problem 12.5).

■ Figure 12.9 illustrates the concepts just discussed. Figure 12.9(a) is $f(x, y)$ and Fig. 12.9(b) is $w(x, y)$. The correlation coefficient $\gamma(x, y)$ is shown as an image in Fig. 12.9(c). The higher (brighter) value of $\gamma(x, y)$ is in the position where the best match between f and w was found. ■

EXAMPLE 12.2:
Object matching via the correlation coefficient.

Although the correlation function can be normalized for amplitude changes via the correlation coefficient, obtaining normalization for changes in size and rotation can be difficult. Normalizing for size involves spatial scaling, a process that in itself adds a significant amount of computation. Normalizing for rotation is even more difficult. If a clue regarding rotation can be extracted from $f(x, y)$, then we simply rotate $w(x, y)$ so that it aligns itself with the degree of rotation in $f(x, y)$. However, if the nature of rotation is unknown, looking for the best match requires exhaustive rotations of $w(x, y)$. This procedure is impractical and, as a consequence, correlation seldom is used in cases when arbitrary or unconstrained rotation is present.

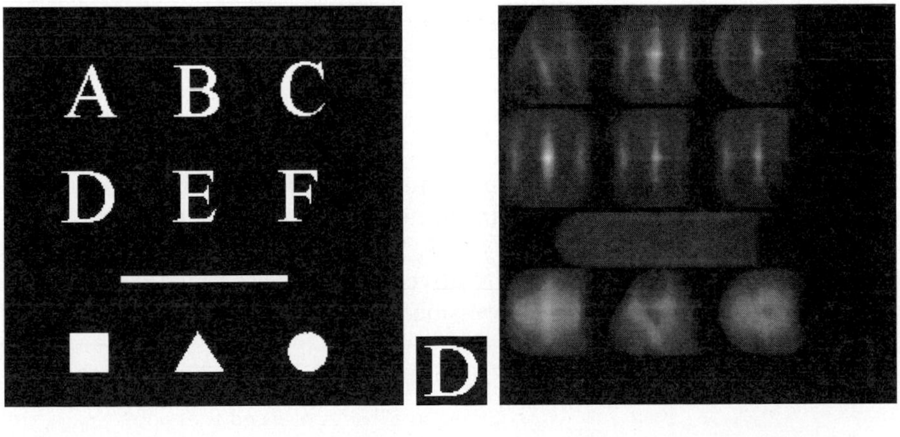

a b c
FIGURE 12.9
(a) Image.
(b) Subimage.
(c) Correlation coefficient of (a) and (b). Note that the highest (brighter) point in (c) occurs when subimage (b) is coincident with the letter "D" in (a).

In Section 4.6.4 we mentioned that correlation also can be carried out in the frequency domain via the FFT. If f and w are the same size, this approach can be more efficient than direct implementation of correlation in the spatial domain. Equation (12.2-7) is used when w is much smaller than f. A trade-off estimate performed by Campbell [1969] indicates that, if the number of nonzero terms in w is less than 132 (a subimage of approximately 13 × 13 pixels), direct implementation of Eq. (12.2-7) is more efficient than the FFT approach. This number, of course, depends on the machine and algorithms used, but it does indicate approximate subimage size at which the frequency domain should be considered as an alternative. The correlation coefficient is more difficult to implement in the frequency domain. It generally is computed directly in the spatial domain.

12.2.2 Optimum Statistical Classifiers

In this section we develop a probabilistic approach to recognition. As is true in most fields that deal with measuring and interpreting physical events, probability considerations become important in pattern recognition because of the randomness under which pattern classes normally are generated. As shown in the following discussion, it is possible to derive a classification approach that is optimal in the sense that, on average, its use yields the lowest probability of committing classification errors (see Problem 12.10).

Foundation

The probability that a particular pattern $\mathbf{x}$ comes from class ω_i is denoted $p(\omega_i/\mathbf{x})$. If the pattern classifier decides that $\mathbf{x}$ came from ω_j when it actually came from ω_i, it incurs a loss, denoted L_{ij}. As pattern $\mathbf{x}$ may belong to any one of W classes under consideration, the average loss incurred in assigning $\mathbf{x}$ to class ω_j is

$$r_j(\mathbf{x}) = \sum_{k=1}^{W} L_{kj} p(\omega_k/\mathbf{x}). \tag{12.2-9}$$

This equation often is called the *conditional average risk* or *loss* in decision-theory terminology.

From basic probability theory, we know that $p(A/B) = [p(A)p(B/A)]/p(B)$. Using this expression, we write Eq. (12.2-9) in the form

$$r_j(\mathbf{x}) = \frac{1}{p(\mathbf{x})} \sum_{k=1}^{W} L_{kj} p(\mathbf{x}/\omega_k) P(\omega_k) \tag{12.2-10}$$

See inside front cover

Consult the book web site for a brief review of probability theory.

where $p(\mathbf{x}/\omega_k)$ is the probability density function of the patterns from class ω_k and $P(\omega_k)$ is the probability of occurrence of class ω_k. Because $1/p(\mathbf{x})$ is positive and common to all the $r_j(\mathbf{x})$, $j = 1, 2, \ldots, W$, it can be dropped from Eq. (12.2-10) without affecting the relative order of these functions from the smallest to the largest value. The expression for the average loss then reduces to

$$r_j(\mathbf{x}) = \sum_{k=1}^{W} L_{kj} p(\mathbf{x}/\omega_k) P(\omega_k). \tag{12.2-11}$$

The classifier has W possible classes to choose from for any given unknown pattern. If it computes $r_1(\mathbf{x}), r_2(\mathbf{x}), \ldots, r_W(\mathbf{x})$ for each pattern $\mathbf{x}$ and assigns the pattern to the class with the smallest loss, the total average loss with respect to all decisions will be minimum. The classifier that minimizes the total average loss is called the *Bayes classifier*. Thus the Bayes classifier assigns an unknown pattern $\mathbf{x}$ to class ω_i if $r_i(\mathbf{x}) < r_j(\mathbf{x})$ for $j = 1, 2, \ldots, W; j \neq i$. In other words, $\mathbf{x}$ is assigned to class ω_i if

$$\sum_{k=1}^{W} L_{ki} p(\mathbf{x}/\omega_k) P(\omega_k) < \sum_{q=1}^{W} L_{qj} p(\mathbf{x}/\omega_q) P(\omega_q) \qquad (12.2\text{-}12)$$

for all $j; j \neq i$. The "loss" for a correct decision generally is assigned a value of zero, and the loss for any incorrect decision usually is assigned the same nonzero value (say, 1). Under these conditions, the loss function becomes

$$L_{ij} = 1 - \delta_{ij} \qquad (12.2\text{-}13)$$

where $\delta_{ij} = 1$ if $i = j$ and $\delta_{ij} = 0$ if $i \neq j$. Equation (12.2-13) indicates a loss of unity for incorrect decisions and a loss of zero for correct decisions. Substituting Eq. (12.2-13) into Eq. (12.2-11) yields

$$r_j(\mathbf{x}) = \sum_{k=1}^{W} \left(1 - \delta_{kj}\right) p(\mathbf{x}/\omega_k) P(\omega_k)$$
$$= p(\mathbf{x}) - p(\mathbf{x}/\omega_j) P(\omega_j). \qquad (12.2\text{-}14)$$

The Bayes classifier then assigns a pattern $\mathbf{x}$ to class ω_i if, for all $j \neq i$,

$$p(\mathbf{x}) - p(\mathbf{x}/\omega_i) P(\omega_i) < p(\mathbf{x}) - p(\mathbf{x}/\omega_j) P(\omega_j) \qquad (12.2\text{-}15)$$

or, equivalently, if

$$p(\mathbf{x}/\omega_i) P(\omega_i) > p(\mathbf{x}/\omega_j) P(\omega_j) \qquad j = 1, 2, \ldots, W; j \neq i. \quad (12.2\text{-}16)$$

With reference to the discussion leading to Eq. (12.2-1), we see that the Bayes classifier for a 0-1 loss function is nothing more than computation of decision functions of the form

$$d_j(\mathbf{x}) = p(\mathbf{x}/\omega_j) P(\omega_j) \qquad j = 1, 2, \ldots, W \qquad (12.2\text{-}17)$$

where a pattern vector $\mathbf{x}$ is assigned to the class whose decision function yields the largest numerical value.

The decision functions given in Eq. (12.2-7) are optimal in the sense that they minimize the average loss in misclassification. For this optimality to hold, however, the probability density functions of the patterns in each class, as well as the probability of occurrence of each class, must be known. The latter requirement usually is not a problem. For instance, if all classes are equally likely to occur, then $P(\omega_j) = 1/M$. Even if this condition is not true, these probabilities generally can be inferred from knowledge of the problem. Estimation of the probability density functions $p(\mathbf{x}/\omega_j)$ is another matter. If the pattern vectors, $\mathbf{x}$, are n dimensional, then $p(\mathbf{x}/\omega_j)$ is a function of n variables, which, if its form is not known, requires methods from multivariate probability theory for its estimation. These methods are difficult to apply in practice, especially if the number of representative patterns from each class is not large or if the underlying form of the probability density functions is not well behaved. For these reasons, use of the Bayes classifier generally is based on the assumption of an analytic expression for the various density functions and then an estimation of the necessary parameters from sample patterns from each class. By far the most prevalent form assumed for $p(\mathbf{x}/\omega_j)$ is the Gaussian probability density function. The closer this assumption is to reality, the closer the Bayes classifier approaches the minimum average loss in classification.

Bayes classifier for Gaussian pattern classes

To begin, let us consider a 1-D problem ($n = 1$) involving two pattern classes ($W = 2$) governed by Gaussian densities, with means m_1 and m_2 and standard deviations σ_1 and σ_2, respectively. From Eq. (12.2-17) the Bayes decision functions have the form

$$d_j(x) = p(x/\omega_j)P(\omega_j)$$

$$= \frac{1}{\sqrt{2\pi}\sigma_j} e^{-\frac{(x-m_j)^2}{2\sigma_j^2}} P(\omega_j) \qquad j = 1, 2 \qquad (12.2\text{-}18)$$

where the patterns are now scalars, denoted by x. Figure 12.10 shows a plot of the probability density functions for the two classes. The boundary between the two classes is a single point, denoted x_0 such that $d_1(x_0) = d_2(x_0)$. If the two classes are equally likely to occur, then $P(\omega_1) = P(\omega_2) = 1/2$, and the decision boundary is the value of x_0 for which $p(x_0/\omega_1) = p(x_0/\omega_2)$. This point is the intersection of the two probability density functions, as shown in Fig. 12.10. Any pattern (point) to the right of x_0 is classified as belonging to class ω_1. Similarly, any pattern to the left of x_0 is classified as belonging to class ω_2. When the classes are not equally likely to occur, x_0 moves to the left if class ω_1 is more likely to occur or, conversely, to the right if class ω_2 is more likely to occur. This result is to be expected, because the classifier is trying to minimize the loss of misclassification. For instance, in the extreme case, if class ω_2 never occurs, the classifier would never make a mistake by always assigning all patterns to class ω_1 (that is, x_0 would move to negative infinity).

FIGURE 12.10
Probability density functions for two 1-D pattern classes. The point x_0 shown is the decision boundary if the two classes are equally likely to occur.

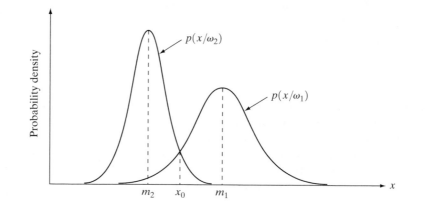

In the n-dimensional case, the Gaussian density of the vectors in the jth pattern class has the form

$$p(\mathbf{x}/\omega_j) = \frac{1}{(2\pi)^{n/2}|\mathbf{C}_j|^{1/2}} e^{-\frac{1}{2}(\mathbf{x}-\mathbf{m}_j)^T \mathbf{C}_j^{-1}(\mathbf{x}-\mathbf{m}_j)} \tag{12.2-19}$$

where each density is specified completely by its mean vector $\mathbf{m}_j$ and covariance matrix $\mathbf{C}_j$, which are defined as

$$\mathbf{m}_j = E_j\{\mathbf{x}\} \tag{12.2-20}$$

and

$$\mathbf{C}_j = E_j\{(\mathbf{x} - \mathbf{m}_j)(\mathbf{x} - \mathbf{m}_j)^T\} \tag{12.2-21}$$

where $E_j\{\cdot\}$ denotes the expected value of the argument over the patterns of class ω_j. In Eq. (12.2-19), n is the dimensionality of the pattern vectors, and $|\mathbf{C}_j|$ is the determinant of the matrix $\mathbf{C}_j$. Approximating the expected value E_j by the average value of the quantities in question yields an estimate of the mean vector and covariance matrix:

$$\mathbf{m}_j = \frac{1}{N_j} \sum_{\mathbf{x} \in \omega_j} \mathbf{x} \tag{12.2-22}$$

and

$$\mathbf{C}_j = \frac{1}{N_j} \sum_{\mathbf{x} \in \omega_j} \mathbf{x}\mathbf{x}^T - \mathbf{m}_j\mathbf{m}_j^T \tag{12.2-23}$$

where N_j is the number of pattern vectors from class ω_j, and the summation is taken over these vectors. Later in this section we give an example of how to use these two expressions.

The covariance matrix is symmetric and positive semidefinite. As explained in Section 11.4, the diagonal element c_{kk} is the variance of the kth element of the pattern vectors. The off-diagonal element c_{jk} is the covariance of x_j and x_k. The multivariate Gaussian density function reduces to the product of the univariate Gaussian density of each element of $\mathbf{x}$ when the off-diagonal elements of the covariance matrix are zero. This happens when the vector elements x_j and x_k are uncorrelated.

According to Eq. (12.2-17), the Bayes decision function for class ω_j is $d_j(\mathbf{x}) = p(\mathbf{x}/\omega_j)P(\omega_j)$. However, because of the exponential form of the Gaussian density, working with the natural logarithm of this decision function is more convenient. In other words, we can use the form

$$\begin{aligned} d_j(\mathbf{x}) &= \ln[p(\mathbf{x}/\omega_j)P(\omega_j)] \\ &= \ln p(\mathbf{x}/\omega_j) + \ln P(\omega_j). \end{aligned} \tag{12.2-24}$$

This expression is equivalent to Eq. (12.2-17) in terms of classification performance because the logarithm is a monotonically increasing function. In other words, the numerical *order* of the decision functions in Eqs. (12.2-17) and (12.2-24) is the same. Substituting Eq. (12.2-19) into Eq. (12.2-24) yields

$$d_j(\mathbf{x}) = \ln P(\omega_j) - \frac{n}{2} \ln 2\pi - \frac{1}{2}\ln|\mathbf{C}_j| - \frac{1}{2}[(\mathbf{x} - \mathbf{m}_j)^T \mathbf{C}_j^{-1}(\mathbf{x} - \mathbf{m}_j)]. \tag{12.2-25}$$

See inside front cover

Consult the book web site for a brief review of vectors and matrices.

The term $(n/2) \ln 2\pi$ is the same for all classes, so it can be eliminated from Eq. (12.2-25), which then becomes

$$d_j(\mathbf{x}) = \ln P(\omega_j) - \frac{1}{2}\ln|\mathbf{C}_j| - \frac{1}{2}\left[(\mathbf{x} - \mathbf{m}_j)^T \mathbf{C}_j^{-1}(\mathbf{x} - \mathbf{m}_j)\right] \quad (12.2\text{-}26)$$

for $j = 1, 2, \ldots, W$. Equation (12.2-26) represents the Bayes decision functions for Gaussian pattern classes under the condition of a 0-1 loss function.

The decision functions in Eq. (12.2-26) are hyperquadrics (quadratic functions in n-dimensional space), because no terms higher than the second degree in the components of $\mathbf{x}$ appear in the equation. Clearly, then, the best that a Bayes classifier for Gaussian patterns can do is to place a general second-order decision surface between each pair of pattern classes. If the pattern populations are truly Gaussian, however, no other surface would yield a lesser average loss in classification.

If all covariance matrices are equal, then $\mathbf{C}_j = \mathbf{C}$, for $j = 1, 2, \ldots, W$. By expanding Eq. (12.2-26) and dropping all terms independent of j, we obtain

$$d_j(\mathbf{x}) = \ln P(\omega_j) + \mathbf{x}^T \mathbf{C}^{-1}\mathbf{m}_j - \frac{1}{2}\mathbf{m}_j^T \mathbf{C}^{-1}\mathbf{m}_j, \quad\quad (12.2\text{-}27)$$

which are linear decision functions (*hyperplanes*) for $j = 1, 2, \ldots, W$.

If, in addition, $\mathbf{C} = \mathbf{I}$, where $\mathbf{I}$ is the identity matrix, and also $P(\omega_j) = 1/W$, for $j = 1, 2, \ldots, W$, then

$$d_j(\mathbf{x}) = \mathbf{x}^T \mathbf{m}_j - \frac{1}{2}\mathbf{m}_j^T \mathbf{m}_j \quad\quad j = 1, 2, \ldots, W. \quad\quad (12.2\text{-}28)$$

These are the decision functions for a minimum distance classifier, as given in Eq. (12.2-5). Thus the minimum distance classifier is optimum in the Bayes sense if (1) the pattern classes are Gaussian, (2) all covariance matrices are equal to the identity matrix, and (3) all classes equally likely to occur. Gaussian pattern classes satisfying these conditions are spherical clouds of identical shape in n dimensions (called *hyperspheres*). The minimum distance classifier establishes a hyperplane between every pair of classes, with the property that the hyperplane is the perpendicular bisector of the line segment joining the center of the pair of hyperspheres. In two dimensions, the classes constitute circular regions, and the boundaries become lines that bisect the line segment joining the center of every pair of such circles.

EXAMPLE 12.3:
A Bayes classifier for three-dimensional patterns.

■ Figure 12.11 shows a simple arrangement of two pattern classes in three dimensions. We use these patterns to illustrate the mechanics of implementing the Bayes classifier, assuming that the patterns of each class are samples from a Gaussian distribution.

Applying Eq. (12.2-22) to the patterns of Fig. 12.11 yields

$$\mathbf{m}_1 = \frac{1}{4}\begin{bmatrix} 3 \\ 1 \\ 1 \end{bmatrix} \quad \text{and} \quad \mathbf{m}_2 = \frac{1}{4}\begin{bmatrix} 1 \\ 3 \\ 3 \end{bmatrix}.$$

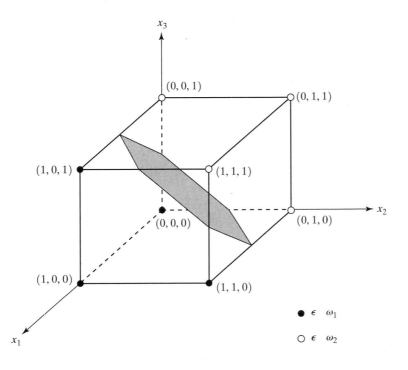

FIGURE 12.11
Two simple
pattern classes
and their Bayes
decision boundary
(shown shaded).

Similarly, applying Eq. (12.2-23) to the two pattern classes in turn yields two covariance matrices, which in this case are equal:

$$\mathbf{C}_1 = \mathbf{C}_2 = \frac{1}{16} \begin{bmatrix} 3 & 1 & 1 \\ 1 & 3 & -1 \\ 1 & -1 & 3 \end{bmatrix}.$$

Because the covariance matrices are equal the Bayes decision functions are given by Eq. (12.2-27). If we assume that $P(\omega_1) = P(\omega_2) = 1/2$, then Eq. (12.2-28) applies, giving

$$d_j(\mathbf{x}) = \mathbf{x}^T \mathbf{C}^{-1} \mathbf{m}_j - \frac{1}{2} \mathbf{m}_j^T \mathbf{C}^{-1} \mathbf{m}_j$$

in which

$$\mathbf{C}^{-1} = \begin{bmatrix} 8 & -4 & -4 \\ -4 & 8 & 4 \\ -4 & 4 & 8 \end{bmatrix}.$$

Carrying out the vector-matrix expansion for $d_j(\mathbf{x})$ provides the decision functions:

$$d_1(\mathbf{x}) = 4x_1 - 1.5 \quad \text{and} \quad d_2(\mathbf{x}) = -4x_1 + 8x_2 + 8x_3 - 5.5.$$

The decision surface separating the two classes then is

$$d_1(\mathbf{x}) - d_2(\mathbf{x}) = 8x_1 - 8x_2 - 8x_3 + 4 = 0.$$

Figure 12.11 shows a section of this surface, where we note that the classes were separated effectively.   ■

One of the most successful applications of the Bayes classifier approach is in the classification of remotely sensed imagery generated by multispectral scanners aboard aircraft, satellites, or space stations. The voluminous image data generated by these platforms make automatic image classification and analysis a task of considerable interest in remote sensing. The applications of remote sensing are varied and include land use, crop inventory, crop disease detection, forestry, air and water quality monitoring, geological studies, weather prediction, and a score of other applications having environmental significance. The following example shows a typical application.

EXAMPLE 12.4:
Classification of multispectral data using the Bayes classifier.

■ As discussed in Sections 1.3.4 and 11.4, a multispectral scanner responds to electromagnetic energy in selected wavelength bands; for example, 0.40-0.44, 0.58-0.62, 0.66-0.72, and 0.80-1.00 microns. These ranges are in the violet, green, red, and infrared bands, respectively. A region on the ground scanned in this manner produces four digital images, one image for each band. If the images are registered, a condition which is generally true in practice, they can be visualized as being stacked one behind the other, as Fig. 12.12 shows. Thus, just as we did in Section 11.4, every point on the ground can be represented by a 4-element pattern vector of the form $\mathbf{x} = (x_1, x_2, x_3, x_4)^T$, where x_1 is a shade of violet, x_2 is a shade of green, and so on. If the images are of size 512×512 pixels, each stack of four multispectral images can be represented by 262,144 4-dimensional pattern vectors.

As noted previously, the Bayes classifier for Gaussian patterns requires estimation of the mean vector and covariance matrix for each class. In remote sensing applications these estimates are obtained by collecting multispectral data for each region of interest and then using these samples, as described in the preceding example. Figure 12.13(a) shows a typical image sensed remotely from an aircraft (this is a monochrome version of a multispectral original). In this

FIGURE 12.12
Formation of a pattern vector from registered pixels of four digital images generated by a multispectral scanner.

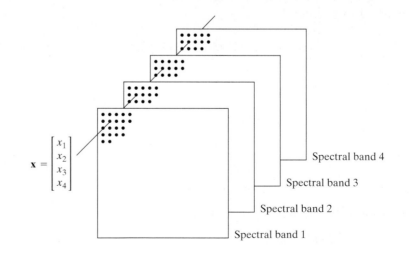

$$\mathbf{x} = \begin{bmatrix} x_1 \\ x_2 \\ x_3 \\ x_4 \end{bmatrix}$$

Spectral band 4

Spectral band 3

Spectral band 2

Spectral band 1

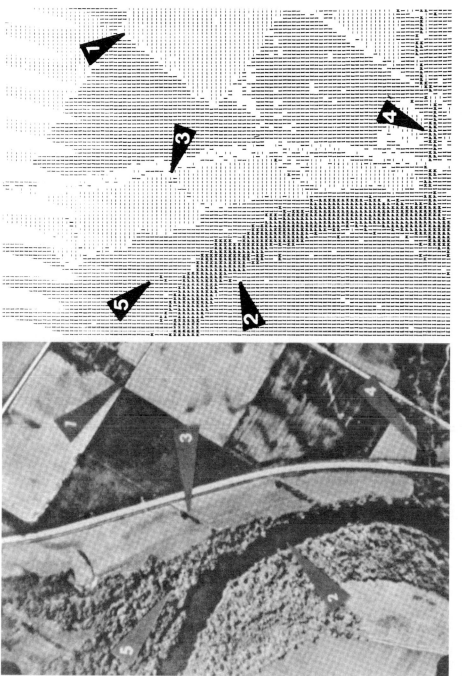

a b

FIGURE 12.13 (a) Multispectral image. (b) Printout of machine classification results using a Bayes classifier. (Courtesy of the Laboratory for Applications of Remote Sensing, Purdue University.)

particular case, the problem was to classify areas such as vegetation, water, and bare soil. Figure 12.13(b) shows the results of machine classification, using a Gaussian Bayes classifier. The arrows indicate some features of interest. Arrow 1 points to a corner of a field of green vegetation, and arrow 2 points to a river. Arrow 3 identifies a small hedgerow between two areas of bare soil. Arrow 4 indicates a tributary correctly identified by the system. Arrow 5 points to a small pond that is almost indistinguishable in Fig. 12.13(a). Comparing the original image with the computer output reveals recognition results that are very close to those that a human would generate by visual analysis.   ■

Before leaving this section, it is of interest to note that pixel-by-pixel classification of an image as described in the previous example actually segments the image into various classes. This approach is like segmentation by thresholding with several variables, as discussed briefly in Section 10.3.7.

12.2.3 Neural Networks

The approaches discussed in the preceding two sections are based on the use of sample patterns to estimate statistical parameters of each pattern class. The minimum distance classifier is specified completely by the mean vector of each class. Similarly, the Bayes classifier for Gaussian populations is specified completely by the mean vector and covariance matrix of each class. The patterns (of *known* class membership) used to estimate these parameters usually are called *training patterns*, and a set of such patterns from each class is called a *training set*. The process by which a training set is used to obtain decision functions is called *learning* or *training*.

In the two approaches just discussed, training is a simple matter. The training patterns of each class are used to compute the parameters of the decision function corresponding to that class. After the parameters in question have been estimated, the structure of the classifier is fixed, and its eventual performance will depend on how well the actual pattern populations satisfy the underlying statistical assumptions made in the derivation of the classification method being used.

The statistical properties of the pattern classes in a problem often are unknown or cannot be estimated (recall our brief discussion in the preceding section regarding the difficulty of working with multivariate statistics). In practice, such decision-theoretic problems are best handled by methods that yield the required decision functions directly via training. Then, making assumptions regarding the underlying probability density functions or other probabilistic information about the pattern classes under consideration is unnecessary. In this section we discuss various approaches that meet this criterion.

Background

The essence of the material that follows is the use of a multitude of elemental nonlinear computing elements (called *neurons*) organized as networks reminiscent of the way in which neurons are believed to be interconnected in the brain. The resulting models are referred to by various names, including *neural networks, neurocomputers, parallel distributed processing* (PDP) *models, neuromorphic systems, layered self-adaptive networks*, and *connectionist models*. Here, we use the name *neural networks*, or *neural nets* for short. We use these networks as vehicles for adaptively developing the coefficients of decision functions via successive presentations of training sets of patterns.

Interest in neural networks dates back to the early 1940s, as exemplified by the work of McCulloch and Pitts [1943]. They proposed neuron models in the form of binary threshold devices and stochastic algorithms involving sudden 0-1 and 1-0 changes of states in neurons as the bases for modeling neural systems. Subsequent work by Hebb [1949] was based on mathematical models that attempted to capture the concept of learning by reinforcement or association.

During the mid-1950s and early 1960s, a class of so-called *learning machines* originated by Rosenblatt [1959, 1962] caused significant excitement among researchers and practitioners of pattern recognition theory. The reason for the great interest in these machines, called *perceptrons*, was the development of mathematical proofs showing that perceptrons, when trained with linearly separable training sets (i.e., training sets separable by a hyperplane), would converge to a solution in a finite number of iterative steps. The solution took the form of coefficients of hyperplanes capable of correctly separating the classes represented by patterns of the training set.

Unfortunately, the expectations following discovery of what appeared to be a well-founded theoretic model of learning soon met with disappointment. The basic perceptron and some of its generalizations at the time were simply inadequate for most pattern recognition tasks of practical significance. Subsequent attempts to extend the power of perceptron-like machines by considering multiple layers of these devices, although conceptually appealing, lacked effective training algorithms such as those that had created interest in the perceptron itself. The state of the field of learning machines in the mid-1960s was summarized by Nilsson [1965]. A few years later, Minsky and Papert [1969] presented a discouraging analysis of the limitation of perceptron-like machines. This view was held as late as the mid-1980s, as evidenced by comments by Simon [1986]. In this work, originally published in French in 1984, Simon dismisses the perceptron under the heading "Birth and Death of a Myth."

More recent results by Rumelhart, Hinton, and Williams [1986] dealing with the development of new training algorithms for multilayer perceptrons have changed matters considerably. Their basic method, often called the *generalized delta rule for learning by backpropagation*, provides an effective training method for multilayer machines. Although this training algorithm cannot be shown to converge to a solution in the sense of the analogous proof for the single-layer perceptron, the generalized delta rule has been used successfully in numerous problems of practical interest. This success has established multilayer perceptron-like machines as one of the principal models of neural networks currently in use.

Perceptron for two pattern classes

In its most basic form, the perceptron learns a linear decision function that dichotomizes two linearly separable training sets. Figure 12.14(a) shows schematically the perceptron model for two pattern classes. The response of this basic

FIGURE 12.14
Two equivalent
representations of
the perceptron
model for two
pattern classes.

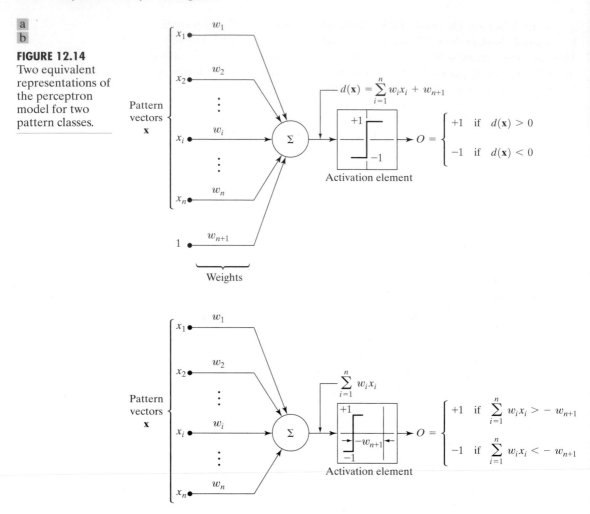

device is based on a weighted sum of its inputs; that is,

$$d(\mathbf{x}) = \sum_{i=1}^{n} w_i x_i + w_{n+1}, \tag{12.2-29}$$

which is a linear decision function with respect to the components of the pattern vectors. The coefficients $w_i, i = 1, 2, \ldots, n, n + 1$, called *weights*, modify the inputs before they are summed and fed into the threshold element. In this sense, weights are analogous to synapses in the human neural system. The function that maps the output of the summing junction into the final output of the device sometimes is called the *activation function*.

When $d(\mathbf{x}) > 0$ the threshold element causes the output of the perceptron to be +1, indicating that the pattern $\mathbf{x}$ was recognized as belonging to class ω_1.

The reverse is true when $d(\mathbf{x}) < 0$. This mode of operation agrees with the comments made earlier in connection with Eq. (12.2-2) regarding the use of a single decision function for two pattern classes. When $d(\mathbf{x}) = 0$, $\mathbf{x}$ lies on the decision surface separating the two pattern classes, giving an indeterminate condition. The decision boundary implemented by the perceptron is obtained by setting Eq. (12.2-29) equal to zero:

$$d(\mathbf{x}) = \sum_{i=1}^{n} w_i x_i + w_{n+1} = 0 \qquad (12.2\text{-}30)$$

or

$$w_1 x_1 + w_2 x_2 + \cdots + w_n x_n + w_{n+1} = 0, \qquad (12.2\text{-}31)$$

which is the equation of a hyperplane in n-dimensional pattern space. Geometrically, the first n coefficients establish the orientation of the hyperplane, whereas the last coefficient, w_{n+1}, is proportional to the perpendicular distance from the origin to the hyperplane. Thus if $w_{n+1} = 0$, the hyperplane goes through the origin of the pattern space. Similarly, if $w_j = 0$, the hyperplane is parallel to the x_j-axis.

The output of the threshold element in Fig. 12.14(a) depends on the sign of $d(\mathbf{x})$. Instead of testing the entire function to determine whether it is positive or negative, we could test the summation part of Eq. (12.2-29) against the term w_{n+1}, in which case the output of the system would be

$$O = \begin{cases} +1 & \text{if } \sum_{i=1}^{n} w_i x_i > -w_{n+1} \\[2mm] -1 & \text{if } \sum_{i=1}^{n} w_i x_i < -w_{n+1}. \end{cases} \qquad (12.2\text{-}32)$$

This implementation is equivalent to Fig. 12.14(a) and is shown in Fig. 12.14(b), the only differences being that the threshold function is displaced by an amount $-w_{n+1}$ and that the constant unit input is no longer present. We return to the equivalence of these two formulations later in this section when we discuss implementation of multilayer neural networks.

Another formulation used frequently is to augment the pattern vectors by appending an additional $(n + 1)$st element, which is always equal to 1, regardless of class membership. That is, an augmented pattern vector $\mathbf{y}$ is created from a pattern vector $\mathbf{x}$ by letting $y_i = x_i, i = 1, 2, \ldots, n$, and appending the additional element $y_{n+1} = 1$. Equation (12.2-29) then becomes

$$\begin{aligned} d(\mathbf{y}) &= \sum_{i=1}^{n+1} w_i y_i \\ &= \mathbf{w}^T \mathbf{y} \end{aligned} \qquad (12.2\text{-}33)$$

where $\mathbf{y} = (y_1, y_2, \ldots, y_n, 1)^T$ is now an *augmented pattern vector*, and $\mathbf{w} = (w_1, w_2, \ldots, w_n, w_{n+1})^T$ is called the *weight vector*. This expression is usually more convenient in terms of notation. Regardless of the formulation used, however, the key problem is to find $\mathbf{w}$ by using a given training set of pattern vectors from each of two classes.

Training algorithms

The algorithms developed in the following discussion are representative of the numerous approaches proposed over the years for training perceptrons.

Linearly separable classes   A simple, iterative algorithm for obtaining a solution weight vector for two linearly separable training sets follows. For two training sets of augmented pattern vectors belonging to pattern classes ω_1 and ω_2, respectively, let $\mathbf{w}(1)$ represent the initial weight vector, which may be chosen arbitrarily. Then, at the kth iterative step, if $\mathbf{y}(k) \in \omega_1$ and $\mathbf{w}^T(k)\mathbf{y}(k) \leq 0$, replace $\mathbf{w}(k)$ by

$$\mathbf{w}(k + 1) = \mathbf{w}(k) + c\mathbf{y}(k) \tag{12.2-34}$$

where c is a positive correction increment. Conversely, if $\mathbf{y}(k) \leq \omega_2$ and $\mathbf{w}^T(k)\mathbf{y}(k) \geq 0$, replace $\mathbf{w}(k)$ with

$$\mathbf{w}(k + 1) = \mathbf{w}(k) - c\mathbf{y}(k). \tag{12.2-35}$$

Otherwise, leave $\mathbf{w}(k)$ unchanged:

$$\mathbf{w}(k + 1) = \mathbf{w}(k). \tag{12.2-36}$$

This algorithm makes a change in $\mathbf{w}$ only if the pattern being considered at the kth step in the training sequence is misclassified. The correction increment c is assumed to be positive and, for now, to be constant. This algorithm sometimes is referred to as the *fixed increment correction rule*.

Convergence of the algorithm occurs when the entire training set for both classes is cycled through the machine without any errors. The fixed increment correction rule converges in a finite number of steps if the two training sets of patterns are linearly separable. A proof of this result, sometimes called the *perceptron training theorem*, can be found in the books by Duda, Hart, and Stork [2001]; Tou and Gonzalez [1974]; and Nilsson [1965].

EXAMPLE 12.5:
Illustration of the perceptron algorithm.

■ Consider the two training sets shown in Fig. 12.15(a), each consisting of two patterns. The training algorithm will be successful because the two training sets are linearly separable. Before the algorithm is applied the patterns are augmented,

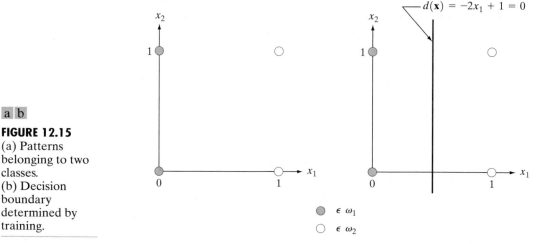

a b

FIGURE 12.15
(a) Patterns belonging to two classes.
(b) Decision boundary determined by training.

yielding the training set $\{(0, 0, 1)^T, (0, 1, 1)^T\}$ for class ω_1 and $\{(1, 0, 1)^T, (1, 1, 1)^T\}$ for class ω_2. Letting $c = 1, \mathbf{w}(1) = \mathbf{0}$, and presenting the patterns in order results in the following sequence of steps:

$$\mathbf{w}^T(1)\mathbf{y}(1) = [0, 0, 0]\begin{bmatrix} 0 \\ 0 \\ 1 \end{bmatrix} = 0 \qquad \mathbf{w}(2) = \mathbf{w}(1) + \mathbf{y}(1) = \begin{bmatrix} 0 \\ 0 \\ 1 \end{bmatrix}$$

$$\mathbf{w}^T(2)\mathbf{y}(2) = [0, 0, 1]\begin{bmatrix} 0 \\ 1 \\ 1 \end{bmatrix} = 1 \qquad \mathbf{w}(3) = \mathbf{w}(2) = \begin{bmatrix} 0 \\ 0 \\ 1 \end{bmatrix}$$

$$\mathbf{w}^T(3)\mathbf{y}(3) = [0, 0, 1]\begin{bmatrix} 1 \\ 0 \\ 1 \end{bmatrix} = 1 \qquad \mathbf{w}(4) = \mathbf{w}(3) - \mathbf{y}(3) = \begin{bmatrix} -1 \\ 0 \\ 0 \end{bmatrix}$$

$$\mathbf{w}^T(4)\mathbf{y}(4) = [-1, 0, 0]\begin{bmatrix} 1 \\ 1 \\ 1 \end{bmatrix} = -1 \qquad \mathbf{w}(5) = \mathbf{w}(4) = \begin{bmatrix} -1 \\ 0 \\ 0 \end{bmatrix}$$

where corrections in the weight vector were made in the first and third steps because of misclassifications, as indicated in Eqs. (12.2-34) and (12.2-35). Because a solution has been obtained only when the algorithm yields a complete error-free iteration through all training patterns, the training set must be presented again. The machine learning process is continued by letting $\mathbf{y}(5) = \mathbf{y}(1)$, $\mathbf{y}(6) = \mathbf{y}(2)$, $\mathbf{y}(7) = \mathbf{y}(3)$, and $\mathbf{y}(8) = \mathbf{y}(4)$, and proceeding in the same manner. Convergence is achieved at $k = 14$, yielding the solution weight vector $\mathbf{w}(14) = (-2, 0, 1)^T$. The corresponding decision function is $d(\mathbf{y}) = -2y_1 + 1$. Going back to the original pattern space by letting $x_i = y_i$ yields $d(\mathbf{x}) = -2x_1 + 1$, which, when set equal to zero, becomes the equation of the decision boundary shown in Fig. 12.15(b).   ■

Nonseparable classes   In practice, linearly separable pattern classes are the (rare) exception, rather than the rule. Consequently, a significant amount of research effort during the 1960s and 1970s went into development of techniques designed to handle nonseparable pattern classes. With recent advances in the training of neural networks, many of the methods dealing with nonseparable behavior have become merely items of historical interest. One of the early methods, however, is directly relevant to this discussion: the original delta rule. Known as the *Widrow-Hoff*, or *least-mean-square* (LMS) *delta rule* for training perceptrons, the method minimizes the error between the actual and desired response at any training step.

Consider the criterion function

$$J(\mathbf{w}) = \frac{1}{2}(r - \mathbf{w}^T\mathbf{y})^2 \tag{12.2-37}$$

where r is the desired response (that is, $r = +1$ if the augmented training pattern vector $\mathbf{y}$ belongs to class ω_1, and $r = -1$ if $\mathbf{y}$ belongs to class ω_2). The task is to adjust **w** incrementally in the direction of the negative gradient of $J(\mathbf{w})$ in order to seek the minimum of this function, which occurs when $r = \mathbf{w}^T\mathbf{y}$; that is, the minimum corresponds to correct classification. If $\mathbf{w}(k)$ represents the weight vector at the kth iterative step, a general gradient descent algorithm may be written as

$$\mathbf{w}(k+1) = \mathbf{w}(k) - \alpha\left[\frac{\partial J(\mathbf{w})}{\partial \mathbf{w}}\right]_{\mathbf{w}=\mathbf{w}(k)} \tag{12.2-38}$$

where $\mathbf{w}(k+1)$ is the new value of **w**, and $\alpha > 0$ gives the magnitude of the correction. From Eq. (12.2-37),

$$\frac{\partial J(\mathbf{w})}{\partial \mathbf{w}} = -(r - \mathbf{w}^T\mathbf{y})\mathbf{y}. \tag{12.2-39}$$

Substituting this result into Eq. (12.2-38) yields

$$\mathbf{w}(k+1) = \mathbf{w}(k) + \alpha\left[r(k) - \mathbf{w}^T(k)\mathbf{y}(k)\right]\mathbf{y}(k) \tag{12.2-40}$$

with the starting weight vector, $\mathbf{w}(1)$, being arbitrary.

By defining the change (delta) in weight vector as

$$\Delta\mathbf{w} = \mathbf{w}(k+1) - \mathbf{w}(k) \tag{12.2-41}$$

we can write Eq. (12.2-40) in the form of a *delta correction algorithm*:

$$\Delta\mathbf{w} = \alpha e(k)\mathbf{y}(k) \tag{12.2-42}$$

where

$$e(k) = r(k) - \mathbf{w}^T(k)\mathbf{y}(k) \tag{12.2-43}$$

is the error committed with weight vector $\mathbf{w}(k)$ when pattern $\mathbf{y}(k)$ is presented.

Equation (12.2-43) gives the error with weight vector $\mathbf{w}(k)$. If we change it to $\mathbf{w}(k+1)$, but leave the pattern the same, the error becomes

$$e(k) = r(k) - \mathbf{w}^T(k+1)\mathbf{y}(k). \tag{12.2-44}$$

The change in error then is

$$\begin{aligned}\Delta e(k) &= \left[r(k) - \mathbf{w}^T(k+1)\mathbf{y}(k)\right] - \left[r(k) - \mathbf{w}^T(k)\mathbf{y}(k)\right] \\ &= -\left[\mathbf{w}^T(k+1) - \mathbf{w}^T(k)\right]\mathbf{y}(k) \\ &= -\Delta\mathbf{w}^T\mathbf{y}(k).\end{aligned} \tag{12.2-45}$$

But $\Delta\mathbf{w} = \alpha e(k)\mathbf{y}(k)$, so

$$\begin{aligned}\Delta e &= -\alpha e(k)\mathbf{y}^T(k)\mathbf{y}(k) \\ &= -\alpha e(k)\|\mathbf{y}(k)\|^2.\end{aligned} \tag{12.2-46}$$

Hence changing the weights reduces the error by a factor $\alpha\|\mathbf{y}(k)\|^2$. The next input pattern starts the new adaptation cycle, reducing the next error by a factor $\alpha\|\mathbf{y}(k+1)\|^2$, and so on.

The choice of α controls stability and speed of convergence (Widrow and Stearns [1985]). Stability requires that $0 < \alpha < 2$. A practical range for α is $0.1 < \alpha < 1.0$. Although the proof is not shown here, the algorithm of

Eq. (12.2-40) or Eqs. (12.2-42) and (12.2-43) converges to a solution that minimizes the mean square error over the patterns of the training set. When the pattern classes are separable, the solution given by the algorithm just discussed may or may not produce a separating hyperplane. That is, a mean-square-error solution does not imply a solution in the sense of the perceptron training theorem. This uncertainty is the price of using an algorithm that converges under both the separable and nonseparable cases in this particular formulation.

The two perceptron training algorithms discussed thus far can be extended to more than two classes and to nonlinear decision functions. Based on the historical comments made earlier, exploring multiclass training algorithms here has little merit. Instead, we address multiclass training in the context of neural networks.

Multilayer feedforward neural networks

In this section we focus on decision functions of multiclass pattern recognition problems, independent of whether or not the classes are separable, and involving architectures that consist of layers of perceptron computing elements.

Basic architecture   Figure 12.16 shows the architecture of the neural network model under consideration. It consists of layers of structurally identical computing nodes (neurons) arranged so that the output of every neuron in one layer feeds into the input of every neuron in the next layer. The number of neurons in the first layer, called layer A, is N_A. Often, $N_A - n$, the dimensionality of the input pattern vectors. The number of neurons in the output layer, called layer Q, is denoted N_Q. The number N_Q equals W, the number of pattern classes that the neural network has been trained to recognize. The network recognizes a pattern vector $\mathbf{x}$ as belonging to class ω_i if the ith output of the network is "high" while all other outputs are "low," as explained in the following discussion.

As the blowup in Fig. 12.16 shows, each neuron has the same form as the perceptron model discussed earlier (see Fig. 12.14), with the exception that the hard-limiting activation function has been replaced by a soft-limiting "sigmoid" function. Differentiability along all paths of the neural network is required in the development of the training rule. The following sigmoid activation function has the necessary differentiability:

$$h_j(I_j) = \frac{1}{1 + e^{-(I_j + \theta_j)/\theta_o}} \tag{12.2-47}$$

where $I_j, j = 1, 2, \ldots, N_J$, is the input to the activation element of each node in layer J of the network, θ_j is an offset, and θ_o controls the shape of the sigmoid function.

Equation (12.2-47) is plotted in Fig. 12.17, along with the limits for the "high" and "low" responses out of each node. Thus when this particular function is used, the system outputs a high reading for any value of I_j greater than θ_j. Similarly, the system outputs a low reading for any value of I_j less than θ_j. As Fig. 12.17 shows, the sigmoid activation function always is positive, and it can reach its limiting values of 0 and 1 only if the input to the activation element is infinitely negative or positive, respectively. For this reason, values near 0 and 1

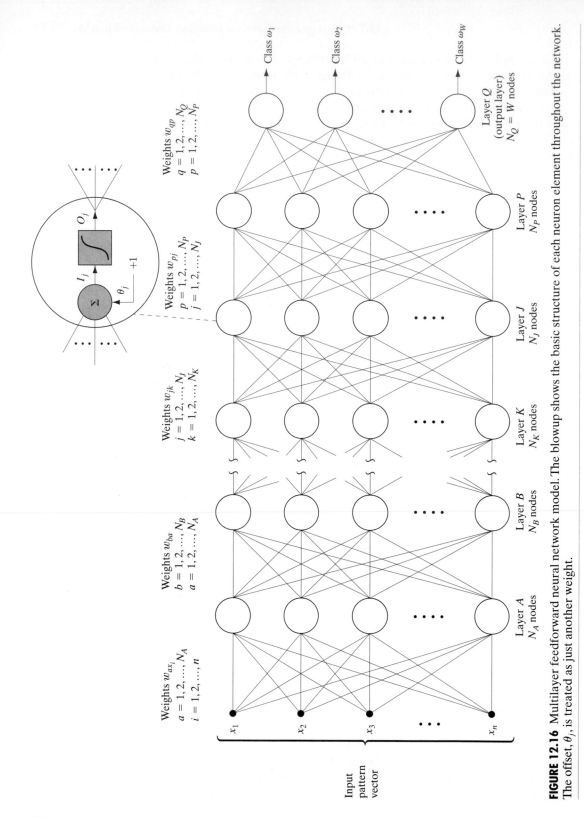

FIGURE 12.16 Multilayer feedforward neural network model. The blowup shows the basic structure of each neuron element throughout the network. The offset, θ_j, is treated as just another weight.

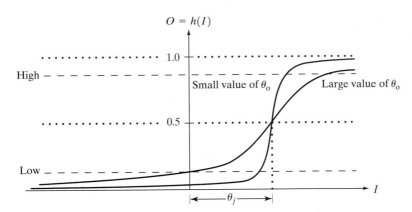

$O = h(I)$

High

Small value of θ_0

Large value of θ_0

Low

θ_j

I

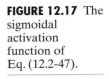

FIGURE 12.17 The sigmoidal activation function of Eq. (12.2-47).

(say, 0.05 and 0.95) define low and high values at the output of the neurons in Fig. 12.16. In principle, different types of activation functions could be used for different layers or even for different nodes in the same layer of a neural network. In practice, the usual approach is to use the same form of activation function throughout the network.

With reference to Fig. 12.14(a), the offset θ_j shown in Fig. 12.17 is analogous to the weight coefficient w_{n+1} in the earlier discussion of the perceptron. Implementation of this displaced threshold function can be done in the form of Fig. 12.14(a) by absorbing the offset θ_j as an additional coefficient that modifies a constant unity input to all nodes in the network. In order to follow the notation predominantly found in the literature, we do not show a separate constant input of +1 into all nodes of Fig. 12.16. Instead, this input and its modifying weight θ_j are integral parts of the network nodes. As noted in the blowup in Fig. 12.16, there is one such coefficient for each of the N_J nodes in layer J.

In Fig. 12.16, the input to a node in any layer is the weighted sum of the outputs from the previous layer. Letting layer K denote the layer preceding layer J (no alphabetical order is implied in Fig. 12.16) gives the input to the activation element of each node in layer J, denoted I_j:

$$I_j = \sum_{k=1}^{N_K} w_{jk} O_k \qquad (12.2\text{-}48)$$

for $j = 1, 2, \ldots, N_J$, where N_J is the number of nodes in layer J, N_K is the number of nodes in layer K, and w_{jk} are the weights modifying the outputs O_k of the nodes in layer K before they are fed into the nodes in layer J. The outputs of layer K are

$$O_k = h_k(I_k) \qquad (12.2\text{-}49)$$

for $k = 1, 2, \ldots, N_K$.

A clear understanding of the subscript notation used in Eq. (12.2-48) is important, because we use it throughout the remainder of this section. First, note that $I_j, j = 1, 2, \ldots, N_J$, represents the input to the *activation element* of the jth node in layer J. Thus I_1 represents the input to the activation element of the first (topmost) node in layer J, I_2 represents the input to the activation element of the second node in layer J, and so on. There are N_K inputs to every node in layer J, but *each* individual input can be weighted differently. Thus the N_K inputs to the first node in layer J are weighted by coefficients $w_{1k}, k = 1, 2, \ldots, N_K$; the inputs to the second node are weighted by coefficients $w_{2k}, k = 1, 2, \ldots, N_K$; and so on. Hence a total of $N_J \times N_K$ coefficients are necessary to specify the weighting of the outputs of layer K as they are fed into layer J. An additional N_J offset coefficients, θ_j, are needed to specify completely the nodes in layer J.

Substitution of Eq. (12.2-48) into (12.2-47) yields

$$h_j(I_j) = \frac{1}{1 + e^{-\left(\sum_{k=1}^{N_K} w_{jk}O_k + \theta_j\right)/\theta_o}}, \tag{12.2-50}$$

which is the form of activation function used in the remainder of this section.

During training, adapting the neurons in the output layer is a simple matter because the desired output of each node is known. The main problem in training a multilayer network lies in adjusting the weights in the so-called *hidden layers*. That is, in those other than the output layer.

Training by back propagation We begin by concentrating on the output layer. The total squared error between the desired responses, r_q, and the corresponding actual responses, O_q, of nodes in (output) layer Q, is

$$E_Q = \frac{1}{2} \sum_{q=1}^{N_Q} (r_q - O_q)^2 \tag{12.2-51}$$

where N_Q is the number of nodes in output layer Q and the $\frac{1}{2}$ is used for convenience in notation for taking the derivative later.

The objective is to develop a training rule, similar to the delta rule, that allows adjustment of the weights in each of the layers in a way that seeks a minimum to an error function of the form shown in Eq. (12.2-51). As before, adjusting the weights in proportion to the partial derivative of the error with respect to the weights achieves this result. In other words,

$$\Delta w_{qp} = -\alpha \frac{\partial E_Q}{\partial w_{qp}} \tag{12.2-52}$$

where layer P precedes layer Q, Δw_{qp} is as defined in Eq. (12.2-42), and α is a positive correction increment.

The error E_Q is a function of the outputs, O_q, which in turn are functions of the inputs I_q. Using the chain rule, we evaluate the partial derivative of E_Q as follows:

$$\frac{\partial E_Q}{\partial w_{qp}} = \frac{\partial E_Q}{\partial I_q} \frac{\partial I_q}{\partial w_{qp}}. \tag{12.2-53}$$

From Eq. (12.2-48),

$$\frac{\partial I_q}{\partial w_{qp}} = \frac{\partial}{\partial w_{qp}} \sum_{p=1}^{N_P} w_{qp}O_p = O_p. \tag{12.2-54}$$

Substituting Eqs. (12.2-53) and (12.2-54) into Eq. (12.2-52) yields

$$\Delta w_{qp} = -\alpha \frac{\partial E_Q}{\partial I_q} O_p$$

$$= \alpha \delta_q O_p \qquad (12.2\text{-}55)$$

where

$$\delta_q = -\frac{\partial E_Q}{\partial I_q}. \qquad (12.2\text{-}56)$$

In order to compute $\partial E_Q/\partial I_q$, we use the chain rule to express the partial derivative in terms of the rate of change of E_Q with respect to O_q and the rate of change of O_q with respect to I_q. That is,

$$\delta_q = -\frac{\partial E_Q}{\partial I_q} = -\frac{\partial E_Q}{\partial O_q}\frac{\partial O_q}{\partial I_q}. \qquad (12.2\text{-}57)$$

From Eq. (12.2-51),

$$\frac{\partial E_Q}{\partial O_q} = -(r_q - O_q) \qquad (12.2\text{-}58)$$

and, from Eq. (12.2-49),

$$\frac{\partial O_Q}{\partial I_q} = \frac{\partial}{\partial I_q} h_q(I_q) = h'_q(I_q). \qquad (12.2\text{-}59)$$

Substituting Eqs. (12.2-58) and (12.2-59) into Eq. (12.2-57) gives

$$\delta_q = (r_q - O_q)h'_q(I_q), \qquad (12.2\text{-}60)$$

which is proportional to the error quantity $(r_q - O_q)$. Substitution of Eqs. (12.2-56) through (12.2-58) into Eq. (12.2-55) finally yields

$$\Delta w_{qp} = \alpha(r_q - O_q)h'_q(I_q)O_p$$

$$= \alpha \delta_q O_p. \qquad (12.2\text{-}61)$$

After the function $h_q(I_q)$ has been specified, all the terms in Eq. (12.2-61) are known or can be observed in the network. In other words, upon presentation of any training pattern to the input of the network, we know what the desired response, r_q, of each output node should be. The value O_q of each output node can be observed as can I_q, the input to the activation elements of layer Q, and O_p, the output of the nodes in layer P. Thus we know how to adjust the weights that modify the links between the last and next-to-last layers in the network.

Continuing to work our way back from the output layer, let us now analyze what happens at layer P. Proceeding in the same manner as above yields

$$\Delta w_{pj} = \alpha(r_p - O_p)h'_p(I_p)O_j$$

$$= \alpha \delta_p O_j \qquad (12.2\text{-}62)$$

where the error term is

$$\delta_p = (r_p - O_p)h'_p(I_p). \qquad (12.2\text{-}63)$$

With the exception of r_p, all the terms in Eqs. (12.2-62) and (12.2-63) either are known or can be observed in the network. The term r_p makes no sense in an internal layer because we do not know what the response of an internal node in terms of pattern membership should be. We may specify what we want the response r to be only at the outputs of the network where final pattern classification takes place. If we knew that information at internal nodes, there would be no need for further layers. Thus we have to find a way to restate δ_p in terms of quantities that are known or can be observed in the network.

Going back to Eq. (12.2-57), we write the error term for layer P as

$$\delta_p = -\frac{\partial E_p}{\partial I_p} = -\frac{\partial E_p}{\partial O_p}\frac{\partial O_p}{\partial I_p}. \tag{12.2-64}$$

The term $\partial O_p/\partial I_p$ presents no difficulties. As before, it is

$$\frac{\partial O_p}{\partial I_p} = \frac{\partial h_p(I_p)}{\partial I_p} = h'_p(I_p), \tag{12.2-65}$$

which is known once h_p is specified because I_p can be observed. The term that produced r_p was the derivative $\partial E_p/\partial O_p$, so this term must be expressed in a way that does not contain r_p. Using the chain rule, we write the derivative as

$$
\begin{aligned}
-\frac{\partial E_P}{\partial O_p} &= -\sum_{q=1}^{N_Q}\frac{\partial E_P}{\partial I_q}\frac{\partial I_q}{\partial O_p} = \sum_{q=1}^{N_Q}\left(-\frac{\partial E_P}{\partial I_q}\right)\frac{\partial}{\partial O_p}\sum_{p=1}^{N_P}w_{qp}O_p \\
&= \sum_{q=1}^{N_Q}\left(-\frac{\partial E_P}{\partial I_q}\right)w_{qp} \\
&= \sum_{q=1}^{N_Q}\delta_q w_{qp}
\end{aligned}
\tag{12.2-66}
$$

where the last step follows from Eq. (12.2-56). Substituting Eqs. (12.2-65) and (12.2-66) into Eq. (12.2-64) yields the desired expression for δ_p:

$$\delta_p = h'_p(I_p)\sum_{q=1}^{N_Q}\delta_q w_{qp}. \tag{12.2-67}$$

The parameter δ_p can be computed now because all its terms are known. Thus Eqs. (12.2-62) and (12.2-67) establish completely the training rule for layer P. The importance of Eq. (12.2-67) is that it computes δ_p from the quantities δ_q and w_{qp}, which are terms that were computed in the layer immediately following layer P. After the error term and weights have been computed for layer P, these quantities may be used similarly to compute the error and weights for the layer immediately preceding layer P. In other words, we have found a way to propagate the error back into the network, starting with the error at the output layer.

We may summarize and generalize the training procedure as follows. For any layers K and J, where layer K immediately precedes layer J, compute the weights w_{jk}, which modify the connections between these two layers, by using

$$\Delta w_{jk} = \alpha\delta_j O_k. \tag{12.2-68}$$

If layer J is the output layer, δ_j is

$$\delta_j = (r_j - O_j)h_j'(I_j). \tag{12.2-69}$$

If layer J is an internal layer and layer P is the next layer (to the right), then δ_j is given by

$$\delta_j = h_j'(I_j) \sum_{p=1}^{N_P} \delta_p w_{jp} \tag{12.2-70}$$

for $j = 1, 2, \ldots, N_j$. Using the activation function in Eq. (12.2-50) with $\theta_o = 1$ yields

$$h_j'(I_j) = O_j(1 - O_j) \tag{12.2-71}$$

in which case Eqs. (12.2-69) and (12.2-70) assume the following, particularly attractive forms:

$$\delta_j = (r_j - O_j)O_j(1 - O_j) \tag{12.2-72}$$

for the output layer, and

$$\delta_j = O_j(1 - O_j) \sum_{p=1}^{N_P} \delta_p w_{jp} \tag{12.2-73}$$

for internal layers. In both Eqs. (12.2-72) and (12.2-73), $j = 1, 2, \ldots, N_J$.

Equations (12.2-68) through (12.2-70) constitute the generalized delta rule for training the multilayer feedforward neural network of Fig. 12.16. The process starts with an arbitrary (but not all equal) set of weights throughout the network. Then application of the generalized delta rule at any iterative step involves two basic phases. In the first phase, a training vector is presented to the network and is allowed to propagate through the layers to compute the output O_j for each node. The outputs O_q of the nodes in the output layer are then compared against their desired responses, r_p, to generate the error terms δ_q. The second phase involves a backward pass through the network during which the appropriate error signal is passed to each node and the corresponding weight changes are made. This procedure also applies to the bias weights θ_j. As discussed earlier in some detail, these are treated simply as additional weights that modify a unit input into the summing junction of every node in the network.

Common practice is to track the network error, as well as errors associated with individual patterns. In a successful training session, the network error decreases with the number of iterations and the procedure converges to a stable set of weights that exhibit only small fluctuations with additional training. The approach followed to establish whether a pattern has been classified correctly during training is to determine whether the response of the node in the output layer associated with the pattern class from which the pattern was obtained is high, while all the other nodes have outputs that are low, as defined earlier.

After the system has been trained, it classifies patterns using the parameters established during the training phase. In normal operation, all feedback paths are disconnected. Then any input pattern is allowed to propagate through the various layers, and the pattern is classified as belonging to the class of the output node that was high, while all the others were low. If more than one output is labeled high, or if none of the outputs is so labeled, the choice is one of declaring a misclassification or simply assigning the pattern to the class of the output node with the highest numerical value.

FIGURE 12.18
(a) Reference
shapes and
(b) typical noisy
shapes used in
training the
neural network of
Fig. 12.19.
(Courtesy of Dr.
Lalit Gupta, ECE
Department,
Southern Illinois
University.)

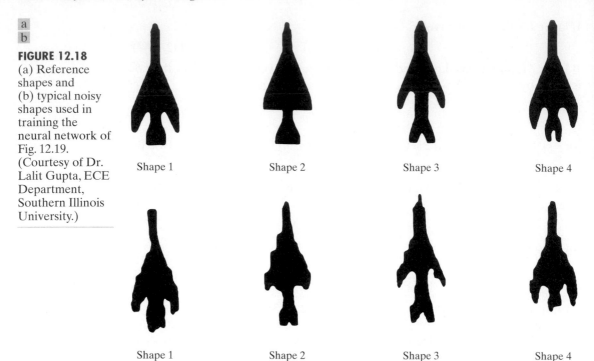

Shape 1   Shape 2   Shape 3   Shape 4

Shape 1   Shape 2   Shape 3   Shape 4

EXAMPLE 12.6:
Shape
classification
using a neural
network.

■ We illustrate now how a neural network of the form shown in Fig. 12.16 was trained to recognize the four shapes shown in Fig. 12.18(a), as well as noisy versions of these shapes, samples of which are shown in Fig. 12.18(b).

Pattern vectors were generated by computing the normalized signatures of the shapes (see Section 11.1.3) and then obtaining 48 uniformly spaced samples of each signature. The resulting 48-dimensional vectors were the inputs to the three-layer feedforward neural network shown in Fig. 12.19. The number of neuron nodes in the first layer was chosen to be 48, corresponding to the dimensionality of the input pattern vectors. The four neurons in the third (output) layer correspond to the number of pattern classes, and the number of neurons in the middle layer was heuristically specified as 26 (the average of the number of neurons in the input and output layers). There are no known rules for specifying the number of nodes in the internal layers of a neural network, so this number generally is based either on prior experience or simply chosen arbitrarily and then refined by testing. In the output layer, the four nodes from top to bottom in this case represent the classes $\omega_j, j = 1, 2, 3, 4$, respectively. After the network structure has been set, activation functions have to be selected for each unit and layer. All activation functions were selected to satisfy Eq. (12.2-50) with $\theta_o = 1$ so that, according to our earlier discussion, Eqs. (12.2-72) and (12.2-73) apply.

The training process was divided in two parts. In the first part, the weights were initialized to small random values with zero mean, and the network was then trained with pattern vectors corresponding to noise-free samples like the

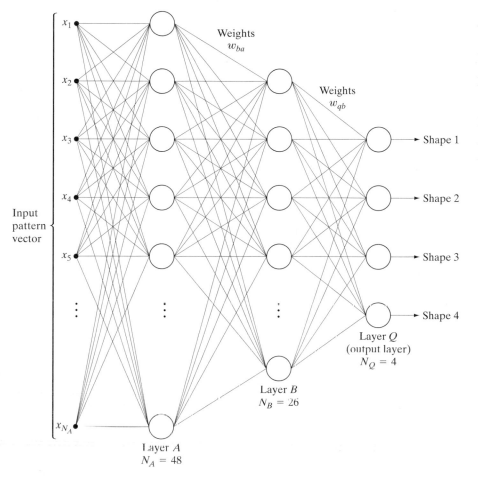

FIGURE 12.19
Three-layer
neural network
used to recognize
the shapes in
Fig. 12.18.
(Courtesy of Dr.
Lalit Gupta, ECE
Department,
Southern Illinois
University.)

shapes shown in Fig. 12.18(a). The output nodes were monitored during training. The network was said to have learned the shapes from all four classes when, for any training pattern from class ω_i, the elements of the output layer yielded $O_i \geq 0.95$ and $O_q \leq 0.05$, for $q = 1, 2, \ldots, N_Q; q \neq i$. In other words, for any pattern of class ω_i, the output unit corresponding to that class had to be high (≥ 0.95) while, simultaneously, the output of all other nodes had to be low (≤ 0.05).

The second part of training was carried out with noisy samples, generated as follows. Each contour pixel in a noise-free shape was assigned a probability V of retaining its original coordinate in the image plane and a probability $R = 1 - V$ of being randomly assigned to the coordinates of one of its eight neighboring pixels. The degree of noise was increased by decreasing V (that is, increasing R). Two sets of noisy data were generated. The first consisted of 100 noisy patterns of each class generated by varying R between 0.1 and 0.6, giving a total of 400 patterns. This set, called the *test set*, was used to establish system a performance after training.

FIGURE 12.20
Performance of
the neural
network as a
function of noise
level. (Courtesy of
Dr. Lalit Gupta,
ECE Department,
Southern Illinois
University.)

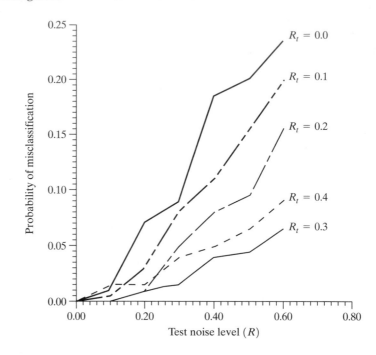

Several noisy sets were generated for training the system with noisy data. The first set consisted of 10 samples for each class, generated by using $R_t = 0$, where R_t denotes a value of R used to generate training data. Starting with the weight vectors obtained in the first (noise-free) part of training, the system was allowed to go through a learning sequence with the new data set. Because $R_t = 0$ implies no noise, this retraining was an extension of the earlier, noise-free training. Using the resulting weights learned in this manner, the network was subjected to the test data set yielding the results shown by the curve labeled $R_t = 0$ in Fig. 12.20. The number of misclassified patterns divided by the total number of patterns tested gives the probability of misclassification, which is a measure commonly used to establish neural network performance.

Next, starting with the weight vectors learned by using the data generated with $R_t = 0$, the system was retrained with a noisy data set generated with $R_t = 0.1$. The recognition performance was then established by running the test samples through the system again with the new weight vectors. Note the significant improvement in performance. Figure 12.20 shows the results obtained by continuing this retraining and retesting procedure for $R_t = 0.2, 0.3$, and 0.4. As expected if the system is learning properly, the probability of misclassifying patterns from the test set decreased as the value of R_t increased because the system was being trained with noisier data for higher values of R_t. The one exception in Fig. 12.20 is the result for $R_t = 0.4$. The reason is the small number of samples used to train the system. That is, the network was not able to adapt itself sufficiently to the larger variations in shape at higher noise levels with the number of samples used. This hypothesis is verified by the results in Fig. 12.21, which show a lower probability of misclassification as the number of training

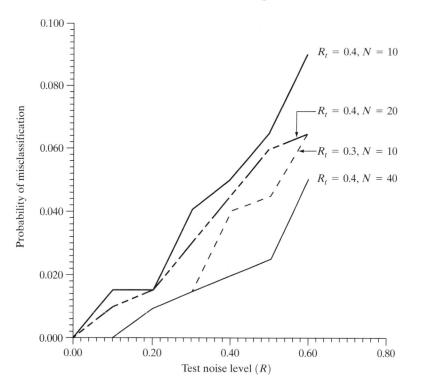

FIGURE 12.21
Improvement in
performance for
$R_t = 0.4$ by
increasing the
number of
training patterns
(the curve for
$R_t = 0.3$ is shown
for reference).
(Courtesy of Dr.
Lalit Gupta, ECE
Department,
Southern Illinois
University.)

samples was increased. Figure 12.21 also shows as a reference the curve for $R_t = 0.3$ from Fig. 12.20.

The preceding results show that a three-layer neural network was capable of learning to recognize shapes corrupted by noise after a modest level of training. Even when trained with noise-free data ($R_t = 0$ in Fig. 12.20), the system was able to achieve a correct recognition level of close to 77% when tested with data highly corrupted by noise ($R = 0.6$ in Fig. 12.20). The recognition rate on the same data increased to about 99% when the system was trained with noisier data ($R_r = 0.3$ and 0.4). It is important to note that the system was trained by increasing its classification power via systematic, small incremental additions of noise. When the nature of the noise is known, this method is ideal for improving the convergence and stability properties of a neural network during learning.   ■

Complexity of decision surfaces   We have already established that a single-layer perceptron implements a hyperplane decision surface. A natural question at this point is, What is the nature of the decision surfaces implemented by a multilayer network, such as the model in Fig. 12.16? It is demonstrated in the following discussion that a three-layer network is capable of implementing arbitrarily complex decision surfaces composed of intersecting hyperplanes.

As a starting point, consider the two-input, two-layer network shown in Fig. 12.22(a). With two inputs, the patterns are two dimensional, and therefore, each node in the first layer of the network implements a line in 2-D space. We

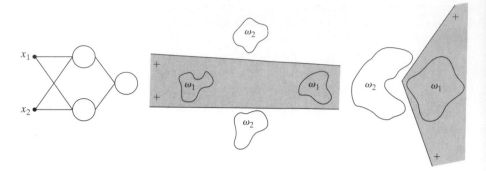

a b c

FIGURE 12.22 (a) A two-input, two-layer, feedforward neural network. (b) and (c) Examples of decision boundaries that can be implemented with this network.

denote by 1 and 0, respectively, the high and low outputs of these two nodes. We assume that a 1 output indicates that the corresponding input vector to a node in the first layer lies on the positive side of the line. Then the possible combinations of outputs feeding the single node in the second layer are $(1, 1)$, $(1, 0)$, $(0, 1)$, and $(0, 0)$. If we define two regions, one for class ω_1 lying on the positive side of both lines and the other for class ω_2 lying anywhere else, the output node can classify any input pattern as belonging to one of these two regions simply by performing a logical AND operation. In other words, the output node responds with a 1, indicating class ω_1, only when both outputs from the first layer are 1. The AND operation can be performed by a neural node of the form discussed earlier if θ_j is set to a value in the half open interval $(1, 2]$. Thus if we assume 0 and 1 responses out of the first layer, the response of the output node will be high, indicating class ω_1, only when the sum performed by the neural node on the two outputs from the first layer is greater than 1. Figures 12.22(b) and (c) show how the network of Fig. 12.22(a) can successfully dichotomize two pattern classes that could not be separated by a single linear surface.

If the number of nodes in the first layer were increased to three, the network of Fig. 12.22(a) would implement a decision boundary consisting of the intersection of three lines. The requirement that class ω_1 lie on the positive side of all three lines would yield a convex region bounded by the three lines. In fact, an arbitrary open or closed convex region can be constructed simply by increasing the number of nodes in the first layer of a two-layer neural network.

The next logical step is to increase the number of layers to three. In this case the nodes of the first layer implement lines, as before. The nodes of the second layer then perform AND operations in order to form regions from the various lines. The nodes in the third layer assign class membership to the various regions. For instance, suppose that class ω_1 consists of two distinct regions, each of which is bounded by a different set of lines. Then two of the nodes in the second layer are for regions corresponding to the same pattern class. One of the output nodes

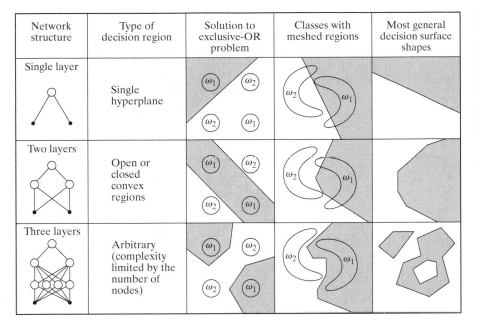

FIGURE 12.23
Types of decision
regions that can
be formed by
single- and
multilayer
feed-forward
networks with
one and two
layers of hidden
units and two
inputs. (Lippman)

Network structure	Type of decision region	Solution to exclusive-OR problem	Classes with meshed regions	Most general decision surface shapes
Single layer	Single hyperplane	ω_1 ω_2 / ω_2 ω_1	ω_2 ω_1	
Two layers	Open or closed convex regions	ω_1 ω_2 / ω_2 ω_1	ω_2 ω_1	
Three layers	Arbitrary (complexity limited by the number of nodes)	ω_1 ω_2 / ω_2 ω_1	ω_2 ω_1	

needs to be able to signal the presence of that class when either of the two nodes in the second layer goes high. Assuming that high and low conditions in the second layer are denoted 1 and 0, respectively, this capability is obtained by making the output nodes of the network perform the logical OR operation. In terms of neural nodes of the form discussed earlier, we do so by setting θ_j to a value in the half-open interval $[0, 1)$. Then, whenever at least one of the nodes in the second layer associated with that output node goes high (outputs a 1), the corresponding node in the output layer will go high, indicating that the pattern being processed belongs to the class associated with that node.

Figure 12.23 summarizes the preceding comments. Note in the third row that the complexity of decision regions implemented by a three-layer network is, in principle, arbitrary. In practice, a serious difficulty usually arises in structuring the second layer to respond correctly to the various combinations associated with particular classes. The reason is that lines do not just stop at their intersection with other lines, and, as a result, patterns of the same class may occur on both sides of lines in the pattern space. In practical terms, the second layer may have difficulty figuring out which lines should be included in the AND operation for a given pattern class—or it may even be impossible. The reference to the exclusive-OR problem in the third column of Fig. 12.23 deals with the fact that, if the input patterns were binary, only four different patterns could be constructed in two dimensions. If the patterns are so arranged that class ω_1 consists of patterns $\{(0, 1), (1, 0)\}$ and class ω_2 consists of the patterns $\{(0, 0), (1, 1)\}$, class membership of the patterns in these two classes is given by the exclusive-OR (XOR) logical function, which is 1 only when one or the other of the two variables is 1, and it is 0 otherwise. Thus an XOR value of 1 indicates patterns of class ω_1, and an XOR value of 0 indicates patterns of class ω_2.

The preceding discussion is generalized to n dimensions in a straightforward way: Instead of lines, we deal with hyperplanes. A single-layer network implements a single hyperplane. A two-layer network implements arbitrarily convex regions consisting of intersections of hyperplanes. A three-layer network implements decision surfaces of arbitrary complexity. The number of nodes used in each layer determines the complexity of the last two cases. The number of classes in the first case is limited to two. In the other two cases, the number of classes is arbitrary, because the number of output nodes can be selected to fit the problem at hand.

Considering the preceding comments, it is logical to ask, Why would anyone be interested in studying neural networks having more than three layers? After all, a three-layer network can implement decision surfaces of arbitrary complexity. The answer lies in the method used to train a network to utilize only three layers. The training rule for the network in Fig. 12.16 minimizes an error measure but says nothing about how to associate groups of hyperplanes with specific nodes in the second layer of a three-layer network of the type discussed earlier. In fact, the problem of how to perform trade-off analyses between the number of layers and the number of nodes in each layer remains unresolved. In practice, the trade-off is generally resolved by trial and error or by previous experience with a given problem domain.

12.3 Structural Methods

The techniques discussed in Section 12.2 deal with patterns quantitatively and largely ignore any structural relationships inherent in a pattern's shape. The structural methods discussed in this section, however, seek to achieve pattern recognition by capitalizing precisely on these types of relationships.

12.3.1 Matching Shape Numbers

A procedure analogous to the minimum distance concept introduced in Section 12.2.1 for pattern vectors can be formulated for the comparison of region boundaries that are described in terms of shape numbers. With reference to the discussion in Section 11.2.2, the *degree of similarity*, k, between two region boundaries (shapes) is defined as the largest order for which their shape numbers still coincide. For example, let a and b denote shape numbers of closed boundaries represented by 4-directional chain codes. These two shapes have a degree of similarity k if

$$
\begin{aligned}
s_j(a) &= s_j(b) && \text{for } j = 4, 6, 8, \ldots, k \\
s_j(a) &\neq s_j(b) && \text{for } j = k + 2, k + 4, \ldots
\end{aligned}
\tag{12.3-1}
$$

where s indicates shape number and the subscript indicates order. The *distance* between two shapes a and b is defined as the inverse of their degree of similarity:

$$
D(a, b) = \frac{1}{k}.
\tag{12.3-2}
$$

This distance satisfies the following properties:

$$D(a, b) \geq 0$$
$$D(a, b) = 0 \quad \text{iff } a = b \qquad (12.3\text{-}3)$$
$$D(a, c) \leq \max\left[D(a, b), D(b, c)\right].$$

Either k or D may be used to compare two shapes. If the degree of similarity is used, the larger k is, the more similar the shapes are (note that k is infinite for identical shapes). The reverse is true when the distance measure is used.

■ Suppose that we have a shape f and want to find its closest match in a set of five other shapes ($a, b, c, d,$ and e), as shown in Fig. 12.24(a). This problem is analogous to having five prototype shapes and trying to find the best match to a given unknown shape. The search may be visualized with the aid of the similarity tree shown in Fig. 12.24(b). The root of the tree corresponds to the lowest possible degree of similarity, which, for this example, is 4. Suppose that the shapes are identical up to degree 8, with the exception of shape a, whose degree of similarity with respect to all other shapes is 6. Proceeding down the tree, we find that shape d has degree of similarity 8 with respect to all others, and so on. Shapes

EXAMPLE 12.7:
Using shape numbers to compare shapes.

FIGURE 12.24
(a) Shapes.
(b) Hypothetical similarity tree.
(c) Similarity matrix. (Bribiesca and Guzman.)

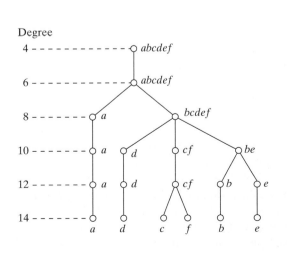

Degree

	a	b	c	d	e	f
a	∞	6	6	6	6	6
b		∞	8	8	10	8
c			∞	8	8	12
d				∞	8	8
e					∞	8
f						∞

f and *c* match uniquely, having a higher degree of similarity than any other two shapes. At the other extreme, if *a* had been an unknown shape, all we could have said using this method is that *a* was similar to the other five shapes with degree of similarity 6. The same information can be summarized in the form of a *similarity matrix*, as shown in Fig. 12.24(c).   ■

12.3.2 String Matching

Suppose that two region boundaries, *a* and *b*, are coded into strings (see Section 11.5) denoted $a_1 a_2 \ldots a_n$ and $b_1 b_2 \ldots b_m$, respectively. Let α represent the number of matches between the two strings, where a match occurs in the *k*th position if $a_k = b_k$. The number of symbols that do not match is

$$\beta = \max(|a|, |b|) - \alpha \tag{12.3-4}$$

where |arg| is the length (number of symbols) in the string representation of the argument. It can be shown that $\beta = 0$ if and only if *a* and *b* are identical (see Problem 12.21).

A simple measure of similarity between *a* and *b* is the ratio

$$R = \frac{\alpha}{\beta} = \frac{\alpha}{\max(|a|, |b|) - \alpha}. \tag{12.3-5}$$

Hence *R* is infinite for a perfect match and 0 when none of the corresponding symbols in *a* and *b* match ($\alpha = 0$ in this case). Because matching is done symbol by symbol, the starting point on each boundary is important in terms of reducing the amount of computation. Any method that normalizes to, or near, the same starting point is helpful, so long as it provides a computational advantage over brute-force matching, which consists of starting at arbitrary points on each string and then shifting one of the strings (with wraparound) and computing Eq. (12.3-5) for each shift. The largest value of *R* gives the best match.

EXAMPLE 12.8:
Illustration of string matching.

■ Figures 12.25(a) and (b) show sample boundaries from each of two object classes, which were approximated by a polygonal fit (see Section 11.1.2). Figures 12.25(c) and (d) show the polygonal approximations corresponding to the boundaries shown in Figs. 12.25(a) and (b), respectively. Strings were formed from the polygons by computing the interior angle, θ, between segments as each polygon was traversed clockwise. Angles were coded into one of eight possible symbols, corresponding to 45° increments; that is, $\alpha_1 : 0° < \theta \le 45°$; $\alpha_2 : 45° < \theta \le 90°; \ldots; \alpha_8 : 315° < \theta \le 360°$.

Figure 12.25(e) shows the results of computing the measure *R* for six samples of object 1 against themselves. The entries correspond to *R* values and, for example, the notation 1.c refers to the third string from object class 1. Figure 12.25(f) shows the results of comparing the strings of the second object class against themselves. Finally, Fig. 12.25(g) shows a tabulation of *R* values obtained by comparing strings of one class against the other. Note that, here, all *R* values are considerably smaller than any entry in the two preceding tabulations, indicating that the *R* measure achieved a high degree of discrimination between the two classes of objects. For example, if the class membership of string 1.a had

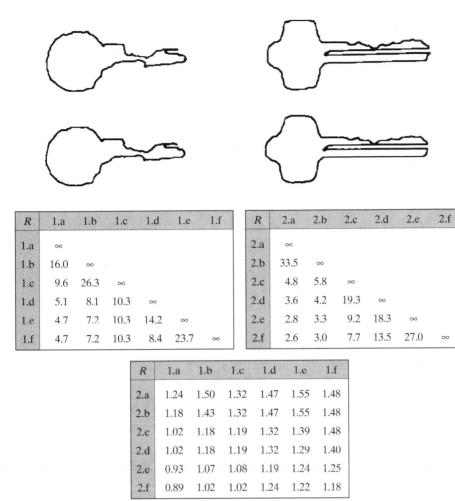

FIGURE 12.25 (a) and (b) Sample boundaries of two different object classes; (c) and (d) their corresponding polygonal approximations; (e)–(g) tabulations of R. (Sze and Yang.)

R	1.a	1.b	1.c	1.d	1.e	1.f
1.a	∞					
1.b	16.0	∞				
1.c	9.6	26.3	∞			
1.d	5.1	8.1	10.3	∞		
1.e	4.7	7.2	10.3	14.2	∞	
1.f	4.7	7.2	10.3	8.4	23.7	∞

R	2.a	2.b	2.c	2.d	2.e	2.f
2.a	∞					
2.b	33.5	∞				
2.c	4.8	5.8	∞			
2.d	3.6	4.2	19.3	∞		
2.e	2.8	3.3	9.2	18.3	∞	
2.f	2.6	3.0	7.7	13.5	27.0	∞

R	1.a	1.b	1.c	1.d	1.e	1.f
2.a	1.24	1.50	1.32	1.47	1.55	1.48
2.b	1.18	1.43	1.32	1.47	1.55	1.48
2.c	1.02	1.18	1.19	1.32	1.39	1.48
2.d	1.02	1.18	1.19	1.32	1.29	1.40
2.e	0.93	1.07	1.08	1.19	1.24	1.25
2.f	0.89	1.02	1.02	1.24	1.22	1.18

been unknown, the *smallest* value of R resulting from comparing this string against sample (prototype) strings of class 1 would have been 4.7 [Fig. 12.25(e)]. By contrast, the *largest* value in comparing it against strings of class 2 would have been 1.24 [Fig. 12.25(g)]. This result would have led to the conclusion that string 1.a is a member of object class 1. This approach to classification is analogous to the minimum distance classifier introduced in Section 12.2.1. ■

12.3.3 Syntactic Recognition of Strings

Syntactic methods provide a unified methodology for handling structural recognition problems. Basically, the idea behind syntactic pattern recognition is the specification of a set of pattern *primitives* (see Section 11.5), a set of rules (in the form of a *grammar*) that governs their interconnection, and a *recognizer* (called an *automaton*), whose structure is determined by the set of rules in the grammar. First we consider string grammars and automata and then extend these ideas in the next section to tree grammars and their corresponding automata.

String grammars

Suppose that we have two classes, ω_1 and ω_2, whose patterns are strings of primitives, generated by one of the methods discussed in Section 11.5. We can interpret each primitive as being a symbol permissible in the *alphabet* of some *grammar*, where a grammar is a set of rules of syntax (hence the name syntactic recognition) that govern the generation of *sentences* formed from symbols of the alphabet. The set of sentences generated by a grammar, G, is called its *language* and is denoted $L(G)$. Here, sentences are strings of symbols (which in turn represent patterns), and languages correspond to pattern classes.

Consider two grammars, G_1 and G_2, whose rules of syntax are such that G_1 only allows generation of sentences that correspond to patterns from class ω_1, and G_2 only allows generation of sentences corresponding to patterns from class ω_2. After two grammars with these properties have been established, the syntactic pattern recognition process, in principle, is straightforward. For a sentence representing an unknown pattern, the task is to decide in which language the pattern represents a valid sentence. If the sentence belongs to $L(G_1)$, we say that the pattern is from class ω_1. Similarly, the pattern is said to be from class ω_2 if the sentence is valid in $L(G_2)$. A unique decision cannot be made if the sentence belongs to both languages. A sentence that is invalid in both languages is rejected.

When there are more than two pattern classes, the syntactic classification approach is the same as described in the preceding paragraph, with the exception that more grammars (at least one per class) are involved in the process. For multiclass classification, a pattern belongs to class ω_i if it is a valid sentence only of $L(G_i)$. As before, a unique decision cannot be made if a sentence belongs to more than one language. A sentence that is invalid over all languages is rejected.

When dealing with strings, we define a grammar as the 4-tuple

$$G = (N, \Sigma, P, S) \tag{12.3-6}$$

where

N is a finite set of variables called *nonterminals,*

Σ is a finite set of constants called *terminals,*

P is a set of rewriting rules called *productions,* and

S in N is called the *starting symbol.*

It is required that N and Σ be disjoint sets. In the following discussion, capital letters, $A, B, \ldots, S, \ldots$, denote nonterminals. Lowercase letters, $a, b, c, \ldots$ at the beginning of the alphabet denote terminals. Lowercase letters, v, w, x, y, z toward the end of the alphabet denote strings of terminals. Lowercase Greek letters $\alpha, \beta, \theta, \ldots$ denote strings of mixed terminals and nonterminals. The *empty sentence* (the sentence with no symbols) is denoted λ. Finally, for a set V of symbols, the notation V^* denotes the set of all sentences composed of elements from V.

String grammars are characterized by the form of their productions. Of particular interest in syntactic pattern recognition are *regular grammars* and *context-free grammars*. Regular grammars have productions only of the form $A \rightarrow aB$ or

$A \rightarrow a$, with A and B in N and a in Σ. Context-free grammars have productions only of the form $A \rightarrow \alpha$, with A in N and α in the set $(N \cup \Sigma)^*$; that is, α can be any string composed of terminals and nonterminals, except the empty string.

■ Before proceeding, it will be useful to consider the mechanics of how grammars generate object classes. Suppose that the object shown in Fig. 12.26(a) is represented by its (pruned) skeleton and that we define the primitives shown in Fig. 12.26(b) to describe the structure of this (and similar) skeletons. Consider the grammar $G = (N, \Sigma, P, S)$, with $N = \{A, B, S\}$, $\Sigma = \{a, b, c\}$ and $P = \{S \rightarrow aA, A \rightarrow bA, A \rightarrow bB, B \rightarrow c\}$, where the terminals a, b, and c correspond to the primitives shown in Fig. 12.26(b). As indicated earlier, S is the starting symbol from which the strings of $L(G)$ are generated. For instance, applying the first production followed by two applications of the second production yields $S \Rightarrow aA \Rightarrow abA \Rightarrow abbA$, where $(\Rightarrow)$ indicates a string derivation starting from S and using productions from the set P. The first production allowed rewriting S as aA, and the second production allowed rewriting A as bA. With a nonterminal in the string $abbA$, we can continue the derivation. For example, applying the second production two more times, followed by one application of the third production and one application of the fourth production, yields the string $abbbbbc$, which corresponds to the structure shown in Fig. 12.26(c). No nonterminals remain after application of the fourth production, so the derivation terminates when this production is used. The language generated by the rules of this grammar is $L(G) = \{ab^nc | n \geq 1\}$, where b^n indicates n repetitions of the symbol b. In other words, G is capable of generating *only* skeletons of the form shown in Fig. 12.26(c) but having arbitrary length. ■

EXAMPLE 12.9:
Object class generation using a regular string grammar.

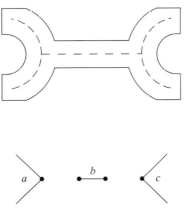

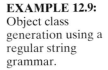

FIGURE 12.26
(a) Object represented by its (pruned) skeleton.
(b) Primitives.
(c) Structure generated by using a regular string grammar.

TABLE 12.1
Example of
semantic
information
attached to
production rules.

Production	Semantic Information
$S \rightarrow aA$	Connections to a are made only at the dot. The direction of a, denoted θ, is given by the direction of the perpendicular bisector of the line joining the end points of the two undotted segments. The line segments are 3 cm each.
$A \rightarrow bA$	Connections to b are made only at the dots. No multiple connections are allowed. The direction of b must be the same as the direction of a. The length of b is 0.25 cm. This production cannot be applied more than 10 times.
$A \rightarrow bB$	The direction of a and b must be the same. Connections must be simple and made only at the dots.
$B \rightarrow c$	The direction of c and a must be the same. Connections must be simple and made only at the dots.

Use of semantics

In the preceding example we assumed that the interconnection between primitives takes place only at the dots shown in Fig. 12.26(b). In more complicated situations the rules of connectivity, as well as information regarding other factors (such as primitive length and direction) and the number of times a production can be applied, must be made explicit. This can be accomplished by using *semantic rules* stored in the *knowledge base* of Fig. 1.23. Basically, the syntax inherent in the production rules establishes the structure of an object, whereas semantics deal with its correctness. For example, in a programming language like C, the statement $A = D/E$ is syntactically correct, but it is semantically correct only if $E \neq 0$.

Suppose that we attach semantic information to the grammar discussed in the preceding example. The information can be attached to the production rules in the form shown in Table 12.1. By using semantic information, we are able to use a few rules of syntax to describe a broad (but limited as desired) class of patterns. For instance, by specifying the direction of θ in Table 12.1, we avoid having to specify primitives for each possible orientation. Similarly, by requiring that all primitives be oriented in the same direction, we eliminate from consideration nonsensical structures that deviate from the basic shapes typified by Fig. 12.26(a).

Automata as string recognizers

So far we have demonstrated that grammars are *generators* of patterns. In the following discussion we consider the problem of recognizing whether a pattern belongs to the language $L(G)$ generated by a grammar G. The basic concepts underlying syntactic recognition may be illustrated by the development of mathematical models of computing machines, called *automata*. Given an input pattern string, an automaton is capable of recognizing whether the pattern belongs to the language with which the automaton is associated. Here, we focus only on *finite automata*, which are the recognizers of languages generated by regular grammars.

A *finite automaton* is defined as the 5-tuple

$$A_f = (Q, \Sigma, \delta, q_0, F) \tag{12.3-7}$$

where Q is a finite, nonempty set of *states*, Σ is a finite input *alphabet*, δ is a *mapping* from $Q \times \Sigma$ (the set of ordered pairs formed from elements of Q and Σ) into the collection of all subsets of Q, q_0 is the *starting state*, and F (a subset of Q) is a set of *final*, or *accepting*, *states*.

▪ Consider an automaton given by Eq. (12.3-7), with $Q = \{q_0, q_1, q_2\}$, $\Sigma = \{a, b\}$, $F = \{q_0\}$, and mappings $\delta(q_0, a) = \{q_2\}$, $\delta(q_0, b) = \{q_1\}$, $\delta(q_1, a) = \{q_2\}$, $\delta(q_1, b) = \{q_0\}$, $\delta(q_2, a) = \{q_0\}$, and $\delta(q_2, b) = \{q_1\}$. If, for example, the automaton is in state q_0 and an a is input, its state changes to q_2. Similarly, if a b is input next, the automaton changes to state q_1, and so on. The initial and final states are the same in this case. ▪

EXAMPLE 12.10: A simple automaton.

Figure 12.27 shows a *state diagram* for the automaton just discussed. The state diagram consists of a node for each state and directed arcs showing the possible transitions between states. The final state is shown as a double circle, and each arc is labeled with the symbol that causes the transition between the states joined by that arc. In this case the initial and final states are the same. A string w of terminal symbols is said to be *accepted* or *recognized* by an automaton if, starting in state q_0, the sequence of symbols (encountered as w is scanned from left to right) causes the automaton to be in a final state after the last symbol from w has been scanned. For example, the automaton in Fig. 12.27 recognizes the string $w = abbabb$ but rejects the string $w = aabab$.

There is a one-to-one correspondence between regular grammars and finite automata. That is, a language is recognized by a finite automaton if and only if it is generated by a regular grammar. The design of a syntactic string recognizer based on the concepts discussed so far is a straightforward procedure, consisting of obtaining a finite automaton from a given regular grammar. Let the grammar be denoted $G = (N, \Sigma, P, X_0)$, where $X_0 \equiv S$, and suppose that N is composed of X_0 plus n additional nonterminals $X_1, X_2, \ldots, X_n$. The set Q for the automaton is formed by introducing $n + 2$ states $\{q_0, q_1, \ldots, q_n, q_{n+1}\}$ such that q_i corresponds

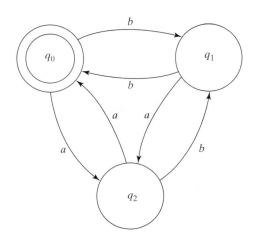

FIGURE 12.27 A finite automaton.

to X_i for $0 \le i \le n$, and q_{n+1} is the final state. The set of input symbols is identical to the set of terminals in G. The mappings in δ are obtained by using two rules based on the productions of G; namely, for each i and j, with $0 \le i \le n, 0 \le j \le n$,

1. If $X_i \to aX_j$ is in P, then $\delta(q_i, a)$ contains q_j.
2. If $X_i \to a$ is in P, then $\delta(q_i, a)$ contains q_{n+1}.

Conversely, given a finite automaton, $A_f = (Q, \Sigma, \delta, q_0, F)$, we obtain the corresponding regular grammar, $G = (N, \Sigma, P, X_0)$ by letting N consist of the elements of Q, with the starting symbol X_0 corresponding to q_0, and the productions of G obtained as follows:

1. If q_j is in $\delta(q_i, a)$, there is a production $X_i \to aX_j$ in P.
2. If a state in F is in $\delta(q_i, a)$ there is a production $X_i \to a$ in P.

The terminal set, Σ, is the same in both cases.

EXAMPLE 12.11:
Finite automaton for recognizing the patterns in Fig. 12.26.

■ The finite automaton for the grammar given in connection with Fig. 12.26 is obtained by writing the productions as $X_0 \to aX_1, X_1 \to bX_1, X_1 \to bX_2$, and $X_2 \to c$. Then $A_f = (Q, \Sigma, \delta, q_0, F)$, with $Q = \{q_0, q_1, q_2, q_3\}$, $\Sigma = \{a, b, c\}$, $F = \{q_3\}$ and mappings $\delta(q_0, a) = \{q_1\}, \delta(q_1, b) = \{q_1, q_2\}, \delta(q_2, c) = \{q_3\}$. For completeness, we write $\delta(q_0, b) = \delta(q_0, c) = \delta(q_1, a) = \delta(q_1, c) = \delta(q_2, a) = \delta(q_2, b) = \varnothing$, where $\varnothing$ is the null set, indicating that these transitions are not defined for this automaton. ■

12.3.4 Syntactic Recognition of Trees

Following a format similar to the preceding discussion for strings, we now expand the discussion to include tree descriptions of patterns. We assume that the image regions or objects of interest have been expressed in the form of trees by using the appropriate primitive elements, as discussed in Section 11.5.

Tree grammars

A *tree grammar* is defined as the 5-tuple

$$G = (N, \Sigma, P, r, S) \tag{12.3-8}$$

where, as before, N and Σ are sets of nonterminals and terminals, respectively; S, contained in N, is the start symbol, which in general can be a tree; P is a set of productions of the form $T_i \to T_j$, where T_i and T_j are trees; and r is a *ranking function* that denotes the number of direct descendants (offspring) of a node whose label is a terminal in the grammar. Of particular relevance to our discussion are *expansive* tree grammars having productions of the form

where $X_1, X_2, \ldots, X_n$ are nonterminals and k is a terminal.

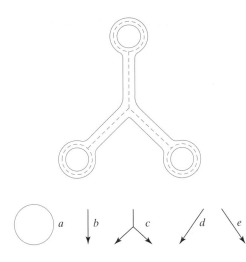

a
b

FIGURE 12.28
(a) An object and
(b) primitives
used for
representing the
skeleton by
means of a tree
grammar.

■ The skeleton of the structure shown in Fig. 12.28(a) can be generated by using a tree grammar with $N = \{X_1, X_2, X_3, S\}$ and $\Sigma = \{a, b, c, d, e\}$, where the terminals represent the primitives shown in Fig. 12.28(b). Assuming head-to-tail connectivity of the line primitives, and arbitrary connections to the circle along its circumference, the grammar under consideration has productions of the form

EXAMPLE 12.12:
A simple tree
grammar.

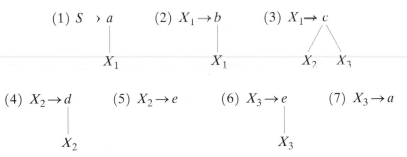

The ranking functions in this case are $r(a) = \{0, 1\}$, $r(b) = r(d) = \{1\}$, $r(e) = \{0, 1\}$, and $r(c) = \{2\}$. Restricting application of productions 2, 4, and 6 to the same number of times would generate a structure in which all three legs have the same length. Similarly, requiring application of productions 4 and 6 the same number of times would produce a structure that is symmetrical about its vertical axis. This type of semantic information is similar to the earlier discussion in connection with Table 12.1 and the knowledge base of Fig. 1.23.   ■

Tree automata

Whereas a conventional finite automaton scans an input string symbol by symbol from left to right, a tree automaton must begin simultaneously at each node on the frontier (the leaves taken in order from left to right) of an input tree and proceed along parallel paths toward the root. Specifically, a *frontier-to-root automaton* is defined as

$$A_t = (Q, F, \{f_k \mid k \in \Sigma\})$$ (12.3-9)

where

Q is a finite set of states,

F, a subset of Q, is a set of final states, and

f_k is a relation in $Q^m \times Q$ such that m is a rank of k.

The notation Q^m indicates the Cartesian product of Q with itself m times: $Q^m = Q \times Q \times Q \times \ldots \times Q$. From the definition of the Cartesian product, we know that this expression means the set of all ordered m-tuples with elements from Q. For example, if $m = 3$, than $Q^3 = Q \times Q \times Q = \{x, y, z \mid x \in Q, y \in Q, z \in Q\}$. Recall that a *relation* R from a set A to a set B is a subset of the Cartesian product of A and B; that is, $R \subseteq A \times B$. Thus a relation in $Q^m \times Q$ is simply a subset of the set $Q^m \times Q$.

For an expansive tree grammar, $G = (N, \Sigma, P, r, S)$, we construct the corresponding tree automaton by letting $Q = N$, with $F = \{S\}$ and, for each symbol a in Σ, defining a relation f_k such that $(X_1, X_2, \ldots, X_m, X)$ is in f_k if and only if there is in G a production

$$X \to k$$
$$X_1 X_2 \ldots X_m$$

For example, consider the tree grammar $G = (N, \Sigma, P, r, S)$, with $N = \{S, X\}$, $\Sigma = \{a, b, c, d\}$, productions

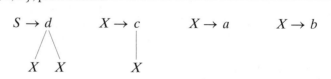

and rankings $r(a) = \{0\}$, $r(b) = \{0\}$, $r(c) = \{1\}$, and $r(d) = \{2\}$. The corresponding tree automaton, $A_t = (Q, F, \{f_k \mid k \in \Sigma\})$, is specified by letting $Q = \{S, X\}$, $F = \{S\}$, and $\{f_k \mid k \in \Sigma\} = \{f_a, f_b, f_c, f_d\}$, where the relations are defined as

$$f_a = \{(\varnothing, X)\}, \text{ arising from production } X \to a$$
$$f_b = \{(\varnothing, X)\}, \text{ arising from production } X \to b$$
$$f_c = \{(X, X)\}, \text{ arising from production } X \to c$$
$$X$$

and

$$f_d = \{(X, X, S)\}, \text{arising from production } S \rightarrow d$$

$$\begin{array}{cc} & \\ X & X \end{array}$$

The interpretation of relation f_a is that a node labeled a with no offspring (hence the null symbol $\varnothing$) is assigned state X. The interpretation of f_c is that a node labeled c, with one offspring having state X, is assigned state X. The interpretation of relation f_d is that a node labeled d with two offspring, each having state X, is assigned state S.

In order to see how this tree automaton goes about recognizing a tree generated by the grammar discussed earlier, consider the tree shown in Fig. 12.29(a). Automaton A_t first assigns states to the frontier nodes a and b via relations f_a and f_b, respectively. In this case, according to these two relations, state X is assigned to both leaves, as Fig. 12.29(b) shows. The automaton now moves up one level from the frontier and makes a state assignment to node c on the basis of f_c and the state of this node's offspring. The state assignment based on f_c again is X, as indicated in Fig. 12.29(c). Moving up one more level, the automaton encounters node d and, as its two offspring have been assigned states, relation f_d, which calls for assigning state S to node d, is used. Because this is the last node and the state S is in F, the automaton accepts (recognizes) the tree as being a valid member of the language of the tree grammar given earlier. Figure 12.29(d) shows the final representation of the state sequences followed along the frontier-to-root paths.

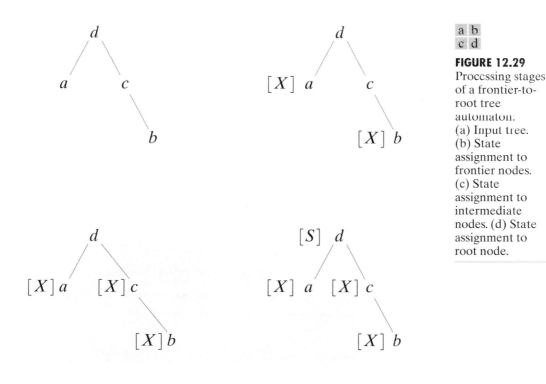

FIGURE 12.29
Processing stages of a frontier-to-root tree automaton.
(a) Input tree.
(b) State assignment to frontier nodes.
(c) State assignment to intermediate nodes. (d) State assignment to root node.

EXAMPLE 12.13:
Use of tree
grammars for
recognizing
events in bubble
chamber images.

■ Images of bubble chamber events are taken routinely during experiments in high-energy physics in which a beam of particles of known properties is directed onto a target of known nuclei. A typical event consists of tracks of secondary particles emanating from the point of collision, such as the example shown in Fig. 12.30. The incoming tracks are the horizontal parallel lines. Note the natural tree structure of the event near the middle of the photograph.

A typical experiment produces hundreds of thousands of photographs, many of which do not contain events of interest. Examining and categorizing these photographs is tedious and time-consuming for a human interpreter, thus creating a need for automatic event recognition techniques.

A tree grammar $G = (N, \Sigma, P, r, S)$ can be specified that generates trees representing events typical of those found in a hydrogen bubble chamber as a result of incoming positively charged particle streams. In this case, $N = \{S, X_1, X_2\}$, $\Sigma = \{a, b\}$, and the primitives a and b are interpreted as follows:

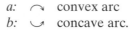

a: ⌒ convex arc
b: ⌣ concave arc.

The productions in P are

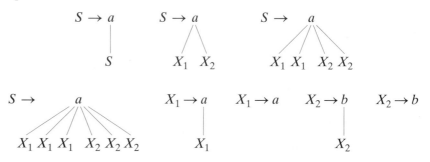

The rankings are $r(a) = \{0, 1, 2, 4, 6\}$ and $r(b) = \{0, 1\}$. The branching productions represent the number of tracks emanating from a collision, which occur

FIGURE 12.30 A
bubble chamber
photograph. (Fu
and Bhargava.)

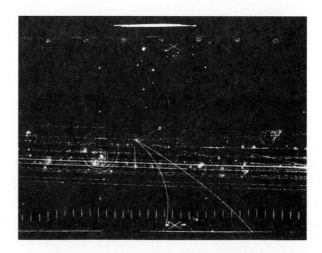

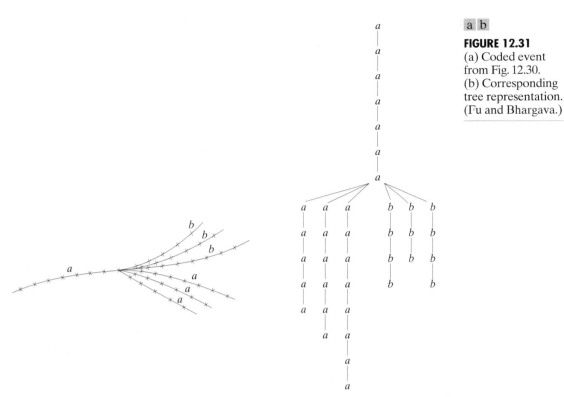

FIGURE 12.31
(a) Coded event
from Fig. 12.30.
(b) Corresponding
tree representation.
(Fu and Bhargava.)

in pairs and usually do not exceed six. Figure 12.31(a) shows the collision event in Fig. 12.30 segmented into convex and concave sections, and Fig. 12.31(b) shows the corresponding tree representation. This tree, as well as variations of it, can be generated by the grammar given above.

The tree automaton needed to recognize the types of trees just discussed is defined by using the procedure outlined in the preceding discussion. Thus, $A_t = (Q, F, \{f_k | k \in \Sigma\})$ is specified by letting $Q = \{S, X_1, X_2\}$, $F = \{S\}$, and $\{f_k | k \in \Sigma\} = \{f_a, f_b\}$. The relations are defined as $f_a = \{(S, S), (X_1, X_2, S), (X_1, X_1, X_2, X_2, S), (X_1, X_1, X_1, X_2, X_2, X_2, S), (X_1, X_1), (\varnothing, X_1)\}$ and $f_b = \{(X_2, X_2), (\varnothing, X_2)\}$. We leave it as an exercise to show that this automaton accepts the tree in Fig. 12.31(b).   ■

Learning

The syntactic recognition approaches introduced in the preceding discussion require specification of the appropriate automata (recognizers) for each class under consideration. In simple situations, inspection may yield the necessary automata. In more complicated cases, an algorithm for learning the automata from sample patterns (such as strings or trees) may be required. Because of the one-to-one correspondence between automata and grammars described previously, the learning problem sometimes is posed in terms of learning grammars directly from sample patterns, a process that usually is called *grammatical inference.* Our focus is on learning finite automata directly from sample pattern strings. The references at the end of this chapter provide a guide to methods for learning tree grammars and automata, as well as other syntactic recognition approaches.

Suppose that all patterns of a class are generated by an *unknown* grammar G and that a finite set of samples R^+ with the property

$$R^+ \subseteq \{v \mid v \text{ in } L(G)\} \tag{12.3-10}$$

is available. The set R^+, called a *positive sample set*, is simply a set of training patterns from the class associated with grammar G. This sample set is said to be *structurally complete* if each production in G is used to generate at least one element of R^+. We want to learn (synthesize) a finite automaton A_f that will accept the strings of R^+ and possibly some strings that resemble those of R^+.

Based on the definition of a finite automaton and the correspondence between G and A_f, it follows that $R^+ \subseteq \Sigma^*$, where Σ^* is the set of all strings composed of elements from Σ. Let z in Σ^* be a string such that zw is in R^+ for some w in Σ^*. For a positive integer k, we define the k *tail* of z with respect to R^+ as the set $h(z, R^+, k)$, where

$$h(z, R^+, k) = \{w \mid zw \text{ in } R^+, |w| \leq k\}. \tag{12.3-11}$$

In other words, the k tail of z is the set of strings w with the properties (1) zw is in R^+, and (2) the length of w is less than or equal to k.

A procedure for learning an automaton $A_f(R^+, k) = (Q, \Sigma, \delta, q_0, F)$ from a sample set R^+ and a particular value of k consists of letting

$$Q = \{q \mid q = h(z, R^+, k) \text{ for } z \text{ in } \Sigma^*\} \tag{12.3-12}$$

and, for each a in Σ,

$$\delta(q, a) = \{q' \text{ in } Q \mid q' = h(za, R^+, k), \text{with } q = h(z, R^+, k)\}. \tag{12.3-13}$$

In addition, we let

$$q_0 = h(\lambda, R^+, k) \tag{12.3-14}$$

and

$$F = \{q \mid q \text{ in } Q, \lambda \text{ in } q\} \tag{12.3-15}$$

where λ is the empty string (the string with no symbols). We note that the automaton $A_f(R^+, k)$ has as states subsets of the set of all k tails that can be constructed from R^+.

EXAMPLE 12.14:
Inferring a finite automaton from sample patterns.

■ Suppose that $R^+ = \{a, ab, abb\}$ and $k = 1$. Then from the preceding discussion,

$$z = \lambda, \qquad h(\lambda, R^+, 1) = \{w \mid \lambda w \text{ in } R^+, |w| \leq 1\}$$
$$= \{a\}$$
$$= q_0$$

$$z = a, \qquad h(a, R^+, 1) = \{w \mid aw \text{ in } R^+, |w| \leq 1\}$$
$$= \{\lambda, b\}$$
$$= q_1$$

$$z = ab, \quad h(ab, R^+, 1) = \{\lambda, b\}$$
$$= q_1$$
$$z = abb, \quad h(abb, R^+, 1) = \{\lambda\}$$
$$= q_2.$$

In this case, other strings z in Σ^* yield strings zw that do not belong to R^+, giving rise to a fourth state, denoted $q_\varnothing$, which corresponds to the condition that h is the null set. The states, therefore, are $q_0 = \{a\}, q_1 = \{\lambda, a\}, q_2 = \{\lambda\}$, and $q_\varnothing$, which give the set $Q = \{q_0, q_1, q_2, q_\varnothing\}$. Although the states are obtained as sets of symbols (k tails), only the state labels $q_0, q_1, \ldots$ are used in forming the set Q. The next step is to obtain the transition functions. Since $q_0 = h(\lambda, R^+, 1)$, it follows that

$$\delta(q_0, a) = h(\lambda a, R^+, 1) = h(a, R^+, 1) = q_1$$

and

$$\delta(q_0, b) = h(\lambda b, R^+, 1) = h(b, R^+, 1) = q_\varnothing.$$

Similarly, $q_1 = h(a, R^+, 1) = h(ab, R^+, 1)$ and it follows that

$$\delta(q_1, a) = h(aa, R^+, 1) = h(aba, R^+, 1) = q_\varnothing.$$

Also, $\delta(q_1, b) \supseteq h(ab, R^+, 1) = q_1$ and $\delta(q_1, b) \supseteq h(abb, R^+, 1) = q_2$; that is, $\delta(q_1, b) = \{q_1, q_2\}$. Following the procedure just described gives $\delta(q_2, a) = \delta(q_2, b) = \delta(q_\varnothing, a) = \delta(q_\varnothing, b) = q_\varnothing$. The set of final states contains those states that have the empty string λ in their k-tail representation. In this case, $q_1 = \{\lambda, b\}$ and $q_2 = \{\lambda\}$, so $F = \{q_1, q_2\}$.

Based on these results, the inferred automaton is given by

$$A_f(R^+, 1) = (Q, \Sigma, \delta, q_0, F)$$

where $Q = \{q_0, q_1, q_2, q_\varnothing\}$, $\Sigma = \{a, b\}$, $F = \{q_1, q_2\}$, and the transition functions are as given above. Figure 12.32 shows the state diagram. The automaton accepts strings of the form $a, ab, abb, \ldots, ab^n$, which are consistent with the given sample set. ■

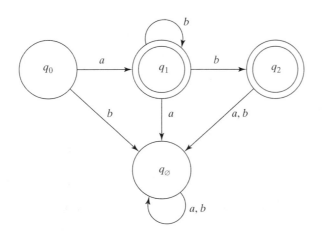

FIGURE 12.32
State diagram for the finite automaton inferred from the sample set $R^+ = \{a, ab, abb\}$.

The preceding example shows that the value of k controls the nature of the resulting automaton. The following properties exemplify the dependence of $A_f(R^+, k)$ on this parameter.

Property 1. $R^+ \subseteq L[A_f(R^+, k)]$ for all $k \geq 0$, where $L[A_f(R^+, k)]$ is the language accepted by $A_f(R^+, k)$.

Property 2. $L[A_f(R^+, k)] = R^+$ if k is equal to, or greater than, the length of the longest string in R^+; $L[A_f(R^+, k)] = \Sigma^*$ if $k = 0$.

Property 3. $L[A_f(R^+, k + 1) \subseteq L[A_f(R^+, k)]$.

Property 1 guarantees that $A_f(R^+, k)$ will, as a minimum, accept the strings in the sample set R^+. If k is equal to, or greater than, the length of the longest string in R^+, then by Property 2 the automation will accept *only* the strings in R^+. If $k = 0$, $A_f(R^+, 0)$ will consist of one state $q_0 = \{\lambda\}$, which will act as both the initial and final states. The transition functions will then be of the form $\delta(q_0, a) = q_0$ for a in Σ. Therefore, $L[A_f(R^+, 0)] = \Sigma^*$, and the automaton will accept the empty string λ and all strings composed of symbols from Σ. Finally, Property 3 indicates that the scope of the language accepted by $A_f(R^+, k)$ decreases as k increases.

These three properties allow control of the nature of $A_f(R^+, k)$ simply by varying the parameter k. If $L[A_f(R^+, k)]$ is a guess of the language L_0 from which the sample R^+ was chosen and if k is very small, this guess of L_0 will constitute a liberal inference that may include most or all of the strings in Σ^*. However, if k is equal to the length of the longest string in R^+, the inference will be conservative in the sense that $A_f(R^+, k)$ will accept only the strings contained in R^+. Figure 12.33 shows these concepts graphically.

FIGURE 12.33
Relationship between $L[A_f(R^+, k)]$ and k. The value of k_m is such that $k_m \geq$ (length of the longest string in R^+).

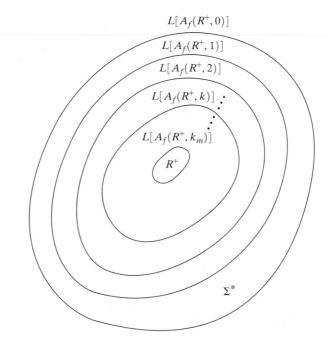

$L[A_f(R^+, 0)]$

$L[A_f(R^+, 1)]$

$L[A_f(R^+, 2)]$

$L[A_f(R^+, k)]$

$L[A_f(R^+, k_m)]$

R^+

Σ^*

■ Consider the set $R^+ = \{caaab, bbaab, caab, bbab, cab, bbb, cb\}$. For $k = 1$,
following the same procedure used in the preceding example gives

EXAMPLE 12.15:
Another example
of inferring an
automaton from a
given set of
patterns.

1. $z = \lambda$, $h(\lambda, R^+, 1) = \{\varnothing\} = q_\varnothing$;
2. $z = c$, $h(z, R^+, 1) = \{b\} = q_1$;
3. $z = ca$, $h(z, R^+, 1) = \{b\} = q_1$;
4. $z = cb$, $h(z, R^+, 1) = \{\lambda\} = q_0$;
5. $z = caa$, $h(z, R^+, 1) = \{b\} = q_1$;
6. $z = cab$, $h(z, R^+, 1) = \{\lambda\} = q_0$;
7. $z = caaa$, $h(z, R^+, 1) = \{b\} = q_1$;
8. $z = caab$, $h(z, R^+, 1) = \{\lambda\} = q_0$;
9. $z = caaab$, $h(z, R^+, 1) = \{\lambda\} = q_0$;
10. $z = b$, $h(z, R^+, 1) = \{\varnothing\} = q_\varnothing$;
11. $z = bb$, $h(z, R^+, 1) = \{b\} = q_1$;
12. $z = bba$, $h(z, R^+, 1) = \{b\} = q_1$;
13. $z = bbb$, $h(z, R^+, 1) = \{\lambda\} = q_0$;
14. $z = bbaa$, $h(z, R^+, 1) = \{b\} = q_1$;
15. $z = bbab$, $h(z, R^+, 1) = \{\lambda\} = q_0$;
16. $z = bbaab$, $h(z, R^+, 1) = \{\lambda\} = q_0$.

The automaton is

$$A_f(R', 1) = (Q, \Sigma, \delta, q_0, F)$$

with $Q - \{q_0, q_1, q_\varnothing\}$, $\Sigma = \{a, b, c\}$, $F = \{q_0\}$, and the transitions shown in
the state diagram in Fig. 12.34. To be accepted by the automaton, a string must
begin with a, b, or c and end with a symbol b. Also, strings with repetitions of a,
b, or c are accepted by $A_f(R^+, 1)$. ■

The principal advantage of the preceding method is simplicity of implementation. The synthesis procedure can be simulated in a digital computer with
a modest amount of effort. The main disadvantage is deciding on a proper value
for k, although this problem is simplified to some degree by the three properties discussed earlier.

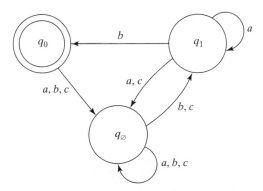

FIGURE 12.34 State diagram for the automaton $A_f(R^+, 1)$ inferred from the sample
set $R^+ = \{caaab, bbaab, caab, bbab, cab, bbb, cb\}$.

Summary

Starting with Chapter 9, our treatment of digital image processing began a transition from processes whose outputs are images to processes whose outputs are attributes about images, in the sense defined in Section 1.1. Although the material in the present chapter is introductory in nature, the topics covered are fundamental to understanding the state of the art in object recognition. As mentioned at the beginning of this chapter, recognition of individual objects is a logical place to conclude this book. To go past this point, we need concepts that are beyond the scope we set for our journey back in Section 1.4. Specifically, the next logical step would be the development of image analysis methods whose proper development requires concepts from machine intelligence.

As mentioned in Sections 1.1 and 1.4, machine intelligence and some areas that depend on it, such as scene analysis and computer vision, still are in their relatively early stages of practical development. Solutions of image analysis problems today are characterized by heuristic approaches. While these approaches are indeed varied, most of them share a significant base of techniques that are precisely the methods covered in this book.

Having concluded study of the material in the preceding twelve chapters, the reader is now in the position of being able to understand the principal areas spanning the field of digital image processing, both from a theoretical and practical point of view. Care was taken throughout all discussions to lay a solid foundation upon which further study of this and related fields could be based. Given the task-specific nature of many imaging problems, a clear understanding of basic principles enhances significantly the chances for their successful solution.

References and Further Reading

Background material for Sections 12.1 through 12.2.2 are the books by Duda, Hart, and Stork [2001], and by Tou and Gonzalez [1974]. The survey article by Jain et al. [2000] also is of interest. The book by Principe et al. [1999] presents a good overview of neural networks. A special issue of *IEEE Trans. Image Processing* [1998] is worth comparing with a similar special issue ten years earlier (*IEEE Computer* [1988]). The material presented in Section 12.2.3 is introductory. In fact, the neural network model used in that discussion is one of numerous models proposed over the years. However, the model we discussed is representative and also is used quite extensively in image processing. The example dealing with the recognition of distorted shapes is adapted from Gupta et al. [1990, 1994]. The paper by Gori and Scarselli [1998] discusses the classification power of multilayer neural networks. An approach reported by Ueda [2000] based on using linear combinations of neural networks to achieve minimum classification error is good additional reading in this context.

For additional reading on the material in Section 12.3.1, see Bribiesca and Guzman [1980]. On string matching see Sze and Yang [1981], Oommen and Loke [1997], and Gdalyahu and Weinshall [1999]. References for Sections 12.3.3 and 12.3.4 are Gonzalez and Thomason [1978], Fu [1982], and Bunke and Sanfeliu [1990]. See also Tanaka [1995], Vailaya et al. [1998], Aizaka and Nakamura [1999], and Jonk et al. [1999].

See inside front cover

Detailed solutions to the problems marked with a star can be found in the book web site. The site also contains suggested projects based on the material in this chapter.

Problems

12.1   **(a)** Compute the decision functions of a minimum distance classifier for the patterns shown in Fig. 12.1. You may obtain the required mean vectors by (careful) inspection.

(b) Sketch the decision surfaces implemented by the decision functions in (a).

★ **12.2**   Show that Eqs. (12.2-4) and (12.2-5) perform the same function in terms of pattern classification.

12.3   Show that the surface given by Eq. (12.2-6) is the perpendicular bisector of the line joining the n-dimensional points $\mathbf{m}_i$ and $\mathbf{m}_j$.

★ **12.4**   Show how the minimum distance classifier discussed in connection with Fig. 12.7 could be implemented by using W resistor banks (W is the number of classes), a summing junction at each bank (for summing currents), and a maximum selector capable of selecting the maximum of W inputs, where the inputs are currents.

12.5   Show that the correlation coefficient of Eq. (12.2-8) has values in the range $[-1, 1]$. *Hint:* Express $\gamma(x, y)$ in vector form.

★ **12.6**   An experiment produces binary images of blobs that are nearly elliptical in shape (see the following figure). The blobs are of three sizes, with the average values of the principal axes of the ellipses being (1.3, 0.7), (1.0, 0.5), and (0.75, 0.25). The dimensions of these axes vary $\pm 10\%$ about their average values. Develop an image processing system capable of rejecting incomplete or overlapping ellipses and then classifying the remaining single ellipses into one of the three size classes given. Show your solution in block diagram form, giving specific details regarding the operation of each block. Solve the classification problem using a minimum distance classifier, indicating clearly how you would go about obtaining training samples and how you would use these samples to train the classifier.

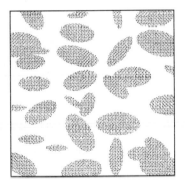

12.7   The following pattern classes have Gaussian probability density functions: ω_1: $\{(0, 0)^T, (2, 0)^T, (2, 2)^T, (0, 2)^T\}$ and ω_2: $\{(4, 4)^T, (6, 4)^T, (6, 6)^T, (4, 6)^T\}$.

(a) Assume that $P(\omega_1) = P(\omega_2) = \frac{1}{2}$ and obtain the equation of the Bayes decision boundary between these two classes.

(b) Sketch the boundary.

★ **12.8**   Repeat Problem 12.7, but use the following pattern classes: ω_1: $\{(-1, 0)^T, (0, -1)^T, (1, 0)^T, (0, 1)^T\}$ and ω_2: $\{(-2, 0)^T, (0, -2)^T, (2, 0)^T, (0, 2)^T\}$. Observe that these classes are not linearly separable.

12.9   Repeat Problem 12.6, but use a Bayes classifier (assume Gaussian densities). Indicate clearly how you would go about obtaining training samples and how you would use these samples to train the classifier.

★ **12.10**   The Bayes decision functions $d_j(\mathbf{x}) = p(\mathbf{x}/\omega_j)P(\omega_j), j = 1, 2, \ldots, W$, were derived using a 0-1 loss function. Prove that these decision functions minimize the probability of error. (*Hint:* The probability of error $p(e)$ is $1 - p(c)$, where $p(c)$ is the probability of being correct. For a pattern vector $\mathbf{x}$ belonging to class ω_i,

$p(c/\mathbf{x}) = p(\omega_i/\mathbf{x})$. Find $p(c)$ and show that $p(c)$ is maximum [$p(e)$ is minimum] when $p(\mathbf{x}/\omega_i)P(\omega_i)$ is maximum.)

12.11 **(a)** Apply the perceptron algorithm to the following pattern classes: ω_1: $\{(0, 0, 0)^T, (1, 0, 0)^T, (1, 0, 1)^T, (1, 1, 0)^T\}$ and ω_2: $\{(0, 0, 1)^T, (0, 1, 1)^T, (0, 1, 0)^T, (1, 1, 1)^T\}$. Let $c = 1$, and $\mathbf{w}(1) = (-1, -2, -2, 0)^T$.

(b) Sketch the decision surface obtained in (a). Show the pattern classes and indicate the positive side of the surface.

★**12.12** The perceptron algorithm given in Eqs. (12.2-34) through (12.2-36) can be expressed in a more concise form by multiplying the patterns of class ω_2 by -1, in which case the correction steps in the algorithm become $\mathbf{w}(k + 1) = \mathbf{w}(k)$, if $\mathbf{w}^T(k)\mathbf{y}(k) > 0$, and $\mathbf{w}(k + 1) = \mathbf{w}(k) + c\mathbf{y}(k)$ otherwise. This is one of several perceptron algorithm formulations that can be derived by starting from the general gradient descent equation

$$\mathbf{w}(k + 1) = \mathbf{w}(k) - c\left[\frac{\partial J(\mathbf{w}, \mathbf{y})}{\partial \mathbf{w}}\right]_{\mathbf{w}=\mathbf{w}(k)}$$

where $c > 0$, $J(\mathbf{w}, \mathbf{y})$ is a criterion function, and the partial derivative is evaluated at $\mathbf{w} = \mathbf{w}(k)$. Show that the perceptron algorithm formulation is obtainable from this general gradient descent procedure by using the criterion function $J(\mathbf{w}, \mathbf{y}) = \frac{1}{2}(|\mathbf{w}^T\mathbf{y}| - \mathbf{w}^T\mathbf{y})$, where |arg| is the absolute value of the argument.

(*Note:* The partial derivative of $\mathbf{w}^T\mathbf{y}$ with respect to $\mathbf{w}$ equals $\mathbf{y}$.)

12.13 Prove that the perceptron training algorithm given in Eqs. (12.2-34) through (12.2-36) converges in a finite number of steps if the training pattern sets are linearly separable. [*Hint:* Multiply the patterns of class ω_2 by -1 and consider a non-negative threshold, T, so that the perceptron training algorithm (with $c = 1$) is expressed as $\mathbf{w}(k + 1) = \mathbf{w}(k)$, if $\mathbf{w}^T(k)\mathbf{y}(k) > T$, and $\mathbf{w}(k + 1) = \mathbf{w}(k) + \mathbf{y}(k)$ otherwise. You may need to use the Cauchy-Schwartz inequality: $\|\mathbf{a}\|^2\|\mathbf{b}\|^2 \geq (\mathbf{a}^T\mathbf{b})^2$.]

★**12.14** Specify the structure and weights of a neural network capable of performing *exactly* the same function as a minimum distance classifier for two pattern classes in n-dimensional space.

12.15 Specify the structure and weights of a neural network capable of performing *exactly* the same function as a Bayes classifier for two pattern classes in n-dimensional space. The classes are Gaussian with different means but equal covariance matrices.

★**12.16** **(a)** Under what conditions are the neural networks in Problems 12.14 and 12.15 identical?

(b) Would the generalized delta rule for multilayer feedforward neural networks developed in Section 12.2.3 yield the particular neural network in (a) if trained with a sufficiently large number of samples?

12.17 Two pattern classes in two dimensions are distributed in such a way that the patterns of class ω_1 lie randomly along a circle of radius r_1. Similarly, the patterns of class ω_2 lie randomly along a circle of radius r_2, where $r_2 = 2r_1$. Specify the structure of a neural network with the minimum number of layers and nodes needed to classify properly the patterns of these two classes.

★**12.18** Repeat Problem 12.6, but use a neural network. Indicate clearly how you would go about obtaining training samples and how you would use these samples to train the classifier. Select the simplest possible neural network that, in your opinion, is capable of solving the problem.

12.19 Show that the expression $h_j'(I_j) = O_j(1 - O_j)$ given in Eq. (12.2-71), where $h_j'(I_j) = \partial h_j(I_j)/\partial I_j$, follows from Eq. (12.2-50) with $\theta_o = 1$.

★**12.20** Show that the distance measure $D(A, B)$ of Eq. (12.3-2) satisfies the properties given in Eq. (12.3-3).

12.21 Show that $\beta = \max(|a|, |b|) - \alpha$ in Eq. (12.3-4) is 0 if and only if a and b are identical strings.

12.22★(a) Specify a finite automaton capable of recognizing pattern strings of the form ab^na.

   (b) Obtain the corresponding regular grammar from your solution in (a). (Do not solve by inspection.)

12.23 Give an expansive tree grammar for generating images consisting of alternating 1's and 0's in both spatial directions (in a checkerboard pattern). Assume that the top left element is a 1 and that all images terminate with a 1 as the bottom left element.

★**12.24** Use the learning procedure specified in Eqs. (12.3-12) through (12.3-15) to learn a finite automaton capable of recognizing strings of the form ab^na, with $n > 0$. Start with the sample set $\{aba, abba, abbba\}$. If this set is insufficient for the algorithm to discover the iterative regularity of symbol b, add more sample strings until it can.

12.25 Show that the tree automaton given in connection with Fig. 12.30 accepts the tree given in Fig. 12.31(b).

12.26 A certain factory mass produces small American flags for sporting events. The quality assurance team has observed that, during periods of peak production, some printing machines have a tendency to drop (randomly) between one and three stars and one or two entire stripes. Aside from these errors, the flags are perfect in every other way. Although the flags containing errors represent a small percentage of total production, the plant manager decides to solve the problem. After much investigation, he concludes that automatic inspection using image processing techniques is the most economical way to handle the problem. The basic specifications are as follows: The flags are approximately 7.5 cm by 12.5 cm in size. They move lengthwise down the production line (individually, but with a ±15° variation in orientation) at approximately 50 cm/s, with a separation between flags of approximately 5 cm. In all cases, "approximately" means ±5%. The plant manager hires you to design an image processing system for each production line. You are told that cost and simplicity are important parameters in determining the viability of your approach. Design a complete system based on the model of Fig. 1.23. Document your solution (including assumptions and specifications) in a brief (but clear) written report addressed to the plant manager.

Bibliography

Abidi, M. A., and Gonzalez, R. C. (eds.) [1992]. *Data Fusion in Robotics and Machine Intelligence*, Academic Press, New York.

Abidi, M. A., Eason, R. O., and Gonzalez, R. C. [1991]. "Autonomous Robotics Inspection and Manipulation Using Multisensor Feedback." *IEEE Computer*, vol. 24, no. 4, pp. 17–31.

Abramson, N. [1963]. *Information Theory and Coding*, McGraw-Hill, New York.

Adiv, G. [1985]. "Determining Three-Dimensional Motion and Structure from Optical Flow Generated by Several Moving Objects." *IEEE Trans. Pattern Anal. Mach. Intell.*, vol. PAMI-7, no. 4, pp. 384–401.

Aggarwal, J. K., and Badler, N. I. (eds.) [1980]. "Motion and Time-Varying Imagery." *IEEE Trans. Pattern Anal. Mach. Intell.*, Special Issue, vol. PAMI-2, no. 6, pp. 493–588.

Aguado, A. S., Nixon, M. S., and Montiel, M. M. [1998], "Parameterizing Arbitrary Shapes via Fourier Descriptors for Evidence-Gathering Extraction," *Computer Vision and Image Understanding*, vol. 69, no. 2, pp. 202–221.

Ahmed, N., Natarajan, T., and Rao, K. R. [1974]. "Discrete Cosine Transforms." *IEEE Trans. Comp.*, vol. C-23, pp. 90–93.

Ahmed, N., and Rao, K. R. [1975]. *Orthogonal Transforms for Digital Signal Processing*, Springer Verlag, New York.

Aizaka, K., and Nakamura, A. [1999]. "Parsing of Two-Dimensional Images Represented by Quadtree Adjoining Grammars," *Pattern Recog.*, vol. 32, no. 2, pp. 277–294.

Alliney, S. [1993]. "Digital Analysis of Rotated Images," *IEEE Trans. Pattern Anal. Machine Intell.*, vol. 15, no. 5, pp. 499–504.

Ando, S. [2000]. "Consistent Gradient Operators," *IEEE Trans. Pattern Anal. Machine Intell.*, vol. 22, no. 3, pp. 252–265.

Andrews, H. C. [1970]. *Computer Techniques in Image Processing*, Academic Press, New York.

Andrews, H. C., and Hunt, B. R. [1977]. *Digital Image Restoration*, Prentice Hall, Englewood Cliffs, N.J.

Anelli, G., Broggi, A., and Destri, G. [1998]. "Decomposition of Arbitrarily-Shaped Morphological Structuring Elements Using Genetic Algorithms," *IEEE Trans. Pattern Anal. Machine Intell.*, vol. 20, no. 2, pp. 217–224.

Ang, P. H., Ruetz, P. A., and Auld, D. [1991]. "Video Compression Makes Big Gains." *IEEE Spectrum*, vol. 28, no. 10, pp. 16–19.

Antonini, M., Barlaud, M., Mathieu, P., and Daubechies, I. [1992]. "Image Coding Using Wavelet Transform." *IEEE Trans. Image Processing*, vol. 1, no. 2, pp. 205–220.

Atchison, D. A., and Smith, G. [2000]. *Optics of the Human Eye*, Butterworth-Heinemann, Boston, Mass.

Baccar, M., Gee, L. A., Abidi, M. A., and Gonzalez, R. C. [1996]. "Segmentation of Range Images Via Data Fusion and Morphological Watersheds," *Pattern Recog.*, vol. 29, no. 10, pp. 1671–1685.

Bajcsy, R., and Lieberman, L. [1976]. "Texture Gradient as a Depth Cue." *Comput. Graph. Image Proc.*, vol. 5, no. 1, pp. 52–67.

Bakir, T., and Reeves, J. S. [2000]. "A Filter Design Method for Minimizing Ringing in a Region of Interest in MR Spectroscopic Images," *IEEE Trans. Medical Imaging*, vol. 19, no. 6, pp. 585–600.

Ballard, D. H. [1981]. "Generalizing the Hough Transform to Detect Arbitrary Shapes." *Pattern Recognition*, vol. 13, no. 2, pp. 111–122.

Ballard, D. H., and Brown, C. M. [1982]. *Computer Vision*, Prentice Hall, Englewood Cliffs, N.J.

Banham, M. R., Galatsanos, H. L., Gonzalez, H. L., and Katsaggelos, A. K. [1994]. "Multichannel Restoration of Single Channel Images Using a Wavelet-Based Subband Decomposition." *IEEE Trans. Image Processing*, vol. 3, no. 6, pp. 821–833.

Banham, M. R., and Katsaggelos, A. K. [1996]. "Spatially Adaptive Wavelet-Based Multiscale Image Restoration." *IEEE Trans. Image Processing*, vol. 5, no. 5, pp. 619–634.

Basart, J. P., and Gonzalez, R. C. [1992]. "Binary Morphology," in *Advances in Image Analysis*, Y. Mahdavieh and R. C. Gonzalez (eds.), SPIE Press, Bellingham, Wash., pp. 277–305.

Basart, J. P., Chacklackal, M. S., and Gonzalez, R. C. [1992]. "Introduction to Gray-Scale Morphology," in *Advances in Image Analysis*, Y. Mahdavieh and R. C. Gonzalez (eds.), SPIE Press, Bellingham, Wash., pp. 306–354.

Bates, R. H. T., and McDonnell, M. J. [1986]. *Image Restoration and Reconstruction*. Oxford University Press, New York.

Battle, G. [1987]. "A Block Spin Construction of Ondelettes. Part I: Lemarié Functions." *Commun. Math. Phys.*, vol. 110, pp. 601–615.

Battle, G. [1988]. "A Block Spin Construction of Ondelettes. Part II: the QFT Connection." *Commun. Math. Phys.*, vol. 114, pp. 93–102.

Baumert, L. D., Golomb, S. W., and Hall, M., Jr. [1962]. "Discovery of a Hadamard Matrix of Order 92." *Bull. Am. Math. Soc.*, vol. 68, pp. 237–238.

Baxes, G. A. [1994]. *Digital Image Processing: Principles and Applications*, John Wiley & Sons, New York.

Baylon, D. M., and Lim, J. S. [1990]. "Transform/Subband Analysis and Synthesis of Signals." *Tech. Report*, MIT Research Laboratory of Electronics. Cambridge. Mass.

Bell, E. T. [1965]. *Men of Mathematics*, Simon & Schuster, New York.

Bengtsson, A., and Eklundh, J. O. [1991]. "Shape Representation by Multiscale Contour Approximation." *IEEE Trans. Pattern Anal. Machine Intell.*, vol. 13, no. 1, pp. 85–93.

Benson, K. B. [1985]. *Television Engineering Handbook*, McGraw-Hill, New York.

Berger, T. [1971]. *Rate Distortion Theory*, Prentice Hall, Englewood Cliffs, N.J.

Beucher, S. [1990]. Doctoral Thesis, Centre de Morphologie Mathématique, École des Mines de Paris, France. (The core of this material is contained in the following paper.)

Beucher, S., and Meyer, F. [1992]. "The Morphological Approach of Segmentation: The Watershed Transformation," in *Mathematical Morphology in Image Processing*, E. Dougherty (ed.), Marcel Dekker, New York.

Bhaskaran, V., and Konstantinos, K. [1997]. *Image and Video Compression Standards: Algorithms and Architectures*. Kluwer, Boston, Mass.

Bhatt, B., Birks, D., Hermreck, D. [1997]. "Digital Television: Making It Work." *IEEE Spectrum*, vol. 34, no. 10, pp. 19–28.

Biberman, L. M. [1973]. "Image Quality." In *Perception of Displayed Information*. Biberman, L. M. (ed.), Plenum Press, New York.

Bichsel, M. [1998]. "Analyzing a Scene's Picture Set under Varying Lightning," *Computer Vision and Image Understanding*, vol. 71, no. 3, pp. 271–280.

Bieniek, A., and Moga, A. [2000]. "An Efficient Watershed Algorithm Based on Connected Components," *Pattern Recog.*, vol. 33, no. 6, pp. 907–916.

Bisignani, W. T., Richards, G. P., and Whelan, J. W. [1966]. "The Improved Grey Scale and Coarse-Fine PCM Systems: Two New Digital TV Bandwidth Reduction Techniques." *Proc. IEEE*, vol. 54, no. 3, pp. 376–390.

Blahut, R. E. [1987]. *Principles and Practice of Information Theory*, Addison-Wesley, Reading, Mass.

Bleau, A. and Leon, L. J. [2000]. "Watershed-Based Segmentation and Region Merging," *Computer Vision and Image Understanding*, vol. 77, no. 3, pp. 317–370.

Blouke, M. M., Sampat, N., and Canosa, J. [2001]. *Sensors and Camera Systems for Scientific, Industrial, and Digital Photography Applications-II*, ISBN 0–8194–3583–X, SPIE Press, Bellingham, Wash.

Blum, H. [1967]. "A Transformation for Extracting New Descriptors of Shape." In *Models for the Perception of Speech and Visual Form*, Wathen-Dunn, W. (ed.), MIT Press, Cambridge, Mass.

Blume, H., and Fand, A. [1989]. "Reversible and Irreversible Image Data Compression Using the S-Transform and Lempel-Ziv Coding." *Proc. SPIE Medical Imaging III: Image Capture and Display*, vol. 1091, pp. 2–18.

Boie, R. A., and Cox, I. J. [1992]. "An Analysis of Camera Noise," *IEEE Trans. Pattern Anal. Machine Intell.*, vol. 14, no. 6, pp. 671–674.

Born, M., and Wolf, E. [1999]. *Principles of Optics: Electromagnetic Theory of Propagation, Interference and Diffraction of Light*, 7th ed., Cambridge University Press, Cambride, UK.

Boulgouris, N. V., Tzovaras, D., and Strintzis, M. G. [2001]. "Lossless Image Compression Based on Optimal Prediction, Adaptive Lifting, and Conditional Arithmetic Coding," *IEEE Trans. Image Processing*, vol. 10, no. 1, pp. 1–14.

Bouman, C., and Liu, B. [1991]. "Multiple Resolution Segmentation of Textured Images." *IEEE Trans. Pattern. Anal. Machine Intell.*, vol. 13, no. 2, pp. 99–113.

Boyd, J. E. and Meloche, J. [1998]. "Binary Restoration of Thin Objects in Multidimensional Imagery," *IEEE Trans. Pattern Anal. Machine Intell.*, vol. 20, no. 6, pp. 647–651.

Bracewell, R. N. [1995]. *Two-Dimensional Imaging*, Prentice Hall, Upper Saddle River, N.J.

Bracewell, R. N. [2000]. *The Fourier Transform and its Applications*, 3rd ed. McGraw-Hill, New York.

Bribiesca, E. [1981]. "Arithmetic Operations Among Shapes Using Shape Numbers." *Pattern Recog.*, vol. 13, no. 2, pp. 123–138.

Bribiesca, E. [1999]. "A New Chain Code," *Pattern Recog.*, vol. 32, no. 2, pp. 235–251.

Bribiesca, E. [2000]. "A Chain Code for Representing 3–D Curves," *Pattern Recog.*, vol. 33, no. 5, pp. 755–765.

Bribiesca, E., and Guzman, A. [1980]. "How to Describe Pure Form and How to Measure Differences in Shape Using Shape Numbers." *Pattern Recog.*, vol. 12, no. 2, pp. 101–112.

Brigham, E. O. [1988]. *The Fast Fourier Transform and its Applications*, Prentice Hall, Upper Saddle River, N.J.

Brinkman, B. H., Manduca, A., and Robb, R. A. [1998]. "Optimized Homomorphic Unsharp Masking for MR Grayscale Inhomogeneity Correction," *IEEE Trans. Medical Imaging*, vol. 17, no. 2, pp. 161–171.

Brummer, M. E. [1991]. "Hough Transform Detection of the Longitudinal Fissure in Tomographic Head Images." *IEEE Trans. Biomed. Images*, vol. 10, no. 1, pp. 74–83.

Brzakovic, D., Patton, R., and Wang, R. [1991]. "Rule-Based Multi-template Edge Detection." *Comput. Vision, Graphics, Image Proc: Graphical Models and Image Proc.*, vol. 53, no. 3, pp. 258–268.

Bunke, H. and Sanfeliu, A. (eds.) [1990]. *Syntactic and Structural Pattern Recognition: Theory and Applications*, World Scientific, Teaneck, N.J.

Burrus, C. S., Gopinath, R. A., and Guo, H. [1998]. *Introduction to Wavelets and Wavelet Transforms*, Prentice Hall, Upper Saddle River, N J., pp. 250–251.

Burt, P. J., and Adelson, E. H. [1983]. "The Laplacian Pyramid as a Compact Image Code." *IEEE Trans. Commun.*, vol. COM-31, no. 4, pp. 532–540.

Campbell, J. D. [1969]. "Edge Structure and the Representation of Pictures." Ph.D. dissertation, Dept. of Elec. Eng., University of Missouri, Columbia.

Candy, J. C., Franke, M. A., Haskell, B. G., and Mounts, F. W. [1971]. "Transmitting Television as Clusters of Frame-to-Frame Differences." *Bell Sys. Tech. J.*, vol. 50, pp. 1889–1919.

Cannon, T. M. [1974]. "Digital Image Deblurring by Non-Linear Homomorphic Filtering." Ph.D. thesis, University of Utah.

Canny, J. [1986]. "A Computational Approach for Edge Detection," *IEEE Trans. Pattern Anal. Machine Intell.*, vol. 8, no. 6, pp. 679–698.

Carey, W. K., Chuang, D. B., and Hamami, S. S. [1999]. "Regularity-Preserving Image Interpolation," *IEEE Trans. Image Processing*, vol. 8, no. 9, pp. 1293–1299.

Caselles, V., Lisani, J.-L., Morel, J.-M., and Sapiro, G. [1999]. "Shape Preserving Local Histogram Modification," *IEEE Trans. Image Processing*, vol. 8, no. 2, pp. 220–230.

Castleman, K. R. [1996]. *Digital Image Processing*, 2nd ed. Prentice Hall, Upper Saddle River, N.J.

Centeno, J. A. S., and Haertel, V. [1997]. "An Adaptive Image Enhancement Algorithm," *Pattern Recog.*, vol. 30, no. 7 pp. 1183–1189.

Chan, R. C., Karl, W. C., and Lees, R. S. [2000]. "A New Model-Based Technique for Enhanced Small-Vessell Measurements in X-Ray Cine-Angiograms," *IEEE Trans. Medical Imaging*, vol. 19, no. 3, pp. 243–255.

Chang, S. G., Yu, B., and Vetterli, M. [2000]. "Spatially Adaptive Wavelet Thresholding with Context Modeling for Image Denoising." *IEEE Trans. Image Processing*, vol. 9, no. 9, pp. 1522–1531.

Chang, S. K. [1989]. *Principles of Pictorial Information Systems Design*, Prentice Hall, Englewood Cliffs, N.J.

Chang, T., and Kuo, C.-C. J. [1993]. "Texture Analysis and Classification with Tree-Structures Wavelet Transforms." *IEEE Trans. Image Processing*, vol. 2, no. 4, pp. 429–441.

Chaudhuri, B. B. [1983]. "A Note on Fast Algorithms for Spatial Domain Techniques in Image Processing." *IEEE Trans. Syst. Man Cyb.*, vol. SMC-13, no. 6, pp. 1166–1169.

Chen, M. C., and Wilson, A. N. [2000]. "Motion-Vector Optimization of Control Grid Interpolation and Overlapped Block Motion Compensation Using Iterated Dynamic Programming," *IEEE Trans. Image Processing.*, vol. 9, no. 7, pp. 1145–1157.

Chen, Y.-S., and Yu, Y.-T. [1996]. "Thinning Approach for Noisy Digital Patterns," *Pattern Recog.*, vol. 29, no. 11, pp. 1847–1862.

Cheng, H. D., and Huijuan Xu, H. [2000]. "A Novel Fuzzy Logic Approach to Contrast Enhancement," *Pattern Recog.*, vol. 33, no. 5, pp. 809–819.

Cheriet, M., Said, J. N., and Suen, C. Y. [1998]. "A Recursive Thresholding Technique for Image Segmentation," *IEEE Trans. Image Processing*, vol. 7, no. 6, pp. 918–921.

Cheung, J., Ferris, D., and Kurz, L. [1997]. "On Classification of Multispectral Infrared Image Data," *IEEE Trans. Image Processing*, vol. 6, no. 10, pp. 1456–1464.

Cheung, K. K. T., and Teoh, E. K. [1999]. "Symmetry Detection by Generalized Complex (GC) Moments: A Closed-Form Solution," *IEEE Trans. Pattern Anal. Machine Intell.*, vol. 21, no. 5, pp. 466–476.

Chow, C. K., and Kaneko, T. [1972]. "Automatic Boundary Detection of the Left Ventricle from Cineangiograms." *Comp., and Biomed. Res.*, vol. 5, pp. 388–410.

Chu, C.-C., and Aggarwal, J. K [1993]. "The Integration of Image Segmentation Maps Using Regions and Edge Information," *IEEE Trans. Pattern Anal. Machine Intell.*, vol. 15, no. 12, pp. 1241–1252.

CIE [1978]. *Uniform Color Spaces—Color Difference Equations—Psychometric Color Terms*, Commission Internationale de L'Eclairage, Publication No. 15, Supplement No. 2, Paris.

Clark, J. J. [1989]. "Authenticating Edges Produced by Zero-Crossing Algorithms." *IEEE Trans. Pattern Anal. Machine Intell.*, vol. 12, no. 8, pp. 830–831.

Clarke, R. J. [1985]. *Transform Coding of Images*, Academic Press, New York.

Cochran, W. T., Cooley, J. W., et al. [1967]. "What Is the Fast Fourier Transform?" *IEEE Trans. Audio and Electroacoustics*, vol. AU-15, no. 2, pp. 45–55.

Cohen, A., and Daubechies, I. [1992]. *A Stability Criterion for Biorthogonal Wavelet Bases and Their Related Subband Coding Schemes*, Technical report, AT&T Bell Laboratories.

Cohen, A., Daubechies, I., and Feauveau, J.-C. [1992]. "Biorthogonal Bases of Compactly Supported Wavelets." *Commun. Pure and Appl. Math.*, vol. 45, pp. 485–560.

Coifman, R. R., and Wickerhauser, M. V. [1992]. "Entropy-Based Algorithms for Best Basis Selection." *IEEE Tran. Information Theory*, vol. 38, no. 2, pp. 713–718.

Cooley, J. W., Lewis, P. A. W., and Welch, P. D. [1967a]. "Historical Notes on the Fast Fourier Transform." *IEEE Trans. Audio and Electroacoustics*, vol. AU-15, no. 2, pp. 76–79.

Cooley, J. W., Lewis, P. A. W., and Welch, P. D. [1967b]. "Application of the Fast Fourier Transform to Computation of Fourier Integrals." *IEEE Trans. Audio and Electroacoustics*, vol. AU-15, no. 2, pp. 79–84.

Cooley, J. W., Lewis, P. A. W., and Welch, P. D. [1969]. "The Fast Fourier Transform and Its Applications." *IEEE Trans. Educ.*, vol. E-12, no. 1. pp. 27–34.

Cooley, J. W., and Tukey, J. W. [1965]. "An Algorithm for the Machine Calculation of Complex Fourier Series." *Math. of Comput.*, vol. 19, pp. 297–301.

Cornsweet, T. N. [1970]. *Visual Perception*, Academic Press, New York.

Cortelazzo, G. M., Lucchese, L., and Monti, C. M. [1999]. "Frequency Domain Analysis of General Planar Rigid Motion with Finite Duration," *J. Opt. Soc. Amer.-A. Optics, Image Science, and Vision*, vol. 16, no. 6, pp. 1238–1253.

Cowart, A. E., Snyder, W. E., and Ruedger, W. H. [1983]. "The Detection of Unresolved Targets Using the Hough Transform." *Comput. Vision Graph Image Proc.*, vol. 21, pp. 222–238.

Croisier, A., Esteban, D., and Galand, C. [1976]. "Perfect Channel Splitting by Use of Interpolation/Decimation/Tree Decomposition Techniques." *Int. Conf. On Inform. Sciences and Systems*, Patras, Greece, pp. 443–446.

Cumani, A., Guiducci, A., and Grattoni, P. [1991]. "Image Description of Dynamic Scenes," *Pattern Recog.*, vol. 24, no. 7, pp. 661–674.

Cutrona, L. J., and Hall, W. D. [1968]. "Some Considerations in Post-Facto Blur Removal." In *Evaluation of Motion-Degraded Images*, NASA Publ. SP-193, pp. 139–148.

Danielson, G. C., and Lanczos, C. [1942]. "Some Improvements in Practical Fourier Analysis and Their Application to X-Ray Scattering from Liquids." *J. Franklin Institute*, vol. 233, pp. 365–380, 435–452.

Daubechies, I. [1988]. "Orthonormal Bases of Compactly Supported Wavelets." *Commun. On Pure and Appl. Math.*, vol. 41, pp. 909–996.

Daubechies, I. [1990]. "The Wavelet Transform, Time-Frequency Localization and Signal Analysis." *IEEE Transactions on Information Theory*, vol. 36, no. 5, pp. 961–1005.

Daubechies, I. [1992]. *Ten Lectures on Wavelets*, Society for Industrial and Applied Mathematics, Philadelphia, Pa.

Daubechies, I. [1993]. "Orthonormal Bases of Compactly Supported Wavelets II, Variations on a Theme." *SIAM J. Mathematical Analysis*, vol. 24, no. 2, pp. 499–519.

Daubechies, I. [1996]. "Where Do We Go from Here?—A Personal Point of View." *Proc. IEEE*, vol. 84, no. 4, pp. 510–513.

Daul, C., Graebling, P., and Hirsch, E. [1998]. "From the Hough Transform to a New Approach for the Detection of and Approximation of Elliptical Arcs," *Computer Vision and Image Understanding*, vol. 72, no. 3, pp. 215–236.

Davis, L. S. [1982]. "Hierarchical Generalized Hough Transforms and Line-Segment Based Generalized Hough Transforms." *Pattern Recog.*, vol. 15, no. 4, pp. 277–285.

Davis, T. J. [1999]. "Fast Decomposition of Digital Curves into Polygons Using the Haar Transform," *IEEE Trans. Pattern Anal. Machine Intell.*, vol. 21, no. 8, pp. 786–790.

Davisson, L. D. [1972]. "Rate-Distortion Theory and Application." *Proc. IEEE*, vol. 60, pp. 800–808.

Delaney, A. H., and Bresler, Y. [1995]. "Multiresolution Tomographic Reconstruction Using Wavelets." *IEEE Trans. Image Processing*, vol. 4, no. 6, pp. 799–813.

Delp, E. J., and Mitchell, O. R. [1979]. "Image Truncation using Block Truncation Coding." *IEEE Trans. Comm.*, vol. COM-27, pp. 1335–1342.

Di Zenzo, S. [1986]. "A Note on the Gradient of a Multi-Image," *Computer Vision, Graphics, and Image Processing*, vol. 33, pp. 116–125.

Dijkstra, E. [1959]. "Note on Two Problems in Connection with Graphs." *Numerische Mathematik*, vol. 1, pp. 269–271.

Djeziri, S., Nouboud, F., and Plamondon, R. [1998]. "Extraction of Signatures from Check Background Based on a Filiformity Criterion," *IEEE Trans. Image Processing*, vol. 7, no. 102, pp. 1425–1438.

Dougherty, E. R. [1992]. *An Introduction to Morphological Image Processing*, SPIE Press, Bellingham, Wash.

Dougherty, E. R. (ed.) [2000]. *Random Processes for Image and Signal Processing*, IEEE Press, New York.

Drew, M. S., Wei, J., and Li, Z.-N. [1999]. "Illumination Invariant Image Retrieval and Video Segmentation," *Pattern Recog.*, vol. 32, no. 8, pp. 1369–1388.

Duda, R. O., and Hart, P. E. [1972]. "Use of the Hough Transformation to Detect Lines and Curves in Pictures." *Comm. ACM*, vol. 15, no. 1, pp. 11–15.

Duda, R. O, Hart, P. E., and Stork, D. G. [2001]. *Pattern Classification*, John Wiley & Sons, New York.

Edelman, S. [1999]. *Representation and Recognition in Vision*, The MIT Press, Cambridge, Mass.

Elias, P. [1952]. "Fourier Treatment of Optical Processes." *J. Opt. Soc. Am.*, vol. 42, no. 2, pp. 127–134.

Elliott, D. F., and Rao, K. R. [1983]. *Fast Transforms: Algorithms and Applications*. Academic Press, New York.

Eng, H.-L., and Ma, K.-K. [2001]. "Noise Adaptive Soft-Switching Median Filter," *IEEE Trans. Image Processing*, vol. 10, no. 2, pp. 242–251.

Equitz, W. H. [1989]. "A New Vector Quantization Clustering Algorithm." *IEEE Trans. Acous. Speech Signal Processing*, vol. ASSP-37, no. 10, pp. 1568–1575.

Etienne, E. K., and Nachtegael, M. (eds.) [2000]. *Fuzzy Techniques in Image Processing*, Springer-Verlag, New York.

Falconer, D. G. [1970]. "Image Enhancement and Film Grain Noise." *Opt. Acta*, vol. 17, pp. 693–705.

Fairchild, M. D. [1998]. *Color Appearance Models*, Addison-Wesley, Reading, Mass.

Federal Bureau of Investigation [1993]. *WSQ Gray-Scale Fingerprint Image Compression Specification*, IAFIS-IC-0110v2, Washington, D. C.

Felsen, L. B., and Marcuvitz, N. [1994]. *Radiation and Scattering of Waves*, IEEE Press, New York.

Ferreira, A., and Ubéda, S. [1999]. "Computing the Medial Axis Transform in Parallel with Eight Scan Operators," *IEEE Trans. Pattern Anal. Machine Intell.*, vol. 21, no. 3, pp. 277–282.

Fischler, M. A. [1980]. "Fast Algorithms for Two Maximal Distance Problems with Applications to Image Analysis." *Pattern Recog.*, vol. 12, pp. 35–40.

Fisher, R. A. [1936]. "The Use of Multiple Measurements in Taxonomic Problems," *Ann. Eugenics*, vol. 7, Part 2, pp. 179–188. (Also in *Contributions to Mathematical Statistics*, John Wiley & Sons, New York, 1950.)

Fortner, B., and Meyer, T. E. [1997]. *Number by Colors*, Springer-Verlag, New York.

Fox, E. A. [1991]. "Advances in Interactive Digital Multimedia Systems." *Computer*, vol. 24, no. 10, pp. 9–21.

Fram, J. R., and Deutsch, E. S. [1975]. "On the Quantitative Evaluation of Edge Detection Schemes and Their Comparison with Human Performance." *IEEE Trans. Computers*, vol. C-24, no. 6, pp. 616–628.

Freeman, A. (translator) [1878]. J. Fourier, *The Analytical Theory of Heat*, Cambridge: University Press, London.

Freeman, C. [1987]. *Imaging Sensors and Displays*, ISBN 0–89252–800–1, SPIE Press, Bellingham, Wash.

Freeman, H. [1961]. "On the Encoding of Arbitrary Geometric Configurations." *IEEE Trans. Elec. Computers*, vol. EC-10, pp. 260–268.

Freeman, H. [1974]. "Computer Processing of Line Drawings." *Comput. Surveys*, vol. 6, pp. 57–97.

Freeman, H., and Shapira, R. [1975]. "Determining the Minimum-Area Encasing Rectangle for an Arbitrary Closed Curve." *Comm. ACM*, vol. 18, no. 7, pp. 409–413.

Freeman, J. A., and Skapura, D. M. [1991]. *Neural Networks: Algorithms, Applications, and Programming Techniques*, Addison-Wesley, Reading, Mass.

Frei, W., and Chen, C. C. [1977]. "Fast Boundary Detection: A Generalization and a New Algorithm." *IEEE Trans. Computers*, vol. C-26, no. 10, pp. 988–998.

Frendendall, G. L., and Behrend, W. L. [1960]. "Picture Quality—Procedures for Evaluating Subjective Effects of Interference." *Proc. IRE*, vol. 48, pp. 1030–1034.

Fu, K. S. [1982]. *Syntactic Pattern Recognition and Applications*, Prentice Hall, Englewood Cliffs, N.J.

Fu, K. S., and Bhargava, B. K. [1973]. "Tree Systems for Syntactic Pattern Recognition." *IEEE Trans. Comput.*, vol. C-22, no. 12, pp. 1087–1099.

Fu, K. S., Gonzalez, R. C., and Lee, C. S. G. [1987]. *Robotics: Control, Sensing, Vision, and Intelligence*, McGraw-Hill, New York.

Fu, K. S., and Mui, J. K. [1981]. "A Survey of Image Segmentation." *Pattern Recog.*, vol. 13, no. 1, pp. 3–16.

Garcia, P. [1999]. "The Use of Boolean Model for Texture Analysis of Grey Images," *Computer Vision and Image Understanding*, vol. 74, no. 3, pp. 227–235.

Gdalyahu, Y., and Weinshall, D. [1999]. "Flexible Syntactic Matching of Curves and Its Application to Automated Hierarchical Classification of Silhouettes," *IEEE Trans. Pattern Anal. Machine Intell.*, vol. 21, no. 12, pp. 1312–1328.

Gegenfurtner, K. R., and Sharpe, L. T. (eds.) [1999]. *Color Vision: From Genes to Perception*, Cambridge University Press, New York.

Geladi, P., and Grahn, H. [1996]. *Multivariate Image Analysis*, John Wiley & Sons, New York.

Geman, D., and Reynolds, G. [1992]. "Constrained Restoration and the Recovery of Discontinuities," *IEEE Trans. Pattern Anal. Machine Intell.*, vol. 14, no. 3, pp. 367–383.

Gentleman, W. M. [1968]. "Matrix Multiplication and Fast Fourier Transformations." *Bell System Tech. J.*, vol. 47, pp. 1099–1103.

Gentleman, W. M., and Sande, G. [1966]. "Fast Fourier Transform for Fun and Profit." *Fall Joint Computer Conf.*, vol. 29, pp. 563–578, Spartan, Washington, D. C.

Gharavi, H., and Tabatabai, A. [1988]. "Sub-Band Coding of Monochrome and Color Images." *IEEE Trans. Circuits Sys.*, vol. 35, no. 2, pp. 207–214.

Giannakis, G. B., and Heath, R. W., Jr. [2000]. "Blind Identification of Multichannel FIR Blurs and Perfect Image Restoration," *IEEE Trans. Image Processing*, vol. 9, no. 11, pp. 1877–1896.

Giardina, C. R., and Dougherty, E. R. [1988]. *Morphological Methods in Image and Signal Processing*, Prentice Hall, Upper Saddle River, N.J.

Gonzalez, R. C. [1985]. "Computer Vision." *Yearbook of Science and Technology*, McGraw-Hill, New York, pp. 128–132.

Gonzalez, R. C. [1985]. "Industrial Computer Vision." In *Advances in Information Systems Science*, Tou, J. T. (ed.), Plenum, New York, pp. 345–385.

Gonzalez, R. C. [1986]. "Image Enhancement and Restoration." In *Handbook of Pattern Recognition and Image Processing*, Young, T. Y., and Fu, K. S. (eds.), Academic Press, New York, pp. 191–213.

Gonzalez, R. C., Edwards, J. J., and Thomason, M. G. [1976]. "An Algorithm for the Inference of Tree Grammars." *Int. J. Comput. Info. Sci.*, vol. 5, no. 2, pp. 145–163.

Gonzalez, R. C., and Fittes, B. A. [1977]. "Gray-Level Transformations for Interactive Image Enhancement." *Mechanism and Machine Theory*, vol. 12, pp. 111–122.

Gonzalez, R. C., and Safabakhsh, R. [1982]. "Computer Vision Techniques for Industrial Applications." *Computer*, vol. 15, no. 12, pp. 17–32.

Gonzalez, R. C., and Thomason, M. G. [1978]. *Syntactic Pattern Recognition: An Introduction*. Addison-Wesley, Reading, Mass.

Gonzalez, R. C., and Woods, R. E. [1992]. *Digital Image Processing*. Addison-Wesley, Reading, Mass.

Good, I. J. [1958]. "The Interaction Algorithm and Practical Fourier Analysis." *J. R. Stat. Soc. (Lond.)*, vol. B20, pp. 361–367; *Addendum*, vol. 22, 1960, pp. 372–375.

Goodson, K. J., and Lewis, P. H. [1990]. "A Knowledge-Based Line Recognition System." *Pattern Recog. Letters*, vol. 11, no. 4, pp. 295–304.

Gordon, I. E. [1997]. *Theories of Visual Perception*, 2nd ed., John Wiley & Sons, New York.

Gori, M., and Scarselli, F. [1998]. "Are Multilayer Perceptrons Adequate for Pattern Recognition and Verification?" *IEEE Trans. Pattern Anal. Machine Intell.*, vol. 20, no. 11, pp. 1121–1132.

Goutsias, J., Vincent, L., and Bloomberg, D. S. (eds) [2000]. *Mathematical Morphology and Its Applications to Image and Signal Processing*, Kluwer Academic Publishers, Boston, Mass.

Graham, R. E. [1958]. "Predictive Quantizing of Television Signals." *IRE Wescon Conv. Rec.*, vol. 2, pt. 2, pp. 147–157.

Graham, R. L., and Yao, F. F. [1983]. "Finding the Convex Hull of a Simple Polygon." *J. Algorithms*, vol. 4, pp. 324–331.

Gray, R. M. [1984]. "Vector Quantization." *IEEE Trans. Acous. Speech Signal Processing*, vol. ASSP-1, no. 2, pp. 4–29.

Gröchenig, K., and Madych, W. R. [1992]. "Multiresolution Analysis, Haar Bases and Self-Similar Tilings of R^n." *IEEE Trans. Information Theory,* vol. 38, no. 2, pp. 556–568.

Grossman, A., and Morlet, J. [1984]. "Decomposition of Hardy Functions into Square Integrable Wavelets of Constant Shape." *SIAM J. Appl. Math.* vol. 15, pp. 723–736.

Guil, N., Villalba, J., and Zapata, E. L. [1995]. "A Fast Hough Transform for Segment Detection," *IEEE Trans. Image Processing,* vol. 4, no. 11, pp. 1541–1548.

Guil, N., and Zapata, E. L. [1997]. "Lower Order Circle and Ellipse Hough Transform," *Pattern Recog.,* vol. 30, no. 10, pp. 1729–1744.

Gunn, S. R. [1998]. "Edge Detection Error in the Discrete Laplacian of a Gaussian," *Proc. 1998 Int'l Conference on Image Processing,* vol. II, pp. 515–519.

Gunn, S. R. [1999]. "On the Discrete Representation of the Laplacian of a Gaussian," *Pattern Recog.,* vol. 32, no. 8, pp. 1463–1472.

Gupta, L., Mohammad, R. S., and Tammana, R. [1990]. "A Neural Network Approach to Robust Shape Classification." *Pattern Recog.,* vol. 23, no. 6, pp. 563–568.

Gupta, L., and Srinath, M. D. [1988]. "Invariant Planar Shape Recognition Using Dynamic Alignment." *Pattern Recog.,* vol. 21, pp. 235–239.

Gupta, L., Wang, J., Charles, A., and Kisatsky, P. [1994]. "Three-Layer Perceptron Based Classifiers for the Partial Shape Classification Problem," *Pattern Recog.,* vol. 27, no. 1, pp. 91–97.

Haar, A. [1910]. "Zur Theorie der Orthogonalen Funktionensysteme." *Math. Annal.,* vol. 69, pp. 331–371.

Habibi, A. [1971]. "Comparison of Nth Order DPCM Encoder with Linear Transformations and Block Quantization Techniques." *IEEE Trans. Comm. Tech.,* vol. COM-19, no. 6, pp. 948–956.

Habibi, A. [1974]. "Hybrid Coding of Pictorial Data." *IEEE Trans. Comm.,* vol. COM-22, no. 5, pp. 614–624.

Haddon, J. F., and Boyce, J. F. [1990]. "Image Segmentation by Unifying Region and Boundary Information." *IEEE Trans. Pattern Anal. Machine Intell.,* vol. 12, no. 10, pp. 929–948.

Hall, E. L. [1979]. *Computer Image Processing and Recognition,* Academic Press, New York.

Hamming, R. W. [1950]. "Error Detecting and Error Correcting Codes." *Bell Sys. Tech. J.,* vol. 29, pp. 147–160.

Hannah, I., Patel, D., and Davies, R. [1995]. "The Use of Variance and Entropy Thresholding Methods for Image Segmentation," *Pattern Recog.,* vol. 28, no. 8, pp. 1135–1143.

Haralick, R. M., and Lee, J. S. J. [1990]. "Context Dependent Edge Detection and Evaluation." *Pattern Recog.,* vol. 23, no. 1–2, pp. 1–20.

Haralick, R. M., and Shapiro, L. G. [1985]. "Survey: Image Segmentation." *Comput. Vision, Graphics, Image Proc.,* vol. 29, pp. 100–132.

Haralick, R. M., and Shapiro, L. G. [1992]. *Computer and Robot Vision,* vols. 1 & 2, Addison-Wesley, Reading, Mass.

Haralick, R. M., Sternberg, S. R., and Zhuang, X. [1987]. "Image Analysis Using Mathematical Morphology." *IEEE Trans. Pattern Anal. Machine Intell.,* vol. PAMI-9, no. 4, pp. 532–550.

Haralick, R. M., Shanmugan, R., and Dinstein, I. [1973]. "Textural Features for Image Classification." *IEEE Trans Syst. Man Cyb.,* vol. SMC-3, no. 6, pp. 610–621.

Harikumar, G., and Bresler, Y. [1999]. "Perfect Blind Restoration of Images Blurred by Multiple Filters: Theory and Efficient Algorithms," *IEEE Trans. Image Processing,* vol. 8, no. 2, pp. 202–219.

Harmuth, H. F. [1970]. *Transmission of Information by Orthogonal Signals*, Springer-Verlag, New York.

Haris, K., Efstratiadis, S. N, Maglaveras, N., and Katsaggelos, A. K. [1998]. "Hybrid Image Segmentation Using Watersheds and Fast Region Merging," *IEEE Trans. Image Processing*, vol. 7, no. 12, pp. 1684–1699.

Hart, P. E., Nilsson, N. J., and Raphael, B. [1968]. "A Formal Basis for the Heuristic Determination of Minimum-Cost Paths." *IEEE Trans. Syst. Man Cyb*, vol. SMC-4, pp. 100–107.

Hartenstein, H., Ruhl, M., and Saupe, D. [2000]. "Region-Based Fractal Image Compression." *IEEE Trans. Image Processing*, vol. 9, no. 7, pp. 1171–1184.

Haskell, B. G., and Netravali, A. N. [1997]. *Digital Pictures: Representation, Compression, and Standards*, Perseus Publishing, New York.

Haykin, S. [1996]. *Adaptive Filter Theory*, Prentice Hall, Upper Saddle River, N.J.

Healy, D. J., and Mitchell, O. R. [1981]. "Digital Video Bandwidth Compression Using Block Truncation Coding." *IEEE Trans. Comm.*, vol. COM-29, no. 12, pp. 1809–1817.

Heath, M. D., Sarkar, S., Sanocki, T., and Bowyer, K. W. [1997]. "A Robust Visual Method for Assessing the Relative Performance of Edge-Detection Algorithms," *IEEE Trans. Pattern Anal. Machine Intell.*, vol. 19, no. 12, pp. 1338–1359.

Heath, M., Sarkar, S., Sanoki, T., and Bowyer, K. [1998]. "Comparison of Edge Detectors: A Methodology and Initial Study," *Computer Vision and Image Understanding*, vol. 69, no. 1, pp. 38–54.

Hebb, D. O. [1949]. *The Organization of Behavior: A Neuropsychological Theory*, John Wiley & Sons, New York.

Heijmans, H. J. A. M., and Goutsias, J. [2000]. "Nonlinear Multiresolution Signal Decomposition Schemes—Part II: Morphological Wavelets." *IEEE Trans. Image Processing*, vol. 9, no. 11, pp. 1897–1913.

Highnam, R., and Brady, M. [1997]. "Model-Based Image Enhancement of Far Infrared Images," *IEEE Trans. Pattern Anal. Machine Intell.*, vol. 19, no. 4, pp. 410–415.

Hojjatoleslami, S. A., and Kittler, J. [1998]. "Region Growing: A New Approach," *IEEE Trans. Image Processing*, vol. 7, no. 7, pp. 1079–1084.

Hoover, R. B., and Doty, F. [1996]. *Hard X-Ray/Gamma-Ray and Neutron Optics, Sensors, and Applications*, ISBN 0-8194-2247-9, SPIE Press, Bellingham, Wash.

Horn, B. K. P. [1986]. *Robot Vision*, McGraw-Hill, New York.

Hotelling, H. [1933]. "Analysis of a Complex of Statistical Variables into Principal Components." *J. Educ. Psychol.*, vol. 24, pp. 417–441, 498–520.

Hough, P. V. C. [1962]. "Methods and Means for Recognizing Complex Patterns." U. S. Patent 3,069,654.

Hsu, C. C., and Huang, J. S. [1990]. "Partitioned Hough Transform for Ellipsoid Detection." *Pattern Recog.*, vol. 23, no. 3–4, pp. 275–282.

Hu. J., and Yan, H. [1997]. "Polygonal Approximation of Digital Curves Based on the Principles of Perceptual Organization," *Pattern Recog.*, vol. 30, no. 5, pp. 701–718.

Hu, M. K. [1962]. "Visual Pattern Recognition by Moment Invariants." *IRE Trans. Info. Theory*, vol. IT-8, pp. 179–187.

Huang, S.-C., and Sun, Y.-N. [1999]. "Polygonal Approximation Using Generic Algorithms," *Pattern Recog.*, vol. 32, no. 8, pp. 1409–1420.

Huang, T. S. [1965]. "PCM Picture Transmission." *IEEE Spectrum*, vol. 2, no. 12, pp. 57–63.

Huang, T. S. [1966]. "Digital Picture Coding." *Proc. Natl. Electron. Conf.*, pp. 793–797.

Huang, T. S., ed. [1975]. *Picture Processing and Digital Filtering*, Springer-Verlag, New York.

Huang, T. S. [1981]. *Image Sequence Analysis*, Springer-Verlag, New York.

Huang, T. S., and Hussian, A. B. S. [1972]. "Facsimile Coding by Skipping White." *IEEE Trans. Comm.*, vol. COM-23, no. 12, pp. 1452–1466.

Huang, T. S., and Tretiak, O. J. (eds.). [1972]. *Picture Bandwidth Compression*, Gordon and Breech, New York.

Huang, T. S., Yang, G. T., and Tang, G. Y. [1979]. "A Fast Two-Dimensional Median Filtering Algorithm." *IEEE Trans. Acoust., Speech, Sig. Proc.*, vol. ASSP-27, pp. 13–18.

Huang, Y., and Schultheiss, P. M. [1963]. "Block Quantization of Correlated Gaussian Random Variables." *IEEE Trans. Commun. Syst.*, vol. CS-11, pp. 289–296.

Hubbard, B. B. [1998]. *The World According to Wavelets-The Story of a Mathematical Technique in the Making*, 2nd ed, A. K. Peters, Ltd., Wellesley, Mass.

Hubel, D. H. [1988]. *Eye, Brain, and Vision*, Scientific Amer. Library, W. H. Freeman, New York.

Huertas, A., and Medione, G. [1986]. "Detection of Intensity Changes with Subpixel Accuracy using Laplacian-Gaussian Masks," *IEEE Trans. Pattern. Anal. Machine Intell.*, vol. PAMI-8, no. 5, pp. 651–664.

Huffman, D. A. [1952]. "A Method for the Construction of Minimum Redundancy Codes." *Proc. IRE*, vol. 40, no. 10, pp. 1098–1101.

Hufnagel, R. E., and Stanley, N. R. [1964]. "Modulation Transfer Function Associated with Image Transmission Through Turbulent Media," *J. Opt. Soc. Amer.*, vol. 54, pp. 52–61.

Hummel, R. A. [1974]. "Histogram Modification Techniques." Technical Report TR-329. F-44620–72C-0062, Computer Science Center, University of Maryland, College Park, Md.

Hunt, B. R. [1971]. "A Matrix Theory Proof of the Discrete Convolution Theorem." *IEEE Trans. Audio and Electroacoust.*, vol. AU-19, no. 4, pp. 285–288.

Hunt, B. R. [1973]. "The Application of Constrained Least Squares Estimation to Image Restoration by Digital Computer." *IEEE Trans. Comput.*, vol. C-22, no. 9, pp. 805–812.

Hunter, R., and Robinson, A. H. [1980]. "International Digital Facsimile Coding Standards." *Proc. IEEE*, vol. 68, no. 7, pp. 854–867.

Hurn, M., and Jennison, C. [1996]. "An Extension of Geman and Reynolds' Approach to Constrained Restoration and the Recovery of Discontinuities," *IEEE Trans. Pattern Anal. Machine Intell.*, vol. 18, no. 6, pp. 657–662.

Hwang, H., and Haddad, R. A. [1995]. "Adaptive Median Filters: New Algorithms and Results," *IEEE Trans. Image Processing*, vol. 4, no. 4, pp. 499–502.

IEEE Computer [1974]. Special issue on digital image processing. vol. 7, no. 5.

IEEE Computer [1988]. Special issue on artificial neural systems. vol. 21, no. 3.

IEEE Trans. Circuits and Syst. [1975]. Special issue on digital filtering and image processing, vol. CAS-2, pp. 161–304.

IEEE Trans. Computers [1972]. Special issue on two-dimensional signal processing, vol. C-21, no. 7.

IEEE Trans. Comm. [1981]. Special issue on picture communication systems, vol. COM-29, no. 12.

IEEE Trans. on Image Processing [1994]. Special issue on image sequence compression, vol. 3, no. 5.

IEEE Trans. on Image Processing [1996]. Special issue on vector quantization, vol. 5, no. 2.

IEEE Trans. Image Processing [1996]. Special issue on nonlinear image processing, vol. 5, no. 6.

IEEE Trans. Image Processing [1997]. Special issue on automatic target detection, vol. 6, no. 1.

IEEE Trans. Image Processing [1997]. Special issue on color imaging, vol. 6, no. 7.

IEEE Trans. Image Processing [1998]. Special issue on applications of neural networks to image processing, vol. 7, no. 8.

IEEE Trans. Information Theory [1992]. Special issue on wavelet transforms and mulitresolution signal analysis, vol. 11, no. 2, Part II.

IEEE Trans. Pattern Analysis and Machine Intelligence [1989]. Special issue on multiresolution processing, vol. 11, no. 7.

IEEE Trans. Signal Processing [1993]. Special issue on wavelets and signal processing, vol. 41, no. 12.

IES Lighting Handbook, 9th ed. [2000]. Illuminating Engineering Society Press, New York.

ISO/IEC [1999]. *ISO/IEC 14495-1:1999: Information technology—Lossless and near-lossless compression of continuous-tone still images: Baseline.*

ISO/IEC JTC 1/SC 29/WG 1 [2000]. *ISO/IEC FCD 15444-1: Information technology—JPEG 2000 image coding system: Core coding system.*

Jahne, B. [1997]. *Digital Image Processing: Concepts, Algorithms, and Scientific Applications,* Springer-Verlag, New York.

Jain, A. K. [1981]. "Image Data Compression: A Review." *Proc. IEEE,* vol. 69, pp. 349–389.

Jain, A. K. [1989]. *Fundamentals of Digital Image Processing,* Prentice Hall, Englewood Cliffs, N.J.

Jain, A. K., Duin, R. P. W., and Mao, J. [2000]. "Statistical Pattern Recognition: A Review," *IEEE Trans. Pattern Anal. Machine Intell.,* vol. 22, no. 1, pp. 4–37.

Jain, J. R., and Jain, A. K. [1981]. "Displacement Measurement and Its Application in Interframe Image Coding." *IEEE Trans. Comm.,* vol. COM-29, pp. 1799–1808.

Jain, R. [1981]. "Dynamic Scene Analysis Using Pixel-Based Processes." *Computer,* vol. 14, no. 8, pp. 12–18.

Jain, R., Kasturi, R., and Schunk, B. [1995]. *Computer Vision,* McGraw-Hill, New York.

Jang, B. K., and Chin, R. T. [1990]. "Analysis of Thinning Algorithms Using Mathematical Morphology." *IEEE Trans. Pattern Anal. Machine Intell.,* vol. 12, no. 6, pp. 541–551.

Jayant, N. S. (ed.) [1976]. *Waveform Quantization and Coding,* IEEE Press, New York.

Jones, R., and Svalbe, I. [1994]. "Algorithms for the Decomposition of Gray-Scale Morphological Operations," *IEEE Trans. Pattern Anal. Machine Intell.,* vol. 16, no. 6, pp. 581–588.

Jonk, A., van den Boomgaard, S., and Smeulders, A. [1999]. "Grammatical Inference of Dashed Lines," *Computer Vision and Image Understanding,* vol. 74, no. 3, pp. 212–226.

Kahaner, D. K. [1970]. "Matrix Description of the Fast Fourier Transform." *IEEE Trans. Audio Electroacoustics,* vol. AU-18, no. 4, pp. 442–450.

Kak, A. C., and Slaney, M. [2001]. *Principles of Computerized Tomographic Imaging,* Published by the Society for Industrial and Applied Mathematics, Philadelphia, Pa.

Karhunen, K. [1947]. "Über Lineare Methoden in der Wahrscheinlichkeitsrechnung." *Ann. Acad. Sci. Fennicae,* Ser. A137. (Translated by I. Selin in "On Linear Methods in Probability Theory." T-131, 1960, The RAND Corp., Santa Monica, Calif.)

Kasson, J., and Plouffe, W. [1992]. "An Analysis of Selected Computer Interchange Color Spaces. " *ACM Trans. on Graphics,* vol. 11, no. 4, pp. 373–405.

Katzir, N., Lindenbaum, M., and Porat, M. [1994]. "Curve Segmentation Under Partial Occlusion," *IEEE Trans. Pattern Anal. Machine Intell.,* vol. 16, no. 5, pp. 513–519.

Khanna, T. [1990]. *Foundations of Neural Networks,* Addison-Wesley, Reading, Mass.

Kim, J. K., Park, J. M., Song, K. S., and Park, H. W. [1997]. "Adaptive Mammographic Image Enhancement Using First Derivative and Local Statistics," *IEEE Trans. Medical Imaging,* vol. 16, no. 5, pp. 495–502.

Kimme, C., Ballard, D. H., and Sklansky, J. [1975]. "Finding Circles by an Array of Accumulators." *Comm. ACM*, vol. 18, no. 2, pp. 120–122.

Kirsch, R. [1971]. "Computer Determination of the Constituent Structure of Biological Images." *Comput. Biomed. Res.*, vol. 4, pp. 315–328.

Kiver, M. S. [1965]. *Color Television Fundamentals*, McGraw-Hill, New York.

Klinger, A. [1976]. "Experiments in Picture Representation Using Regular Decomposition." *Comput. Graphics Image Proc.*, vol. 5, pp. 68–105.

Knowlton, K. [1980]. "Progressive Transmission of Gray-Scale and Binary Pictures by Simple, Efficient, and Lossless Encoding Schemes." *Proc. IEEE*, vol. 68, no. 7, pp. 885–896.

Kohler, R. J., and Howell, H. K. [1963]. "Photographic Image Enhancement by Superposition of Multiple Images." *Photogr. Sci. Eng.*, vol. 7, no. 4, pp. 241–245.

Kokaram, A. [1998]. *Motion Picture Restoration*, Springer-Verlag, New York.

Kramer, H. P., and Mathews, M. V. [1956]. "A Linear Coding for Transmitting a Set of Correlated Signals." *IRE Trans. Info. Theory*, vol. IT-2, pp. 41–46.

Langdon, G. C., and Rissanen, J. J. [1981]. "Compression of Black—White Images with Arithmetic Coding." *IEEE Trans. Comm.*, vol. COM-29, no. 6, pp. 858–867.

Lantuéjoul, C. [1980]. "Skeletonization in Quantitative Metallography." In *Issues of Digital Image Processing*, Haralick, R. M., and Simon, J. C. (eds.), Sijthoff and Noordhoff, Groningen, The Netherlands.

Latecki, L. J., and Lakämper, R. [1999]. "Convexity Rule for Shape Decomposition Based on Discrete Contour Evolution," *Computer Vision and Image Understanding*, vol. 73, no. 3, pp. 441–454.

Le Gall, D., and Tabatabai, A. [1988]. "Sub-band Coding of Digital Images Using Symmetric Short Kernel Filters and Arithmetic Coding Techniques." *IEEE International Conference on Acoustics, Speech, and Signal Processing*, New York, NY, pp. 761–765.

Ledley, R. S. [1964]. "High-Speed Automatic Analysis of Biomedical Pictures." *Science*, vol. 146, no. 3461, pp. 216–223.

Lee, J.-S., Sun, Y.-N., and Chen, C.-H. [1995]. "Multiscale Corner Detection by Using Wavelet Transforms." *IEEE Trans. Image Processing*, vol. 4, no. 1, pp. 100–104.

Lee, S. U., Chung, S. Y., and Park, R. H. [1990]. "A Comparative Performance Study of Several Global Thresholding Techniques for Segmentation." *Comput. Vision, Graphics, Image Proc.*, vol. 52, no. 2, pp. 171–190.

Lehmann, T. M., Gönner, C., and Spitzer, K. [1999]. "Survey: Interpolation Methods in Medical Image Processing," *IEEE Trans. Medical Imaging*, vol. 18, no. 11, pp. 1049–1076.

Lema, M. D., and Mitchell, O. R. [1984]. "Absolute Moment Block Truncation Coding and Its Application to Color Images." *IEEE Trans. Comm.*, vol. COM-32, no. 10. pp. 1148–1157.

Levine, M. D. [1985]. *Vision in Man and Machine*, McGraw-Hill, New York.

Liang, K.-C., and Kuo, C.-C. J. [1991]. "Waveguide: A Joint Wavelet-Based Image Representation and Description System." *IEEE Trans. Image Processing*, vol. 8, no. 11, pp. 1619–1629.

Liang, Q., Wendelhag, J. W., and Gustavsson, T. [2000]. "A Multiscale Dynamic Programming Procedure for Boundary Detection in Ultrasonic Artery Images," *IEEE Trans. Medical Imaging*, vol. 19, no. 2, pp. 127–142.

Lillesand, T. M., and Kiefer, R. W. [1999]. *Remote Sensing and Image Interpretation*, John Wiley & Sons, New York.

Lim, J. S. [1990]. *Two-Dimensional Signal and Image Processing*, Prentice Hall, Upper Saddle River, N.J.

Limb, J. O., and Rubinstein, C. B. [1978]. "On the Design of Quantizers for DPCM Coders: A Functional Relationship Between Visibility, Probability, and Masking." IEEE Trans. Comm., vol. COM-26, pp. 573–578.

Lindblad, T., and Kinser, J. M. [1998]. *Image Processing Using Pulse-Coupled Neural Networks*, Springer-Verlag, New York.

Linde, Y., Buzo, A., and Gray, R. M. [1980]. "An Algorithm for Vector Quantizer Design." *IEEE Trans. Comm.*, vol. COM-28, no. 1, pp. 84–95.

Lippmann, R. P. [1987]. "An Introduction to Computing with Neural Nets." *IEEE ASSP Magazine*, vol. 4, pp. 4–22.

Liu, J., and Yang, Y.-H. [1994]. "Multiresolution Color Image Segmentation," *IEEE Trans Pattern Anal. Machine Intell.*, vol. 16, no. 7, pp. 689–700.

Liu-Yu, S., and Antipolis, M. [1993]. "Description of Object Shapes by Apparent Boundary and Convex Hull," *Pattern Recog.*, vol. 26, no. 1, pp. 95–107.

Lo, R.-C., and Tsai, W.-H. [1995]. "Gray-Scale Hough Transform for Thick Line Detection in Gray-Scale Images," *Pattern Recog.*, vol. 28, no. 5, pp. 647–661.

Loncaric, S. [1998]. "A Survey of Shape Analysis Techniques," *Pattern Recog.*, vol. 31, no. 8, pp. 983–1010.

Lu, H. E., and Wang, P. S. P. [1986]. "A Comment on 'A Fast Parallel Algorithm for Thinning Digital Patterns.' " *Comm. ACM*, vol. 29, no. 3, pp. 239–242.

Lu, N. [1997]. *Fractal Imaging*, Academic Press, New York.

Lu, W.-S and Antoniou, A. [1992]. "Two-Dimensional Digital Filters," Marcel Dekker, New York.

MacAdam, D. L. [1942]. "Visual Sensitivities to Color Differences in Daylight," *J. Opt. Soc. Am.*, vol. 32, pp. 247–274.

MacAdam, D. P. [1970]. "Digital Image Restoration by Constrained Deconvolution." *J. Opt. Soc. Am.*, vol. 60, pp. 1617–1627.

Maki, A., Nordlund, P., and Eklundh, J.-O. [2000]. "Attentional Scene Segmentation: Integrating Depth and Motion," *Computer Vision and Image Understanding*, vol. 78, no. 3, pp. 351–373.

Malacara, D. [2001]. *Color Vision and Colorimetry: Theory and Applications*, SPIE Optical Engineering Press, Bellingham, Wash.

Mallat, S. [1987]. "A Compact Multiresolution Representation: The Wavelet Model." *Proc. IEEE Computer Society Workshop on Computer Vision*, IEEE Computer Society Press, Washington, D. C., pp. 2–7.

Mallat, S. [1989a]. "A Theory for Multiresolution Signal Decomposition: The Wavelet Representation." *IEEE Trans. Pattern Anal. Mach. Intell.*, vol. PAMI-11, pp. 674–693.

Mallat, S. [1989b]. "Multiresolution Approximation and Wavelet Orthonormal Bases of L^2." *Trans. American Mathematical Society*, vol. 315, pp. 69–87.

Mallat, S. [1989c]. "Multifrequency Channel Decomposition of Images and Wavelet models." *IEEE Trans. Acoustics, Speech, and Signal Processing*, vol. 37, pp. 2091–2110.

Mallat, S. [1998]. *A Wavelet Tour of Signal Processing*, Academic Press, Boston, Mass.

Mallat, S. [1999]. *A Wavelet Tour of Signal Processing*. 2nd ed., Academic Press, San Diego, Calif.

Mallot, A. H. [2000]. *Computational Vision*, The MIT Press, Cambridge, Mass.

Mamistvalov, A. [1998]. "n-Dimensional Moment Invariants and Conceptual Mathematical Theory of Recognition [of] n-Dimensional Solids," *IEEE Trans. Pattern Anal. Machine Intell.*, vol. 20, no. 8, pp. 819–831.

Maragos, P. [1987]. "Tutorial on Advances in Morphological Image Processing and Analysis." Optical Engineering, vol. 26, no. 7, pp. 623–632.

Marchand-Maillet, S., and Sharaiha, Y. M. [2000]. *Binary Digital Image Processing: A Discrete Approach*, Academic Press, New York.

Maren, A. J., Harston, C. T., and Pap, R. M. [1990]. *Handbook of Neural Computing Applications*, Academic Press, New York.

Marr, D. [1982]. *Vision*, Freeman, San Francisco.

Marr, D., and Hildreth, E. [1980]. "Theory of Edge Detection." *Proc. R. Soc. Lond.*, vol. B207, pp. 187–217.

Martelli, A. [1972]. "Edge Detection Using Heuristic Search Methods." *Comput. Graphics Image Proc.*, vol. 1, pp. 169–182.

Martelli, A. [1976]. "An Application of Heuristic Search Methods to Edge and Contour Detection." *Comm. ACM*, vol. 19, no. 2, pp. 73–83.

Martin, M. B., and Bell, A. E. [2001]. "New image compression techniques using multiwavelets and multiwavelet packets," *IEEE Trans. on Image Proc.*, vol. 10, no. 4, pp. 500–510.

Mather, P. M. [1999]. *Computer Processing of Remotely Sensed Images: An Introduction*, John Wiley & Sons, New York.

Max, J. [1960]. "Quantizing for Minimum Distortion." *IRE Trans. Info. Theory*, vol. IT-6, pp. 7–12.

McClelland, J. L., and Rumelhart, D. E. (eds.) [1986]. *Parallel Distributed Processing: Explorations in the Microstructures of Cognition*, vol. 2: *Psychological and Biological Models*, The MIT Press, Cambridge, Mass.

McCulloch, W. S., and Pitts, W. H. [1943]. "A Logical Calculus of the Ideas Imminent in Nervous Activity." *Bulletin of Mathematical Biophysics*, vol. 5, pp. 115–133.

McFarlane, M. D. [1972]. "Digital Pictures Fifty Years Ago." *Proc. IEEE*, vol. 60, no. 7, pp. 768–770.

McGlamery, B. L. [1967]. "Restoration of Turbulence-Degraded Images." *J. Opt. Soc. Am.*, vol. 57, no. 3, pp. 293–297.

Meijering, E. H. W., Niessen, W. J., and Viergever, M. A. [1999]. "Retrospective Motion Correction in Digital Subtraction Angiography: A Review," *IEEE Trans. Medical Imaging*, vol. 18, no. 1, pp. 2–21. [chap 3]

Memon, N., Neuhoff, D. L., and Shende, S. [2000]. "An Analysis of Some Common Scanning Techniques for Lossless Image Coding." *IEEE Trans. Image Processing*, vol. 9, no. 11, pp. 1837–1848.

Mesarović, V. Z. [2000]. "Iterative Linear Minimum Mean-Square-Error Image Restoration from Partially Known Blur," *J. Opt. Soc. Amer.-A. Optics, Image Science, and Vision*, vol. 17, no. 4, pp. 711–723.

Meyer, Y. [1987]. "L'analyses par Ondelettes." *Pour la Science*.

Meyer, Y. [1990]. *Ondelettes et ope'rateurs*, Hermann, Paris.

Meyer, Y. (ed.) [1992a]. *Wavelets and Applications: Proceedings of the International Conference, Marseille, France*, Mason, Paris, and Springer-Verlag, Berlin.

Meyer, Y. (translated by D. H. Salinger) [1992b]. *Wavelets and Operators*, Cambridge University Press, Cambridge, UK.

Meyer, Y. (translated by R. D. Ryan) [1993]. *Wavelets: Algorithms and Applications*, Society for Industrial and Applied Mathematics, Philadelphia.

Meyer, F. G., Averbuch, A. Z., and Strömberg, J.-O. [2000]. "Fast Adaptive Wavelet Packet Image Compression," *IEEE Trans. Image Processing*, vol. 9, no. 7, pp. 792–800.

Meyer, F., and Beucher, S. [1990]. "Morphological Segmentation." *J. Visual Comm., and Image Representation*, vol. 1, no. 1, pp. 21–46.

Meyer, H., Rosdolsky, H. G., and Huang, T. S. [1973]. "Optimum Run Length Codes." *IEEE Trans. Comm.*, vol. COM-22, no. 6, pp. 826–835.

Minsky, M., and Papert, S. [1969]. *Perceptrons: An Introduction to Computational Geometry*, the MIT Press, Cambridge, Mass.

Mirmehdi, M., and Petrou, M. [2000]. "Segmentation of Color Textures," *IEEE Trans. Pattern Anal. Machine Intell.*, vol. 22, no. 2, pp. 142–159.

Misiti, M., Misiti, Y., Oppenheim, G., and Poggi, J.-M. [1996]. *Wavelet Toolbox User's Guide*, The MathWorks, Inc., Natick, Mass.

Mitiche, A. [1994]. *Computational Analysis of Visual Motion*, Perseus Publishing, New York.

Mitra, S. K., and Sicuranza, G. L. (eds.) [2000]. *Nonlinear Image Processing*, Academic Press, New York.

Murase, H., and Nayar, S. K. [1994]. "Illumination Planning for Object Recognition Using Parametric Eigen Spaces," *IEEE Trans. Pattern Anal. Machine Intell.*, vol. 16, no. 12, pp. 1219–1227.

Murino, V., Ottonello, C., and Pagnan, S.]1998]. "Noisy Texture Classfication: A Higher-Order Statistical Approach," *Pattern Recog.*, vol. 31, no. 4, pp. 383–393.

Nagao, M., and Matsuyama, T. [1980]. *A Structural Analysis of Complex Aerial Photographs*, Plenum Press, New York.

Najman, L., and Schmitt, M. [1996]. "Geodesic Saliency of Watershed Contours and Hierarchical Segmentation," *IEEE Trans. Pattern Anal. Machine Intell.*, vol. 18, no. 12, pp. 1163–1173.

Narendra, P. M., and Fitch, R. C. [1981]. "Real-Time Adaptive Contrast Enhancement." *IEEE Trans. Pattern Anal. Mach. Intell.*, vol. PAMI-3, no. 6, pp. 655–661.

Netravali, A. N. [1977]. "On Quantizers for DPCM Coding of Picture Signals." *IEEE Trans. Info. Theory*, vol. IT-23, no. 3, pp. 360–370.

Netravali, A. N., and Limb, J. O. [1980]. "Picture Coding: A Review." *Proc. IEEE*, vol. 68, no. 3, pp. 366–406.

Nevatia, R. [1982]. *Machine Perception*, Prentice Hall, Upper Saddle River, N.J.

Ngan, K. N., Meier, T., and Chai, D. [1999]. *Advanced Video Coding: Principles and Techniques*. Elsevier, Boston.

Nilsson, N. J. [1965]. *Learning Machines: Foundations of Trainable Pattern-Classifying Systems*, McGraw-Hill, New York.

Nilsson, N. J. [1971]. *Problem Solving Methods in Artificial Intelligence*, McGraw-Hill, New York.

Nilsson, N. J. [1980]. *Principles of Artificial Intelligence*, Tioga, Palo Alto, Calif.

Noble, B., and Daniel, J. W. [1988]. *Applied Linear Algebra*, 3rd ed., Prentice Hall, Upper Saddle River, N.J.

Odegard, J. E., Gopinath, R. A., and Burrus, C. S. [1992]. "Optimal Wavelets for Signal Decomposition and the Existence of Scale-Limited Signals." *Proceedings of IEEE Int. Conf. On Signal Proc.*, ICASSP-92, San Francisco, CA, vol. IV, 597–600.

Olson, C. F. [1999]. "Constrained Hough Transforms for Curve Detection," *Computer Vision and Image Understanding*, vol. 73, no. 3, pp. 329–345.

O'Neil, J. B. [1971]. "Entropy Coding in Speech and Television Differential PCM Systems." *IEEE Trans. Info. Theory*, vol. IT-17, pp. 758–761.

Oommen, R. J., and Loke, R. K. S. [1997]. "Pattern Recognition of Strings with Substitutions, Insertions, Deletions, and Generalized Transpositions," *Pattern Recog.*, vol. 30, no. 5, pp. 789–800.

Oppenheim, A. V., and Schafer, R. W. [1975]. *Digital Signal Processing*, Prentice Hall, Englewood Cliffs, N.J.

Oppenheim, A. V., Schafer, R. W., and Stockham, T. G., Jr. [1968]. "Nonlinear Filtering of Multiplied and Convolved Signals." *Proc. IEEE*, vol. 56, no. 8, pp. 1264–1291.

Oyster, C. W. [1999]. *The Human Eye: Structure and Function*, Sinauer Associates, Sunderland, Mass.

Paez, M. D., and Glisson, T. H. [1972]. "Minimum Mean-Square-Error Quantization in Speech PCM and DPCM Systems." *IEEE Trans. Comm.*, vol. COM-20, pp. 225–230.

Pao, Y. H. [1989]. *Adaptive Pattern Recognition and Neural Networks*, Addison-Wesley, Reading, Mass.

Papamarkos, N., and Atsalakis, A. [2000]. "Gray-Level Reduction Using Local Spatial Features," *Computer Vision and Image Understanding*, vol. 78, no. 3, pp. 336–350.

Papoulis, A. [1991]. *Probability, Random Variables, and Stochastic Processes*, 3rd ed., Mc-Graw-Hill, New York.

Park, H., and Chin, R. T. [1995]. "Decomposition of Arbitrarily-Shaped Morphological Structuring," *IEEE Trans. Pattern Anal. Machine Intell.*, vol. 17, no. 1, pp. 2–15.

Parker, J. R. [1991]. "Gray Level Thresholding in Badly Illuminated Images." *IEEE Trans. Pattern Anal. Machine Intell.*, vol. 13, no. 8, pp. 813–819.

Pattern Recognition [2000]. Special issue on Mathematical Morphology and Nonlinear Image Processing, vol. 33, no. 6, pp. 875–1117.

Pavlidis, T. [1977]. *Structural Pattern Recognition*, Springer-Verlag, New York.

Pavlidis, T. [1982]. *Algorithms for Graphics and Image Processing*, Computer Science Press, Rockville, Md.

Pavlidis, T., and Liow, Y. T. [1990]. "Integrating Region Growing and Edge Detection." *IEEE Trans. Pattern Anal. Mach. Intell.*, vol. 12, no. 3, pp. 225–233.

Peebles, P. Z. [1993]. *Probability, Random Variables, and Random Signal Principles*, 3rd ed., McGraw-Hill, New York.

Pennebaker, W. B., and Mitchell, J. L. [1992]. *JPEG: Still Image Data Compression Standard*. Van Nostrand Reinhold, New York.

Pennebaker, W. B., Mitchell, J. L., Langdon, G. G., Jr., and Arps, R. B. [1988]. "An Overview of the Basic Principles of the Q-coder Adaptive Binary Arithmetic Coder." *IBM J. Res. Dev.*, vol. 32, no. 6, pp. 717–726.

Perez, A., and Gonzalez, R. C. [1987]. "An Iterative Thresholding Algorithm for Image Segmentation." *IEEE Trans. Pattern Anal. Machine Intell.*, vol. PAMI-9, no. 6, pp. 742–751.

Perona, P., and Malik, J. [1990]. "Scale-Space and Edge Detection Using Anisotropic Diffusion." *IEEE Trans. Pattern Anal. Machine Intell.*, vol. 12, no. 7, pp. 629–639.

Persoon, E., and Fu, K. S. [1977]. "Shape Discrimination Using Fourier Descriptors." *IEEE Trans. Systems Man Cyb.*, vol. SMC-7, no. 2, pp. 170–179.

Petrou, M., and Bosdogianni, P. [1999]. *Image Processing: The Fundamentals*, John Wiley & Sons, UK.

Petrou, M., and Kittler, J. [1991]. "Optimal Edge Detector for Ramp Edges." *IEEE Trans. Pattern Anal. Machine Intell.*, vol. 13, no. 5, pp. 483–491.

Piech, M. A. [1990]. "Decomposing the Laplacian." *IEEE Trans. Pattern Anal. Machine Intell.*, vol. 12, no. 8, pp. 830–831.

Pitas, I., and Vanetsanopoulos, A. N. [1990]. *Nonlinear Digital Filters: Principles and Applications*, Kluwer Academic Publishers, Boston, Mass.

Plataniotis, K. N, and Venetsanopoulos, A. N. [2000]. *Color Image Processing and Applications*, Springer-Verlag, New York.

Pokorny, C. K., and Gerald, C. F. [1989]. *Computer Graphics: The Principles Behind the Art and Science*, Franklin, Beedle & Associates, Irvine, Calif.

Poynton, C. A. [1996]. *A Technical Introduction to Digital Video*, John Wiley & Sons, New York.

Prasad, L., and Iyengar, S. S. [1997]. *Wavelet Analysis with Applications to Image Processing*, CRC Press, Boca Raton, Fla.

Pratt, W. K. [1991]. *Digital Image Processing*, 2nd ed., John Wiley & Sons, New York.

Pratt, W. K. [2001]. *Digital Image Processing*, 3rd ed., John Wiley & Sons, New York.

Preparata, F. P., and Shamos, M. I. [1985]. *Computational Geometry: An Introduction*, Springer-Verlag, New York.

Preston, K. [1983]. "Cellular Logic Computers for Pattern Recognition." *Computer*, vol. 16, no. 1, pp. 36–47.

Prewitt, J. M. S. [1970]. "Object Enhancement and Extraction." in *Picture Processing and Psychopictorics*, Lipkin, B. S., and Rosenfeld, A. (eds.), Academic Press, New York.

Principe, J. C., Euliano, N. R., and Lefebre, W. C. [1999]. *Neural and Adaptive Systems: Fundamentals through Simulations*, John Wiley & Sons, New York.

Pritchard, D. H. [1977]. "U. S. Color Television Fundamentals—A Review." *IEEE Trans. Consumer Electronics*, vol. CE-23, no. 4, pp. 467–478.

Proc. IEEE [1967]. Special issue on redundancy reduction, vol. 55, no. 3.

Proc. IEEE [1972]. Special issue on digital picture processing, vol. 60, no. 7.

Proc. IEEE [1980]. Special issue on the encoding of graphics, vol. 68, no. 7.

Proc. IEEE [1985]. Special issue on visual communication systems, vol. 73, no. 2.

Qian, R. J., and Huang, T. S. [1996]. "Optimal Edge Detection in Two-Dimensional Images," *IEEE Trans. Image Processing*, vol. 5, no. 7, pp. 1215–1220.

Rabbani, M., and Jones, P. W. [1991]. *Digital Image Compression Techniques*, SPIE Press, Bellingham, Wash.

Rajala, S. A., Riddle, A. N., and Snyder, W. E. [1983]. "Application of One-Dimensional Fourier Transform for Tracking Moving Objects in Noisy environments." *Comp., Vision, Image Proc.*, vol. 21, pp. 280–293.

Reddy, B. S., and Chatterji, B. N. [1996]. "An FFT-Based Technique for Translation, Rotation, and Scale Invariant Image Registration," *IEEE Trans. Image Processing*, vol. 5, no. 8, pp. 1266–1271.

Regan, D. D. [2000]. *Human Perception of Objects: Early Visual Processing of Spatial Form Defined by Luminance, Color, Texture, Motion, and Binocular Disparity*, Sinauer Associates, Sunderland, Mass.

Ritter, G. X., and Wilson, J. N. [2001]. *Handbook of Computer Vision Algorithms in Image Algebra*, CRC Press, Boca Raton, Fla.

Roberts, L. G. [1965]. "Machine Perception of Three-Dimensional Solids." In *Optical and Electro-Optical Information Processing*, Tippet, J. T. (ed.), MIT Press, Cambridge, Mass.

Robertson, A. R. [1977]. "The CIE 1976 Color Difference Formulae," *Color Res. Appl.*, vol. 2, pp. 7–11.

Robinson, G. S. [1976]. "Detection and Coding of Edges Using Directional Masks." University of Southern California, Image Processing Institute, Report no. 660.

Robinson, J. A. [1965]. "A Machine-Oriented Logic Based on the Resolution Principle." *J. ACM*, vol. 12, no. 1, pp. 23–41.

Rock, I. [1984]. *Perception*, W. H. Freeman, New York, New York.

Roese, J. A., Pratt, W. K., and Robinson, G. S. [1977]. "Interframe Cosine Transform Image Coding." *IEEE Trans. Comm.*, vol. COM-25, pp. 1329–1339.

Rosenblatt, F. [1959]. "Two Theorems of Statistical Separability in the Perceptron." In *Mechanisation of Thought Processes: Proc. of Symposium No. 10*, held at the National Physical Laboratory, November 1958, H. M. Stationery Office, London, vol. 1, pp. 421–456.

Rosenblatt, F. [1962]. *Principles of Neurodynamics: Perceptrons and the Theory of Brain Mechanisms*, Spartan, Washington, D. C.

Rosenfeld, A. (ed.) [1984]. *Multiresolution Image Processing and Analysis*, Springer-Verlag, New York.

Rosenfeld, A. [1999]. "Image Analysis and Computer Vision: 1998," *Computer Vision and Image Understanding*, vol. 74, no. 1, pp. 36–95.

Rosenfeld, A. [2000]. "Image Analysis and Computer Vision: 1999," *Computer Vision and Image Understanding*, vol. 78, no. 2, pp. 222–302.

Rosenfeld, A., and Kak, A. C. [1982]. *Digital Picture Processing*, vols. 1 and 2, 2nd ed., Academic Press, New York.

Rosin, P. L. [1997]. "Techniques for Assessing Polygonal Approximations of Curves," *IEEE Trans. Pattern Anal. Machine Intell.*, vol. 19, no. 6, pp. 659–666.

Rudnick, P. [1966]. "Note on the Calculation of Fourier Series." *Math. Comput.*, vol. 20, pp. 429–430.

Rumelhart, D. E., Hinton, G. E., and Williams, R. J. [1986]. "Learning Internal Representations by Error Propagation." In *Parallel Distributed Processing: Explorations in the Microstructures of Cognition*, vol. 1: *Foundations*, Rumelhart, D. E., et al. (eds.), MIT Press, Cambridge, MA., pp. 318–362.

Rumelhart, D. E., and McClelland, J. L. (eds.) [1986]. *Parallel Distributed Processing: Explorations in the Microstructures of Cognition*, vol. 1: *Foundations*, MIT Press, Cambridge, Mass.

Runge, C. [1903]. *Zeit. für Math., and Physik*, vol. 48, p. 433.

Runge, C. [1905]. *Zeit. für Math., and Physik*, vol. 53, p. 117.

Runge, C., and König, H. [1924]. "Die Grundlehren der Mathematischen Wissenschaften." *Vorlesungen über Numerisches Rechnen*, vol. 11, Julius Springer, Berlin.

Russ, J. C. [1999]. *The Image Processing Handbook*, 3rd ed., CRC Press, Boca Raton, Fla.

Sahni, S., and Jenq, J.-F. [1992]. "Serial and Parallel Algorithms for the Medial Axis Transform," *IEEE Trans. Pattern Anal. Machine Intell.*, vol. 14, no. 12, pp. 1218–1224.

Sahoo, P. K., Soltani, S., Wong, A. K. C., and Chan, Y. C. [1988]. "A Survey of Thresholding Techniques." *Comput. Vision, Graphics, Image Proc.*, vol. 4, pp. 233–260.

Saito, N., and Cunningham, M. A. [1990]. "Generalized E-Filter and its Application to Edge Detection." *IEEE Trans. Pattern Anal. Machine Intell.*, vol. 12, no. 8, pp. 814–817.

Sakrison, D. J., and Algazi, V. R. [1971]. "Comparison of Line-by-Line and Two-Dimensional Encoding of Random Images." *IEEE Trans. Info. Theory*, vol. IT-17, no. 4, pp. 386–398.

Salari, E., and Siy, P. [1984]. "The Ridge-Seeking Method for Obtaining the Skeleton of Digital Images." *IEEE Trans. Syst. Man Cyb.*, vol. SMC-14, no. 3, pp. 524–528.

Salinas, R. A., Abidi, M. A., and Gonzalez, R. C. [1996]. "Data Fusion: Color Edge Detection and Surface Reconstruction Through Regularization," *IEEE Trans. Industrial Electronics*, vol. 43, no. 3, pp. 355–363, 1996.

Sato, Y. [1992]. "Piecewise Linear Approximation of Plane Curves by Perimeter Optimization," *Pattern Recog.*, vol. 25, no. 12, pp. 1535–1543.

Sauvola, J., and Pietikainen, M. [2000]. "Adaptive Document Image Binarization," *Pattern Recog.*, vol. 33, no. 2, pp. 225–236.

Schalkoff, R. J. [1989]. *Digital Image Processing and Computer Vision*, John Wiley & Sons, New York.

Schonfeld, D., and Goutsias, J. [1991]. "Optimal Morphological Pattern Restoration from Noisy Binary Images." *IEEE Trans. Pattern Anal. Machine Intell.*, vol. 13, no. 1, pp. 14–29.

Schowengerdt, R. A. [1983]. *Techniques for Image Processing and Classification in Remote Sensing*, Academic Press, New York.

Schreiber, W. F. [1956]. "The Measurement of Third Order Probability Distributions of Television Signals." *IRE Trans. Info. Theory*, vol. IT-2, pp. 94–105.

Schreiber, W. F. [1967]. "Picture Coding." *Proc. IEEE* (Special issue on Redundancy Reduction), vol. 55, pp. 320–330.

Schreiber, W. F., and Knapp, C. F. [1958]. "TV Bandwidth Reduction by Digital Coding." *Proc. IRE National Convention*, pt. 4, pp. 88–99.

Schwartz, J. W., and Barker, R. C. [1966]. "Bit-Plane Encoding: A Technique for Source Encoding." *IEEE Trans. Aerosp. Elec. Systems*, vol. AES-2, no. 4, pp. 385–392.

Serra, J. [1982]. *Image Analysis and Mathematical Morphology*, Academic Press, New York.

Serra, J. (ed.) [1988]. *Image Analysis and Mathematical Morphology*, vol. 2, Academic Press, New York.

Sezan, M. I., Rabbani, M., and Jones, P. W. [1989]. "Progressive Transmission of Images Using a Prediction/Residual Encoding Approach." *Opt. Eng.*, vol. 28, no. 5, pp. 556–564.

Shack, R. V. [1964]. "The Influence of Image Motion and Shutter Operation on the Photographic Transfer Function." *Appl. Opt.*, vol. 3, pp. 1171–1181.

Shafarenko, L., Petrou, M., and Kittler, J. [1998]. "Histogram-Based Segmentation in a Perceptually Uniform Color Space." *IEEE Trans. Image Processing*, vol. 7, no. 9, pp. 1354–1358.

Shaked, D., and Bruckstein, A. M. [1998]. "Pruning Medial Axes," *Computer Vision and Image Understanding*, vol. 69, no. 2, pp. 156–169.

Shannon, C. E. [1948]. "A Mathematical Theory of Communication." *The Bell Sys. Tech. J.*, vol. XXVII, no. 3, pp. 379–423.

Shapiro, L. G., and Stockman, G. C. [2001]. *Computer Vision*, Prentice Hall, Upper Saddle River, N.J.

Shapiro, V. A. [1996]. "On the Hough Transform of Multi-Level Pictures," *Pattern Recog.*, vol. 29, no. 4, pp. 589–602.

Shariat, H., and Price, K. E. [1990]. "Motion Estimation with More Than Two Frames." *IEEE Trans. Pattern Anal. Machine Intell.*, vol. 12, no. 5, pp. 417–434.

Sheppard, J. J., Jr., Stratton, R. H., and Gazley, C., Jr. [1969]. "Pseudocolor as a Means of Image Enhancement." *Am. J. Optom. Arch. Am. Acad. Optom.*, vol. 46, pp. 735–754.

Shih, F. Y., and Wong, W.-T. [1994]. "Fully Parallel Thinning with Tolerance to Boundary Noise," *Pattern Recog.*, vol. 27, no. 12, pp. 1677–1695.

Shih, F. Y. C., and Mitchell, O. R. [1989]. "Threshold Decomposition of Gray-Scale Morphology into Binary Morphology." *IEEE Trans. Pattern Anal. Machine Intell.*, vol. 11, no. 1, pp. 31–42.

Sid-Ahmed, M. A. [1995]. *Image Processing: Theory, Algorithms, and Architectures*, McGraw-Hill, New York.

Sikora, T. [1997]. "MPEG Digital Video-Coding Standards." *IEEE Signal Processing*, vol. 14, no. 5, pp. 82–99.

Simon, J. C. [1986]. *Patterns and Operators: The Foundations of Data Representations*, McGraw-Hill, New York.

Sklansky, J., Chazin, R. L., and Hansen, B. J. [1972]. "Minimum-Perimeter Polygons of Digitized Silhouettes." *IEEE Trans. Comput.*, vol. C-21, no. 3, pp. 260–268.

Smirnov, A. [1999]. *Processing of Multidimensional Signals*, Springer-Verlag, New York.

Smith, A. R. [1978]. "Color Gamut Transform Pairs." *Proc. SIGGRAPH* '78, published as *Computer Graphics*, vol. 12, no. 3, pp. 12–19.

Smith, M. J. T., and Barnwell, T. P. III [1984]. "A Procedure for Building Exact Reconstruction Filter Banks for Subband Coders." *Proc. IEEE Int. Conf. Acoust., Speech,
and Signal Proc.*, San Diego, Calif.

Smith, M. J. T., and Barnwell, T. P. III [1986]. "Exact Reconstruction Techniques for Tree-
Structured Subband Coders." *IEEE Trans. On Acoust., Speech, and Signal Proc.*, vol. 34,
no. 3, pp. 434–441.

Sobel, I. E. [1970]. "Camera Models and Machine Perception," Ph.D. dissertation, Stan-
ford University, Palo Alto, Calif.

Sonka, M., Hlavac, V., and Boyle, R. [1999]. *Image Processing, Analysis, and Machine Vision*, 2nd ed., PWS Publishing, New York.

Stark, H. (ed.) [1987]. *Image Recovery: Theory and Application*, Academic Press, New
York.

Stark, J. A. [2000]. "Adaptive Image Contrast Enhancement Using Generalizations of
Histogram Equalization," *IEEE Trans. Image Processing*, vol. 9, no. 5, pp. 889–896.

Stockham, T. G., Jr. [1972]. "Image Processing in the Context of a Visual Model." *Proc.
IEEE*, vol. 60, no. 7, pp. 828–842.

Storer, J. A., and Reif, J. H., eds. [1991]. *Proceedings of DDC '91*, IEEE Computer Society Press, Los Alamitos, Calif.

Strang, G., and Nguyen, T. [1996]. *Wavelets and Filter Banks*, Wellesley-Cambridge Press
Wellesley, Mass.

Stumpff, K. [1939]. *Tafeln und Aufgaben zur Harmonischen Analyse und Peri-
odogrammrechnung*, Julius Springer, Berlin.

Sussner, P., and Ritter, G. X. [1997]. "Decomposition of Gray-Scale Morphological Templates Using the Rank Method," *IEEE Trans. Pattern Anal. Machine Intell.*, vol. 19,
no. 6, pp. 649–658.

Swets, D. L., and Weng, J. [1996]. "Using Discriminant Eigenfeatures for Image Retrieval,"
IEEE Trans. Pattern Anal. Machine Intell., vol. 18, no. 8, pp. 1831–1836.

Symes, P. D. [2001]. *Video Compression Demystified*. McGraw-Hill, New York.

Sze, T. W., and Yang, Y. H. [1981]. "A Simple Contour Matching Algorithm." *IEEE Trans.
Pattern Anal. Mach. Intell.*, vol. PAMI-3, no. 6, pp. 676–678.

Tanaka, E. [1995]. "Theoretical Aspects of Syntactic Pattern Recognition," *Pattern Recog.*,
vol. 28, no. 7 pp. 1053–1061.

Tanimoto, S. L. [1979]. "Image Transmission with Gross Information First." *Comput.
Graphics Image Proc.*, vol. 9, pp. 72–76.

Tasto, M., and Wintz, P. A. [1971]. "Image Coding by Adaptive Block Quantization."
IEEE Trans. Comm. Tech., vol. COM-19, pp. 957–972.

Tasto, M., and Wintz, P. A. [1972]. "A Bound on the Rate-Distortion Function and Application to Images." *IEEE Trans. Info. Theory*, vol. IT-18, pp. 150–159.

Teh, C. H., and Chin, R. T. [1989]. "On the Detection of Dominant Points on Digital
Curves." *IEEE Trans. Pattern Anal. Machine Intell.*, vol. 11, no. 8, pp. 859–872.

Thévenaz, P., and Unser, M [2000]. "Optimization of Mutual Information for Multiresolution Image Registration." *IEEE Trans. Image Processing*, vol. 9, no. 12,
pp. 2083–2099.

Thomas, L. H. [1963]. "Using a Computer to Solve Problems in Physics." *Application of
Digital Computers*, Ginn, Boston.

Thomason, M. G., and Gonzalez, R. C. [1975]. "Syntactic Recognition of Imperfectly
Specified Patterns." *IEEE Trans. Comput.*, vol. C-24, no. 1, pp. 93–96.

Thompson, W. B. (ed.) [1989]. "Special Issue on Visual Motion." *IEEE Trans. Pattern
Anal. Machine Intell.*, vol. 11, no. 5, pp. 449–541.

Thompson, W. B., and Barnard, S. T. [1981]. "Lower-Level Estimation and Interpretation of Visual Motion." *Computer*, vol. 14, no. 8, pp. 20–28.

Thorell, L. G., and Smith, W. J. [1990]. *Using Computer Color Effectively*, Prentice Hall, Upper Saddle River, N.J.

Tian, J., and Wells, R. O., Jr. [1995]. *Vanishing Moments and Wavelet Approximation*, Technical Report CML TR-9501, Computational Mathematics Lab., Rice University, Houston, Texas.

Tomita, F., Shirai, Y., and Tsuji, S. [1982]. "Description of Texture by a Structural Analysis." *IEEE Trans. Pattern Anal. Mach. Intell.*, vol. PAMI-4, no. 2, pp. 183–191.

Topiwala, P. N. (ed.) [1998]. *Wavelet Image and Video Compression*, Kluwer Academic Publishers, Boston, Mass.

Tou, J. T., and Gonzalez, R. C. [1974]. *Pattern Recognition Principles*, Addison-Wesley, Reading, Mass.

Toussaint, G. T. [1982]. "Computational Geometric Problems in Pattern Recognition." In *Pattern Recognition Theory and Applications*, Kittler, J., Fu, K. S., and Pau, L. F. (eds.), Reidel, New York, pp. 73–91.

Tsai, J.-C., Hsieh, C.-H, and Hsu, T.-C. [2000]. "A New Dynamic Finite-State Vector Quantization Algorithm for Image Compression," *IEEE Trans. Image Processing*, vol. 9, no. 11, pp. 1825–1836.

Tsujii, O., Freedman, M. T., and Mun, K. S. [1998]. "Anatomic Region-Based Dynamic Range Compression for Chest Radiographs Using Warping Transformation of Correlated Distribution," *IEEE Trans. Medical Imaging*, vol. 17, no. 3, pp. 407–418.

Udpikar, V. R., and Raina, J. P. [1987]. "BTC Image Coding Using Vector Quantization." *IEEE Trans. Comm.*, vol. COM-35, no. 3, pp. 352–356.

Ueda, N. [2000]. "Optimal Linear Combination of Neural Networks for Improving Classification Performance," *IEEE Trans. Pattern Anal. Machine Intell.*, vol. 22, no. 2, pp. 207–215.

Ullman, S. [1981]. "Analysis of Visual Motion by Biological and Computer Systems," *IEEE Computer*, vol. 14, no. 8, pp. 57–69.

Umbaugh, S. E. [1998]. *Computer Vision and Image Processing: A Practical Approach Using CVIPtools*, Prentice Hall, Upper Saddle River, N.J.

Umeyama, S. [1988]. "An Eigendecomposition Approach to Weighted Graph Matching Problems." *IEEE Trans. Pattern Anal. Machine Intell.*, vol. 10, no. 5, pp. 695–703.

Unser, M. [1995]. "Texture Classification and Segmentation Using Wavelet Frames." *IEEE Trans. on Image Processing*, vol. 4, no. 11, pp. 1549–1560.

Unser, M., Aldroubi, A., and Eden, M. [1993]. "A Family of Polynomial Spline Wavelet Transforms." *Signal Proc.*, vol. 30, no. 2, pp. 141–162.

Unser, M., Aldroubi, A., and Eden, M. [1993]. "B-Spline Signal Processing, Parts I and II." *IEEE Trans. Signal Proc.*, vol. 41, no. 2, pp. 821–848.

Unser, M., Aldroubi, A., and Eden, M. [1995]. "Enlargement or Reduction of Digital Images with Minimum Loss of Information," *IEEE Trans. Image Processing*, vol. 4, no. 5, pp. 247–257.

Vaidyanathan, P. P., and Hoang, P.-Q. [1988]. "Lattice Structures for Optimal Design and Robust Implementaion of Two-Channel Perfect Reconstruction Filter Banks." *IEEE Trans. Acoust., Speech, and Signal Proc.*, vol. 36, no. 1, pp. 81–94.

Vailaya, A., Jain, A., and Zhang, H. J. [1998]. "On Image Classification: City Images vs. Landscapes," *Pattern Recog.*, vol. 31, no. 12, pp. 1921–1935.

Vetterli, M. [1986]. "Filter Banks Allowing Perfect Reconstruction Signal Proc., vol. 10, no. 3, pp. 219–244".

Vetterli, M., and Kovacevic, J. [1995]. *Wavelets and Suband Coding*, Prentice Hall, Englewood Cliffs, N.J.

Voss, K., and Suesse, H. [1997]. "Invariant Fitting of Planar Objects by Primitives," *IEEE Trans. Pattern Anal. Machine Intell.*, vol. 19, no. 1, pp. 80–84.

Vuylsteke, P., and Kittler, J. [1990]. "Edge-Labeling Using Dictionary-Based Relaxation." *IEEE Trans. Pattern Anal. Machine Intell.*, vol. 12, no. 2, pp. 165–181.

Walsh, J. W. T. [1958]. *Photometry*, Dover, New York.

Wang, G., Zhang, J., and Pan, G.-W. [1995]. "Solution of Inverse Problems in Image Processing by Wavelet Expansion." *IEEE Trans. Image Processing*, vol. 4, no. 5, pp. 579–593.

Wang, Y.-P., Lee, S. L., and Toraichi, K. [1999]. "Multiscale Curvature-Based Shape Representation Using β-Spline Wavelets." *IEEE Trans. Image Processing*, vol. 8, no. 11, pp. 1586–1592.

Wang, Z., Rao, K. R., and Ben-Arie, J. [1996]. "Optimal Ramp Edge Detection Using Expansion Matching," *IEEE Trans. Pattern Anal. Machine Intell.*, vol. 18, no. 11, pp. 1092–1097.

Watt, A. [1993]. *3D Computer Graphics*, 2nd ed., Addison-Wesley, Reading, Mass.

Wechsler [1980]. "Texture Analysis—A Survey." *Signal Process*, vol. 2, pp. 271–280.

Wei, D., Tian, J., Wells, R. O., Jr., and Burrus, C. S. [1998]. "A New Class of Biorthogonal Wavelet Systems for Image Transform Coding." *IEEE Trans. Image Processing*, vol. 7, no. 7, pp. 1000–1013.

Weinberger, M. J., Seroussi, G., and Sapiro, G. [2000]. "The LOCO-I Lossless Image Compression Algorithm: Principles and Standardization into JPEG-LS," *IEEE Trans. Image Processing*, vol. 9, no. 8, pp. 1309–1324.

Westenberg, M. A., and Roerdink, J. B. T. M. [2000]. "Frequency Domain Volume Rendering by the Wavelet X-ray Transform." *IEEE Trans. Image Processing*, vol. 9, no. 7, pp. 1249–1261.

Weszka, J. S. [1978]. "A Survey of Threshold Selection Techniques." *Comput. Graphics Image Proc.*, vol. 7, pp. 259–265.

White, J. M., and Rohrer, G. D. [1983]. "Image Thresholding for Optical Character Recognition and Other Applications Requiring Character Image Extraction." *IBM J. Res. Devel.*, vol. 27, no. 4, pp. 400–411.

Widrow, B. [1962]. "Generalization and Information Storage in Networks of 'Adaline' Neurons." In *Self-Organizing Systems 1962*, Yovitz, M. C., et al. (eds.), Spartan, Washington, D. C., pp. 435–461.

Widrow, B., and Hoff, M. E. [1960]. "Adaptive Switching Circuits." *1960 IRE WESCON Convention Record*, Part 4, pp. 96–104.

Widrow, B., and Stearns, S. D. [1985]. *Adaptive Signal Processing*, Prentice Hall, Englewood Cliffs, N.J.

Wiener, N. [1942]. *Extrapolation, Interpolation, and Smoothing of Stationary Time Series*, the MIT Press, Cambridge, Mass.

Wilburn, J. B. [1998]. "Developments in Generalized Ranked-Order Filters," *J. Opt. Soc. Amer.-A. Optics, Image Science, and Vision*, vol. 15, no. 5, pp. 1084–1099.

Windyga, P. S. [2001]. "Fast Impulsive Noise Removal," *IEEE Trans. Image Processing*, vol. 10, no. 1, pp. 173–179.

Wintz, P. A. [1972]. "Transform Picture Coding." *Proc. IEEE*, vol. 60, no. 7, pp. 809–820.

Witten, I. H., Neal, R. M., and Cleary, J. G. [1987]. "Arithmetic Coding for Data Compression." *Comm. ACM*, vol. 30, no. 6, pp. 520–540.

Wolff, R. S., and Yaeger, L. [1993]. *Visualization of Natural Phenomena*, Springer-Verlag, New York.

Woods, J. W., and O'Neil, S. D. [1986]. "Subband Coding of Images." *IEEE Trans. Acous. Speech Signal Proc.*, vol. ASSP-35, no. 5, pp. 1278–1288.

Woods, R. E., and Gonzalez, R. C. [1981]. "Real-Time Digital Image Enhancement." *Proc. IEEE*, vol. 69, no. 5, pp. 643–654.

Xu, Y., Weaver, J. B., Healy, D. M., Jr., and Lu, J. [1994]. "Wavelet Transform Domain Filters: A Spatially Selective Noise Filtration Technique." *IEEE Trans. Image Processing*, vol. 3, no. 6, pp. 747–758.

Yachida, M. [1983]. "Determining Velocity Maps by Spatio-Temporal Neighborhoods from Image Sequences." *Comput. Vis. Graph. Image Proc.*, vol. 21, no. 2, pp. 262–279.

Yamazaki, Y., Wakahara, Y., and Teramura, H. [1976]. "Digital Facsimile Equipment 'Quick-FAX' Using a New Redundancy Reduction Technique." *NTC '76*, pp. 6.2-1–6.2-5.

Yang, X., and Ramchandran, K. [2000]. "Scalable Wavelet Video Coding Using Aliasing-Reduced Hierarchical Motion Compensation," *IEEE Trans. Image Processing*, vol. 9, no. 5, pp. 778–791.

Yates, F. [1937]. "The Design and Analysis of Factorial Experiments." Commonwealth Agricultural Bureaux, Farnam Royal, Burks, England.

Yitzhaky, Y., Lantzman, A., and Kopeika, N. S. [1998]. "Direct Method for Restoration of Motion Blurred Images," *J. Opt. Soc. Amer.-A. Optics, Image Science, and Vision*, vol. 15, no. 6, pp. 1512–1519.

You, J., and Bhattacharya, P. [2000]. "A Wavelet-Based Coarse-to-Fine Image Matching Scheme in a Parallel Virtual Machine Environment." *IEEE Trans. Image Processing*, vol. 9, no. 9, pp. 1547–1559.

Yu, D., and Yan, H. [2001]. "Reconstruction of Broken Handwritten Digits Based on Structural Morphology," *Pattern Recog.*, vol. 34, no. 2, pp. 235–254.

Yu, S. S., and Tsai, W. H. [1990]. "A New Thinning Algorithm for Gray-Scale Images." *Pattern Recog.*, vol. 23, no. 10, pp. 1067–1076.

Yuan, M., and Li, J. [1987]. "A Production System for LSI Chip Anatomizing." *Pattern Recog. Letters*, vol. 5, no. 3, pp. 227–232.

Zahn, C. T., and Roskies, R. Z. [1972]. "Fourier Descriptors for Plane Closed Curves." *IEEE Trans. Comput.*, vol. C-21, no. 3, pp. 269–281.

Zhang, T. Y., and Suen, C. Y. [1984]. "A Fast Parallel Algorithm for Thinning Digital Patterns." *Comm. ACM*, vol. 27, no. 3, pp. 236–239.

Zhu, H., Chan F. H. Y., and Lam, F. K. [1999]. "Image Contrast Enhancement by Constrained Local Histogram Equalization," *Computer Vision and Image Understanding*, vol. 73, no. 2, pp. 281–290.

Zhu, P., and Chirlian, P. M. [1995]. "On Critical Point Detection of Digital Shapes," *IEEE Trans. Pattern Anal. Machine Intell.*, vol. 17, no. 8, pp. 737–748.

Zimmer, Y., Tepper, R., and Akselrod, S. [1997]. "An Improved Method to Compute the Convex Hull of a Shape in a Binary Image," *Pattern Recog.*, vol. 30, no. 3, pp. 397–402.

Ziou, D. [2001]. "The Influence of Edge Direction on the Estimation of Edge Contrast and Orientation," *Pattern Recog.*, vol. 34, no. 4, pp. 855–863.

Ziv, J., and Lempel, A. [1977]. "A Universal Algorithm for Sequential Data Compression." *IEEE Trans. Info. Theory*, vol. IT-23, no. 3, pp. 337–343.

Ziv, J., and Lempel, A. [1978]. "Compression of Individual Sequences Via Variable-Rate Coding." *IEEE Trans. Info. Theory*, vol. IT-24, no. 5, pp. 530–536.

Zucker, S. W. [1976]. "Region Growing: Childhood and Adolescence." *Comput. Graphics Image Proc.*, vol. 5, pp. 382–399.

Zugaj, D., and Lattuati, V. [1998]. "A New Approach of Color Images Segmentation Based on Fusing Region and Edge Segmentation Outputs," *Pattern Recog.*, vol. 31, no. 2, pp. 105–113.

Index